Particles in Contact

Sergiy Antonyuk
Editor

Particles in Contact

Micro Mechanics, Micro Process Dynamics and Particle Collective

Springer

Editor
Sergiy Antonyuk
Institute of Particle Process Engineering
Technische Universität Kaiserslautern
Kaiserslautern, Germany

ISBN 978-3-030-15898-9 ISBN 978-3-030-15899-6 (eBook)
https://doi.org/10.1007/978-3-030-15899-6

This Springer imprint is published by the registered company Springer Nature Switzerland AG
The registered company address is: Gewerbestrasse 11, 6330 Cham, Switzerland

This book is dedicated in memory of Professor Jürgen Tomas—an outstanding scientist, colleague and friend.

Preface

The properties and behaviour of particulate products, such as powder materials, agglomerated or coated materials, fillers, active ingredients or pigments and natural substances, like soils, sludge or aerosols, are significantly determined by interactions between fine solid particles in granular matter. The production, handling and transport processes of the solid products are also heavily influenced by the inter-particle interactions. For the optimization of existing solid processes and development of novel particulate products, the understanding and description of physicochemical microprocesses, which take place during approaching, direct contact and detachment of adhesive fine particles (from a few nanometers to 100 micrometers) are needed. Therefore, the description of macroscopic phenomena and properties of particulate matter consisting of many particles needs a basic understanding of the micromechanical and adhesive interactions of single particles in contact. They can be characterized by stress–strain, force–displacement, torgue–angle and potential–distance relationships for six degrees of freedom for the translation and rotation of particles in contact. New scientific developments in the field of simulations as well as measuring technique make it possible to understand and describe these relationships in the form of contact or potential models. These interaction models based on the physical parameters of individual particles can be used in modern particle-based simulation methods, such as Discrete Element Method or Molecular Dynamics, to provide an accurate prediction of the morphology, microstructure, deformation and flow behaviour and distribution of the properties of particulate systems in industrial processes.

The German Research Foundation (Deutsche Forschungsgemeinschaft, DFG) has supported an international research program in the form of Priority Program (SPP 1486) "Particles in Contact—Micro Mechanics, Micro Process Dynamics and Particle Collective" from 2010 to 2017. This book summarizes the research results of this joint research project obtained by 27 cooperating institutes from Germany, Austria, The Netherlands and Switzerland. The coordinator of this program was Prof. Jürgen Tomas from the Otto von Guericke University Magdeburg (Germany). This program was created and successfully carried out thanks to his innovative ideas, his enthusiasm for powder technology that has brought the scientists from different research areas together into this research project. Unfortunately, he passed away too early. Since 2015, Prof. Sergiy Antonyuk from the Technische Universität Kaiserslautern has taken over the coordination of this program.

We want to thank DFG for the financial support, and specially Dr.-Ing. Bernd Giernoth, Dr.-Ing. Georg Bechtold, Dr. Simon Jörres and Ms. Anja Kleefuß for their excellent coordination and support for the research activities of the SPP.

I am grateful to my Ph.D. students Vanessa Puderbach, Robert Hesse, Philipp Lösch, Dominik Weis, and Oleg Urazmetov, who caught misprints in the proofs of this book.

Kaiserslautern, Germany
December 2018

Sergiy Antonyuk

Contents

Analysis of Adsorbates and Interfacial Forces at Metal Oxide Interfaces at Defined Environmental Conditions

A. G. Orive, C. Kunze, B. Torun, T. de los Arcos and G. Grundmeier

Introduction

Adhesion forces between ensembles of micro- and nanoparticles are central in processes such as fluidization, agglomeration and sintering [1]. These effects find applications in an extensive range of technical fields that include the development of pharmaceutical powders, paints, and solar cells [2, 3].

Contact forces between oxide nanoparticles are strongly dependent on the surface chemistry, the presence of adsorbate layers and the environmental conditions [4]. High-energy oxide surfaces under ambient conditions spontaneously adsorb organic molecules and water. This adsorption leads to reduced surface energies. While the existence of atmospheric contaminations on oxide surfaces is routinely observed in surface analytical studies, their surface concentration and composition is often not perfectly controlled.

For most technological applications, assemblies of micro and nanoparticles will be necessarily utilized under normal ambient experimental conditions. As a consequence, the prevalence of water layers adsorbed onto the particle surfaces which are in contact with humid air needs to be taken into account. Indeed, capillary forces due to capillary condensation are particularly interesting since, when present, they usually dominate the inter-particle adhesion interactions. Over the last few years, both theoretical and experimental approaches have been carried out to propose a unified model which might account for the relevance of contact forces between particles under humid conditions [5, 6]. In particular the high-energy surfaces of micro- and nano-metal oxide particles experience the spontaneous adsorption of molecule

A. G. Orive · C. Kunze · B. Torun · T. de los Arcos · G. Grundmeier (✉)
Institute of Technical and Macromolecular Chemistry, University Paderborn, Paderborn, Germany
e-mail: g.grundmeier@tc.upb.de; guido.grundmeier@uni-paderborn.de

S. Antonyuk (ed.), *Particles in Contact*,
https://doi.org/10.1007/978-3-030-15899-6_1

adsorbates and water, which gives rise to a reduced surface energy. At this point, it is worth noting that although the afore-mentioned molecular adsorptions due to atmospheric exposure are well-known and easily identified by analytical techniques in interfacial surface studies, their quantitative surface concentration and composition is often not stated or it is poorly characterized. In particular, nano-contact force studies carried out at gas/solid interfaces, e.g. by means of atomic force microscopy (AFM), are often performed upon rather ill-defined conditions.

The quantification of the contact forces occurring between micro- and nanoparticles can be straightforwardly assessed by force spectroscopy measurements by means of AFM [7]. In a seminal contribution, Harrison et al. [8] collected experimental and theoretical results coming from 28 different approaches dealing with the assessment of capillary forces in the nanoparticle regime. It was clearly stated that the impact of capillary forces on particle interfacial adhesion depends on a nontrivial combination of the particle shape, size, deformability, adsorbate layers, surface roughness and hydrophilicity [6, 9]. Indeed, Farshchi-Tabrizi et al. [7] showed that the contact geometry has a critical influence on the humidity dependence of the adhesion force. They addressed the contact forces occurring between regular AFM tips and also the result of attaching particles to them and different solid samples, smooth and rough surfaces. Interestingly, the dependence between adhesion forces and humidity register for hydrophilic surfaces exhibited two distinct behaviors: a maximum or an increasing linear growing behavior. A new numerical models considering both surface roughness and cantilever shape that accurately estimates the quantification of the meniscus force was successfully proposed.

Regarding the presence of molecule adsorbate layers, Gojzewski et al. [10] elegantly introduced the measurement of the contact forces arising between the surfaces of silicon nitride AFM cantilevers and self-assembled monolayers (SAM) on gold as a function of relative humidity by means of dynamic force spectroscopy. Methylene- and hydroxyl-terminated SAMs were used as model hydrophobic and hydrophilic surfaces, respectively. The adhesion forces registered for the hydrophilic SAMs were noticeably higher than those obtained for the hydrophobic ones, even at low humidity values. The latter cannot be thus not only attributed to the formation of hydrogen bridges between hydroxyl groups from the SAM and silanol functionalities of the AFM tip, but also to the strength of the capillary forces.

Taking all the afore-mentioned facts into account, it is clear that AFM measurements carried out under ultra-high vacuum or clean (pure) electrolytes offer the advantage of working with a precisely defined interface chemistry. The aim of the work presented here was to analyze (characterize) the impact of poorly defined ambient surface contamination such as water and organic molecule adsorbates by directly removing them or replacing them by model adsorbates. In the following, TiO_2, ZnO and Al_2O_3 surfaces have been considered as model metal oxides with high surface energy as promising candidates/platforms for fundamental adhesion force studies. In particular, AFM-based methods such as nanoshaving (AFM scratching), UHV-based nanocontact force studies and single molecule force spectroscopy were applied. In addition, in situ FTIR surface spectroscopy was performed to analyze the behavior of some molecule adsorbates under humid conditions.

Materials and Methods

Materials and Surface Preparation

Materials and Chemicals

Rutile TiO_2 single crystals with crystallographic (100) and (110) orientations in the size of 10 mm × 10 mm × 0.5 mm were obtained from *Mateck GmbH* (Juelich, Germany) and *CrysTec* (Germany), respectively. Al_2O_3(11–20) single crystalline substrates of 20 mm × 20 mm × 0.5 mm were also purchased from *Mateck GmbH* (Juelich, Germany).

All chemicals were of p.a. grade (analytical reagent grade) and used without any further purification. Nonadecanoic acid (97%) was purchased from *Fluka*. Octadecylphosphonic acid (97%) and hexylphosphonic acid were obtained from *Alfa Aesar*. α-Mercapto-ω-amino PEG hydrochloride (HS–PEG–NH_2 × HCl, PEG-MW: 10,000 Da), α-Hydroxy-ω-mercapto PEG (HO–PEG–NHCO–CH_2–CH_2–SH, PEG-MW: 10,000 Da), α-Methoxy-ω-mercapto PEG (CH_3O–PEG–SH, PEG-MW: 10,000 Da) were acquired form *Rapp Polymere GmbH*. Absolute ethanol, used as a solvent and for general cleaning purposes, was purchased from *VWR International*. Hydrofluoric acid (40%) was obtained from Merck *KGa*. Ultrapure water was obtained from an *Ultraclear TWF* (SG Wasseraufbereitung, Barsbüttel, Germany) system with a maximum electrical conductance of 0.055 $\mu S\ cm^{-1}$. RCA-1 solution, i.e. 5:1:1; purified water, ammonia solution 30%, hydrogen peroxide 30%, was used for both etching and cleaning purposes.

Preparation of Rutile TiO_2(110) and (110) Single Crystalline Surfaces [11, 12]

In order to prepare surfaces stable under ambient conditions, TiO_2(100) single crystalline surfaces were prepared from commercial TiO_2 rutile single crystals. The procedure followed consisted of a pre-cleaning step, where the TiO_2 crystals were sonicated in ultrapure water, followed by wet chemical etching, and finally an annealing step according to the methodology developed by Nakamura et al. [13] and Yamamoto et al. [14]. In order to avoid possible contamination of the crystals during the preparation, the etching was performed in a PTFE vessel. (100) and (110) oriented TiO_2 crystals were etched in 20% HF solution for 20 min. After the etching process, the crystals were intensively rinsed with ultrapure water and dried in a stream of pure nitrogen gas. The crystals were then deposited into a tube furnace with Al_2O_3 walls and annealed at 680 °C for 48–72 h in ambient air. A slow heating and cooling rate of 3 °C/min (ramping time of 3.5 h) was selected in order to minimize both mechanical stress in the crystal and in the furnace tube. The quality of the so-prepared TiO_2(100) and (110) single crystals was finally assessed by means of AFM and LEED. A nonconventional 1 × 3 superstructure was obtained in the LEED pattern registered for the as-prepared $TiO_2(100)$ surfaces.

Preparation of $Al_2O_3(0001)$ and (11–20) Single Crystals [11]

In order to produce well-defined $Al_2O_3(0001)$ and $Al_2O_3(11–20)$ single crystalline surfaces, alkaline and acidic etching procedures were applied to the as-received commercial substrates according to a well-established methodology published elsewhere [11]. Briefly, the as-received substrates were immersed into RCA-1 (5:1:1; purified water, ammonia solution 30%, hydrogen peroxide 30%) solution for 2 min at 70 °C to remove organic contaminants. Then, they were subsequently immersed into a phosphoric acid (85%) aqueous solution for 2 min, rinsed with ultra-pure water, and dried with nitrogen. The as-treated crystals were then annealed under an ambient atmosphere at 1450 °C for 36 h.

Additionally, in the case of $Al_2O_3(11–20)$ the single crystals were again immersed into phosphoric acid (85%, 2 min), rinsed with ultra-pure water, and dried with nitrogen. The goal was to simulate the acidic treatment conventionally used in the conditioning Al-alloy surfaces before being exposed to adhesive compounds. A representative AFM image of such prepared surfaces is shown in Fig. 1.

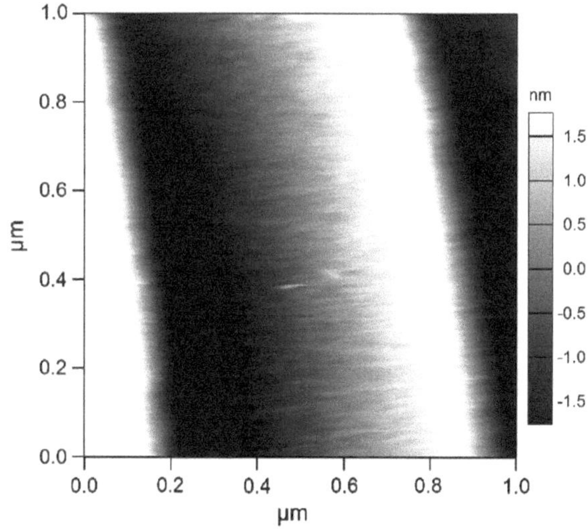

Fig. 1 AFM topography image of $Al_2O_3(11–20)$ single crystal cleaned with RCA-1, etched with concentrated phosphoric acid, heat treated at 1450 °C, and finally etched in concentrated phosphoric acid. Reprinted from [15] with permission from Elsevier

Adsorption of Organic Monolayers on TiO_2 and Al_2O_3 Single Crystalline Surfaces [11]

In order to carry out fundamental studies about the interaction between carboxylic acids with well-defined TiO_2 surfaces, nonadecanoic acid (NDCA) was selected as a model system to characterize the adsorbate formation process onto TiO_2(100) single crystalline substrates [11]. In this regard, freshly prepared TiO_2 single crystals were immersed into a 1 mM NDCA ethanolic solution for 12–16 h. After that, the as-modified substrates were rinsed in ethanol and dried in a nitrogen stream.

Additionally, with the aim to remove the scarcely defined carbon contaminations due to atmospheric contamination, TiO_2(110) substrates were chemically modified by self-assembly (SA) with octadecylphosphonic acid (ODPA) molecule layer. In this regard, freshly prepared and annealed TiO_2(110) single crystals were immersed into a 1 mM ODPA ethanolic solution for 16 h. The as-modified substrates were thoroughly rinsed in ethanol and dried in a nitrogen stream. This approach was carried out to create a defect-free surface covered by a well-defined aliphatic adsorbate layer as a reference surface to assess the impact of such aliphatic-based film on interfacial contact forces [12].

ODPA and hexyl phosphonic acid (HPA)-based organic films onto Al_2O_3(0001) single crystal were prepared by self-assembly. The freshly chemically etched and thermally annealed aluminum oxide substrates were immersed into 1 mM molecule (ODPA or HPA)-containing ethanolic solution for 12 h. The as-modified crystals were thoroughly rinsed with ultrapure ethanol, dried in nitrogen stream, and immediately used.

AFM Instrumentation

UHV-AFM

UHV AFM imaging and Force-distance curve (F-D) spectroscopy were carried out by using a VT-AFM XA (*Oxford Instruments*, Taunusstein, Germany) microscope (Fig. 2). AFM images were collected in contact mode. CSC17 cantilevers (resonant frequency of 13 kHz, spring constant of 0.18 N/m, and tip radius of 8 nm) purchased from *Mikromasch* (Bulgaria) were used. For AFM-based F-D spectroscopy measurements, data obtained from 16 × 16-point grids over 2.0 × 2.0 μm^2 areas were collected at room temperature for statistical purposes. The AFM tip was approached and retracted over the sample surface at 1 μm/s for a 500 nm tip-sample separation. The cantilevers were calibrated according to their deflection sensitivity, obtained from the slope of the repulsive regime, and the spring constant, which was obtained by means of the thermal noise method [16].

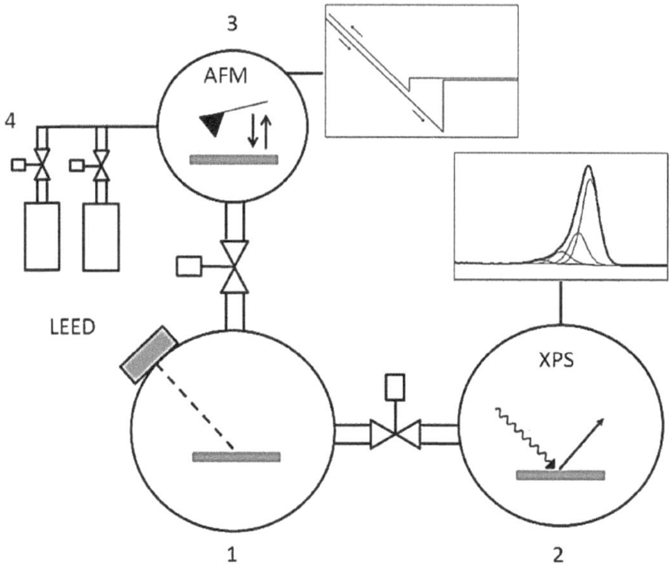

Fig. 2 Illustration of the UHV analytical setup allowing the measurement of surface chemistry and contact forces. (1) preparation chamber; (2) analysis chamber; (3) AFM chamber; (4) gas and water dosing system

Ambient Pressure AFM

AFM imaging was carried out by using a JPK Nanowizard II Ultra (*JPK Instruments AG*, Berlin, Germany) operating in air and at room temperature with a scan rate of 1 Hz. The topographical features exhibited by pristine $TiO_2(100)$ single crystal surfaces were characterized in contact mode in a window loading force range of 2–5 nN. In this case, DP17 silicon cantilevers (12 kHz, 0.17 N/m), purchased form *Mikromasch*, were used. However, for the AFM characterization of NDCA-modified $TiO_2(100)$ substrates, alternating contact mode (AC) was chosen. For the latter, an amplitude corresponding to a 90% of the free-air amplitude was maintained constant during the profiling. Both AC AFM imaging and nanoshaving experiences were carried out using NSC15 AFM tips (320 kHz, 40 N/m), acquired from *Mikromasch*. In the nanoshaving and nanografting processes, an organic molecule monolayer (typically prepared by self-assembly) is removed/altered as a consequence of AFM tip ploughing. Essentially, when high enough loading forces are applied in a pure solvent with the AFM tip, the latter is expected to scratch the modified surface and subsequently remove the organic film, unveiling thus the underlying substrate surface. This procedure is known as nanoshaving. If this process is carried out when other secondary molecules are also present in the solvent, then a fast chemisorption of the latter onto the shaved area can be obtained, i.e. nanografting [17]. It is worth noting that although the organic layer is removed as a consequence of AFM tip ploughing, no noticeable damage on the single crystalline surfaces can be detected for the loading force window range considered.

Single molecule force spectroscopy experiences on Al_2O_3(11–20) single crystals were carried out using an MFP-3D-SA (*Asylum Research*) microscope equipped with an anti-vibration system. A fluid cell (Fluid Cell Lite from *Asylum Research*) was employed to obtain the corresponding force-distance curves in 10 mM $NaClO_4$ aqueous solution at pH 6. In order to achieve good reproducibility, several hundred force curves were recorded in each experiment. The rupture events were evaluated with Igor-Pro software. Note that only a fraction (10–30%) of all curves showed rupture events. In those cases, the force recorded at the rupture point was noted and used to build an additional set of statistical data. A Gaussian fit was applied to determine the mean force value.

Modification of AFM-Tips

TiO_2 Coated AFM-Tips

NSC19 AFM cantilevers (65 kHz, 0.5 N/m), acquired from *Mikromasch*, were modified with 150 nm thick titanium coating by means of plasma-enhanced PVD. A titanium target was sputtered by a DC plasma in pure Argon atmosphere at 0.2×10^{-3} mPa at a mean power of 2 kW and 300 V bias voltage. Under these experimental conditions, a deposition rate of 14 nm/min was obtained. In order to completely oxidize the titanium coating of the cantilevers, the latter modified AFM tips were annealed in air at 600 °C for 16 h.

Macromolecular Modification of AFM Tips

Gold-coated silicon cantilevers HQ:CSC17 (13 kHz, 0.18 N/m), acquired from *Mikromasch*, were cleaned according to a well-stablished procedure published elsewhere [15]. Briefly, as-received cantilevers were immersed into an RCA-1 solution for 2 min. at 70 °C in order to remove organic contaminants. After that, the tips were rinsed with ultrapure water and subsequently dried in a nitrogen stream.

OH-Functionalization

The solvent cleaned gold-coated cantilevers were introduced into a 1 mM α-Hydroxy-ω-mercapto PEG-containing ultrapure water solution for $t = 4$ h.

NH_2 x HCl-Functionalization

The solvent cleaned gold-coated cantilevers were introduced into a 1 mM α-Mercapto-ω-amino PEG hydrochloride-containing ultrapure water solution for $t = 4$ h.

OCH_3-Functionalization

The solvent cleaned gold-coated cantilevers were introduced into a 1 mM α-Methoxy-ω-mercapto PEG-containing ultrapure water solution for $t = 4$ h.

XPS Analysis

XPS measurements were performed in an Omicron ESCA$^+$ System (Omicron NanoTechnology GmbH, Germany). The element spectra were recorded at pass energies of 25 eV. A monochromated Al Kα (1486.3 eV) X-ray source with a spot diameter of 600 μm was used. The calibration of the spectra was done by using the position of the C 1s core level peak of adventitious carbon (binding energy, BE = 285 eV) as a reference. For data evaluation, the CasaXPS software was used [18].

For the quantitative analysis of the surface chemistry of the prepared crystals, angle-resolved XPS experiments were performed. XPS spectra were recorded at take-off angles of the photoelectrons from 70° to 10° with respect to the surface plane. The probing depth d can be calculated as a function of the take-off angle Θ [19] according to $d = \lambda \sin\Theta$, where λ is the inelastic mean free path of the photoelectrons.

LEED Instrumentation

Omicron Spectra LEED optics were employed at an incident angle close to 90° with respect to the surface for LEED experiences. With the aim to obtain high-quality LEED images, the beam energy was adjusted from 80 to 120 eV. The as-registered LEED patterns of the corresponding surfaces were simulated and compared using the software LEEDpat30.

Results

Analysis of Disordered Molecular Adsorbates by Means of Nanoshaving and XPS Analysis

The molecular adsorbate formation of alkanehydroxamic and carboxylic acids on various metal oxides was investigated by Folkers et al. [20]. They observed the formation of self-assembled hydroxamic monolayers, which they attributed to the high complexation constant of the hydroxamic acids with metal ions as opposite to carboxylic acids. Thus, in the case of carboxylic acids adsorbed onto amorphous TiO_2 oxide films on Ti no self-assembly process was observed. The authors proposed that this behaviour was mainly governed by the acid to neutral behavior of the TiO_2 oxide film. Therefore, in this section we address the adsorption of model molecules with the aim of getting a good understanding of particle–wall interactions in the presence of a well-defined adsorbate under ambient conditions. Nonadecanoic acid (NDCA) was chosen as a model molecule to simulate typical adsorbates present in ambient atmosphere. The reasoning behind was that (i) the carboxylic acid group can mimic the chemical functionalities of molecules present at atmospheric conditions, (ii) the well-defined composition of the aliphatic chain results in an intense carbon signal in XPS experiments and (iii) it leads to a measurable height contrast in AFM based nanoshaving investigations. NDCA monolayers were prepared on a $TiO_2(100)$ single crystal by self-assembly from NDCA-containing ethanolic solutions [21].

Nanoshaving

The topographical features exhibited by adsorption of NDCA molecules onto $TiO_2(100)$ single crystal surfaces were investigated by AFM operating in air in AC mode [22]. By comparison to the smooth atomically flat terraces of pristine $TiO_2(100)$, rounded nanostructured islands or domains decorating the terraces can be observed, which are attributable to a sub-monolayer coverage of NDCA molecules. The ploughing of a $500 \times 500\ nm^2$ area with the AFM tip in contact mode with a significant high value of loading force, leads to the removal of the adsorbed NDCA molecules takes place (see Fig. 3).

As a consequence of the nanoshaving process, pristine areas of $TiO_2(100)$ terraces are uncovered. This is an indication that the NDCA adsorbate layer is easily removed from the surface. By carrying out cross section profiles along both the outer NDCA coated and the nanoshaved areas, a corrugation of 120 pm could be measured. The latter value is noticeably lower than that expected for a tightly packed monolayer of NDCA molecules arranged in a slightly tilted standing up configuration, since the theoretical length of the NDCA molecules is in in the range of 2 nm. Consequently, the formation of a disordered film together with a low fractional surface coverage of NDCA molecules can be concluded from the analysis of the presented AFM data. These results hint at a poor interaction between the anchoring carboxylate

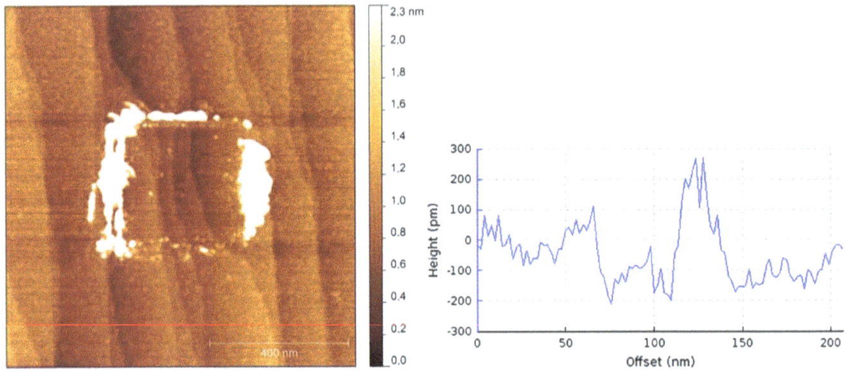

Fig. 3 *Left hand side*: Nanoshaving on an NDCA covered $TiO_2(100)$ surface shows the disordered structure of the adsorbate layer, the scan size is $1 \times 1\ \mu m^2$. *Right hand side*: cross section of the area marked at the left. Reprinted from [22] with permission from Elsevier

groups and the $TiO_2(100)$ surface. The result is that NDCA molecules exhibiting carboxylate functional/anchoring groups, reversible adsorbed at active $TiO_2(100)$ adsorption sites, stay rather mobile due to relatively weak interactions.

XPS Analysis

The TiO_2 substrate was characterized by XPS before and after NDCA adsorption. The results of measurements of the pristine $TiO_2(100)$ surface measured at different take-off angles is shown in Table 1. Titanium, oxygen and carbon were found at the surface. The atomic percentage of carbon increases when the photoelectron take-off angle is decreased, which indicates that the carbon is associated to an adsorbate film formed at the crystalline surface by exposure to air (adventitious carbon).

Most importantly, in comparison to the spectrum obtained for $TiO_2(100)$ single crystals prepared in UHV conditions where the O 1s signal is exclusively originated from O^{-2} species, [23] the high resolution O 1s AR-XP core level peaks exhibit a profile with different components that can be assigned to the presence of hydroxides, adsorbed water and carboxylic groups. Indeed, the increase in the O/Ti ratio when the take-off angle is reduced hints at an oxygen enrichment of the surface supported by the surface R_{O-1}/O^{2-} and R_{O-2}/O^{2-} contributions [11], indicating a

Table 1 Surface composition of $TiO_2(100)$ surface as measured by XPS at varying take-off angles with respect to the surface plane (from [11])

Angle (°)	O 1s (at.–%) at 530.2 eV	Ti 2p 3/2 (at.–%) at 458.8 eV	C 1s (at.–%) at 285 eV
70	63.2 ± 0.5	27.9 ± 0.5	8.9 ± 0.5
40	62.5 ± 0.5	24.7 ± 0.5	12.8 ± 0.5
20	61.3 ± 0.5	19.0 ± 0.5	19.7 ± 0.5
10	51.2 ± 0.5	13.8 ± 0.5	35.0 ± 0.5

Table 2 Surface composition of the NDCA covered $TiO_2(100)$ crystal as measured by XPS at varying take-off angles with respect to the surface plane (from [11])

Angle (°)	O 1s (at.–%) at 530.2 eV	Ti 2p 3/2 (at.–%) at 458.8 eV	C 1s (at.–%) at 285 eV
70	54.1 ± 0.5	22.6 ± 0.5	23.2 ± 0.5
40	52.6 ± 0.5	18.2 ± 0.5	29.2 ± 0.5
20	41.7 ± 0.5	13.9 ± 0.5	44.4 ± 0.5
10	29.1 ± 0.5	8.9 ± 0.5	62.0 ± 0.5

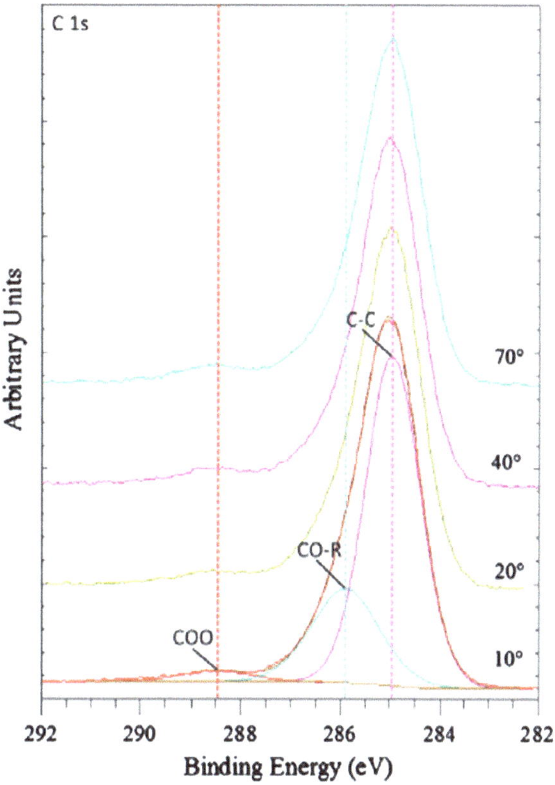

Fig. 4 C 1s spectrum of NDCA covered TiO_2 crystal at varying take-off angles. For ease of comparison, the spectra were normalized and shifted vertically. The vertical dashed lines mark the positions of the different components of the C 1s peak. Reprinted from [11] with permission from Elsevier

hydroxide-terminated surface. Consequently, it could be concluded that a layer of surface hydroxides is ubiquitously present on the $TiO_2(100)$ single crystals prepared under ambient conditions.

The adsorption of NDCA on TiO_2 could also be proven by XPS by the significantly increased carbon concentration in comparison to that exhibited by the bare $TiO_2(100)$ substrate (see Table 2).

An extensive XPS analysis of the NDCA-coated TiO_2 crystals exhibited a C 1s signal with a peak width of 1.25 eV of the C–C component at a takeoff angle of 10° as depicted in Fig. 4.

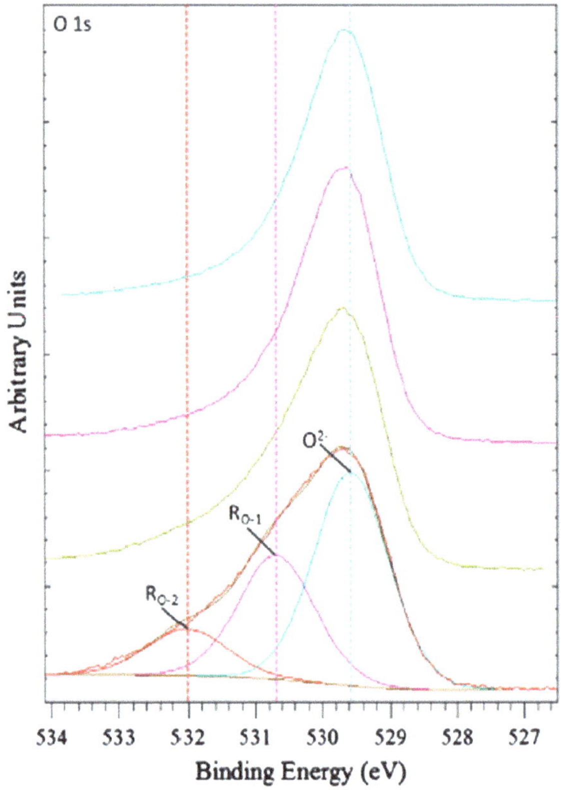

Fig. 5 O 1s spectra of the NDCA covered $TiO_2(100)$ surface measured at the same take-off angles as in Fig. 4. For ease of comparison, the spectra were normalized and shifted vertically. Reprinted from [11] with permission from Elsevier

The as-prepared NDCA adsorbate film exhibits the characteristic C 1s spectrum expected for a long chain carboxylic acid, as can be seen in Fig. 4. The registered ratio of C–C/COO is close to 16–18 range, which corresponds to the expected stoichiometry of NDCA. Thus, the undefined original layer of adventitious carbon has been replaced in this case by a well-defined layer of NDCA molecules, most likely due to the stabilizing effect of intermolecular long aliphatic chain van der Waals interactions.

Additionally, the detailed assessment of the components present in the O 1s signal, displayed in Fig. 5, together with the corresponding atomic fraction shown in Table 3, allows to conclude that the underlying $TiO_2(100)$ single crystalline surface still exhibits the same $OH:O^{2-}$ ratio as the pristine crystalline surface (Table 1). A slightly reduced value of the O/Ti ratio is seen in this case, which is attributed to the displacement of adsorbed water molecules by NDCA.

XPS characterization reinforces the observation that NDCA molecules chemisorb porly onto the $TiO_2(100)$ single crystalline surfaces due to weak interactions occurring between the hydroxyl-stabilized TiO_2 surface and anchoring carboxylate groups. Indeed, taking into account the presented data, no condensation between the latter and hydroxyl groups present in the substrate surfaces can be detected. Consequently,

Table 3 Detailed composition of the O 1s element spectra of the NDCA covered surface at varying take-off angles (from [11])

Angle (°)	$\underline{O}_{\Sigma}$/Ti	$\underline{O^{2-}}$/Ti	$\underline{O}H/O^{2-}$	$(H_2\underline{O}, CO\underline{O}H)/O^{2-}$
70	2.4	2.0	0.16	0.05
40	2.8	2.1	0.24	0.08
20	3.0	2.0	0.32	0.16
10	3.3	1.8	0.60	0.24

the ubiquitous presence of these hydroxyl groups, together with water adsorbate layers, on the TiO_2 surface (prepared in ambient conditions) is tentatively proposed to be responsible for the formation of the poorly packed and disordered NDCA layer initially observed by AFM and XPS and lately supported by AFM nanoshaving. This argument is additionally supported by Henderson et al. who reported the dissociation of water molecules on $TiO_2(100)$ surfaces to give rise to two hydroxide groups adsorbed onto the surface: the first one to Ti^{4+} adsorption sites, whilst the second one would be bound at bridging O^{2-} [24]. It has also been proposed that the adsorption of volatile molecules could also take place by forming Ti–O bonds [25]. The latter, together with the above-mentioned hydroxide groups and the water adlayers, would strongly inhibit the anchoring of carboxylate groups to this $TiO_2(100)$ surface prepared in ambient conditions.

Conclusions

A new methodology for the preparation of rutile $TiO_2(100)$ crystalline surfaces in water and oxygen partial pressures at high temperature is reported. Apart from small amounts of carbon-based contamination bearing carboxylate moieties (i.e. organocarboxilate adsorbates coming from the laboratory atmosphere), angle-resolved XPS shows the presence of a hydroxyl layer terminating the single crystal surface. Hydroxyl groups are proposed to chemisorb preferentially at defects present on the $TiO_2(100)$ surface lowering thus the surface energy.

The adsorption of NDCA molecules onto the as-prepared bare $TiO_2(100)$ surfaces results in the displacement of the adventitious carbon layer typically present at surfaces exposed to air and resulted in a noticeable rise in the contact angle from $18 \pm 3°$ to $45 \pm 3°$ [12]. The combination of angle-resolved-XPS and AFM nanoshaving allows to conclude that a disordered adsorbate layer of NDCA molecules essentially lying flat on the substrate terraces is formed (as opposed to the formation of a self-assembled monolayer). Consequently, nanoshaving has been shown to be a reliable method for the fundamental characterization of adsorbate layers. In the particular case of NDCA-coated $TiO_2(100)$ substrates, nanoshaving unveiled disordered organic layer exhibiting a fractional surface coverage consistent with a submonolayer.

The presence of terminating hydroxyl groups, together with water adlayers, on the as-prepared $TiO_2(100)$ crystalline surface strongly inhibits the formation of well-defined tightly packed and ordered NDCA monolayers. Consequently, weak interactions between the anchoring carboxylate groups and the hydroxyl-stabilized $TiO_2(100)$ terraces, namely hydrogen bridging, are proposed to be responsible for the observed behaviour.

Nanoshaving and Nanografting of ODPA Monolayers on Al_2O_3 Surfaces [22]

The combination of nanografting and nanoshaving techniques can be considered a valuable tool to assess the quality of self-assembled monolayers. The characterization of a vast variety of spontaneously adsorbed molecules on noble metals has been carried out by means of AFM-based nanografting over the last few years [26–28]. In particular, the impact exerted by different distal functional groups and chain lengths on frictional coefficients has been assessed by nanografting. In addition, the differential adsorption of thiol molecules by means of either nanografting or self-assembly has been compared [27, 28]. Most importantly, it has been stated that as a consequence of the reduced spatial confinement, fewer pinholes and defects can be obtained in the grafted patches in comparison to those obtained in the outer self-assembled molecule layer regions [28]. In this regard, nanografting appears as a very reliable and useful technique for the characterization and modelling of the molecular adsorption and self-assembly processes at the nanoscale. Indeed, when the spontaneous chemisorption of certain molecules exhibiting enhanced affinities for specific surfaces is considered (for instance aliphatic thiols on gold), then the suitability of using in situ-AFM measurements for the monitoring of the monolayer assembly decreases. In the following, nanografting and nanoshaving processes have been carried out to characterize the adsorption of phosphonate-terminated aliphatic molecules onto $Al_2O_3(0001)$ single crystals. The study was supplemented with XPS measurements of the surfaces.

High-resolution AFM images registered in both contact and AC mode for the bare Al_2O_3 single crystals measured in ambient conditions are displayed in Fig. 6. The solvent-cleaned substrates exhibited a relatively smooth surface which turned into characteristic atomically flat terraces after etching and thermal annealing. Similar results to these presented in this chapter have been reported elsewhere [29, 30]. These terraces are in the range of ~1 μm wide. However, the methodology employed for the preparation of this multistep terraced morphology brought about some inhomogeneities, even when analogue $Al_2O_3(0001)$ substrates were submitted to identical experimental conditions. These particular results can be summarized as follows: (i) typical morphological features, involving multistep ~1 μm-wide atomically flat terraces, were obtained for all the samples measured, as can be seen in the topographical AFM images depicted in Fig. 6a, c, e; however, (ii) noticeable heterogeneities were then unveiled by both friction (Fig. 6b, f), and phase contrast images (Fig. 6d).

These results can be tentatively interpreted as follows. When working in ambient conditions, water adlayers adsorbed onto the surface are expected to modify signifi-

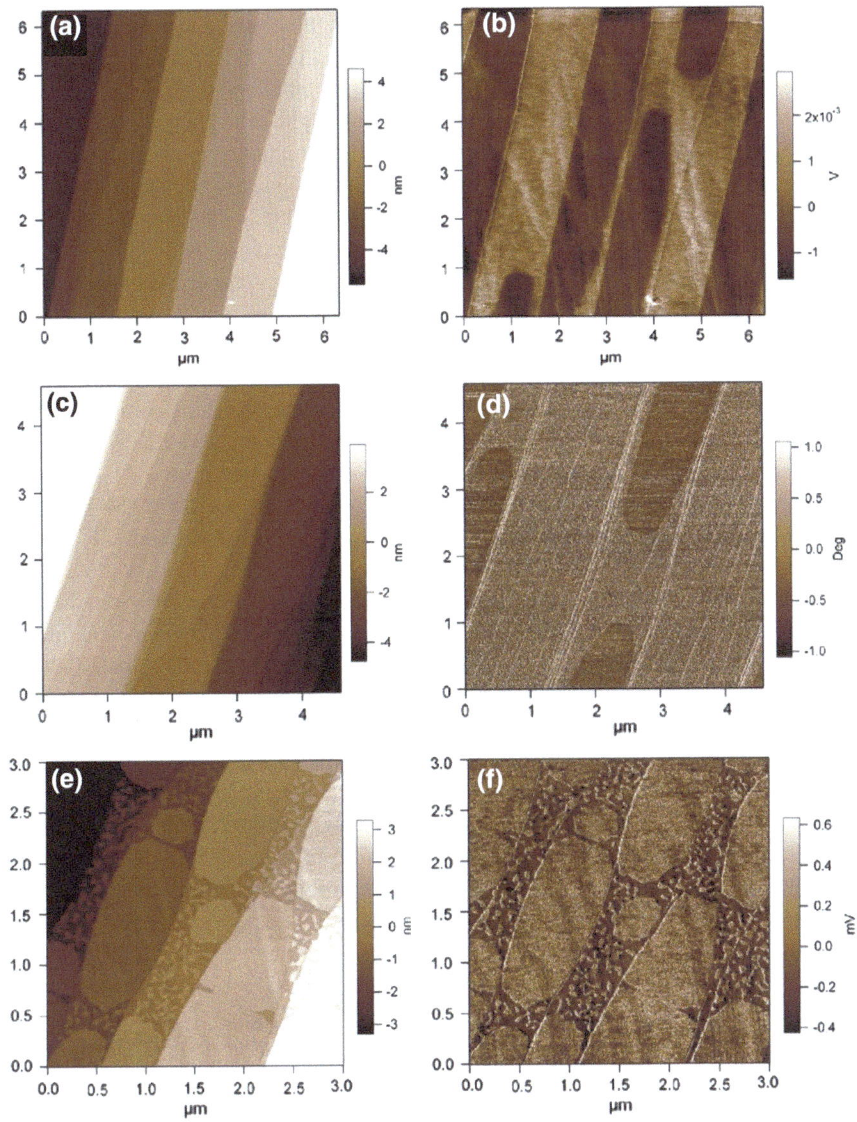

Fig. 6 AFM images showing morphological differences and heterogeneities on $Al_2O_3(0001)$ surfaces. **a** and **c** Topography images of typical atomically flat terraces. **b** Friction and **d** phase images reveal different terminations on the same terrace. **e** and **f** Similarly prepared single crystals showing domains (**e** topography image; **f** friction image). The measurements were performed under ambient conditions. Reprinted from [22] with permission from Elsevier

cantly the tip-to-surface interaction contact forces. In this context, although the height differences, which can be obtained from topographical AFM images by performing cross section profiles, are in the range of ~100 pm, noticeable phase and friction contrast showing well-defined domains can be observed in Fig. 6b, d, f. The latter could then be attributed to the presence of water adsorbate layers on the crystalline surfaces. Interestingly, large rounded and elongated domains seem to coexist with smaller and amorphous islands (see Fig. 6e). Consequently, the presence of these domains definitely accounts for a differential friction behaviour which is eventually indicative of a different surface termination. However, AFM experiments carried out in UHV at 140 °C showed the persistence of these friction-contrast domains even at high temperature, i.e. presumably in the absence of physisorbed water [22]. The origin of the heterogeneities revealed by phase and friction contrast AFM imaging is unclear. In the following, fot the sake of simplification, the morphologies shown in Fig. 6b and d will be identified as *type 1*, whilst those displayed in Fig. 6f will be named *type 2*. It is worth noting that previous studies carried out onto $Al_2O_3(0001)$ single crystals in UHV conditions pointed out to two distinct surface terminations, i.e. Al and O [31]. Additionally, AFM-based studies, also performed in UHV, have demonstrated a hexagonal conformation for two layers of Al atoms for these crystalline facets [32]. However, in contrast to the surfaces obtained in UHV conditions, single crystalline $Al_2O_3(0001)$ substrates prepared in ambient experimental conditions exhibit a differential surface composition since they are exposed to both oxygen activity and water adlayers. Indeed, it has been widely stated that hydroxide-coated alumina crystalline surfaces are the most stable configurations under ambient conditions since hydroxylation is expected to decrease the surface energy of aluminium oxide [33]. Thus, in the absence of water, Al and O surface terminations would be expected. However, aluminium oxide surfaces are expected to suffer hydroxylation in water-containing ambient condition as published elsewhere [29]. In fact, the existence of hydroxyl-stabilized $Al_2O_3(0001)$ terraces in the alumina substrates etched and annealed in ambient (water containing) conditions has been proved by XPS, shown in Fig. 7.

Once the adsorption of octadecylphosphonic acid (ODPA) molecules (from ethanol) on the as-treated $Al_2O_3(0001)$ terraces (after etching and annealing in ambient conditions) has taken place, AFM nanoshaving was performed onto the ODPA-modified alumina single crystal. In this context, AFM imaging carried out onto the latter modified substrate and displayed in Fig. 8a shows the formation of a very homogeneous and closely packed ODPA layer. A representative cross section profile taken in the AFM image depicted in Fig. 8a point at two terraces, indicated with stars, exhibiting a rougher surface, in contrast with the other much smoother surrounding terraces. This fact has been tentatively attributed to weak interactions between the phosphonate anchoring groups and the $Al_2O_3(0001)$ terraces giving rise to a disordered and poorly bound ODPA layer. Conversely, a stronger interaction with the $Al_2O_3(0001)$ surface together with the formation of a compact and tightly packed ODPA film on the smoother terraces is proposed.

Taking this fact into account, for the following an ODPA-modified $Al_2O_3(0001)$ substrate exhibiting the afore-mentioned two types of organic films (disordered and

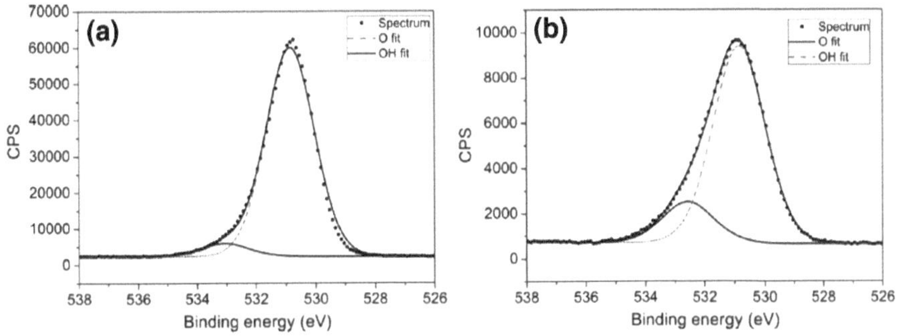

Fig. 7 XPS measurements of the O 1s region of the $Al_2O_3(0001)$ single crystal at **a** 70° and **b** 20° take-off angle. The spectra were fitted with two components assigned to oxide (at lower binding energy) and hydroxide species (at higher binding energy positions). A clear increase of the hydroxide contribution was found for the spectrum acquired with a lower take-off angle, indicating that the hydroxide lies on top of the surface. Reprinted from [22] with permission from Elsevier

compactly packed) is considered, as can be seen in the white rectangle area indicated in Fig. 8b. By applying with the AFM tip a high enough loading force in contact mode to the ODPA-modified aluminum oxide single crystal, the organic layer is completely removed unveiling thus the underlying substrate surface, i.e. nanoshaving, see the ploughed squared area in Fig. 8c. The as-detached ODPA molecules rapidly diffuses to the bulk ethanolic solution and no readsorption is detected at all. Cross section profiles carried out through the "shaved" and surrounding areas in the same terrace allows to accurately quantify the thickness of the ODPA layer. Thus, the as-measured thicknesses in both differentiated regions (1 and 2 as identified in Fig. 8b) reaches a value of ~2.0 nm which is quite close to the theoretical length calculated for the ODPA molecule. The latter is therefore indicative of ODPA monolayer film formation in a slightly tilted standing-up configuration in the two different areas shown in Fig. 8b. Most importantly, after successive AFM scanning on the same area, only ODPA molecules strongly adsorbed on the ordered and well-packed domains on the $Al_2O_3(0001)$ terraces remain adsorbed, while those weakly adsorbed in disordered ODPA regions are subsequently removed as can be observed in Fig. 8d. This result hints at the fact that ODPA molecules can presumably bind strongly to just one single type of surface termination. Conversely, physisorption is most likely to occur on the other crystalline surface termination.

Nanografting experiments have been carried out on hexylphosphonic acid (HPA)-modified $Al_2O_3(0001)$ single crystals by using longer aliphatic chain ODPA molecules. More stable self-assembled monolayers can be obtained for phosphonate-terminated molecules incorporating longer aliphatic chain due to the enhanced intermolecular van der Waals interactions. After nanoshaving the squared area displayed in Fig. 9a of an HPA-modified $Al_2O_3(0001)$ in a 1 mM ODPA-containing ethanolic solution, the grafting of ODPA molecules in the "shaved" area, where HPA molecules have been desorbed by AFM ploughing, has taken place. The latter can be justified in terms of the brighter squared area corresponding to higher regions observed in

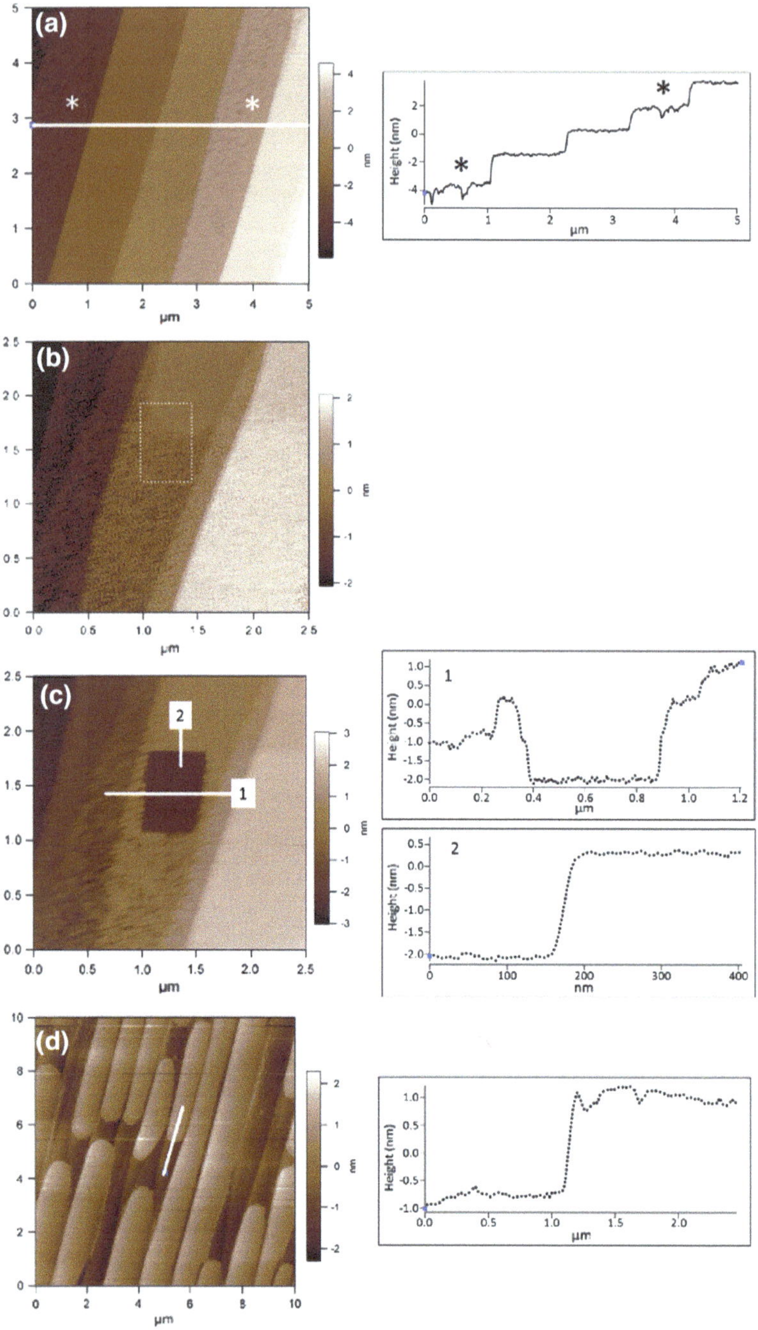

Fig. 8 **a** AFM topography of ex-situ adsorbed ODPA on $Al_2O_3(0001)$. Topography **b** before and **c** after nanoshaving on an area containing two different surface terminations. **d** Same crystal after two AFM scans. ODPA stays stable only on one type of surface termination. The measurements were performed in ethanol. Reprinted from [22] with permission from Elsevier

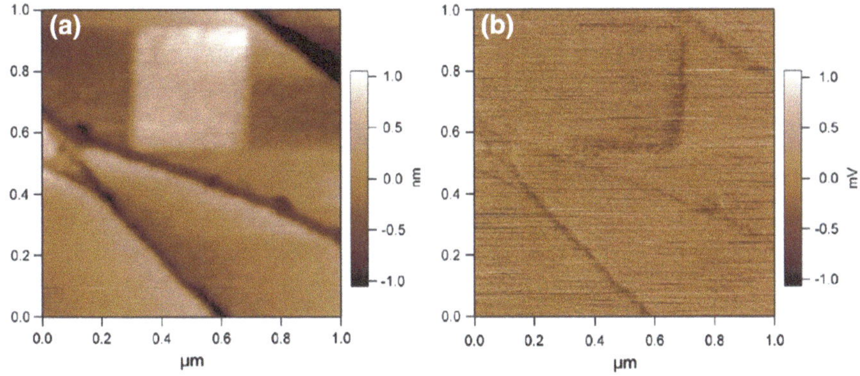

Fig. 9 Nanografting on Al_2O_3 single-crystal surfaces. **a** Topography and **b** friction images after nanografting in the presence of ethanolic ODPA solution on the HPA-covered (0001) surface. Reprinted from [22] with permission from Elsevier

Fig. 9a. Since both molecules, i.e. HPA and ODPA, are CH_3-terminated, no remarkable differences could be found in the AFM friction image (see Fig. 9b). Cross section profiles performed through the squared and surrounding areas exhibit a positive height difference of 0.8 nm which is slightly lower but reasonably consistent with the difference between the theoretical lengths of HPA and ODPA [22].

The results of the afore-mentioned experiments accounts for the systematic characterization of the adsorption processes and self-assembly kinetics of phosphonate-terminated aliphatic molecules on model aluminum oxide crystalline surfaces. The preparation (chemical etching and thermal annealing) of $Al_2O_3(0001)$ single crystalline surfaces upon exposure to ambient conditions brought about surface heterogeneities corresponding to a differential surface termination. The latter determines the nature of the bonding interaction between the adsorbed molecule and the aluminum oxide surface and, consequently, both the ordering and density of the packing arrangement into a film, as reliably demonstrated by nanoshaving. Additionally, nanografting has been shown to be a useful and reliable technique for studying both adsorption and self-assembly kinetics of phosphonate-based moieties at the liquid/metal oxide interface.

Investigation of TiO_2–TiO_2-Contact Forces Under UHV Conditions

Contact forces between micro- and nanoparticles can be accurately estimated by force spectroscopy measurements using an atomic force microscope (AFM). As previously stated, under ambient conditions particle-to-particle interactions are significantly influenced by the formation of capillary forces. Indeed, Farshchi-Tabrizi et al. [7]

showed that the contact geometry exerts a significant influence on the dependence between relative humidity and adhesion forces. Additionally, the superficial chemical composition of TiO_2 can be modified by water adlayers and by exposure to UV-light, which may result in remarkable modifications of adhesion behavior [34]. However, under ultrahigh vacuum conditions (UHV) experimental conditions, capillary bridges do not form and particle-particle contact force interactions can be then evaluated for well-defined surface chemistries. To the best of our knowledge only a few works based on the study of contact forces via an UHV AFM have been reported so far [35].

TiO_2 single crystalline surfaces provide an excellent platform as a model system for studying adhesion phenomena due to their well-defined atomic surface structure. Within this section, the contact forces between a TiO_2-coated AFM tip and a rutile $TiO_2(110)$ single crystalline surface in different chemical states were investigated under UHV-conditions in order to address particle-wall interactions under well-defined experimental conditions [12]. Additionally, particle-wall contact forces were measured between TiO_2-coated AFM tips and $TiO_2(110)$ single crystal modified with an octadecylphosphonic acid (ODPA) monolayer. The coating with ODPA provided a defect-free reference system, and also prevented from ulterior additional uncontrolled adsorption by exposure to air, due to the low surface energy of the ODPA-coated $TiO_2(110)$.

Both the surface chemistry of the TiO_2 films grown on the AFM cantilevers and $TiO_2(110)$ single crystalline surfaces were modified in situ by argon ion bombardment at low energy, water dosing at 3×10^{-6} Pa, and an annealing step at 200 °C. These treatments led to measurable differences in the density of Ti(III) defect states by XPS (note that the transfer between the AFM setup and the XPS measurement position were done without breaking UHV conditions). The measured contact forces could be correlated to the surface defect density. The results are discussed as function of electronic structure changes and their influence on the Hamaker constant.

XPS and UPS Investigation of the Surface

High-resolution Ti 2p spectra registered for a $TiO_2(110)$ single crystal before and after the Ar^+ bombardment are displayed in Fig. 10a, b, respectively. It should be noted at this point that the XPS characterization of TiO_2-coated AFM cantilevers and $TiO_2(110)$ single crystals showed no large differences between the two oxide surfaces. Most importantly, after Ar^+ bombardment treatment, both types of TiO_2 showed an increase of surface defect density, identifiable by the appearance of Ti(III) states (around 20% of the original titanium changed to the Ti(III) oxidation state after bombardment). This effect has been attributed to the transfer of one electron to Ti 3d empty levels during the formation of surface defects created by Ar^+ bombardment [36].

The Ti 2p core level spectra displayed in Fig. 10b corresponds to the successive surface treatment experienced by the TiO_2 single crystal, namely Ar^+ bombardment, water dosing at 3×10^{-6} Pa, and annealing at 200 °C. The as-registered spectra

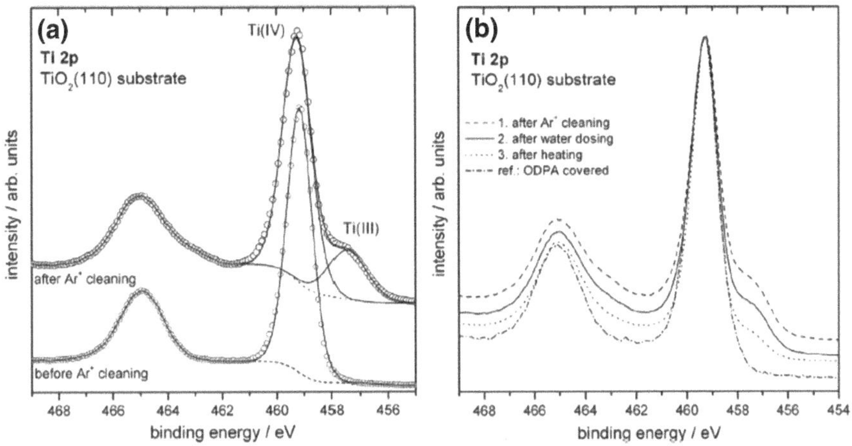

Fig. 10 **a** Ti 2p core level spectra of the $TiO_2(110)$ substrate before and after Ar^+ cleaning. **b** Ti 2p core level spectra of the $TiO_2(110)$ substrate after Ar^+ cleaned surface, after water dosing at 3 $\times$ 10^{-6} Pa for 30 min and after annealing the sample to 473 K for 30 min. For comparison the Ti 2p core level spectrum of an ODPA adsorbed monolayer onto TiO_2 bare crystal which reflects a defect-free surface is also shown. Reprinted from [12] with permission from Elsevier

exhibit a decrease in Ti(III) component after water dosing, indicating thus a reduced defect density in the $TiO_2(110)$ surface (drop in the intensity of the peak shoulder at 457.2 eV). The subsequent annealing of $TiO_2(110)$ led to a further decrease of Ti(III) surface defects as shown in Fig. 10b. For comparison purposes, the ODPA-modified $TiO_2(110)$ surface was also analyzed as a Ti(III) defect-free reference. From these results, a saturation of surface defects by means of a reaction with H_2O and thermal annealing could be concluded.

Changes in the valence orbitals occurring during the different treatments were observed by UPS (HeI, 21.2 eV). UPS spectra registered for the $TiO_2(110)$ substrates after the successive surface treatments, i.e. Ar^+ bombardment, water dosing, and thermal annealing, are displayed in Fig. 11a. For comparison, the defect-free ODPA-modified $TiO_2(110)$ surface has been included as well. The main contribution can be attributed to O 2p band exhibiting a main peak located at 5.5 eV with shoulders located at 7 and 11 eV [37]. Most importantly, the peak at 0.9 V corresponds to Ti 3d transitions. The normalized area of the latter is straightforwardly associated with the presence of defects in the surface-near region [38]. As expected, the profile of the UPS spectrum points out that the ODPA-modified TiO_2 surface is essentially free of Ti(III) defects. Interestingly, in comparison to the results obtained for the $TiO_2(110)$ substrates treated with Ar^+ ion bombardment, the samples after water dosing shows a significant decrease in the intensity of the Ti 3d transition. The latter clearly indicates that water exposure noticeably reduces the surface defect density. Additionally, after thermal annealing at 200 °C, the intensity of the Ti 3d transition is further reduced, confirming thus that an additional surface repair step is occurring.

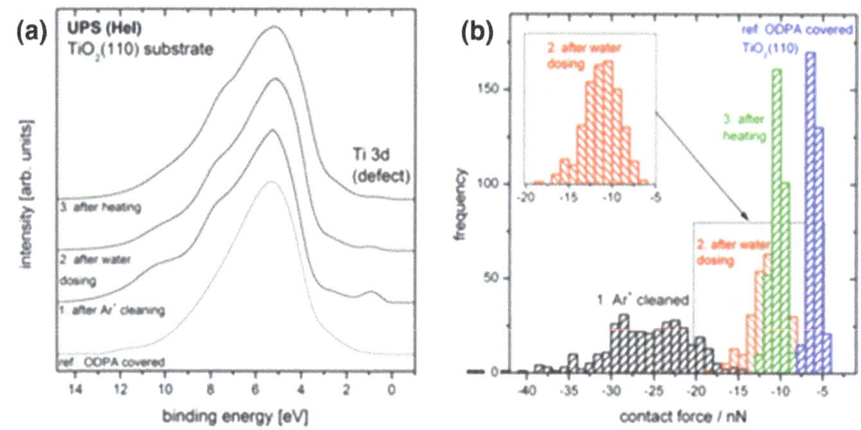

Fig. 11 **a** UPS data of the $TiO_2(110)$ substrate surfaces after different step procedure. **b** Histograms of contact forces for different surface chemistries of the TiO_2–TiO_2 nanocontact. Both figures (**a**, **b**) are reprinted from [12] with permission from Elsevier

UHV AFM Force-Distance Spectroscopy

The histograms with the contact force values extracted from the AFM force-distance curves measurements are presented in Fig. 11b. The $TiO_2(110)$ surface after Ar^+ bombardment displays a broad distribution and also exhibits the highest adhesion force values. Therefore, these high values can be attributed to the as-obtained defective $TiO_2(110)$ surface state. In this case, the strong contact forces measured between these two TiO_2 surfaces can be justified in terms of the occupied Ti $3d^1$ orbital configurations that provide electrons for the formation of additional bonds. Yong et al. [39] thoroughly characterized in a seminal contribution contact forces between TiO_2-based surfaces by means of molecular dynamics simulations. The authors concluded that the subsequent deformation of TiO_2 in the contact area gives rise to defects that would bring about supplementary bonds, together with enhanced adhesion forces. In our approach, TiO_2 defective surfaces are created by means of the cleaning process, namely Ar^+ ion bombardment. Consequently, an analogue mechanism of interaction can be expected. According to DFT modeling, an electron gradient between the surface and the bulk can be expected due to the electron excess originated in the oxygen vacancies delocalized through the surface [40]. Thus, as a consequence of the strong gradient in electron density due to the arising TiO_2 defects, the Hamaker constant is increased. The registered broad distribution of the contact force histogram, Fig. 11b, hints at a chemically inhomogeneous defect-rich TiO_2 surface. The latter would be related to the presence of multiple defects in the interaction contact area. After the water dosing treatment, a noticeable narrower histogram together with remarkable lower adhesion force values could be measured as well. This effect has been tentatively attributed to the decrease of the surface defect density which has already been shown by XPS and UPS, see Figs. 10b and 11a. As previously stated, after thermal annealing of the $TiO_2(110)$ single crystalline surface, the defect density is further decreased. This effect would be accompanied by a narrower contact force value distribution which would definitely hint at a more homogeneous surface composition. Finally, the ODPA-modified $TiO_2(110)$ surface exhibits the smallest adhesion force and the narrowest histogram. This effect is mainly originated from the variations in the surface chemistry, which changes from pure TiO_2 to an aliphatic carbon.

Estimation of Interaction Forces Based on Hamaker Constants

Based on the current literature, the non-retarded Hamaker constant A_{11} expected for the interaction TiO_2/vacuum/TiO_2 should be normally in the range of 150×10^{-21} J [41]. Likewise, for the interaction of (hydro)carbon chains, i.e. CxHy/vacuum/CxHy, the Hamaker constant A_{22} has been estimated to be around 28×10^{-21} J [42]. According to the equation:

$$A_{12} \approx \sqrt{A_{11} \cdot A_{22}}$$

allows to obtain a combined Hamaker constant, A_{12}, for the TiO_2/vacuum/C_nH_{2n} system of 65×10^{-21} J. This result reduces the interaction force by a factor of ~2 in comparison to those values registered for the TiO_2-modified AFM cantilever in contact with the $TiO_2(110)$ single crystalline surface. The latter would be definitely in very good agreement with the measured adhesion forces when the AFM results obtained for $TiO_2(110)$ substrates after the healing of defects (water dosing and thermal annealing) are compared with those registered for the monolayer-coated surface. Therefore, it can be concluded that the rise in the interaction adhesion forces detected for the defect-rich, i.e. Ti(III)-rich, surfaces can be explained by the enhanced donor and electron density in the conduction band of TiO_2.

Analysis of Acid-Base Properties of Single Crystalline Al_2O_3 Surfaces

Single-Molecule Force Spectroscopy (SMFS) appears as a valuable tool to accurately characterize the complex de-adhesion processes occurring between macromolecules and metal oxide interfaces in a corrosive environment induced by well-defined external forces upon a precise control of the fundamental dynamics of desorption processes. This technique is essentially based on the analytical analysis of the force-distance curves which can be obtained from the result of the approach-retraction cycles of a polymer-modified AFM cantilever over a specific substrate.

In the following, a systematic investigation on the interaction forces occurring between aluminum oxide substrates and aliphatic amino, hydroxyl and ether group-

Table 4 High resolution AR-XPS data of the measurements of the Al_2O_3 annealed and etched surface [46] (Table from [15])

Take-off angle/°	Peak	Chemical state	Position/eV	at.% Conc.
20	C 1s	C–C	284.8	11.4
20	C 1s	C–O	286.8	1.1
20	C 1s	C=O	288.9	0.8
20	O 1s	O^{2-}	530.7	39.5
20	O 1s	O–H, OR	532.0	8.8
20	Al 2p	Al^{3+}	74.0	38.4
75	C 1s	C–C	284.8	5.2
75	C 1s	C–O	286.0	1.0
75	C 1s	C=O	289.0	0.5
75	O 1s	O^{2-}	531.2	46.4
75	O 1s	O–H, OR	532.5	7.7
75	Al 2p	Al^{3+}	73.9	39.2

terminated surfaces has been carried out by means of SMFS in an electrolyte environment at pH 6 [43]. These force-distance curve-based studies were based on the variation of the terminal group of a polyethylene glycol-based polymer, which was covalently bound via thiol chemistry to the gold-coated AFM tip. The chemical characterization of both substrates and modified AFM cantilevers as well as the acid-base interactions due to hydroxyl and amino groups was carried out by XPS [44, 45]. The experimental approach presented herein consisted of measuring force-distance curves originated form the approach-retract cycles of a PEG-NH_2, –OH and –OCH_3 functionalized atomic force microscope (AFM) tip over the non-polar single crystalline Al_2O_3(11–20) surface. Interestingly, non-equilibrium peel-off desorption forces which point out to acid-base interactions occurring between the electron donating hydroxyl and amino groups with the Al-ions present in the oxide surface. Desorption forces in the range of 100–200 pN could be measured for the NH_2, OH/Al_2O_3(11–20) combinations.

Non-polar Al_2O_3(11–20) single crystalline surfaces have been chosen since those crystallographic planes exhibit more accessible Al ions at the oxide interface which allows for the formation of strong interfacial bonds with electron donating hydroxyl and amine groups present at the modified AFM cantilevers. Al_2O_3(11–20) single crystals were prepared by acidic and alkaline etching according to the procedure stablished in the experimental section. By means of AFM imaging it could be shown that the surface of the as-treated aluminum oxide single crystal surfaces consisted of atomically flat terraces which were ~600 nm wide. An optimum clean surface is definitely obtained after the final acidic etching process as has been shown by AR-XPS. The surface compositions at two different take-off angles (20° and 75°) are displayed in Table 4.

From the high-resolution XP spectra it could be unambiguously concluded that the Al_2O_3(11–20) is not covered with a hydroxyl monolayer and, consequently Al-atoms should be accessible for the interaction with electron donating groups $–NH_2$ and –OH. The latter is based on the measured OH/O ratio of 15:100.

SMFS Measurements

As previously stated, the functionalized AFM tips consisted of a PEG spacer covalently attached to the tip via thiol-gold chemistry. Thus, the terminating functional groups, namely –OH, $–NH_2$, and $–OCH_3$, would get in contact with the crystalline aluminum oxide surface during the approach-retract cycles of the AFM cantilever. As a consequence of hindrance effects, it is assumed that the density of PEG molecules attached to the surface should be considerably lower than that exhibited by the result of a self-assembled monolayer formed onto a smooth substrate.

Representative force-distance curves registered for the Al_2O_3(11–20) surface with the modified cantilevers exhibiting different terminating functional groups have been displayed in Fig. 12. As expected, the bare Au-coated cantilever exhibited one-single featureless unspecific desorption event as can be seen in Fig. 12a. Interestingly, when the PEG linker is used, the latter acts as a polymer spacer and, consequently, non-specific interactions, due to the gold-coated AFM tip surface, are significantly decreased. Indeed, the plot depicted in Fig. 12b obtained by using the $–OCH_3$-modified AFM cantilever does not exhibit any specific interaction with the Al_2O_3(11–20) single crystalline surface. What it is more, an entropic repulsion can be detected in the approach regime. A typical force-curve registered for the amine-modified AFM tip has been displayed in Fig. 12c. The latter exhibits several non-equilibrium rupture force events indicating that terminating molecules are adsorbed at specific sites on the surface due to strong interactions (most likely coordinative bonding) [47, 48]. Interestingly, OH-terminated AFM cantilevers exhibited analogue results with the presence of multiple rupture force events. Consequently, both electron-donating NH_2 and OH-functionalized PEG gave rise to non-equilibrium rupture force events associated with specific interactions, Fig. 12c, d, in contrast to the plane profile registered for the force-distance curve measured for the $–OCH_3$ groups, see Fig. 12b.

The enhanced performance shown for both NH_2 and OH groups in terms of adhesion force interactions has been tentatively attributed to the fact that the latter terminating moieties can definitely behave as Lewis bases interacting thus with Lewis acidic groups present in the Al_2O_3(11–20) single crystalline surface. These acidic groups in the aluminum oxide surface have their origin in surface Al ions not coordinated by ending oxygen atoms.

The afore-mentioned adhesion results unambiguously indicate that interfacial water adlayers present at Al_2O_3(11–20) surfaces can be removed by the $–NH_2$ and –OH organic Lewis bases present in the modified AFM cantilevers bringing about to coordinative bonds between the latter and the aluminum oxide crystalline. Dashed lines in Fig. 12c, d correspond to the fitting based on worm-like chain model (WLC)

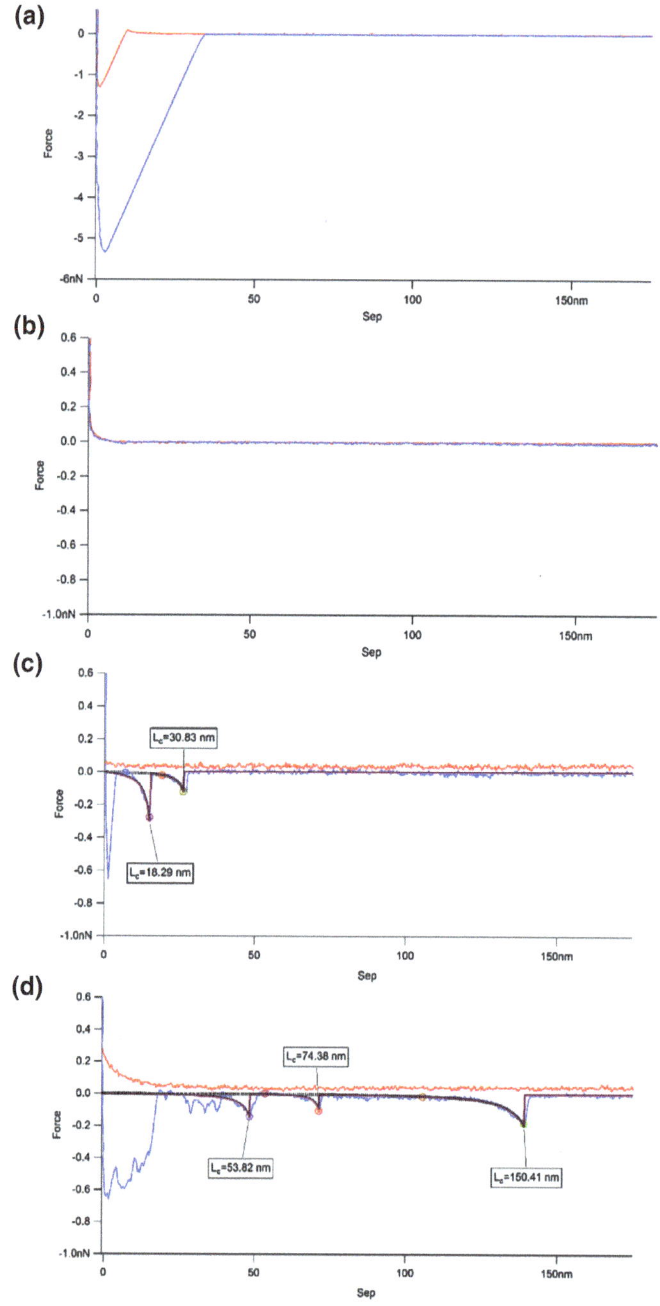

Fig. 12 **a–d** Force distance curves obtained with **a** Au coated cantilever, **b** CH_3O-PEG modified cantilever, **c** NH_2-PEG modified cantilever and **d** OH-PEG-modified cantilever (red: approach-curve, blue: retraction curve, dotted line: fitting according to the WLC model). Reprinted from [15] with permission from Elsevier

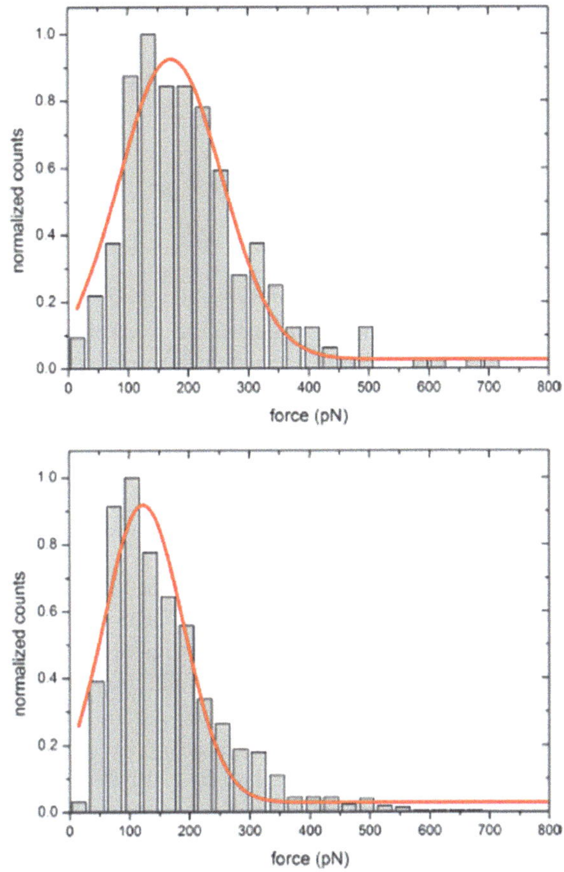

Fig. 13 *Up*: Force histogram PEG–NH_2 versus Al_2O_3(11–20); *down*: Force histogram PEG–OH versus Al_2O_3(11–20). Reprinted from [15] with permission from Elsevier

[48]. This analysis allows to obtain the contour length (Lc) of the tethered PEG molecule. The latter turned out to be is 70 nm. The measured contour length is consistent with the PEG spacer used in the present study. The corresponding force values were then recorded at the rupture point. Accordingly, a thorough set of statistical data was collected for every modified AFM tip. A Gaussian fit was then applied to the adhesion force value distribution, presented in Fig. 13 as histograms, in order to establish the mean force value.

The averaged measured forces (also displayed in Fig. 13) and the calculated desorption energies are presented in Table 5. These results show consistently lower adhesion forces for the ending hydroxyl moieties in comparison to the amino groups.

The reported trend is assigned to the well-known strong interaction between the lone electron pair of the amino group with metal cations, namely Al^{3+} accessible in the Al_2O_3(11–20) surface. Additionally, it could be stated that the adsorption of amino groups at neutral pH would involve a deprotonation of the amino group according to:

Table 5 Adhesion forces PEG-NH_2 and PEG-OH versus Al_2O_3 (table from [15])

	Peak maximum	
	Force (pN)	Error (pN)
−OH	123.7	4.8
$-NH_2$	172.4	4.7

$$Al_2O_3|Al\cdots OH_2 + NH_3^+ - PEG \rightarrow Al_2O_3|Al\cdots NH_2 - PEG + H_3O^+$$

The reported pK_b-value of the amino group is proposed to support this statement. Accordingly, the amino groups in the electrolyte medium would be protonated at $pH = 6$. Thus, the observed acid-base interaction can then be explained in terms of the deprotonation of the terminating amino moieties occurring during their adsorption.

Conclusions

AFM and XPS-based studies allowed for the correlation of surface chemistry and nano-contact forces as well as molecular forces. It has been stated that for semi-conducting TiO_2 the adhesion properties of oxide properties are characterized by the defective structure and the electronic properties. In this case, UHV-based AFM studies in combination with ion beam modification allows for the analysis of high energy surfaces under clean conditions.

Nevertheless, monomolecular films with thicknesses of less than two nanometers lead to a diminishing impact of the underlying oxide film. Additionally, the strong binding of hydroxyls and water adlayers on the TiO_2 surface leads to rather weak adsorption of organic adsorbates such as carboxylic acids. On the contrary, Al_2O_3 surfaces show enhanced acid-base interactions due to the Al^{3+} ion-centers acting as Lewis acids. This property has been analyzed for the adsorption of organophosphonate monolayers as well as amino and hydroxyl-terminated macromolecules. In this case, it can be assumed the molecular monolayers are chemisorbed and rather immobile. Nanoshaving of pre-adsorbed organic monolayers proved to be a versatile tool for the assessment of the adsorbate thickness and the adsorption rate. Based on these results, this technique could also serve as a reliable tool to study the mobility of adsorbates at elevated temperatures in the near future. Single molecule force spectroscopy could also be used as a complementary methodology for the quantification of atomic defects at oxide surfaces.

In general, the afore-mentioned studies show that the formation of molecular organic adsorbates on oxide surface is an important factor that has to be necessarily considered for the better understanding of the interfacial contact-force interactions occurring when particles are in contact.

References

1. Salameh, S., Schneider, J., Laube, J., Alessandrini, A., Facci, P., Seo, J.W., Ciacchi, L.C., Mädler, L.: Langmuir ACS J. Surf. Colloids **28**, 11457–11464 (2012)
2. Weller, H.: Adv. Mater. **5**, 88 (1993)
3. Park, N.-G., van de Lagemaat, J., Frank, A.J.: J. Phys. Chem. B **104**, 8989–8994 (2000)
4. Laube, J., Salameh, S., Kappl, M., Mädler, L., Colombi Ciacchi, L.: Langmuir ACS J. Surf. Colloids **31**, 1288–11295 (2015)
5. Leroch, S., Wendland, M.: Langmuir ACS J. Surf. Colloids **29**, 12410–12420 (2013)
6. Xiao, X., Qian, L.: Langmuir **16**, 8153–8158 (2000)
7. Farshchi-Tabrizi, M., Kappl, M., Cheng, Y., Gutmann, J., Butt, H.-J.: Langmuir **22**, 2171–2184 (2006)
8. Harrison, A.J., Corti, D.S., Beaudoin, S.P.: Part. Sci. Technol. **33**, 526–538 (2015)
9. Kralchevsky, P.A., Nagayama, K.: Langmuir **10**, 23–36 (1994)
10. Gojzewski, H., Kappl, M., Ptak, A., Butt, H.-J.: Langmuir ACS J. Surf. Colloids **26**, 1837–1847 (2010)
11. Kunze, C., Torun, B., Giner, I., Grundmeier, G.: Surf. Sci. **606**, 1527–1533 (2012)
12. Kunze, C., Giner, I., Torun, B., Grundmeier, G.: Chem. Phys. Lett. **597**, 134–138 (2014)
13. Nakamura, R., Ohashi, N., Imanishi, A., Osawa, T., Matsumoto, Y., Koinuma, H., Nakato, Y.: J. Phys. Chem. B **109**, 1648–1651 (2005)
14. Yamamoto, Y., Nakajima, K., Ohsawa, T., Matsumoto, Y., Koinuma, H.: Jpn. J. Appl. Phys. **44**, L511 (2005)
15. Mosebach, B., Ozkaya, B., Giner, I., Keller, A., Grundmeier, G.: Appl. Surf. Sci. **420**, 296–302 (2017)
16. Levy, R., Maaloum, M.: Nanotechnology **13**, 33–37 (2002)
17. Xu, S., Liu, G.-Y.: Langmuir **13**, 127–129 (1997)
18. CasaXPS Version 2.3, Casa Software Ltd (Computer Software). Available from http://www.casaxps.com
19. Mitchell, D.F., Clark, K.B., Bardwell, J.A., Lennard, W.N., Massoumi, G.R., Mitchell, I.V.: Surf. Interface Anal. **21**, 44–50 (1994)
20. Folkers, J.P., Gorman, C.B., Laibinis, P.E., Buchholz, S., Whitesides, G.M., Nuzzo, R.G.: Langmuir **11**, 813–824 (1995)
21. Love, J.C., Estroff, L.A., Kriebel, J.K., Nuzzo, R.G., Whitesides, G.M.: Chem. Rev. **105**, 1103–1169 (2005)
22. Torun, B., Oezkaya, B., Grundmeier, G.: Langmuir **28**, 6919–6927 (2012)
23. Johansson, E.M.J., Plogmaker, S., Walle, L.E., Schölin, R., Borg, A., Sandell, A., Rensmo, H.: J. Phys. Chem. C **114**, 15015–15020 (2010)
24. Henderson, M.A., Otero-Tapia, S., Castro, M.E.: Faraday Disc. **114**, 313–329 (1999)
25. Diebold, U.: Surf. Sci. Rep. **48**, 53–229 (2003)
26. Liu, G.Y., Xu, S., Qian, Y.L.: Acc. Chem. Res. **33**, 457–466 (2000)
27. Yu, J.H., Ngunjiri, J.N., Kelley, A.T., Gano, J.C.: Langmuir **24**, 11661–11668 (2008)
28. Te Riet, J., Smit, T., Gerritsen, J.W., Cambi, A., Elemans, J.A.A.W., Figdor, C.G., Speller, S.: Langmuir **26**, 6357–6366 (2010)
29. Liakos, I.L., McAlpine, E., Chen, X., Newman, R., Alexander, M.R.: Appl. Surf. Sci. **255**, 3276–3282 (2008)
30. Isono, T., Ikeda, T., Aoki, R., Yamazaki, K., Ogino, T.: Surf. Sci. **604**, 2055–2063 (2010)
31. Toofan, J., Watson, P.R.: Surf. Sci. **401**, 162–172 (1998)
32. Barth, C., Reichling, M.: Nature **414**, 54–57 (2001)
33. Gan, Y., Franks, G.V.: J. Phys. Chem. B **109**, 12474–12479 (2005)
34. Takeuchi, M., Martra, G., Coluccia, S., Anpo, M.: J. Phys. Chem. C **111**, 9811–9817 (2007)
35. Tadepalli, R., Thompson, C.V.: Appl. Phys. Lett. **90**, 151919 (2007)
36. Wendt, S., Sprunger, P.T., Lira, E., Madsen, G.K.H., Li, Z., Hansen, J.Ø., Matthiesen, J., Blekinge-Rasmussen, A., Lægsgaard, E., Hammer, B., Besenbacher, F.: The role of interstitial sites in the Ti3d defect state in the band gap of titania. Science **320**, 1755 (2008)

37. Krischok, S., Günster, J., Goodman, D.W., Höfft, O., Kempter, V.: Surf. Interface Anal. **37**, 77–82 (2005)
38. Yim, C.M., Pang, C.L., Thornton, G.: Phys. Rev. Lett. **104**, 36806 (2010)
39. Yong, C.W., Smith, W., Kendall, K.: J. Mater. Chem. **12**, 2807–2815 (2002)
40. Chrétien, S., Metiu, H.: J. Phys. Chem. C **115**, 4696–4705 (2011)
41. Bergström, L.: Adv. Colloid Interface Sci. **70**, 125–169 (1997)
42. Drummond, C.J., Georgaklis, G., Chan, D.Y.C.: Langmuir **12**, 2617 (1996)
43. Mosebach, B., Bayer, F.M., Fels, C.-C., Voigt, M., Oezkaya, B., Pomorska, A., Torun, B., Keller, A., Grundmeier, G.: Surf. Interface Anal. **48**, 1017–1025 (2016)
44. van den Brand, J., Blajiev, O., Beentjes, P.C.J., Terryn, H., de Wit, J.H.W.: Langmuir ACS J. Surf. Colloids **20**, 6308–6317 (2004)
45. van den Brand, J., Snijders, P.C., Sloof, W.G., Terryn, H., de Wit, J.H.W.: J. Phys. Chem. B **108**, 6017–6024 (2004)
46. Kloprogge, J.T., Duong, L.V., Wood, B.J., Frost, R.L.: J. Colloid Interface Sci. **296**, 572–576 (2006)
47. Ozkaya, B., Ozcan, O., Thissen, P., Grundmeier, G.: Langmuir ACS J. Surf. Colloids **26**, 8155–8160 (2010)
48. Valtiner, M., Grundmeier, G.: Langmuir ACS J. Surf. Colloids **26**, 815–820 (2010)

Understanding and Manipulation of Nanoparticle Contact Forces by Capillary Bridges

Hans-Joachim Schmid, Guido Grundmeier, Michael Dörmann, Alejandro González Orive, Teresa de los Arcos and Boray Torun

Nomenclature

ff_c	Flowability index (–)
D	Particle diameter (m)
d	Distance between particle surfaces (m)
f	Frequency (Hz)
F	Force (N)
h	Size of capillary bridge (m)
k	Stiffness of capillary bridge contact (m^{-1})
n	Overtone number
p	Pressure (Pa)
r	Radius of curvature (m)
R	Universal gas constant (8,314 J mol^{-1} K^{-1})
T	Temperature (K)
V_M	Molar volume (m^3 mol^{-1})
V	Volume (m^3)
β	Filling angle of capillary bridge (°)
γ	Surface tension (N m^{-1})
ΔΓ	Dissipation in (Hz)
φ	Relative humidity (–)
λ	Surface enhancement factor (–)
Θ	Contact angle (°)
ρ	Density (kg m^{-3})
σ	Stress (Pa)

H.-J. Schmid (✉) · M. Dörmann
Particle Technology Group, University of Paderborn, Paderborn, Germany
e-mail: hans-joachim.schmid@upb.de

G. Grundmeier · A. G. Orive · T. de los Arcos · B. Torun
Technical and Macromolecular Chemistry Group, University of Paderborn, Paderborn, Germany
e-mail: g.grundmeier@tc.upb.de

S. Antonyuk (ed.), *Particles in Contact*,
https://doi.org/10.1007/978-3-030-15899-6_2

Indices

C	Capillary
eff	Effective
K	Kelvin
max	Maximum
min	Minimum
mod	Modified
P	Pressure related
Pl	Plate
S	Saturation
S	Surface tension related
Sp	Sphere

Abbreviations

BET	Determination of specific surface area according to Brunauer, Emmett, and Teller
FE-SEM	Field emission scanning electron microscopy
FTIR	Fourier transform infrared (spectroscopy)
MD	Molecular dynamics
ODS	Octadecyltriethoxysiloxane
ODT	1-octadecanthiol
PE-CVD	Plasma enhanced chemical vapour deposition
PM-IRRAS	Photoelastic-modulated Fourier transform infrared absorption spectroscopy
QCM	Quartz crystal microbalance
QCM-D	QCM with dissipation monitoring
RH	Relative humidity
SEM	Scanning electron microscopy
DEM	Discrete Element Modeling

Introduction

The adhesion forces between particles are crucial for the handling of powders and bulk solids as they can change the flowability considerably [1–4]. On the one hand these adhesion forces can be desirable, for example in case of granulation, but on the other hand they can be undesirable, e.g. when a freely flowing powder is required. The theory of the principal interacting forces, which in most cases are based on electrostatic, van der Waals and capillary bridge forces, and the force distance relationships been reviewed by Israelachvili in his well-known monograph on "Intermolecular and Surface Forces" [5].

The measurement of interaction forces between microscopic particles attracted much attention in the last years and was reviewed by Butt and Kappl [6]. Butt and co-workers also studied the contact mechanics of TiO_2 microparticles in atmospheres of defined humidity and distinguished the contribution of capillary forces from van der Waals interactions [7]. However, only very little attention had been paid on particle interactions on the nanoscale. Therefore, this work has two major aims:

1. Understanding capillary bridge formation and resulting interparticle forces for nanoscaled particles.
2. Understanding the interaction of surface chemistry, water adsorption and consequently the formation of capillary bridges due to humidity.

Therefore, the studies presented here consider the influence of surface chemistry on the water adsorption isotherms and capillary bridge formation in particle ensembles. Such particle-particle interactions determine the macroscopic flow behavior. To bridge the gap between the interface chemistry of particles and the flow behavior under ambient conditions, a multi-scale approach is proposed as depicted in Fig. 1. Experiments will be performed ranging from the individual particle level (AFM) and small particle ensembles (QCM-D/FTIR) over to macroscopic powders (shear tester). In this context, the combined in situ QCM-D/FTIR experiments will bridge the gap between experiments on an individual particle level and macroscopic shear test. Since the surface chemistry is of outmost importance for such a fundamental study, particle preparation and especially surface modification on site is mandatory.

Specific variation of surface chemistry by means of adsorption of functional organic molecules will facilitate the correlation of macroscopic particle behavior like water adsorption isotherms and flow properties to nanoscopic effects like the presence and structure of adsorbate layers as well as the formation of capillary bridges.

An important aspect of the approach is the detailed experimental analysis and simulation of water adsorbate films and capillary bridges including the formation of rather ice-like or liquid like layers as a function of the surface chemistry of the particles.

In addition, the experimental data obtained by this multi-scale approach will be directly transferred to corresponding simulations facilitating the development of a model allowing for the prediction of inter-particle forces as a function of surface chemistry and environmental factors.

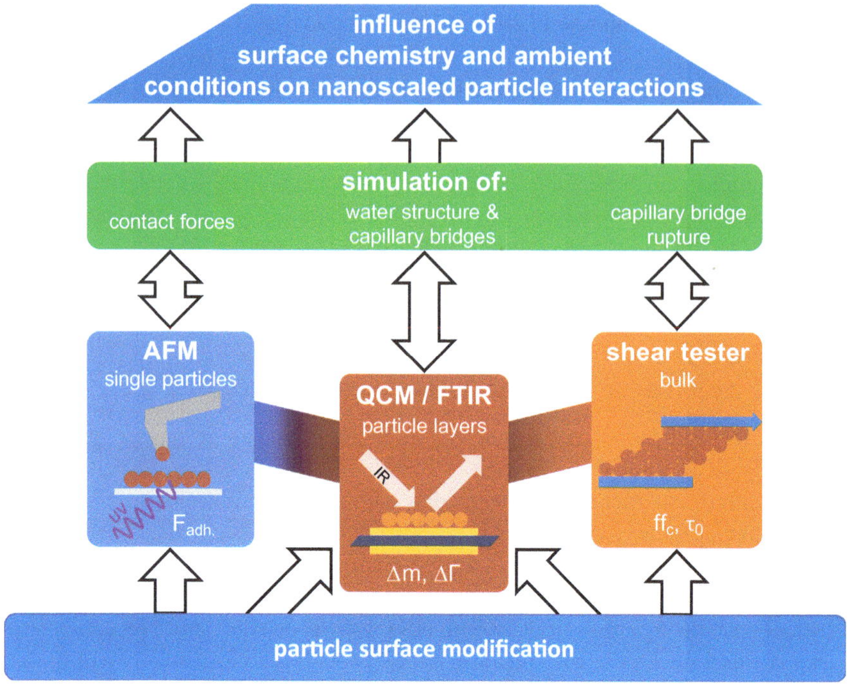

Fig. 1 Structure of the scientific approach bridging the gap between individual contact formation and the behavior of particle ensembles

Experimental Investigations on Water Adsorption and Capillary Bridge Formation

Ambient Water Adsorption on Oxide Surfaces Covered with Organic Monolayers as Studied by In Situ PM-IRRAS

As oxide surfaces are characterised by high surface energies, they spontaneously adsorb organic contaminations from ambient atmospheres [8]. Thereby, relevant adsorbates formed under such conditions are dominated by the interdependence between water adsorption and the adsorbed molecules. Grundmeier et al. have previously studied model contamination layers on TiO_2 based on the controlled adsorption of organic acids [9]. However, the aim of the here presented studies is to analyze the interplay between the structure of model organic adsorbates and the adsorption of water from humid atmospheres on reference oxide surfaces. In this case, nanocrystalline ZnO films on a reflecting Au substrate were used. The latter allowed for the analysis of adsorption processes by means of monolayer sensitive FTIR reflection absorption spectroscopy under grazing incidence. Experimental details can be found in [10].

The interfacial stability of 1-octadecanthiol (ODT) and octadecyltriethoxysiloxane (ODS) monolayers on polycrystalline zinc oxide thin films in presence of high water activities, i.e. 80% of relative humidity (RH), in comparison to low water activities (0% RH) were investigated by means of in situ Fourier transform infrared reflection absorption spectroscopy (PM-IRRAS) [11]. The assumption is that ODS binds covalently to the ZnO surface via S–Zn bonds while organosilanes form water-sensitive hydrogen bonds between Si–OH groups and surface hydroxyls.

The measurements were then performed at 0% RH, after exposure to 80% RH, and after coming back to 0% RH. The PM-IRRAS spectra of the ZnO surface modified with ODT (displayed in Fig. 2a, b) indicated no formation of water layers after exposure to 80% RH and thus a great stability at high water activities. The peak position of the symmetric and asymmetric methylene stretching modes of the aliphatic chain at 2920 and 2850 cm^{-1} are characteristic for all-trans configuration of methylene groups and a high level of molecular ordering [12]. In comparison, the peak positions for the asymmetric and symmetric methylene contributions appeared at 2925 and 2855 cm^{-1} in the PM-IRRAS spectra registered for the ODS-modified ZnO surface (Fig. 2c). Such peak positions indicate a certain disorder in the film [12]. The PM-IRRAS spectrum measured at 0% RH on the ODS-modified ZnO surface showed in the fingerprint region (Fig. 2d) a large peak at 1154 cm^{-1} exhibiting a little hump at 1100 cm^{-1}. These signals are attributed to the longitudinal (LO) and transverse optical (TO) modes of the asymmetric Si–O–Si stretching vibration confirming thus the successful modification of the substrate with a cross-linked siloxane network [13–17]. Moreover, a noticeable increase of the ν(OH) and δ(H_2O) signals at high water activities (80% RH) assigned to simultaneous adsorption of water molecule adlayers and an enhanced surface hydroxylation can be observed in the PM-IRRAS spectrum depicted in Fig. 2d. The extended tail of the H_2O-related stretching vibration peak toward lower wavenumbers is indicative of firmly bound and ordered ice-like water species [10]. These signals can then be mostly attributed to either trapped water molecules between ODS molecule chains or accessible non-blocked active adsorption sites on the ZnO surface as can be seen in Fig. 2f. This tight confinement of water molecules would be responsible for the observed ordering. A broadening of the asymmetric stretching vibration bands corresponding to Si–O–Si groups together with a displacement to lower wavenumbers (Fig. 2d), namely 1100 cm^{-1}, can be observed. While the TO contribution is markedly increased, the LO component has mostly disappeared. Interestingly, the significant changes occurred when the system was exposed to 80% RH, i.e. the adsorption of water molecule adlayers, an enhanced surface hydroxylation, the Si–O–Si peak shifting, and the decrease in the methylene stretching intensities, turned out be fully reversible, as can be observed in Fig. 2d when the RH is reversed to 0%. This reversible process can be justified on the basis of the adsorption of a compact and cross-linked siloxane network via either hydrogen bonding and/or exhibiting just a few covalent bonding points with the ZnO-based pristine surface (Fig. 2e).

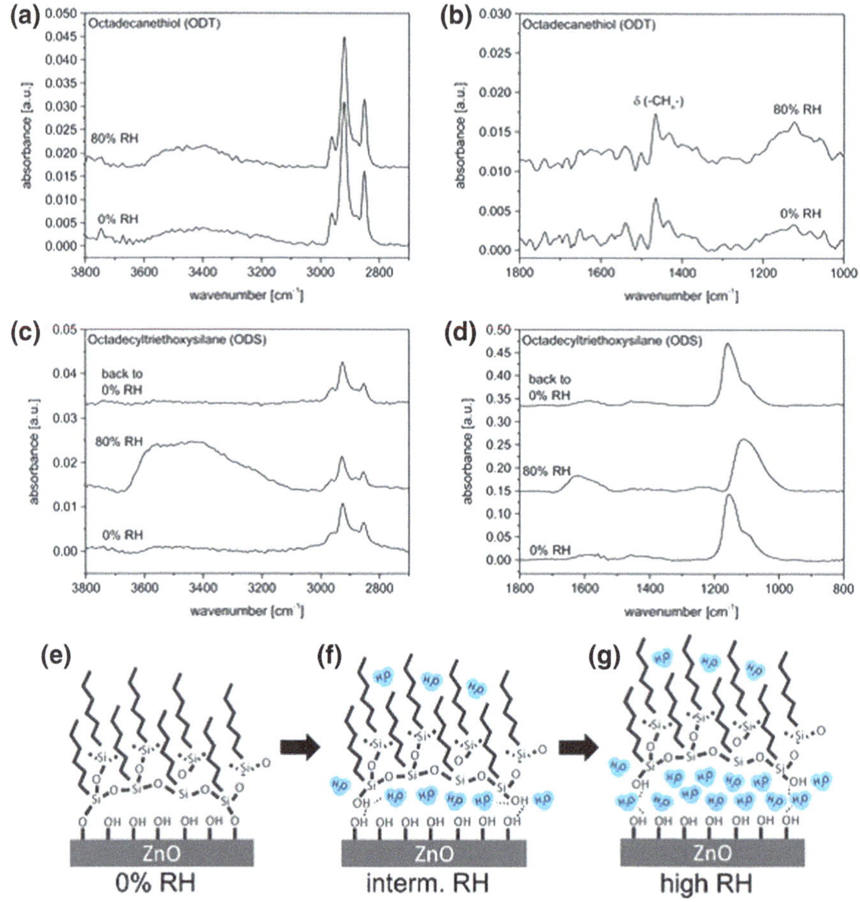

Fig. 2 **a–g** In situ PM-IRRAS spectra of ODT (**a–b**) and ODS (**c–d**) modified surfaces at 0% RH, after exposure to 80% RH and after coming back to dry conditions. Proposed mechanisms of water induced changes at the alkyl polysiloxane film/ZnO interface (**e–g**). Reprinted with permission from [11]

The studied zinc oxide films are semiconducting in nature, however, provide Lewis acid centers as well. It could be shown for this oxide, that the adsorption of adsorbates is affected by the simultaneous adsorption of water from ambient atmospheres. While organothiols act as Lewis bases leading to adsorbates that are not affected by water adsorption, organosilanols lead to adsorbates that are dominantly bound via hydrogen bonding. In this case the water adsorption produces changes in the adsorbate structure. The crosslinked film which is only partially coordinated to the surfaces is mechanically stressed due to the adsorption of water in the nanoscopic free volumes at the oxide/film interface. Overall, the presented results indicate that the structure of an organic adsorbate and the formation of water is highly interrelated.

Functionalization of particles based on organosilane chemistry is thereby strongly affected by the surrounding water activity.

Combined In Situ FTIR Reflection Absorption/QCM Studies on Smooth SiO_2-Surfaces and SiO_2-Particle Layers

Quartz crystal microbalance (QCM) measurements as well as FTIR spectroscopy allow for the in situ analysis of water adsorbate formation under ambient conditions with monolayer sensitivity [18]. FTIR spectroscopy in the mid-infrared region is sensitive to the coordination of the water molecules and enables to distinguish liquid-like and ice-like structures [19].

QCM studies including dissipation analysis (QCM-D) moreover provide information on the viscoelastic properties of the adsorbed film [20]. The hypothesis of our studies was that the transition from ice-like ultra-thin adsorbates to rather liquid-like capillary bridges can be monitored by both techniques as a function of the relative humidity and the surface chemistry of the oxides. For the variation of the surface energy, even perfluorinated organosilane monolayers will be adsorbed on model oxide substrates [21].

In the here presented studies, water adsorption was analyzed on two different substrates: smooth silica-like plasma polymer films and such silica-like plasma films coated with SiO_2 particle layers. Figure 3 illustrates the corresponding sample design:

The QCM-D data presented in Fig. 4a shows the evolution of frequency as a function of humidity. As a consequence of the mass change associated with adsorbed water, a noticeable shifting to smaller frequencies is observed. It should be noted that an appreciable increase on the effective surface area is expected due to the modification of the silica-like plasma polymer film with the SiO_2 particle layer as shown in Fig. 3.

Consequently, with the aim to quantitatively compare the water uptake taking place onto the SiO_2 particle layer with that registered for the smooth silica-like plasma polymer, this enhancement in the effective surface area must be necessarily considered. By taking into account simple geometric considerations described in detail in [21] this surface roughness factor can be calculated.

Then, with the aim to normalize the registered QCM-D data with the effective surface area, all the frequency shifts have been divided by a surface increase factor,

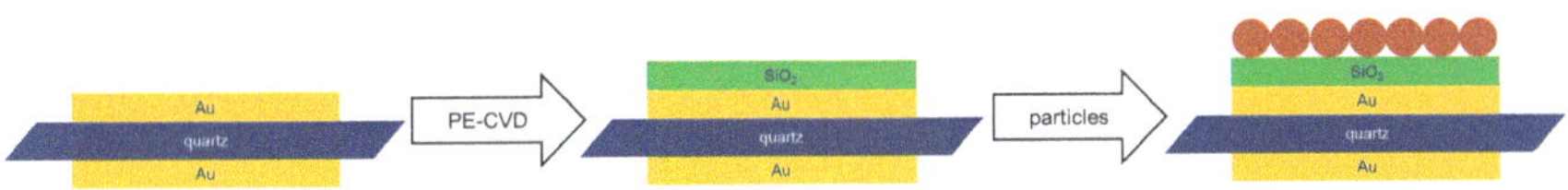

Fig. 3 Schematic of sample preparation procedure. After cleaning, the samples were coated with a thin SiO_2 film by means of PE-CVD. Subsequently, the sample was coated with SiO_2 particles with a diameter of about 250 nm (in analogy to [10])

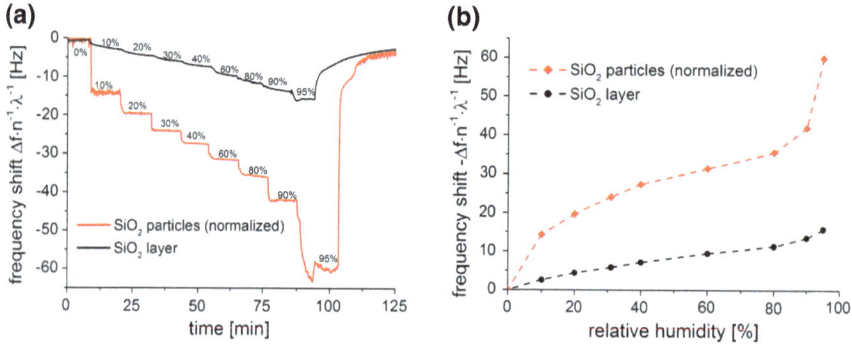

Fig. 4 QCM-D measurement (**a**) of a smooth SiO_2 plasma polymer coated crystal (black) and the particle layer coated sample (red). Frequency shifts were divided by the order n of overtone observed and the surface increase factor λ. The surface increase factor for the correction was found to be 5.3 based on the SEM and BET data. Adsorption isotherm (**b**) of water on the SiO_2 like plasma polymer (black) and SiO_2 particle covered sample (red). Although the isotherm has been derived from the area corrected QCM-D data the total amount of water uptake is significantly higher than on the reference sample. This indicates that additional water is absorbed forming capillaries in the contact regions (Reprinted with permission from [11]—Published by the PCCP Owner Societies.)

λ, calculated according to [10]. Before carrying out the QCM-D measurements, the SiO_2 nanoparticle-surface coverage has been determined by FE-SEM measurements. For the experiments presented here, a surface coverage of 1.1 monolayer could be estimated. By also considering the BET registered of SiO_2 particles, a surface increase factor of $\lambda = 5.3$ has been measured.

The water uptake exhibited by the SiO_2 particle layer is noticeably higher than that registered for the smooth plasma polymer film, as can be deduced from the comparison of the corrected frequency shifts displayed in Fig. 4a. This additional adsorbed water is expected to be involved in the formation of capillary bridges in contact areas between particles and particle-to-substrate. The QCM-D data depicted in Fig. 4a allows us to obtain a water adsorption isotherm. For statistical purposes the frequency shift values corresponding to every relative humidity plateau step in Fig. 4a were averaged. These mean values have been subsequently plotted versus RH and displayed in Fig. 4b.

Three different regimes can be distinguished in the water-adsorption isotherm of Fig. 4b for the SiO_2 nanoparticle-modified substrate. For the segment up to 20% of RH a pronounced slope is observed. As can be observed in Fig. 4b a significant amount of the total water layer has already been adsorbed at this point. A highly favored water adsorption can then be concluded on this hydrophilic surface. The second section, from 20 to 80% RH, exhibits however a smoother transient pointing out to an almost saturated surface. For RH > 80% a steep slope is again observed which can be attributed to the formation of water-adsorbed multilayers together with capillary condensation taking place on the SiO_2 nanoparticle layer. As a control reference, the smooth plasma polymer substrate shows a smooth slope for water uptake which spreads out almost linearly over the whole RH range.

The dissipation data collected for both surfaces have been displayed in Fig. 5a. An almost linear increase in dissipation has been registered for the SiO_2 particle layer until a maximum in dissipation has been reached for the segment of 60–80% RH. An analogue effect has been reported by Dybwad [22] and Du [23]. Indeed, it has been proposed that contact forces taking place between particles and the resonator increases as a consequence of the occurrence of capillaries. This maximum in dissipation should be subsequently related to an evolution from elastic to inertial loading. A similar mechanism is tentatively proposed to further explain our dissipation results. Accordingly, capillaries would be mostly originated in a range of RH between 60 and 80% giving rise thus to a change in the substrate-to-particle contact stiffness. The high figures registered for water uptake at large RH (e.g. >80%) can be related to capillary growth due to the ongoing condensation. As a matter of comparison, the smooth plasma polymer surface does not show any maximum, but a reasonable linear increase in dissipation. Consequently, dramatic modifications of the mechanical properties dependent on RH can be ruled out in this case.

By means of an experimental setup explained in further detail in [11], simultaneous collection of QCM-D data together with the corresponding FT-IR spectra as a function of the RH have been carried out. A reference measurement using a densely packed octadecanethiol (ODT) self-assembled monolayer (SAM) onto a gold-coated quartz crystal has been employed for these FT-IR experiments. As a consequence of its high hydrophobicity, very little amount of water can be adsorbed then on the ODT SAM as verified by QCM-D. It has been found for the SiO_2 particle layer an increase of the OH peak (~3300 cm^{-1}), Fig. 5b, for growing RH corroborating thus the QCM-D data which previously showed an enhanced water uptake on the particle-modified substrate. Similarly to that observed in the QCM-D experiments,

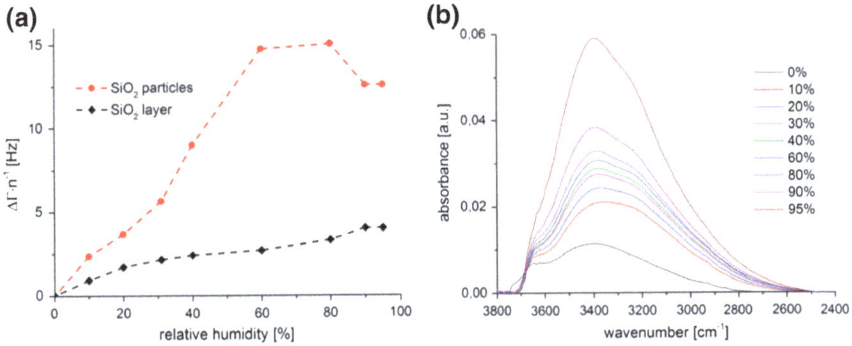

Fig. 5 Evolution of dissipation quantified by $\Delta\Gamma$ during the experiment (**a**). FT-IR data of the SiO_2-particle coated QCM-crystal (**b**). Presented spectra where measured simultaneously to the QCM-D data. All spectra are referenced to an ODT SAM covered QCM-crystal measured under identical conditions (Reprinted with permission from [11]—Published by the PCCP Owner Societies.)

the intensity of the OH peak for the smooth plasma polymer surface is significantly lower.

Most importantly, the FT-IR spectrum collected for the particle-modified substrate at 0% RH exhibits a broad peak centered at ~3400 cm^{-1} related to "free" or "dangling" OH groups of the surface silanols and/or of the adsorbed water monolayer [24, 25]. In addition, the presence of a characteristic hump at ~3700 cm^{-1} can also be inferred from Fig. 5b. This shoulder can be assigned to a single monolayer of water molecules adsorbed at the silanol groups of the silica surface [26] and essentially vanishes when the substrate is exposed to any RH > 0%. The subsequent adsorption of water multilayers observed for high values of RH is related to the occurrence of the "ice-like" peak at ~3200 cm^{-1}. An even more detailed analysis of the water structures is provided in [10].

The comparison of bare high energy SiO_2 surfaces and the same surface covered with perfluorinated organosilane monolayers showed that the behaviour of particle layers already at medium relative humidity is dominated by the contact regions of the particles [21]. The gas-phase adsorption of organosilanes on particle layers revealed that the contact regions remained adsorbate-free due to the confined dimensions of the contact region which hindered the adsorption of the organosilanes [21].

In Situ AFM Studies of Contact Forces

The measurement of interaction contact forces between individual nanoscopic and microscopic particles under ambient conditions, mainly based on atomic force microscopy (AFM) methods, attracted a great deal of attention over the last few years and were extensively reviewed [7].

In a very recent contribution by Ciacchi et al. a comprehensive study dealing with the impact exerted by solvation, capillaries and dispersion forces on the interactions taking place between TiO_2 particles of less than 10 nm [27] is reported. Both the structure of adsorbed water layers and the surface roughness strongly influence the RH dependent interparticle forces. However, the influence of UV-light exposure was not considered in these studies. A new experimental setup was designed allowing for simultaneous control of both relative humidity and UV irradiation during AFM imaging and force distance spectroscopy. The cell (Fig. 6) features a sample compartment which can be purged with gas of the desired composition and relative humidity.

Both the temperature and relative humidity are measured within the cell in direct proximity to the specimen. Moreover, the particles deposited on a special UV-transparent glass slide could be irradiated with UV light from underneath to clean and activate the surface.

Nanosized colloidal probes were prepared by an annealing method similar to the procedure proposed by Huttl et al. [29]. Briefly, as received Si based cantilevers were initially cleaned by means of RCA-1 cleaning. Subsequently, cleaned probes were annealed at 1150 °C for approximately 1 h under ambient atmosphere resulting in a

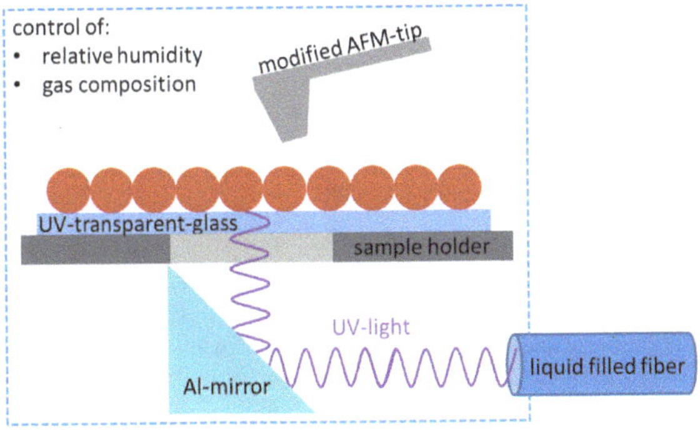

Fig. 6 Schematic of the new experimental setup allowing for AFM imaging and force distance spectroscopy under environmental control and UV irradiation [28]

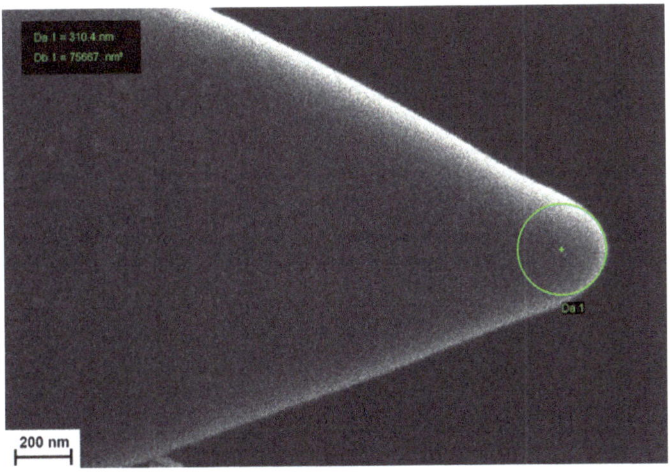

Fig. 7 High resolution FE-SEM image of an annealed AFM tip featuring a diameter of 310 nm [28]

blunt tip with spherical geometry and a typical diameter of ≈300 nm as characterized by high resolution FE-SEM (see Fig. 7).

In addition, atomic layer deposition was performed to cover the cone-like tips and smooth plane silica substrates with an ultra-thin TiO_2-like film (film thickness: about 30 nm).

Both the plane substrate and the cantilever were annealed at 500 °C for 5 h in order to trigger the crystallization process from the as received amorphous phase to an anatase crystalline phase. The experiment consisted of force-distance spectroscopy without as well as with UV-light exposure, each at various relative humidity values. The as-obtained results are presented in Figs. 8 and 9 [28].

It was found that in the absence of UV-light the contact force increased immediately in the presence of humidity in the system (Fig. 8-left, red box). In addition,

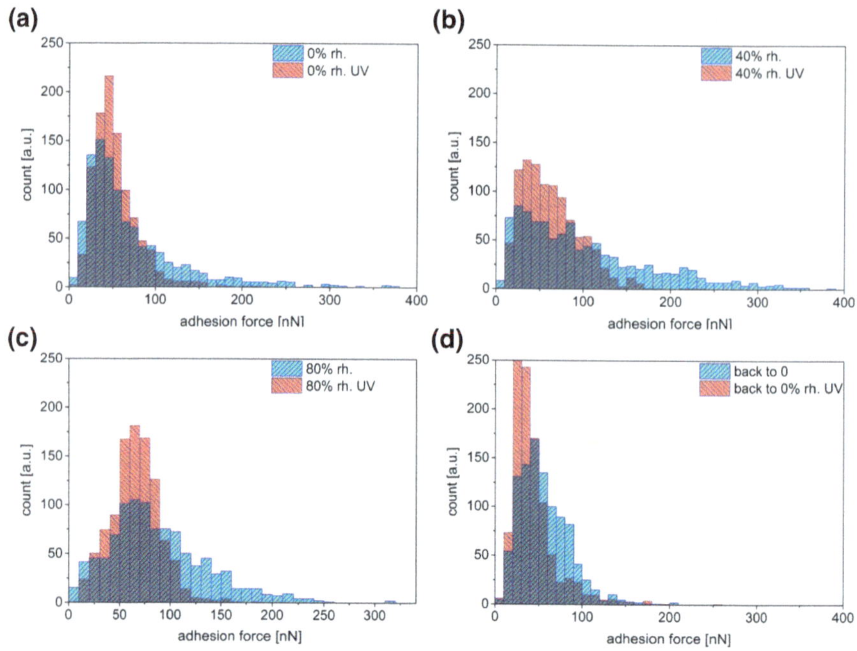

Fig. 8 Force histograms of the experiments performed on an anatase sphere wall contact at different relative humidity values both with (violet) and without (blue) UV-light exposure [28]

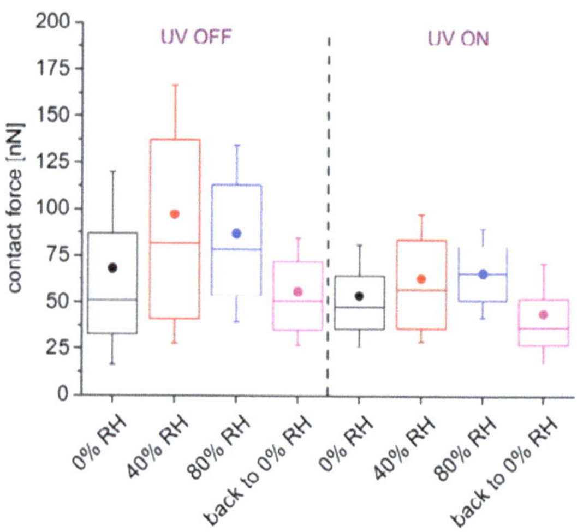

Fig. 9 Boxplot of the histograms presented in Fig. 8. Data sets acquired without the influence of UV-light (left) and during ongoing UV-light irradiation (right)

a significant broadening of the histogram was noticed at 40% RH indicating many different contact events. This behavior could be attributed to the formation of capillary bridges with a broader size distribution. This is further supported by the fact that the distribution narrowed considerably at higher relative humidity values (Fig. 9-left, blue box) due to the fact that the critical radius predicted by the Kelvin equation again surpassed the length scale of the surface roughness. As a result, instead of multiple smaller capillaries, one single, much larger capillary with more uniform properties was formed. Moreover, the studied contacts also exhibited the characteristic decrease of the contact force at high water activities as predicted by the analytical solution. Presented results are further supported by the fact that the experimentally derived values of the contact force are found to feature the same order of magnitude as the theoretical values as predicted by the analytical solution. However, the experimental values are generally slightly lower than expected which again is attributed to the roughness effectively reducing the overall force due to the lower effective contact area.

Force histograms acquired in presence of UV-light were found to exhibit a narrower distribution, as can be seen in Fig. 9-right. This can be explained with the assumption that surface contaminations were oxidized and desorbed upon UV-light exposure, leading to a more uniform surface chemistry. Moreover, water desorbs from the surface during UV-light exposure. This leads to a diminishing contribution of capillary bridges.

Consequently, the contact events tended to be much more similar resulting in a significantly narrower force distribution. In addition, the histograms mean value was slightly shifted toward higher values in the case of dry and very humid (80% RH) conditions. The histogram at 40% lacked the many multi capillary events present in the dataset without UV-light exposure. This finding is explained by the photooxidative desorption of water and organic contaminations on TiO_2-surface in the presence of UV-light [30]. A similar, but less pronounced trend was also observed in the high humidity region.

In summary, the here presented in situ AFM experiment allows for the analysis of nanocontact forces under controlled ambient conditions. The strong influence of the UV-light exposure on the capillary forces could be demonstrated.

Bulk Powder Flow

Shear experiments on commercially available TiO_2 nanoparticles (Aeroxide P25, company of Evonik, Hanau) with a primary particle diameter of about 20 nm were performed on a Jenike shear tester. The experiments were conducted inside a glove box with controlled relative air humidity. Inside the glove box there is a humidity sensor which measures the actual relative humidity and two feed streams of dry and humid air, respectively, are controlled to obtain the target relative humidity. The powder and all devices, which are needed for the experiments, are located inside the glove box in order to maintain constant conditions throughout the experiments. A moisture equilibrium is established between the air in the glove box and the powder by placement in a container with a porous bottom and an air flow passing through the powder. After 10 h of humidification the moisture in the powder was found to be in equilibrium with the air humidity. Therefore, this humidification procedure was used in all experiments.

The experiments at a set relative humidity are reproducible, as it can be seen exemplarily in Fig. 10 for a relative humidity of 55%. The experiments were performed three times at each shear level for three shear levels at one pre-shear stress. The average pre-shear stress of all nine experiments is shown in Fig. 10 as a red square with the standard deviation as error bar. The standard deviation is so small that it cannot be seen. The shear stress for each experiment is shown as a blue diamond and it can be seen that the measured stresses coincide very well, showing a good reproducibility.

During the experiments, an increase of the bulk density of the powder could be determined when it was used repeatedly, as it can be seen in Fig. 11a. Due to the external forces during the shear experiments, the powder gets compacted and the bulk density increases. This compaction cannot be reversed to obtain a defined state of the powder with any mechanical method. Of course, this increase in bulk density also influences the measured flow properties as presented in Fig. 11b. At the beginning and at a very low relative humidity, the powder has a flowability index

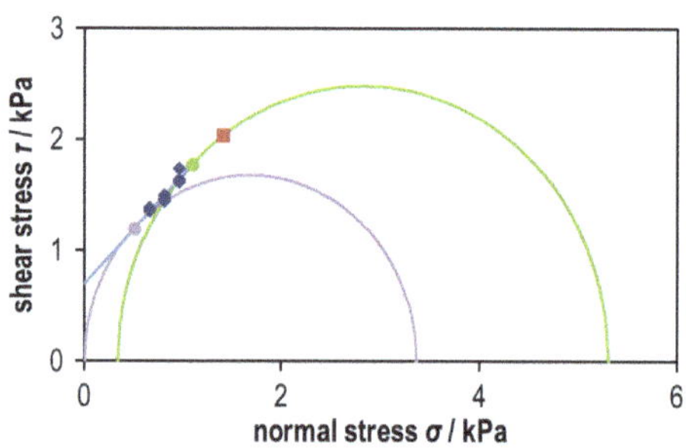

Fig. 10 Yield locus of TiO_2 nanoparticles at a relative humidity of 55%

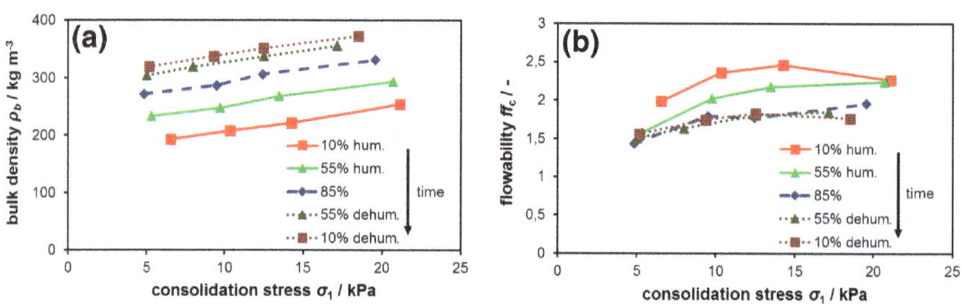

Fig. 11 Bulk density (**a**) and flowability index (**b**) depending on consolidation stress and number of experiments

ff_c of about 2–2.5, which corresponds to a strongly cohesive behavior [2]. This is an expected result caused by the small particle size and the corresponding high ratio between adhesion and gravitational forces. With higher relative humidity and a higher number of performed experiments, the flowability decreases even more. Primarily, the increasing relative humidity causes higher adhesion forces due to capillary bridges. Unfortunately, this behavior is masked by the effect that the higher bulk density causes a higher number of contact points between the particles and therefore leads to a more cohesive behavior as well. The powder then shows a very cohesive behavior with a flowability index of about 1.5. When the relative humidity is decreased subsequently, the flowability index is more or less stagnating which reflects the two—now compensating—effects of reduced capillary forces and increasing bulk density, respectively. As only the effect of the humidity shall be examined conditioned powder with a defined bulk density has to be used. The use of new material for each experiment is not advisable due to ecological and economic reasons. Particularly it cannot be ensured that powder properties (i.e. bulk density) are identical for each experiment and therefore, reproducibility would be not sufficient either.

Follow-up shear experiments will be performed using a Schulze ring shear tester. Due to the rotatory movement, the powder can be sheared as long as it is needed to ensure stationary behavior. This is expected to result in a stationary bulk density as well, which means that performing shear experiments will not lead to a further compaction.

Simulation of Capillary Bridges

If particles are in contact and a liquid is present capillary bridges are formed at the contact points between the particles and they induce an adhesion force which is, in general, stronger than other adhesion forces, like van der Waals forces or electrostatic forces. The liquid can either be added to the powder separately or alternatively capillary condensation occurs when humidity of the air condenses into the gap between particles and the liquid forms a meniscus there. In either case the meniscus induces an adhesion force due to the pressure difference between liquid and gaseous phase and due to the surface tension on the liquid-gas interface, respectively. Especially for nanoscale particles capillary condensation is important.

Method

The method presented here allows numerical calculation of capillary bridges in case of condensing conditions between two perfectly smooth, spherical particles (Fig. 1a) or a spherical particle and a plate (Fig. 1b), respectively. Only rotationally-symmetrical capillary bridges are considered and therefore, also the particles have to be rotationally-symmetrical with respect to the rotation axis of the capillary bridge.

This assumption allows the two-dimensional modeling of the meniscus shape due to the symmetry. The particles are described by the particle diameter D and the contact angle Θ, the plate is described only by the contact angle Θ. The distance between the surfaces of the two objects is defined by the parameter d. All parameters can be set independently from each other. As this model is based on condensing conditions, the ratio of the actual vapor pressure p to the saturated vapor pressure over a planar surface p_s has to be considered, too. In the case of water as liquid phase and air as continuous phase this ratio p/p_s can be interpreted as relative humidity φ.

Early works on capillary bridges were done by Haines [31], Fisher [32], Orr [33] by solving the corresponding elliptical integrals. This allowed the determination of the forces of a capillary bridge. Schubert [34] calculated the forces for different configurations regarding particle shape, particle size, distance, and contact angle and represented his results in dimensionless diagrams. Hence, they can be used for various particle sizes and fluids. In all cases the capillary forces were determined depending on the bridge size, characterized by the so-called filling angle β (compare Fig. 12). However, in the case of condensing conditions, the filling angle cannot be determined directly from the existing relative humidity. Furthermore, the solutions cannot be adapted to parameter configurations that were not covered by the authors. However, these results are used to validate our simulation method for a variety of parameter sets (cf. section "Comparison to Classical Theory").

Other modeling approaches that are often used make assumptions regarding the shape of the meniscus. The exact shape of the meniscus can be described as a nodoid [33, 35, 36] but for an easier calculation the profile is often assumed to be a circle, i.e. one curvature radius is constant at every point of the surface. This assumption

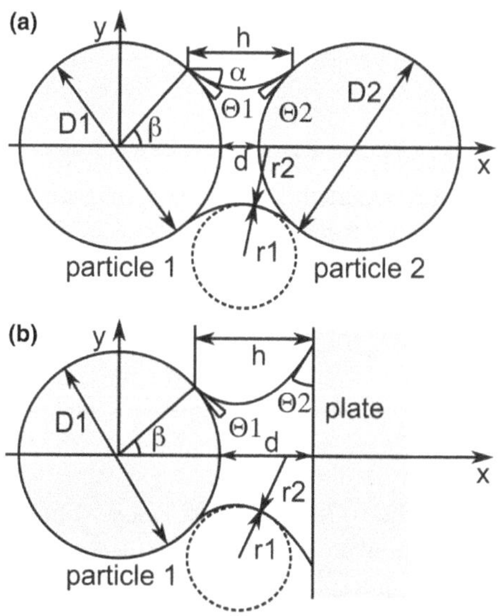

Fig. 12 Sketch of capillary bridge between **a** two spherical particles and **b** a sphere and a smooth plate

was developed for micrometer sized particles, and therefore, one must check if it can be used also for the determination of capillary bridges between nanoscale particles.

Pakarinen et al. [37] developed a method to calculate numerically the exact meniscus shape between an AFM tip and a planar surface based on the Young–Laplace equation. As only contacts to planes are considered, it is not possible to determine the forces between two particles which is of great interest to characterize bulk properties. This method will be extended here to determine the volume and the forces of capillary bridges between arbitrary sphere–sphere or sphere–plate contacts with regard to relative humidity.

Another approach is to calculate the energy-minimal surface of the liquid phase and to determine the resulting force as the derivative of the energy [35]. With this approach it is even possible to model nonsymmetrical bridges, but the volume has to be prescribed and the capillary force cannot be determined dependent on the relative humidity in thermodynamic equilibrium directly. Lambert et al. [38] showed that both approaches, based on the Young–Laplace equation and the minimization of energy, respectively, are equivalent.

Both aforementioned modeling approaches treat the capillary bridge as a continuum. In contrast to this assumption it is also possible to use molecular dynamics and calculate the interactions between all molecules [39, 40]. Here the molecular properties of the materials have to be known in contrast to macroscopic properties in continuum mechanical approaches. The computational costs are very high, so that only very small particles with diameters up to about 10 nm can be described with reasonable effort. Furthermore, this simulation method is not deterministic, so the results show stochastic fluctuations, and they have to be evaluated statistically. Nevertheless, section "Ultrasmall Particles" will show some comparison of MD simulations with our continuum fluid model.

Our approach assumes a continuous fluid phase forming the liquid capillary bridge by condensation. Therefore, the overall surface curvature is identical to the Kelvin radius which can be calculated depending on the relative humidity by

$$r_K = \frac{\gamma V_M}{RT \ln\left(\frac{p}{p_S}\right)} \tag{1}$$

if following requirements are valid [41]: (1) the liquid phase is in thermodynamic equilibrium with the gaseous phase and (2) gravity can be neglected. The second requirement is valid for sufficiently small capillary bridges and consequently can be assumed valid for nanoscaled particles in general [42]. Thermodynamic equilibrium is assumed as only stationary capillary bridges are calculated in this model and the transient state is not considered. The overall curvature radius (i.e. the Kelvin radius r_K) is determined by the two curvature radii r_1 and r_2 with

$$\frac{1}{r_K} = \frac{1}{r_1} + \frac{1}{r_2} \tag{2}$$

The center of the curvature radius r_2 is on the rotation axis of the capillary bridge, because the capillary bridge is rotationally symmetrical. Furthermore, both radii are perpendicular to the capillary bridge surface and they have opposite signs corresponding to a convex and concave curvature, respectively.

As the surface of the meniscus is curved, the pressure inside the capillary may be different from the pressure of the gaseous phase and the pressure difference, the so-called Laplace pressure [5], can be calculated by

$$\Delta p = \frac{\gamma}{r_K} \tag{3}$$

Usually the pressure inside the capillary is lower than the pressure in the gas phase and hence an adhesion force is induced between the particles by the capillary pressure force F_P. Another force component arises from the surface tension of the liquid-gas interface with the surface tension force F_S being attractive in case the wetting angle is smaller than 90°. These two forces sum up to the total capillary force F_C. If the shape of the meniscus is known the capillary bridge force can be calculated at every cross-section of the meniscus. The most straightforward method however is to calculate the force at the neck of the capillary bridge by

$$F_C = F_P + F_S = r^2 \Delta p \pi + 2 r \gamma \pi cos(\alpha) \tag{4}$$

The radius r is the radius of the meniscus at this cross-section and α is the angle the interface deviates from normal direction in this cross-section. The first term on the right-hand side of Eq. (4) represents the capillary pressure force and the second term represents the surface tension force, respectively.

In order to determine the exact bridge shape the meniscus is composed of many discrete points that are iteratively determined and interpolated linearly: At the beginning a filling angle β of the liquid bridge at the first particle is chosen and the first point is set on the surface of the first particle, corresponding to this filling angle. The slope of the meniscus in this point can be determined with the curvature of the particle and the contact angle. If the slope is known the next point of the meniscus can be calculated. In the next step, the slope in the new point has to be determined. Therefore, the radius r_2 is computed as the perpendicular distance of the point to the rotation axis and the radius r_1 is computed with Eq. (2) under consideration of the Kelvin radius. The radius r_1 is connected to the curvature of the meniscus with

$$r_1 = \frac{\left[1 + \left(\frac{dy}{dx}\right)^2\right]^{3/2}}{\frac{d^2y}{dx^2}} \tag{5}$$

which allows the determination of the slope of the meniscus surface in this point. Afterwards, the position of the next point can be determined. This procedure is continued until either the second particle is reached or it is not possible to reach the second particle anymore. In case of a contact or intersection of liquid bridge and particle surface the contact angle between meniscus and particle is calculated

Fig. 13 Sketch of the iteration sequence (Reprinted with permission from [43].)

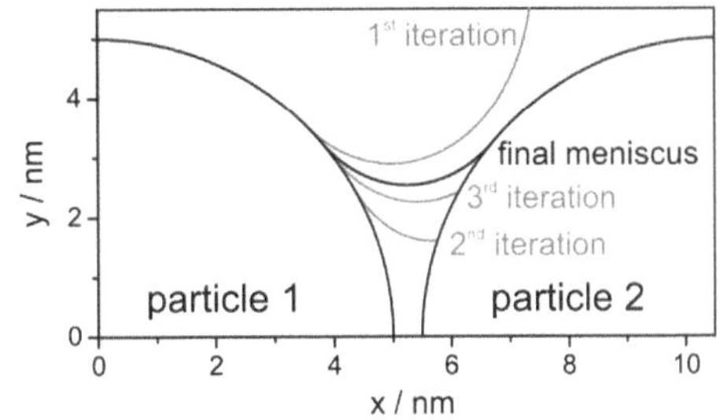

at the contact point and compared to the desired contact angle given by the material properties. If they agree within numerical precision the result is stored. If this is not the case, the filling angle on the first particle will be varied and a new meniscus is determined iteratively until there is an agreement between the simulated and the desired contact angle. A schematic of the iteration sequence is shown in Fig. 13. The grey lines show menisci calculated at incorrect filling angles and the black line is the correct capillary bridge.

When the meniscus is determined, the forces and the volume of the capillary bridge are calculated. As additional second convergence criteria force and volume of an iteration step are compared to force and volume of the previous iteration step, respectively. If any of the relative differences is greater than a chosen threshold another iteration of the calculation will be performed.

Comparison to Classical Theory

As a first step the results provided by this simulation were validated with the exact calculated results by Orr et al. [33], Schubert [34] and Pakarinen et al. [37].

Figure 14a shows the very good agreement between the simulation presented in this paper and the calculation by Orr et al. for a sphere-plate configuration with

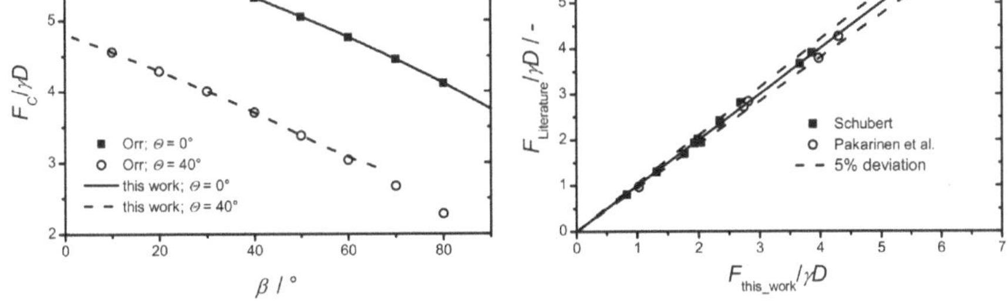

Fig. 14 Comparison of the capillary force from the presented method with the results **a** by Orr et al. [33] and **b** Schubert [34] and Pakarinen et al. [37] (Reprinted with permission from [43].)

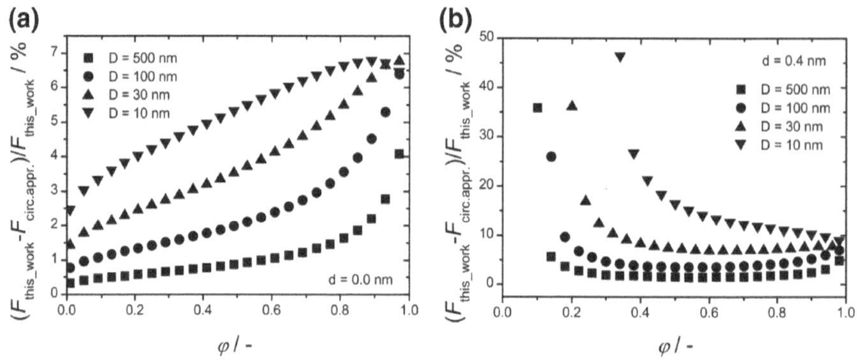

Fig. 15 Relative deviation of capillary forces obtained with circular approximation $F_{\text{circ.appr.}}$ and the full shape simulation of this work $F_{\text{this\_work}}$, respectively. Two spheres of identical diameter D, contact angle $\Theta = 0°$ at distance **a** $d = 0$ nm and **b** $d = 0.4$ nm (Reprinted with permission from [43].)

deviations in all cases of less than 1%. A comparison with the results by Schubert and by Pakarinen et al. for various sphere-sphere and sphere-plate configurations in Fig. 14b shows deviations smaller than 5% which is a very good agreement considering reading accuracy from available graphical data. Consequently, the new method shows very good agreement with classical, macroscopic solutions.

In order to reduce the calculation complexity several approximations and simplifications are used frequently in literature. The most popular assumption is that the profile of the capillary bridge can be described as a circle and hence the curvature radius r_1 remains constant for every point of the meniscus profile. In that case the capillary force can be calculated depending on the filling angle analytically. The forces between two spherical particles were calculated with this approximation with equations given by Butt and Kappl [44] and compared to the solutions of this work. The difference in the total capillary force between both methods is shown in Fig. 15. For relatively big particles in direct contact, e.g. with a diameter of 500 nm or more, the difference between both methods is small, in this case under 5%. This substantiates again, that the circular approximation might well be used for macroscopic particles. However, if the particle size decreases or the distance increases only slightly the difference grows especially at relative humidities near the minimum humidity for capillary bridge formation. In these examples the deviations might become even larger than 40%. It should be noted, that the approximation originally was developed for capillary bridges between micron sized particles where one curvature radius is much greater than the other one. In this case this assumption is reasonable. But for nanoscale particles both curvature radii are at about the same order of magnitude. Thus, it is not possible that both, the Kelvin radius and one curvature radius, can be constant while the second curvature radius varies. The deviation increases with a greater distance as the length of the capillary bridge increases and hence the difference between the real profile and an assumed circular profile is larger. Furthermore, the capillary bridges calculated with this work can exist at a longer distance assuming the same humidity, compared to using the circular approximation. The distance of

0 nm was chosen because it is used very often if particles are in contact. The distance of 0.4 nm is the classical distance for calculations of the van der Waals force and is still extremely small and could be easily caused e.g. by a small roughness. Despite the question, which is the correct minimum contact distance, Fig. 15 clearly shows that the circular approximation should not be used for nanoscale particles.

Interparticle Capillary Forces for Capillary Bridges at Rest

For capillary bridges under condensing conditions of course relative humidity φ is one of the most important parameters for the capillary force. This humidity determines the curvature of the capillary bridge in thermodynamic equilibrium with the environment. This curvature in turn determines the capillary pressure and therefore influences the capillary forces as well as the liquid volume of the bridge. With increasing humidity, the volume of the meniscus is continuously growing by successive condensation of humidity to the capillary bridge. Furthermore, the Kelvin radius and hence the surface radius of curvature and the bridge cross section are increasing as well.

At increasing humidity two counteractive effects must be considered: The increasing radius of curvature will lead to a smaller capillary underpressure inside the bridge leading to a decrease of the capillary force. However, this pressure is acting on an increasing bridge cross section leading to a larger pressure force F_{p} (comp. Eq. (4)). Simultaneously the surface tension force F_{S} is increasing as well due to the increasing perimeter of the capillary bridge. As shown in Fig. 16, this leads to an almost constant overall capillary force between two particles in close contact, i.e. $d = 0$ nm. However, only a marginal increase in the particle distance to $d = 0.2$ nm leads to

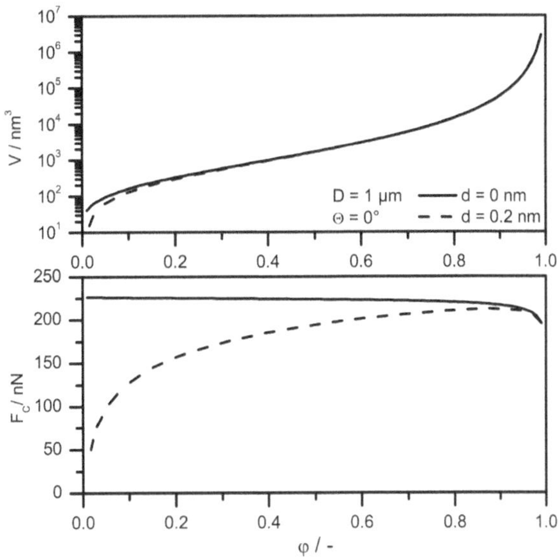

Fig. 16 Volume V (top) and capillary force F_C (bottom) between two identical particles of diameter $D = 1$ µm and a contact angle $\Theta = 0°$. Distance of particles d = 0 nm and 0.2 nm respectively (Reprinted with permission from [45].)

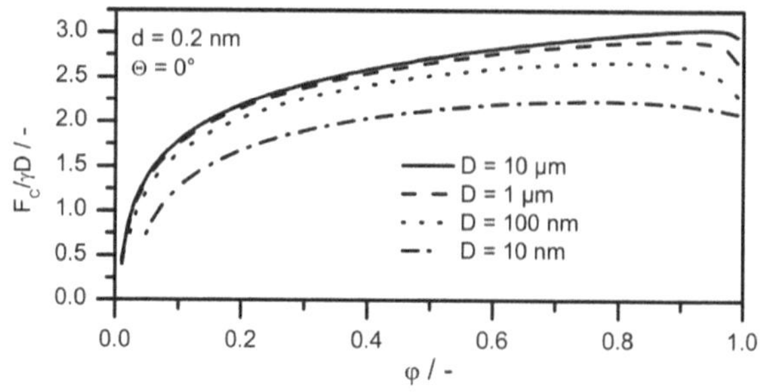

Fig. 17 Dimensionless capillary force between two identical spherical particles with different diameters D (Reprinted with permission from [45].)

a drastically decreasing capillary force for decreasing humidities. While the interaction force is reduced to about 1/5 compared to two particles in touch, the bridge volume is almost unchanged. In case of the small distance the capillary force shows a maximum at around 90% humidity. For even larger humidities the force is decreasing again despite the increasing bridge volume due to the strongly decreasing capillary pressure.

The size of particles is indeed a very important parameter. Figure 17 shows the dimensionless capillary force between two identical spherical particles with respect to particle size. From classical, microscopic theory, one would expect no influence since the capillary force scales linearly with the particle diameter. However, it can be clearly seen that for smaller particles there are substantial negative deviations, i.e. decreasing dimensionless forces particularly at higher humidity levels. This can be explained by the fact that in those cases the particle size and the Kelvin radius are of the same order. Therefore, the filling angle at a given humidity is larger for smaller particles leading to a further decrease of the capillary force.

Furthermore, in many cases a critical relative humidity must be exceeded in order to form a stable capillary bridge in thermodynamic equilibrium. In Fig. 18 this minimum humidity φ_{min} is shown depending on particle size D for different distances d and contact angles Θ. For particles in close contact (i.e. $d = 0$ nm) or at a distance of d $= 0.2$ nm and a contact angle of $\Theta = 0°$ there is no minimum humidity, i.e. a capillary bridge is formed for all humidities larger than 0. In general, φ_{min} is rising for increasing distance as well as for increasing contact angle. For particles larger than $D = 1$ μm the minimum humidity φ_{min} appears to be independent of particle size. However, for nanoscaled particles there is a considerable increase in φ_{min} for decreasing particle sizes.

In many cases, not only the forces between particles but also between particles and walls are very important. For macroscopic particles the capillary force between a particle and a wall is twice as large compared to the force between two identical spheres. Figure 19 depicts the ratio of capillary force between a particle of size D with a wall F_{C,Sp_Pl} and between two particles of size D F_{C,Sp_sp} with respect to humidity. While for particles of 100 nm only minor deviations to this macroscopic theory occur, i.e. the force ratio is close to 2.0 for the whole range of humidities, for particles of size $D = 10$ nm a completely different behavior can be observed.

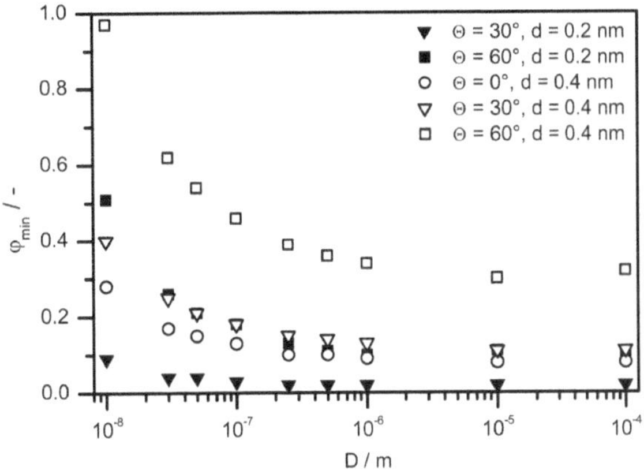

Fig. 18 Minimum relative humidity required in order to get a capillary bridge in thermodynamic equilibrium (Reprinted with permission from [45].)

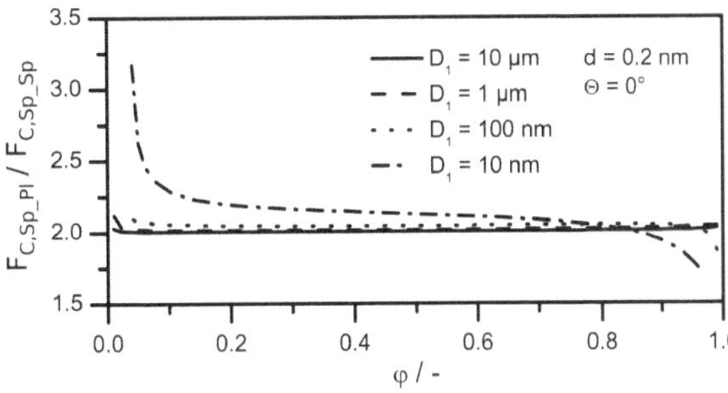

Fig. 19 Ratio of capillary force between two identical spheres $F_{C,SpSp}$ and one sphere and a plane wall F_{C,Sp_Pl} (Reprinted with permission from [45].)

For small humidities it turns out that the capillary forces at the wall is up to 3 times larger than between two spheres. However, at large humidity the capillary force is even smaller than 2 times the force between spheres. This behavior can be explained by the volume of the capillary bridge which is significantly larger for nanoscaled particles in contact with a wall compared to a sphere-sphere contact. This leads to stronger forces at small humidities due to significantly larger bridges and to a decrease at high humidities due to the decreasing surface curvature.

In case of particles with different diameters D_1 and D_2, respectively, Fig. 20 shows the resulting capillary force with respect to relative humidity. For all curves one particle diameter is fixed at $D_1 = 100$ nm while the second particle diameter is allowed to change from $D_2 = 10$ nm to $D_2 = 50$ μm. All combinations of particle sizes show a similar dependency on relative humidity. However, the absolute value of the capillary force is quite different. In general, the capillary force depends on both particle sizes with increasing capillary force with increasing one of both particle sizes. From macroscopic theory an appropriate way to combine both particle sizes as an effective particle is given by Eq. (6):

$$D_{eff} = \frac{D_1 \cdot D_2}{D_1 + D_2} \tag{6}$$

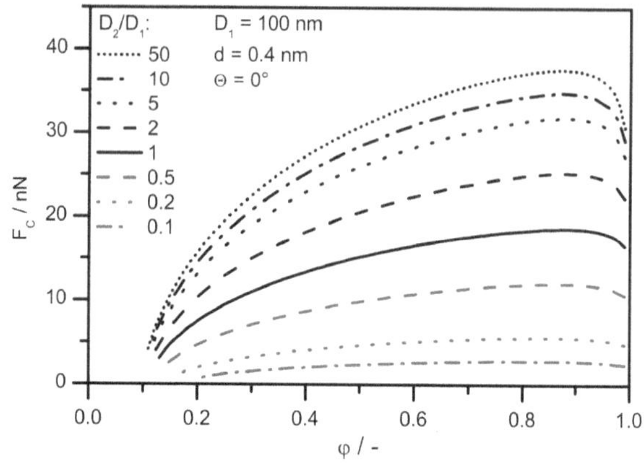

Fig. 20 Capillary forces between particles with different sizes (Reprinted with permission from [45].)

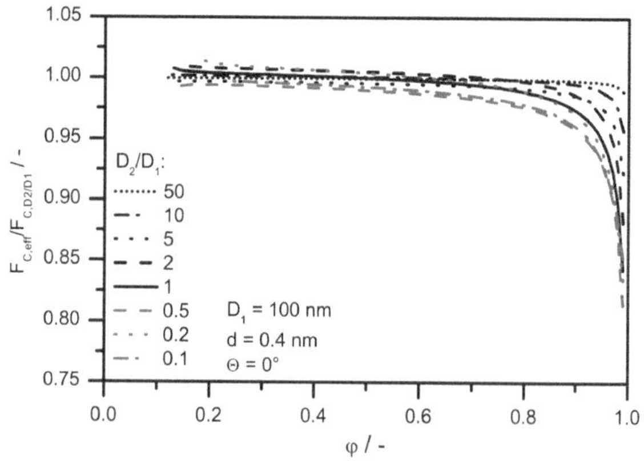

Fig. 21 Ratio of capillary force between two spherical particles of different diameters D_1 and D_2, calculated with the 'effective diameter' approximation and the full simulation, respectively (Reprinted with permission from [45].)

For macroscopic particles two identical particles of size D_{eff} are expected to experience a similar capillary force as two particles of size D_1 and D_2. As shown by Willet et al. [46], for millimeter sized spheres a deviation of typically less than 5% and in all cases less than 20% can be achieved using the effective particle size as an approximation. Figure 21 examines this approximation for nanoscaled particles. For all combinations of particle sizes the effective diameter appears to be a good approximation for small to moderate humidities. However, for very large humidities using D_{eff} leads to a substantial under-estimation of the capillary forces with deviations up to 20%.

Since the contact angle Θ between the solid particle surface and the wetting liquid strongly influences the shape of the meniscus and therefore the capillary pressure inside the bridge, it is expected that capillary bridge forces are decreasing with increasing contact angle. As shown in Fig. 22, a scaling of the capillary force with $\cos\Theta$ is a reasonable way to consolidate the results: For two spheres in close contact with the contact angle varying between 0° (perfect wetting) to 60° (poor wetting) the consolidated forces almost coincide in one 'master curve'. For some distance between

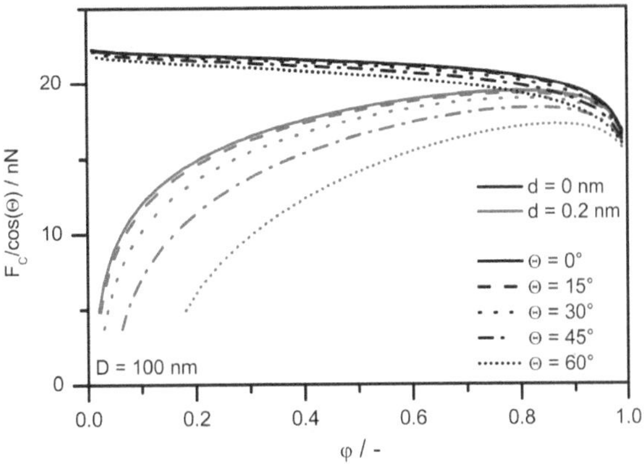

Fig. 22 Capillary force scaled by cosine of contact angle Θ for two spheres with diameter $D = 100$ nm, and surface distance $d = 0.0$ nm and $d = 0.2$ nm, respectively (Reprinted with permission from [45].)

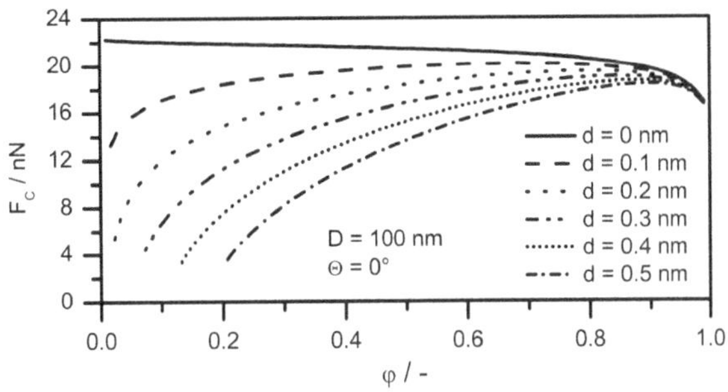

Fig. 23 Capillary Forces between two spherical particles of size D for varying distance d with a perfectly wetting fluid (Reprinted with permission from [45].)

the particles the scaling with cos Θ works reasonably well for contact angles up to 30° (moderate wetting), whereas for larger contact angles even the scaled capillary force is significantly smaller, i.e. the reduction of the capillary force is larger than expected by the scaling with cos Θ.

Finally, a systematic evaluation of the influence of the distance between two particles is shown in Fig. 23. It is obvious that particularly at small humidities there is a very strong influence of the distance with rapidly decreasing capillary forces for only slightly increasing distances. However, at larger humidities small variations of the distance—in this case only distances up to 0.5 nm were considered—have only negligible influence on the capillary bridge forces. The strong influence of the distance at small humidities can be explained by the strong curvature of the surface which leads to rapidly shrinking capillary bridges if the distance of the particles is increased. This implies as well that for a given distance below some threshold value of the relative humidity φ_{min} no capillary bridge can be formed at all. Figure 24 shows this minimum relative humidity with respect to particle distance for different particle sizes and contact angles. For the same reason as given above capillary bridges between smaller particles are much more sensitive to changes in distance leading to a significantly stronger increase in the threshold humidity φ_{min} required to form

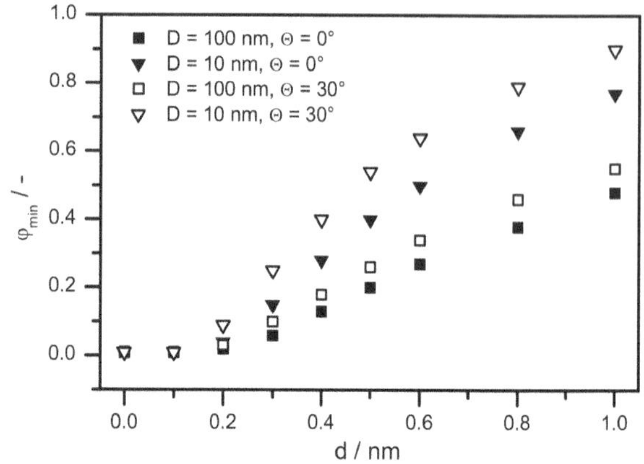

Fig. 24 Minimum relative humidity φ_{min} required to form any capillary bridge (Reprinted with permission from [45].)

a capillary bridge at all. Furthermore, a decrease of wettability, i.e. an increasing contact angle, leads to a strong increase of the distance sensitivity as well.

Capillary Forces upon Bridge Rupture

An important aspect in handling of bulk particulate solids is the separation of two particles that are connected by a capillary bridge. While separating the surfaces two fundamental limiting cases are imaginable. On the one hand, the capillary bridge stays in thermodynamic equilibrium with the surrounding air resulting in a constant curvature of the meniscus corresponding to the actually given relative humidity. However, for changing interparticle distances this would result in a variation of the capillary bridge volume. On the other hand, the volume of the capillary bridge may be assumed to stay constant and therefore a varying distance would result in curvature changes. When the surfaces are separated slowly there is time that liquid can evaporate or condense and the curvature might be considered constant. During faster separation processes the kinetics of evaporation and condensation, respectively, may become relevant up to very fast separation processes, where the second limiting case applies with a constant bridge volume [44]. The rate of condensation and evaporation of liquids is mainly determined by the diffusion of molecules to or from the interface, respectively, although heat transfer of latent heat may have a minor influence as well [45, 47]. Since capillary bridge volumes for nanoscaled particles are extremely small, a rough estimation leads to typical times to achieve equilibrium of well below 1 ms. Therefore, it is expected that for typical powder handling processes the kinetics of distance variation is slow enough in order to use the thermodynamic equilibrium assumption. Figure 25 exemplarily shows the difference in capillary bridge forces for both limiting cases. The simulations for constant volume have been performed with the public domain software Surface Evolver [48] and for some points iteratively using our method as presented above. It can be clearly seen, that for small distances

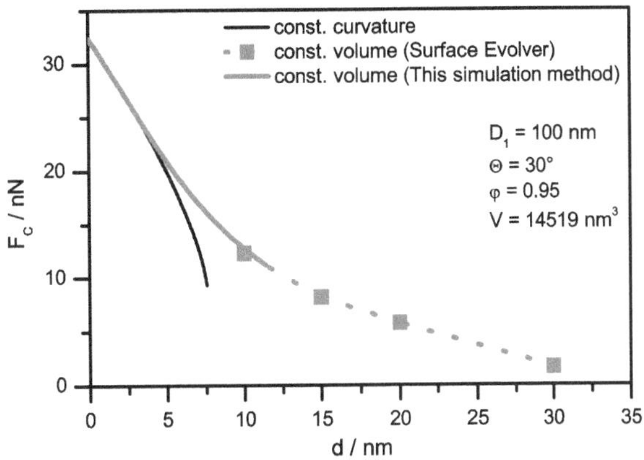

Fig. 25 Comparison of capillary force for distance variation with assumption of thermodynamic equilibrium (i.e. const. curvature) and infinitely fast change (i.e. const. volume), respectively (Reprinted with permission from [49].)

the deviations are negligible. Examining the actual bridge volume for thermodynamic equilibrium reveals that the volume stays almost constant for small distances which explains the correspondence to simulations with prescribed constant volume. However, the capillary bridge in equilibrium ruptures much earlier than for constant bridge volume—in this case at around 7 nm compared to 30 nm. Furthermore, there is a steeper decrease of the capillary bridge force in case of thermodynamic equilibrium.

The dependence of capillary bridge force on distance for different particle sizes, contact angles and relative humidities is shown in Fig. 26. The capillary force decreases with increasing distance between the particles. It appears that the gradient of the force is nearly constant over a wide range of distances. Only near the maximum distance shortly before rupture of the capillary bridge there is some stronger decrease. With a higher relative humidity, the slope of the force decreases. At a distance of 0 nm the force is highest at a low relative humidity and at an increasing distance this changes to a higher force at higher humidity. The maximum distance is also higher if the relative humidity is higher, which will be explained later in more detail. At higher contact angles the force is lower because the volume of the capillary bridge is also lower. However, the contact angle has obviously no significant influence on the gradient of the force. Combining these two results, the maximum distance is smaller for a higher contact angle. Only at very small particle sizes differences in the gradients due to the contact angle can be detected. Furthermore, the force scales nearly linear with particle size for micrometer-sized particles, for nanoscaled particles the force decreases more. The maximum distance of the capillary bridge increases for bigger particles.

The gradient of the force over the distance between two particles is called the contact stiffness [50]. As the force is depending nearly linearly on the distance and there is only a slight curvature in the force-distance curves the contact stiffness of a capillary bridge can be assumed to be almost independent of the separation distance. Therefore, a linear regression was done over all distances for capillary bridges between particles with various diameters and contact angles at different

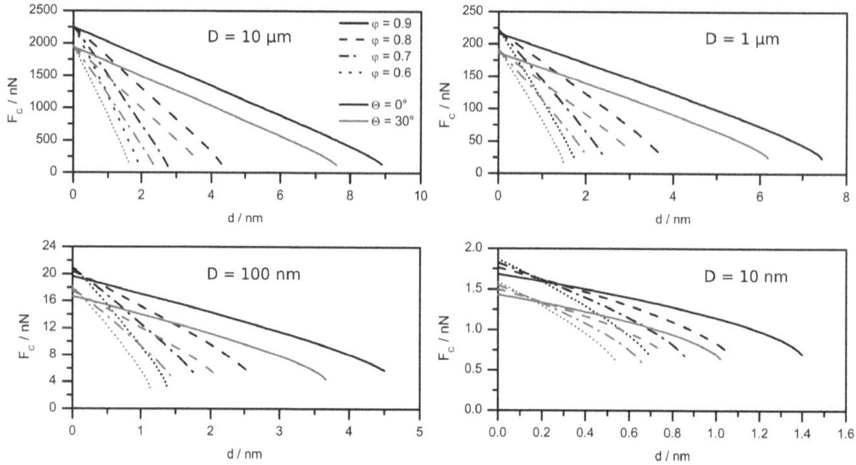

Fig. 26 Distance dependency of capillary bridge forces for different particle sizes D, relative humidities φ and contact angles Θ, respectively (Reprinted with permission from [49].)

Fig. 27 Exemplary comparison of simulated capillary bridge forces and linear regression in order to determine the contact stiffness (Reprinted with permission from [49].)

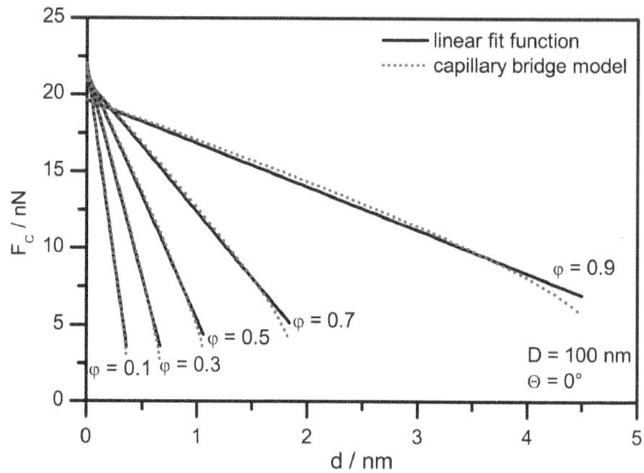

relative humidities. For the regression the force at direct contact was fixed with the value from the simulation and all distances were considered to determine the slope of the regression curve. In Fig. 27 the original and the reconstructed force-distance curves are compared for some exemplary cases. The excellent correspondence of the linear fit function is clearly demonstrated.

The contact stiffness k^* is finally determined from linear regression to the dimensionless capillary bridge force:

$$k^* = \frac{\partial \frac{F}{\gamma D}}{\partial d} \qquad (7)$$

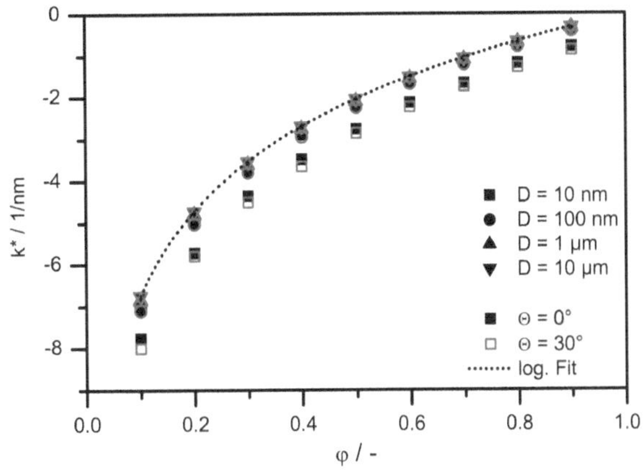

Fig. 28 Contact stiffness of capillary bridge forces with respect to relative humidity φ, for different particle sizes *D* and contact angle Θ, respectively (Reprinted with permission from [49].)

The determined contact stiffness k^* for different particle sizes, contact angles and relative humidities is shown in Fig. 28. In general, the absolute value of the contact stiffness decreases for higher relative humidity. The contact angle obviously has only a small influence on the contact stiffness as the values for different contact angles are almost identical. For particles with diameters in the micrometer range the contact stiffness does not change with particle size. At a particle diameter of 100 nm there is a small increase in stiffness and for even smaller particles the influence of particle size increases and the contact stiffness becomes higher.

The dependence of the contact stiffness on relative humidity can be described very well with a logarithmic fit function. In Fig. 28 the fit function given by Eq. (8) for a particle diameter of 10 µm and a contact angle of 0° is shown.

$$k^* = 2.931\frac{1}{\text{nm}} \cdot \ln(\varphi) \quad (8)$$

Furthermore, the capillary bridge has a maximum distance of existence (cf. above). At higher distances a capillary bridge between these particles cannot exist at the respective relative humidity. In theoretical considerations for micronsized particles the maximum bridge length $h_{\max}$ (cf. Fig. 12) of particles before capillary bridge rupture occurs can be assessed. It turns out that in these cases $h_{\max}$ of a capillary bridge in thermodynamic equilibrium is two times the Kelvin radius scaled with the cosine of the contact angle [51–53]. Since bridge size h and particle separation d are closely correlated, in order to compare the results of our simulations with theoretical considerations, the bridge size h of the capillary bridge was calculated from the simulation results quite easily. In Fig. 29 the maximum size h_{max} of capillary bridges between particles with various diameters are shown. With increasing relative humidity, the maximum bridge size becomes higher as the Kelvin radius increases. Furthermore, the maximum size agrees quite will with the theoretical maximum $2 \cdot r_K$. However, the smaller the particles are the lower is the maximum size as the curvature of the particle surface increases. Therefore, the achievable maximum height depends

Fig. 29 Maximum bridge size h_{max} (cf. Fig. 12) until rupture of capillary bridge occurs (Reprinted with permission from [49].)

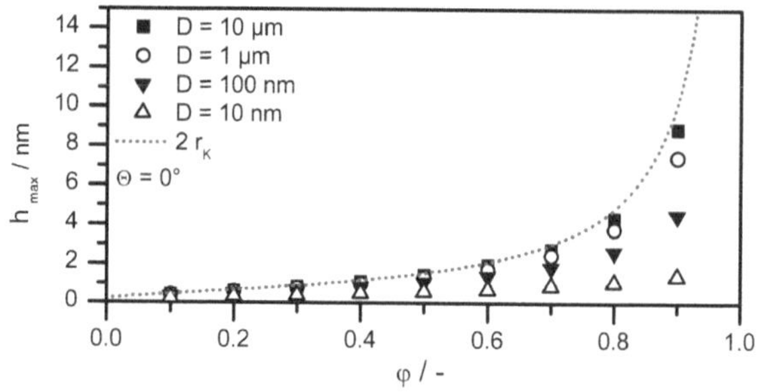

on the curvature of the particle and the curvature of the capillary bridge. Furthermore, it turns out that the influence of the contact angle can be described very well with the cosine of the angle as it was proposed before [53].

Ultrasmall Particles

All simulations presented in this work are based on the continuum assumption of the liquid bridge. However, it is not obvious that this assumption is still acceptable for particles in the ultrafine particle size regime, i.e. particle sizes in the range of several nm. On the other hand, Molecular Dynamics (MD) simulations provide a physical model for phenomena at very small length scales. However, MD is quite limited in achievable particle sizes due to a very high computational effort. Therefore, results from MD simulations of Laube et al. [54] have been compared to continuum simulations of the capillary bridge for two spherical particles of $D_0 = 4$ nm. In the MD simulations the fluid properties have been adjusted to give macroscopic values for the surface tension of $\gamma_0 = 52.3$ N/m and for the contact angle of $\Theta_0 = 10°$. Figure 30 shows that the continuum simulation using the original parameters for γ_0, Θ_0, D_0 (dotted line) results in about 3 times smaller capillary bridge forces compared to MD simulations (point cloud). Furthermore, the slope of the force-distance curve is slightly different. However, due to a closed water absorption layer on the particles the effective contact angle which turns out from MD simulations is close to 0°. Furthermore, from MD simulations one can deduce the actual surface tension. Although the force field had been chosen in order to give good agreement with the macroscopic surface tension, in case of the capillary bridge it turns out that the surface tension is affected by the high curvature of the surface. Therefore, a modified value of the surface tension at the actual surface curvature taken from MD simulations has to be used. Finally, the closed water adsorption layer on top of the particles leads to an effective increase of the radius of each particle of about 0.3 nm. Changing the contact angle and the surface tension appropriately leads to some increase in the bridge forces. Although the force from continuum simulations is still about two times smaller compared to the MD simulations the slope of the curve with respect to

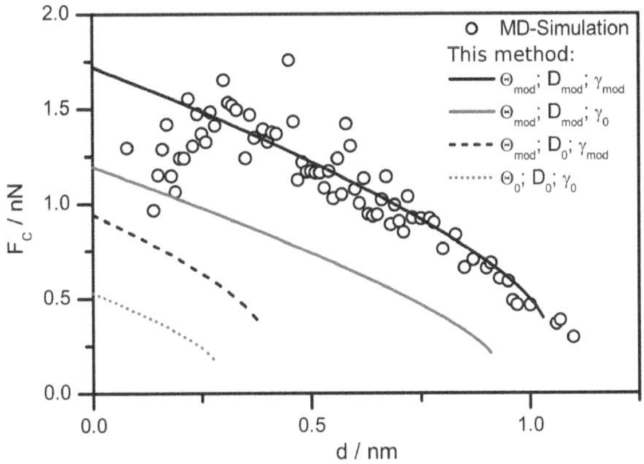

Fig. 30 Comparison of Molecular Dynamics simulations with continuum model of this work for particles of $D_0 = 4$ nm, $\Theta_0 = 10°$, $\gamma_0 = 52.3$ N/m at $\varphi = 40.2\%$. Continuum simulations done with various modifications of particle size (due to adsorption layer), contact angle and surface tension in order to comply with MD simulations (Reprinted with permission from [45].)

distance now agrees quite well with MD simulations. If the effective particle diameter of $D = 4.6$ nm—i.e. including a 0.3 nm adsorption layer—is used in the continuum simulations, a considerable increase in the capillary bridge forces can be observed. If all three effects are considered in the capillary bridge simulations an excellent agreement between MD and continuum simulations is obtained. Yet it should be emphasized, that these modifications are not based on an arbitrary parameter fitting but only on physically sound adjustments which can be reasoned to become effective for such ultrasmall particles.

Summary and Conclusions

This chapter describes experimental and theoretical studies of water surface adsorption and the formation of capillary bridges between nanoscaled particles. First, experimental studies of water adsorption depending on surface chemistry using PM-IRRAS and QCM-D FTIR are described. It is shown that only small changes in surface chemistry either by adsorption of monolayer or by UV-irradiation lead to substantial changes in both total amount of adsorbed humidity and structure of the adsorbed molecules. Using QCM it was even possible to quantitatively measure the substantial amount of water adsorbing into capillary bridges between nanoscaled particles. Furthermore, the contact force between nanoscaled spherical TiO_2 particles has been investigated using AFM. In this case a strong dependence on humidity and surface chemistry can be ob served as well. Particularly for low humidities even extremely small roughness leads to a formation of multiple very small bridges instead of one capillary bridge. Therefore, a wide distribution of contact forces is obtained.

Due to the photocatalytic effect, the formation of capillary bridges is strongly suppressed upon UV irradiation. Measurements of the bulk powder flow behavior clearly show a reduced flowability due to capillary condensation, although the quantitative measurement is biased by a growing compaction of the powder during the measurements.

Second, simulations of capillary bridges of nanoscaled particles have been investigated in detail. It is clearly demonstrated, that for nanoscaled particles classical approximations do not work well and that capillary forces behave quite different due to the curvature of the particles which might be in the range of the curvature of the capillary bridge. In contrast to microscopic particles a very strong dependency on particle distance can be observed since capillary bridges between nanoscaled particles formed by condensation are extremely small. This explains the wide variation of contact forces found in the AFM measurements in case of capillary bridge formation. Furthermore, the maximum separation of particles until the capillary bridge ruptures decreases with decreasing particle size and increasing contact angle. A comparison with Molecular Dynamics simulations revealed that the continuum approach may still be applied for particle sizes down to a few nanometers to calculate the capillary bridges. However, in that case changes of surface tension due to the surface curvature, changes in contact angle due to an adsorbed layer of water and particularly an enlargement of the particle size due to the adsorption layer (even if it is as small as 0.3 nm) must be taken into account in order to get a quantitative prediction of the capillary forces.

Finally, in order to include these results in DEM simulations of humid powder flow, a force-distance relationship based on a stiffness approach has been formulated. It appears that the stiffness is only weakly affected by particle size and contact angle. However, the maximum separation until bridge rupture occurs is strongly depending on both, particle size and contact angle.

Acknowledgements We would like to thank the Deutsche Forschungsgemeinschaft (DFG) for supporting this work by the projects SCHM1429/7 and GR1709/12.

We thank Prof. Anjana Devi and her co-workers at the Ruhr University in Bochum for the atomic layer deposition of ultra-thin TiO_2 films.

References

1. Hann, D., Stražišar, J.: Influence of particle size distribution, moisture content, and particle shape on the flow properties of bulk solids. Instrum. Sci. Technol. **35**(5), 571–584 (2007). https://doi.org/10.1080/10739140701540453
2. Schulze, D.: Powders and Bulk Solids: Behavior, Characterization, Storage and Flow. Springer, Berlin (2008)
3. Rabinovich, Y., Esayanur, M., Johanson, K., et al.: The flow behavior of the liquid/powder mixture, theory and experiment. I. The effect of the capillary force (bridging rupture). Powder Technol. **204**(2–3), 173–179 (2010). https://doi.org/10.1016/j.powtec.2010.07.035
4. Schubert, H.: Capillary forces—modeling and application in particulate technology. Powder Technol. **37**(1), 105–116 (1984). https://doi.org/10.1016/0032-5910(84)80010-8

5. Israelachvili, J.N.: Intermolecular and Surface Forces: Revised Third Edition, 3rd edn. Elsevier Science, Burlington (2011)
6. Butt, H.-J., Cappella, B., Kappl, M.: Force measurements with the atomic force microscope: technique, interpretation and applications. Surf. Sci. Rep. **59**(1–6), 1–152 (2005). https://doi.org/10.1016/j.surfrep.2005.08.003
7. Farshchi-Tabrizi, M., Kappl, M., Cheng, Y., et al.: On the adhesion between fine particles and nanocontacts: an atomic force microscope study. Langmuir **22**(5), 2171–2184 (2006). https://doi.org/10.1021/la052760z
8. Keller, A., Fritzsche, M., Ogaki, R., et al.: Tuning the hydrophobicity of mica surfaces by hyperthermal Ar ion irradiation. J. Chem. Phys. **134**(10), 104705 (2011). https://doi.org/10.1063/1.3561292
9. Kunze, C., Torun, B., Giner, I., et al.: Surface chemistry and nonadecanoic acid adsorbate layers on TiO_2(100) surfaces prepared at ambient conditions. Surf. Sci. **606**(19–20), 1527–1533 (2012). https://doi.org/10.1016/j.susc.2012.05.025
10. Torun, B., Kunze, C., Zhang, C., et al.: Study of water adsorption and capillary bridge formation for SiO(2) nanoparticle layers by means of a combined in situ FT-IR reflection spectroscopy and QCM-D set-up. Phys. Chem. Chem. Phys. **16**(16), 7377–7384 (2014). https://doi.org/10.1039/c3cp54912g
11. Torun, B., Giner, I., Grundmeier, G., et al.: In situ PM-IRRAS studies of organothiols and organosilane monolayers-ZnO interfaces at high water activities. Surf. Interface Anal. **49**(1), 71–74 (2017). https://doi.org/10.1002/sia.6058
12. Rajca, A.: An Introduction to Ultrathin Organic Films: From Langmuirblodgett to Self-assembly, By Abraham Ulman. Academic Press, London (1991), Xxiii, 442 pp. $65. ISBN 0-12-708230-1. Adv. Mater. **4**(4), 309 (1992). https://doi.org/10.1002/adma.19920040424
13. Marrone, M., Montanari, T., Busca, G., et al.: A Fourier transform infrared (FTIR) study of the reaction of triethoxysilane (TES) and bis[3-triethoxysilylpropyl]tetrasulfane (TESPT) with the surface of amorphous silica. J. Phys. Chem. B **108**(11), 3563–3572 (2004). https://doi.org/10.1021/jp036148x
14. Riccio, M., Montanari, T., Castellano, M., et al.: An IR study of the chemistry of triethoxysilane at the surface of metal oxides. Colloids Surf. A **294**(1–3), 181–190 (2007). https://doi.org/10.1016/j.colsurfa.2006.08.010
15. Yen, Y.S., Wong, J.S.: Infrared reflectance properties of surface thin films. J. Phys. Chem. **93**(20), 7208–7216 (1989). https://doi.org/10.1021/j100357a036
16. Calzaferri, G., Imhof, R.: In situ attenuated total reflection FTIR investigations of H_2O, $HSiCl_3$ and $Co_2(CO)8$ on ZnSe in the range 600–4000 cm^{-1}. Spectrochim. Acta Part A Mol. Biomol. Spectrosc. **52**(1), 23–28 (1996). https://doi.org/10.1016/0584-8539(95)01533-7
17. Nyanhongo, G.S., Nugroho Prasetyo, E., Herrero Acero, E., et al.: Engineering strategies for successful development of functional polymers using oxidative enzymes. Chem. Eng. Technol. **35**(8), 1359–1372 (2012). https://doi.org/10.1002/ceat.201100590
18. Grundmeier, G., Matheisen, E., Stratmann, M.: Formation and stability of ultrathin organosilane polymers on iron. J. Adhes. Sci. Technol. **10**(6), 573–588 (1996). https://doi.org/10.1163/156856196X00599
19. Asay, D.B., Kim, S.H.: Evolution of the adsorbed water layer structure on silicon oxide at room temperature. J. Phys. Chem. B **109**(35), 16760–16763 (2005). https://doi.org/10.1021/jp053042o
20. Dixon, M.C.: Quartz crystal microbalance with dissipation monitoring: enabling real-time characterization of biological materials and their interactions. J. Biomol. Tech. **19**, 151–158 (2008)
21. Giner, I., Torun, B., Han, Y., et al.: Water adsorption and capillary bridge formation on silica micro-particle layers modified with perfluorinated organosilane monolayers. Appl. Surf. Sci. **475**, 873–879 (2019). https://doi.org/10.1016/j.apsusc.2018.12.221
22. Dybwad, G.L.: A sensitive new method for the determination of adhesive bonding between a particle and a substrate. J. Appl. Phys. **58**(7), 2789–2790 (1985). https://doi.org/10.1063/1.335874

23. Du, B., König, A.M., Johannsmann, D.: On the role of capillary instabilities in the sandcastle effect. New J. Phys. **10**(5), 53014 (2008). https://doi.org/10.1088/1367-2630/10/5/053014
24. Kühne, T.D., Pascal, T.A., Kaxiras, E., et al.: New insights into the structure of the vapor/water interface from large-scale first-principles simulations. J. Phys. Chem. Lett. **2**(2), 105–113 (2011). https://doi.org/10.1021/jz101391r
25. Kühne, T.D., Krack, M., Parrinello, M.: Static and dynamical properties of liquid water from first principles by a novel car–parrinello-like approach. J. Chem. Theory Comput. **5**(2), 235–241 (2009). https://doi.org/10.1021/ct800417q
26. Sulpizi, M., Gaigeot, M.-P., Sprik, M.: The silica-water interface: how the silanols determine the surface acidity and modulate the water properties. J. Chem. Theory Comput. **8**(3), 1037–1047 (2012). https://doi.org/10.1021/ct2007154
27. Laube, J., Salameh, S., Kappl, M., et al.: Contact forces between TiO_2 nanoparticles governed by an interplay of adsorbed water layers and roughness. Langmuir **31**(41), 11288–11295 (2015). https://doi.org/10.1021/acs.langmuir.5b02989
28. Boray Barış Torun: In situ analysis of particles in contact under ambient conditions. Doctoral Thesis, University of Paderborn (2016)
29. Hüttl, G., Klemm, V., Popp, R., et al.: Tailored colloidal AFM probes and their TEM investigation. Surf. Interface Anal. **33**(2), 50–53 (2002). https://doi.org/10.1002/sia.1160
30. Sawunyama, P., Fujishima, A., Hashimoto, K.: Photocatalysis on TiO_2 surfaces investigated by atomic force microscopy: photodegradation of partial and full monolayers of stearic acid on TiO_2 (110). Langmuir **15**(10), 3551–3556 (1999). https://doi.org/10.1021/la9814440
31. Haines, W.B.: Studies in the physical properties of soils: II. A note on the cohesion developed by capillary forces in an ideal soil. J. Agric. Sci. **15**(04), 529 (1925). https://doi.org/10.1017/S0021859600082460
32. Fisher, R.A.: On the capillary forces in an ideal soil; correction of formulae given by W. B. Haines. J. Agric. Sci. **16**(03), 492 (1926). https://doi.org/10.1017/S0021859600007838
33. Orr, F.M., Scriven, L.E., Rivas, A.P.: Pendular rings between solids: Meniscus properties and capillary force. J. Fluid Mech. **67**(04), 723 (1975). https://doi.org/10.1017/S0022112075000572
34. Schubert, H.: Kapillaritat in Porosen Feststoff systemen. Springer, Berlin (1982)
35. Chau, A., Rignier, S., Delchambre, A., et al.: Three-dimensional model for capillary nanobridges and capillary forces. Model. Simul. Mater. Sci. Eng. **15**(3), 305–317 (2007). https://doi.org/10.1088/0965-0393/15/3/009
36. Iinoya, K., Asakawa, S., Hotta, K., et al.: Liquid surface profiles in contact with symmetrical solid surfaces. Powder Technol. **1**(1), 28–32 (1967). https://doi.org/10.1016/0032-5910(67)80005-6
37. Pakarinen, O.H., Foster, A.S., Paajanen, M., et al.: Towards an accurate description of the capillary force in nanoparticle-surface interactions. Model. Simul. Mater. Sci. Eng. **13**(7), 1175–1186 (2005). https://doi.org/10.1088/0965-0393/13/7/012
38. Lambert, P., Chau, A., Delchambre, A., et al.: Comparison between two capillary forces models. Langmuir **24**(7), 3157–3163 (2008). https://doi.org/10.1021/la7036444
39. Salameh, S., Schneider, J., Laube, J., et al.: Adhesion mechanisms of the contact interface of TiO_2 nanoparticles in films and aggregates. Langmuir **28**(31), 11457–11464 (2012). https://doi.org/10.1021/la302242s
40. Leroch, S., Wendland, M.: Influence of capillary bridge formation onto the silica nanoparticle interaction studied by grand canonical Monte Carlo simulations. Langmuir **29**(40), 12410–12420 (2013). https://doi.org/10.1021/la402002f
41. Megias-Alguacil, D., Gauckler, L.J.: Capillary forces between two solid spheres linked by a concave liquid bridge: regions of existence and forces mapping. AIChE J. **55**(5), 1103–1109 (2009). https://doi.org/10.1002/aic.11726
42. Adams, M.J., Johnson, S.A., Seville, J.P.K., et al.: Mapping the influence of gravity on pendular liquid bridges between rigid spheres. Langmuir **18**(16), 6180–6184 (2002). https://doi.org/10.1021/la011823k

43. Dörmann, M., Schmid, H.-J.: Simulation of capillary bridges between particles. Procedia Eng. **102**, 14–23 (2015). https://doi.org/10.1016/j.proeng.2015.01.102
44. Butt, H.-J., Kappl, M.: Surface and Interfacial Forces. Wiley-VCH Verlag GmbH & Co. KGaA, Weinheim, Germany (2010)
45. Dörmann M (2018) Zur Modellierung von Kapillarbrücken zwischen nanoskaligen Partikeln. Doctoral, University of Paderborn
46. Willett, C.D., Adams, M.J., Johnson, S.A., et al.: Capillary bridges between two spherical bodies. Langmuir **16**(24), 9396–9405 (2000). https://doi.org/10.1021/la000657y
47. Langmuir, I.: The evaporation of small spheres. Phys. Rev. **12**(5), 368–370 (1918). https://doi.org/10.1103/PhysRev.12.368
48. Kohonen, M.M., Maeda, N., Christenson, H.K.: Kinetics of capillary condensation in a nanoscale pore. Phys. Rev. Lett. **82**(23), 4667–4670 (1999). https://doi.org/10.1103/PhysRevLett.82.4667
49. Dörmann, M., Schmid, H.-J.: Distance-dependency of capillary bridges in thermodynamic equilibrium. Powder Technol. **312**, 175–183 (2017). https://doi.org/10.1016/j.powtec.2017.01.012
50. Brakke, K.A.: The surface evolver. Exp. Math. **1**(2), 141–165 (1992). https://doi.org/10.1080/10586458.1992.10504253
51. Tomas, J.: Mechanics of nanoparticle adhesion—a continuum approach. In: Particles on Surfaces 8: Detection, Adhesion and Removal, vol. 8, pp. 1–47 (2003)
52. Andrienko, D., Patricio, P., Vinogradova, O.I.: Capillary bridging and long-range attractive forces in a mean-field approach. J. Chem. Phys. **121**(9), 4414–4423 (2004). https://doi.org/10.1063/1.1778154
53. Petrov, P., Olsson, U., Wennerström, H.: Surface forces in bicontinuous microemulsions: water capillary condensation and lamellae formation. Langmuir **13**(13), 3331–3337 (1997). https://doi.org/10.1021/la962085g
54. Laube, J., Dörmann, M., Schmid, H.-J., et al.: Dependencies of the adhesion forces between TiO_2 nanoparticles on size and ambient humidity. J. Phys. Chem. C **121**(28), 15294–15303 (2017). https://doi.org/10.1021/acs.jpcc.7b05655

Microwave Emission During the Impact Compaction of Particle Bed

Sergej Aman, Alexander Aman and Werner Hintz

Nomenclature

g	Gravity constant (m s^{-2})
p	Dipole moment (m C)
ε_0	Vacuum permittivity (As/v/m)
c	Speed of light (m/s)
t_{dp}	Depolarization time (s)
q	Electrical charge (C)
I	Emission current (A)
φ	Work function (v)
β	Field enhancement factor (–)
p	Mechanical impulse of pen (kg m/s)
m	Mass of the pen (g)
v	Velocity of the pen (m/s)
t_i	Time interval (s)

S. Aman (✉) · W. Hintz
Institute of Mechanical Process Engineering, Otto von Guericke University Magdeburg, Kaiserslautern, Germany
e-mail: sergej.aman@ovgu.de

W. Hintz
e-mail: werner.hintz@ovgu.de

A. Aman
Vollmann Group, Research and Development Department, Gevelsberg, Germany
e-mail: a.aman@vollmann-group.com

S. Antonyuk (ed.), *Particles in Contact*,
https://doi.org/10.1007/978-3-030-15899-6_3

Introduction

Cost, quality and productivity in comminution processes are all dependent on distribution of consumed energy. In an inefficient process, the consumed energy leads not to particle size reduction but to heat dissipation and temperature rise in the grinding equipment. Excessive energy consumption by reduced grinding efficiency increases energy costs and demands the grinding equipment. Especially a particle bed grinding is an inefficient process and has great potential to achieve high efficiency.

The relationship between the forces acting on the particle bed and the deformation behavior is the main influencing variable regarding to comminution. The modelling of the stress behaviour based on compaction models used a number of simplifications and lead, nonetheless, to very complex solutions [1, 2]. In this context, it remains to develop any experimental technique, which can be used to increase the grinding efficiency.

At the one hand, the stressing of the dry particle beds leads to energy absorption due to micro processes caused by compaction [3]:

- energy loss due to friction between the particles and the wall of stressing chamber,
- energy loss due to irreversible structural changes of particles (plastic deformation work, generation of micro cracks),
- energy loss caused by the wear of the confinements of the particle,
- energy loss due to thermoplastic effects, sound wave propagation and oscillation of elastically deformed fragments.

At the other hand the energy absorption leads to breakage i.e. size reduction of particles. The ratio between the energies applied to compaction and to size reduction is about 10–50. However, it is difficult to estimate the stored energy without any real time information about micro processes occurring during stressing events. Especially it is actual for impact stressing determination. However, it becomes possible to optimize the comminution by monitoring the crack building in particles. Based on correlation between intensity of microwave emission and rates of cracks building one can estimate the rate of size reduction i.e. efficiency of commination.

There are a number of fracture analysis methods based on fracto-emission [6–16], which usually appears during crack formation in materials. This fracto-emission includes the emission of electrons, ions, free radicals, electromagnetic and acoustic waves. Regarding to application, electromagnetic and acoustic emission are most interesting. This type of emissions are used to monitor crack generation in massive specimens of different materials, such as rocks, wood, metal, ceramic and carbon-fibre/epoxy composites [4–17].

The electromagnetic emission (EME) exhibits over a broad frequency range—from radio wave and microwave up to X-ray [9, 10, 13, 14]. The intensity of the EME depends on the tested material and on the crack sizes. With increasing crack size, the intensity of the EME increases. However, a special type of capacitance sensor was used to detect an extremely low EME signal from the composite material 'extren' [8]. Capacitance sensors are well-suited for studying the EME from small-sized specimens in laboratory experiments.

The level of fracture-EME is usually comparable to the electromagnetic noise of the environment. Because of this, the practical application of poor, i.e., not enhanced, microwave emission for fracture monitoring of large grinding machinery components is problematically. However, it was observed that the stressing of Lead zirconium titanate $Pb[Zr_xTi_{1-x}]O_3$ (PZT) caused by impact of a sharp tungsten pin on the surface of the PZT-ceramic produced sequences of high intensive microwave impulses. The amplitude of an EME-impulse is proportional to the amplitude of discharge current impulse i.e. charge that is induced by crack generation. In terms of impulse generation, the PZT domain is Hertzian dipole with charge that is proportional to the amplitude of impulse. Due to intensive charge generation by crack generation in PZT and other materials, the EME emission from this material becomes detectable by a horn antenna [5, 8, 18, 19].

Processes responsible for EME such as particle deformation, breakage at the contact between the particles, friction, creating of new contacts are caused due to mechanical loading i.e. stressing of particle beds. The basic physical micro-processes, which correspond to these processes, are different. The characteristic duration of these micro processes is in the micro and nanosecond range. It was necessary to develop a new method to record these processes with a corresponding temporal resolution.

The numerous breakage processes take place in the particle contacts by impact stressing of particle beds. The electrons emitted during the breakage are accelerated in the electrical field of newly created charged PZT surface. By the collision with air molecules, which penetrate into the cracks from the atmosphere, the molecules become ionized, accelerated and micro electrical discharge occurs.

Numerous micro gas discharges occur during compression of particle beds and strong friction between the particles takes place. The micro gas discharges generate microwave impulses in a wide-band frequency range with a characteristic duration of few nanoseconds.

Emission of PZT-Ceramic by Impact Stressing with a Sharp Tungsten Indenter

PZT Preparation

The PZT ceramic was formed by pressing process. In the first step, commercial PZT powder (Arburg GmbH) with a mean grain size of 1–5 μm and 91.5 vol.% was mixed with main binder Licomont EK 583 (Clariant International Ltd.)—6.0 vol.% by use of a dual screw extruder. In the second step, Polyethylene glycol 35000 (PEG 35000)—1.5 vol.% was mixed with a special resin 1.0 vol.% at the feedstock and pressed. After pressing process, the specimen has undergone the deboning process with temperature profile, recommended by the manufacturer. In the third step, the specimens were sintered at 450 °C for 40 h. Before polarization, the both seeds of specimen ware coved with silver of about 100 μm thickness. The polarization of specimen occurs by direct voltage with maximum amplitudes of 4.5 kV for 120 s.

Domain Switching in Ferroelectric as Source of Microwave Emission

According to Herzian dipole model, the full energy of microwave emitted from the dipole in all directions per unit time, i.e., the intensity of electromagnetic emission, is given by [20]

$$I_d = \frac{2 \cdot \ddot{\mathbf{p}}^2}{3 \cdot \varepsilon_0 \cdot c^3} \tag{1}$$

where $\mathbf{p}$, ε_0 and c are the dipole moment, the vacuum permittivity and the speed of light. In terms of this model the revolution, polarization or depolarization of dipole causes the emission of electromagnetic waves. By the depolarization the second derivative of the dipole moment can be estimated as

$$\ddot{\mathbf{p}} \approx \frac{\mathbf{d}q}{t_{dp}^2} \tag{2}$$

where t_{dp} is the depolarization time and **d** is distance vector between electrical charges q and –q in dipole. The Eq. (3) is valid for dipole with size d that can be neglected compared with size of receiver and distance between dipole (transmitter) and receiver. This condition can be realized by domain breakage or switching in ferroelectric.

The generation of microwave by stressing of PZT-particles occurs due to reorientation and depolarization of domains. The movement of domains lead the formation of micro crack with charged walls, see Fig. 1. An electron tunneling through an energy barrier takes place from inside of PZT into between of cracks walls [4, 12–14]. It is a very fast process with characteristic times of few picosecond. The emission current depends on the external electric field and structure of materiel but not of temperature. The Fowler–Nordheim equation [21] represents the emission current

$$I = a \cdot F^2 \cdot \exp(-b\varphi^{3/2}/\beta \cdot F) \tag{3}$$

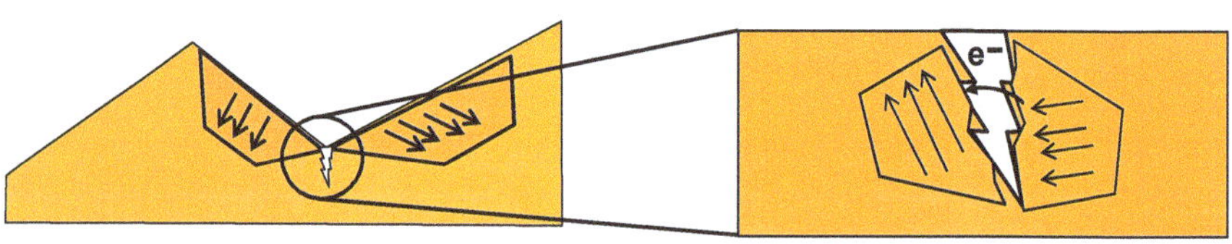

Fig. 1 The charging of micro cracks walls occurs due to reorientation of PZT domains

where I, φ and β are the emission current, the work function and the field enhancement factor, respectively.

F is usually taken as $F = \beta\ V/a$, where V is the applied potential and a the distance between cathode and anode. The initial charge configuration between cracks walls changes due to field emission. That leads to the creation of dipole moment and generation of microwaves. Electron tunneling does not occur for low applied voltages and current becomes exhibited after reaching the threshold voltage, which depends on the type of emitter. For example, conventional metal-emitting surfaces have threshold fields of ~100 V/μm [22]. These values are comparable with field fluctuations, which appear by stressing of PZT ceramics [23].

Experimental Set Up

The measurement set-up (Fig. 2) allowed simultaneous measurement of force, electric impulses and microwave emissions during mechanical impact on PZT. A tungsten pin (1) with a conical 60 μm/30° edging induced the mechanical impact on the PZT. The PZT-specimen (2) was positioned under the pin on a precision table (3). Movement of precision table can change the position of indenter on ceramic surface.

The data acquisition was carried out by means of a four channel oscilloscope "Pico 6403" (4) with a bandwidth of 350 MHz and a sampling rate up to 5 giga-samples per second. After the polarization, the conducting layer of the upper surface was removed from surface of ceramic to achieve direct contact between PZT-surface and pin. The upper surface of specimen was used as source of microwave emission. The

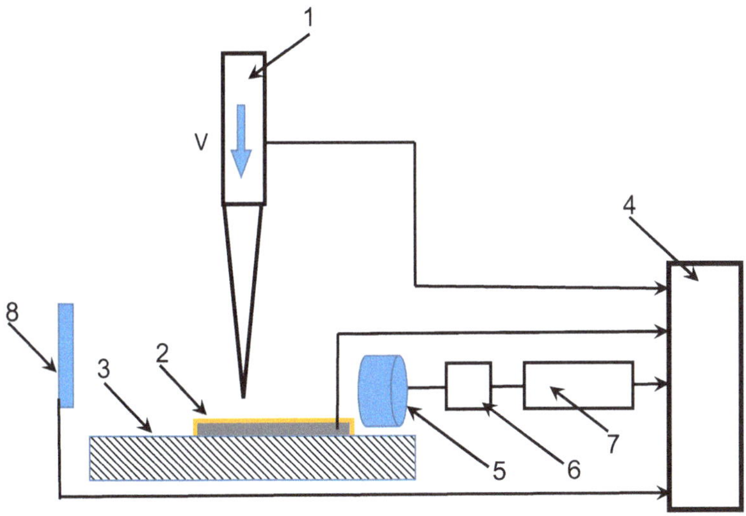

Fig. 2 Experimental set up

remaining metalized layer of specimen was connected to first channel of oscilloscope directly and was used as force sensor. In this way, it was possible to detect the applied mechanical force on the surface of specimen during the contact between specimen and pen. The direct connection between specimen and oscilloscope avoided the time delay caused by application of amplifier or other electronic circuits between signal source and receiving devices.

To measure the current impulses, which appear between the conical ending of tungsten indenter and the electrical ground, the indenter was connected to 50 Ω input of the second channel of oscilloscope. During the contact between the indenter and specimen the electric circuit was closed and current impulses were detected.

The detection of the microwave signal was realized by third channel of oscilloscope. A spiral antenna (5) with a receiving bandwidth from 4 to 18 GHz was fixed on a distance of 0.05 m from the specimen. This spiral antenna allowed the detection of microwaves from dipoles i.e. domains with random orientation. The amplification of signal took place by means of a broadband microwave amplifier 83017A (6). This type of amplifiers with 26.5 GHz bandwidth exhibit a gain of 25 dB. The coaxial diode 8473C (7) with negative polarity output converted the amplified high frequency signals to video signals. A rise time of 8–12 ns and a fall time of approximately 200 ns are typical for this type of diode. Detectable video signals with temporal resolution of about 10 ns were obtained.

The fourth channel was used to acquire a low frequency microwave signal. A WLAN-antenna (8) with carrier frequency of 2.3 GHz was mounted in a distance of 0.07 m from the specimen and connected to the 50 Ω alternating current input of oscilloscope.

Micro Discharge as Source of Microwave Emission from PZT

During the compression of PZT a potential difference induced in the PZT appeared. That led to generation of voltage, with amplitude that was proportional to impact force. Figure 3 shows a typical sequence of force impulses caused by repeated collisions of pen with specimen.

After first impact on the PZT-ceramic, the pin sprang back and repeated its impact with reduced velocity v. One can see, that the signal conceits of positive and negative components. The duration of positive signal was around 320 μs, see Fig. 4. This time corresponds to contact time between pin and specimen. It was measured by means of voltage applied to pin and remained metallically layer on the PZT-surface.

One can conclude that the high amplitude positive signal corresponds to mechanical compression of ceramic by pin and the negative decompression signal appeared after rebounding of the pin. Typical for these impulses is the amplitude modulation by ceramic oscillations with characteristic period of about 30 μs. This time corresponds to the time of sound wave propagation through the pin from the contact point to the end of pin and back. The oscillations are reproducible and the form of signal did not change by repeated impact. That means that time needed to achieve the

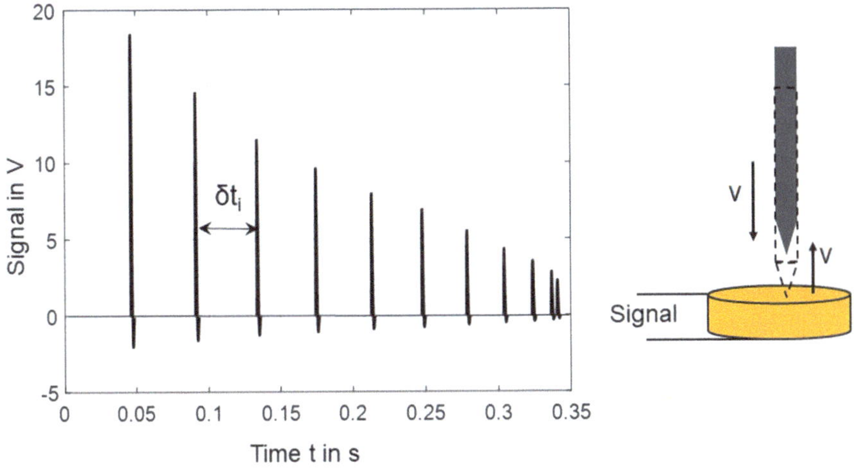

Fig. 3 Sequence of voltage impulses caused by repeated impact of pin

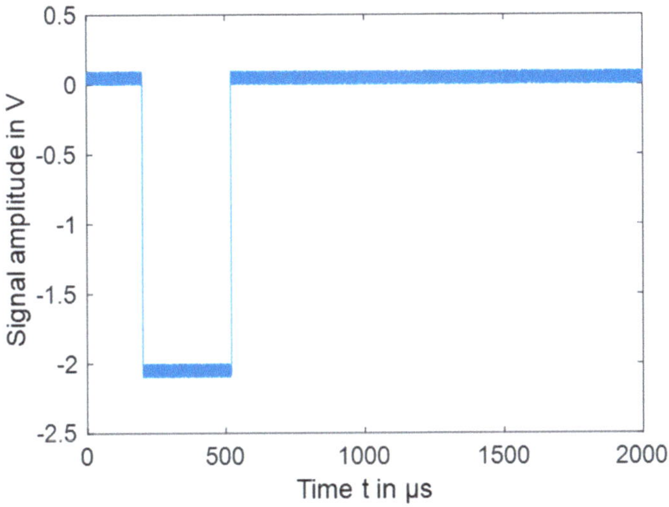

Fig. 4 Voltage applied to pen and remained metallically layer on the PZT-surface. Contact time corresponds to duration of negative impulse

maximum of signal does not dependent on collision number. For the description of ceramic behavior under impact stressing sensor signal it is necessary to describe the correlation between maximum of voltage impulse and mechanical impulse of pen

$$p = mv \tag{4}$$

where m is the mass (150 mg) and v is the velocity of the pin. The velocity was calculated from the time interval t between two impulses by

$$v_i = \frac{g\delta t_i}{2} \quad (5)$$

The linear relationship

$$p = kv_{max} \quad (6)$$

between the maximum voltage of sensor v_{max} and pin impulse p is shown in Fig. 5. The linear fitting of data was carried out with a coefficient of correlation $R = 0.96$. No saturation of signal was observed. That means that the deformation of specimen can be considered as elastic. In this case, the plastic deformation occurs only inside thin contact area between pin and surface of ceramic. In terms of elastic deformation, the impact generates an acoustic wave inside of specimen that is proportional to applied force and propagates with sound velocity. As result, the impact force behavior inside contact point becomes detectable by means of signal obtained from metalized surface of specimen that is located at distance of 10 mm to contact point. That means that the metalized surface of specimen can be used as force sensor.

It is important to consider the correlation between the impact force and generation of high frequency microwave impulses (HFMI), see Fig. 6. The intensive HFMI appear in the initial and finale phase of contact and corresponds to maximal change of contact surface during compression and decompression of indenter. The impulse amplitude caused by decompression is higher in comparison to the impulse amplitude during the initial compression phase. However, the impulse frequency is higher during the initial phase. The same behavior was observed for LFMI, see Fig. 7. Similar to HFMI the LFMI appear in the initial and final phase of contact and the impulse

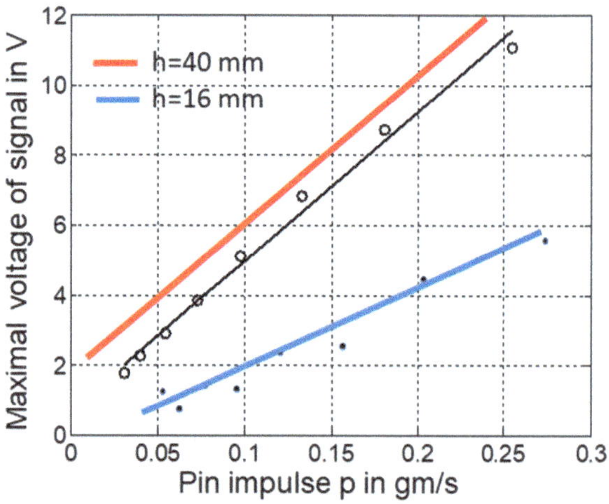

Fig. 5 Linear dependence of maximum sensor voltage on mechanic impulse of pen at different initial high

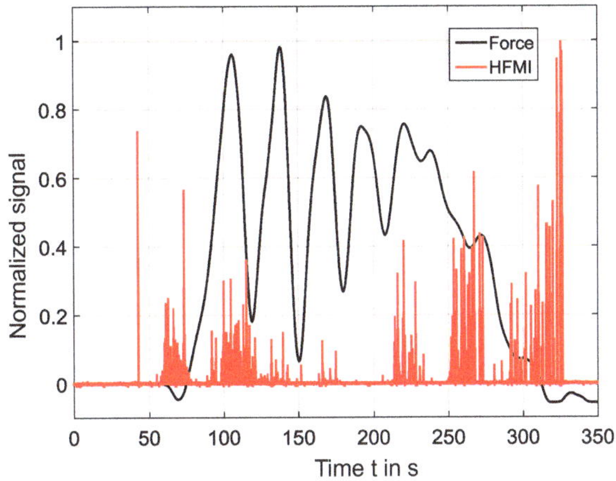

Fig. 6 Correlation between the high frequency microwave impulse and compression force

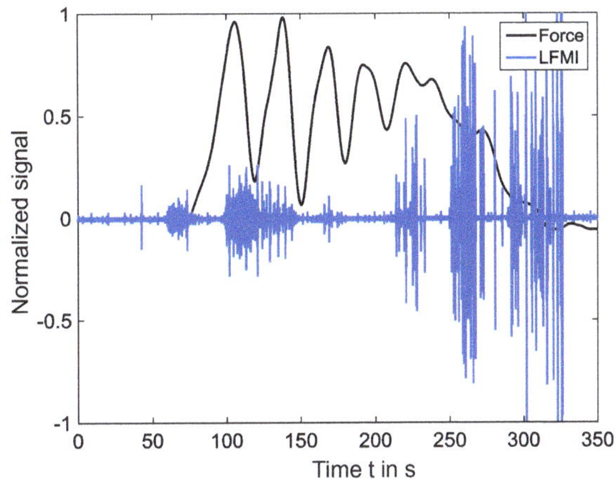

Fig. 7 Low frequency microwave impulse and compression force

amplitude increases during the final phase of contact. Based on this difference in amplitude, it becomes reasonable to assume that there are two different mechanisms of microwave emission. The first one occurs by compression and second one by decompression of PZT. During the compression of PZT at initial contact phase, field emission occurs from pin to positive charged sites on the PZT. This is enough for successful acceleration of ions and creation of micro gas discharges, which lead to generation of microwave impulses in a frequency range of few GHz. It is a very short process with duration in sub-nanosecond range [24].

The emission of electrons in air leads to generation of ions, which become accelerated in the electric field between two charged sites. Taking the tungsten-emitting threshold fields of ~100 V/μm [22] and site potential of about few kV into account one can estimate the distance between indenter surface and charged sites as few tens of micrometers.

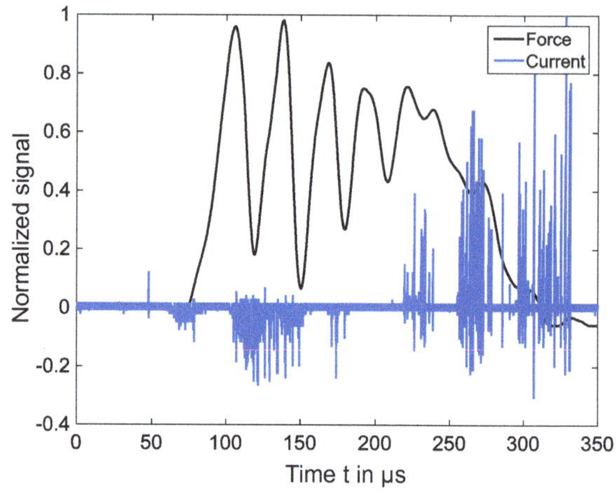

Fig. 8 Correlation between current impulse and force

The ferroemission is a result of spontaneous switching of domains caused by stress relaxation by decompression of PZT. In difference to field emission, ferroemission occurs in opposite direction i.e. from negative charged sites on PZT surface to indenter. That means that the sign of current impulses will be changed by transition from compression to decompression, see Fig. 8.

Because of the polarity of signals is changed by transition from compression to decompression stressing phase, one can used the amplitude distributions of HFMI or LFMI to calculate the contact time. By use of sensitive sensors, it becomes possible to obtain force behavior during the contact of solids from HFMI/LFMI signals. In this context, the HFMI impulses can be used to characterize the free impact of metallic particle on ceramic surface. For example, the contact time can be used to calculate the particle size by Herzian impact of spherical particle. Additionally, the amplitudes of impulses force allows the calculation of the particle velocity.

The generation of LFMI occurs by incoming of indenter to the positive charged site. In this case, the negative charge induces on the indenter. After discharge that occurs by means of field- or ferroemission, the induced charge remains uncompensated and the indenter becomes charged. The discharge of indenter occurs by means of currents flow from indenter to ground. These currents i.e. current impulses are responsible for generation of LFMI. In this case, the indenter serves as transmitter of LFMI. Without contact between indenter and ground, the LFMI becomes negligible.

It should be noted that very short microwave impulses are not directly detectable as real-time signal. It is possible to detect microwave emission only as a form of response of the used devices as part of fast microwave impulse.

The microwave impulse irradiation occurs in subnanosecond time scale. This duration is shorter than the rise time of the coaxial detector and oscilloscope used in this study. However, the amplitude of initial microwave is sufficient to generate a detectable response signal. The amplitude of response signal is proportional to

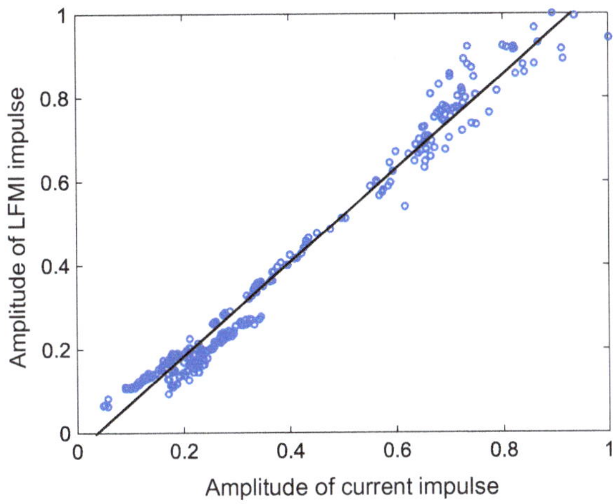

Fig. 9 Correlation between current impulse and force

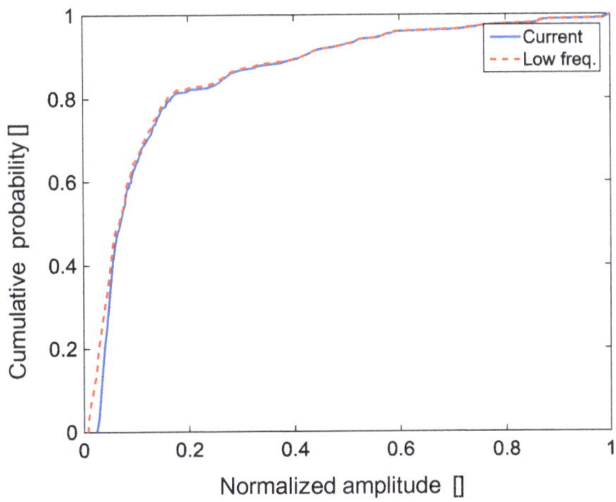

Fig. 10 Cumulative distributions of LFMI and current impulse

microwave amplitude. Taking this fact into account, it is reasonable to use signal amplitudes to characterize the microwave emission.

Figure 9 represents the linear relationships between the amplitudes current impulse and LFMI. The both type of impulses appears simultaneously and exhibit the similar distribution, see Fig. 10.

In terms of this, the pen indenter can be conceded as a transmitter with a size, comparable with the size of receiver (WLAN antenna). The electric field that is responsible for charge (signal) induction on the WLAN antenna can be calculated according to Eq. (2).

The received LFMI signal is proportional to electric field i.e. to induced charge q. On the other hand, the discharge time remains the same during measurement. Here, the amplitude of current impulse is proportional to charge q and, consequently, to amplitude of LFMI, too.

Conclusions

Test of microwave emission caused by impact of a sharp tungsten indenter on the surface of PZT, was carried out. It was found that the microwave impulses appears during the plastic deformation of PZT surface and formation of micro cracks. The impulse amplitude is proportional to deformation rate i.e. charge generated on surface or walls of micro cracks. Thus it becomes possible to monitor the deformation and breakage inside of particles. Based on the technique of microwave monitoring, it becomes possible to develop a wireless stress sensor with high temporal and spatial resolution for impact measurements.

Microwave Response on Particular Bed Stressing

Materials and Methods

The initial PZT ceramic specimens had the form of circular rings. In the first preparation step the rings were cleaned using a short acetone bath. In the second step, the PZT rings were ground to powder in a disk mill. The different sized PZT particles were produced and separated in a sieve stack with different sieve openings. The grain boundaries d of the sieve opening were as follows: 500 μm > d > 315 μm, 315 μm > d > 225 μm, 225 μm > d > 100 μm and d < 100 μm.

The first test particles were spherical glass particles with density of 2500 kg/m^3, E-Modulus of 73 GPA and $d_{3,90} = 7.8$ μm. These glass particles, called soda lime glass microspheres (float glass), were produced by Cospheric Innovations Microtechnology. The mechanical properties of glass and PZT particles are listed in Table 1. It can be assumed that the hardness of the glass spheres exceeds the hardness of the PZT particles tenfold.

Table 1 Properties of glass and PZT-particles

	**Glass** E-Modulus: 73 GPA Density: ϱ : 2500 kg/m³	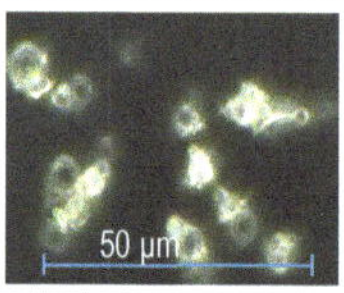	spherical, smooth, $d_{3,90}$=7,8 μm
	**PZT** E-Modulus: 60 GPA Density ϱ : 7800 kg/m³	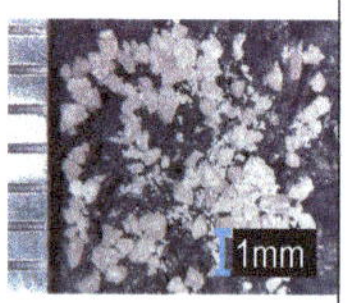	irregularly shaped, $d_{3,90}$=1000 μm

Experimental Set Up

It is necessary to find out the conditions for a reproducible and adequate electromagnetic radiation of the PZT. For this purpose, a plexiglass pipe (Fig. 11) was placed over a basis made of hardened steel, which represented the measurement cell. Because of the transparent wall of pipe, it was possible to film and observe the compression processes directly. At first, the stressing chamber was filled up to 3 mm high with PZT particles, compressed and then filled with 15 mm glass powder. Before start of impact stressing, glass and PZT particles were compressed with force of about 10 N. The compression step was necessary to avoid whole layer shifting and destruction. It avoided also mixing of glass and PZT-particles. The movement of air that remains in uncompressed particle bed causes this effect. The weight of tested particles was measured before the impact compression in order to obtain a specific weight at certain height and to ensure an approximately similar compression rate in the experiments.

A pressure sensor with a measuring range of up to 50 kN and a high sensitivity of −4.1 pC/N from HBM was located on the ram cylinder, see Fig. 12. The operating principle of this sensor based on the crystal structure of a semiconductor (Si). The crystal structure of semiconductor was influenced by compression and the specific resistance of the element changes strongly. A resistance displacement sensor was mounted at the ram cylinder (Fig. 3) with 0.5% accuracy. This sensor is characterized by its high sensitivity and temporal resolution. Due to this, it is ideally suited for displacement registration during impact compression within the measuring cell.

Figure 13 represent the whole experimental setup. The falling weight (1) with a mass of 1118 g slid over a linear guide (2) from the falling heights 120, 150 and 180 mm on the particle bed. A linear guide make it possible to avoid possible slipping and twisting effects that could adversely affect the measurement. The displacement sensor (3) was fixed to the ram cylinder. At the metal plate (4) locates the stressing

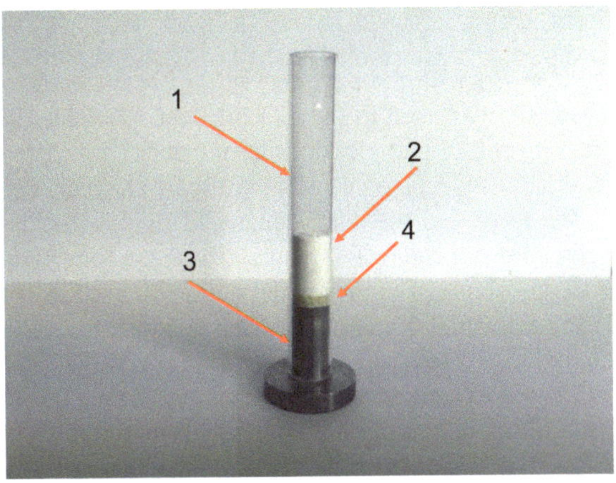

Fig. 11 Stressing chamber; 1-plexiglass pipe, 2-glass particles, 3-hart metal basis, 4-PZT particles

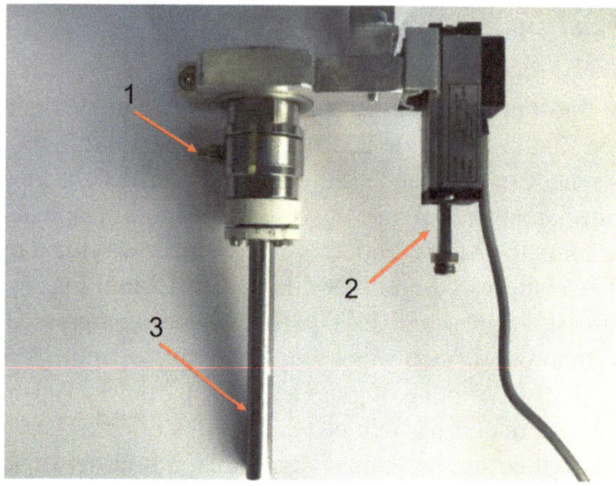

Fig. 12 Stressing chamber; 1-plexiglas pipe, 2-glass particles, 3-piston, 4-PZT particles

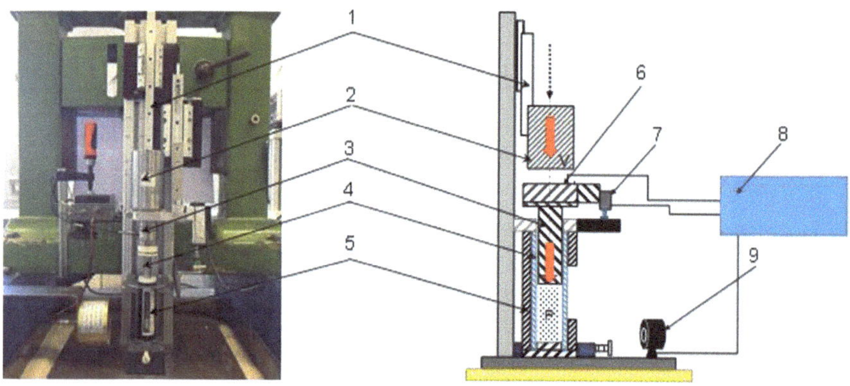

Fig. 13 Stressing equipment; 1-falling weight, 2-linear guidance, 3-displacement sensor, 4-metallic plate, 5-stressing chamber, 6-piston, 7-force sensor, 8-oscilloscope, 9-antenna

chamber (5) with the piston rod (6). The diameter of rod is 9.75 mm. The force sensor (7) was fixed to the ram cylinder. The electromagnetic signal generated by stressing of PZT particles during compaction was detected by antenna (8). The signal i.e. microwave data accusation occurs by channel A of oscilloscope (9).

The force and the displacement of the particle beds during compression were recorded using channels B and C. Here, the time signal on the channel A was used as trigger signal. The measuring scale was used in two ways; firstly for orientation during the initial filling of stressing chamber and, secondary, for acquisition of the approximate displacement during impact compression.

The test experiments have shown that the particle size of approximately 500 μm and a filling height of approximately 3 mm corresponded to the highest amplitude of the microwave pulses. In terms of this, it was assumed that during the mechanical stress, the best conditions for microcracks at the contact points between the particles and deformation of particles were present. The characteristic duration of the physical micro processes was in the micro and nano-seconds range.

Deformation Behavior of Particle Bed by Impact Stressing

In order to study the compaction process precisely, a high-speed camera (Fig. 14) was placed in front of the stressing cell. The high-speed EoSens CL MC1362 camera was produced by Tokina Macro. Due to the technology of the CMOS chip, it is better to minimize the image height to increase the frame rates per second. Thus, the image height was adjusted to obtain a maximum number of recordable images. The images were acquired with a resolution of 320 × 516 pixel and a recording speed of 2704 frames per second (time interval of 365 μs). In the settings of camera parameters, the resolution and the recording speed depend on each other directly.

The stressing cell was filled with PZT and glass powder for recording. An approximately 2 mm thick layer of glass particles colored with blue ink was incorporated in the middle of glass layer to obtain the reference points. The end of the stamp was marked by means of blue color ink, too. It was assumed, that there is no friction between the punch and the vessel wall of the measuring cell. The recorded images show that the previously described initial compression of particle bed was successful and the marked layer moves without mixing with others particles. One can see the upper layer shift versus middle layer (color marked). This shift was achieved

Fig. 14 High-speed camera EoSens CL MC1362

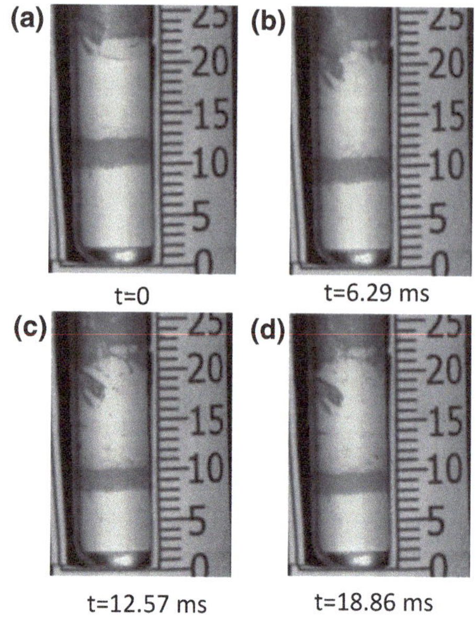

Fig. 15 Images of particulate bed obtained by high-speed camera EoSens CL MC1362

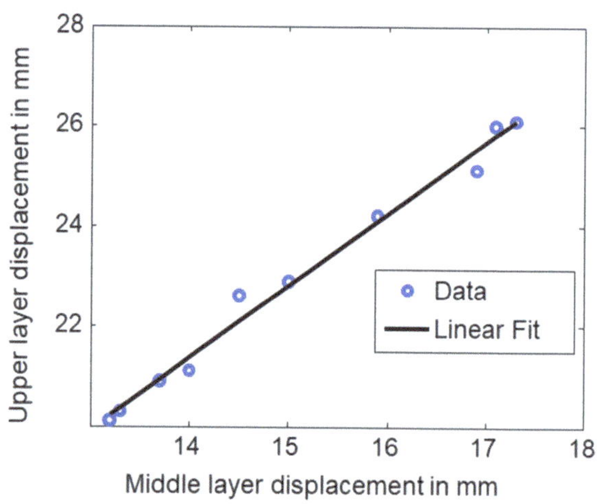

Fig. 16 The upper/middle shift dependence

by means of analyses of images obtained at different times. The upper/middle shift dependence is linear. That means that wall friction is low and their contribution to resistance force by impact stressing can be neglected (Figs. 15, 16 and 17).

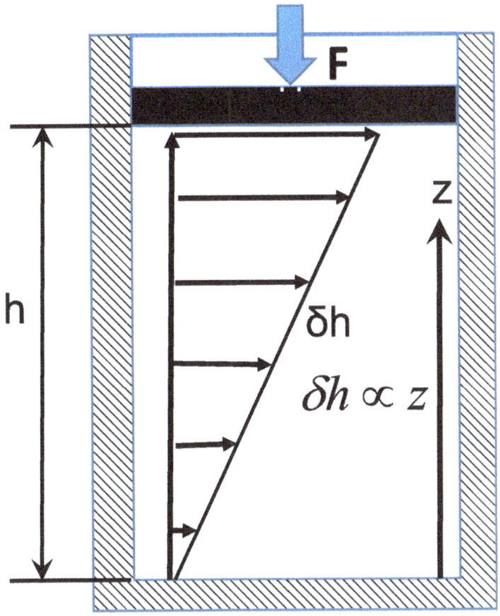

Fig. 17 The shift of particle layer increases linear with high of layer

Static and Dynamic Resistance of Particle Bed

Under applied impact loading, the particle bed moves and its compression take place. Due to regular spherical form and smooth surface of tested glass particles, it can be assumed, that the contribution of van der Waals surface adhesion in the static friction was dominant. To start the movement the applied force have to be overcome particle adhesion within the bed.

This means that the force and displacement are no time-synchronous and the start of bed movement occurs with delay regarding to applied force. Figure 18 shows the force and bed displacement as function of time. For the purpose of better characterization of the temporally behavior of the measured quantities the signals of force and displacement sensors are presented in normalized form. The falling height in this case was 120 mm.

The red, yellow and green rectangles in the figure mark the regions of static, mixing and dynamic friction. Approximately 2.6 ms after starting force signal appearance, the movement of particle bed begins. One can see that the movement of bed started after reaching the first locale force maximum. During static friction an elastic and plastic deformation of particles occurs [25–28]. The form of particles changed not considerably due to high E-module of glass particles. The shift of particle layers was not detectable with applied displacement sensor. The magnitude of resisting force increases due to particle deformation until it reaches the threshold value. The bed consolidation started after reaching the threshold force with increasing packing density and relative movement of particles.

In this moment the reorientation and depolarization of domains in PZT ceramic was caused by reduction of mechanical stress at individual contact points. Because of

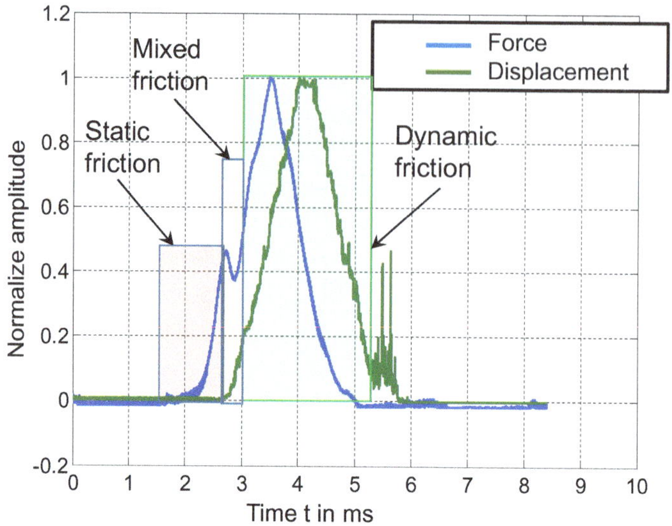

Fig. 18 Signals of force and displacement sensors as a function of time. Falling high is 120 mm

Fig. 19 Microwave emission by impact stressing of particle bed

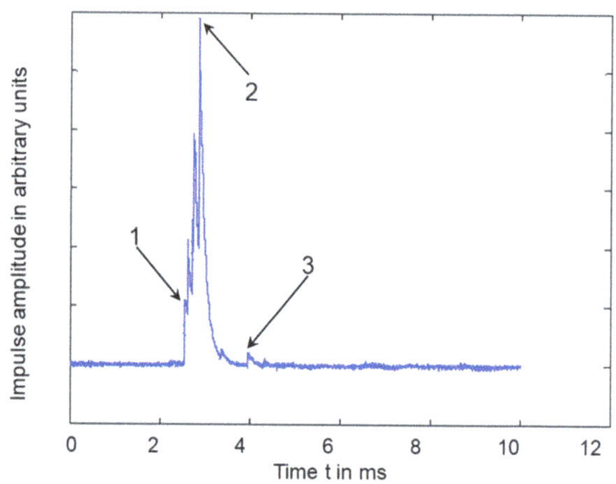

this, the generation of voltage differences took place. An electron emission occurred between the differently charged PZT particles generated at individual highly charged sites. Thereafter, the surrounding gas was ionized, and subsequent impact ionizations generated micro gas discharges, which resulted in impulsive microwave emission. Figure 19 shows an electromagnetic pulse that corresponds to the force and displacement curves shown in Fig. 18. The first impulse (1) that overlapped many small fracture processes started at time of the first locale maximum of resistance force i.e. at the beginning of the bed consolidation.

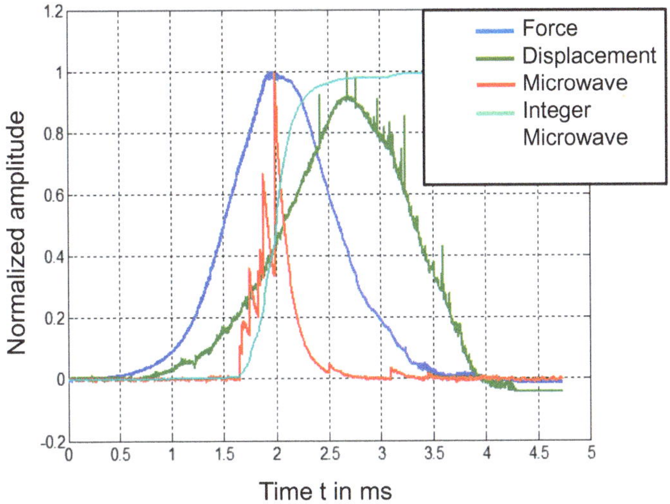

Fig. 20 Signals of force, displacement and microwave sensors as function of time. Falling height is 180 mm

The maximum of microwave impulse (2) corresponds to maximum force. In this time, the displacement changed with high speed and the maximum of breakage rates reached. The subsequent pulses (3) show the stress relaxation in bed. They exhibit only a small deflection in relation to the main load phase. The subsequent pulses (3) show the stress relaxation in bed. They exhibit only a small deflection in relation to the main load phase. The maximum of microwave impulse (2) corresponds to maximum force. In this time, the displacement changed with high speed and maximum breakage rates reached.

Similar behavior of microwave impulse was observed during high energy impact for falling heights of 180 mm, see Fig. 20 However, increase of falling height i.e. applied energy changes the stressing behavior of particle bed. The first local maximum of resistance force disappears and displacement started with low delay of about 0.2 ms regarding to force. The microwave emission started not simultaneously with displacement but with delay of about 0.8 ms. Similar to low energy impact, the maximum of microwave impulse corresponds the maximum of applied force. Based on results shown in Figs. 18 and 19, it becomes reasonable to assume that the consolidation of beds started with microwave emission. Indeed, the intensity of microwave emission by particle deformation is very low and not detectable. However, it becomes detectable by relative movement of particles.

Conclusion

It was shown that microwave-based method is applicable for characterization breakage processes by impact stressing of particle beds. The experiments confirm that the movement of particles by bed consolidation is detectable. The microwave emission started with consolidation of particles i.e. by formation of micro cracks. The time interval where the particle deformation occurs can be obtained by means of combination of force, displacement and microwave sensors. Based on this, the impact energy stored in plastic deformation can be calculated from force-displacement curve. Indeed, the intensity of the microwave emission by particle deformation is very low and is not detectable.

Moreover, in the first part of this paper it was shown that the integrated intensity of microwave emission is proportional to charge generated by particle breakage i.e. to newly created surface of particles. The microwave energy varies depending on stressing conditions. In this way, one can optimize the stressing parameters: layer thickness, particle size, velocity and energy of impact to reduce the energy consummation by grinding in particle bed.

References

1. Müller, F.: Hochdruckzerkleinerung im Gutbett bei Variation von Feuchte und Beanspruchungsgeschwindigkeit. Technische Universitat Clausthal (1989)
2. Schonert, K., Müller, F., Schwechten, D.: Comperssion an energy absorption at interparticle breakage. ZEMENT-KALK-GIPS **43**(2), 65–70 (1990)
3. Schubert, H.: Zu einigen Fragen der Kollektivzerkleinerung. Chem. Techn. **19**(10), 595–598 (1967)
4. Aman, S., Aman, A., Morgner, W.: Monitoring of carbon fibre breakage in composites based on microwave emission. Compos. Sci. Technol. **84**, 58–64 (2013)
5. Aman, A., Majcherek, S., Schmidt, M.-P., Hirsch, S.: Microwave sensor for mechanical stress measurement based on ferroelectric graphene nanosheet composites. Procedia Eng. **87**, 124–127 (2014)
6. Aman, A., Majcherek, S., Hirsch, S.: Microwave emission of carbon fibres during electrical breakdown. Sens. Lett. **13**(1), 98–101 (2015)
7. Aman, A., Aman, S., Martin, M.: Sensoreinrichtung und Verfahren zur Detektierung und Lokalisierung von Rissen in Bauteilen, DE102012006155A1
8. Koktavy, P.: Experimental study of electromagnetic emission signals generated by crack generation in composite materials. Meas. Sci. Technol. **20**(1), 15704 (2008)
9. Srilakshmi, B., Misra, A.: Secondary electromagnetic radiation during plastic deformation and crack propagation in uncoated and tin coated plain-carbon steel. J. Mater. Sci. **40**(23), 6079–6086 (2005)
10. Ogawa, T., Oike, K., Miura, T.: Electromagnetic radiations from rocks. J. Geophys. Res. Atmos. **90**(D4), 6245–6249 (1985)
11. Aman, S., Tomas, J., Molitor, M., Aman, A.: Investigation of a deformation-breakage-parameter by the use of microscopic discharges of gas during the propagation of micro cracks and breakage of particles. Materialwiss. Werkstofftech. **43**(12), 1001–1005 (2012)
12. Dickinson, J.T., Park, M.K., Donaldson, E.E., Jensen, L.C.: Fracto-emission accompanying adhesive failure. J. Vac. Sci. Technol. **20**(3), 436–439 (1982)

13. Dickinson, J.T., Jahan-Latibari, A., Jensen, L.C.: Electron emission and acoustic emission from the fracture of graphite/epoxy composites. J. Mater. Sci. **20**(1), 229–236 (1985)
14. Dickinson, J.T., Donaldson, E.E., Park, M.K.: The emission of electrons and positive ions from fracture of materials. J. Mater. Sci. **16**(10), 2897–2908 (1981)
15. Chandra, B.P.: Mechanoluminescence. Springer Science & Business Media (1998)
16. Barré, S., Benzeggagh, M.L.: On the use of acoustic emission to investigate damage mechanisms in glass-fibre-reinforced polypropylene. Compos. Sci. Technol. **52**(3), 369–376 (1994)
17. Morgner, W.: On-line application of acoustic emission analysis. J. Acoust. Emiss. **8**(1–2), 570 (1994)
18. Aman, S., Aman, A., Majcherek, S., Hirsch, S., Schmidt, B.: Microwave based method of monitoring crack formation. Meas. Sci. Technol. **25**(2), 25014 (2014)
19. Aman, A., Majcherek, S., Hirsch, S., Schmidt, B.: Microwave emission from lead zirconate titanate induced by impulsive mechanical load. J. Appl. Phys. **118**(16), 164105 (2015)
20. Landau, L.D., Lifšic, E.M.: Lehrbuch der theoretischen Physik: In 10 Bändenn 1, 14th edn. Verlag Europa-Lehrmittel, Nourney, Vollmer GmbH et Co. KG, Haan-Gruiten (2016)
21. Cheng, Y., Zhou, O.: Electron field emission from carbon nanotubes. C. R. Phys. **4**(9), 1021–1033 (2003)
22. Gomer, R.: Field Emission and Field Ionization (AVS Classics of Vacuum Science and Technology). AIP, College Park, MD (1993)
23. Flechtner, D., Golkowski, C., Ivers, J.D., Kerslick, G.S., Nation, J.A., Schächter, L.: Electron emission from lead–zirconate–titanate ceramics. J. Appl. Phys. **83**(2), 955–961 (1998)
24. Sud'enkov, Y.V.: Electromagnetic radiation induced by the failure of piezoelectrics under the action of submicrosecond stress pulses. Tech. Phys. **46**(12), 1588–1590 (2001)
25. Müller, P., Antonyuk, S., Tomas, J.: Simulation des Druck-und Stoßvorgangs von Zeolith 4A-Granulaten. Chem. Ing. Tech. **83**(5), 643–651 (2011)
26. Antonyuk, S., Heinrich, S., Tomas, J., Deen, N.G., van Buijtenen, M.S., Kuipers, J.A.: Energy absorption during compression and impact of dry elastic-plastic spherical granules. Granul. Matter **12**(1), 15–47 (2010)
27. Antonyuk, S., Heinrich, S., Deen, N., Kuipers, H.: Influence of liquid layers on energy absorption during particle impact. Particuology **7**(4), 245–259 (2009)
28. Tykhoniuk, R., Tomas, J., Luding, S., Kappl, M., Heim, L., Butt, H.-J.: Ultrafine cohesive powders: from interparticle contacts to continuum behaviour. Chem. Eng. Sci. **62**(11), 2843–2864 (2007)

Formation, Deformation, Rolling and Sliding of Particles and Particle Aggregates: Mechanisms and Applications

Laurent Gilson, Jennifer Wenzl, Maxime Paven, Michael Kappl, Hans-Jürgen Butt, Doris Vollmer and Günter K. Auernhammer

Introduction

The mechanical properties of particles depend on a variety of parameters, including their size, polydispersity, surface modification and surrounding medium [1]. To investigate the formation of particular aggregates, we focused on small particles, whereas for the investigation of the rotational motion of individual particles and their neighbors, micrometer sized particles are well-suitable [2].

Different forces and mechanisms determine the static and kinetic properties of particles and their aggregates [3]. Typically, particles experience an attractive van der Waals force [4]. Van der Waals forces can be minimized or even excluded if the particles are dispersed in a medium of identical refractive index [5]. Now, cohesive interactions can dominate aggregation. Depending on the polydispersity, particles can form crystals, although most common are amorphous structures [6]. Particles can also be stabilized by sintering the particles together or by coating particles with a thin shell, resulting in the formation of porous colloidal networks. The porosity of the colloidal networks can vary greatly and even be as low as a few percent [7]. As a consequence, the shear stress is not equally distributed over the entire surface area, like in the case of a flat surface. The real contact area between a porous colloidal network and an abradant is greatly reduced. Therefore, the stress acting on the topmost particles is much higher than the stress acting on a flat surface instead. Once a critical stress is exceeded, individual colloids can break off. This reduces the surface roughness and can change the chemical composition of the topmost surface

L. Gilson · J. Wenzl · M. Paven · M. Kappl · H.-J. Butt · D. Vollmer (✉) · G. K. Auernhammer
Max Planck Institute for Polymer Research, Physics at Interfaces, Ackermannweg 10, 55128 Mainz, Germany
e-mail: vollmerd@mpip-mainz.mpg.de

G. K. Auernhammer (✉)
Leibniz Institute for Polymer Research, Institute of Physical Chemistry and Polymer Physics, Hohe Straße 6, 01069 Dresden, Germany
e-mail: auernhammer@ipfdd.de

S. Antonyuk (ed.), *Particles in Contact*,
https://doi.org/10.1007/978-3-030-15899-6_4

[8]. Therefore, the mechanical durability results from the interplay of several minute details: Beside of the porosity of the network, elasticity, hardness and thickness of the coating are essential. The interplay of these factors leads to surfaces spanning a broad range in terms of mechanical durability. Unfortunately, this hinders the adoption of a unique and generalized testing procedure to evaluate the mechanical strength and durability of porous colloidal networks. To investigate their mechanical properties, different qualitative methods are commonly used, including sandpaper abrasion, sand abrasion, tape peeling, pencil scratching [9], as discussed in section "Fabrication, Characterization and Applications of Porous Colloidal Networks". Qualitative tests have the advantage that they are generally cheap and quick. The drawback is that they do not offer detailed insights into the mechanical properties of samples. Quantitative measurements include atomic force microscopy (AFM) [4, 10] and nanoindentation [11]. Unfortunately, these measurements are typically challenging because of the low mechanical strength of porous colloidal networks. To understand the advantages and challenges of different strategies, we discuss the mechanical properties of highly porous surfaces using pencil scratching, atomic force microscopy, the colloidal probe method [12, 13], and nanoindentation.

While in the second part of this chapter we analyzed rigid networks of particles, in section "Dynamic Microscopic Investigations with Single Particle Resolution" we are focusing on particle agglomerates which are only physically linked and can react to external stimulus such as shear. The goal of the second part is to discuss possibilities of observing single particle motion by laser scanning confocal microscopy. To resolve the motion of single particles with high special resolution, the particle size should be in the sub-micrometer range or even larger. Therefore, granular matter is used. Granular matter covers a whole range of common materials, from fine sands, soil, table salt or sugar up to material systems with larger particles size like charcoal or iron ore.

Handling granular matter at industrial scale is a complex topic, as there is no universal understanding of the processes during the motion of granular matter under external mechanical stress. Transport with band conveyors or down chutes involves inertia driven motion, shear, compression, fluidization and related stress modes. Particle size, size distribution, densities, surface morphology, surface roughness and the surrounding fluid phase all influence flow and stresses. At small length scales, inter-particular forces have to be considered. One important mechanism for inter-particular forces are liquid bridges between the particles. Liquid bridges give rise to capillary forces, which can be dominant in the μm regime.

In order to correlate the interactions between individual particles and the macroscopic behavior of granular matter, various steps on different time and length scales are necessary [1]. The challenges in three-dimensional systems include the exact knowledge of movement and deformation of the particles, the visualization of the force lines and the determination of the rolling and sliding friction coefficients.

Despite the enormous number of papers published in the field, only few experiments are able to resolve processes like shear in granular matter at single particle level for small, hard particles [14–21]. Even fewer groups [17–21] are working with wet granular matter and at single particle resolution. Correct calibration of DEM simulations is still an issue, especially for systems including liquid bridges [22]. To our knowledge, if at all, only very few experimental datasets of such systems with sufficient quality are available. We aim to provide such datasets for hard sphere systems and for wet systems. Our chosen method (confocal imaging) will provide base data, while detailed analyses and chemistry will fill in the missing data.

With a combination of piezo-mechanical nano manipulators and confocal microscopy, the trajectories of each particle can be tracked during mechanical deformation. Additionally, it is possible to determine their rotational motion. How do the particles and particle agglomerates deform under shear or compression? Are relative movements of the particles dominated by rolling or sliding? Results will be compared with other macroscopic measurements (e.g. on shear rates) and simulations.

Fabrication, Characterization and Applications of Porous Colloidal Networks

The poor mechanical stability of highly porous colloidal networks limits their use for industrial applications. Particles easily break off as soon as a critical stress is exceeded [8, 9, 23, 24]. Therefore, a better understanding of the mechanical stability of nanoporous colloidal networks is desirable. However, measuring and comparing mechanical quantities is not trivial. Significant differences in surface roughness, thickness, bulk material, and mechanical stability exist, which hinders the adoption of a single generalized testing procedure. To investigate the mechanical properties, different qualitative methods are used [9]. Qualitative tests have the advantage that they are generally cheap and quick. The drawback is that they do not offer detailed insights into the mechanical properties of samples. It is also difficult to compare samples of similar mechanical stability.

Quantitative measurements include atomic force microscopy (AFM) [4, 10] and nanoindentation [11]. Unfortunately, these measurements are typically challenging because of the low mechanical strength and the high porosity of the layer. To neglect contributions of the substrate to the measured forces, the penetration depth should not exceed 10% of the film thickness [25, 26]. In case of high surface roughness, the size ratio of the indenter to the surface features influences the results. After an introduction in the preparation of our strategy to fabricate porous colloidal networks, we discuss the advantages and challenges of different strategies of different measurement strategies, including pencil scratching, atomic force microscopy, the colloidal probe method [12, 13], and nanoindentation.

Fabrication of Soot Templated Networks

A simple and versatile technique to fabricate highly porous surface of nanoparticles is based on the collection of carbon soot from a paraffin candle [27–29]. The surface to be coated, in our case a glass slide or silicon wafer, is held in the flame of the candle (Fig. 1). Investigation of the surface by scanning electron microscopy reveals that the collected soot layer is composed of carbon beads with diameters of 30–50 nm, forming a fractal-like network (Fig. 1). The thickness can be tuned by the sooting period [28]. The structure is fragile because the network formed out of particle chains are held together by Van der Waals forces. Even a water droplet, rolling over the surface carries soot particles away. To increase the mechanical stability of the soot network, it was stabilized by coating the particles with a silica shell making use of chemical vapor deposition (CVD) of tetraethoxysilane [30]. Therefore, the soot-coated substrates were placed in a desiccator together with two open glass vessels containing about 2 ml of tetraethoxysilane and aqueous ammonia solution, respectively. Chemical vapor deposition of tetraethoxysilane was carried out for 1–3 days, Fig. 1. Similar to a Stöber reaction [31], silica is formed by hydrolysis and condensation of tetraethoxysilane catalyzed by ammonia [32]. The shell thickness increases with the duration of chemical vapor deposition. After 24 h the particles are coated by a 25 ± 10 nm thick silica shell. The shell thickness increases to approximately 110 nm after three days of chemical vapor deposition [9].

To investigate the effect of temperature on the mechanical properties of the porous network, the samples were annealed at a temperature of up to 1150 °C. During annealing, the carbon cores thermally degraded and carbon diffused through the silica shell. Annealing the carbon/silica network at temperatures above 500 °C in air results in a network formed out of hollow silica shells, Fig. 2. Annealing goes in

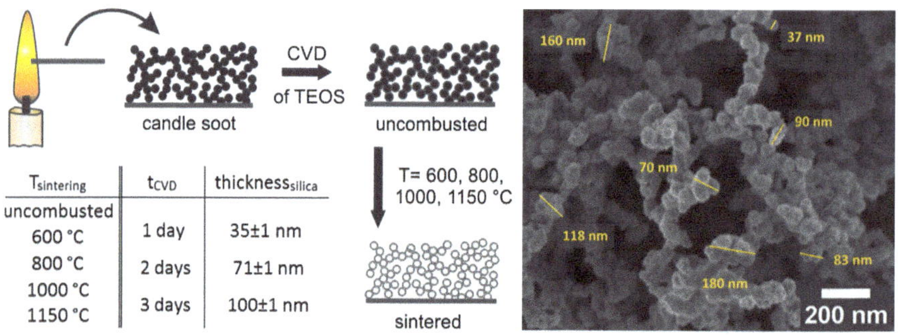

Fig. 1 (left) Sketch of preparing a porous network based on the collection of candle soot. Deposition of silica by chemical vapor deposition (CVD) of tetraethoxysilane. Combustion of carbon at different temperatures. Hydrophobization with a fluorosilane (PFOTS). Inset: Summary of reaction parameters, i.e. sintering temperature $T_{sintering}$, time of exposure to chemical vapor deposition of silica t_{CVD} in days and the respective silica shell thickness measured by ellipsometry on a silicon wafer coated under identical conditions. Right: SEM image of a templated silica network. After chemical vapor deposition of tetraethoxysilane for 24 h, the sample was annealed at 600 °C

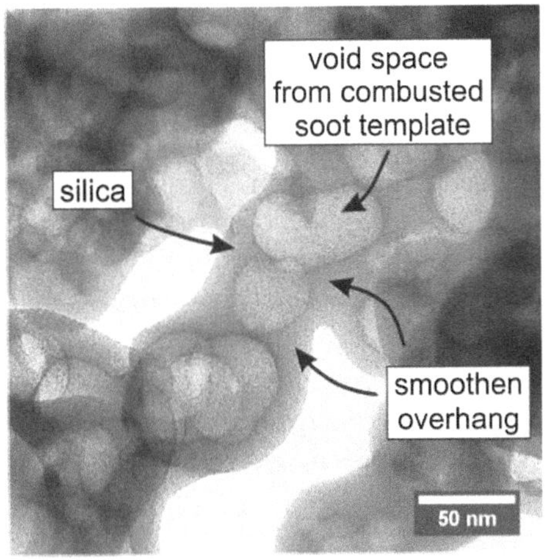

Fig. 2 TEM image of a soot-templated silica surface exposed to 72 h CVD annealed at 600 °C

hand with a decrease of the shell thickness. However, the shells remained intact, and the layer kept its roughness and network texture, Fig. 2. The silica shell started to sinter when annealing the soot layer at temperatures above approximately 1100 °C [9, 33]. This is reflected in a smoothening of the silica strings and a decrease of the thickness of the coating.

Soot-templated surfaces have the nice property that their wetting properties can be widely tuned by changing the chemical properties of the surface. Whereas the just prepared soot network shows super-water-repellent properties [29], it becomes super-hydrophilic after coating the soot with silica. After fluorination, the soot templated network shows super-water and super-oil repellent properties [27]. This implies that a deposited water and oil droplet roll off if the surface is tilted by less than 10°. Both water and oil droplets partially rest on the fluorinated network and partially on air, which is entrapped within the network.

Hydrophobization of the soot-templeated network is based on the reaction between the Si–Cl groups on the silane molecule and –OH groups on the silica surface:

$$\mathrm{Cl\ SiR_3 + HOR' \rightarrow R'OSiR_3 + HCl}$$

In our case $\boldsymbol{R'}$ represents the silicon oxide surface and $\boldsymbol{R}$ the fluoralkyl chain. After the reaction silica surface is covered by the semi-fluorinated silane which lowers the surface energy of the surface. Chlorosilanes can also react with water (R′ = H), so it is favorable to carry out the reaction in aprotic solvents, in an inert gas atmosphere or under reduced pressure. For example, the surfaces to be coated were put in a desiccator for approximately 1 h together with an open glass vessel containing about 0.1 ml of the semi-fluorinated silane and the pressure was reduced to 0.150 bar [27].

Mechanical Characterization

Pencil Scratching

To quantify the mechanical strength of the soot templated network on a macroscopic scale, we scratched the surfaces with a pencil [9]. Pencil scratch test is a common and quick test in industry. 6 B pencil was moved over the surface at an angle of 45° while applying a load of approximately $\mathbf{10^{-2}\,kg}$. This corresponds to a pressure P of approximately $\boldsymbol{P} = \frac{F}{A} \approx \frac{10^{-2}\,\mathrm{kg}\cdot 10\,\mathrm{ms}^{-2}}{\pi\cdot 0.5^2\cdot 10^{-6}\,\mathrm{m}^2} \approx \mathbf{10^5\,N\,m^{-2}}$, assuming a contact diameter of **1 mm**. To characterize the damage, the optical and electron microscopy images of the left trace were compared. Under these conditions, the width and depth of the damage stripe decreased with increasing sintering temperature.

Nanoscale Mechanical Test Using the Atomic Force Microscope

By scaling down the size of the impacting tip, details about the breaking mechanism of such soot-templated surfaces can be revealed. Using an atomic force microscope (AFM, JPK Nanowizard3) with tips as small as $r_{tip} \approx 10$ nm, mechanical properties at the nanoscale were investigated The tip radius is much smaller than the characteristic size of the silica protrusions and the average distance between silica protrusions (about 1 μm).

After the tip made contact with the network, force jumps of up to 10 nN show up in the approach curve and of 2–5 nN in the retract curve (Fig. 3). The small jumps in the approach part are likely caused by the force required to move or deform a single string (Fig. 3b). The larger jumps are likely caused by breaking single necks. Attractive surfaces forces keep broken pieces in contact with the tip or with the porous layer.

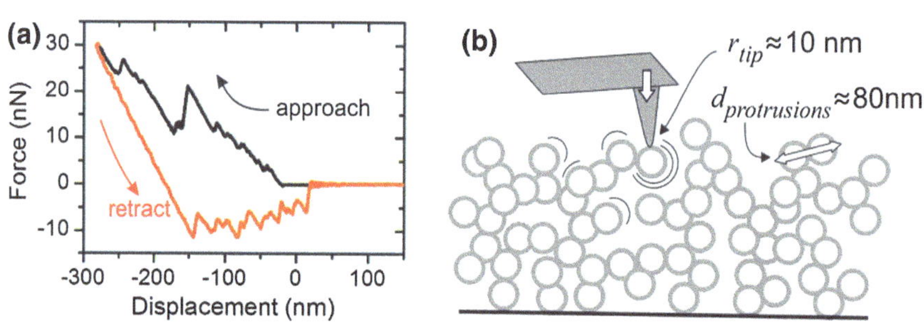

Fig. 3 **a** Force-displacement curve of a soot-templated silica surface. The soot was exposed for 24 h to chemical vapor deposition of tetraethoxysilane and after that annealed at 600 °C. **b** Schematic indentation of a soot-templated surface by an AFM tip. The probe radius is about 10 nm and thus much smaller than the typical size of silica protrusions. Reprinted with permission from [28]

At a scale of 100 nm, the relevant surface forces are Van der Waals forces, capillary forces, and electrostatic forces [34]. Capillary forces can arise because the silica-coated soot particles are hydrophilic. Water can condense into the necks and the gap between two pieces, forming a meniscus. Breaking can charge up the pieces, resulting in electrostatic attraction of the pieces. Note, gravitational forces can be neglected because of the small length scales.

The force jumps observed during retraction are most likely caused by rearrangements of single pieces or their agglomerates. The last jump corresponds to the detachment of the tip from the topmost protrusion [35]. Assuming that the jump is determined by van der Waals interactions, the force would be $F = A_H r_{tip}/6D^2 \approx 4$ nN, which is in the same order of magnitude as the measured peak heights. Here, $A_H \approx 6 \times 10^{-20}$ J is the Hamaker constant and $D \approx 0.16$ nm the distance at contact [36].

In summary, the advantage of AFM is that it provides local information. However, it cannot discern independently whether a peak result from breaking or from deformation of the network, i.e. the underlying process cannot be assigned.

Colloidal probe:
To obtain information on the elastic and plastic modulus of the network, larger areas need to be probed. This can be done by replacing the sharp AFM tip with a microsphere (colloidal probe technique) [9]. Again, force-distance curves were recorded. To have a well-defined reference point, the silica sphere, having a radius of 24 μm, approached the surface until a force of 300 nN was reached. After that, the colloidal probe was indented into the network by another 2 μm. To avoid an influence of the substrate, the initial thickness of the soot layer exceeded 20 μm. Because the measured force results from an average over many individual events (Fig. 4), the force curves are much smoother than those measured by AFM.

From force curves we extracted different parameters, which characterize the stability of the porous layer: (i) the effective elastic modulus E_{eff}, (ii) the maximum indentation force F_{max}, (iii) the work of plastic deformation $W_{plastic}$ and (iv) the effective work of adhesion $W_{adhesive}$. The elastic modulus quantifies the elastic energy that is stored in the network upon compression. E_{eff} is obtained from the slope of the onset of the retract curve (Fig. 4a). F_{max} is a measure of the hardness of the network, i.e. the force that needs to be applied to reach a certain compression. The work of plastic deformation is a measure of the energy required to break chains. It is given by the area between the force curve with the x-axis (displacement) on approach (Fig. 4b): $W_{plastic} = \int_{\infty}^{d} F_a dz$. Here, d is the final position of the colloidal probe and z is a coordinate normal to the surface. The work of adhesion gives the energy that is needed to break physical contacts upon retract. It is given by the area of the force with the x-axis upon retracting (Fig. 4c): $W_{adhesive} = -\int_{d}^{\infty} F_r dz$. F_a and F_r are the forces on approach and retraction, respectively.

Figure 5 shows the averaged quantities. E_{eff} and F_{max} show very similar behavior. Both increase with increasing sintering temperature and shell thickness. In agreement with electron microscopy images, the silica shell becomes more compact at the higher sintering temperature. In agreement with data on silica shells prepared by

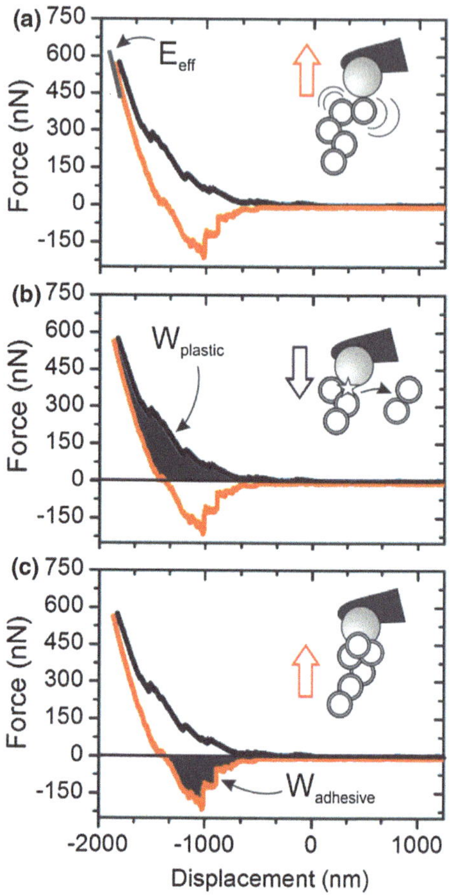

Fig. 4 Mechanical properties extracted from typical force-displacement curves: **a** effective elastic modulus E_{eff}, **b** work of plastic deformation $W_{plastic}$ and **c** effective work of adhesion, $W_{adhesive}$. Reprinted with permission from [28]

the Stöber method, the elastic modulus increases strongly for temperatures above 800 °C. An increase of the shell thickness by a factor of approximately three (1 day CVD compared to three days of CVD) caused an increase of E_{eff} by more than a factor of five. F_{max} increased by a factor (400–900 nN) for samples exposed to 24 h CVD to samples exposed to 72 h CVD and sintered at 1000 °C (Fig. 5b). With increasing hardness of the network, the plastic work of adhesion decreased strongly for temperatures above 800 °C (Fig. 5c). Both $W_{plastic}$ and $W_{adhesive}$ decreased with increasing shell thickness (Fig. 5d). Samples sintered at 1150 °C could not be measured with the colloidal probe technique, because F_{max} exceeded the force measurement range.

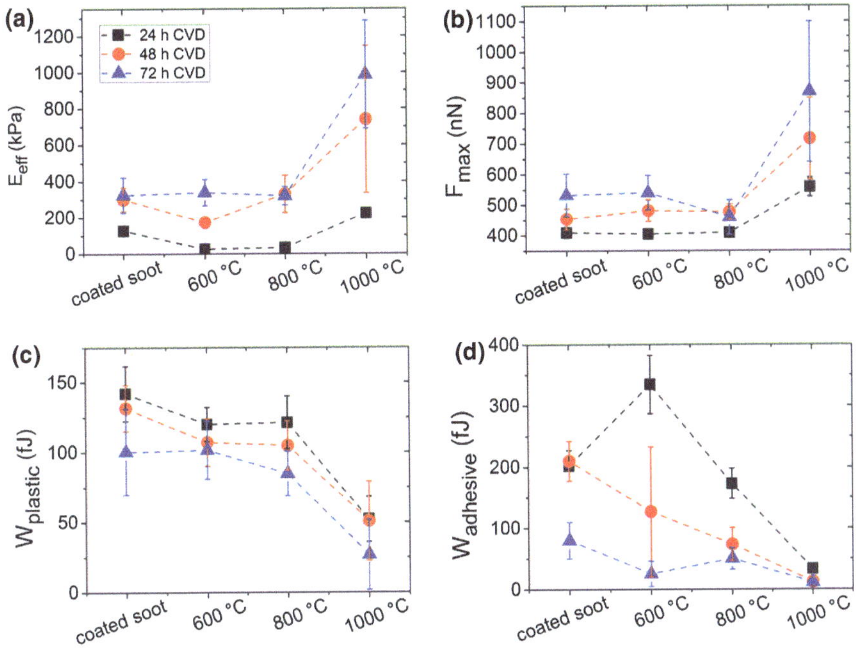

Fig. 5 Mechanical properties of soot-templated surfaces. The black, red and blue data points represent surfaces exposed to 24 h, 48 h and 72 h of CVD, respectively. Coated soot and soot-templated silica surfaces sintered at 600 °C, 800 °C, 1000 °C were investigated, respectively. **a** Effective elastic modulus E_{eff}. **b** Maximum force F_{max}. **c** Plastic work $W_{plastic}$. **d** Effective adhesive work $W_{adhesive}$. The errors are the standard deviations of the respective values obtained from different force curves at different positions on multiple samples. Reprinted with permission from [28]

Nanoindentation

To gain insight into the mechanical properties of networks sintered at 1150 °C, nanoindentation was used (MFP NanoIndenter, Asylum Research) [9]. Unfortunately, nanoindentation and the colloidal probe technique suffered overlap, because the nanoindenter requires a minimum elastic modulus of 10 MPa. A typical force-displacement curve for a network exposed to 72 h CVD is given in Fig. 6. The approach part (black) shows a few jumps in the order of a few up to 40 μN. The jumps hint that part of the network collapsed. Again, the effective elastic modulus increased with increasing shell thickness (Fig. 6b).

In conclusion, because of the large range of length scales and forces, different methods need to be used to characterize the mechanical properties of porous networks. Pencil scratching provides a first idea of the mechanical robustness of a network. The yield force of single chains can be measured by AFM. However, due to the irregular arrangement of the chains, the height and position of the peaks vary from scan to scan. To gain effective values for the elastic modulus, the colloidal probe technique is well suited, assuming that the modulus does not exceed 1 MPa.

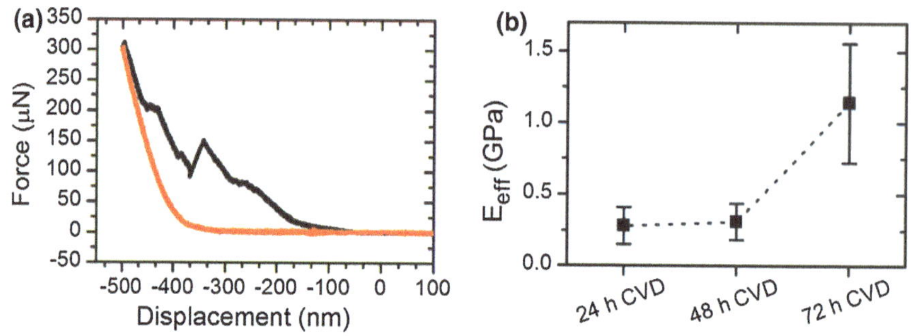

Fig. 6 Nanoindentation of superamphiphobic surfaces sintered at 1150 °C with a conical diamond tip. **a** Typical force-displacement curve of a surface exposed to 72 h CVD indented for 500 nm. Black curve: approach, red curve: retraction **b** E_{eff} of surfaces exposed to 24, 48, and 72 h CVD. Reprinted with permission from [28]

A requirement is that the radius of the colloidal probe must be much larger than the spacing of neighboring chains and the thickness of the coating needs to exceed 10 times the indentation depth. Harder networks can be investigated by nanoindentation. Thicker shells lead to higher effective elastic moduli as do higher sintering temperatures.

Adhesion Test

Soot-templated surfaces have the nice property that their wetting properties can be widely tuned by changing the chemical properties of the surface. After fluorination, the soot templated network shows super-water and super-oil repellent properties [27]. To be able to understand the interactions between the fluorinated porous networks with water, we measured the adhesion force using the colloidal probe technique [37]. Even better would have been to use a water droplet as the colloidal probe and a soot-templated surface as the substrate. Such a procedure, however faces the problem that water evaporates on the time scale of the measurement. Therefore, we "inverted" the experiment: we used super-water-repellent microspheres as the colloidal probe and measured the adhesion towards water (Fig. 7). To prepare the colloidal probe, we coated 40 μm sized glass sphere with a few μm thick layer of soot, deposited silica and fluorinated the silica surface.

Figure 8a shows a typical force-distance curve (Nanowizard 3, JPK). After the coated sphere jumped ("zero distance") into contact, the force increased almost linearly. The slope of the linear part of the force curve is dominated by the deformation of the interface between water and the sphere [38, 39]. The small peaks during advancing and retraction reflect the discontinuous movement of the contact line over the porous network. The adhesion force can be determined from the retraction part, i.e. it is the force just before the water-sphere meniscus ruptured, Fig. 8a. For small

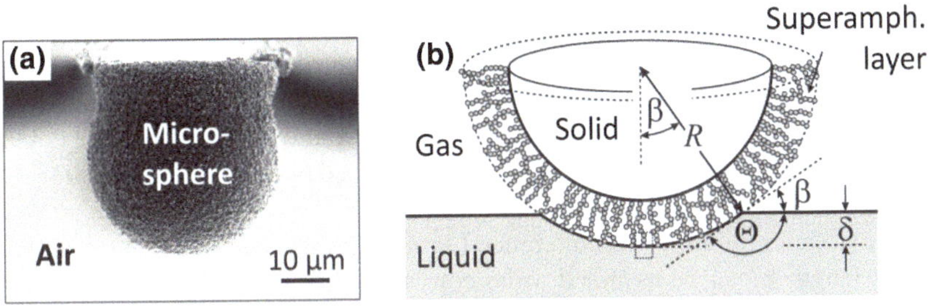

Fig. 7 Scanning electron microscope image of a microsphere coated with a porous network of fluorinated nanoparticles. **a** The microspheres were sintered or glued at the end of tipless, rectangular silicon cantilevers (2 μm thickness, 130 μm or 250 μm length) by using a micromanipulator and two-component epoxy glue. **b** Schematic of a coated microsphere in contact with water. The top particle creates the overhang that prevents the water from penetrating into the hydrophobic soot templated layer. Adapted from [37]

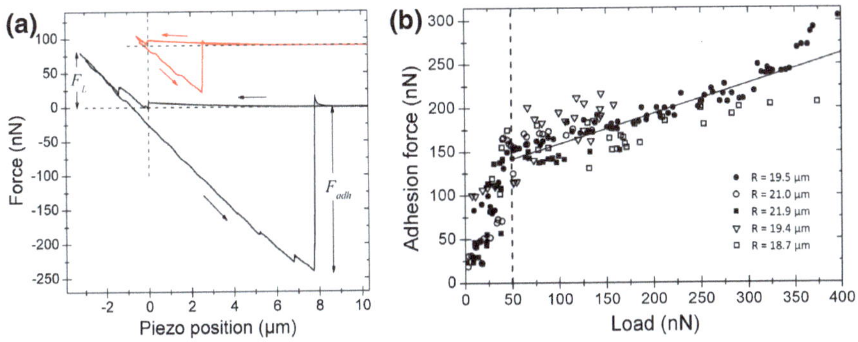

Fig. 8 **a** Two typical force-versus-distance curves for a microsphere coated with a porous network of nanoparticles ($R = 18.7\,\mu$m) interacting with a water drop (volume of 130 μl). Repulsive forces are positive, attractive forces negative. F_{adh} is the adhesion force and F_L the load, i.e. the maximal applied force. The top curve, which was recorded with a lower load, was shifted vertically by 90 nN for better visibility. The tiny increase of the force just before the sphere jumped into contact is caused by electrostatic repulsion. **b** Adhesion force versus load for five experiments with different microspheres interacting with water. The continuous line is a guide to the eye. The dashed vertical line separates two regimes. Adapted from [37]

sphere where buoyancy and gravity can be neglected, the adhesion force is given by $F_{adh} = 2\pi\gamma R cos^2\frac{\Theta}{2}$ [40, 41], where γ is the surface tension of water, R the radius of the coated microsphere and Θ is the Young contact angle. For the derivation of F_{adh}, it is assumed that the contact line is free to move over the surface of the particle while the contact angle remains constant.

Surprisingly, the adhesion force depends on the applied load. For loads below 50 nN, the adhesion force increases steeply with increasing load. This steep part is followed up by a shallow slope. The steep regime reflects the sensitive dependence of the force on the number of contacts between the fluorinated silica coated soot particles and the water interface. The force per nanoparticle can be estimated as 8 nN, assuming a particle radius of $r = 60\,\mathrm{nm}$ and a local contact angle of 115°. This implies that up to 20 coated soot particles contribute to the adhesion force. For higher loads, the motion of the contact line over the rough surface needs to be considered explicitly. Now, even more particles contribute to the force. If the contact line advances, for the force the *cos* needs to be replaced by a sinus ($\Theta \rightarrow 180° - \Theta$): $F^{single}_{advancing} = 2\pi\gamma r sin^2\frac{\Theta}{2}$ [42].

For a sufficiently large circumference ($2\pi R sin\beta$), the number of nanoparticles per unit length κ can be assumed to be constant. This yields for the load: $F_L = 4\pi^2\gamma\kappa r R sin^2\frac{\Theta}{2}$, assuming that the contact line slides to a position given by β if a load F_L is applied. When the contact line recedes, all contacts along the circumference need to rupture, giving for the adhesion force: $F_L = 4\pi^2\gamma\kappa r R cos^2\frac{\Theta}{2} = \frac{F_L}{tan^2(\Theta/2)}$. Indeed, the adhesion force increases with the applied load for a rough surface made out of hydrophobic nanoparticles.

Applications

The special wetting properties of fluorinated soot-templated surfaces make them ideally suitable as substrates for the synthesis of mesoporous supraparticles [43]. Supraparticles are porous particles having a diameter of a few hundred micrometers consisting of smaller particles. Supraparticles can be prepared by placing a drop of an aqueous suspension of (here 0.1 vol.% of 21 nm sized TiO_2 NP particles, P25 Degussa Evonik) on a soot-templated surface. The suspension forms a high contact angle with the surface, greatly reducing the adhesion force between the substrate and the suspension drop. While the water evaporates, the nanoparticle dispersion keeps its high contact angle with the substrate (Fig. 9). During drying, the TiO_2 nanoparticles assemble by capillary forces. After the water has evaporated, almost spherical mesoporous supraparticles remained on the surface. The particles could be easily removed from the surface because of the low contact area with the fluorinated soot-templated surface. The mesoporous supraparticles did not fall apart after redispersion in a solvent if no strong external mechanical force was applied. The strategy is not limited to fabricate mesoporous TiO_2 supraparticles [43]. Depending on the application, the type of nanoparticle can be varied. Characteristic for this strategy is that almost spherical particles can be synthesized without the use of any emulsifiers or stabilizer [44].

Dynamic Microscopic Investigations with Single Particle Resolution

While in section "Fabrication, Characterization and Applications of Porous Colloidal Networks" we analyzed rigid networks of particles, here we are more interested in particle agglomerates which are only physically linked and can react to external

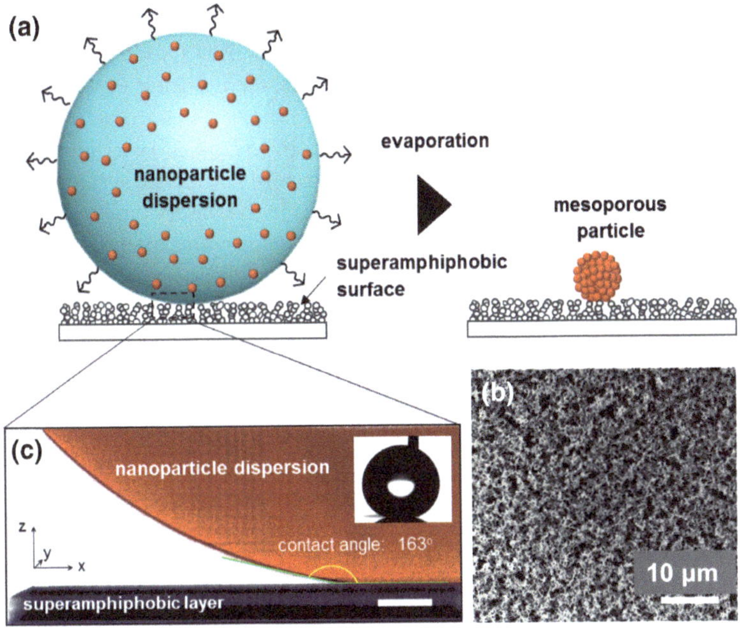

Fig. 9 Drop of an aqueous nanoparticle dispersion on a fluorinated soot-templated surface. **a** A schematic illustration of mesoporous supraparticle formation on the soot-templated surface. Due to the high liquid repellency, the shape of the evaporating drop remains spherical throughout the process. **b** SEM image of the highly porous structure of the soot-templated surface prepared (scale bar: 10 μm). **c** Interface between the drop and soot-templated surface imaged by confocal microscopy. Water (red) was fluorescently labeled with the Alexa488 dye for illustration. Reflection (gray) shows the porous surface and the drop–air interface (scale bar: 50 μm). The inset is a side view of the drop of the nanoparticle dispersion. Reprinted with permission from [43]

stimuli such as shear. The goal of this part is to discuss possibilities of observing 3D single particle motion in fine granular powders. For a detailed understanding of the underlying physics of flowing powders, we aim at the observation of all particle trajectories in the observation volume while a mechanical deformation is applied to the powder. We will end with a first comparison of observed particle motion to simulations.

Experimental Considerations

To obtain detailed information on the internal deformation of particulate systems, we use confocal microscopy. Among 3D imaging methods that are able to image dense particulate systems confocal microscopy has the best chance for imaging (at least slow) dynamics because of its resolution. Following an approach developed earlier by various authors [14, 45–53], we use fluorescently labeled particles that are embedded in an index matching carrier fluid. This allows, for example, studying the properties of the system in thermal equilibrium, the influence of weak attractive forces, and the colloidal glass transition.

For the work presented here, confocal microscopy [52, 54, 55] was identified as the best-suited method for quasi-dynamic investigations of particulate systems that allow for tracking of the motion of (almost all) particles in the observation volume. Using confocal microscopy allows for tracking many particles with a high accuracy in their particle positions [45, 46, 52, 53, 56, 57]. However, confocal microscopy also sets some limitation to the sample system. The most severe limitation being the matching of the refractive index of the particles and the carrier fluid. This excludes gases as carrier fluids. To develop a good model system for fine granular powders which is suitable for confocal microscopy, a couple of additional modifications have to be done to the sample system used in the colloidal studies discussed above: Typical fine powders are neither monodisperse spheres, nor are they subjected to thermal motion. Before describing the used model system, we briefly explain the general strategy for this approach.

Particles of fine granular powders are still large enough so that thermal motion can be neglected. A quantitative measure for the importance of thermal motion is the ratio of the characteristic time scales for Brownian diffusion τ_B and sedimentation τ_S of the particles. These time scales can be estimated by $\tau_B = R_p^2/D$ and $\tau_S = R_p/v_S$, with the particle radius R_p, the particle diffusion constant $D = (k_B T)/(6\pi\eta R_p)$, and the sedimentation velocity $v_S = 2\Delta\varrho g R_p/9$. Here k_B is the Boltzmann constant, T the absolute temperature, η the viscosity, $\Delta\varrho$ the density difference between the particles and the carrier fluid, and g the acceleration due to gravity. The ratio

$$Pe = \frac{\tau_B}{\tau_S} = \frac{4}{3}\pi\frac{R_p^4}{k_B T}\Delta\varrho g \tag{1}$$

is often called the Peclet number. The larger Pe is, the faster sedimentation is relative to thermal motion, i.e., the more sedimentation is dominating and the more thermal motion can be neglected. Additionally, the comparison to cohesive forces between the particles is relevant. As we will show below, the cohesive forces can be changed over a wide range.

Fine Polydisperse Granular Particles for Confocal Microscopy

The 3D observation of particles in confocal microscopy is best realized using fluorescent particles. We have chosen commercial spherical particles (Kromasil 100 Sil, average diameter 7 µm, Akzo Nobel, Sweden, purchased from MZ-Analysentechnik GmbH, Germany) that have nanopores which can easily be used to load the particles with fluorescent dye [57]. These particles consist of silica and allow for a covalent binding of the fluorescent dye using the chemistry described previously by other authors [58, 59]. This results in homogeneously dyed particles that can easily be observed in confocal microscopes [57]. The size of the particles is a compromise between competing requirements. Large particles are needed for a good imaging of the particles and their internal structure (section "Rotation of Spherical Particles") and a high Peclet number. On the other hand, a relevant cohesion is easier to realize with small particles. The purchased particles are close to spherical with a small roughness and a certain polydispersity. These properties are preserved during the covalent attachment of the fluorescent dye.

Using silica-based particles has the advantage of a broad flexibility of possible surface modifications. After synthesis the particles are hydrophilic and can be easily dispersed in aqueous environments. For the adjustment of the refractive index, sodium thiocyanate (NaSCN) was dissolved in the water. This leads to a rather high ionic strength of the solutions that efficiently screens all electrostatic interactions between particles. Effectively, this system has only a weak cohesion but a hard-sphere-like repulsion [60]. Alternatively, the silica particles can be coated with hydrophobic molecules to allow for a dispersion in organic solvents. By finely tuning the concentration of hydrophobic molecules on the surface of the particles, the particles can stabilized in the interface between water and organic solvents, e.g., decaline. Particles at the water-decaline interface were partially hydrophobized with hexamethyl-disilazane (HDMS) [60]. These particles are perfectly index-matched to the carrier liquid, but they are not density-matched. In the following experiments, they sediment to the bottom glass slide during the preparation of the sample. A typical solid volume fraction of such sedimented bed is in the order of 40%.

The original algorithms for the automatic detection of particle positions and trajectories [45, 46, 52, 56] are optimized for monodisperse particles. For the polydisperse silica particles used as model systems for fine granular powders, a couple of modifications to the detections algorithm were necessary (Fig. 10). The original algorithm by Crocker and Grier relies on the numerical convolution of the measured 3D image ($I_{i,j,k}$ in Fig. 10) with a single mask $M_{i,j,k}$ that perfectly matches the particle size [57]. This leads to strong peaks at the particle positions in the convoluted image $C_{i,j,k}$. To detect polydisperse particles, a set of masks $M_{i,j,k}$ with different sizes is used. Depending on the particle size some masks will be oversized or undersized. The strongest peak in the convoluted image will be observed at matching sizes of the mask with the particle. This gives a first estimate of the particle size and position. A further refinement of the size and position is obtained by searching the particle surface along a number of arbitrarily drawn lines through the particle. Fitting a sphere to the detected points on the particle surface gives a precise positioning of the particles and measures the particle diameter [57].

The observation of fine granular powders was realized in a home built confocal microscope that was previously described [52]. Briefly, this is a laser scanning microscope with a diffraction limited resolution (laterally, about 300 and 700 nm in beam direction). The scanning is realized by galvanic mirrors and the detection uses single photon counting by an avalanche diode with a maximum photon frequency of about 17 MHz. Two separate color channels can be used in parallel for imaging

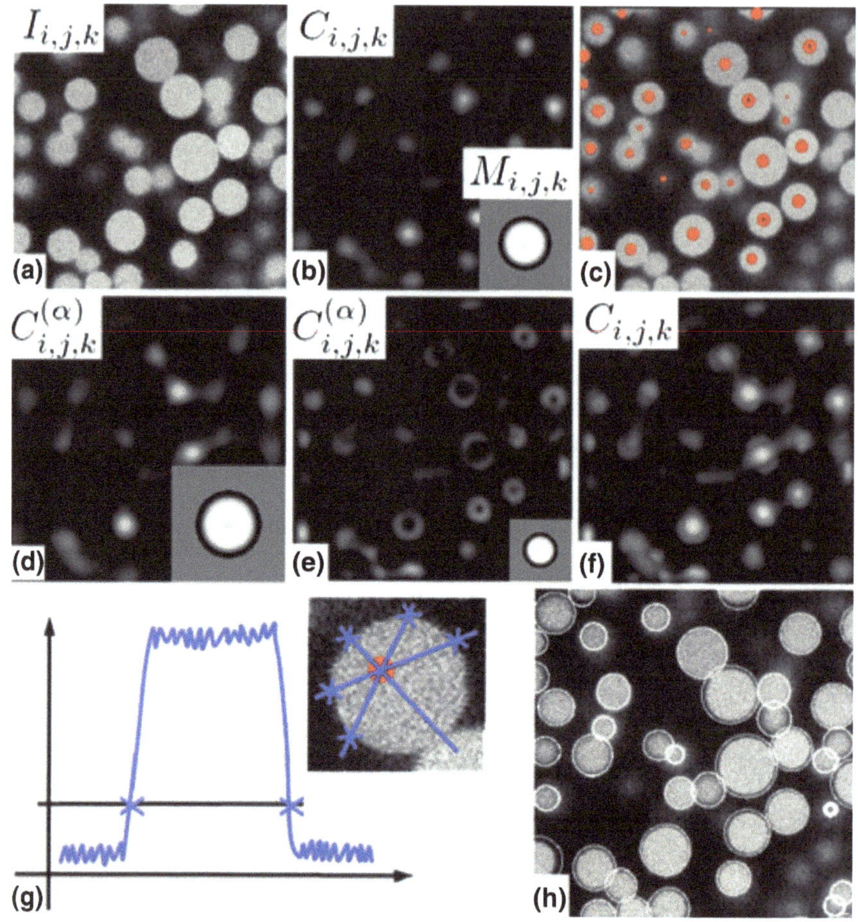

Fig. 10 Confocal images $I_{i,j,k}$ at different steps of the automatic position detection of polydisperse spherical particles. Image, as measured with a confocal microscope (**a**), after the convolution with the mask $M_{i,j,k}$ (**b**), confocal image with obtained particle positions marked with *red dots* (**c**), after the convolution with an oversized (**d**) and undersized (**e**) mask $M_{i,j,k}$, merged data after convolution with different mask sizes (**f**). Intensity profile of a localized particle with obtained points at the rim (**g**), particles with obtained diameter after refinement step (**h**). Reprinted with permission from Granular Matter **15**: 391 [57]

(excitation 457 and 532 nm, detection wavelengths can be chosen by suitable filters). Depending on the number of pixels and the fluorescent intensity of the sample a 2D image takes about one to a few seconds to aquire. The advantage of this setup is its high flexibility. This especially implies that additional methods can be applied to the sample. Figure 11 depicts a nanomanipulation setup that can be used to shear and/or compress the sample. Here a shear cell is realized between the glass slide that carries

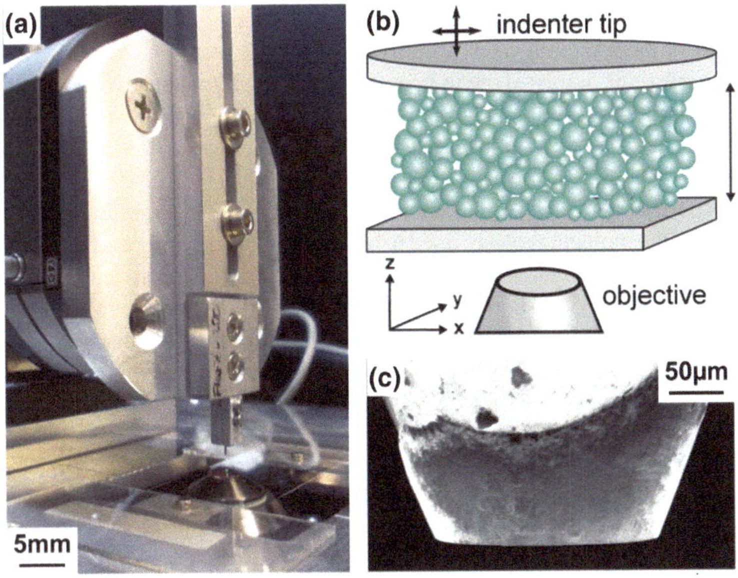

Fig. 11 Nano-manipulator setup consisting of a piezo-actuator and a mounted diamond indenter tip (**a**). Schematic view of the plate-plate geometry (**b**). Scanning electron microscope image of the indenter tip (**c**). Reprinted with permission from Granular Matter **15**: 391 [57]

the sample and a flat-end diamond tip. With this setup shear deformations with a maximum amplitude of 160 μm can be applied to the sample.

Rotation of Spherical Particles

The methods described in the previous section allow the reliable tracking of the position of thousands of particles in a series of 3D confocal images. However, particles are solid bodies that have six degrees of freedom for their motion: Three translational and three rotational degrees of freedom. The methods presented so far resolve the translational degrees of freedom. In the following methods for the rotational degrees of freedom are discussed.

The optical detection of the rotation of particles depends on some optical anisotropy of the particles. The probably simplest method is using particles with a non-spherical shape [61–65]. Also, Janus particles have been used [66, 67] to observe the rotation of single particles. These methods, however, necessarily break the spherical symmetry of the particle interaction, either through the particle shape or through changes in the surface chemistry. For a model system, a fully isotropic interaction with the surrounding particles is intended. Any internal structure should be uncoupled from the particle interactions. In the following, we describe methods that use bleaching of the fluorescent dyes in the interior of the particles to give each particle one or more internal axes. To fully decouple this internal structuring from the particle interactions, it turned out to be necessary to surround the dye laden particles with a non-fluorescent shell. This shell decouples all changes done to the interior of the particles from the surface chemistry (and thus the interaction potential) of the particles.

Since fluorescent dye molecules interact with the light field almost like dipoles, the orientation of the molecules can be used to bleach molecules with specific orientations relative to an incident polarized light filed, either on a rather macroscopic scale [68], or using a confocal microscope [51]. This polarized bleaching results in particles with an orientation-dependent efficiency of the fluorescence. Originally, this method has been applied to particle ensembles [68]. It has also been shown that this method works on a single particle level [51]. The advantage of this method is that it can be applied to particles that have a diameter in the order of the optical resolution of the used microscope. The disadvantage of this method is, however, that only the rotation of the particles around one axis can be determined. Additionally, the interaction of the dye molecules is in general not strongly polarization dependent and might be different for the absorption and the emission process. In summary, these effects lead to a rather weak dependence of the fluorescence efficiency on the particle orientation [51].

A more efficient way of inscribing the orientation in the interior of the particle is to use structured bleaching [57]. The goal is to obtain all rotational degrees of freedom of the particle, i.e., to follow all three rotation angles. For this purpose, the internal structure of the particle has to be complex enough. The following method is applicable when the optical resolution of the confocal microscope is better than the size of the particles. The excitation laser that is used for the imaging can also bleach the fluorescent dye molecules, if the intensity is high enough. For good imaging situations (i.e., when multiple imaging of the sample is possible) there is a significant difference between the light intensity used for imaging and the intensity needed to bleach the dye molecules. We used this intensity difference for structured bleaching of the particles (Fig. 12). In this specific example a line was "written" into the particle (Fig. 12 top row). Due to the focusing of the laser by the objective of the microscope, the bleaching in the interior of the particles was not only along the written line, but also above and below it (Fig. 12 bottom row). Due to this feature actually two axes were written into the particle: the line $\vec{f}$ and the propagation direction of the light $\vec{a}$. For a full set of coordinate axis, the vector product of $\vec{f}$ and $\vec{a}$ results in the third axis $\vec{n}$. We note that a simplified version of this method (having only one axis) was later on also used to generate shape anisotropic particles [65]. From the two axes bleached in the interior of the particle the full rotational motion of the particle can be measured [57].

Particles with an internal fluorescent structure can be used to determine the rotational motion of particles in deformed fine granular powders. Figure 13 gives a simple example of such an experiment. A sedimented bed (section "Fine Polydisperse Granular Particles for Confocal Microscopy") was compressed and the resulting restructuring of the particle arrangement was observed. The particle highlighted on

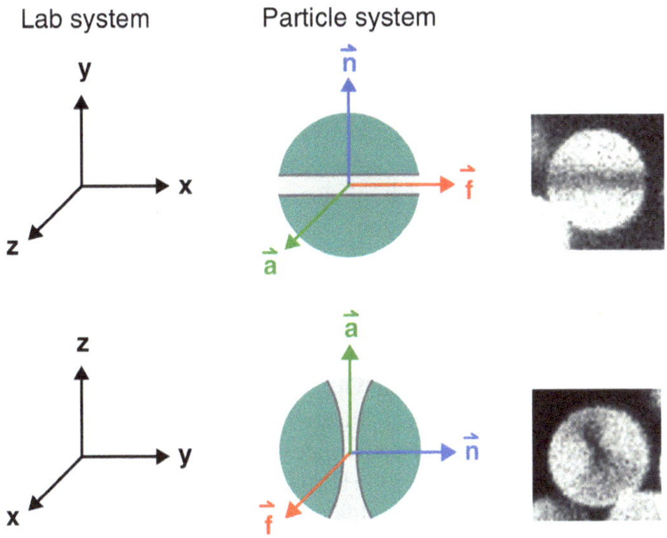

Fig. 12 Geometry of the bleaching process in the lab and particle system (left) and real images of particles after bleaching (right). The diameter of the particle was 7.3 μm. Reprinted with permission from Granular Matter **15**: 391 [57]

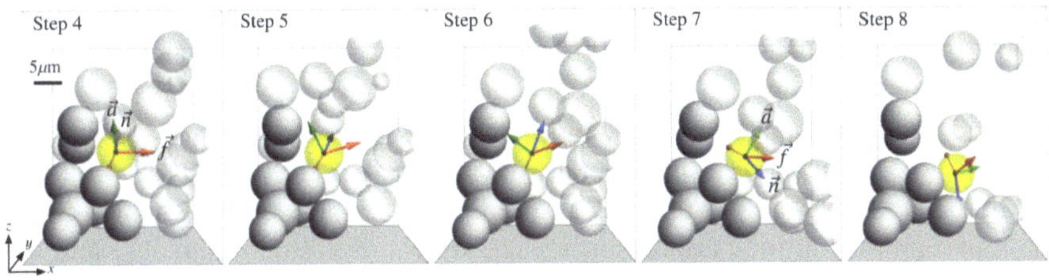

Fig. 13 Visualization of rotation and translation of one particle as measured during the experiment. Resting particles are shown by *opaque gray*, the others *translucent*, the *red* and *blue arrows* ($\vec{f}$ and $\vec{n}$, respectively) indicate the focal and normal vectors, the *green arrow* ($\vec{a}$) their vector product. The *brown line* indicates the original contact point to the neighboring particle. Reprinted with permission from Granular Matter **15**: 391 [57]

yellow first exhibited a torsional motion around the contact point of its neighbor (brown line). During this torsional motion the contact point was unchanged. Only after a finite torsional motion, the particle rolled over its neighbor. This experiment illustrates the importance of all possible modes in particle interaction [1]. All of them are actually realized in fine granular powders.

Wet Granular Powders Under External Deformation

In the previous sections, situations with only one fluid phase were discussed. This corresponds to dry powders (or to completely immersed powders). In situations, with two fluid phases present (e.g. wet granular powders with capillary bridges between the particles), we used partially hydrophobized particles that were immersed in a mixture of an aqueous (NaSCN solution) and an organic phase (cis-decaline) [60].

To prepare clusters that were composed of drops and particles, partially hydrophobized particles (section "Fine Polydisperse Granular Particles for Confocal Microscopy") were dispersed in an aqueous NaSCN-solution. The cis-decaline was fluorescently dyed with the hydrophobic dye Nile Red at the saturation concentration. To produce a liquid-liquid mixture the fluorescenty labelled cis-decaline was added to the aqueous particle suspension and manually shaken. For enhancing the contrast of the liquid-liquid interface, the refractive index of the surrounding NaSCN-solution was adjusted to the refractive index of the HMDS-modified particles (index of refraction approximately 1.443), Fig. 14a. In this specific example, the contact angle of the liquid-liquid interface on the particles was rather high (about 130°). Consequently, the particles were connected by drops rather through capillary bridges. The amount of shaking was chosen to have a drop size that is approximately similar to the particle size.

Whereas the algorithms describe above for the position of the particles can also be applied in this case, the identification of the drop coordinates and sizes needs some more attention. For drops that are connected only to a few particles the same algorithm results in good coordinates and sizes. For drops connected to several particles, the direct analysis is not possible, but an indirect one gives good results (Fig. 14b). When the coordinates and radii $r_{p,i}$ of the connected particles and the contact angle of the two liquid phases on the particle surface are known, the drop radius $r_{D,i}$ and the coordinates of the drop can be identified through an optimization procedure. This optimization procedure uses points on the drop surface that can be calculated from the drop coordinates, the drop radii $r_{p,i}$ and the part of the particle inside the drop $sub_{p,i}$ (Fig. 14b). Note that $sub_{p,i}$ is a function of the drop radius and the contact angle.

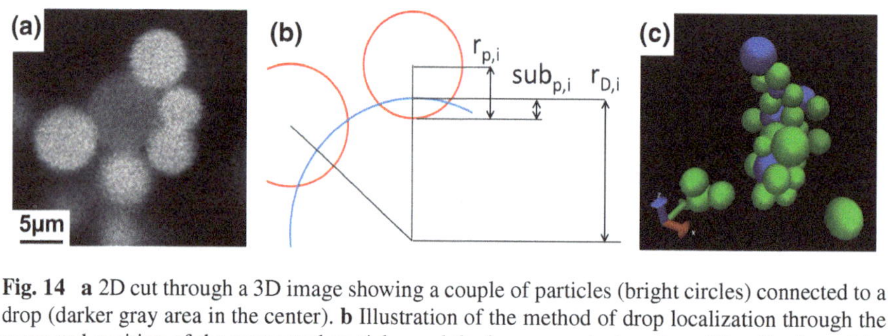

Fig. 14 **a** 2D cut through a 3D image showing a couple of particles (bright circles) connected to a drop (darker gray area in the center). **b** Illustration of the method of drop localization through the measured position of the connected particles and the known contact angle. **c** Visualization of the measured position and sizes of all drops and particles in the cluster

Additionally, the assumption of a spherical drop increases the reliability of the results. This assumption is justified, if the forces acting on the drops are small compared to the restoring force due to the surface tension of the drop. The combination of these methods gives a reliable representation of the measured system that can be used for further analysis Fig. 14c. Especially, the measured drop diameter did not change significantly during the investigated time frame.

In a system of particles connected by capillary bonds to drops two different interactions are relevant. The partially hydrophobized particle exhibit a small attractive van-der-Waals-type interaction. The binding force of the particles in the drops due to the reduction of the surface energy is larger by orders of magnitudes.

To test these ideas, we investigated a cluster consisting of 23 particles and four drops of cis-decaline in an aqueous continuous phase [60]. Eight particles were attached to more than one droplet. One of the drop was sticking to the side of the indenter tip surface, i.e. at the cone surface of the tip (Fig. 11c). Several particles were in contact with the glass slide at the bottom.

The mechanical load was applied stepwise onto the sample. After each step, a 3D image of the sample was taken. Effectively, in each step the cluster was shaken by changing the distance between the indenter and the glass slide by a few μm back and forth. Before starting a 3D scan any motion in the system had stopped. In total 21 steps were made. No drift of the sample was observed in the images. The whole cluster was imaged in an observation volume of $100 \times 100 \times 40\ \mu m^3$ with a resolution of 512 pixels × 512 pixels 96 pixels [60]. The trajectories of all visible droplets and particles could be followed over the whole measurement series.

In the analysis of the system dynamics a few key features were identified. First, all particles stayed connected to their drops and move only within the surface of these drops (Fig. 15a). Second, particle-particle connections are weaker and not stable over time. Third, particles connected to two drops actually bind the cluster together. The cluster was connected to the supporting glass slide with one drop (drop 1 in Fig. 15b) and to the indenter with one drop (drop 4 in Fig. 15b). The particles can be classified in particles that are involved in the direct connection between the glass slide and the indenter (the backbone of the cluster) and particles that are just on the surface of one of the drops. The analysis of the dynamics of the particles allows extracting the backbone of the cluster (Fig. 15b). All rearrangements of the cluster keep the connections within this backbone unchanged but change the relative positions of the all particles and drops in the cluster (including those of the backbone).

Comparing X-Ray Tomography Data to Simulations

The methods for data analysis discussed so far are not limited to series of 3D images obtained with confocal microscopy. As an example for further analysis of data, we include here the analysis of X-ray micro tomography data [69]. Grey polyethylene microspheres (PE-particles, Cospheric, type "GRYPMS-1.00 250–300 μm") in the size region of 250–300 μm and a density of approximately 1 g/cm^3 were used in this

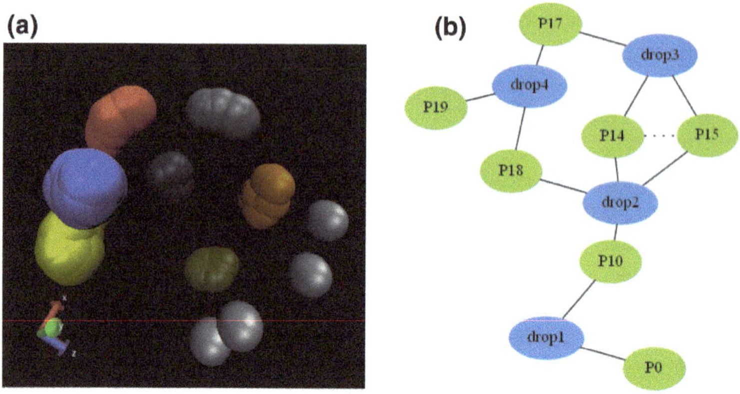

Fig. 15 **a** The motion of particles in a drop surface (in the frame of reference co-moving with the drop center). **b** Minimal reduced graph of the network forming the cluster. Solid lines show permanent connections and dotted lines show reversible connections

study. The particles were filled in a custom-made mixer. The purpose of the study was to combine a detailed data analysis of a series of 3D images and suitably adapted simulations. This combination allowed to identify simulation parameters that were otherwise hard to measure, e.g. sliding and rolling friction [2].

The X-ray tomography images include a much larger volume than typical confocal images, however at a lower resolution. In total, the trajectories of almost 33,000 particles were extracted from a series of 23 3D images of the sample, taken over a 22° rotation of the mixer [69]. A representative selection of results of this analysis is shown in Fig. 16. For the comparison of experimental and numerical results a discrepancy function D was introduced. It was calculated depending on the summarized difference between relative angular motion ω_{rel} of particles in different zones according to Eq. (2).

$$D = \frac{\sum \left| \omega_{rel,exp,i} - \omega_{rel,sim,i} \right| \times N_{sim,i}}{N_{sim,tot}} \tag{2}$$

Here the index i counts the different zones, $N_{sim,tot}$ and $N_{sim,i}$ are the number of particles in zone i and the total number of simulated particles.

A minimization of the discrepancy D was used to identify the most probably coefficient of the rolling friction. In this specific example the rolling friction coefficient between the particles was found to be $\mu_{rol,pp} = 0.04$ and between the particles and the wall $\mu_{rol,pw} = 0.002$. For these parameters the discrepancy function showed a minimum [69]. For this interpretation of these numbers is important to consider that these coefficients are obtained through a minimization. To finally conclude the values a more in-depth study would be necessary, e.g. by experiments and simulations with different geometries and external conditions.

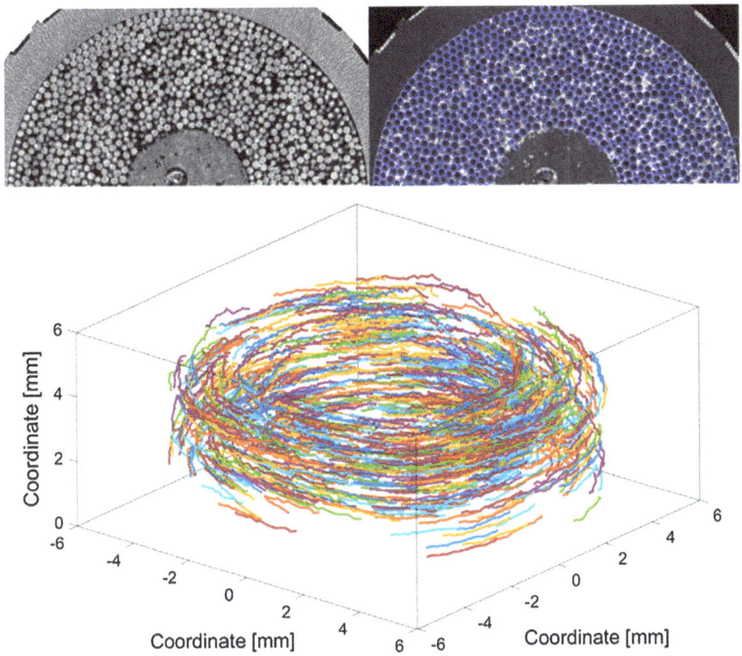

Fig. 16 Top: Extraction of particle coordinates from 3D micro tomography images dataset: original image (left), extracted coordinates (right). Bottom: Trajectories of 1150 particles extracted from experimental measurements. Reprinted with permission from Chem. Engineer. Res. Design **135**: 121–128 [69]

Conclusion

The mechanical porous colloidal networks were investigated using force-sensitive measurements. However, the method of choice depends on the porosity and hardness of the network and the question to be addressed. Atomic force microscopy reveals information on the break-off force of a single particle. The colloidal probe technique provides intrinsic material properties like the effective elastic modulus. If the case of soot templated networks, the thickness of the shell and sintering temperatures allow to tune the effective elasticity and hardness of the network in a wide range. In contrast to smooth surfaces, for fluorinated soot-templated layer the roughness gives rise to an adhesion force which depends on the load.

Furthermore, we reported on recent developments in the experimental study of fine granular powder. We emphasized the possibilities arising from single particle tracking in series of 3D images of granular systems under mechanical deformation. Exemplarily, the insights coming from the detection of the rotation (additional to the translation) of the particles is shown. For wet granular systems the force due to the capillary bridges induces strong binding in the particles and changes drastically the motion of the particles. In a detailed comparison between experiments and simulations, it is possible to estimate material parameters that are hard to measure. To judge the possibilities of this combination of methods more work has to be invested.

References

1. Tomas, J.: Adhesion of ultrafine particles—a micromechanical approach. Chem. Eng. Sci. **62**, 1997–2010 (2007)
2. Fuchs, R., et al.: Rolling, sliding and torsion of micron-sized silica particles: experimental, numerical and theoretical analysis. Granul. Matter **16**, 281–297 (2014). https://doi.org/10.1007/s10035-014-0481-9
3. Goncu, F., Luding, S.: Effect of particle friction and polydispersity on the macroscopic stress-strain relations of granular materials. Acta Geotech. **8**, 629–643 (2013). https://doi.org/10.1007/s11440-013-0258-z
4. Butt, H.J., Cappella, B., Kappl, M.: Force measurements with the atomic force microscope: Technique, interpretation and applications. Surf. Sci. Rep. **59**, 1–152 (2005). https://doi.org/10.1016/j.surfrep.2005.08.003
5. Yethiraj, A., van Blaaderen, A.: A colloidal model system with an interaction tunable from hard sphere to soft and dipolar. Nature **421**, 513–517 (2003). https://doi.org/10.1038/nature01328
6. Pusey, P.N., Vanmegen, W.: Phase-behavior of concentrated suspensions of nearly hard colloidal spheres. Nature **320**, 340–342 (1986). https://doi.org/10.1038/320340a0
7. Koylu, U.O., Faeth, G.M., Farias, T.L., Carvalho, M.G.: Fractal and projected structure properties of soot aggregates. Combust. Flame **100**, 621–633 (1995). https://doi.org/10.1016/0010-2180(94)00147-k
8. Verho, T., et al.: Mechanically durable superhydrophobic surfaces. Adv. Mater. **23**, 673–678 (2011). https://doi.org/10.1002/adma.201003129
9. Paven, M., et al.: Mechanical properties of highly porous super liquid-repellent surfaces. Adv. Func. Mater. **26**, 4914–4922 (2016). https://doi.org/10.1002/adfm.201600627
10. Binnig, G., Quate, C.F., Gerber, C.: Atomic force microscope. Phys. Rev. Lett. **56**, 930–933 (1986)
11. Oliver, W.C., Pharr, G.M.: An improved method for determining hardness and elastic modulus using load and displacement sensing indentation experiments. J. Mater. Res. **7**, 1564–1583 (1992)
12. Ducker, W.A., Senden, T.J., Pashley, R.M.: Direct measurement of colloidal forces using an atomic force microscope. Nature **353**, 239–241 (1991)
13. Kappl, M., Butt, H.-J.: The colloidal probe technique and its application to adhesion force measurements. Part. Part. Syst. Charact. **19**, 129–143 (2002)
14. Kwade, A., et al.: Micromechanical properties of colloidal structures. In: AIP Conference Proceedings, vol. 1542, 939–942 (2013)
15. Strege, S., et al.: Approach to structural anisotropy in compacted cohesive powder. Granul. Matter **16**, 401–409 (2014). https://doi.org/10.1007/s10035-013-0454-4
16. MiDi, G.: On dense granular flows. Eur. Phys. J. E **14**, 341–365 (2004)
17. Scheel, M., et al.: Morphological clues to wet granular pile stability. Nat. Mater. **7**, 189–193 (2008)
18. Herminghaus, S.: Wet Granular Matter: A Truly Complex Fluid. World Scientific Publishing Co. Pte. Ltd. (2013)
19. Mani, R., et al.: Role of contact-angle hysteresis for fluid transport in wet granular matter. Phys. Rev. E **91** (2015). https://doi.org/10.1103/physreve.91.042204
20. Koos, E., Willenbacher, N.: Capillary forces in suspension rheology. Science **331**, 897–900 (2011)
21. Bossler, F., Koos, E.: Structure of particle networks in capillary suspensions with wetting and nonwetting fluids. Langmuir **32**, 1489–1501 (2016). https://doi.org/10.1021/acs.langmuir.5b04246

22. Roy, S., Singh, A., Luding, S., Weinhart, T.: Micro–macro transition and simplified contact models for wet granular materials. Comput. Part. Mech. **3**, 449–462 (2016). https://doi.org/10.1007/s40571-015-0061-8
23. Greenwood, J.A., Williamson, J.B.P.: The contact of two nominally flat surfaces. Proc. R. Soc. Lond. **295**, 300–319 (1966)
24. Groten, J., Ruhe, J.: Surfaces with combined microscale and nanoscale structures: a route to mechanically stable superhydrophobic surfaces? Langmuir **29**, 3765–3772 (2013). https://doi.org/10.1021/la304641q
25. Zhang, L., et al.: Hollow silica spheres: synthesis and mechanical properties. Langmuir **25**, 2711–2717 (2009). https://doi.org/10.1021/la803546r
26. Reissner, E.: Stresses and small displacements of shallow spherical shells 2. J. Math. Phys. **25**, 279–300 (1946)
27. Deng, X., Mammen, L., Butt, H.-J., Vollmer, D.: Candle soot as a template for a transparent robust superamphiphobic coating. Science **335**, 67–70 (2012). https://doi.org/10.1126/science.1207115
28. Paven, M., et al.: Optimization of superamphiphobic layers based on candle soot. Pure Appl. Chem. **86**, 87–96 (2014). https://doi.org/10.1515/pac-2014-5015
29. Callies, M., Quere, D.: On water repellency. Soft Matter **1**, 55–61 (2005). https://doi.org/10.1039/b501657f
30. Deng, X., et al.: Transparent, thermally stable and mechanically robust superhydrophobic surfaces made from porous silica capsules. Adv. Mater. **23**, 2962–2965 (2011)
31. Stober, W., Fink, A., Bohn, E.: Controlled growth of monodisperse silica spheres in micron size range. J. Colloid Interface Sci. **26**, 62– (1968)
32. D'Acunzi, M., et al.: Superhydrophobic surfaces by hybrid raspberry-like particles. Faraday Discuss. **146**, 35–48 (2010). https://doi.org/10.1039/b925676h
33. Zhang, L.J., et al.: Tuning the mechanical properties of silica microcapsules. Phys. Chem. Chem. Phys. **12**, 15392–15398 (2010). https://doi.org/10.1039/c0cp00871k
34. Butt, H.-J., Kappl, M.: Surface and Interfacial Forces. Wiley-VCH (2010)
35. Heim, L.O., Blum, J., Preuss, M., Butt, H.-J.: Adhesion and friction forces between spherical micrometer-sized particles. Phys. Rev. Lett. **83**, 3328–3331 (1999)
36. Heim, L., Butt, H.-J., Schräpler, R., Blum, J.: Analyzing the compaction of high-porosity microscopic agglomerates. Aust. J. Chem. **58**, 1–3 (2005)
37. Ye, M., et al.: Superamphiphobic particles: how small can we go? Phys. Rev. Lett. **112** (2014). https://doi.org/10.1103/physrevlett.112.016101
38. Ducker, W.A., Xu, Z., Israelachvili, J.N.: Measurement of hydrophobic and DLVO forces in bubble-surface interactions in aqueous solutions. Langmuir **10**, 3279–3289 (1994)
39. Fielden, M.L., Hayes, R.A., Ralston, J.: Surface and capillary forces affecting air bubble-particle interactions in aqueous electrolyte. Langmuir **12**, 3721–3727 (1996)
40. Preuss, M., Butt, H.-J.: Direct measurement of forces between particles and bubbles. Int. J. Miner. Process. **56**, 99–115 (1999)
41. Scheludko, A., Toshev, B.V., Bojadjiev, D.T.: Attachment of particles to a liquid surface (Capillary theory of flotation). J. Chem. Soc. Faraday Trans. I **72**, 2815–2828 (1976)
42. Butt, H.J., et al.: Design principles for superamphiphobic surfaces. Soft Matter **9**, 418–428 (2013). https://doi.org/10.1039/c2sm27016a
43. Wooh, S., et al.: Synthesis of mesoporous supraparticles on superamphiphobic surfaces. Adv. Mater. **27**, 7338–+ (2015). https://doi.org/10.1002/adma.201503929
44. Deng, X., et al.: Solvent-free synthesis of microparticles on superamphiphobic surfaces. Angew. Chem. Int. Ed. **52**, 11286–11289 (2013). https://doi.org/10.1002/anie.201302903
45. Weeks, E.R., Crocker, J.C., Levitt, A.C., Schofield, A., Weitz, D.A.: Three-dimensional direct imaging of structural relaxation near the colloidal glass transition. Science **287**, 627–631 (2000)
46. Crocker, J.C., Grier, D.G.: Methods of digital video microscopy for colloidal studies. J. Colloid Interface Sci. **179**, 298–310 (1996)
47. Sedgwick, H., Egelhaaf, S.U., Poon, W.C.K.: Clusters and gels in systems of sticky particles. J. Phys.: Condens. Matter **16** (2004)

48. Zausch, J., et al.: From equilibrium to steady state: the transient dynamics of colloidal liquids under shear. J. Phys.: Condens. Matter **20**, 404210 (2008)
49. Van Blaaderen, A., Imhof, A., Hage, W., Vrij, A.: Three-dimensional imaging of submicrometer colloidal particles in concentrated suspensions using confocal scanning laser microscopy. Langmuir **8**, 1514–1517 (1992). https://doi.org/10.1021/la00042a005
50. van Blaaderen, A., Wiltzius, P.: Real-space structure of colloidal hard-sphere glasses. Science **270**, 1177–1179 (1995)
51. Roth, M., Franzmann, M., D'Acunzi, M., Kreiter, M., Auernhammer, G.K.: Experimental analysis of single particle deformations and rotations in colloidal and granular systems. arXiv.org **1106**, 1106.3623 [cond-mat.soft] (2011)
52. Roth, M., Schilde, C., Lellig, P., Kwade, A., Auernhammer, G.: Colloidal aggregates tested via nanoindentation and quasi-simultaneous 3D imaging. Eur. Phys. J. E **35**, 1–12 (2012). https://doi.org/10.1140/epje/i2012-12124-8
53. Roth, M., Schilde, C., Lellig, P., Kwade, A., Auernhammer, G.K.: Simultaneous nanoindentation and 3D imaging on semicrystalline colloidal films. Chem. Lett. **41**, 1110–1112 (2012)
54. Minsky, M.: Microscopy Apparatus. USA patent (1957)
55. Minski, M.: Double focussing stage scanning microscope. Scanning **10**, 128–138 (1988)
56. Jenkins, M.C., Haw, M.D., Barker, G.C., Poon, W.C.K., Egelhaaf, S.U.: Finding bridges in packings of colloidal spheres. Soft Matter **7**, 684–690 (2011)
57. Wenzl, J., Seto, R., Roth, M., Butt, H.-J., Auernhammer, G.: Measurement of rotation of individual spherical particles in cohesive granulates. Granul. Matter **15**, 391–400 (2013). https://doi.org/10.1007/s10035-012-0383-7
58. Van Blaaderen, A., Vrij, A.: Synthesis and characterization of colloidal dispersions of fluorescent, monodisperse silica spheres. Langmuir **8**, 2921–2931 (1992). https://doi.org/10.1021/la00048a013
59. Verhaegh, N.A.M., Blaaderen, A.V.: Dispersions of rhodamine-labeled silica spheres: synthesis, characterization, and fluorescence confocal scanning laser microscopy. Langmuir **10**, 1427–1438 (1994). https://doi.org/10.1021/la00017a019
60. Wenzl, J.: Wet and Dry Model Granulates Under Mechanical Load : A Confocal Microscopy Study. Johannes Gutenberg-Universität (2014)
61. Sacanna, S., Rossi, L., Kuipers, B.W.M., Philipse, A.P.: Fluorescent monodisperse silica ellipsoids for optical rotational diffusion studies. Langmuir **22**, 1822–1827 (2006). https://doi.org/10.1021/la052484o
62. Th, G., Odenbach, S.: Investigation of the motion of particles in magnetorheological elastomers by X-μ CT. Smart Mater. Struct. **23**, 105013 (2014)
63. Gundermann, T., Cremer, P., Löwen, H., Menzel, A.M., Odenbach, S.: Statistical analysis of magnetically soft particles in magnetorheological elastomers. Smart Mater. Struct. **26**, 045012 (2017)
64. Schutter, S., Roller, J., Kick, A., Meijer, J.-M., Zumbusch, A.: Real-space imaging of translational and rotational dynamics of hard spheres from the fluid to the crystal. Soft Matter **13**, 8240–8249 (2017). https://doi.org/10.1039/C7SM01400G
65. Harrington, M., Lin, M., Nordstrom, K.N., Losert, W.: Experimental measurements of orientation and rotation of dense 3D packings of spheres. Granul. Matter **16**, 185–191 (2014). https://doi.org/10.1007/s10035-013-0474-0
66. Anthony, S.M., Hong, L., Kim, M., Granick, S.: Single-particle colloid tracking in four dimensions. Langmuir **22**, 9812–9815 (2006)
67. Hsiao, L.C., Saha-Dalal, I., Larson, R.G., Solomon, M.J.: Translational and rotational dynamics in dense suspensions of smooth and rough colloids. Soft Matter **13**, 9229–9236 (2017). https://doi.org/10.1039/C7SM02115A
68. Lettinga, M.P., Koenderink, G.H., Kuipers, B.W.M., Bessels, E., Philipse, A.P.: Rotational dynamics of colloidal spheres probed with fluorescence recovery after photobleaching. J. Chem. Phys. **120**, 4517–4529 (2004)
69. Dosta, M., et al.: Application of micro computed tomography for adjustment of model parameters for discrete element method. Chem. Eng. Res. Des. **135**, 121–128 (2018). https://doi.org/10.1016/j.cherd.2018.05.030

Contact Models and DEM Simulation of Micrometer-Sized Particles and Agglomerates at Static Loading Based on Experimental Characterization

Philipp Grohn, Dominik Weis, Ulrich Bröckel, Stefan Heinrich and Sergiy Antonyuk

Introduction

Many products of the chemical, food, and pharmaceutical industry, e.g. tablets, briquettes, catalysts, adsorbents, fertilizers etc., are often produced as granules, which consist of fine particles. In comparison to powders, the obvious advantages of granules are higher packing density, better flow behaviour as well as less dust formation. Desirable properties such as a narrow particle size distribution, porosity, internal surfaces, regular shape and chemical composition can be achieved. Some problems in handling, like segregation or long-time consolidation of the bulk materials in bunkers and transport containers can also be avoided due to granulation [1].

During the processing, handling and transportation the agglomerates are subjected to mechanical stressing because of agglomerate-agglomerate and agglomerate-apparatus wall contacts. This mechanical stressing can lead to plastic deformation, attrition and breakage, which may reduce the product quality. In addition, dust may arise. For this reason, the strength is an important property of the granules [2]. In 1962 Rumpf published a significant work to describe the tensile strength of agglomerates depending on the adhesion forces and coordination number between primary particles [3]. Further systematic investigations of the agglomerate strength performed by

P. Grohn · D. Weis · S. Antonyuk (✉)
Institute of Particle Process Engineering, Technische Universität Kaiserslautern, Kaiserslautern, Germany
e-mail: sergiy.antonyuk@mv.uni-kl.de

U. Bröckel
Institute for Micro Process Engineering and Particle Technology, Environmental Campus Birkenfeld, Birkenfeld, Germany
e-mail: u.broeckel@umwelt-campus.de

S. Heinrich
Institute of Solids Process Engineering and Particle Technology, Hamburg University of Technology, Hamburg, Germany
e-mail: stefan.heinrich@tuhh.de

S. Antonyuk (ed.), *Particles in Contact*,
https://doi.org/10.1007/978-3-030-15899-6_5

Schubert [4, 5], Kendall et al. [6–8] confirmed the micro-macro correlation of particle bonds [3, 9, 10]. A correlation between the strengths of dry granules and their internal structure consisting of the primary particles bonded with the solid bridges was found by Bika et al. [11]. Delenne et al. [12] modelled and simulated the deformation and breakage behaviour of a couple of cylindrical rods by applying the Mohr-Coulomb failure criterion.

The strength of granules can be investigated experimentally by using uniaxial compression tests. In uniaxial compression tests a granule is placed on a fixed plate and loaded by a punch, which moves towards this plate. During the loading, the force and displacement are continuously measured. The compression tests are carried out either with stress control (up to a predefined stress) or strain control (up to a predefined strain) at constant loading rates. In various works, different granules, such as aluminium oxide produced by fluidized bed spray granulation [13], enzymes [14], polymer bound Al_2O_3 granules [15], detergents [16], calcium carbonate [17] were tested using compression test.

During processing and handling of particulate materials the single particles are repeatedly loaded in contact with their neighbours, wall of apparatus or components. The deformation behaviour of particles and agglomerates can significantly change during cyclic loading since softening or hardening of material can take place. This effect can occur locally in the small volume in the range of the contact area so that particle shape and the material properties, such as volume density, can be assumed to be unchanged. In other cases the cyclic loading can influence the whole volume of the particle resulting in the shape change, for example, as it is the case in a spheronization process, where spherical pellets are produced by rounding of cylindrical extrudates [18]. The breakage of granules which are subjected to cyclic loading can occur at stresses that are substantially lower than the strength of the material. The fatigue can be attributed to the formation and propagation of shear zones and micro cracks in each loading cycle. Riehle and Simmchen [19] described this effect with the Wöhler curve. Tavares and King [20] found a decrease of the elastic-plastic stiffness of particles during repeated impact and explained the fatigue behaviour with the formation and propagation of damages. On the contrary, Antonyuk [21] found that the stiffness of spherical granules increases with increasing number of loading/unloading compression cycles until the saturation of the plastic deformation is reached. Despite the increase in the stiffness the breakage strength decreases. The resistance of a material against cyclic loading is dependent on the loading intensity and the frequency as well as on the particle size and the microstructure. For agglomerates, this effect was confirmed by [14] and [22]. In order to describe real processes, where the particles are loaded very often, this effects need to be taken into account since they may significantly affect the behaviour of the entire system. So far, there are only limited works in the literature which model and describe this effect. Therefore, it is important to measure the cyclic behaviour and describe it with suitable models.

The Discrete Element Method (DEM), firstly introduced in the work of Cundall and Strack [23], can be used for the study of the mechanical interactions between particles as well as between particles and apparatus walls. This method calculates the trajectory of each individual particle by solving the equations of motion considering the forces which act on the particle. Therefore, in addition to the kinetics of particulate system, the important data about the dynamics concerning force and stress distributions within the bulk material are obtained. Thus, it is possible to conclude from the microscopic particle-particle or particle-wall interactions on the macroscopic state of the particle collective. In the last twenty years the method was applied to model many applications in the mechanics of bulk materials, i.e. filling a packing [24, 25], shearing [26–28] flow behaviour and discharge of silos [29–31], mixing and transportation [18, 32–34], fluidized bed processes [35–39], compression behaviour [16, 40] and milling processes [41, 42]. With the help of DEM simulations the influence of process parameters on the microprocesses occurring on the scale of single particle interactions can be determined. These data can be used for the modelling of the macro process using a multiscale approach [43, 44].

The basis for accurate DEM simulations is a suitable contact model which should predict the force-displacement relationship for contact loads in normal and tangential directions. This kind of contact model depends particularly on the material deformation behaviour of particles and the contribution of the adhesion interactions. However, the particle shape, size, surface morphology and deformation rate can also influence the contact behaviour. In order to choose and validate a contact model, a precise characterization of the particles involved is necessary. Especially for particles in the micrometre scale, which is typical for a large number of different particulate products, it is challenging to determine the required parameters. For this size range of particle only few methods and devices are available, for example the nanoindentation tests [45] and compression tests [46]. These measuring techniques are not designed for the investigation of the irregularly shaped particles. Therefore, the examination of several samples can be very time-consuming. For this reason, we developed a novel device, which enables the measurement of mechanical properties of micrometre-sized particles and agglomerates like contact stiffness, Young's modulus, yield point as well as restitution coefficient. It allows the investigation of particles during uniaxial compression and shear by static or cyclic loading of single particles. In this work, the loading-unloading and cyclic behaviour of fine maltodextrin and titanium dioxide particles is examined with the developed device. In addition, the influence of moisture on amorphous particles is studied by performing the experiments under climatic conditions.

In order to calculate the mechanical parameters of the particles and the energy absorption from the determined force-displacement curves, different contact models are adapted. The accordance of the selected contact models and the experimental data is examined. Since the used particles have an irregular shape the force-displacement behaviour is strongly dependent on the shape of the particle and the orientation to the loading axis. In this case, the description with the models assuming a spherical particle shape can lead to significant errors. Therefore, another aim of this study is to develop a method that allows the determination of the material parameters from the force-displacement curves measured for irregular shaped particles. For this reason, a microscope was integrated into the setup to determine the shape and orientation of the particle relative to the loading axis during the compression test.

Furthermore, the compression test of the irregularly shaped titanium dioxide agglomerates is numerically reproduced using DEM and FEM simulations. For this purpose, the real size and shape of the particles were obtained from X-ray microtomography (XRMT) measurements. The irregular particle shape is modelled in DEM with the multi-sphere approach as well as with bonded-particle models. The results are compared with the experimental results and results from the FEM. By using bonded-particle models, important influences of the microstructure concerning porosity, density, primary particle distribution, stiffness of the primary particles, their elasticity or plasticity as well as binder properties of the agglomerate on deformation and breakage behaviour can be examined. In addition to many interesting results from single agglomerate simulations [2, 16, 21, 40, 44, 47–50], a new stage of modelling is the simulation of an assembly of agglomerates subjected to different stressing conditions [51–56]. Grof et al. [57] performed the simulations of uniaxial compression tests of needle-shaped agglomerates within a random packed bed. However, no simulation of a bed with agglomerates containing the real number of primary particles in the size range of micrometres or nanometres has been provided. First reason is the large computational effort that is needed to simulate a very high number of small particles. The time step of DEM calculations that ensures numerical stability is proportional to the size of the smallest particle in the system [21]. Therefore, in the DEM models the influence of the microstructure of the agglomerates is often neglected and they are represented as homogeneous particles [16, 58–60]. Moreover, the material behaviour of the micrometre-sized particles is difficult to obtain experimentally. This work focuses on the development of the measurement setup for the study of micrometre-sized particles, estimation of the model parameters and validation of contact models for the DEM simulation.

Theory and Models

The deformation behaviour of the agglomerates obtained experimentally was described using different contact models known from literature. Moreover the compression of agglomerates was reproduced and analysed by applying different simulation methods.

Contact Modelling

Various models to describe the force-displacement relation during the deformation of spherical particles can be found in the literature for different material behaviour. In addition to the Hertz model of an ideal elastic deformation, models to describe elastic-plastic behaviour are shown.

Hertz Model

For normal loading of ideal elastic and spherical particles the contact force F can be calculated from the normal displacement s using Eq. (1) according to Hertz theory [61]. The non-linear equation is derived from the integration of the stress distribution in the circular contact area of a homogeneous and isotropic sphere.

$$F_{\text{Hertz}}(s) = \frac{4}{3} \cdot E^* \cdot \sqrt{R^* \cdot s^3} \tag{1}$$

where R^* and E^* are the effective curvature radius and the effective Young's modulus of the contact partners (for example between a spherical particle-index p and wall-index w) given as:

$$E^* = \left(\frac{1-\nu_p^2}{E_p} + \frac{1-\nu_w^2}{E_w}\right)^{-1} \approx \frac{E_p}{1-\nu_p^2} \tag{2}$$

$$R^* = \left(\frac{1}{R_p} + \frac{1}{R_w}\right)^{-1} \approx R_p \tag{3}$$

In Eq. (2) ν_p and ν_w are the Poisson's ratios of the pellet and wall material respectively. Assuming that the Young's modulus of the particle E_p is significantly smaller than the Young's modulus of the rigid wall (flat surface of a steel punch in the developed experimental setup) E_w the Eq. (1) considers the elastic properties and size only for the particle as:

$$F_{\text{Hertz}}(s) = \frac{4}{3} \cdot \frac{E_p}{1-\nu_p^2} \cdot \sqrt{R_p \cdot s^3} \tag{4}$$

Due to the non-linear (parabolic) curvature $F_{\text{Hertz}}(s)$, the contact stiffness $k_{\text{el}}(s)$ of the particle in normal direction increases with increasing displacement s and particle size radius R_p:

$$k_{\text{el}}(s) = \frac{dF_{\text{Hertz}}}{ds} = \frac{2 \cdot E_p}{1-\nu_p^2}\sqrt{R_p \cdot s} \tag{5}$$

Since the stiffness is not a constant, in many simulation works described in the literature the constant coefficient $\frac{4}{3}\frac{E_p}{1-\nu_p^2}\sqrt{R_p}$ in Eq. (4) (1) is called as the elastic stiffness. For the calculation of the elastic stiffness from the measured force-displacement curve, the number of contacts that took place during the measurement should be taken into account. In the uniaxial compression test considered in this work, the particle is loaded between two punches. The slope of the measured curve corresponds to the deformation in two contacts and represents only a half of the contact stiffness of the particle.

Tsuji Model

The viscoelastic model of Tsuji extends the ideal elastic model of Hertz with a damping term to account for viscous effects [62]. Thus, the non-linear viscoelastic model of Tsuji also accounts for the energy dissipation during contact.

$$F_{\text{Tsuji}}(s) = F_{\text{Hertz}}(s) + \beta \cdot \sqrt[4]{s} \cdot \dot{s} \tag{6}$$

The damping coefficient β is a function of the coefficient of restitution of the particle and therefore considers the energy dissipation during the interaction. $\dot{s}$ is the normal component of the relative velocity of the contact partners and corresponds to the deformation rate.

Walton and Braun Model

Walton and Braun developed a linear model to predict elastic-plastic deformation behaviour [63]. The stiffness as a main parameter in the model is different for loading k_L and unloading case k_U:

$$F_{\text{WB}}(s) = \begin{cases} k_L \cdot s & if\ s \in [0, s_{max}] \\ k_U \cdot (s - s_0) & if\ s \in [s_0, s_{max}] \end{cases} \tag{7}$$

The stiffnesses for loading and unloading are related via the equivalent coefficient of restitution e_{eq} as given in Eq. (8). From the maximum force F_{max} level the plastic displacement s_0 can be calculated as a function of the unloading stiffness and the equivalent coefficient of restitution:

$$e_{eq} = \sqrt{\frac{k_L}{k_U}} = const \tag{8}$$

$$s_0 = \frac{F_{max}}{k_U} \cdot \left(\frac{1}{e_{eq}^2} - 1\right) \tag{9}$$

In the case of cyclic loading, the contact behaviour changes due to hardening as well as the increasing flattening of the contact area. Therefore, the existing model needs to be extended in order to account for those effects [64]. To achieve this, the equivalent coefficient of restitution is expressed as a saturation function:

$$e_{eq}(N) = A \cdot \left(1 - e^{-b \cdot N}\right) \tag{10}$$

where N is the number of cycles and A and b are fitting parameters. The unloading stiffness remains constant. This, in turn, leads to an increase of the loading stiffness with the increasing number of cycles. In our previous work [1] for different granulated industrial products with the size of particles in mm range this behaviour was shown. Since the plastic displacement is a function of the equivalent coefficient of restitution it is also dependent on the cycle number.

Thornton Model

In the model of Thornton [65], the Hertz model for ideal elastic behaviour is extended to cover plastic deformation, once the yield limit is reached. This is realized by a linear approximation of the force-displacement data when the maximum elastic displacement s_{el-max} is exceeded.

$$F_{\text{Thornton}}(s) = \begin{cases} F_{\text{Hertz}}(s) & if\ s \in \left[0, s_{el-max}\right] \\ F_{\text{Hertz}}(s_{el-max}) + \pi \cdot p_Y \cdot R^*(s - s_{el-max}) & if\ s \in \left[s_{el-max}, s_{max}\right] \end{cases} \tag{11}$$

The constant stiffness during plastic deformation corresponds to the stiffness according to Hertz at the yield point. By using the Hertz force at yield point, the following relation can be derived for the yield pressure p_Y:

$$p_Y = \frac{6 \cdot F_{\text{Hertz}}(s_{el-max})}{\pi \cdot d_{p,load} \cdot s_{el-max}} \tag{12}$$

The model uses two different diameters for loading and unloading phase, respectively, which describe the curvature of the surface in the contact area. $d_{p,load}$ is the particle diameter in the loading case, while the diameter increases for the unloading phase to account for an increased contact area due to plastic deformation. For the unloading phase, the Hertz model is applied again using the enlarged particle diameter $d_{p,unload}$, which dependents on the maximum deformation s_{max}:

$$d_{p,unload} = d_{p,load} \cdot \frac{F_{\text{Hertz}}(s_{max})}{F_{\text{Thornton}}(s_{max})} \tag{13}$$

Zener Model

The model of Zener [66] combines the description of the viscoelastic behaviour according to models of Maxwell and Kelvin-Voigt. This model, also known as Standard Linear Solids (SLS) model, was the first approach to describe creep of the material under load as well as recovery after removal of load and stress relaxation. The equation of the Zener model is:

$$\sigma + \frac{\eta}{E_1 + E_2}\dot{\sigma} = \frac{E_1 E_2}{E_1 + E_2}\varepsilon + \frac{\eta E_1}{E_1 + E_2}\dot{\varepsilon} \tag{14}$$

where σ and ε are the stress and strain, $\dot{\sigma}$ and $\dot{\varepsilon}$ are their time derivatives, respectively. E_1 and E_2 are elastic spring constants of the material, where E_1 corresponds to the Young's modulus for instantaneous loads (initial elastic response at time $t = 0$). The Young's modulus in equilibrium is given as:

$$E(t \to \infty) = \left(\frac{1}{E_1} + \frac{1}{E_2}\right)^{-1} \tag{15}$$

The ratio α of both moduli in Eq. (16) shows the contribution of elastic to viscous behaviour of material. If this ratio equals 1, the material behaves purely elastically, its limit $\alpha = 0$ corresponds to a purely viscous material response.

$$\alpha = \frac{E(t \to \infty)}{E(t = 0)} \tag{16}$$

The relaxation time t_{rel} for the system can be calculated as:

$$t_{rel} = \frac{\eta}{E_2} \tag{17}$$

where η describes the viscosity (dashpot component of the model).

Simulation of Deformation of the Irregular Shaped Particles

Basically there are two different approaches for modelling the deformation behaviour of agglomerates. On the one hand there is the Finite Element Method (FEM) which is a well-known continuum approach and on the other hand the particle based Discrete Element Method (DEM). Since the DEM is based on the models for the spherical objects, an additional approach for the shape representation of irregular shaped agglomerates is needed. There are two techniques which are commonly used to model a non-spherical particle shape in DEM. The multi-sphere approach and bonded-particle models. In the multi-sphere approach a cluster of overlapping spheres, which are not interacting with each other, is used for the shape representation. In bonded-particle models the agglomerate consists of spherical smaller primary particles, which are connected with solid bonds, representing a solid binder with specific mechanical properties. In this case, the interactions between primary particles takes place through the solid bonds. For each of these approaches the data about the real agglomerate shape are needed. For this reason, the real shape of studied agglomerates was determined by X-ray microtomography (XRMT) measurements with XRMT Skyscan 1072. The 3D image of the titanium dioxide agglomerate shape obtained with a resolution of 1.7 μm was extracted by a tomographic software as STL (standard triangulation language) file containing triangulated surface determined by the unit normal and vertices of the triangles. Finally, in order to reduce the number of sur-

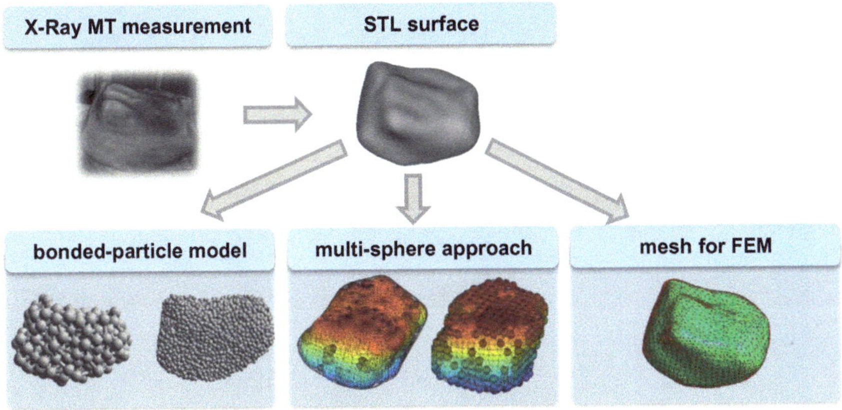

Fig. 1 Representation of the agglomerate shape from tomographic 3D measurement for simulations

face facets, the mesh was simplified. Figure 1 shows a comparison of the shape representation of an agglomerate for the different approaches.

Discrete Element Method

The Discrete Element Method (DEM) was firstly introduced by Cundall and Strack [67, 68] in 1971. In the DEM the trajectory of each individual particle in a particulate system is calculated by solving the equations of motion:

$$m_{\mathrm{p,i}}\frac{\mathrm{d}\vec{v}_{\mathrm{p,i}}}{\mathrm{d}t} = \vec{F}_{\mathrm{g,i}} + \sum_{\mathrm{j=0}}^{\mathrm{k}} \vec{F}_{\mathrm{c,ij}} \tag{18}$$

$$J_{\mathrm{p,i}}\frac{\mathrm{d}\vec{\omega}_{\mathrm{p,i}}}{\mathrm{d}t} = \sum_{\mathrm{j=0}}^{\mathrm{k}} (\vec{M}_{\mathrm{t,ij}} + \vec{M}_{\mathrm{r,ij}}) \tag{19}$$

To obtain the velocity $\vec{v}_{\mathrm{p,i}}$ and rotation $\vec{\omega}_{\mathrm{p,i}}$ of a particle i with mass $m_{\mathrm{p,i}}$ and moment of inertia $J_{\mathrm{p,i}}$, the contact forces $\vec{F}_{\mathrm{c,ij}}$ which act on the particle due to interactions with other particles j or the apparatus walls need to be modeled. For the contact force in normal direction $\vec{F}_{\mathrm{c,n,ij}}$ the models described in the section "Contact Modelling" might be used. The tangential component $\vec{F}_{\mathrm{c,t,ij}}$ can be calculated using the model of Mindlin and Deresiewicz [69]. The tangential force causes the torque $\vec{M}_{\mathrm{t,ij}}$ which affects the rotation of particle i. $\vec{r}_{\mathrm{c,ij}}$ runs from the center of particle i to the center of the contact area with particle j. In addition to $\vec{M}_{\mathrm{t,ij}}$ there is a torque $\vec{M}_{\mathrm{r,ij}}$ due to rolling friction which acts in the opposite direction of the rotation.

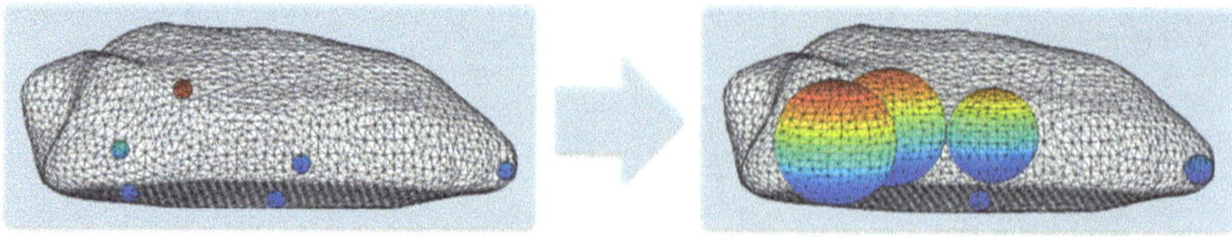

Fig. 2 Creation of overlapping particle cluster as shape representation

$$\vec{M}_{\mathrm{t,ij}} = \vec{r}_{\mathrm{c,ij}} \times \vec{F}_{\mathrm{c,t,ij}} \tag{20}$$

$$\vec{M}_{\mathrm{r,ij}} = \frac{\vec{\omega}_{\mathrm{p,i}}}{\left|\vec{\omega}_{\mathrm{p,i}}\right|} \cdot \mu_{\mathrm{r,i}} \cdot R^*_{ij} \cdot \left|\vec{F}_{\mathrm{c,n,ij}}\right| \tag{21}$$

where $\mu_{\mathrm{r,i}}$ is the coefficient of rolling friction.

Multi-sphere Approach

One possibility to represent a non-spherical particle shape in DEM is the multi-sphere approach, which provides numerical stability and efficient contact detection capabilities [70, 71]. In this approach a cluster of overlapping spheres which are not interacting with each other is used to represent the real irregular shaped particle surface. To create such particle clusters a STL representation of the real particle surfaces was obtained by XRMT measurements. Based on this surface information the initial cluster-spheres are randomly placed on the surface. Afterwards, the initial spheres start to grow, until another point on the particle surface is reached as exemplarily shown in Fig. 2.

In DEM simulations with such multi-sphere particles, the contact forces are calculated for every sphere which is in contact with another particle or wall. Therefore, the accuracy as well as the computational cost of the DEM simulation is highly dependent on the number of spheres used for the shape representation.

Bonded-Particle Model

A more detailed representation of the agglomerate in the DEM can be achieved by generation of its microstructure with non-overlapping spheres connected by solid bonds. The bonded-particle model has been successfully used to simulate the deformation as well as the breakage behaviour of different materials, such as rock [72], ceramic [73], fibres [74], agglomerates [40, 49], crystals [2] and composites [75]. In Fig. 3 a schematic representation of two particles connected with a cylindrical solid bond is shown.

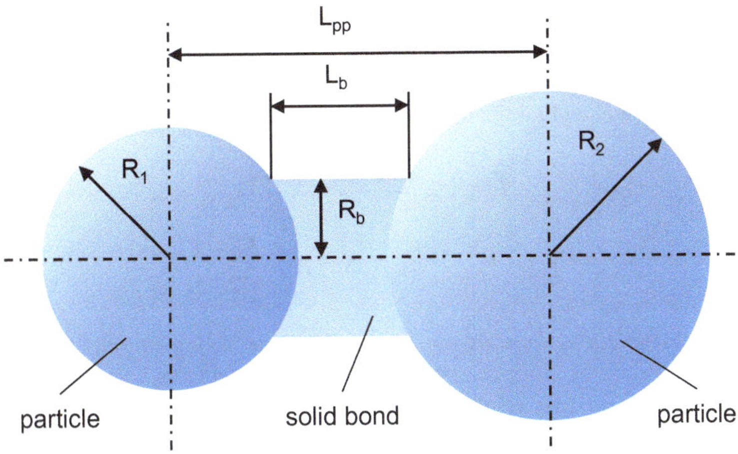

Fig. 3 Schematic representation of two particles connected with a solid bond

From geometrical considerations, the length of the bridge L_b can be derived from the distance of the particles L_{pp} and their radii R_1 and R_2 as well as the radius of the bridge R_b:

$$L_b = L_{pp} - \sqrt{R_1^2 - R_b^2} - \sqrt{R_2^2 - R_b^2} \tag{22}$$

The maximum value of the solid bond radius R_b is limited by the radius of the smaller bonded particle. During the deformation of the whole agglomerate, the particles can move relative to each other and deform their bond. The forces and moments acting in the bond because of this deformation are calculated with an appropriate model depending on the material behaviour of solid binder, as described in the following sections. For each time step the current length of the bond $L_{b,cur}$ is compared with the so-called recovery length $L_{b,rec}$, which describes the length the bond would have if there would be no external forces acting on both contact partners. Thus, the deformation of the bond in the normal direction ξ_b in each time step is calculated as:

$$\xi_b = L_{b,rec} - L_{b,cur} \tag{23}$$

The value of ξ_b is positive for compression of the bond and negative for bonds tension.

Elastic Deformation in Solid Bonds

For pure elastic behaviour of the bond the recovery length $L_{b,rec}$ is constant and equals to the initial bond length $L_{b,init}$, at the creation time of the bond in the agglomerate.

This means, the bond always aspires to recover its initial length, independently on previous deformation. The forces and moments in the bond for the current time step at $(t + \Delta t)$ are calculated based on the values from the previous time step at (t) [44]. The forces in normal and tangential direction are calculated as:

$$\vec{F}_{b,n}^{t+\Delta t} = \frac{A_b \cdot E_b \cdot \xi_b \cdot \vec{r}_n}{L_{b,rec}} \tag{24}$$

$$\vec{F}_{b,t}^{t+\Delta t} = T \cdot \vec{F}_{b,t}^{t} - \vec{v}_{rel,t} \cdot \Delta t \cdot A_b \cdot \frac{E_b \cdot (1 + \nu_b)}{2 \cdot L_{b,rec}} \tag{25}$$

A_b is the cross-cut surface area of the bond, E_b its Young's modulus and ν_b its Poisson's ratio. $\vec{r}_n$ is the unit vector between the particle centers, $\vec{v}_{rel,t}$ the tangential component of the relative velocity at the contact points. T is the transformation matrix, which describes the movement of the contact partners in space and Δt is the numerical time step. The moments acting in the bonds calculated as:

$$\vec{M}_{b,n}^{t+\Delta t} = T \cdot \vec{M}_{b,n}^{t} - \vec{\omega}_{rel,n} \cdot \Delta t \cdot A_b \cdot \frac{E_b \cdot (1 + \nu_b)}{2 \cdot L_{b,rec}} \cdot J \tag{26}$$

$$\vec{M}_{b,t}^{t+\Delta t} = T \cdot \vec{M}_{b,t}^{t} - \vec{\omega}_{rel,t} \cdot \Delta t \cdot A_b \cdot \frac{E_b}{L_{b,rec}} \cdot I \tag{27}$$

where $\vec{\omega}_{rel,n}$ and $\vec{\omega}_{rel,t}$ are vectors of the relative rotational velocities of the particles in normal and tangential direction. I and J are the moment of inertia and the polar moment inertia of the solid bond cross-section, respectively.

Elastic Deformation in Solid Bonds with Brittle Fracture

This model extends the pure elastic model by implementation of breakage criteria for the bond according to classical beam theory. If the stresses in a bond reach or exceed some predefined limits, the bridge is destroyed and consequently removed from the system. Assuming brittle material behaviour the breakage criteria for fracture due to bending/tension or shearing/torsion can be written as:

$$\frac{\left|\vec{F}_{b,n}\right|}{A_b} + \frac{\left|\vec{M}_{b,t}\right| \cdot R_b}{I} \geq \sigma_{max} \tag{28}$$

$$\frac{\left|\vec{F}_{b,t}\right|}{A_b} + \frac{\left|\vec{M}_{b,n}\right| \cdot R_b}{J} \geq \tau_{max} \tag{29}$$

where σ_{max} and τ_{max} are tensile and shear strength of the bond material, respectively. The solid bond breakage takes place when at least one of the breakage criteria is satisfied [72]. Due to breakages of bonds energy dissipation as well as an irreversible plastic deformation is received, even if the load is completely removed.

Fig. 4 Schematic dependency of the recovery bond length on the bond deformation

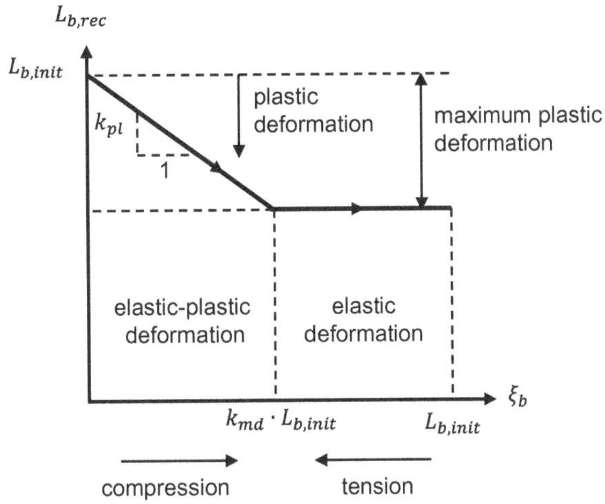

Inelastic Deformations in Solid Bonds

In order to account for plastic deformation, in this model the recovery length of the bond $L_{b,rec}$, modelled as a constant in elastic case, is assumed to be decreasing after the loading-unloading of the bond. It is calculated as a function of the deformation ξ_b and two additional model parameters k_{md} and k_{pl}. The model parameter $k_{md} \in [0; 1]$ corresponds to the ratio of the maximum bond deformation. The decrease recovery length of the bond $L_{b,rec}$ with the bond deformation is limited by a maximum $L_{b,rec} = k_{md} \cdot L_{b,init}$. The second parameter $k_{pl} \in [0; 1]$ defines the bond plasticity. A plasticity parameter of $k_{pl} = 0$ means pure elastic behaviour, while $k_{pl} = 1$ results in a pure plastic behaviour. The dependency of $L_{b,rec}$ on the normal deformation of the bond ξ_b is schematically shown in Fig. 4.

Maxwell Solid Bond Model

The basic viscoelastic Maxwell model, consisting of a viscous damper and an elastic spring connected in series, can also be used to calculate the forces and moments acting in a solid bond. The elastic material properties will be described with the Young's modulus E_b for the normal deformation and the shear modulus $G_b = \frac{E_b}{2\cdot(1+\nu_b)}$ for the tangential deformation of the bond, respectively, as well as with the Poisson's ratio of the bond ν_b. The damping term is a function of the viscosity η, which characterizes the response on the deformation rate. The forces and moments for the current time step $(t + \Delta t)$ are obtained from the values on the previous time step (t) as:

$$\vec{F}_{b,n}^{t+\Delta t} = T \cdot \vec{F}_{b,n}^{t} \cdot \left(1 - \frac{\Delta t \cdot E_b}{\eta}\right) + \vec{v}_{rel,n} \Delta t \cdot \frac{A_b \cdot E_b}{L_{b,rec}} \tag{30}$$

$$\vec{F}_{b,t}^{t+\Delta t} = T \cdot \vec{F}_{b,t}^{t} \cdot \left(1 - \frac{\Delta t \cdot G_b}{\eta}\right) + \vec{v}_{rel,t} \cdot \Delta t \cdot \frac{A_b \cdot E_b}{2 \cdot L_{b,rec} \cdot (1 + \nu_b)} \tag{31}$$

In this model the moments are calculated using the Eqs. (26) and (27).

Finite Element Method

The macroscopic deformation of agglomerates can be depicted using the Finite Element Method (FEM). Hereby the agglomerate is considered as a homogeneous material. Since the FEM is a continuum approach a spatial discretization of the agglomerate volume is necessary. The single finite elements are connected via the nodes of the mesh. Based on a stress-strain law this nodes can be displaced, what leads to a deformation of the agglomerate. The accuracy depends on the quality of the mesh. With higher spatial resolution the accuracy but also the computational effort increases. Adams [76] performed FEM analysis of the impact test of a spherical particle with elastic-plastic material behaviour. In the works of Kienzler et al. [77] the compressive stress of particles with different material behaviour (from elastic to viscoplastic) was simulated. Also Khanal and Schubert [78] applied the FEM to describe the compression and impact behaviour of particle compounds using an elastic material law.

Materials and Methods

Developed Experimental Setup for Mechanical Tests of Microparticles

For the study of the contact behaviour of micrometer-sized particles (minimum size of about 20 μm) at different loading conditions (compression, tension, shear, torsion) a novel experimental setup (Fig. 5) was designed at the Institute for Micro-Process-Engineering and Particle Technology (IMiP). This setup was used to obtain the experimental data about the forces acting on the single particle in contact with a rigid wall. The measured data were analysed at the Institute of Solids Process Engineering and Particle Technology (Hamburg University of Technology) and compared with DEM simulations of mechanical tests performed with different contact models described in section "Contact Modelling". Also the shape of the particle and its position relatively to the loading direction were considered in the simulation according to approaches given in section "Simulation of Deformation of the Irregular Shaped Particles". These data were obtained before starting a measurement. The particle was glued to the polished load pin with an average roughness of 6.2 nm (measured by AFM) and scanned with X-ray microtomography as described in section "Simulation

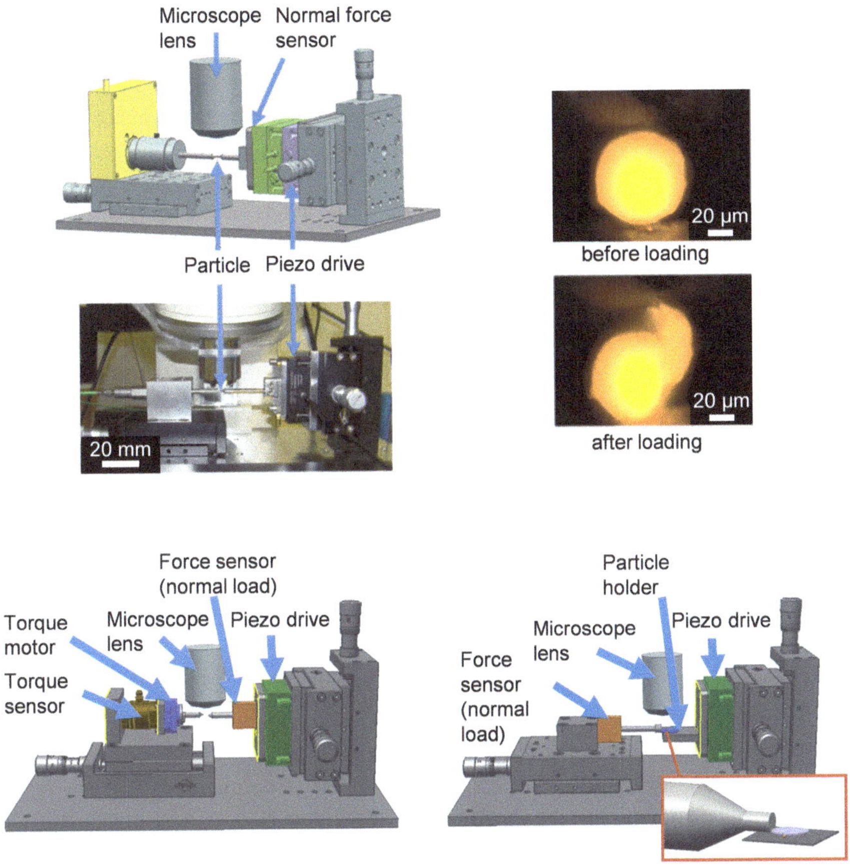

Fig. 5 Developed experimental setup for the mechanical test of single micrometer sized particles under different loading conditions: compression and tension (top), torsion (bottom right), shear (bottom left) [64]

of Deformation of the Irregular Shaped Particles". After the X-ray microtomography, the pin with the particle was mounted into the setup shown in Fig. 5.

During a compression test, the pin with the fixed particle moves to the other fixed pin and the particle will be loaded uniaxially between them. A piezo drive (Physik Instrumente PIHera P-622.Z) is used to move the pin and load the particle with a constant deformation rate of 50 µm/s. The maximal displacement of the piezo drive is 250 µm at a repeatability of 1 nm. The force in contact is measured with a force sensor LSB200 Futek. This force sensor has a maximal force of 200 mN and a non-linearity of 0.1% of the rated output. Using the laser-vibrometer (SIOS SPS 120), the displacement can be measured with a resolution of 0.2 nm. It is equipped with temperature and pressure sensors to compensate the influence of environmental conditions. With the help of the laser-scanning-microscope (Leica TCS NT) the samples can be monitored during the experiments. Figure 5 shows two images of a particle before loading and after unloading captured by the microscope. With an additional climatic chamber on the object stage, the relative humidity can be adjusted from 10% to 85% ± 2%. A thermostat was installed for temperature control (±1 °C). The self-written control software allows to perform different loading cycles of the particle.

The developed setup also allows performing shear and torsion tests (Fig. 5, bottom left and right, respectively). In a shear test, three particles were glued in a triangular arrangement on a glass pane and placed on the surface of a test substrate. The glass pane is pushed with the load pin by piezo drive and the acting shear force and the corresponding displacement are measured at a constant normal load. The applied normal load is determined from the defined weights put on the glass pane considering its mass. The course of the force over time is measured in order to be able to determine sliding and static friction. The normal force was set by corresponding weights. During the torsion test of the particle, first the pin with the solid particle is brought with the particle. It is moved onto the fixed pin and rotated by a motor at a defined speed. Meanwhile the torsional moment is measured by a torque sensor.

Validation of Experimental Setup

For direct comparison of the new setup with established measuring systems, results were compared to nanoindentation experiments performed at the Max-Planck Institute for Polymer Research in Mainz (group of Dr. Auernhammer). As a reference substance, monodisperse, spherical polystyrene particles from micro Particle GmbH (Berlin, Germany) were used. The particles have a well-defined diameter of 48.15 ± 0.67 μm and mechanical properties are independent of relative humidity. To create a wall-particle-wall configuration during the nanoindentation experiments an indenter with a flat tip was used. Particles were loaded up to a maximal force of 4 mN in force controlled mode. Five measurements were made where temperature and loading rate were adapted to the conditions of the experiments in Mainz. The experiment was replicated four times at the same temperature in our new setup. Only the maximal force was raised to 20 mN to account for the lower sensitivity sensor. The results of both experiments show a purely elastic deformation behaviour with elastic modulus of 3.12 ± 0.17 GPa with the new setup and of 2.99 ± 0.18 GPa with the nanoindentation. The results are in very good agreement and confirm the accuracy of the system.

Material

In the performed experiments, the behaviour of two different materials is examined. Besides spherical maltodextrin particles, agglomerates of titanium dioxide are used.

Maltodextrin and Production of Test Particles

Maltodextrin, an amorphous polysaccharide, is an ingredient in many food products. It is flavourless to mildly sweet, mostly used to raise the nutritious value of foods. Maltodextrin is commercially available in different dextrose equivalents (DE), which stand for the average molecule chain length, ranging from 2 up to 20. Materials with a DE above 20 are classified as glucose syrup. For this work, Maltodextrin DE6 (MD6) provided by Cargill Holding Germany GmbH was used. The material exhibits a highly irregular particle shape (Fig. 6) and a mean particle diameter of 160 μm.

At high relative humidities (RH) the particle liquefies at room temperature when the glass transition temperature T_g is exceeded [79, 80]. To characterize the behaviour of amorphous materials regarding the climatic conditions the value $T - T_g$ is used [79]. Glass transition can be achieved by either raising the temperature or the water content of the material. Gorden and Taylor [81] developed a model in Eq. (32), which described the glass transition point depending on glass transition temperatures of the dry material $T_{g,s}$ (216 °C, for MD6 according to [80]), water $T_{g,w}$ (−138 °C according to [82]), the water and solid content (index *ww* resp. *ws*) of the material and the experimentally measured material parameter k.

$$T_g = \frac{w_w T_{g,w} + k w_s T_{g,s}}{w_w + k w_s} \tag{32}$$

During the experiments, the temperature was kept constant at 25 °C and only RH was modified. The water content of the test particles can be estimated by using the sorption isotherm data available in literature [80] and the Guggenheim-Anderson-De Boer model [83]. For 25 °C and MD6, T_g corresponds to 70.9% RH.

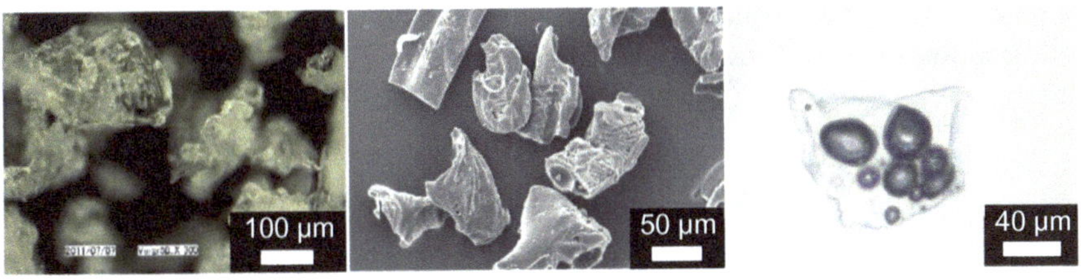

Fig. 6 Left to right: reflected and transmitted light microscopic image of MD6, SEM image, image of the pores inside the particle (were made visible by an index matching fluid) [64]

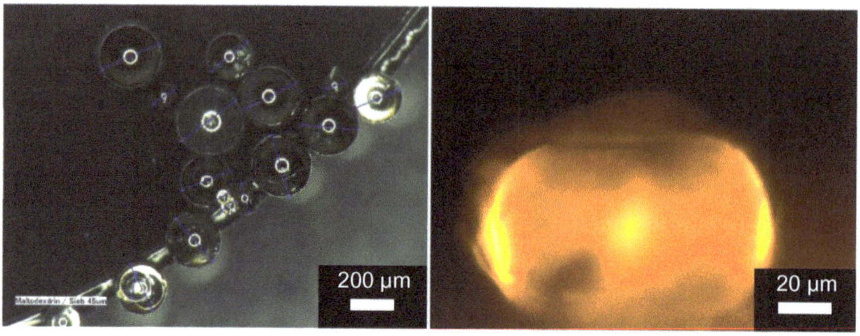

Fig. 7 MD6 spheres (left) and a single, approximately cylindrical MD6 particle (right) [64]

In order to minimize the influence of the particle shape on the test results, we refused on the utilization of commercially available maltodextrin particles which are usually irregular shaped. An evaluation according to Hertz is limited in the case of irregular particles, since the contact zone does not correspond to the Hertzian hemisphere. This makes it very difficult to measure their material properties like Young's modulus or yield point, directly. For the performed tests we produced the spherical maltodextrin particles according to the method described in [84]. A water solution of MD6 is dropped into a hydrophobic liquid. The formed emulsion is stirred with a magnetic stirrer resulting in the size reduction of the individual droplets. At elevated temperatures, the water content of the droplets is slowly decreasing and solid spheres of MD6 are formed (Fig. 7 left). Different particle diameters can be achieved by varying temperature, initial water content and the intensity of the dispersion. For this study particles in the range 80–120 μm were produced.

In addition to the spherical particles, cylindrical particles were formed using the testing equipment developed at the IMiP. Spherical MD6 particles were softened by storage at the conditions above glass transition point and then after equilibration, a small load was applied. Thus, the particles get flattened on top and bottom (Fig. 7 right). By changing the climatic environment, the material hardens again. The resulting particles are approximately cylindrical. Due to irregularities in the flattened surface of the individual particles, precise measurement of contact area was not possible. Therefore, values for E_1, E_2 and ν could not be calculated. However, the relaxation time t_{rel} (Eq. 17) and the ratio α (Eq. 16) are not dependent on the contact area.

Mechanical Characterization of Maltodextrin Particles

For the characterization of the MD6 particles, tests were carried out under constant and cyclic loading with the experimental setup described in section "Developed Experimental Setup for Mechanical Tests of Microparticles" and run under controlled climatic conditions. As already described, the temperature in the experiments was

Table 1 Comparing relative humidity and water content of the particles and the resulting $T - T_g$ value

RH/%	Water content/g/g	$T - T_g$/°C
40	0.089	−53.4
50	0.105	−37.7
60	0.126	−19.6
70	0.145	−2.3
80	0.186	20.5

held constant at 25 °C. In order to calibrate the material model, 94 cylindrical MD6 particles were stressed at a constant load of 50 mN for 500 ms and the deformation was recorded. The tests were run at a relative humidity of 50% and from 70 to 83%. For the cyclic tests, 53 spherical particles were exposed 40 times to a predefined load (33.7 mN ± 0.47 mN). The duration of each loading cycle corresponds exactly to the holding period of the constant load used for the cylindrical particles, that means 500 ms. The relative humidity was varied from 40 to 80% in 10% increments. The corresponding water content of the particles and $T-T_g$ can be found in Table 1. $T - T_g$ was calculated using the Gordon and Taylor equation [81] (Eq. 32) and data from [80]. The equation of Guggenheim-Anderson-De Boer [83] was used to estimate the water content [64].

In [85] it was shown that the glass transition of maltodextrin depends on the storage history of the material. This was partly taken into account by using exsiccators loaded with silica gel to store the particles at least one week before the test.

Tested Titanium Dioxide Particles

Titanium dioxide (TiO_2), also known as titanium (IV) oxide or titania, is the naturally occurring oxide of titanium. Titanium dioxide can exist in three crystalline modifications (brookite, rutile and anatase) with different crystal lattice structures and as an amorphous material. In this study, amorphous irregular shaped titanium dioxide particles were used. The particles were produced at the Institute of Mechanical Process Engineering (Otto von Guericke University of Magdeburg, group of Prof. J. Tomas$^+$) via the sol-gel process without calcination [64]. The SEM images of derived particles indicated that titanium dioxide particles had a typical agglomerate structure Fig. 8.

Titania was analysed by combined differential scanning calorimetry (DSC) (NETZSCH DSC 204 F1 Phoenix) and thermogravimetric analysis (TGA) (Q500, TA Instruments), which detect the mass loss and the heat flow as a function of temperature changes. At a temperature of 170 °C the measurements showed a height loss of approximately 17.74 ± 2.62% (Table 2). As pointed out in [86], chemical bonding $TiO_2 \cdot H_2O$ can arise during a sol-gel process, whereby H_2O has a mass fraction of 18.4% and is evaporated at temperatures from 170 to 200 °C. Due to the manufacturing process, the used sample can contain not only water residue but also sodium dodecyl sulfate (SDS). Therefore, that the decomposition of SDS usually takes place

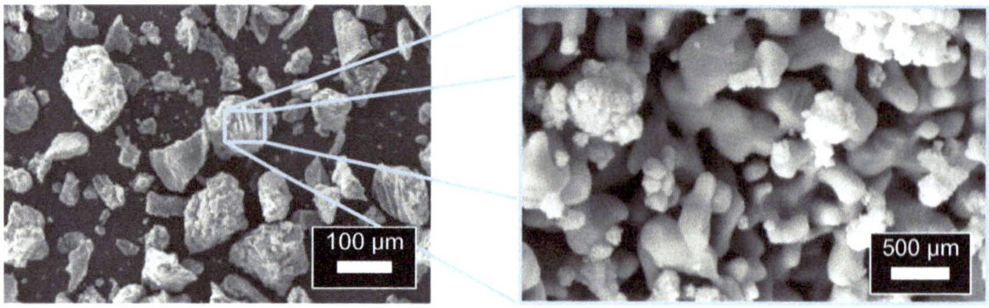

Fig. 8 SEM images: titanium dioxide agglomerates (left); image with higher magnification showing agglomerate inner structure (right)

Table 2 Results of the combined thermal DSC/TGA analysis. Mean values with standard deviation [87]

Sample weight/mg: 22.82 ± 8.42						
Temperature/°C	150	170	250	600	1000	After cooling
Weight loss/%	17.54 ± 3.2	17.74 ± 2.6	20.98 ± 1.3	23.07 ± 1.2	23.13 ± 1.5	23.59 ± 0.6
Sample weight/mg: 19.70 ± 2.12						
Temperature of the endothermic peak/°C (dehydration)			Temperature of the first exothermic peak/°C (amorphous–anatase)		Temperature of the second exothermic peak/°C (anatase–rutile)	
100 ± 34			402 ± 35		776 ± 51	

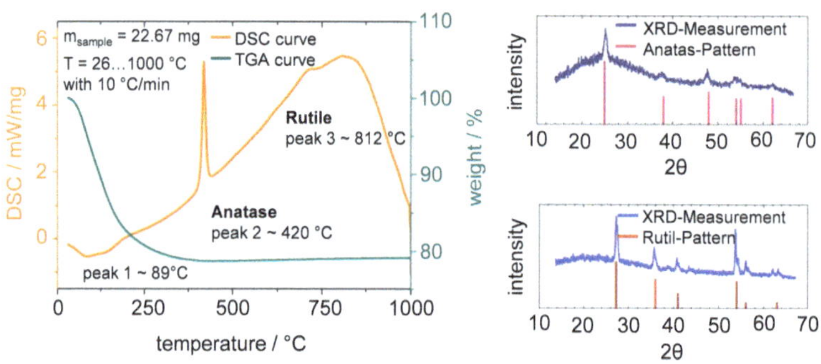

Fig. 9 Diagram of TGA/DSC analysis (left); XRD for titania agglomerates after 600 °C (top right) and after 1000 °C (down right) [87]

in a range between 200 and 400 °C, SDS plays only a minor role in the mass loss at temperatures up to 170 °C (Fig. 9).

With a helium pycnometer the agglomerate density was measured and was found to be 2.45 g/cm^3. Since in the literature [88, 89] the titanium dioxide density is about 4 g/cm^3, this suggests that the lower measured agglomerate density can be explained by the presence of pores and water residues in the internal structure of the agglomerates.

Mechanical Characterization of Titanium Dioxide Agglomerates

Single particle uniaxial compression tests up to breakage were performed with the help of the setup described above to define the mechanical properties of titania agglomerates. By analysing images of the particles taken by laser-scanning-microscope from above and from the side before compression, the equivalent sizes were determined. The maximum distance between two loading pins at first contact was determined by the free image processing program ImageJ. This maximum particle height in the loading position was assumed to be the particle equivalent size $h_{eq,side}$. The equivalent cross-section area was calculated as $A_{eq} = d_{eq,top} \cdot h_{eq,side}$, where $d_{eq,top}$ is the diameter of a circle with the same area as the particle's projection.

In total, 34 particles with equivalent sizes in the range of 70–350 μm were tested, the average value of the breakage force F_{break} was found to be 40 ± 37 mN. The engineering compressive strength, $\sigma_{eng} = A_{break}/A_{eq}$, was estimated as 1.5 ± 0.9 MPa. It must be taken into account that due to the shape irregularity an excessively complex state of stress in the agglomerates during uniaxial compression occur. Therefore, it is important not to confuse this parameter with a true compressive strength of titanium dioxide. In earlier works of Rumpf [3] as well as in many recent investigations [90], was pointed out that the strength of an agglomerated structure is affected by interparticle bonds, not by the strength of constituent primary particles. The results of performed experiments also confirm this assumption for TiO_2 agglomerates, since the tensile strength of pure titanium dioxide is at least two orders of magnitude higher than the strength obtained for titanium dioxide agglomerates at compression [89].

The breakage probability P of TiO_2 agglomerates stressed by compression were described by a three-parameter Weibull distribution function (Eq. 33) in Fig. 10. The following distribution parameters were obtained: Weibull modulus $m = 1.49$, minimum compressive stress $\sigma_{min} = 0.15$ MPa and a characteristic value of the compressive stress $\sigma_0 = 1.6$ MPa. The distribution of defects, the homogeneity of the internal structure, the shape and morphology of the agglomerates depend on the particle size. The smaller the agglomerates, the more homogeneous their structure and less large defects occur on the surface of the agglomerates. As shown in our previous work for larger agglomerates (in the mm size range) [21, 91], the failure stress increases with the decrease in the size of the agglomerates. The finer agglomerates show higher strength. This is confirmed here for finer TiO_2 agglomerates. The obtained broad fracture probability distribution (Fig. 10) described by a small Weibull modulus is partly the result of the wide distribution of particle sizes of the tested TiO_2 agglomerates.

$$P = 1 - exp\left[-\left(\frac{\sigma - \sigma_{min}}{\sigma_0 - \sigma_{min}}\right)^m\right] \tag{33}$$

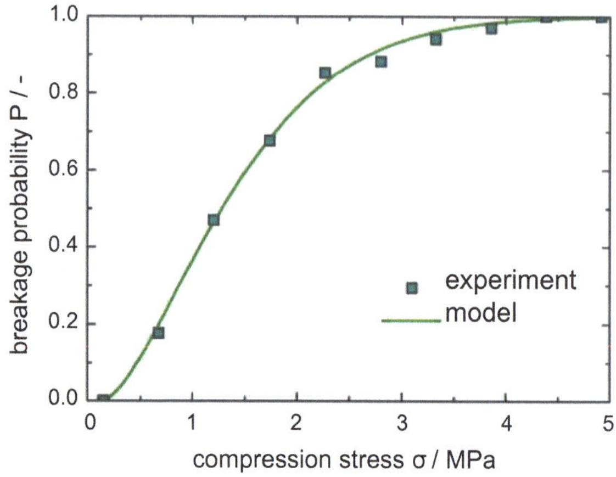

Fig. 10 Measured breakage probability of TiO_2 agglomerates described by Weibull distribution function [87]

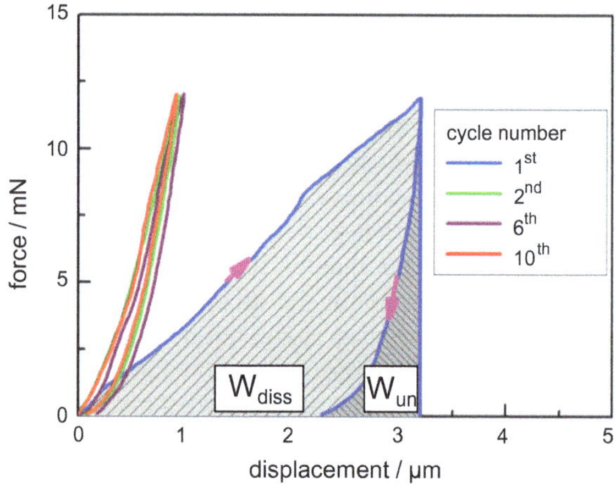

Fig. 11 Typical force-displacement curves of cyclic loading of the titania agglomerate at cyclic compression with constant amplitude of 12 mN. Each cycle is shown as separate curve starting from origin of coordinates. The hitched area under loading and unloading curves of first loading cycle indicates the estimation procedure of the equivalent restitution coefficient [87]

Additionally, compression tests with agglomerates under cyclic loading–unloading at two different levels of maximum force of the cycle were carried out. The typical force-displacement curves of the cyclic loading of a single titania agglomerate are shown in Fig. 11. It can be seen that the agglomerates demonstrated plastic shakedown response at cyclic loading since the value of residual plastic displacement decreased with the number of cycle. To calculate the equivalent restitution coefficient (34), that describes the energy loss during particle loading and unloading [92], the obtained force-displacement curves are numerical integrated (see hatched areas in Fig. 11. The value of the restitution coefficient must lie between zero (pure inelastic impact without rebounding) and one (pure elastic impact). For each tested particle, the equivalent restitution coefficient was calculated as:

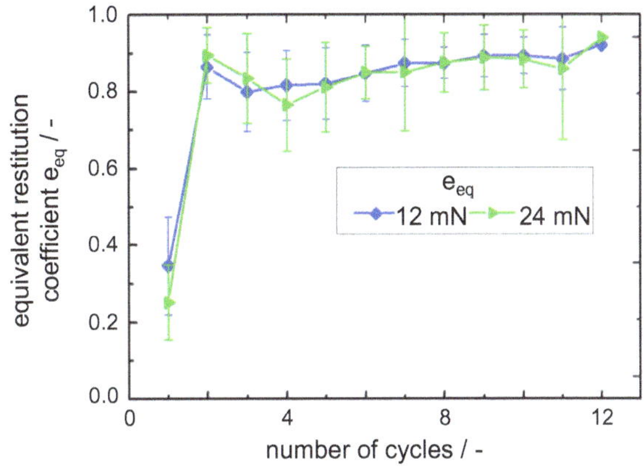

Fig. 12 Equivalent restitution coefficient as function of number of cycles for titania agglomerates during cyclic compression test [87]

$$e_{eq} = \sqrt{\frac{W_{un}}{W_l}} = \sqrt{1 - \frac{W_{diss}}{W_l}} \tag{34}$$

where W_{un} is the area under the force-displacement curve at unloading; W_l is the area under the loading curve during the compression phase and $W_{diss} = W_l - W_{un}$ is the energy dissipation.

The obtained mean values of restitution coefficient as a function of the number of cycles for the tested agglomerates are shown in Fig. 12. The mean value of equivalent coefficient of restitution at first loading cycle was determined to be equal to 0.3. In the second cycle the restitution coefficient highly increases up to a value of about 0.9 and remains nearly constant during next cycles. No influence of the force level in the range of 12–24 mN during the cycling loading was shown.

Experimental Results and Description with Contact Models

Maltodextrin

Experimental Results of Cylindrical Particles

Using a constant normal load, 94 cylindrical MD6 particles were stressed and analysed using least square fits to the Zener model (Eq. 14). In Table 3 the ratio α between $E(t \to \infty)$ and $E(0)$ and t_{rel} for the different RH tested are listed. With increasing RH the material approaches the glass transition point and transfers to the viscous state at high RH. This results in the rapidly decrease of the ratio α after RH of 80%. Despite a large standard deviation, the results for t_{rel} show a roughly linear relationship between glass transition temperature T_g and relaxation time t_{rel}. The Zener model predicts decreasing sample height for tests at a constant load. Equilibrium

Table 3 Results from compression test with cylindrical MD6 particles [64]

RH/%	Ratio α/–	t_{rel}/s
50	0.92 ± 0.04	2.38 ± 1.40
70	0.87 ± 0.05	1.79 ± 0.75
80	0.71 ± 0.08	0.87 ± 0.40
83	0.19 ± 0.08	0.17 ± 0.06

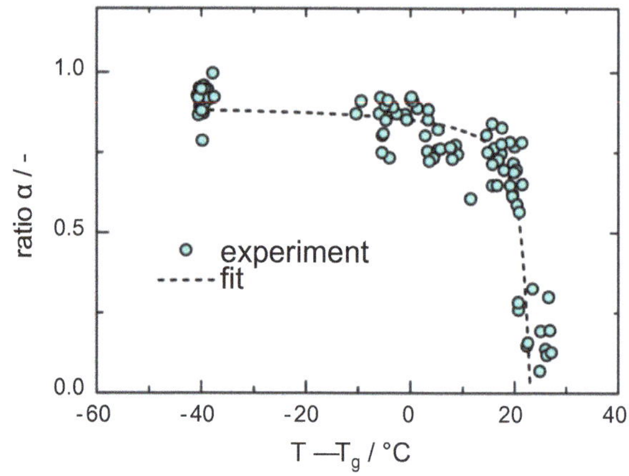

Fig. 13 Measurement results (dots) and fit (curve) of parameter α as function of $T - T_g$. Obtained values for the fit parameters (Eq. 13) are a = 0.90, b = 0.79 and c = −18.50 [64]

must be reached after approximately five t_{rel}, which corresponds to 99.3% of the end value. Due to the restrictions of the measurement equipment, the constant load could only be applied for 500 ms. As a consequence, equilibrium was not reached for the majority of samples and, additionally to the natural variation within the particles, leads to a higher standard deviation.

The results for the ratio α are fitted using the following relation:

$$\alpha = ae^{\frac{1}{bT_g+c}} \tag{35}$$

as shown in Fig. 13. RH strongly influences this parameter and leads to a sharp drop in α above 80% RH. As t_{rel} and α are independent of contact area and particle height, the significant particle deformation under load during these experiments has limited impact on the results.

Model for Spherical MD6 Particles

In the experiments with cylindrical MD6 particles, it was shown that the MD6 material parameters are time- and load-dependent and can be explained using the Zener model. To apply this model to spherical particles, Thornton and Zener models were combined to form our new Thornton and Zener model. From experimental data, the global particle strain can be calculated using:

$$\varepsilon(t) = \frac{s(t)}{2R} \tag{36}$$

where $s(t)$ is the deformation at time t and R the particle radius. The differential equation describing Zener (Eq. 14) was solved numerically using $\varepsilon(t)$ as input, resulting in a time depended Young's modulus $E(t)$. $E(t)$ was used to calculate the force predicted by the new model, instead of using a constant E in Thornton's model (Eq. 11). A least square fit using the initial Young's modulus $E(t = 0)$, the yield point s_{el-max} and the end-deformation of unloading curve s_0 as free parameters was used to characterize experimental data. α and t_{rel} were set to values derived from fits described in section "Experimental Results of Cylindrical Particles". In all experiments with spherical MD6 particles, the new model was used.

Experimental Results of Spherical Particles

53 spherical MD6 particles with a mean diameter of 90.5 ± 12.9 μm were loaded in uniaxial compression tests. In Fig. 14 typical force-displacement curves for particles below and above glass transition T_g are shown, which indicate the decrease of the contact stiffness and increase of plastic deformation after glass transition. The approximation curves according to Thornton and Zener model are also shown in Fig. 14. In both cases, model and experiment are in very good agreement. The initial Young's modulus for MD6 was estimated by Zener model. Figure 15 demonstrates a relationship between T_g and initial Young's modulus $E(t = 0)$.

In addition, the spherical MD6 particles were stressed 40 times in uniaxial cyclic compression tests under different climatic conditions, with a wait time of 1 min between each loading. Based on the results of relaxation from section "Experimental Results of Cylindrical Particles", MD6 will reach an equilibrium between loadings. Therefore, the reloading of the particles should not be affected by deformation due to

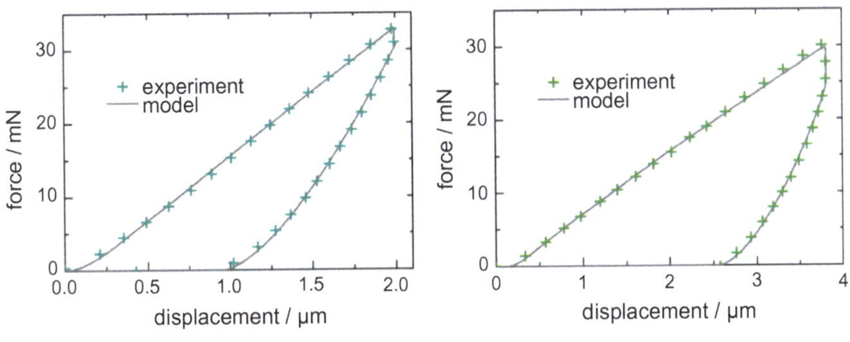

Fig. 14 Typical force-deformation curves for spherical MD6, experimental and least square fit of Thornton model. Example 1 was run at $T - T_g = -40.86$ °C (left), example 2 at $T - T_g = 19.96$ °C (right) [64]

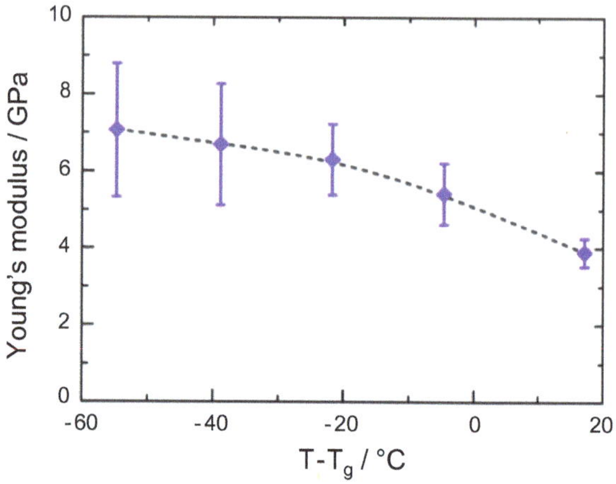

Fig. 15 Young's modulus for different RH during the first loading. Error bars indicate the standard deviation from 9 to 13 particles per RH [64]

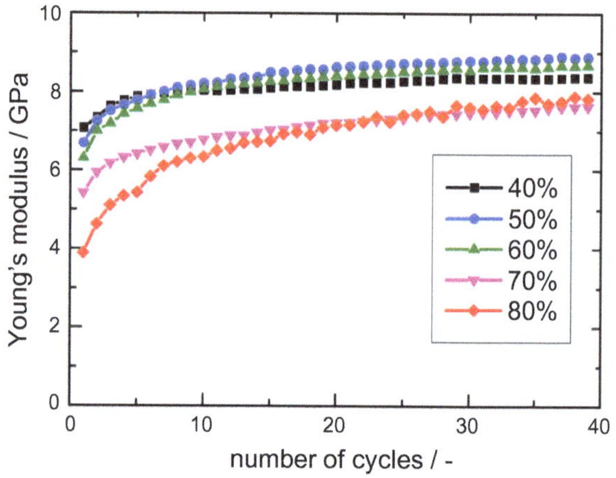

Fig. 16 Young's modulus for different RH over loading cycles

viscous behaviour from the previous cycle. Young's modulus according to Thornton and Zener model is presented in Fig. 16.

The cyclic hardening follows an exponential curve, which shows smaller values of the Young's modulus at higher RH just before (70%) and after glass transition (80%). Following the argumentation of Thornton in [93], an increase in the Young's modulus can be explained by a modification of particle curvature in the contact zone due to plastic deformation in each cycle. By using an optical microscopy, it was not possible to reliably measure changes in the contact zone. For this reason, a specially prepared sample was examined using a SEM. The expected deformation can be seen in Fig. 17. The increase in contact area diameter after 40 loadings for each RH is listed in Table 4.

In Fig. 18 the yield pressure calculated from Thornton and Zener is presented. It also demonstrates the influence of moisture content on the mechanical properties. The lower yield pressure at higher RH (after 70%) corresponds well with the higher

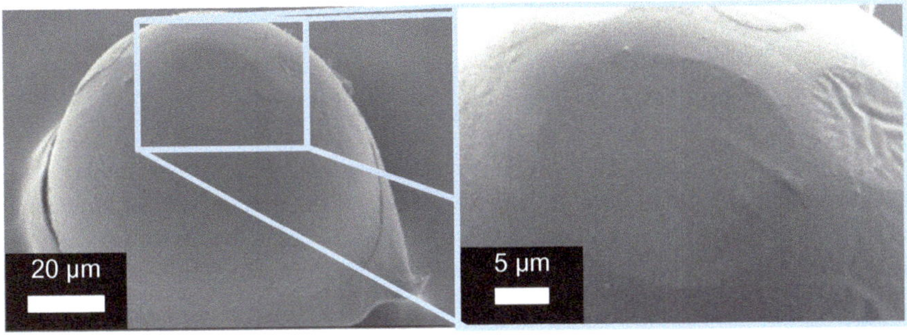

Fig. 17 SEM images showing a single spherical MD6 particle after 40 cyclic loadings. The circular region in the middle is assumed to be the plastically deformed contact zone. Due to production, process and handling there are other bumps around the particle [64]

Table 4 Equivalent increase in contact zone diameter after 40 loadings [64]

RH/%	Contact zone diameter increase/–
40	1.40
50	1.76
60	1.89
70	1.98
80	4.04

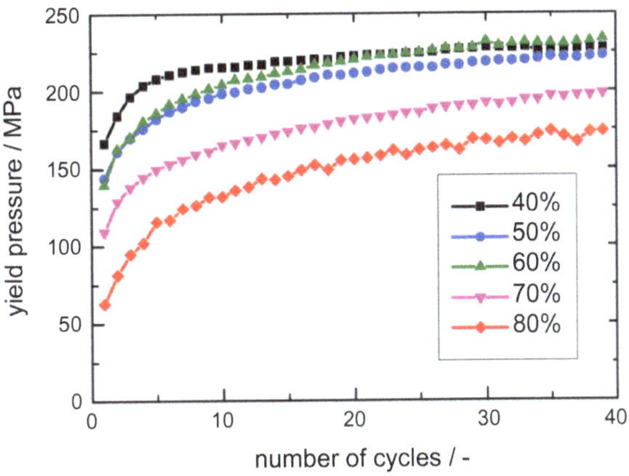

Fig. 18 Yield pressure for different RH over loading cycles [64]

increase in contact zone diameter. Lower yield pressure leads to a higher plastic deformation during cyclic loading.

Titan Dioxide

Cyclic Compression Tests

In addition to the compression test until a constant maximum load, cyclic loading/unloading tests of TiO_2 agglomerates were carried out at room temperature using the own tester described in section "Developed Experimental Setup for Mechanical Tests of Microparticles". The different loading regimes are listed in Table 5.

Loading Regimes A, B and C

During the cyclic loading A, B and C (Table 5), the particle were repeatedly loaded until a constant maximum force and then completely unloaded. Each particle was subjected up to 20 compression cycles, although some particles were broken at lower numbers of cycles. The applied maximum forces were equal to 13, 28 and 54% of the average breakage force obtained from previous compressive tests (section "Mechanical Characterization of Titanium Dioxide Agglomerates"). The typical force-displacement curves of cyclic loading of a TiO_2 agglomerate at such cyclic compression can be seen in Fig. 19. For further estimation of the measurement results, it was assumed that the equivalent particle diameter equals to the obtained

Table 5 Parameters of the chosen loading regimes for cyclic compression tests with TiO_2 agglomerates [64]

<table>
<tr><td colspan="3">Loading regime A, B and C</td></tr>
<tr><td>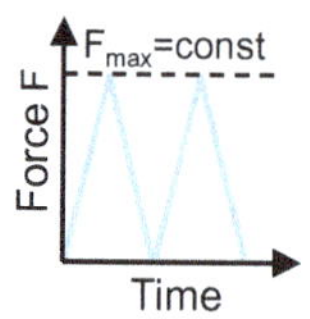
</td><td>Loading parameter:
A: $F_{max} = 10$ mN
B: $F_{max} = 21$ mN
C: $F_{max} = 40$ mN</td><td>Number tested particles:
13
16
10</td></tr>
<tr><td colspan="3">Loading regime D</td></tr>
<tr><td>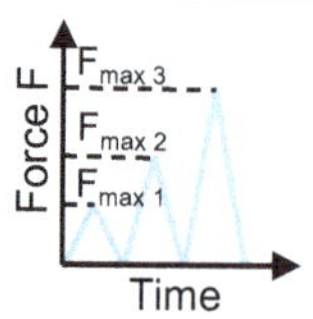
</td><td>Loading parameter:
$F_{max,i} = 5; \ldots;$
100 mN
Increase in 5 mN increments</td><td>Number tested particles:
10</td></tr>
<tr><td colspan="3">Loading regime E</td></tr>
<tr><td>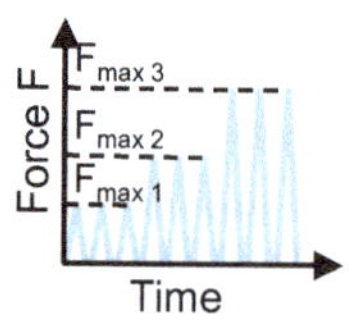
</td><td>Loading parameter:
$F_{max,1} = 10$ mN
$F_{max,2} = 20$ mN
$F_{max,3} = 40$ mN</td><td>Number tested particles:
15</td></tr>
</table>

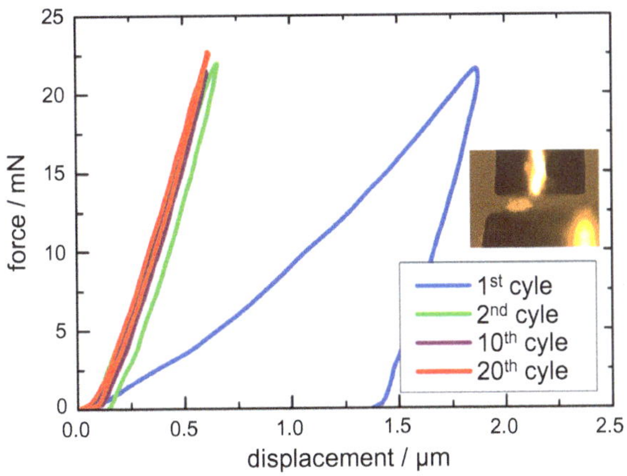

Fig. 19 Typical force-displacement curves for a TiO_2 agglomerate at cyclic compression with a constant amplitude (regime B). Each curve starting from origin of coordinates represents a loading-unloading cycle [64]

height of the particle. The force-displacement curve of each cycle was approximated in the elastic deformation range up to the yield point by the Hertz model (Eq. 4).

Table 6 shows the values of Young's modulus and displacements obtained from loading-unloading curves. The Young's modulus at unloading was calculated on the assumption that the particle behaves pure elastically during unloading path. The cyclic stressing of the particle leads to a cyclic hardening effect. This effect can be explained on the basis of microstructure change within the contact zone of the agglomerate, where the highest compressive stresses are generated and the surface irregularities and the porous structure inside of a contact zone is greatly deformed plastically. Therefore, that a stable contact zone is formed, the normal force cannot cause large irreversible deformations in the following loading cycles with the same load amplitude. The increasing slope of the loading curve from the first to the second cycle (Fig. 19) indicates that the largest changes occur at the first cycle. For all three regimes A, B and C, the residual displacement in each cycle decreased until a saturation value with increasing number of cycles as can be seen in Fig. 20. In the

Table 6 Mechanical properties of TiO_2 agglomerates with mean equivalent diameter of 100.5 ± 23.6 μm obtained by cyclic compression tests at loading regimes A, B and C [64]

Loading regime	Young's modulus/GPa		Displacement at maximum force/μm		Residual displacement/μm	
	At loading	At unloading	In first cycle	Average value	In first cycle	Average value
A	9.5 ± 3.2	11.1 ± 2.8	1.08 ± 0.65	0.51 ± 0.07	0.50 ± 0.20	0.08 ± 0.007
B	11.1 ± 4.5	15.6 ± 6.8	2.27 ± 1.32	0.72 ± 0.02	1.75 ± 1.53	0.11 ± 0.007
C	14.1 ± 6.2	18.0 ± 7.8	2.74 ± 1.96	0.98 ± 0.02	2.03 ± 2.01	0.12 ± 0.009

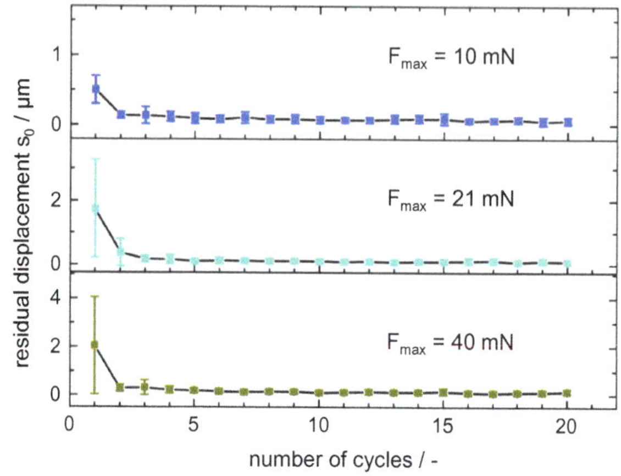

Fig. 20 Residual displacement of TiO_2 particles during cyclic compression at regimes A, B and C (constant amplitudes of 10, 21 and 40 mN) [64]

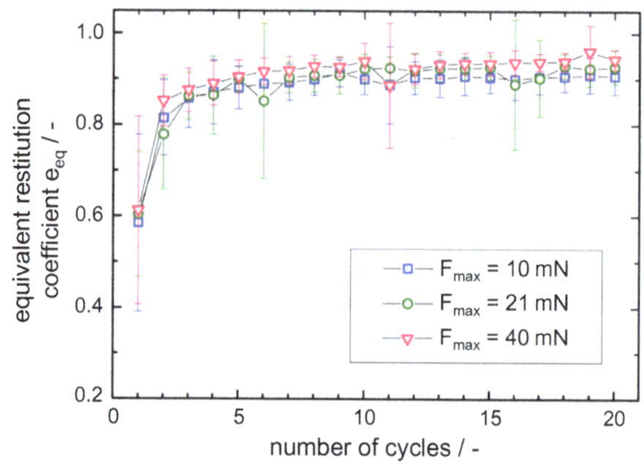

Fig. 21 Equivalent restitution coefficient as function of number of cycles for TiO_2 agglomerates during cyclic compression (regimes A, B and C) [64]

first cycle the highest value of residual displacement at the end of unloading was observed.

By using of Eq. (34) the equivalent restitution coefficient was estimated from force-displacement curves. In Fig. 21 the obtained mean values of restitution coefficient as the function of the number of cycles for the loading regimes A, B and C are given. The saturation values of equivalent coefficient of restitution for loading regimes A, B and C were determined to be equal to 0.90, 0.91 and 0.93.

Loading Regime D

The particles were stressed under cyclic loading with increasing amplitude of the compressive force starting at 5 mN up to 100 mN with a increment of 5 mN (Table 6). If the maximum compressive force amplitude of 100 mN was reached without particle breakage, the particle was further cyclic stressed at this amplitude. No particle

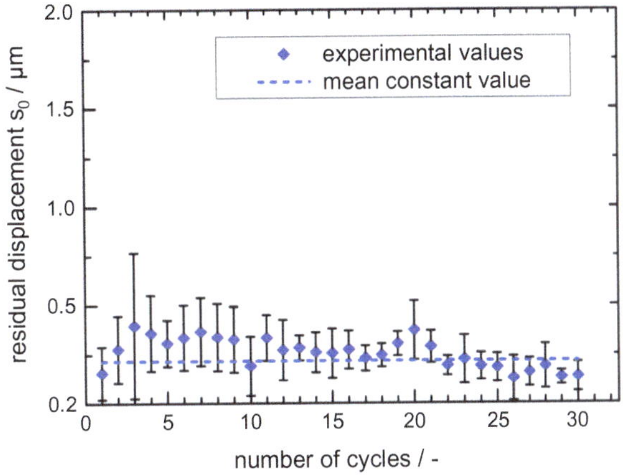

Fig. 22 Residual displacement as function of number of cycles for TiO_2 particles during cyclic compression (loading regime D) [64]

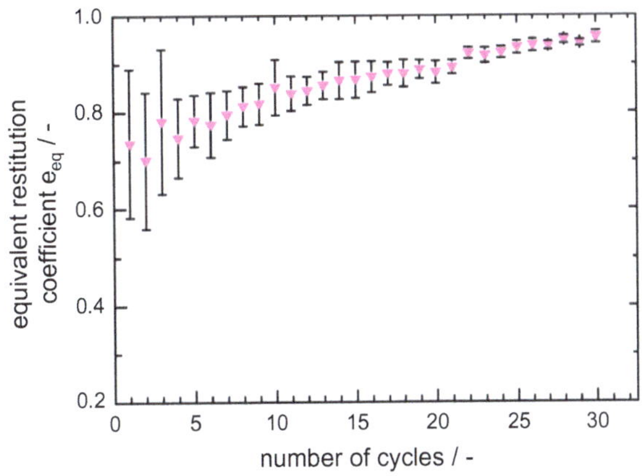

Fig. 23 Equivalent restitution coefficient as function of number of cycles for TiO_2 particles during cyclic compression (loading regime D) [64]

withstood more than 30 cycles for such a loading regime. As shown in Fig. 22, apart from the first cycle, the residual displacement remains nearly constant by increasing number of cycles. Thus, the average constant value of 0.22 μm (corresponds to the stain of about 0.22%) was estimated.

The mean equivalent coefficient of restitution obtained for each loading cycle increases incrementally with the increase of load amplitude as can be clearly seen in Fig. 23. In Table 7 the Young's modules at loading and unloading paths averaged over 30 cycles are listed.

Loading Regime E

In Table 5 this loading regime is shown schematically. Using three different blocks of constant amplitude cycles of 10, 21 and 40 mN respectively, 5 cyclic compression tests were performed in each block. The evolution of the mean residual displacement

Table 7 Average characteristics of TiO_2 particles from cyclic compression tests (loading regimes D and E) [64]

	Mean equivalent diameter/μm	Young's modulus at loading/GPa	Young's modulus at unloading/GPa	Equivalent coefficient of restitution/–
D	97.2 ± 21.6	13.7 ± 2.3	16.5 ± 1.0	0.92 ± 0.01
E	98.5 ± 15.1	11.4 ± 2.3	14.9 ± 2.0	0.89 ± 0.01

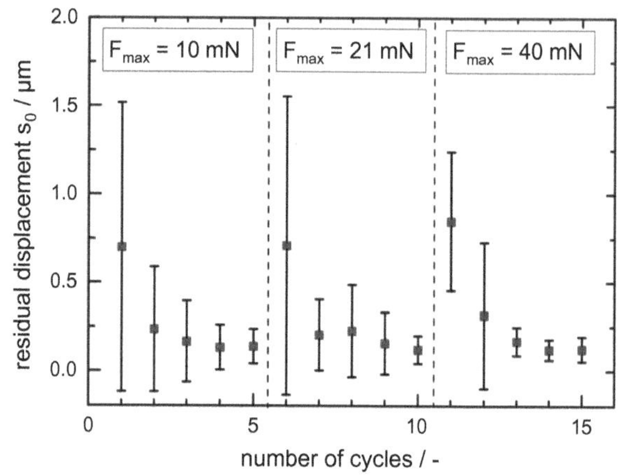

Fig. 24 Residual displacement as function of number of cycles of TiO_2 agglomerates during cyclic compression (loading regime E) [64]

and equivalent restitution coefficient during cyclic loading can be seen in Fig. 24 and Fig. 25, respectively. Similar to the observed agglomerate behaviour during the cyclic compression tests with loading regimes A, B and C, the residual displacement has a maximum value during the first cycle of each loading block and decreases in following four cycles. This maximum value is nearly constant for each loading block. Moreover, the decrease of these residual displacement and increase of the restitution coefficient are quite similar in each block. This indicates that the highest residual displacement and dominantly plastic behaviour measured in the first cycle occurs not due to deformation of asperity structure of the particle surface, but in particular due to the inelastic deformation of the material in contact zone. This depends on the intensity of cyclic load and the number of cycles. The cyclic hardening is thus a function of the load history of the material.

Average Loading-Unloading Curve

The average loading-unloading force-displacement curve can be used to describe the obtained experimental results of all cyclic uniaxial compression tests with theoretical contact models. At first, the mean experimental values of peak force as a function of total displacement assessed from all five loading regimes are illustrated in Fig. 26, where the experimental points of regimes A, B, C, and E correspond to

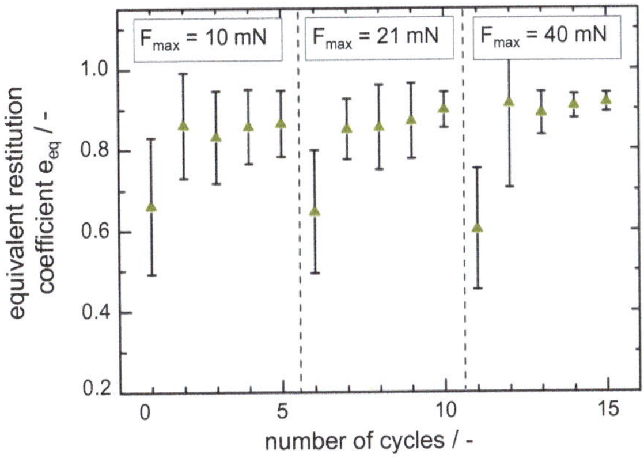

Fig. 25 Equivalent restitution coefficient as function of number of cycles of TiO_2 agglomerates during cyclic compression (loading regime E) [64]

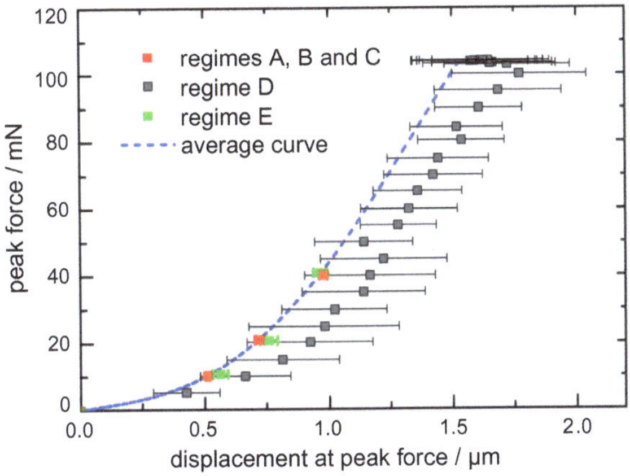

Fig. 26 Average curve (dash line) estimated from all cyclic compression tests of TiO_2 agglomerates [64]

saturation values. Under the assumption that the curve must lie within the range of all experimental tests, the loading part of the average curve was created (dash line in Fig. 26).

Using contact models for normal force described in section "Contact Modelling" the average curve was fitted. In order to approximate the elastic part by the Hertz model with the mean value of Young's module determined from experiments, the average curve was shifted to the left. In Table 8 the comparison of the results is listed. The following values of the equivalent particle diameter, yield point and Young's modulus for average loading-unloading curve were found as the average of all experiments: 99.5 μm, 0.66 μm and 15.7 GPa, respectively.

The results of average loading curve approximated with different contact models are shown in Fig. 27. One can conclude that the analytical contact models of Thornton and Ning and Tomas do not fit the averaged experimental curve since they predict not only the conservative values of the normal force and the equivalent restitution coefficient but also the overestimated value of the dissipated energy. The approximation with the simple model of Walter and Braun fits the average curve best. Although

Table 8 Comparison of values predicted by analytical contact models with the values of average loading-unloading curve of TiO_2 agglomerates with the mean equivalent particle diameter of 99.5 μm [64]

Parameter	Average loading-unloading curve	Tomas	Thornton	Walton and Braun
Peak force/mN	104	37	43	104
Residual displacement/μm	0.22	0.35	0.33	0.22
Energy at loading, 10^{-2}/μJ	5.29	2.63	3.93	7.95
Energy at unloading, 10^{-2}/μJ	4.48	1.74	2.04	6.81
Dissipated energy, 10^{-2}/μJ	0.81	0.88	0.89	1.14
Equivalent restitution coefficient/–	0.92	0.82	0.84	0.93

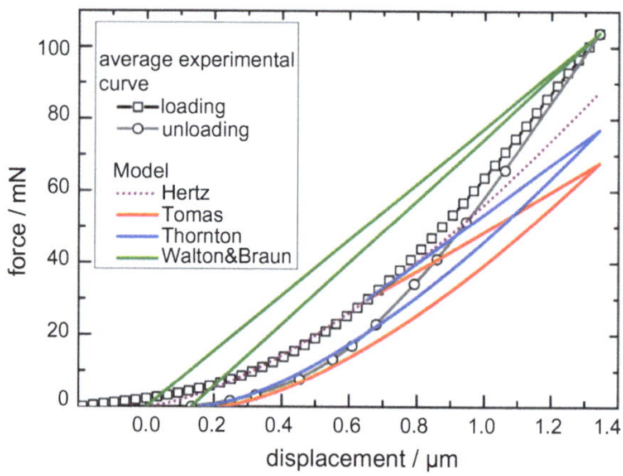

Fig. 27 Average loading-unloading curve of TiO_2 agglomerates in the saturation range approximated with contact models [64]

this model ignores the nonlinear character of force-displacement curve in the elastic range, it can be employed for the determination of plastic contact model parameters that are required for DEM simulations.

Walter and Braun Model Extended by the Saturation

In order to investigate the extended model of Walter and Braun (section "Walton and Braun Model"), it was applied to predict the mean experimental results of the cyclic tests for regime A, B and C. The comparison shown in Fig. 28 indicates the good agreement between the experimental and calculated data. The evolution of the equivalent restitution coefficient in the following cycles was described with the

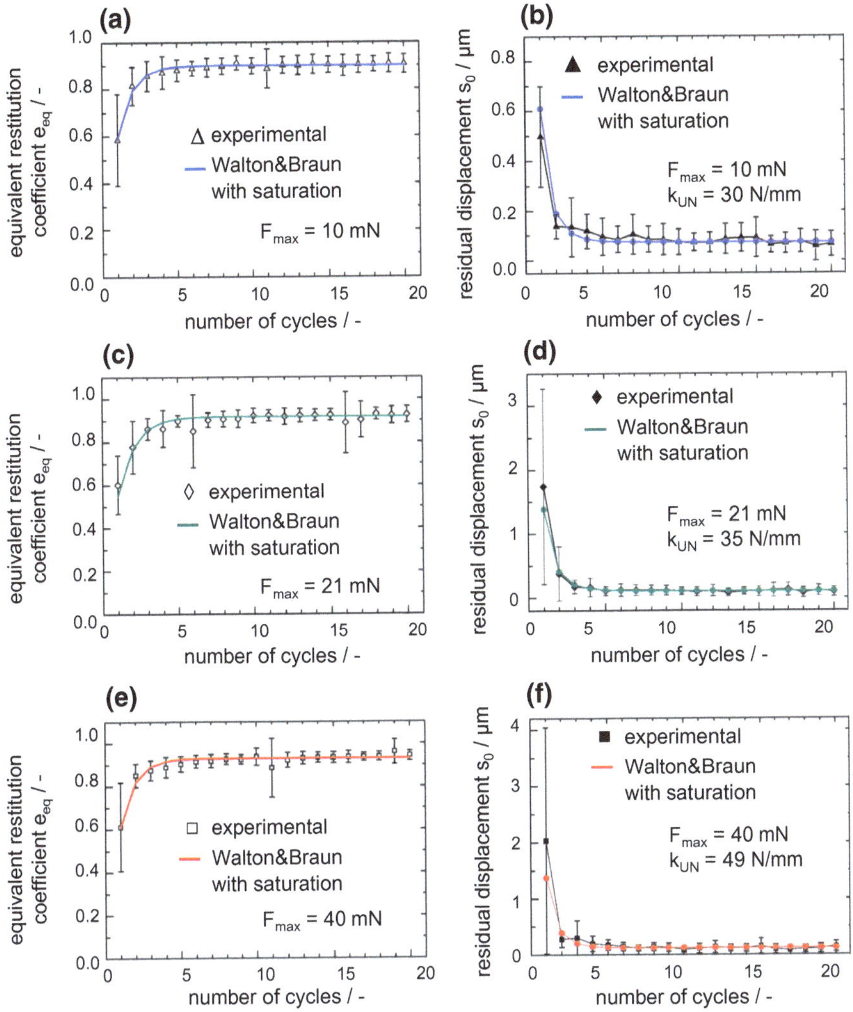

Fig. 28 Equivalent restitution coefficient of TiO_2 agglomerates versus number of cycles at loading regimes A (**a**), B (**c**) and C (**e**) and corresponding residual displacement at loading regimes A (**b**), B (**d**) and C (**f**) described by Walton and Braun model with the saturation [64]

exponential saturation function in Eq. (10). The values of the residual displacement were determined as a function of number of cycles using Eq. (9).

In addition, the particular experiment of one single particle subjected to constant amplitude of 21 mN (see Fig. 19) was reproduced and calculated by using the proposed model. The results are shown in Figs. 29 and 30. The total residual displacement in the cyclic test is equal to 2.3 µm and the value estimated by the model was found to be to 2.5 µm. Again, the theoretical and experimental curves are found to be in general agreement.

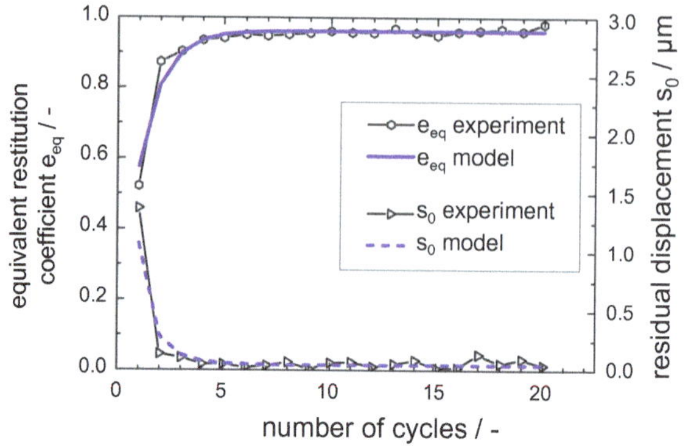

Fig. 29 Equivalent restitution coefficient versus number of cycles and corresponding residual displacement described by Walton and Braun model with the saturation for cyclic loading test of the TiO_2 agglomerate from Fig. 19 [64]

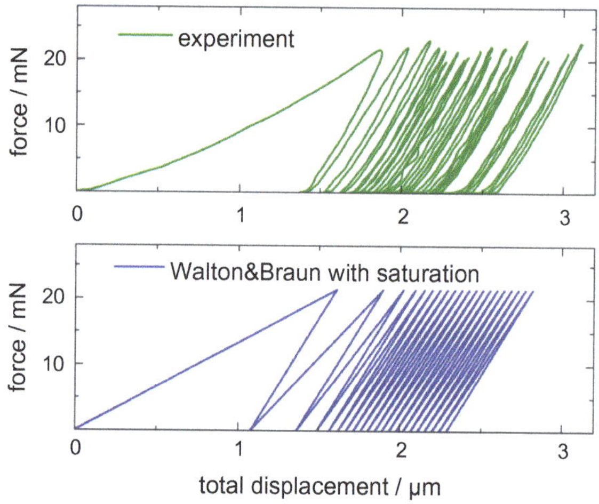

Fig. 30 Comparison of the cyclic loading test of the TiO_2 agglomerate shown in from Fig. 19 with the proposed modified contact model of Walton and Braun [64]

Shear Tests

The measurements of the friction between a glass plate and the TiO_2 agglomerates were conducted with own setup described in section "Developed Experimental Setup for Mechanical Tests of Microparticles". The experiment was repeated 17 times. During each shear test, three non-spherical particles were loaded with five different increasing normal forces in the range of 0.60–3.23 N. In Fig. 31, the typical course of the shear force during the shear test is shown. From the shear curves, a coefficient of static friction of 0.74 ± 0.13 and sliding friction coefficient of 0.49 ± 0.06 were calculated. The obtained shear stiffness was 0.50 ± 0.10 mN/mm.

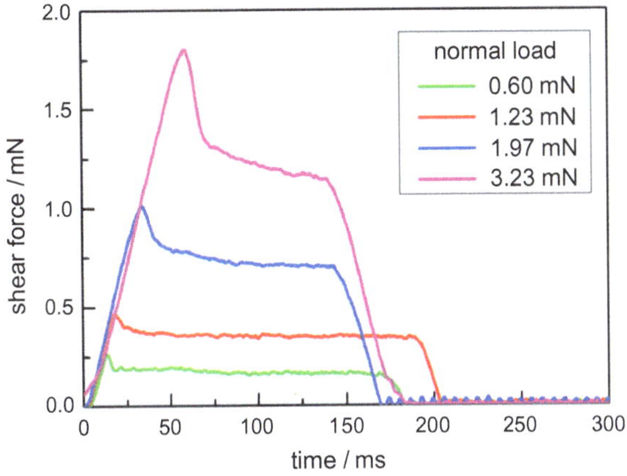

Fig. 31 Results from one measurement of the shear force over the time

Simulation of the Compression Test

DEM Simulations with Multi-sphere Approach

The scanned 3D agglomerate shape (bounding box 220 μm × 258 μm × 88 μm) was filled with 110 overlapping, polydisperse spheres with a minimal diameter of 15 μm as described in section "Multi-sphere Approach". After importing the cluster into the commercial code EDEM 2.5 (DEM Solutions), the simulation of the compression test with the multi-sphere particle was conducted. It must be noted that the influence of the number of the composite spheres in the agglomerate on the simulation results was not studied. The reason is that the used algorithm only had the number of initial surface points for sphere growth and the value of the minimum allowed sphere radius. Therefore, the quality of the shape presentation does not improve by a higher number of starting surface points as well as the usage of smaller initial radii of spheres but drastically enlarges the simulation time.

In the simulations of compression tests the contacts models of Tsuji (see section "Tsuji Model") and Walton and Braun (see section "Walton and Braun Model") for normal contact force between particle and rigid wall were used. In accordance with the experimental compression tests, the upper surface moved down at 50 μm/s strain rate up to the distance of 4.17 μm. The loading pins were represented by two non-deformable STL surfaces by setting of a high value of Young's modulus for wall material. Furthermore, the experimentally measured values of equivalent restitution coefficient, density and coefficient of friction were used as material parameters of the contact model for the multi-sphere particle. The damping coefficient β in Eq. (6) was evaluated from the particle mass and the equivalent restitution coefficient. The particle mass was calculated by software under the assumption that the filled particle volume had the experimentally determined density value. The equivalent restitution coefficient was previously obtained by cyclic compression experiments.

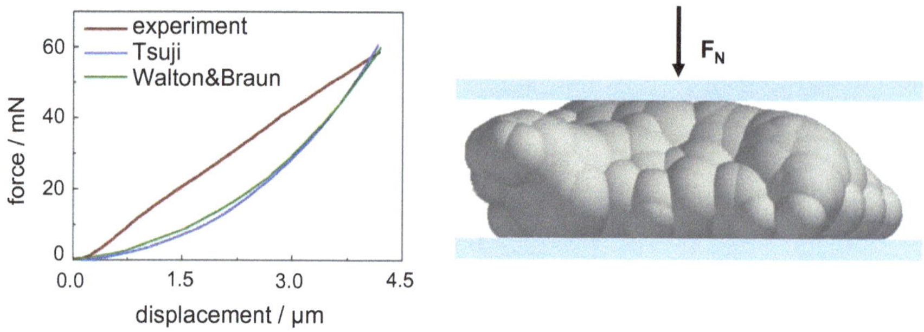

Fig. 32 Comparison of the simulation results obtained using the multi-sphere approach with experimental data

The influence of the second term in the Tsuji model on the overall agglomerate response is assumed to be insignificant, since the low stress rate classifies the pressure load as quasistatic loading. For impact loading, however, this dissipative part of the Tsuji model should not be neglected, because it influences the particle behaviour under dynamic loading, where the impact velocities are relatively high.

The value of 0.28 for Poisson's ratio ν from literature [88] was used for the elastic term of contact models. The other contact model parameters, such as Young's modulus E and yield strengths Y, were varied during simulations to fit the last point of the experimental loading curve (s_{max}; F_{max}). For the Tsuji model a Young's modulus of E $=$ 0.61 GPa was obtained. The yield strengths was found to be $Y = 40$ MPa. The loading stiffness was calculated from the yield strength using: $k_{load} = 5 \cdot R \cdot Y$.

In Fig. 32 the experimentally obtained force-displacement curve is compared with the simulations performed using the multi-sphere approach. It turns out, that both used contact models cannot reproduce the nearly linear course of the experimental curve. This can be explained with the dependency of force on the normal overlap in Tsuji model, which is not linear. The total force-displacement behaviour calculated with the multi-sphere approach shows nonlinearity for both contact models, although the Walton and Braun model uses a linear dependency between force and overlap for the single particles.

The reason is the presence of many contact points between the agglomerate and the punch, which differ in the stiffness due to the different values of equivalent curvature in contact area. In addition, the number of points participating in contact with walls increases during compression. Therefore, the total stiffness of the agglomerate changes during loading phase. This phenomenon leads to the non-linear force-displacement relation in the DEM simulations.

DEM Simulations with Bonded-Particle Model

The scanned agglomerate was rebuilt with 2280 non-overlapping primary particles with a diameter of 10 μm within the volume bounded by STL surface. 6634 solid

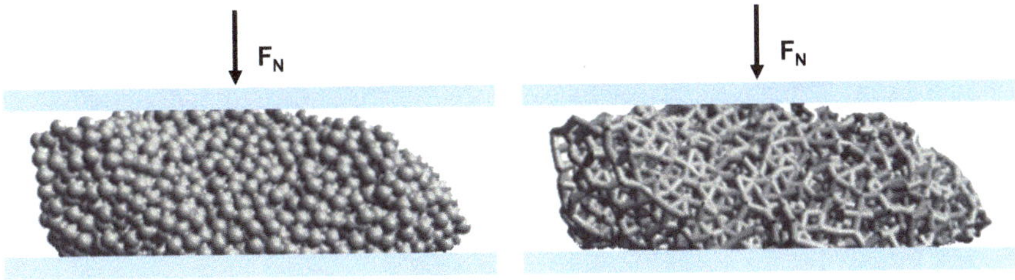

Fig. 33 Bonded-particle model of the agglomerate. Left: primary particles; right: bonds

bonds with a diameter of 5 μm were generated between the particles. The agglomerate structure and bond networks are shown in Fig. 33.

The numerical results of the compression simulation calculated with the bonded-particle model are affected by the values of radii of the primary particle as well as bond radii and bond length. The optimum choice of the primary particle size is always a compromise between an acceptable shape representation and the calculation time. While smaller primary particles increase the shape quality, the numerical time step reduces. The usage of bigger primary particles leads to an unsatisfactory shape representation although the time step becomes larger. Therefore in this work, the diameter of primary particles in the agglomerate model was set to 10 μm.

In accordance with the DEM simulations of compression test described above, the DEM simulations with bonded-particle model were performed on conditions equal to the conducted experiments. The material density, equivalent restitution coefficient and frictions coefficients used in contact models were determined experimentally. The other contact and bond model parameters were estimated, in a way that the last point of the force-displacement curve in the simulation approximately coincides with experimental value. The obtained contact model parameters are listed in Table 9.

In order to investigate the influence of the deformation model of solid bonds on the agglomerate behaviour during compression, four different models were compared: bonds deform only elastically (case 1); bonds behave like brittle solids with a breakage phenomenon (case 2); bonds can show a residual deformation due to the used elastic–plastic model (case 3) and bond material response corresponds to the

Table 9 Parameters for the used boded-particle models

Case	Model parameter
1: Elastic behaviour	$E_{particle} = 1.9$ GPa; $E_{bond} = 1.9$ GPa
2: Elastic behaviour with bond breakage	$E_{particle} = 1.9$ GPa; $E_{bond} = 8.0$ GPa
3: Elastic-plastic behaviour	$E_{particle} = 3.0$ GPa; $E_{bond} = 2.0$ GPa; $k_{md} = 0.9$; $k_{pl} = 0.9$
4: Maxwell viscoelastic behaviour	$E_{particle} = 1.9$ GPa; $E_{bond} = 6$ GPa; $\eta = 22 \times 10^6$ Pa·s

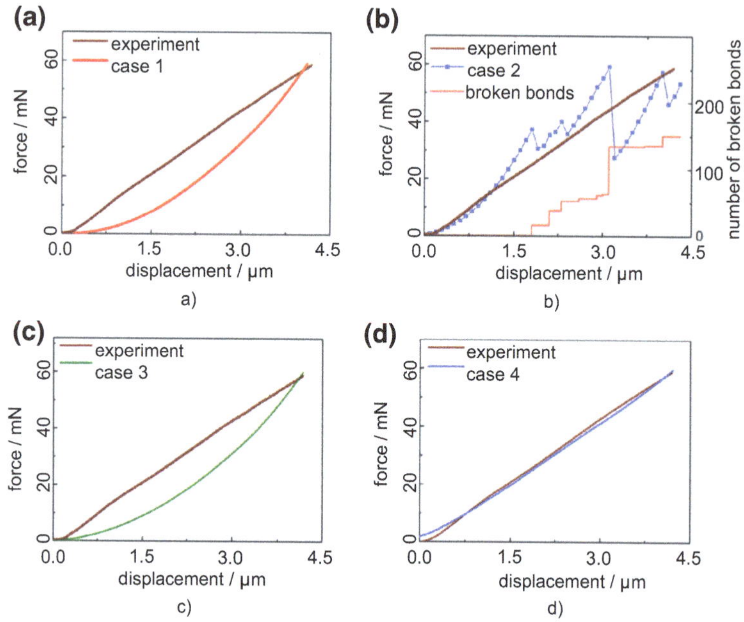

Fig. 34 Comparison of force-displacement curves obtained by bonded-particle simulations with the experimental curve

Maxwell dashpot-damper model (case 4). Details to the applied models can be found in section "Bonded-Particle Model".

Figure 34 shows a comparison of the experimental curve and the force-displacement curves obtained with the different bonded-particle models.

In case 1, the used elastic bond model exhibits a power-law dependency and cannot predict the nearly linear development of the experimental force-displacement curve (Fig. 34a). The same pure elastic behaviour as in case 1 is seen for case 2 except that the solid bonds can be destroyed by reaching either critical normal stress or critical shear stress values. After the breakage of a bond, the bond was removed from the DEM simulation. As a consequence a sudden decrease in the force-displacement curve occurs (Fig. 34b). With the additional plotted curve of the number of broken bonds in the agglomerate, it can be clearly seen that the peaks of the force-displacement curve obtained by DEM simulation correlate with the breakage of solid bonds. At displacement values lower than 1.5 µm, the values of the force before first breakage of a bond are higher than in case 1, although the Young's moduli of the primary particles are the same in case 1 and case 2. This can be explained by the lower Young's modulus for the bonds in the case 1 compared to the value of 8 GPa in the case 2. The elastic solid bond model with breakage cannot depict the course of the experimental force-displacement curve but can be used for DEM simulations where attrition of agglomerates has to be studied.

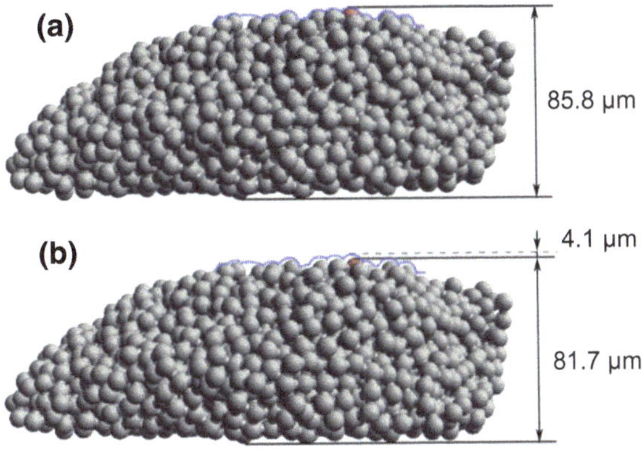

Fig. 35 Comparison of the vertical position of a primary particle in upper contact surface before (**a**) and after (**b**) compression of the agglomerate (simulated with Maxwell viscoelastic bond model)

Also the elastic-plastic deformation model of solid bonds (case 3) cannot reproduce the slope of the experimental curve (Fig. 34c). Nevertheless, this model is suitable for simulation of agglomerates aiming to an accurate prediction of the energy dissipation as well as plastic deformation during collision events.

Figure 34d shows the results of the implemented Maxwell model for solid bonds. Using the viscoelastic solid bond model, the non-linearity of the force-displacement curve obtained by pure elastic bond model was significantly reduced. Therefore, it is possible to represent the energy dissipation during contact and precisely depict the experimental force-displacement curve. Moreover, the plastic deformation of bonds in the contact with the upper wall was also verified. In Fig. 35 the displacement of the primary particle with the highest vertical position before and after unloading is evaluated and found to be 4.1 μm. Thus, changes in the agglomerate shape due to plastic deformation will be taken into account during next contact events in the DEM simulation. This phenomenon cannot be reproduced with the multi-sphere approach, since it does not consider contact plastic deformation. In other words, the agglomerate shape does not change because the radius of curvature in a contact before and after impact is the same.

FEM Simulation of the Compression Test

In order to mimic the inelastic deformation during compression, a FEM simulation of the compression of the agglomerate was carried out. For this purpose a volumetric tetrahedral mesh was created based on the STL surface representing the real agglomerate shape by using the open software GMSH. The resulting solid FE model contained 17,892 ten-node tetrahedral elements. For the simulation, the FE model of the agglomerate is assumed to be the isotropic and homogeneous. Voids and other irregularities in the agglomerate structure are therefore not considered. The simulation was performed with the commercial software ABAQUS using the

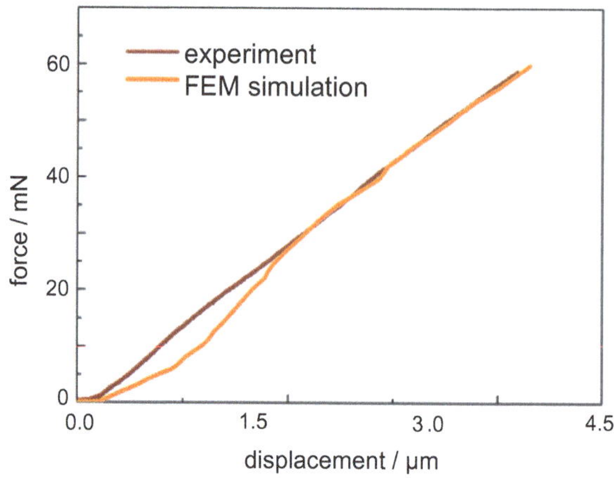

Fig. 36 Comparison of the force-displacement curves obtained from experiment and by FE simulation

surface-to-surface contact pairs. The surfaces of the loading pins were assumed to be rigid. The material law in the FE model was set to elastic-ideal plastic. In accordance with the experiment, the upper rigid surface moved down up to 4.17 μm. Since the results of FE analysis with assumed material model are time-independent, a deformation rate of 1 mm/s was used to increase the simulation speed. The lower surface of the particle was constrained, what corresponds to the boundary condition of the real agglomerate in the bottom region. By varying of Young's modulus and yield stress in the simulations, a maximum contact force of about 59 mN was achieved. The force-displacement relation found in the FEM simulation is compared to the experimental data in Fig. 36.

The estimated values of Young's modulus and yield stress are found to be 3.8 GPa and 9 MPa, respectively. Additionally, the flattening of contact zone was observed which mimics the real agglomerate behaviour in compression test (Fig. 37) and can explain the linearity of the experimental loading curve [94, 95].

Conclusion

A novel setup for the measurement of the contact force and deformation during static loading of single micrometre-sized particles in normal and shear directions was developed, validated and used for the performed particle tests. In the first part of this work, maltodextrin (MD6) and titanium dioxide particles were experimentally investigated. Compression tests with these particles are conducted and the different force-displacement behaviour was analysed in order to obtain mechanical contact properties like contact stiffness, Young's modulus, yield point as well as the restitution coefficient. The spherical and cylindrical MD6 particles were subjected to normal and cyclic loading under different climatic conditions. A clear dependence of the particle properties on the climate conditions was found. To calibrate the param-

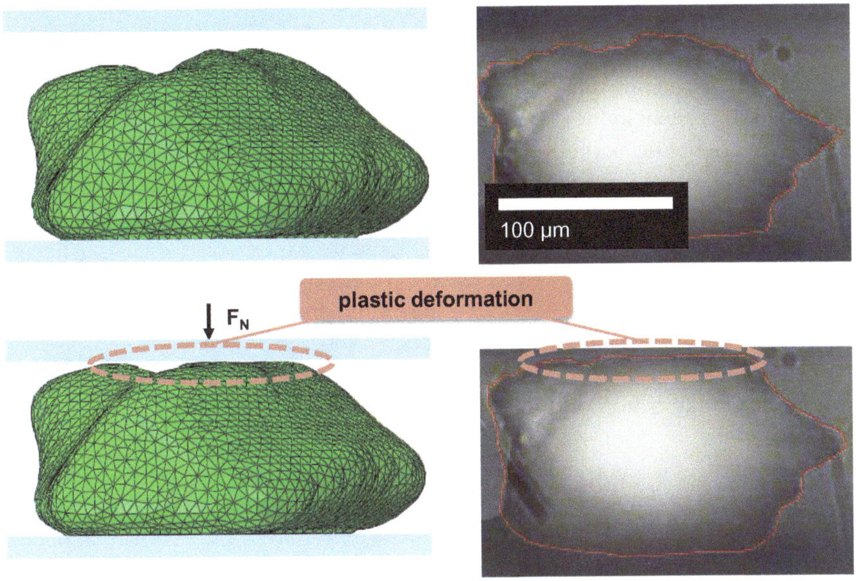

Fig. 37 Comparison of flattening of contact zone at the beginning and the end of loading phase. Left: FEM; right: laser-scanning-microscope images from experiment

eters of the used material model of Zener, experiments with cylindrical particles were performed. In order to take into account the viscous effects observed in the experiments, the Thornton model was applied. This model enables a good prediction of the experimental data.

Uniaxial compression tests up to breakage were carried out with the titanium dioxide agglomerates. The failure properties of these agglomerates, such as breakage force and strength, as well as the breakage probability. The titanium dioxide agglomerates were also subjected to cyclic loading with constant load amplitudes as well as at incrementally increasing load amplitudes. The analysis of the force-displacement curves showed the elasto-plastic behaviour of the material. With increasing number of cycles the residual displacement decreases due to cyclic hardening. The average loading-unloading curve was estimated in order to describe the obtained experimental results with contact models for normal force. Comparing the different models, the best agreement was obtained with the model of Walton and Braun. In addition, the cyclic hardening at a constant force amplitude was approximated adequately by extending this model with an exponential saturation function.

In the second part of this study DEM simulations of the compression test with irregularly shaped titanium dioxide agglomerates were performed. The irregular shape hinders a direct determination of the contact properties from experimental force-displacement curves. Therefore, the contact model parameters were determined in numerical simulations by using the data from experimentally obtained curves. For that purpose, the agglomerates were scanned by X-ray computer microtomography and the obtained 3D shape was used for the DEM and FEM simulations. The minor disadvantages of the proposed experimental method are some losses of the geometry data during scanning, the dependence of the surface quality on the tomographic resolution and a small influence of the glued volume on the entire material response, since the bottom of the particle was fixed. In order to represent the irregular shape of the tested titania agglomerates in the numerical DEM simulations, the multi-sphere approach and bonded-particle models with different deformation behaviour of the solid bonds were used. The simulation results and the experimental data of the compression tests were compared to estimate the contact model parameters in a way that the maximum normal load at maximum normal displacement coincides. It must be noted that the estimated parameters cannot be considered as real material properties in terms of materials science but as model parameters, which influence the stiffness during contact interactions in DEM simulations.

Although the linear plastic contact model of Walton and Braun was used in the simulations with the multi-sphere approach, linear behaviour of the experimental force-displacement curve could not be reproduced. The DEM simulations with the bonded-particle models are more precise, since they consider the particle structure which approximates the real microstructure of the agglomerate with primary particles and solid bonds. Comparing the simulation results with the experimental data, the best agreement was found for the solid bonds which deform according to the Maxwell viscoelastic material law. Based on all simulation results, the elastic-plastic model and Maxwell viscoelastic solid bond model should be preferred for DEM simulations of titania agglomerates. In addition to the DEM simulations, a FEM simulation of the compression test was performed. The obtained loading curve also confirms the dominantly plastic contact behaviour. Moreover, the flattening of the contact zone achieved in the FEM simulation matches the real agglomerate behaviour observed in the compression tests.

Acknowledgements The authors gratefully acknowledge the German Research Foundation DFG (priority programme 1486 "Particles in contact: micromechanics, microprocess dynamics and particle collectives") for supporting the project HE 4526/7 and BR 1151/4. The authors also thank Dr.-Ing. Sergiy Kozhar, Prof. Dr.-Ing. Maksym Dosta, and M.Sc. Laurent Gilson for the help with the experiments and simulations presented in the manuscript.

References

1. Antonyuk, S., Heinrich, S., Tomas, J., et al.: Energy absorption during compression and impact of dry elastic-plastic spherical granules. Granul. Matter **12**(1), 15–47 (2010). https://doi.org/10.1007/s10035-009-0161-3
2. Antonyuk, S., Palis, S., Heinrich, S.: Breakage behaviour of agglomerates and crystals by static loading and impact. Powder Technol. **206**(1–2), 88–98 (2011). https://doi.org/10.1016/j.powtec.2010.02.025
3. Rumpf, H.: The strength of granules and agglomeration. In: Knepper, W.A. (ed.) Agglomeration, pp. 379–418. Interscience, New York (1962)

4. Schubert, H.: Tensile strength of agglomerates. Powder Technol. **11**(2), 107–119 (1975). https://doi.org/10.1016/0032-5910(75)80036-2
5. Schubert, H.: Grundlagen des Agglomerierens. Chem. Ing. Tech. **51**(4), 266–277 (1979). https://doi.org/10.1002/cite.330510404
6. Johnson, K.L., Kendall, K., Roberts, A.D.: surface energy and the contact of elastic solids. Proc. R. Soc. A: Math. Phys. Eng. Sci. **324**(1558), 301–313 (1971). https://doi.org/10.1098/rspa.1971.0141
7. Kendall, K.: Relevance of contact mechanics to powders-elasticity, friction and agglomerate strength. Tribol. Part. Technol. 110–122 (1987)
8. Kendall, K., Alford, N.M., Birchall, J.D.: The strength of green bodies. Spec. Ceram. **8**, 255–265 (1986)
9. Rumpf, H.: Grundlagen und Methoden des Granulierens. Chem. Ing. Tech. **30**(3), 144–158 (1958). https://doi.org/10.1002/cite.330300307
10. Rumpf, H.: Die Einzelkornzerkleinerung als Grundlage einer technischen Zerkleinerungswissenschaft. Chem. Ing. Tech. **37**(3), 187–202 (1965). https://doi.org/10.1002/cite.330370303
11. Bika, D., Tardos, G.I., Panmai, S., et al.: Strength and morphology of solid bridges in dry granules of pharmaceutical powders. Powder Technol. **150**(2), 104–116 (2005). https://doi.org/10.1016/j.powtec.2004.11.024
12. Delenne, J.-Y., El Youssoufi, M.S., Cherblanc, F., et al.: Mechanical behaviour and failure of cohesive granular materials. Int. J. Numer. Anal. Meth. Geomech. **28**(15), 1577–1594 (2004). https://doi.org/10.1002/nag.401
13. Agniel, Y.: Rôle des propriétés des granules pour la fabrication de pièces de poudres céramiques granulées sans défauts de compaction. Ph.D. thesis, University Karlsruhe (1992)
14. Beekman, W.J., Meesters, G.M.H., Becker, T., et al.: Failure mechanism determination for industrial granules using a repeated compression test. Powder Technol. **130**(1–3), 367–376 (2003). https://doi.org/10.1016/S0032-5910(02)00238-3
15. Sheng, Y., Briscoe, B.J., Maung, R., et al.: Compression of polymer bound alumina agglomerates at the micro deformation scale. Powder Technol. **140**(3), 228–239 (2004). https://doi.org/10.1016/j.powtec.2004.01.016
16. Samimi, A., Hassanpour, A., Ghadiri, M.: Single and bulk compressions of soft granules: experimental study and DEM evaluation. Chem. Eng. Sci. **60**(14), 3993–4004 (2005). https://doi.org/10.1016/j.ces.2005.02.036
17. Mangwandi, C., Cheong, Y.S., Adams, M.J., et al.: The coefficient of restitution of different representative types of granules. Chem. Eng. Sci. **62**(1–2), 437–450 (2007). https://doi.org/10.1016/j.ces.2006.08.063
18. Weis, D., Evers, M., Thommes, M., et al.: DEM simulation of the mixing behavior in a spheronization process. Chem. Eng. Sci. **192**, 803–815 (2018). https://doi.org/10.1016/j.ces.2018.07.057
19. Riehle, M., Simmchen, E.: Grundlagen der Werkstofftechnik: Mit 33 Tabellen. Dt. Verl. für Grundstoffindustrie, Stuttgart (1997)
20. Tavares, L.M., King, R.P.: Modeling of particle fracture by repeated impacts using continuum damage mechanics. Powder Technol. **123**(2–3), 138–146 (2002). https://doi.org/10.1016/S0032-5910(01)00438-7
21. Antonyuk, S., Khanal, M., Tomas, J., et al.: Impact breakage of spherical granules: experimental study and DEM simulation. Chem. Eng. Process. **45**(10), 838–856 (2006). https://doi.org/10.1016/j.cep.2005.12.005
22. Pitchumani, R., Strien, S.A., Meesters, G.M.H., et al.: Breakage of sodium benzoate granules under repeated impact conditions. Powder Technol. **140**(3), 240–247 (2004). https://doi.org/10.1016/j.powtec.2004.01.011
23. Cundall, P.A., Strack, O.D.L.: A discrete numerical model for granular assemblies. Géotechnique **29**(1), 47–65 (1979). https://doi.org/10.1680/geot.1979.29.1.47
24. Zhao, S., Zhang, N., Zhou, X., et al.: Particle shape effects on fabric of granular random packing. Powder Technol. **310**, 175–186 (2017). https://doi.org/10.1016/j.powtec.2016.12.094

25. Deng, X.L., Davé, R.N.: Dynamic simulation of particle packing influenced by size, aspect ratio and surface energy. Granul. Matter **15**(4), 401–415 (2013). https://doi.org/10.1007/s10035-013-0413-0
26. Salazar, A., Sáez, E., Pardo, G.: Modeling the direct shear test of a coarse sand using the 3D discrete element method with a rolling friction model. Comput. Geotech. **67**, 83–93 (2015). https://doi.org/10.1016/j.compgeo.2015.02.017
27. Wang, Z., Jing, G., Yu, Q., et al.: Analysis of ballast direct shear tests by discrete element method under different normal stress. Measurement **63**, 17–24 (2015). https://doi.org/10.1016/j.measurement.2014.11.012
28. Jing, X.-Y., Zhou, W.-H., Li, Y.: Interface direct shearing behavior between soil and saw-tooth surfaces by DEM simulation. Procedia Eng. **175**, 36–42 (2017). https://doi.org/10.1016/j.proeng.2017.01.011
29. Wang, Y., Ooi, J.Y.: A study of granular flow in a conical hopper discharge using discrete and continuum approach. Procedia Eng. **102**, 765–772 (2015). https://doi.org/10.1016/j.proeng.2015.01.183
30. Baserinia, R., Sinka, I.C.: Mass flow rate of fine and cohesive powders under differential air pressure. Powder Technol. **334**, 173–182 (2018). https://doi.org/10.1016/j.powtec.2018.04.041
31. Magalhães, F.G.R., Atman, A.P.F., Moreira, J.G., et al.: Analysis of the velocity field of granular hopper flow. Granul. Matter **18**(2), 208 (2016). https://doi.org/10.1007/s10035-016-0636-y
32. Pantaleev, S., Yordanova, S., Janda, A., et al.: An experimentally validated DEM study of powder mixing in a paddle blade mixer. Powder Technol. **311**, 287–302 (2017). https://doi.org/10.1016/j.powtec.2016.12.053
33. Alian, M., Ein-Mozaffari, F., Upreti, S.R.: Analysis of the mixing of solid particles in a plowshare mixer via discrete element method (DEM). Powder Technol. **274**, 77–87 (2015). https://doi.org/10.1016/j.powtec.2015.01.012
34. Ferellec, J.-F., McDowell, G.R.: A method to model realistic particle shape and inertia in DEM. Granul. Matter **12**(5), 459–467 (2010). https://doi.org/10.1007/s10035-010-0205-8
35. Breuninger, P., Weis, D., Behrendt, I., et al.: CFD–DEM simulation of fine particles in a spouted bed apparatus with a Wurster tube. Particuology (2018). https://doi.org/10.1016/j.partic.2018.03.015
36. Mahmoodi, B., Hosseini, S.H., Olazar, M., et al.: CFD-DEM simulation of a conical spouted bed with open-sided draft tube containing fine particles. J. Taiwan Inst. Chem. Eng. **81**, 275–287 (2017). https://doi.org/10.1016/j.jtice.2017.09.051
37. Xu, H., Zhong, W., Yuan, Z., et al.: CFD-DEM study on cohesive particles in a spouted bed. Powder Technol. **314**, 377–386 (2017). https://doi.org/10.1016/j.powtec.2016.09.006
38. Pietsch, S., Kieckhefen, P., Heinrich, S., et al.: CFD-DEM modelling of circulation frequencies and residence times in a prismatic spouted bed. Chem. Eng. Res. Des. **132**, 1105–1116 (2018). https://doi.org/10.1016/j.cherd.2018.01.013
39. Pietsch, S., Heinrich, S., Karpinski, K., et al.: CFD-DEM modeling of a three-dimensional prismatic spouted bed. Powder Technol. **316**, 245–255 (2017). https://doi.org/10.1016/j.powtec.2016.12.046
40. Dosta, M., Dale, S., Antonyuk, S., et al.: Numerical and experimental analysis of influence of granule microstructure on its compression breakage. Powder Technol. **299**, 87–97 (2016). https://doi.org/10.1016/j.powtec.2016.05.005
41. Jiang, Y., Zhang, J., He, Y., et al.: Discrete element simulation of the stress wave in high speed milling. In: Volume 1: Processes. ASME, V001T02A033 (2017)
42. Wu, J., Li, D., Zhu, B., et al.: Milling process simulation of old asphalt mixture by discrete element. Constr. Build. Mater. **186**, 996–1004 (2018). https://doi.org/10.1016/j.conbuildmat.2018.08.015
43. Heinrich, S., Dosta, M., Antonyuk, S.: Multiscale analysis of a coating process in a Wurster fluidized bed apparatus. In: Mesoscale Modeling in Chemical Engineering Part I, vol. 46, pp. 83–135. Elsevier (2015)
44. Dosta, M., Antonyuk, S., Heinrich, S.: Multiscale simulation of agglomerate breakage in fluidized beds. Ind. Eng. Chem. Res. **52**(33), 11275–11281 (2013). https://doi.org/10.1021/ie400244x

45. Fuchs, R., Weinhart, T., Meyer, J., et al.: Rolling, sliding and torsion of micron-sized silica particles: experimental, numerical and theoretical analysis. Granul. Matter **16**(3), 281–297 (2014). https://doi.org/10.1007/s10035-014-0481-9
46. Mader-Arndt, K., Aman, S., Fuchs, R., et al.: Contact properties determination of macroscopic fine disperse glass particles via compression tests in normal direction. Adv. Powder Technol. **28**(1), 101–114 (2017). https://doi.org/10.1016/j.apt.2016.09.014
47. Schubert, W., Khanal, M., Tomas, J.: Impact crushing of particle–particle compounds—experiment and simulation. Int. J. Miner. Process. **75**(1–2), 41–52 (2005). https://doi.org/10.1016/j.minpro.2004.01.006
48. Thornton, C., Ciomocos, M.T., Adams, M.J.: Numerical simulations of diametrical compression tests on agglomerates. Powder Technol. **140**(3), 258–267 (2004). https://doi.org/10.1016/j.powtec.2004.01.022
49. Spettl, A., Dosta, M., Antonyuk, S., et al.: Statistical investigation of agglomerate breakage based on combined stochastic microstructure modeling and DEM simulations. Adv. Powder Technol. **26**(3), 1021–1030 (2015). https://doi.org/10.1016/j.apt.2015.04.011
50. Ge, R., Ghadiri, M., Bonakdar, T., et al.: Experimental study of the deformation and breakage of 3D printed agglomerates: effects of packing density and inter-particle bond strength. Powder Technol. **340**, 299–310 (2018). https://doi.org/10.1016/j.powtec.2018.09.029
51. Hassanpour, A., Antony, S.J., Ghadiri, M.: Effect of size ratio on the behaviour of agglomerates embedded in a bed of particles subjected to shearing: DEM analysis. Chem. Eng. Sci. **62**(4), 935–942 (2007). https://doi.org/10.1016/j.ces.2006.10.026
52. Jiménez-Herrera, N., Barrios, G.K.P., Tavares, L.M.: Comparison of breakage models in DEM in simulating impact on particle beds. Adv. Powder Technol. **29**(3), 692–706 (2018). https://doi.org/10.1016/j.apt.2017.12.006
53. Cleary, P.W., Delaney, G.W., Sinnott, M.D., et al.: Inclusion of incremental damage breakage of particles and slurry rheology into a particle scale multiphase model of a SAG mill. Miner. Eng. **128**, 92–105 (2018). https://doi.org/10.1016/j.mineng.2018.08.026
54. Cleary, P.W., Sinnott, M.D.: Simulation of particle flows and breakage in crushers using DEM: Part 1—Compression crushers. Miner. Eng. **74**, 178–197 (2015). https://doi.org/10.1016/j.mineng.2014.10.021
55. Delaney, G.W., Morrison, R.D., Sinnott, M.D., et al.: DEM modelling of non-spherical particle breakage and flow in an industrial scale cone crusher. Miner. Eng. **74**, 112–122 (2015). https://doi.org/10.1016/j.mineng.2015.01.013
56. Cabiscol, R., Finke, J.H., Kwade, A.: Calibration and interpretation of DEM parameters for simulations of cylindrical tablets with multi-sphere approach. Powder Technol. **327**, 232–245 (2018). https://doi.org/10.1016/j.powtec.2017.12.041
57. Grof, Z., Kohout, M., Štěpánek, F.: Multi-scale simulation of needle-shaped particle breakage under uniaxial compaction. Chem. Eng. Sci. **62**(5), 1418–1429 (2007). https://doi.org/10.1016/j.ces.2006.11.033
58. Ning, Z., Ghadiri, M.: Distinct element analysis of attrition of granular solids under shear deformation. Chem. Eng. Sci. **61**(18), 5991–6001 (2006). https://doi.org/10.1016/j.ces.2006.03.056
59. Ye, M., van der Hoef, M.A., Kuipers, J.A.M.: A numerical study of fluidization behavior of Geldart A particles using a discrete particle model. Powder Technol. **139**(2), 129–139 (2004). https://doi.org/10.1016/j.powtec.2003.10.012
60. Moreno-Atanasio, R., Xu, B.H., Ghadiri, M.: Computer simulation of the effect of contact stiffness and adhesion on the fluidization behaviour of powders. Chem. Eng. Sci. **62**(1–2), 184–194 (2007). https://doi.org/10.1016/j.ces.2006.08.036
61. Hertz, H.: Über die Berührung fester elastischer Körper. Journal für die reine und angewandte Mathematik (92), 156–171 (1881)
62. Tsuji, Y., Tanaka, T., Ishida, T.: Lagrangian numerical simulation of plug flow of cohesionless particles in a horizontal pipe. Powder Technol. **71**(3), 239–250 (1992). https://doi.org/10.1016/0032-5910(92)88030-L

63. Walton, O.R., Braun, R.L.: Viscosity, granular-temperature, and stress calculations for shearing assemblies of inelastic, frictional disks. J. Rheol. **30**(5), 949–980 (1986). https://doi.org/10.1122/1.549893
64. Gilson, L., Kozhar, S., Antonyuk, S., et al.: Contact models based on experimental characterization of irregular shaped, micrometer-sized particles. Granul. Matter **16**(3), 313–326 (2014). https://doi.org/10.1007/s10035-013-0464-2
65. Thornton, C., Yin, K.K.: Impact of elastic spheres with and without adhesion. Powder Technol. **65**(1–3), 153–166 (1991). https://doi.org/10.1016/0032-5910(91)80178-L
66. Zener, C.: Elasticity and Anelasticity of Metals. University of Chicago Press (1948)
67. Cundall, P.A.: A computer model for simulating progressive, large-scale movement in blocky rock system. In: Proceedings of the International Symposium on Rock Mechanics (1971)
68. Cundall, P.A.: A Computer model for rock-mass behavior using interactive graphics for the input and output of geometrical data. Report MDR-2-74 (1974)
69. Mindlin, R.D., Deresiewicz, H.: Elastic spheres in contact under varying oblique force. Trans. ASME J. Appl. Mech. **20**, 327–344 (1953)
70. Lu, G., Third, J.R., Müller, C.R.: Discrete element models for non-spherical particle systems: from theoretical developments to applications. Chem. Eng. Sci. **127**, 425–465 (2015). https://doi.org/10.1016/j.ces.2014.11.050
71. Höhner, D., Wirtz, S., Kruggel-Emden, H., et al.: Comparison of the multi-sphere and polyhedral approach to simulate non-spherical particles within the discrete element method: influence on temporal force evolution for multiple contacts. Powder Technol. **208**(3), 643–656 (2011). https://doi.org/10.1016/j.powtec.2011.01.003
72. Potyondy, D.O., Cundall, P.A.: A bonded-particle model for rock. Int. J. Rock Mech. Min. Sci. **41**(8), 1329–1364 (2004). https://doi.org/10.1016/j.ijrmms.2004.09.011
73. Kudryavtsev, O.A., Sapozhnikov, S.B.: Numerical simulations of ceramic target subjected to ballistic impact using combined DEM/FEM approach. Int. J. Mech. Sci. **114**, 60–70 (2016). https://doi.org/10.1016/j.ijmecsci.2016.04.019
74. Nan, W., Wang, Y., Tang, H.: A viscoelastic model for flexible fibers with material damping. Powder Technol. **276**, 175–182 (2015). https://doi.org/10.1016/j.powtec.2015.02.037
75. Wolff, M.F.H., Salikov, V., Antonyuk, S., et al.: Three-dimensional discrete element modeling of micromechanical bending tests of ceramic–polymer composite materials. Powder Technol. **248**, 77–83 (2013). https://doi.org/10.1016/j.powtec.2013.07.009
76. Adams, M.J., Lawrence, C.J., Urso, M.E.D., et al.: Modelling collisions of soft agglomerates at the continuum length scale. Powder Technol. **140**(3), 268–279 (2004). https://doi.org/10.1016/j.powtec.2004.01.013
77. Kienzler, R., Schmitt, W.: On single-particle comminution; numerical analysis of compressed spheres. Powder Technol. **61**(1), 29–38 (1990). https://doi.org/10.1016/0032-5910(90)80063-5
78. Khanal, M., Schubert, W., Tomas, J.: Ball impact and crack propagation?: Simulations of particle compound material. Granul. Matter **5**(4), 177–184 (2004). https://doi.org/10.1007/s10035-003-0149-3
79. Palzer, S.: The effect of glass transition on the desired and undesired agglomeration of amorphous food powders. Chem. Eng. Sci. **60**(14), 3959–3968 (2005). https://doi.org/10.1016/j.ces.2005.02.015
80. Renzetti, S., Voogt, J.A., Oliver, L., et al.: Water migration mechanisms in amorphous powder material and related agglomeration propensity. J. Food Eng. **110**(2), 160–168 (2012). https://doi.org/10.1016/j.jfoodeng.2011.07.005
81. Gordon, M., Taylor, J.S.: Ideal copolymers and the second-order transitions of synthetic rubbers. I. Non-crystalline copolymers. J. Appl. Chem. **2**(9), 493–500 (1952). https://doi.org/10.1002/jctb.5010020901
82. Sugisaki, M., Suga, H., Seki, S.: Calorimetric study of the glassy state. IV. Heat capacities of glassy water and cubic ice. BCSJ **41**(11), 2591–2599 (1968). https://doi.org/10.1246/bcsj.41.2591

83. Silalai, N., Roos, Y.H.: Mechanical relaxation times as indicators of stickiness in skim milk—maltodextrin solids systems. J. Food Eng. **106**(4), 306–317 (2011). https://doi.org/10.1016/j.jfoodeng.2011.05.018
84. Haider, C.I., Althaus, T., Niederreiter, G., et al.: A micromanipulation particle tester for agglomeration contact mechanism studies in a controlled environment. Meas. Sci. Technol. **23**(10), 105904 (2012). https://doi.org/10.1088/0957-0233/23/10/105904
85. Descamps, N., Palzer, S., Zuercher, U.: The amorphous state of spray-dried maltodextrin: Sub-sub-Tg enthalpy relaxation and impact of temperature and water annealing. Carbohydr. Res. **344**(1), 85–90 (2009). https://doi.org/10.1016/j.carres.2008.06.033
86. Stojanović, B.D., Marinković, Z.V., Branković, G.O., et al.: Evaluation of kinetic data for crystallization of TiO_2 prepared by hydrolysis method. J. Therm. Anal. Calorim. **60**(2), 595–604 (2000). https://doi.org/10.1023/A:1010107423825
87. Kozhar, S., Dosta, M., Antonyuk, S., et al.: DEM simulations of amorphous irregular shaped micrometer-sized titania agglomerates at compression. Adv. Powder Technol. **26**(3), 767–777 (2015). https://doi.org/10.1016/j.apt.2015.05.005
88. Shackelford, J.F., Alexander, W. (eds.): CRC Materials Science and Engineering Handbook, 3rd edn. CRC Press, Boca Raton, FL (2001)
89. Martienssen, W., Warlimont, H.: Springer Handbook of Condensed Matter and Materials Data. Springer Handbook. Springer, Berlin (2005)
90. Salman, A.D. (ed.): Granulation, 1st ed., reprint. In: Williams, J.C., Allen, T. (eds.) Handbook of Powder Technology, vol. 11. Elsevier, Amsterdam (2007)
91. Antonyuk, S., Tomas, J., Heinrich, S., et al.: Breakage behaviour of spherical granulates by compression. Chem. Eng. Sci. **60**(14), 4031–4044 (2005). https://doi.org/10.1016/j.ces.2005.02.038
92. Stronge, W.J.: Impact Mechanics. Cambridge University Press, Cambridge (2000)
93. Thornton, C., Ning, Z.: A theoretical model for the stick/bounce behaviour of adhesive, elastic-plastic spheres. Powder Technol. **99**(2), 154–162 (1998). https://doi.org/10.1016/S0032-5910(98)00099-0
94. Jackson, R.L., Green, I.: Experimental investigation of fully plastic contact of a sphere against a hard flat. J. Tribol. **129**(3), 700 (2007). https://doi.org/10.1115/1.2749229
95. Popov, V.L.: Contact Mechanics and Friction: Physical Principles and Applications, 1st edn. Springer, Berlin, Berlin (2010)

DEM Analysis of Breakage Behavior of Bicomponent Agglomerates

Maksym Dosta, Matthias Weber, Volker Schmidt and Sergiy Antonyuk

Introduction

In this chapter, we investigate the breakage behavior of bicomponent agglomerates under uni-axial compressive load. Knowledge about breakage mechanisms is important for different fields of application. For example, production processes such as granulation and grinding depend on breakage characteristics of the used agglomerates. In recent years, increasing computational resources permit the investigation of mechanical properties based on simulation experiments. Various research has already been conducted to examine the relationship of agglomerate properties and their breakage behavior. For instance, in the case of impact breakage, the effect of agglomerate's shape and impact angle on breakage behavior has been investigated [5].

The aim of this chapter is to quantify the relationship between the microstructure of agglomerates and their macroscopic breakage behavior. We consider agglomerates made up of a relatively dense packing of spherical primary particles connected by cylindrical bonds. We study the effect of different material properties of primary particles and bonds on breakage characteristics of these agglomerates. On the one hand, we consider agglomerates made up from equally-sized spheres and compare them to agglomerates made from two types of particles with different diameters.

M. Dosta (✉)
Institute of Solids Process Engineering and Particle Technology, Hamburg University of Technology, Hamburg, Germany
e-mail: dosta@tuhh.de

M. Weber · V. Schmidt
Institute of Stochastics, Ulm University, Ulm, Germany
e-mail: volker.schmidt@uni-ulm.de

S. Antonyuk
Institute of Particle Process Engineering, Technische Universität Kaiserslautern, Kaiserslautern, Germany
e-mail: sergiy.antonyuk@mv.uni-kl.de

S. Antonyuk (ed.), *Particles in Contact*,
https://doi.org/10.1007/978-3-030-15899-6_6

On the other hand, we take cylindrical agglomerates composed of a single type of spherical primary particles and vary the Young's modulus of the bonds. Again, we compare breakage characteristics of agglomerates made up from bonds with the same Young's modulus to those of agglomerates made up from two types of bonds with different Young's moduli.

For the simulation, we apply the MUSEN framework [1] which uses the discrete element method (DEM) and the bonded particle model (BPM) to simulate the time-resolved behavior of an agglomerate under compressive load. A general description of DEM and a detailed summary of literature dealing with the application of BPM for microscale modeling of agglomerates can be found in Chap. 5. The considered agglomerates are created using a stochastic model [9] which is able to generate densely packed agglomerates of spherical primary particles with varying sizes. Furthermore, it incorporates the placement of cylindrical bonds between primary particles.

The simulation experiments considered in this chapter show a clear dependency of breakage properties such as breakage force and deformation on the composition of bicomponent spherical agglomerates. Homogeneous mixtures of both types of primary particles tend to exhibit a lower breakage force, but a higher deformation, indicating higher elasticity. For cylindrical agglomerates, we show that the Young's modulus of bonds plays an important role in the breakage behavior.

Methods

Agglomerate Structure Generation Algorithm

The agglomerates under study consist of spherical primary particles and cylindrical solid bounds. This resembles typical structures obtained by industrial wet agglomeration processes such as pelletizing discs and tablet presses. For usage in DEM simulations, agglomerate structures are created in two stages [9]. In the first stage, a packing of primary particles is generated; in the second stage, a set of solid bonds connecting the primary particles is added. To generate a packing of primary particles, the force-biased algorithm [7] has been applied. In the first step of this algorithm, primary particles are randomly placed into a specified volume. In further iterations contacts between particles are detected and virtual forces in normal directions are calculated. Afterwards, the particles are moved according to the forces acting on them. Thereby, the inner stresses are reduced. Using this procedure, the overlaps between primary particles are reduced. The stiffness which is used to calculate the inter-particle forces is not directly related to material properties. It is calculated automatically in the algorithm and varies depending on the maximal particle velocity in the system. In order to avoid size or material dependent segregation of particles, their masses during generation are scaled. Additional conditions ensure that all particles are totally contained in the specified volume. For this purpose, virtual walls representing the generation domain is considered. Interactions between particles and walls are taken into account when calculating the virtual forces acting on the particles. The rearrangement algorithm is stopped when the maximal overlap between particles and between particles and walls is smaller than some threshold value. In Fig. 1a, an initial particle placement and the corresponding final packing are shown schematically.

The proposed generation approach allows us to model homogeneous and isotropic structures for packings with high packing densities. This means that there is no preferred direction for e.g. contacts between particles and there is no spatial gradient in the packing density. These are reasonable assumptions for spherical agglomerates produced by e.g. spray granulation and we restricted ourselves to this case. However, non-homogeneous agglomerates occur in reality, too, e.g., agglomerates with spatially varying packing density or agglomerates having a core-shell structure. These could be modeled by extending the present approach, see Weber et al. [14]. Application of the developed algorithm for low packing densities will lead to the formation of spatially distributed particle clusters. In the case of very high packing densities the algorithm will result in ordered structures (see Fig. 1b) and can lead to formation of fcc, bcc or hcp-like structures.

In the second stage, bonds are generated between primary particles. For this purpose, two different algorithms are used for the creation of cylindrical and spherical agglomerates. For the creation of cylindrical agglomerates, a simple algorithm based on a distance threshold is used. A bond between two primary particles is generated if the distance between those is smaller than a specific threshold. For case studies with spherical agglomerates, an advanced approach based on the volume fraction of binder material is applied. The aim is to connect the primary particles by bonds such that the union of all bonds has a given total volume. This volume is obtained as a volume fraction with respect to the total volume of primary particles. The idea is to construct an agglomerate where all particles being sufficiently close to each other are connected and further bonds are only inserted where necessary to obtain

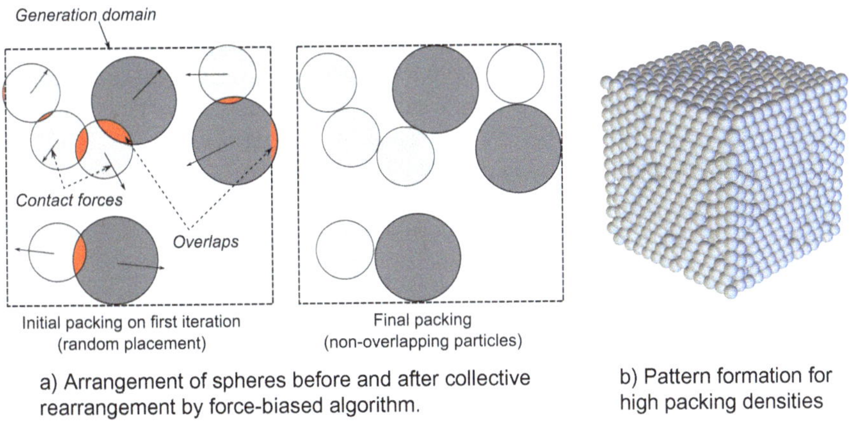

Fig. 1 Generation of particle packing using the force-biased algorithm

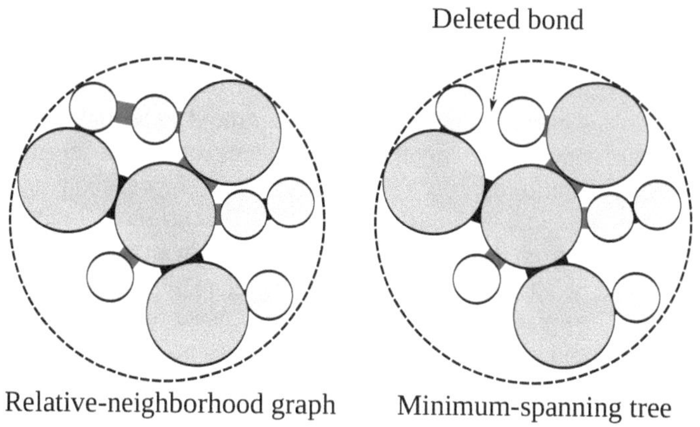

Fig. 2 Illustration of RNG for $l = 0$ and MST for the same graph [9]. Direct contacts are shown in solid black, further connectivity bonds are gray

connectivity. Then, the volume of bonds can be controlled by the thresholding value used to connect particles. If this threshold is zero, only bonds necessary to obtain connectivity are generated. For larger thresholds, the number of bonds (and their total volume) is increased.

This can be implemented as an algorithm using ideas from graph theory [4]. For a given distance threshold $l \geq 0$, we will construct a graph $G^{(l)} = \left(V, E^{(l)}\right)$ that connects the set of vertices (particles) V by the set of edges (bonds) $E^{(l)}$. A bond between two particles is only put if the smallest distance between their surfaces is less than or equal to l or if the bond is required for complete connectivity, i.e., every primary particle has be connected directly or indirectly to all others via the bond system. Such a graph $G^{(l)}$ can be obtained by the following algorithm, which is illustrated in Fig. 2.

1. Compute the relative-neighborhood graph (RNG) [12] $G^{(l)}_{RNG}$ of the primary particles, using a suitable distance function $d : V \times V \to [0, \infty)$ (see below). In the RNG, an edge is put between two vertices when there is no third vertex closer to both vertices (with respect to d). Using this technique, the vertices (primary particles) are always completely connected by the edges in the RNG [12]. A suitable distance function d is given by

$$d\big((x, r), (x', r')\big) = \max\big\{0, \left\|x - x'\right\| - r - r' - l\big\}, \tag{1}$$

where x and x' are the centroids of the considered particles and r and r' are their respective radii. As d is zero if the distance of the surfaces of two particles is at most l, the RNG contains all edges that connect two particles within a given distance l and further edges ensuring complete connectivity. Therefore, $G^{(l)}_{RNG}$ fulfills most properties desired for $G^{(l)}$, but it may contain more edges than necessary for connectivity. More precisely, it is a superset of the graph we are interested in, whose construction is explained in the next step.

2. Compute the minimum spanning tree (MST) [4] $G^{(l)}_{MST}$ of $G^{(l)}_{RNG}$, again using the same distance function d given in (1), now applied to the existing edges. In the MST, as many edges as possible are deleted while minimizing the sum of distances and preserving connectivity.
3. The result $G^{(l)}$ is given by the edges of the graph $G^{(l)}_{MST}$ combined with all edges in $G^{(l)}_{RNG}$ having 'length' zero, i.e., the surfaces of the corresponding primary particles having at most a distance of l from each other.

Recall that for a given threshold value $l \geq 0$, $G^{(l)}$ is the graph connecting all particles that are within distance l to each other, plus further bonds that are required for connectivity. For a given set of primary particles, the binder volume fraction (which is given as a model parameter) can be converted into the absolute binder volume. The aim is to find the 'optimal' threshold value l^* such that the absolute binder volume obtained from the connectivity graph $G^{(l)}$ is closest to the target value. This leads to a minimization problem

$$l^* = \arg\min_{l \geq 0} \left| \sum_{e \in E^{(l)}} \mathrm{vol} B(e) - b_{abs} \right|, \tag{2}$$

where b_{abs} is the absolute binder volume targeted and $\mathrm{vol} B(e)$ denotes the volume of the bond specified by edge e.

Contact Models

For modeling of quasi-static compression of agglomerates, a bonded particle model (BPM) has been used [2, 8]. Here, three rheological contact models have been employed. The soft-sphere visco-elastic Hertz-Tsuju [13] model for normal force and the Mindlin model for tangential force [6] were used to describe particle-particle and particle-wall interactions [3]. For modeling of the solid bonds, a rheological model of the ideally elastic material was applied. The forces and moments caused by bonds have been added to the interparticle interactions. One of the main limitations when applying BPM is a lack of detection and calculation of physical contacts between primary particles and solids bonds. This leads to the reduction of accuracy when the distance between surfaces of bonded particles is relatively large.

The relative motion of connected particles leads to a deformation of bonds and causes corresponding forces and moments. The force in the normal direction F_n is calculated directly from the current positions of primary particles. The force depends on the bond deformation ΔL and is calculated as

$$F_n(t) = \frac{\Delta L(t)}{L_{init}} \cdot A_b \cdot E, \tag{3}$$

where A_b is the bond cross-cut surface, L_{init} the initial bond length, and E the Young's modulus. The distance between particle centers is taken as bond length. For the calculations of force in tangential direction F_t as well as for torsional and bending moments M_n and M_t, respectively, an iterative approach is used. In each DEM simulation time step, the increment of force is calculated and added to the transformed previous value. For example, the tangential force in the global coordinate system for the new time step is computed as

$$F_t(t + \Delta t) = T \cdot F_t(t) + \frac{\Delta t \cdot v_{rel,t}}{L_{init}} \cdot A_b \cdot \frac{E}{2(1 + \nu)}, \tag{4}$$

where ν is the Poisson ratio, $v_{rel,t}$ the relative velocity in tangential direction and T the rotation matrix to consider motion of contact partners in three-dimensional space.

In order to simulate agglomerate breakage, the stresses acting in each individual bond are analyzed and compared with material properties for tension strength σ_{max} and shear strength τ_{max}. If one of the conditions given in Eqs. 5 and 6 is fulfilled, the bond breaks and is removed from the simulation domain. Moreover, it is assumed that there is no compressive breakage of single bonds. Therefore, the condition in Eq. 5 is only taken into account when the bond is elongated. Then, a bond breaks if either of the following conditions holds

$$\frac{F_n(t)}{A_b} + M_t \cdot \frac{R_b}{I} > \sigma_{max} \text{ or} \tag{5}$$

$$\frac{F_t(t)}{A_b} + M_n \cdot \frac{R_b}{J} > \sigma_{max}, \tag{6}$$

where R_b is the bond radius, I and J the area and polar moments of inertia, respectively.

For DEM calculations, the MUSEN simulation framework has been used [1]. The calculations in this system are parallelized for computation on a graphic processor unit (GPU) using the CUDA library. This reduces computation time significantly, thus allowing to perform a large number of simulation studies. For parallelization of computations, atomic functions like *atomicAdd()* are used. Due to the formation of the internal queue of operands and the limited number of digits in floating point operands, the application of such functions leads to non-deterministic results. This means that two runs of the program on the same initial data will lead to varying results. However, the overall variations caused due to the usage of atomic operations are negligible in the investigated case studies.

Spherical Agglomerates

Material Parameters and Agglomerate Geometry

To begin with, we briefly recall the stochastic microstructure model introduced in [9]. We denote the radius of a spherical agglomerate by r. The primary particles are also spheres and their radii are drawn from (independent copies of) a random particle radius R_p which takes values r_1 and r_2 with probabilities p and $1 - p$, respectively. Thus, the agglomerate consists of particles of two different sizes. The volume fraction a (mixing ratio) of particles of radius r_1 is given by

$$a = \frac{p \cdot r_1^3}{p \cdot r_1^3 + (1 - p) \cdot r_2^3}. \tag{7}$$

The targeted volume fraction of primary particles simulated inside the agglomerate is given by the packing density $\eta \in (0; \eta_{max}]$, where the maximal possible packing density η_{max} depends on the radius distribution of the primary particles (e.g., $\eta_{max} = 0.74$ for identical spheres).

The total volume of bonds is given as a ratio $b > 0$ with respect to the total volume of the primary particles. The bonds are cylindrical objects connecting two primary particles. The bond radius depends on the radius of the smaller adjacent particle: For a bond connecting two particles with radii r and r', the bond radius $\mathrm{r_b}$ is given by $\mathrm{r_b} = k_b \cdot \min\{r, r'\}$ for some factor $k_b \in (0, 1]$.

All parameters except the mixing ratio were fixed. The mixing ratio a was randomly chosen from the interval [0, 1]. The agglomerate radius was equal to 4 mm, radii of primary particles r_1 and r_2 were equal to 0.5 mm and 0.25 mm accordingly. The binder volume fraction was equal to 0.1 and bond radius factor k_b equal to 0.4.

Before investigating the breakage behavior of the agglomerates generated by the proposed stochastic model, it is interesting to look at some characteristics of the simulated microstructures. The agglomerate diameter as well as the porosity are always the same, but it is useful to analyze the primary particle radii, coordination numbers and bond radii in dependence of the mixing ratio. For an optical impression, three agglomerates with different mixing ratios are shown in Fig. 3. The larger primary particles are visualized in blue, the smaller primary particles in yellow. For all three agglomerates one eighth has been cut out for visualization of the interior microstructure. For all evaluations, 300 agglomerates have been generated according to the stochastic microstructure model.

The mean radius of primary particles (unweighted, i.e., not weighted by their volume or mass) clearly depends on the mixing ratio a. The average radius of particles in the agglomerate is calculated as

$$\bar{r} = \frac{N_1 r_1 + N_2 r_2}{N_1 + N_2}, \tag{8}$$

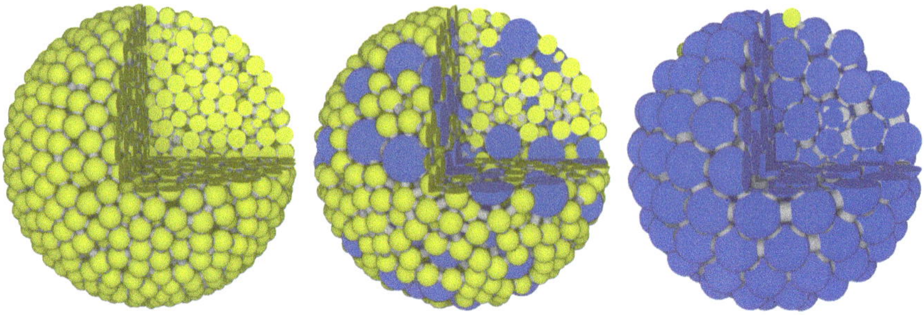

Fig. 3 Three realizations of agglomerates for mixing ratios $a = 0$ (left), $a = 0.45$ (middle), and $a = 0.99$ (right) [9]

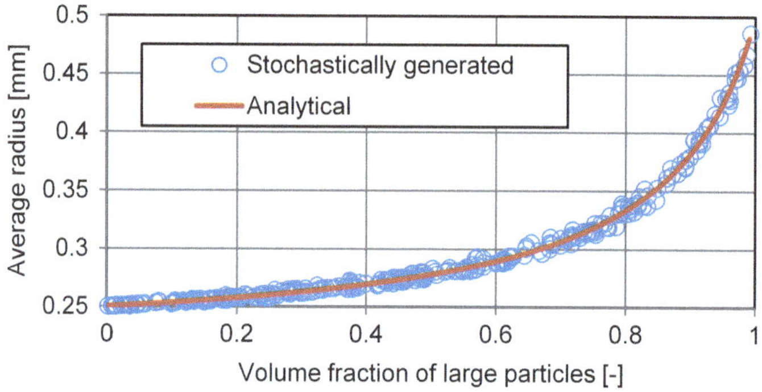

Fig. 4 Effect of mixing ratio on the average radius of primary particles in agglomerate

where N_1 and N_2 is the number of large and small particles respectively.

From the definition of volume fraction and considering the fact that $r_1 = 2r_2$ it can be derived that

$$a = \frac{N_1 r_1^3}{N_1 r_1^3 + N_2 r_2^3} = \frac{8N_1}{8N_1 + N_2}. \tag{9}$$

After substitution of Eq. 9 into Eq. 8, it follows that

$$\bar{r} = \frac{8 - 6a}{8 - 7a} r_2. \tag{10}$$

In Fig. 4, the dependency of average diameter on volume fraction of larger particles is shown. The data obtained from the stochastically generated structure is visualized with circles and the theoretical approximation with a solid line. It can be observed that, due to the relatively large number of primary particles in agglomerate, the mean radius in probability-based generated structure is in good agreement with theoretical calculations.

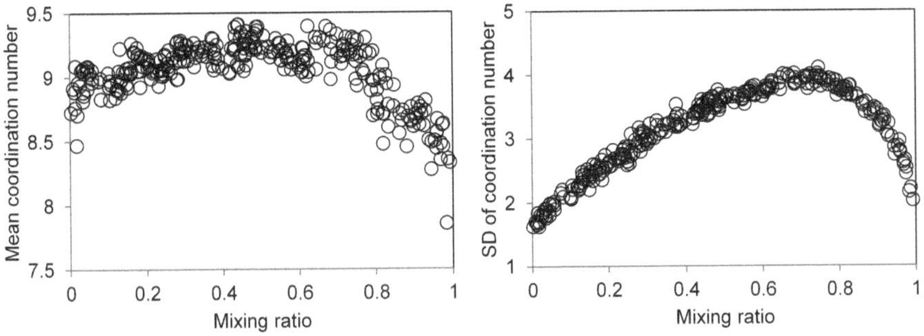

Fig. 5 Effect of mixing ratio on the coordination numbers of primary particles: mean coordination number (left), standard deviation of coordination numbers (right) [9]

However, the behavior of the coordination numbers, i.e., the number of bonds per primary particle, is more interesting. A larger value of the mean coordination number implies that the system of particles is better connected.

Figure 5 shows that the mean coordination number is a bit smaller for very small and very large mixing ratios a than for intermediate 'mixing' scenarios. On the other hand, the standard deviation (SD) is obviously largest in the range of $a \in [0.6, 0.8]$, where it achieves a value of almost 4 in comparison to about 2 for very low/high mixing ratios. This can be interpreted as a higher variability of the microstructure.

In combination with DEM simulations, the presented microstructure model can be used to investigate the relationship of structural and mechanical properties of agglomerates. The first step in order to achieve this goal is the description of real agglomerates using the given model, which has been done in [10]. In this way, we obtain model parameters representing real agglomerates. Further research leads to the exploration and statistical analysis of breakage properties [11]. However, one limitation of the presented model is the restriction to homogeneous and isotropic structures. While there are many types of particles for which these are reasonable assumptions, our model can easily be extended to overcome these shortcomings. An extension to agglomerates which can be spatially separated into core and shell can be found in [14].

Agglomerate strength can be described partly by two important characteristics, see Fig. 6 (top). The maximum force applied before breakage is smaller for mixing scenarios (scenarios where both fractions of particle sizes are relevant) than in the uniform cases. A possible reason is the higher variability in agglomerates with both small and large particles. This leads to more weak points in the microstructure, where forces are not evenly spread and individual bonds are heavily loaded. Therefore, local fracturing occurs before a high (global) force is applied. This implicates that even for almost only small primary particles the force is considerably higher.

On the other hand, the (mass-related) primary breakage energy does not show this behavior. It seems to increase almost linearly with the mixing ratio, i.e., agglomerates consisting of larger primary particles and thicker bonds are more stable. This means that even though the maximum force is smaller for mixing scenarios, this does

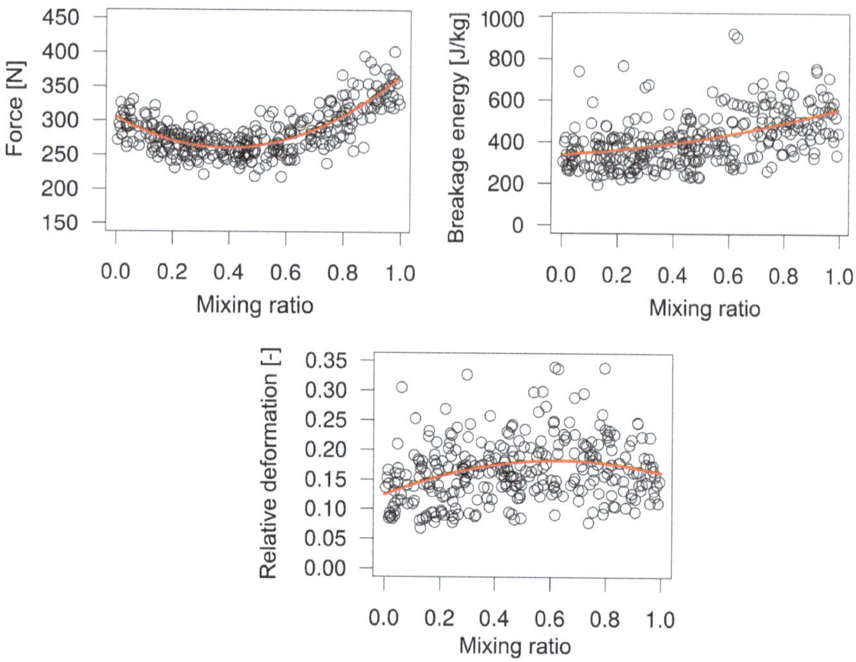

Fig. 6 Effect of mixing ratio on breakage behavior, i.e., on breakage force (top left), breakage energy (top right) and relative deformation (bottom) [9]

not influence the energy required for breakage. Therefore, the average force before breakage has to be similar or the related deformation has to be larger. Figure 6 (bottom) shows the deformation of the agglomerates relative to their original diameter. It can be seen that the deformation at the breakage point really tends to be larger for mixing scenarios, but the fluctuations are in a wide range for all mixing ratios.

Cylindrical Agglomerates

Agglomerate Geometry and Material Properties

The radius and height of the investigated cylindrical agglomerates are equal to 10 mm and 20 mm, respectively. Each agglomerate consists of 7200 primary particles with radius 0.5 mm connected by almost 29,000 solid bonds with radius 0.25 mm. In Fig. 7, an agglomerate and its internal structure are shown. The previously described force-biased packing algorithm (see the section "Agglomerate Structure Generation Algorithm") has been used to create the agglomerates. In order to minimize the error

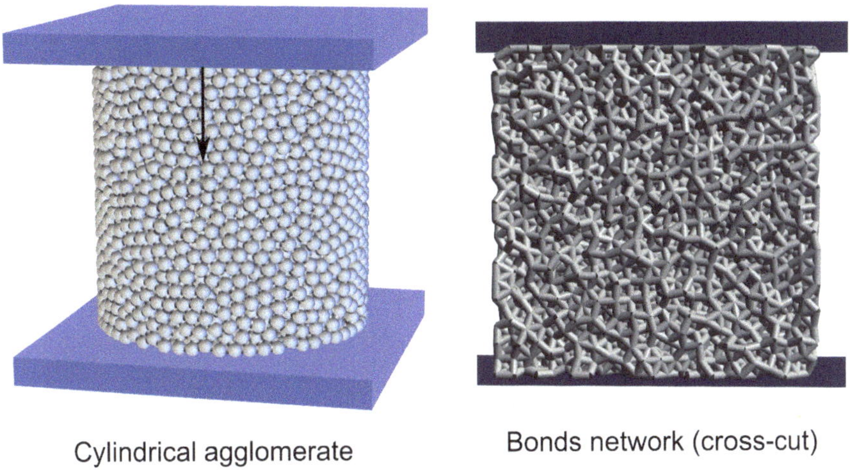

Fig. 7 Cylindrical agglomerate under uniaxial compression test

which is caused due to the stochastic particle placement, ten different agglomerates have been generated and analyzed.

The initial packing density of primary particles was equal to 0.63 which is almost equal to the density of a random close packing. The friction coefficient between primary particles and particles as well as between walls was set to 0.3. To analyze the influence of the distribution of particles and solid bonds on mechanical behavior, four model materials have been defined: two materials M_A and M_B for primary particles and two materials M_C and M_D for solid bonds. Most relevant material parameters are listed in Table 1. The parameters of materials M_A and M_C were kept constant whereas the parameters of materials M_B and M_D have been varied. For these, the range of variation is given in Table 1.

First, we studied the influence of the Young's modulus of bond materials on the deformation and breakage behavior for agglomerates containing only one type of binder M_D and one type of particles M_A. In Fig. 8, typical force-displacement characteristics obtained for two different Young's moduli of solid bonds ($E = 40$ MPa and $E = 10$ MPa) are illustrated. One can observe a linear elastic deformation

Table 1 Basic material parameters of materials M_A and M_B for particles, M_C and M_D for bonds

	Particles		Bonds	
	M_A	M_B	M_C	M_D
Young's modulus (GPa)	2.5	[0.25–2]	0.02	[0.002–0.08]
Poisson ratio	0.2	0.2	0.2	0.2
Normal and shear strength (MPa)	–	–	0.5	0.5
Density (kg/m^3)	2500	2500	–	–

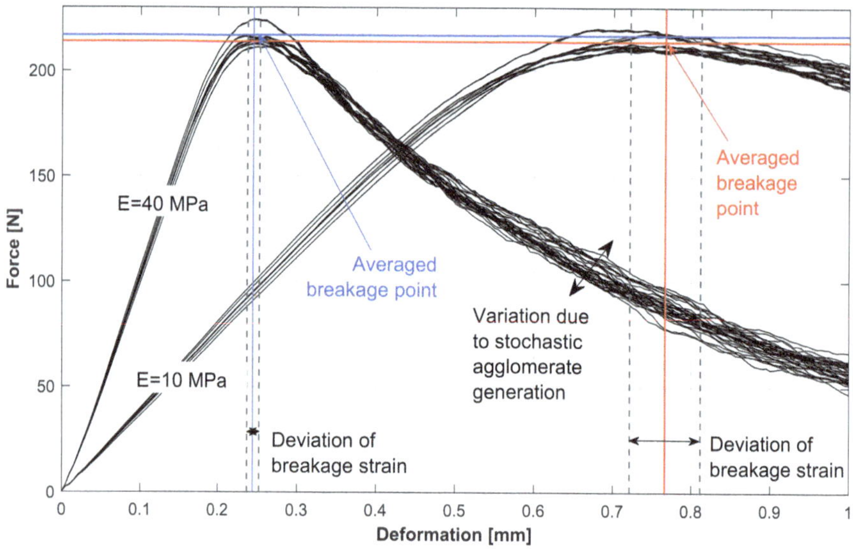

Fig. 8 Force displacement characteristics for agglomerates with varying bond stiffness

behavior of the agglomerate almost until the breakage point. However, the force does not decrease suddenly after the primary breakage, indicating a viscous failure. From the obtained results it can be observed that the increase of the Young's modulus of the bonds decreases the agglomerate stiffness by factor four and the breakage displacement by about factor three. This non-linear dependency between the stiffness of binder material and final stiffness of agglomerate is caused by the agglomerate's internal structure. A more detailed analysis is given in the section "Stiffness Ratio of Bond to Particle" below.

The difference in failure strain is caused by the fact that the strength values of the bonds were kept constant. Thus for the case of the higher Young's modulus of bonds $E = 40$ MPa, the same axial strain of the agglomerate leads to larger stresses in the bonds and as a consequence to earlier breakage. Because of the materials' strength being equal, the breakage force in both cases had an almost identical value of about 215 N.

The obtained results show that variations due to the stochastic generation of the microstructure of agglomerates are relatively small. The large number of primary particles, their high connectivity and high packing density result in a similar microstructure for all agglomerates.

Multicomponent Bonds

In order to investigate agglomerates with multicomponent binder, heterogeneous bond networks consisting of two bond materials M_C and M_D have been generated. The fraction of bonds of material M_D in the total agglomerate is given by

$$K_{DC,N} = \frac{N_D}{(N_C + N_D)}, \tag{11}$$

where N_D and N_C are the total numbers of bonds of component M_C and M_D respectively.

Recall that in this case study it is supposed that all material parameters of bonds except for the Young's moduli are equal. The ratio of Young's moduli of materials M_D (E_D) and M_C (E_C) is described by

$$K_{DC,E} = \frac{E_D}{E_C}. \tag{12}$$

According to the parameter $K_{DC,N}$, bonds of materials M_D and M_C have been created between primary particles. Both components have been distributed at random within all 10 initial agglomerates. For reduction of the statistical error caused by the random spatial positioning of bonds, five random agglomerates have been created for each initial agglomerate and value of $K_{DC,N}$. Therefore, the total number of investigated agglomerates for each specified set of parameters $K_{DC,E}$ was equal to 50.

The internal structure of agglomerates with different numbers of bonds of different types is illustrated in Fig. 9. Here, the cross sections of cylindrical agglomerates are shown and all particles are hidden.

In Fig. 10, the main mechanical characteristics obtained from the simulations of cylindrical agglomerates with varied content of bonds of type M_D (i.e., with

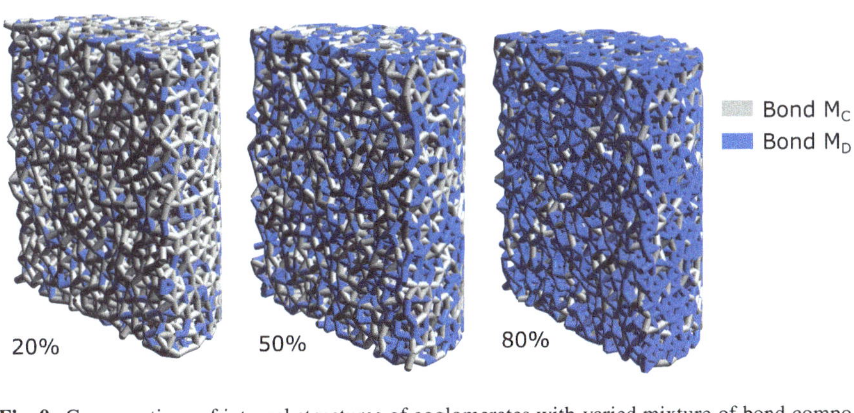

Fig. 9 Cross sections of internal structures of agglomerates with varied mixture of bond components. For better perceptibility of the bond network, primary particles are hidden

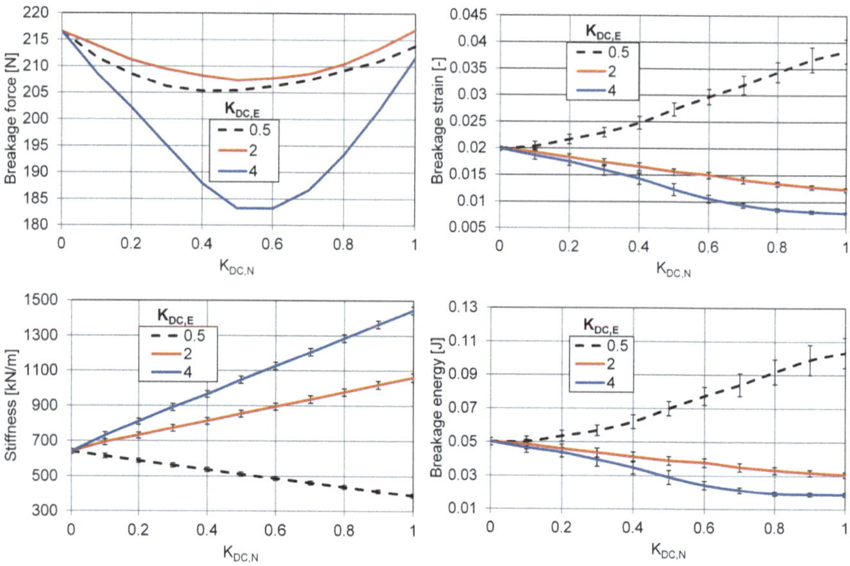

Fig. 10 Mechanical properties of multicomponent agglomerates with varied binder content and Young's modulus

different ratios $K_{DC,N}$) and varied Young's moduli (i.e., with different ratios $K_{DC,E}$) are shown. The values are averaged over force-displacement data of about 1500 simulated compression tests.

The upper left graph in Fig. 10 shows the dependency of breakage force on bond composition and Young's modulus. By the fact that the breakage strength of materials M_C and M_D was equal, it can be seen that the breakage force is almost independent on the stiffness of bonds. In the cases when all bonds consist of material M_C ($K_{DC,N} = 0$) with Young's modulus equal to 20 MPa or when all bonds consisting of material M_D ($K_{DC,N} = 1$) with Young's moduli 10 MPa, 40 MPa or 80 MPa, the breakage force is almost equal to about 215 N. A significant decrease of the breakage force is observed when both types of bonds are mixed together. This happens due to the non-homogeneous stress distribution in the agglomerates. The largest part of total mechanical stress in the agglomerates is distributed over the bonds with higher stiffness, which are subject to higher loads in comparison to bonds in agglomerates consisting of only one material. Agglomerates with the largest ratio of Young's moduli $K_{DC,E} = 4$ consisting of the same amount of bonds of both types M_C and M_D ($K_{DC,N} \approx 0.5$) exhibit the minimal breakage force of about 183 N.

The stiffness of agglomerates with multicomponent bonds depends almost linearly on the amount of bonds $K_{DC,N}$ (lower left graph of Fig. 10). A larger amount of bonds with higher stiffness results in increased overall stiffness of the total agglomerate. As a consequence, the breakage strain as well as the breakage energy decreases (right column of Fig. 10).

Stiffness Ratio of Bond to Particle

In this section, the mechanical properties of cylindrical agglomerates consisting of particles of type M_A and bonds of material M_D are investigated. We analyzed the influence of the ratio between Young's moduli of bonds (E_D) and particles (E_A) on the mechanical properties. This ratio is denoted as $K_{DA,E} = E_D/E_A$. For the simulations, the Young's modulus of particles was kept constant at 25 GPa and the modulus of bonds was varied between 0.2 and 80 MPa. The average coordination number (i.e., the number of bonds per particle) was equal to 4.02. In Fig. 11, the main mechanical characteristics obtained from simulation of quasi-static uni-axial compression of cylindrical agglomerates are shown. On the left-hand side, the influence of the Young's modulus of bonds on agglomerate stiffness and breakage strain is shown. On the right-hand side, the typical force-displacement characteristics for two specific values of $K_{DA,E}$ are illustrated.

The increase of the Young's modulus of bonds leads to a non-linear increase of the stiffness of the whole agglomerate. When the bonds become stiffer, fewer contacts between primary particles are formed under compression, resulting in a reduced effect of particle contacts on the total agglomerate stiffness. In the investigated parameter domain, the stiffness of the agglomerate K_{agg} can be effectively described using the potential function

$$K_{agg} = 15.721[\mathrm{m}] \cdot E_D^{0.729}. \tag{13}$$

Moreover, from Fig. 11 it can be observed that an increase of the Young's modulus of solid bonds E_D leads to a significant decrease of the breakage strain ε. Using the potential function, this dependency can be approximated by

$$\varepsilon = 2078.5\left[\frac{1}{\mathrm{Pa}}\right] \cdot E_D^{0.789}. \tag{14}$$

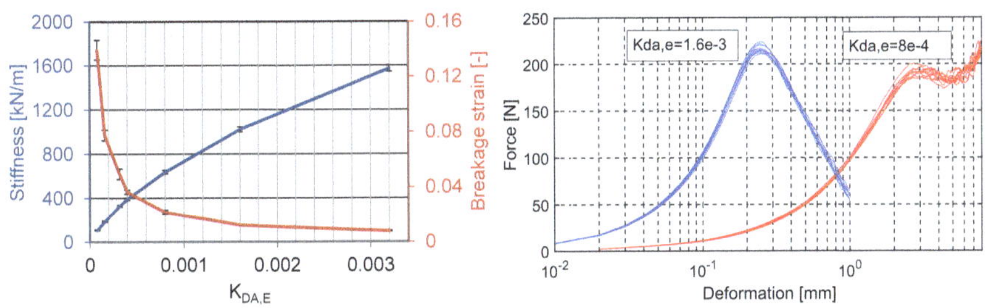

a) Averaged mechanical characteristics b) Typical force-displacement curves

Fig. 11 Simulation results for case studies with varied Young's modulus of bonds

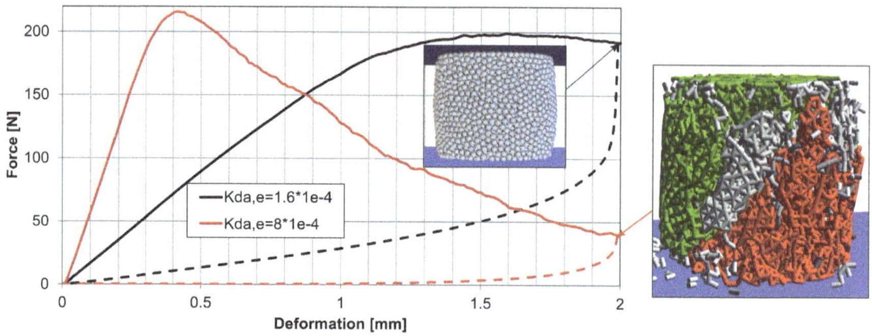

Fig. 12 Left: Influence of the bond stiffness on the force-displacement characteristics of agglomerates. Right: Bond network of sample agglomerate with Young's modulus 20 MPa after breakage. The largest and second largest fragments are shown in green and red, respectively; other fragments (debris) are shown in gray

Higher stiffness of solid bonds results in a more brittle breakage pattern. In Fig. 12 (left), force displacement characteristics for two typical agglomerates with parameters $K_{DA,E} = 1.6 \times 10^{-4}$ and $K_{DA,E} = 8 \times 10^{-4}$ are shown. The loading and unloading phases are shown using solid lines and dashed lines, respectively.

In the case of agglomerates with higher Young's modulus of bonds, a primary breakage point at a deformation of about 450 μm can be observed. Further increase of the load on the sample leads to the destruction of a large amount of solid bonds and consequently to a rapid decrease of the force. Only a small elastic recovery during unloading phase of the broken agglomerate (Fig. 12, left) can be observed. On the right-hand side of Fig. 12, an agglomerate structure colored according to the size of fragments is shown. The deformation behavior during failure indicates shear dominated fracture pattern.

In contrast, the simulated force-displacement curve of the agglomerate with softer bonds did not show a clear primary breakage point with a rapid drop of the force after the point of maximum force. Only a relatively small amount of bonds is destroyed. The sample exhibits a large elastic unloading. Therefore, even after a large deformation of about 2 mm (strain of 0.1), the general structure of the agglomerate remains stable and the sample recovers to its initial height.

Bicomponent Particle Distribution

In this section, the mechanical characteristics of agglomerates which consist of two types of primary particles are analyzed. The primary particles have a constant diameter of 2 mm and differ only in their stiffness. The number fraction of particles of material M_B in the agglomerate is described by

$$K_{BA,N} = \frac{N_B}{(N_B + N_A)}, \tag{15}$$

where N_A and N_B are the total numbers of particles of materials M_A and M_B, respectively. Since the sizes of primary particles are equal, the parameter $K_{BA,N}$ also describes the volume fraction of particles of material M_B.

It was supposed that all material parameters of particles apart from Young's moduli were equal. The ratio of Young's moduli of material M_B (E_B) and M_A (E_A) was described by

$$K_{BA,E} = \frac{E_B}{E_A}. \tag{16}$$

In our simulations, the Young's modulus of material M_A was kept constant while material M_B was chosen softer than material M_A, resulting in $K_{BA,E} < 1$. The (number) fraction of particles of material B was varied between 0 and 1 for five different values of $K_{BA,E}$. In Fig. 13, the approximated overall stiffness of the agglomerate is shown. As expected, the volume fraction and stiffness of the softer component directly influence the agglomerate stiffness. Higher volume fraction and decreased stiffness of the softer primary particles, both result in a reduced overall stiffness of the agglomerate. When the difference between Young's moduli of both components is relative small, the dependency between stiffness of the agglomerate and volume fraction of the second component can be roughly approximated by a linear function. However, a further decrease of E_B leads to non-linear dependency.

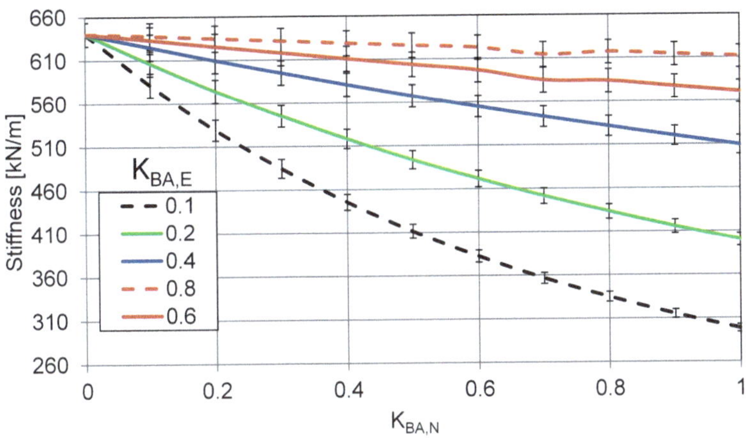

Fig. 13 Influence of the volume fraction $K_{BA,N}$ and stiffness $K_{BA,E}$ of softer primary particles on the agglomerate's total stiffness

Conclusion

In this chapter, the discrete element method and the bonded-particle model have been applied to investigate the mechanical properties of cylindrical and spherical bicomponent agglomerates under quasi-static compression. Four cases have been analyzed:

- large and small particles of the same material (spherical agglomerates)
- solid bonds of two different materials (cylindrical agglomerates)
- varied stiffness ratio of bonds to particles (cylindrical agglomerates)
- primary particles of two different materials (cylindrical agglomerates)

The obtained results show that the microstructure and composition of bicomponent aggregates have a significant influence on their mechanical properties such as stiffness or strength. For example, the breakage force of cylindrical agglomerates with two different bond materials reveals a distinctive minimum for specific mixtures. This occurs due to the large heterogeneity of the stresses inside the bond network. Furthermore, a complex non-linear behavior can be also observed in the case when agglomerates consist of primary particles of different sizes. Here, the breakage force strongly depends on the mixing ratio and reaches its minimum for ratios of about 50%.

It should be mentioned that correlations and conclusions given in this chapter are valid only for a strictly limited parameter domain. It is expected that significant changes of material properties can result in a different material behavior.

References

1. Dosta, M., Antonyuk, S., Heinrich, S.: Multiscale simulation of agglomerates breakage in fluidized beds. Ind. Eng. Chem. Res. **52**, 11275–11281 (2013)
2. Dosta, M., Dale, S., Antonyuk, S., Wassgren, C.R., Heinrich, S., Litster, J.D.: Numerical and experimental analysis of influence of granule microstructure on its compression breakage. Powder Technol. **299**, 87–97 (2016)
3. Dosta, M., Bröckel, U., Gilson, L., Kozhar, S., Auernhammer, G.K., Heinrich, S.: Application of micro computed tomography for adjustment of model parameters for discrete element method. Chem. Eng. Res. Des. **135**, 121–128 (2018)
4. Jungnickel, D.: Graphs, Networks and Algorithms, 3rd edn. Springer, Berlin (2008)
5. Liu, L., Kafui, K., Thornton, C.: Impact breakage of spherical, cuboidal and cylindrical agglomerates. Powder Technol. **199**(2), 189–196 (2010)
6. Mindlin, R.D., Deresiewicz, H.: Elastic spheres in contact under varying oblique force. Trans. ASME J. Appl. Mech. **20**, 327–344 (1953)
7. Mościński, J., Bargieł, M., Rycerz, Z.A., Jacobs, P.W.M.: The force-biased algorithm for the irregular close packing of equal hard spheres. Mol. Simulat. **3**(4), 201–212 (1989)
8. Potyondy, D.O., Cundall, P.A.: A bonded-particle model for rock. Int. J. Rock Mech. Min. Sci. **41**, 1329–1364 (2004)
9. Spettl, A., Dosta, M., Antonyuk, S., Heinrich, S., Schmidt, V.: Statistical investigation of agglomerate breakage based on combined stochastic microstructure modeling and DEM simulations. Adv. Powder Techol. **26**, 1021–1030 (2015)

10. Spettl, A., Bachstein, S., Dosta, M., Goslinska, M., Heinrich, S., Schmidt, V.: Bonded-particle extraction and stochastic modeling of internal agglomerate structures. Adv. Powder Techol. **27**, 1761–1774 (2016)
11. Spettl, A., Klingner, F., Dosta, M., Heinrich, S., Schmidt, V.: Copula-based approximation of particle breakage as link between DEM and PBM. Comput. Chem. Eng. **99**, 158–170 (2017)
12. Toussaint, G.T.: The relative neighbourhood graph of a finite planar set. Pattern Recognit. **12**, 261–268 (1980)
13. Tsuji, Y., Tanaka, T., Ishida, T.: Lagrangian numerical simulation of plug flow of cohesionless particles in a horizontal pipe. Powder Technol. **71**, 239–250 (1992)
14. Weber, M., Spettl, A., Dosta, M., Heinrich, S., Schmidt, V.: Simulation-based investigation of core-shell agglomerates: influence of spatial heterogeneity in particle sizes on breakage characteristics. Comput. Mater. Sci. **137**, 100–107 (2017)

Contact Mechanisms in Ultrasound-Agitated Particulate Systems

Claas Knoop, Tobias Wollborn and Udo Fritsching

Introduction

Dynamic stresses and force effects in a gaseous or liquid environment, that occur for instance in sound fields, can be used to the contactless influence of powder systems containing solid primary particles. For instance, acoustic noise can be utilized for separation of fluidized particles [1] or for the dosage of packed beds [2]. Depending on the process control, agglomeration of single particles [3] or the precipitation of particles on solid surfaces [4] can be induced, but also the dispersion/fragmentation of agglomerated particles [5] and the detachment of wall adhered particles [6]. These examples are common methods that are applied for the dispersion of nanoparticles and the removal of adherences in cleaning baths. Here, the cohesion of particle-particle and particle-wall contacts is repealed by the impact of the sound field, in which each single binding force is overcome by sound induced drag forces.

In this work the potential of the sound induced drag forces on single bonds in agglomerated particles and on wall adhered particles is derived within a gaseous environment. In principle the possibility of this approach has been shown in the contributions of Fuhrman et al. [7] and Chen and Wu [8].

The possibility of agglomerate dispersion is investigated on two different levels. The potential of fragmentation of single agglomerates by the direct application of

C. Knoop · T. Wollborn · U. Fritsching (✉)
Leibniz Institute for Materials Engineering IWT, Bremen, Germany
e-mail: ufri@iwt.uni-bremen.de

C. Knoop
e-mail: knoop@iwt.uni-bremen.de

T. Wollborn
e-mail: wollborn@iwt.uni-bremen.de

U. Fritsching
Particles and Process Engineering, University of Bremen, Bremen, Germany

MAPEX Center for Materials and Processes, University Bremen, Bremen, Germany

S. Antonyuk (ed.), *Particles in Contact*,
https://doi.org/10.1007/978-3-030-15899-6_7

acoustic drag forces is analyzed first. Secondly, the fragmentation of agglomerate collectives is described for the indirect application of acoustic drag forces. Here, agglomerates are specifically accelerated by the acoustic drag forces. Fragmentation occurs by agglomerate-agglomerate and agglomerate-wall impacts. In the third part of the investigation the sound-induced detachment of wall adhered particles is analyzed. Finally the mechanical stability of different types of agglomerated particles is described in tensile and compressive tests. The ratio between tensile and compressive forces is analyzed and compared to the possibility of agglomerate dispersion.

Materials and Methods

Particle Systems

For the experimental investigation soda-lime glass particles from Potters (Spheriglass) have been used (Fig. 1). Particles of different size distributions were agglomerated in a controlled way to analyze particle size effects on the dispersion efficiency.

Different binding processes for particle agglomeration were used to specifically vary the adhesion force at the contact points of the primary particles and hence

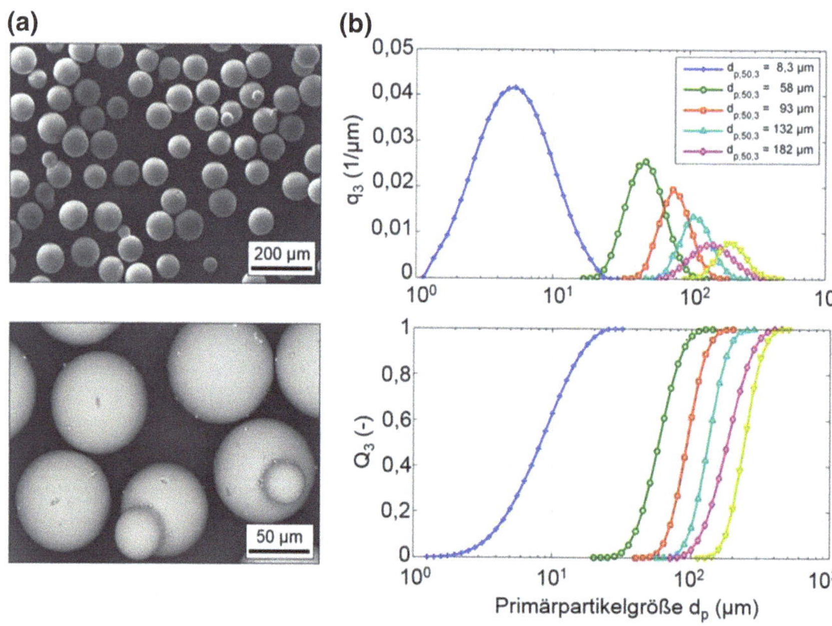

Fig. 1 Scanning electron microscope images of glass particles (**a**) and particle size distribution $q_3(d_P)$ and $Q_3(d_P)$ (**b**)

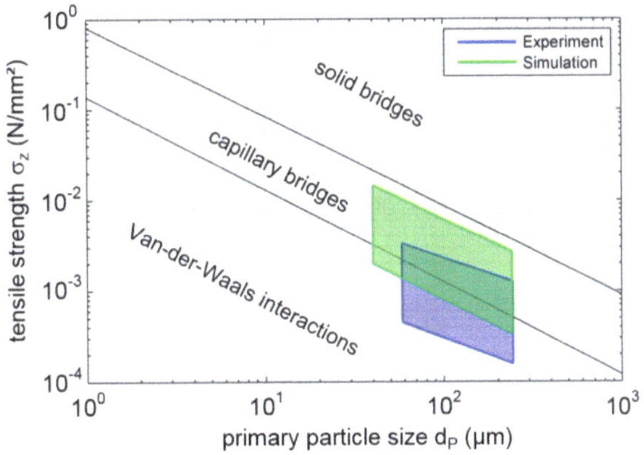

Fig. 2 Tensile strength σ_Z of the investigated agglomerate structures as a function of the primary particle size d_P

the stability of the produced agglomerate structures. Therefore, liquid agglomerates, bonded by capillary bridges and van der Waals forces, and solid bonded agglomerates with essential higher strength were considered. In Fig. 2 the range of the tensile strength σ_Z as a function of the primary particle size d_P for the different binding mechanisms is shown. Strength and particle size ranges considered in experiment and simulation are indicated as well. The tensile strength σ_Z is calculated by Rumpf's equation:

$$\sigma_Z = \frac{1-\varepsilon}{\varepsilon} \frac{1}{d_P^2} F_{Adh}. \quad (1)$$

Here, d_P is the primary particle diameter and F_{Adh} is the adhesion force between primary particles [9]. The agglomerates are considered as homogeneous. The value of porosity ε is estimated to 0.58 based on [10].

The solid bonded agglomerates were produced in a spray agglomeration process. An adhesive agent was dissolved in water and sprayed over the fluidized particles. After drying, only the adhesive agent remains between the particles as solid bond. To vary the bonding strength, different types of agents have been used, as carbomethyl cellulose (CMC—3 wt%), polyvinylpyrolidon (PVP—5 wt%) and sacchorose (SAC—20 wt%). The adhesion forces of the different types of binders were determined via compression tests to calculate the stability of the agglomerates. In Fig. 3 scanning electron microscope (SEM) images of the agglomerate structures are shown.

The stability of liquid bonded agglomerates has been changed by variation of the relative humidity φ and by variation of the contact angle Θ of the primary particles. Thereby it was possible to determine the parameters relevant for adhesion and their influence on the agglomerate stability. This allows conclusion on the dispersibility of the produced structures. For liquid bonded agglomerates the adhesion force is the sum of the capillary forces of each capillary bond between the primary particles and the humidity and contact angle dependent van der Waals force.

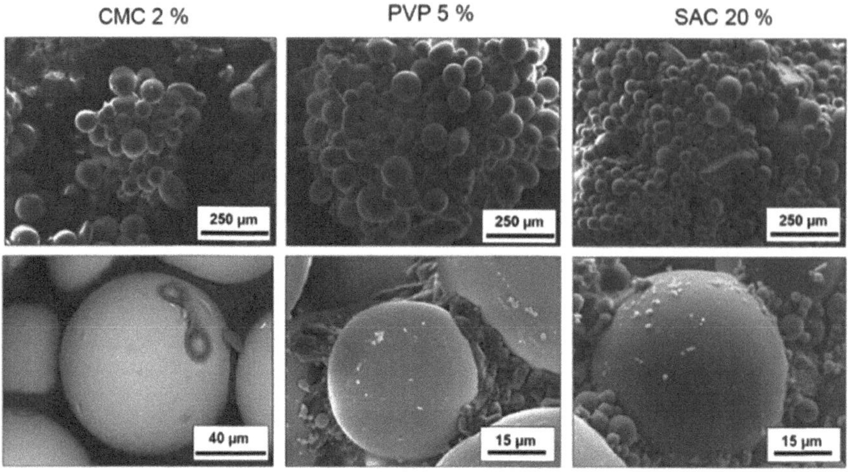

Fig. 3 Scanning electron microscope images of CMC-, PVP- and SAC-bonded agglomerates at different magnification

For the calculation of the capillary force the shape of the meniscus of the capillary bridge is determined via a circular approximation with constant radius. In this case it is possible to calculate the capillary force dependent from contact angle and relative humidity via analytical equations [11]. The calculation of the van der Waals is described as a function of the relative humidity as shown in [12].

Ultrasonic Device

For experimental investigation of single agglomerates an ultrasonic sonotrode (Branson) was used to generate a high intensive standing ultrasound wave. A schematic representation is illustrated in Fig. 4.

The generator frequency is controlled from 50 Hz to 20 kHz. A piezoelectric element converts the electric signal into a mechanical movement. The frequency of the oscillation corresponds to the resonant frequency of the connected components (amplifier and sonotrode). Thus, the components are excited to oscillate and because of the geometry amplification of the amplitude results [13]. The maximum amplitude is reached at the front surface of the sonotrode. After a distance x_{sr} the wave, transferred to the gas phase, attains a passive reflector. The wave is reflected and interact with itself. A standing ultrasound wave occurs. The distance x_{sr} is set to $x_{res} = 3/2\lambda$ to receive a resonant field.

For the investigation of the liquid bonded agglomerates the experimental setup is encased by a climate chamber, which allows the adjustment the atmospheric con-

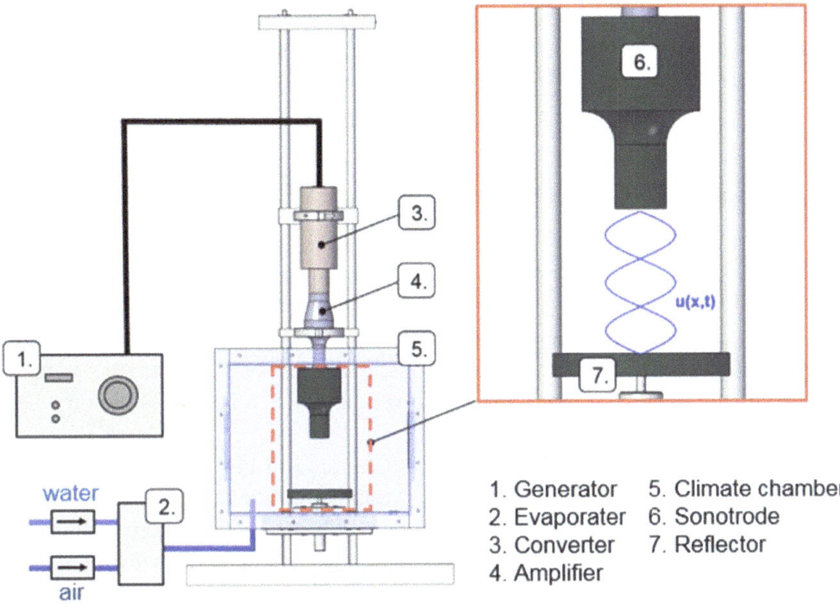

Fig. 4 Schematic setup to generate a high intensive ultrasound field (f = 20 kHz)—vertical arrangement

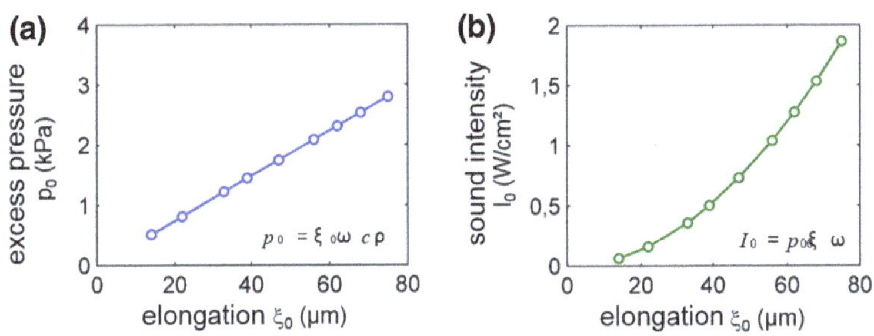

Fig. 5 Excess pressure p_0 and sound intensity I_0 as a function of the elongation ζ_0 of the front surface of the sonotrode

ditions. In this chamber relative humidity Φ and temperature T are controlled by regulated water evaporation.

The periodic deflection of the sonotrode front surface correlates with the elongation ζ_0 of the sound wave emitted to the gas phase. The index 0 indicates a parameter directly at the sonotrode. This indication is also valid for all following acoustic parameters.

The experimental setup enables elongations between 14 and 75 μm. The resulting trends of the amplitude of the excess pressure p_0 and the sound intensity I_0 over the elongation are shown in Fig. 5.

The amplitude of the excess pressure increases to a value of 2.7 kPa, which is a sound pressure level of 142.6 dB. At maximum elongation the sound intensity reaches value of 2 W/cm^2.

Simulation of Acoustic Drag Forces

For the derivation of the distributions of excess pressure and sound velocity of the acoustic field as well as the resulting acoustic drag forces on solids CFD simulations were performed. For these numerical investigations the open source platform OpenFOAM is used. The three dimensional and time dependent equations of mass, momentum and energy conservation are solved, in which a compressible flow field is considered ($\rho \neq$ const.).

The geometry of the numerical setup is illustrated in Fig. 6. It shows the field between the active sonotrode and the passive reflector of the device. The distance in direction of propagation of the ultrasound wave is indicated with x_d and perpendicular to this direction with y_d and z_d. In case of the resonant standing ultrasound field distance x_d is set to $x_d = x_{res}$. This corresponds to the distance used in the experimental investigations and is calculated to $x_{res} = 25.76$ mm ($x_{res} = 3/2\lambda$).

Two different sound field configurations have been investigated. In the first case a standing wave surrounded without undisturbed gas atmosphere (Rayleigh configuration) is considered. In the second case, the Langevin configuration, a standing wave surrounded by an undisturbed gas atmosphere is considered ($y_d \neq y_s$ and $z_d \neq$

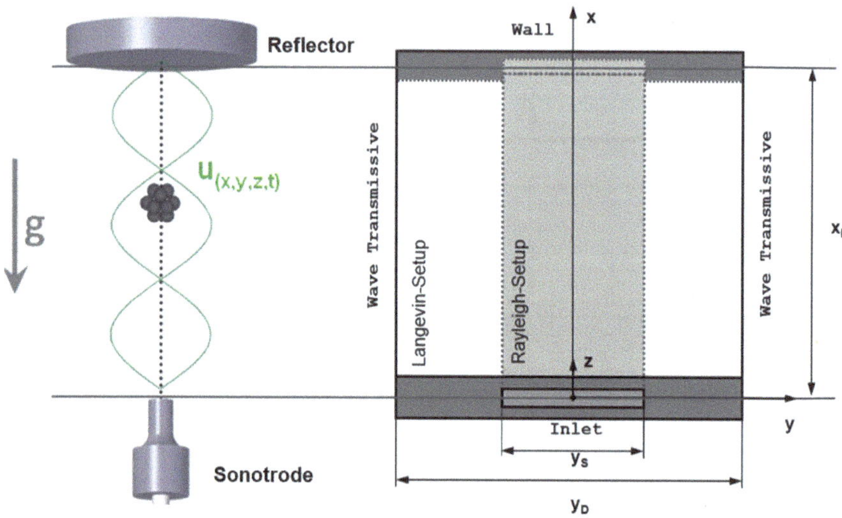

Fig. 6 Geometry of the numerical setup (CFD) of the ultrasound field in Rayleigh and Langevin configuration

z_s). Most of the performed calculations were performed in Rayleigh configuration, because of the lower computational costs.

The ultrasound wave is generated via an inlet boundary condition representing the sonotrode of the experimental setup and is expressed as periodically oscillating velocity:

$$u(t) = u_0 \sin(\omega t) \tag{2}$$

The generated wave traverses the computational domain and attains the reflector, which is considered as solid wall (no slip boundary condition). After reflection a standing wave is established. A wave transmissive boundary condition is used in radial direction of propagation of the ultrasound wave and is applied for the Rayleigh configuration as well as for the Langevin configuration. This boundary condition is necessary to prevent a reflection of the pressure and velocity waves back into the computational domain.

The mesh for the computational domain was designed by using the snappyHexMesh routine of OpenFOAM (see [14]). This enables the construction of a block structured mesh. The resolution of the mesh is chosen to represent each antinode by approximately 30 cells. The wave length of the sound field is in the millimeter range, but the investigated primary particles are in the micrometer range. Therefore a higher mesh resolution for the area around the particle surface is necessary. In each refinement step the cells are divided in eight equal sized smaller cells. Up to 7 refinement steps are necessary. This leads to a final cell number of $nz = 2 \times 10^6$ to 3×10^6 cells depending on the configuration. The mesh quality has been checked to guarantee a mesh independent solution.

For the calculation of transient flow fields the time step width is of essential importance for numerical stability. The dimensionless CFL number (also Courant number) is commonly used as stability criteria [15]. It shows how many cells are traversed per time step. In case of explicit solvers no more than one cell should be traversed per time step. Also for implicit solvers, as used in the numerical calculation, the number should be small. Therefore a time step width of $\Delta t = 2 \times 10^{-7}$ s is set. This guarantees small Courant numbers even for the smallest cells. Also the Nyquist-Shannon-sampling-theorem is fulfilled, which implies that the sampling frequency must be twice as high as the maximum frequency of the waveform [16].

For a full developed standing ultrasound field a simulated time auf $t_s = 0.002$ s is necessary. Therefore a number of $n_{\Delta t} = 10{,}000$ time steps are calculated per simulation case.

The primary particles exposed to the sound field are represented as wall (no slip). Hence they are defined as solid and fixed bodies. Different particle configurations are considered. Besides wall adhered particles also a two-particle agglomerate (doublet) as well as a thirteen particle agglomerate as hexagonal densest packing. The surface

Table 1 Primary particle configurations and parameter variations used in the CFD simulation of the acoustic drag forces on wall adhered and levitated agglomerates

Configuration	Particle-wall	Doublet	Agglomerate
n_p (–)	1	2	13
d_p (μm)	40–240	40–240	80
f (kHz)	20	20	18.7
u_0 (m/s)	2–14	2–14	0.25–2.00
I_0 (W/cm^2)	0.08–4.2	0.08–4.2	0.001–0.08
x_{us} (mm)	25.76	25.76	27.54
x_{ag} (mm)	25.76	12.88–17.17	15.03
α (°)	0–90	0–90	0
n_{pk} (–)	0	1	36

mesh and the relevant geometric and acoustic parameters considered in the simulation are summarized in Table 1.

For the derivation of the acoustic drag forces the simulated pressure and shear stress fields are integrated over the surface S of the considered particles [17]:

$$F_{pi} = \oint_S p_s n_i dS; \quad i = x; y; z; \tag{3}$$

$$F_{ti} = \oint_S \tau_s n_i dS; \quad i = x; y; z; \tag{4}$$

This results to the time dependent components of the applied acoustic drag forces of the ultrasound field on the primary particles. The pressure distribution at the surface leads to the pressure part F_p whereas the friction part F_t results from the wall shear stress calculated from the velocity gradient at the particle surface. The acoustic drag force is the sum of both parts relating to the direction in space.

Dispersion of Agglomerated Particles

Direct Agglomerate Dispersion

Representation of the Standing Ultrasound Field

For the derivation of the acoustic drag forces on primary particles in ultrasound exposed agglomerates the ultrasound field used in the experiment is reproduced by means of CFD. In Fig. 7 the excess pressure p as well as the velocity u_x of the sound field over distance is shown at different time steps with an initial amplitude of $u_0 = 8$ m/s. The trends are related to the amplitude of the incident sound wave p_0 and u_0, respectively. Thus, conclusions about the degree of amplification in the resonant flow field are possible.

The result show the behavior of a standing sound wave with its characteristic nodes and antinodes. The trends of pressure and velocity are phase-delayed, so that velocity antinodes and pressure nodes are arranged at the same position and vice versa. The resonance leads to an increase of the amplitude, around five times of the initial amplitude. Furthermore, the influence of nonlinear effects within the high intensive sound field is identified. Due to the fact that the local phase velocity is different to the average sound velocity, a shift of the maxima of the antinodes against each other can be observed [18]. Moreover, the pressure nodes are blurred because of the dissipative damping. This leads to small differences of the amplitudes of the incident and reflecting sound wave.

Sound Induced Drag Forces

The pressure and velocity distributions are used to calculate the dynamic forces on a solid particle in the sound field. In this case, the pressure as well as the velocity gra-

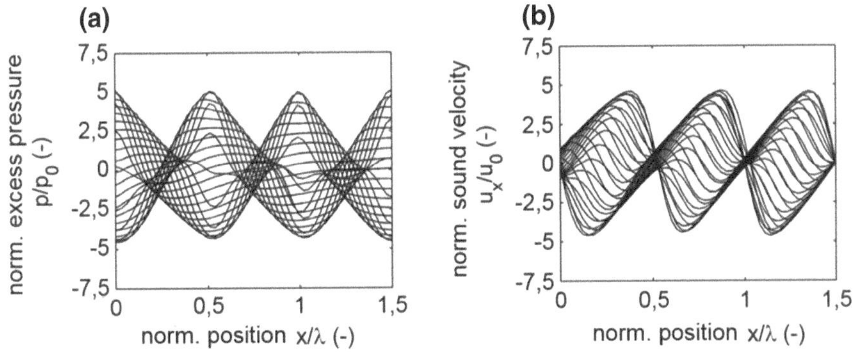

Fig. 7 Excess pressure and sound velocity over distance at different time steps; $u_0 = 8$ m/s

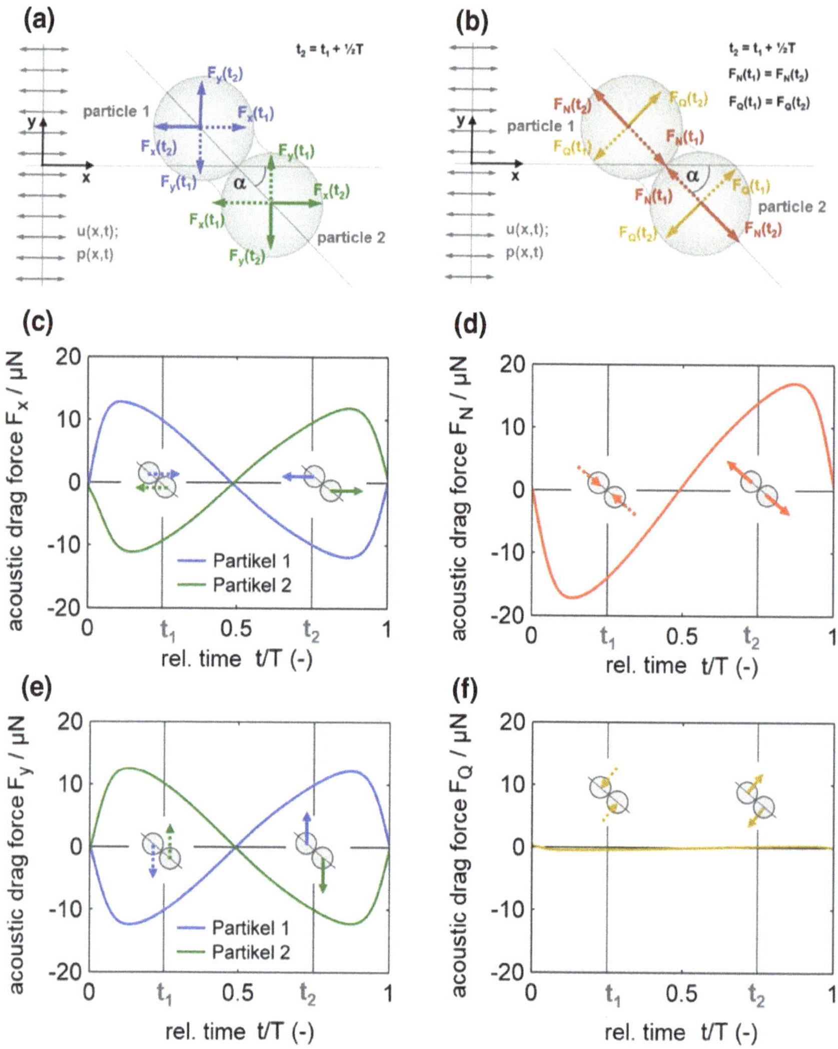

Fig. 8 Acoustic drag forces on a two-particle-configuration for different time steps t_1 and t_2 in x- and y-direction (**a**) and normal and tangential direction at the contact point as well as the time dependent development of the contact point (**b**), force components on the single primary particles (**c** + **e**) and at the contact point (**d** + **f**) at one periodic time

dients are integrated over the particle surface to receive the pressure and the friction component of the acoustic drag force (Eqs. 3 and 4). In Fig. 8 the resulting acoustic drag forces on a two-particle-configuration for one periodic time and hence for one stress cycle is represented. The configuration is placed at a pressure node at x = 17.2 mm. It is rotated by an angle of $\alpha = 45°$. The primary particle diameter is d_p = 80 μm. The left part of the figure shows the acoustic drag forces on the primary particles in x- and y-direction. The right part shows the resulting normal and tangential

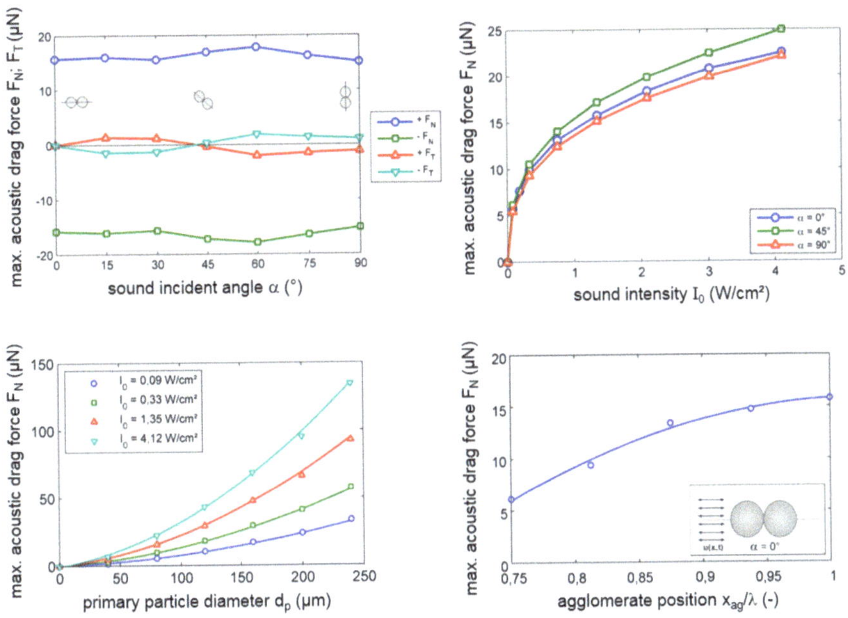

Fig. 9 Amplitudes of the normal (F_N) and tangential (F_T) force on a two-particle-configuration with variation of the contact angle (top left), the sound intensity I_0 (top right), the primary particle diameter d_p (bottom left) and the agglomerate position x_{ag} (bottom right)

force at the contact point of the particles. The primary particles experience alternating stress conditions in direction of propagation of the sound field (x-direction) as well as perpendicular to this direction. Within one period, in a first step the particles are pressed against each other and in a second step they are pulled apart. The same observation is valid for the consideration of normal and tangential force component. Here, the maximum resulting tangential force, thus the force component responsible for a sliding between the primary particles, is obviously lower than the normal component. In the considered case the normal force component reaches values up to about 20 μN.

In Fig. 9 (top left) the development of the positive and the negative amplitudes (positive: tensile force, negative: compressive force) is considered for a variation of the doublet orientation with respect to the sound wave propagation direction (angle α). With rotation of the particle-configuration no significant changes in the resulting drag forces are identified. Maximum normal forces are received at a rotation angle of 60°, but with a value of 18 μN the forces are only slightly higher than the minimum value. The significantly lower tangential force changes its prefix with increasing rotation angle.

Compared to the orientation α the sound intensity I_0, the agglomerate position x_{ag} and the primary particle diameter d_p have an significant influence on the normal force F_N responsible for a detachment of the primary particles (Fig. 9b–d). With increasing

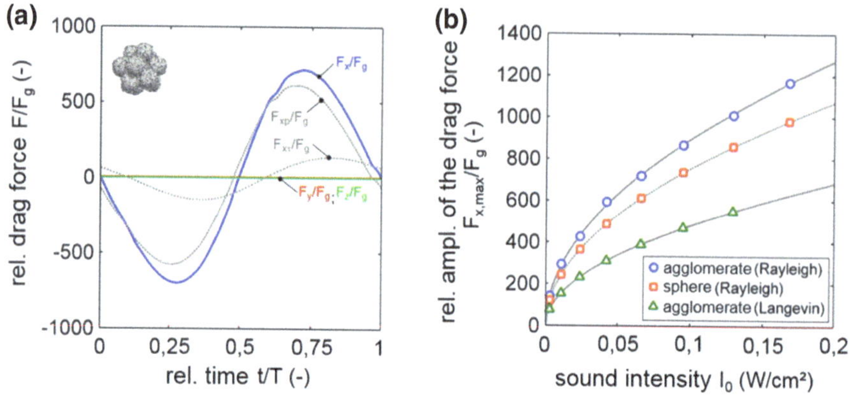

Fig. 10 Time dependent acoustic drag force F_x over one periodic time T (**a**) and the development of the amplitude of $F_{x,max}$ over the sound intensity I_0 for different geometries (**b**) (Reprinted with permission from [17].)

sound intensity, thus with increasing application of energy, also the resulting drag forces increase. This leads to an increase of normal force up to 25 μN within the considered interval. This trend can also be observed by comparing the single curves in Fig. 9c. Here, also an influence of the primary particle diameter is identified. With increasing particle diameter an increase of the pressure and friction component of the acoustic drag force occurs, resulting in an increase of the normal force.

The normal force also increases with increasing agglomerate distance to the velocity anti-node ($x_{ag}/\lambda = 0.75$) and with decreasing distance to the pressure anti-node ($x_{ag}/\lambda = 1.0$). Therefore the pressure is, compared to the friction, the dominant component of the acoustic drag force. This becomes apparent by considering the trend of the dimensionless acoustic drag force on the model agglomerate in Fig. 10a. The integrated values of the drag force on the agglomerate structure in x-direction F_x are related to the gravitational force F_g. The values are also represented as their pressure component F_{px} and friction component F_{vx}. The friction component is significantly lower for an agglomerate placed in the center between pressure and velocity node ($x_{ag} = 15.0$ mm). The drag force in y- and z-direction is small compared to the force component in direction of propagation of the sound wave because of the symmetry conditions of the compact agglomerate structure. Also the gravitational force is several orders of magnitudes lower than F_x.

In Fig. 10b the acoustic drag force in dependence of the sound intensity I_0 is shown for the model agglomerate and a sphere of the same volume. With increasing sound intensity the drag also increases. The trend of the agglomerate is always above the trend of the sphere.

The observed phenomena of alternating compressive and tensile forces at the particle bonds for the two-particle-configuration is also observed for the primary particle contact points of the agglomerate structure (Fig. 11). According to that, dynamic stresses at the contact points in all three dimensions in space exist, even for the y- and z-direction with (very low drag forces), resulting in the alternating

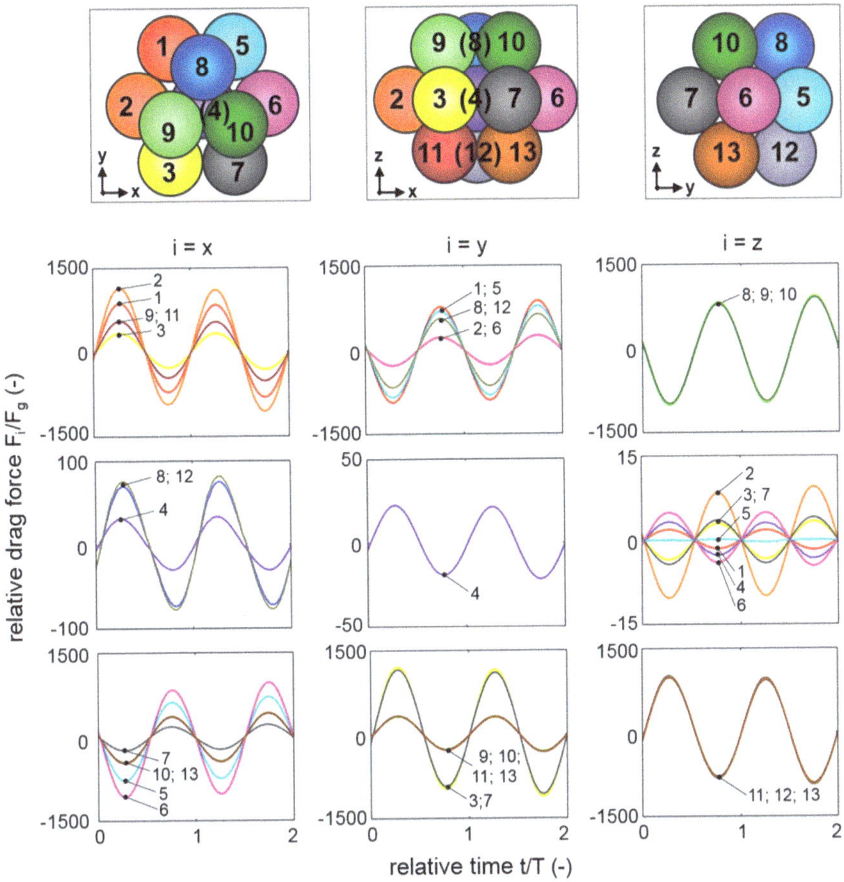

Fig. 11 Time dependent acoustic drag force F_i on the primary particles in x-, y- and z-direction for two time periods T (Reprinted with permission from [17].)

compressive and tensile forces. For example in x-direction this can be observed for the primary particles #1; #2; #3; #9 and #11. The time dependent drag force on these particles is phase-delayed compared to the drag force on the particles #5; #6; #7; #12 and #13 by 180°. The described tensile stress is a unique characteristic of the ultrasound dispersion of particulate structures compared to other dispersion processes.

Particle Bond Forces

The previous considerations show that separation forces in the μN range can be achieved by acoustic impingement, depending on the primary particle size. In order to overcome particle-particle contact, the adhesive forces at the contact points of the particles must be overcome. In the following, the adhesive force between two particles of the same size is calculated analytically. It is assumed that the adhesive force is composed of the capillary force of the condensed humidity and the van der Waals force. Assuming a circular approximation, the capillary force is calculated according to [11] and the van der Waals according to [12].

Figure 12a shows the adhesion force F_{Adh} between two particles with a diameter of 80 μm as a function of humidity φ, while Fig. 12b varies the primary particle diameter. The individual curves also illustrate the influence of the contact angle Θ.

As the humidity is increased, also the adhesive strength increases. However, a critical amount of moisture must be present in the ambient air in order to form a capillary bond. If this is not the case, only the van der Waals forces between the particles are present. The critical humidity level φ_{krit} depends on the contact angle. The higher the contact angle (the more hydrophobic the particle surface is), the higher the critical humidity. At the same time, the maximum adhesive force that can be achieved decreases as the contact angle increases. With increasing particle diameter, however, the adhesive force increases linearly.

For 80 μm diameter particles adhesion forces range up to the 20–30 μN range, as well as acoustic separation forces. It can therefore be assumed that there is potential to overcome the cohesion in moist agglomerates with primary particle sizes in the micrometer range by direct dispersion.

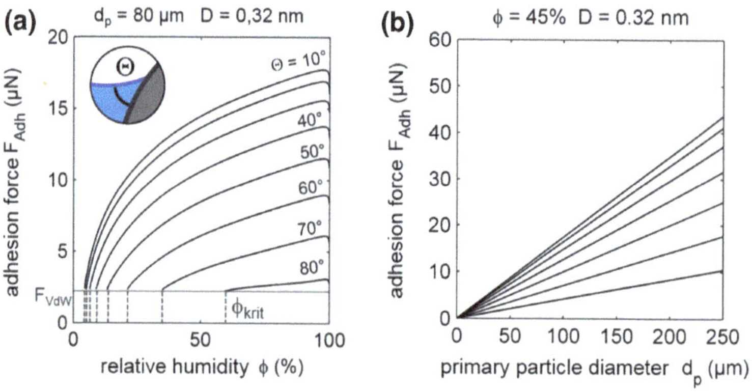

Fig. 12 Adhesion force F_{Adh} depending on relative humidity φ (**a**) and primary particle diameter d_p (**b**) with variation of the contact angle Θ (Reprinted with permission from [28].)

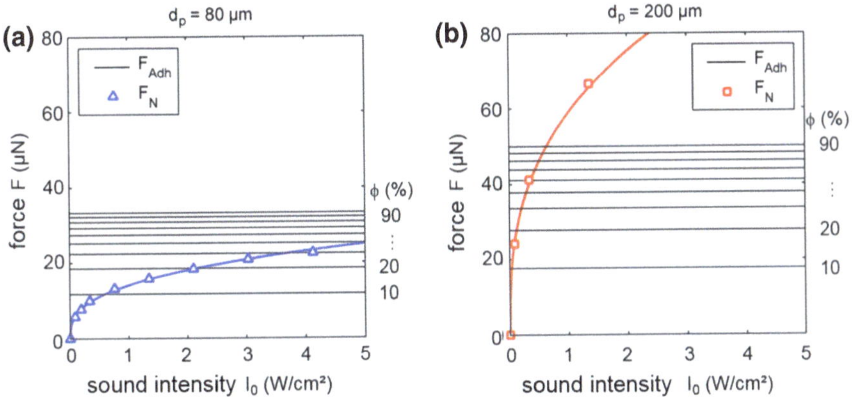

Fig. 13 Comparison of the maximum acoustic drag force in normal direction F_N with analytically calculated adhesion forces F_{Adh} of a two-particle-configuration for different primary particle diameter d_p

Comparison of Analytically Calculated Adhesion Forces and Sound Induced Drag Forces

Figure 13 compares the sound-intensity dependent normal force F_N, which strives to overcome the particle cohesion, with the analytically calculated adhesion force F_{Adh}, which is responsible for the cohesion, for two different primary particle sizes.

In the performance interval considered here, the sound-induced force exceeds the adhesive force only for the larger particles over the entire moisture range, while this is the case for the smaller particles only up to a moisture content below 40%. The theoretical dispersion intensity I_0* can be derived from the intersections of the curve of F_N with the moisture-dependent adhesion forces.

The result of the dispersion intensity as a function of air humidity and the Helmholtz number ($He = d_p/\lambda$) is shown in Fig. 14. The individual graphs also show a variation of the contact angle from $\Theta = 15°$ (hydrophilic) to $\Theta = 70°$ (hydrophobic).

The intensity required to overcome the adhesive force increases with increasing humidity and decreasing Helmholtz number, i.e. smaller particles (assuming a constant frequency) and decreases for increasingly hydrophobic particles.

Simulation of the Breakage Behavior of Agglomerates with the Direct Ultrasound Dispersion

To investigate the breakage behavior of agglomerated solid particles in the ultrasonic field, the drag forces of the standing wave field on the primary particles were derived numerically, as described in section "Dispersion of Agglomerated Particles". The comparison between the sound induced separation forces and the particle binding forces has shown that overcoming the adhesive forces acting between the particles is

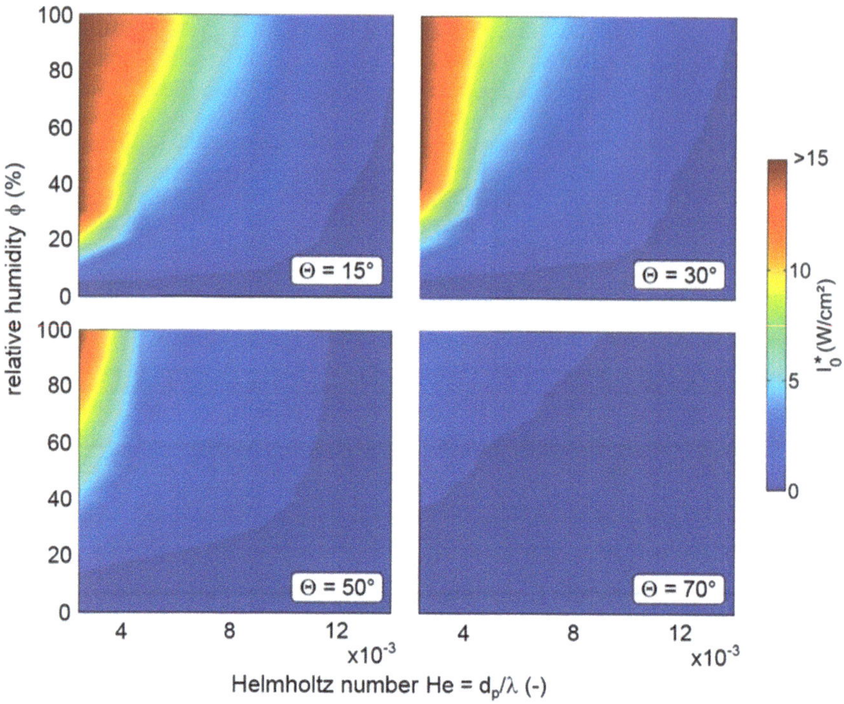

Fig. 14 Dispersion intensity I_0* depending on Helmholtz number He and relative humidity φ for different contact angles Θ = 15°–70°

basically plausible. To confirm this statement, the breakage behavior of agglomerated solid particles in the ultrasonic field was investigated in detail using DEM simulation.

As an extension of section "Sound Induced Drag Forces", an irregularly shaped agglomerate of 12 primary particles was created using a ballistic algorithm (see Table 2).

The applied acoustic drag forces on the primary particles were calculated via CFD simulation of a resonant standing ultrasound wave, in which the agglomerate was placed within the levitation area. The simulation parameters are shown in Table 2. In a next step the agglomerate structure was imported in the DEM simulation. Here, the calculated periodic drag force vectors were applied on each single primary particle. The binding forces between the primary particles were realized by using a stiff-bond-model. This model does not include the physical behavior of the present capillary bridges, but in fact that breakage occurs very rapid, these properties can be neglected in this estimation. The required breakage force of a particle bond was adjusted based on the calculation in section "Particle Bond Forces" for primary particles with d_p = 100 μm and a relative humidity of ϕ = 50%. The breakage distance was adjusted based on [27]:

Table 2 Particle configuration and parameter variation of the calculation (CFD) of the acoustic drag forces on irregular shaped and levitated agglomerates

Parameter	Irregularly shaped agglomerate
n_P (–)	12
d_P (μm)	100
f (kHz)	20
u_0 (m/s)	1–4
I_0 (W/cm^2)	0.04–0.66
x_{us} (mm)	25.76
x_{ag} (mm)	19.32
n_{pk} (–)	15

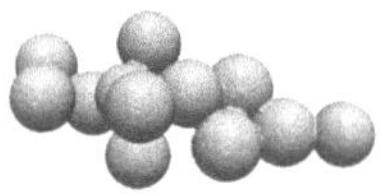

$$D_{rupture} = (1 + 0.5\theta)V^{1/3} \tag{5}$$

Here θ is he contact angle. The bond volume is represented by V and calculated by [19]:

$$V_{sp/sp} = \pi R^2\alpha^2 H + 0.5\pi R^3\alpha^4 \tag{6}$$

The particle radius is described by R, the particle distance by H and the saturation angle by α. The simulation for the calculation of the resulting drag forces has been performed for different amplitudes u_0 (Table 2) within the intensity spectrum of the experimental setup. For low amplitudes no breakage or detachment of single primary particles is observed. With increasing the amplitude, the acoustic drag forces are increased. This leads to higher separation forces on the particle bonds. The result is an increasing degree of dispersion up to full breakage of the agglomerate structure (Fig. 15). This supports the assumption based on the comparison between simulated separation forces and analytically calculated binding forces, that it is possible to overcome capillary bonds by ultrasound induced drag forces.

Experimental Investigations

A schematic representation of the used experimental setup is shown in Fig. 16. In the acoustic levitator, surrounded by a climatic chamber for controlled atmosphere, a vertical standing ultrasound field is generated, in which a single agglomerate is levitated. For this a certain sound intensity I_{lev} is required. From the time of levitation t_0 the intensity is continually increased until the breakage intensity I_B is reached and

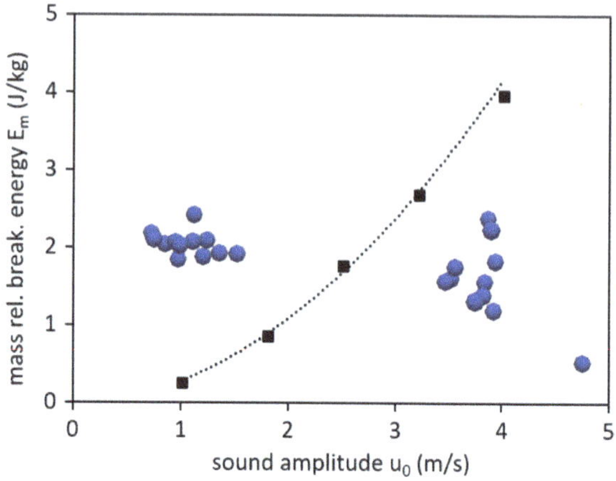

Fig. 15 Breakage behavior of an agglomerate depending on the mass related breakage energy (black curve, CFD) and examples of agglomerate dispersion (blue particles) investigated in the DEM simulation

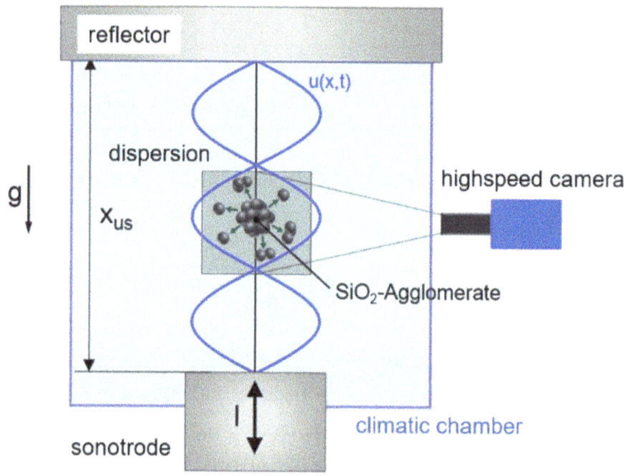

Fig. 16 Schematic representation of the experimental setup for the direct ultrasound dispersion of liquid bonded agglomerates (Reprinted with permission from [28].)

Table 3 Parameters used in the experimental investigation of the direct dispersion of liquid bonded agglomerates

I_0 (W/cm^2)	f (kHz)	x_{us} (mm)	d_{ag} (μm)	$d_{p;50,3}$ (μm)	ϕ (%)	Θ (°)
0.07–1.85	20	25.76	1000–3000	58–242	15–75	14.5–103.4

agglomerate breakage is visually observed. The intensity I_B is proportional to the breakage energy of the agglomerate.

The breakage process was recorded via a highspeed camera (Photron Fastcam SA4). The images have been used to describe the trajectories of the agglomerate and agglomerate fragments as well. Furthermore, agglomerate specific parameters, i.e. the agglomerate diameter, were calculated. An overview of the parameter set used in the experiments is shown in Table 3.

Agglomerate Movement in the Standing Ultrasound Field

An example of the breakage of an agglomerate is shown Fig. 17 for a diameter of d_{ag} = 1.9 mm and an average particle size of $d_p = 93$ μm. The image sequence in the left part, recorded with a frame rate of 1000 fps, shows a levitated agglomerate, which rotates before breakage into several fragments. The right part shows the trajectories of the agglomerate and its fragments.

Before the particulate structure breaks apart, the agglomerate obviously swings around a stable position. This corresponds to the levitation position of a sphere of the same mass and is also referred to here as the levitation level. This up-and-down motion is a typical characteristic of non-spherical bodies during acoustic levitation. It results from the body rotation and the associated change of the projection surface in the direction of sound propagation. This constant change results in a variation of the acoustic drag forces, which counteracts the weight force and levitates the body. Since the weight force remains constant at the same time, the position of the body changes continuously.

The sinusoidal course of the agglomerate movement of the agglomerate before fracture shown in Fig. 18a reveals an amplitude of displacement $X_{max} = 1.22$ mm at a period of T = 38.4 ms. According to this the agglomerate oscillates around the levitation level at a frequency of $f_{ag} = 26.0$ Hz. This frequency is three orders of magnitude lower than the frequency of the ultrasound. The frequency of the movement of the agglomerate under consideration is quite representative, as the evaluation of the frequency of the vertical oscillation movement of more than one hundred agglomerates of different size and circularity confirms. The median of the frequencies is therefore $f_{ag;50} = 39.8$ Hz and the span value $\Delta = \left(f_{ag;90} - f_{ag;10}\right)/f_{ag;50} = 0.30$.

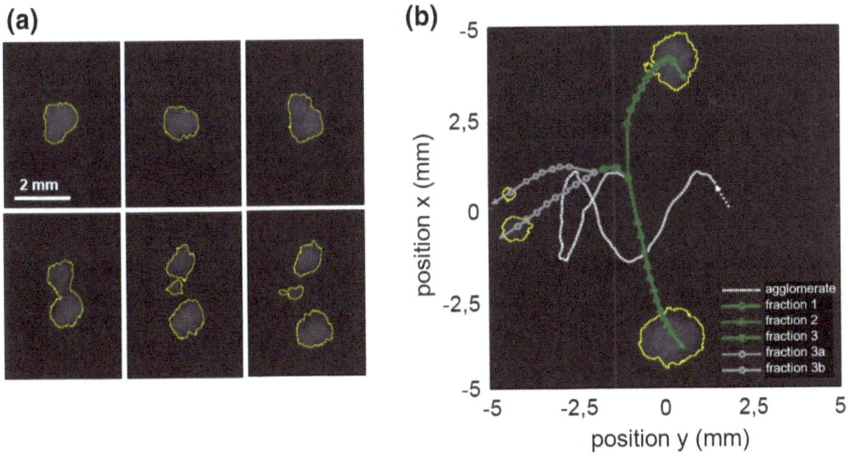

Fig. 17 Breakage process of a liquid bonded agglomerate caused by ultrasound induced drag forces (**a**) and the trajectories of the fragments (**b**); $d_{ag} = 1.9$ mm; $\zeta_{ag} = 0.63$; $\Delta t = 1$ ms (Reprinted with permission from [28].)

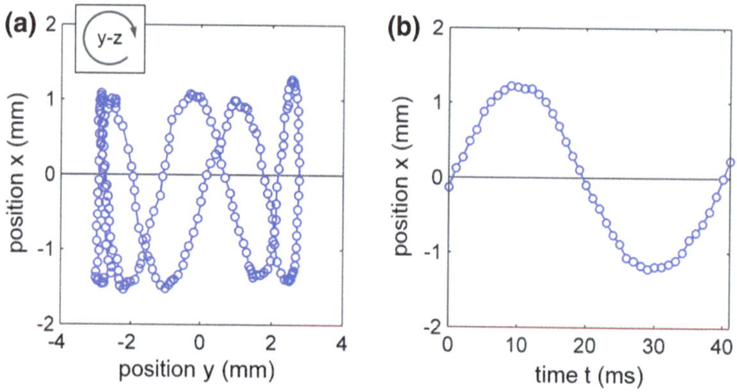

Fig. 18 Time dependent movement of a levitating agglomerate (**a**) (Reprinted with permission from [28].) and the area of agglomerate movement in the sound pressure field (**b**); $d_{ag} = 1.93$ mm; $\zeta_{ag} = 0.63$

The levitation level of the agglomerate examined is $x_0 = 12.3$ mm and is located below the central pressure node. According to Fig. 18b, in which the agglomerate position is plotted as a function of time, the agglomerate also continues to move vertically in the direction of the pressure anti-node, which is located below the zero position, and the pressure node above it. Due to the significantly lower frequency of the fluctuation movement, a levitated agglomerate experiences dynamic loads at every position along its trajectory.

The levitation level of a body x_0 depends above all on its mass. In contrast, the amplitude of the fluctuation movement x_{max} is largely determined by the geometry of the body. The more the shape of a body deviates from the ideal spherical sphere, the more pronounced is its fluctuation. This impression is confirmed by the graph shown in Fig. 19a. This shows the dependence of the fluctuation amplitude x_{max} on the circularity ζ_{ag} of various agglomerates. With increasing deviation of the agglomerates from the spherical shape, i.e. with increasing circularity, the magnitude of the amplitudes and the scattering of the individual values also increase. A comparison with the amplitudes of spherical glass particles (with a mass in the same range as the agglomerates used) with a circularity of $\zeta_{ag} > 0.99$ shows virtually no scattering and only extremely low fluctuation amplitudes.

For spherical bodies, the fluctuation movement and the scattering show minimum values. The amplitudes of the glass particles are at a median of $X_{max;50} = 99.8$ µm, while the span value $\Delta = 0.34$. For agglomerates with significantly higher circularities ($\zeta_{ag} = 0.5$–0.7) the median distribution increases to a value of $X_{max;50} = 1683.2$ µm with a span of $\Delta = 1.10$ (Fig. 19b).

A classification of the examined particles according to the ratio of the minimum and maximum Feret diameter can be seen as a parameter for the slenderness of a body. The lower the value, the slimmer or longer a particle makes it clear that slender bodies tend to have less pronounced fluctuations than blunt bodies of equal circularity.

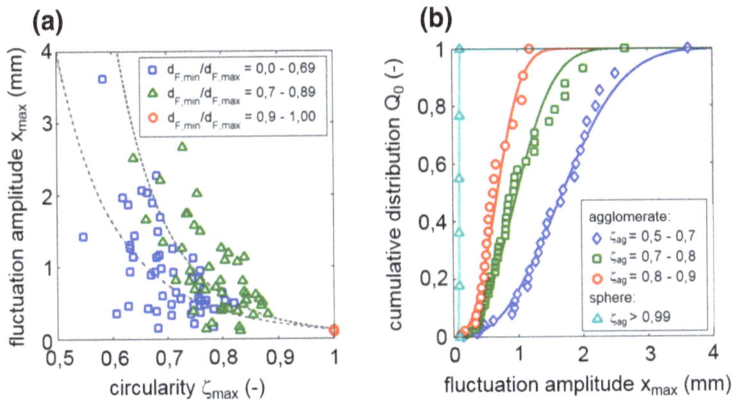

Fig. 19 Dependence of the amplitude of the fluctuation movement x_{max} of levitating agglomerates on circularity ζ_{ag} (**a**) and the distribution of x_{max} for different circularities (**b**) (Reprinted with permission from [28].)

Slender bodies tend to align in the sound field without rotating. Accordingly, the agglomerate movement decreases.

Breakage Probability as a Function of Mass Related Breakage Energy

To characterize the potential of the direct dispersion of agglomerated solid particles, a series of fracture events was recorded. For each parameter variation a total of more than one hundred fracture processes was analyzed. Figure 20 shows the required sound intensity I_b to cause a fracture for agglomerates with different primary particle distributions. Each point of the graph represents a fracture event, while the curve shows the development of the mean values of the fracture intensities.

Fig. 20 Dependence of the experimentally determined acoustic fracture intensity I_b of levitated glass agglomerates on the primary particle size $d_{p;50;3}$

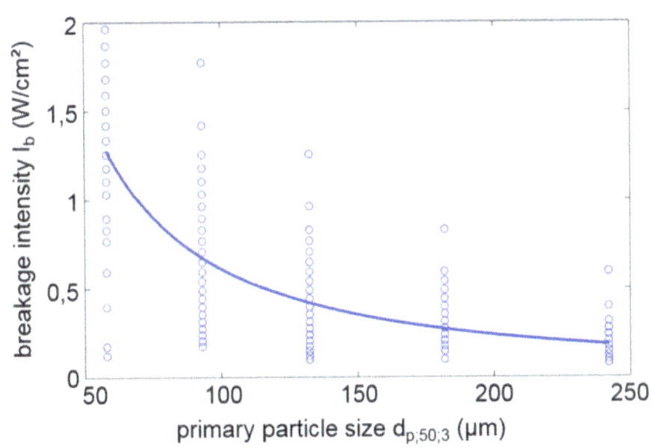

The average fracture intensity decreases significantly with increasing primary particle size. Since the fracture of an agglomerate is a complex process that is never the same even with the same agglomerate configuration, the values show a high standard deviation around the respective mean values. Thus, for an agglomerate structure of a certain shape and size, only the statistical probability of a fracture event depending on the applied ultrasound energy can be predicted. Based on this finding, an evaluation of the fracture probability ψ_{BP} depending on the mass-specific fracture energy W_m is carried out.

For an agglomerate of the mass m_{ag} the fracture energy is calculated by:

$$W_m = \int \frac{I_{lev} A_{ag}}{m_{ag}} dt. \tag{7}$$

Here, A_{ag} describes the cross-sectional area of the agglomerate in direction towards the sound propagation and I_{lev} describes the sound intensity at the point of levitation x_{lev}. The intensity I in a standing wave is not constant along the spatial propagation, but depends on the position in the sound field. The maximum values of the intensity in a resonant standing wave reach significantly higher values than the intensity I_0 of the incident wave.

Based on the sound intensity I_0 and the levitation position, which results from the recordings of the high-speed camera, the intensity I_{lev} can be derived from the diagram given in Fig. 21. This was derived from the distributions of sound pressure and sound velocity obtained from the CFD simulations of the resonant sound field. It shows the intensity of the resonant standing wave field I as a function of the position in the direction of sound propagation x and the intensity of the incident wave I_0.

The graph shows that the intensity in the resonant standing wave is significantly higher than the intensity of the incident wave. The difference is about one order of

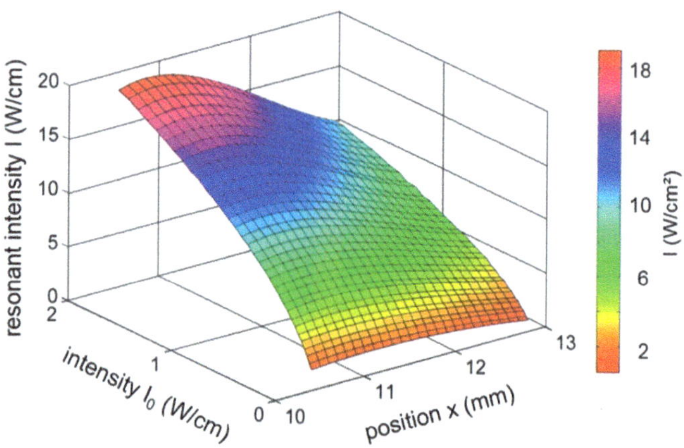

Fig. 21 Dependence of the resonant sound intensity I of a standing wave on the position x and the intensity of the incident wave I_0 (from simulation) (Reprinted with permission from [28].)

magnitude. The more the intensity I decreases, the further the position moves away from the pressure node at x = 12.9 mm.

The breakage probability ψ_{BP} derived from the sum of the individual experiments depending on the mass related fracture energy is given in Fig. 22. The breakage energy has been calculated over the duration of a period. The required quantities of mass and the flowed-on cross-sectional area A_{ag} of the individual agglomerates result from the images of the high-speed camera. The mass was calculated according to the following relationship:

$$m_{ag} = \frac{\pi}{4}\rho_P(1-\varepsilon)d_{ag}^2 \tag{8}$$

The porosity of the agglomerates was estimated according to [10] by ε = 0.58, while the agglomerate diameter d_{ag} was averaged from at least one hundred individual images of the high-speed camera.

The influence of the primary particle size on the fracture probability is given in Fig. 22a. Accordingly, the acoustic energy required to break an agglomerate with a certain probability decreases with increasing primary particle diameter. The larger the diameter of a particle, the lower the adhesive forces related to the weight force.

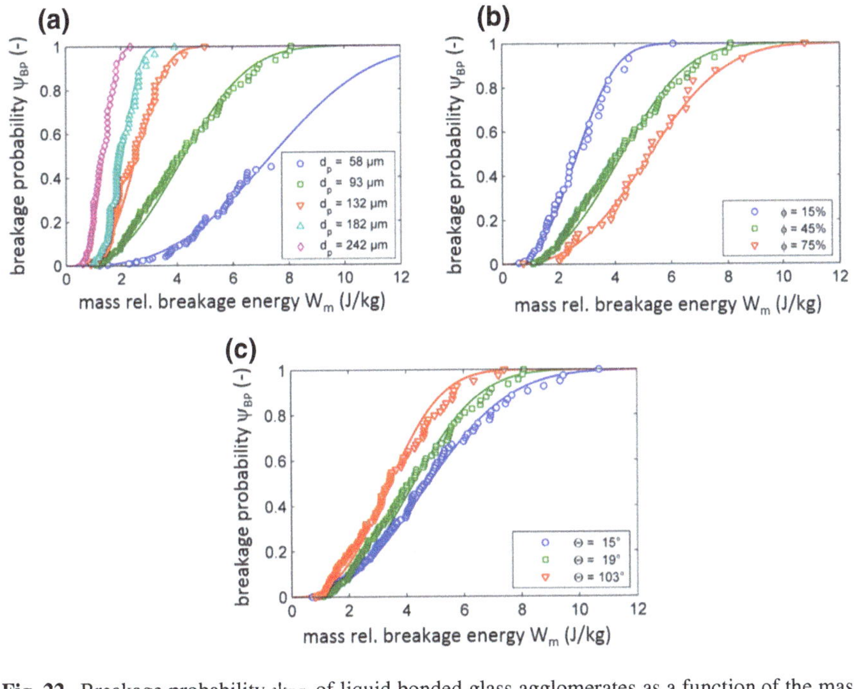

Fig. 22 Breakage probability ψ_{BP} of liquid bonded glass agglomerates as a function of the mass related breakage energy W_m for different primary particle diameters d_p (**a**), relative ambient humidity φ (**b**) and contact angle Θ (**c**) (Reprinted with permission from [28].)

Hence, the strength of the interparticulate bonds decrease. As a result, agglomerates of larger primary particles can be dispersed by a lower specific energy input.

Figure 12 showed that the adhesive force between two particles increases with increasing humidity of the surrounding gas atmosphere. Figure 22b shows the influence of the humidity on the breakage probability. Due to the increasing strength, the probability of breakage with constant energy input decreases with an increase in relative humidity. Accordingly, as the amount of humidity increases, the curves shift to higher fracture energies, although the degree of displacement varies. There are smaller differences between the breakage probabilities for $\varphi = 45\%$ and $\varphi = 75\%$ than between the gradients of $\varphi = 15\%$ and $\varphi = 45\%$. This is due to the larger gradients of the adhesive force $\partial F_{Adh}/\partial\varphi$ in the area of low humidity.

Figure 22c shows the influence of the contact angle on the breakage probability. As a result, the probability of breakage increases with increasing contact angle for the same specific energy input, i.e. particles with a hydrophobic surface ($\Theta = 103°$) can be dispersed with significantly lower acoustic energy densities than more hydrophilic particles. This observation is also in line with the trend illustrated in Fig. 12.

A comparison between the experimentally determined median values of the mass related breakage energies $W_{m;50}$ and the analytically determined curves of the related adhesion force F_{Adh}/F_g is shown in Fig. 23 for variation of the primary particle size d_p (a), the relative humidity φ (b) and the contact angle Θ (c). The vertical bars also indicate the values of $W_{m,10}$ and $W_{m,90}$ for a 10 and 90% breakage probability.

The analytical trends correspond to the experimentally determined points in all three diagrams with regard to the qualitative development. The fact that agglomerates of larger primary particles require a lower specific energy input to cause dispersion is supported by the curve of the adhesive force (Fig. 23a). This also applies to the case of variation in humidity shown in Fig. 23b. As the humidity increases, both the acoustic energy required for the breakage and the interparticulate adhesive force increase. When the contact angle changes (Fig. 23c), the specific energy requirement decreases with increasing contact angle, i.e. with an increasingly hydrophobic particle surface. If the critical contact angle Θ_{krit} is reached, no capillary is formed as the contact angle increases and the adhesive force results solely from the van der Waals force.

Agglomerate Dispersion (Indirect)

Principle Process

The mechanism of indirect dispersion has been investigated using the experimental setup schematically shown in Fig. 24. In this process, a certain quantity of solid bonded agglomerates, which are initially located at the bottom of the process chamber, is fluidized by the application of the high-intensity ultrasound field.

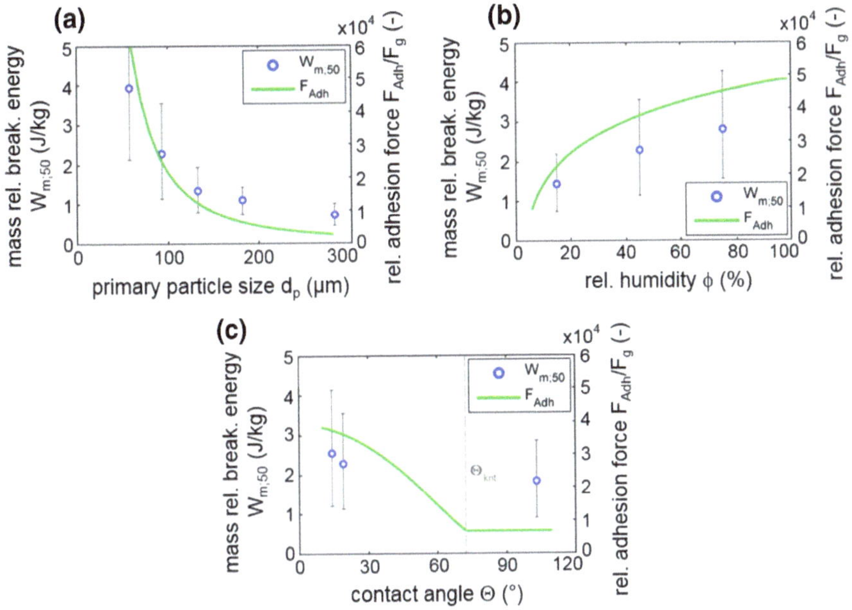

Fig. 23 Development of the mass related breakage energy $W_{m:50}$ and the relative adhesive force F_{Adh}/F_g as a function of the primary particle size d_p (**a**), the relative humidity φ (**b**) and the contact angle Θ (**c**) (Reprinted with permission from [28].)

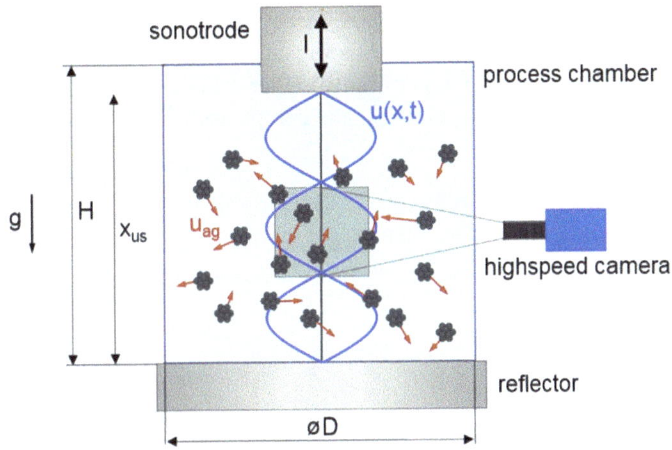

Fig. 24 Schematic representation of the experimental setup for the investigation of the indirect dispersion of solid bonded agglomerates

Table 4 Parameter variation for the experimental investigation of the indirect dispersion of solid bonded agglomerates

I_0 (W/cm²)	f (kHz)	x_{us} (mm)	d_{ag} (μm)	$d_{p;50,3}$ (μm)	t_{us} (s)	*Binder*
0.07–1.87	20	25.76 (3/2λ)	1000	8.3–182	2–30	CMC/PVP/SAC

The size distribution of the agglomerates is determined before the test and after sonication with the Malvern Morphologi G3 analytical microscope. In addition, the high-speed camera is used to derive collision speeds. Spherical particles whose weight corresponds to that of the agglomerates are used for this purpose. This is necessary because the agglomerates decompose and the resulting fragments prevent image analysis. Table 4 shows the parameters for the investigations of indirect dispersion.

Collision Velocity

The velocity distributions as a function of the sound intensity I_0 derived from the image evaluation are shown in Fig. 25a.

As the sound intensity increases, also the particle velocity increases. Thus, the median values in the range of the considered sound intensity change from 3.2 m/s to 4.3 m/s. For high sound intensities, the particles can reach speeds in the double-digit range. A comparison of the kinetic energy density E_m of the particles calculated from the median values (Fig. 25b) shows that the energy level for both the particle-particle impact (for which the relative velocity is assumed to be twice the median value) and for the collision with the wall is significantly higher than the static breakage energies of the generated agglomerates determined in compression tests.

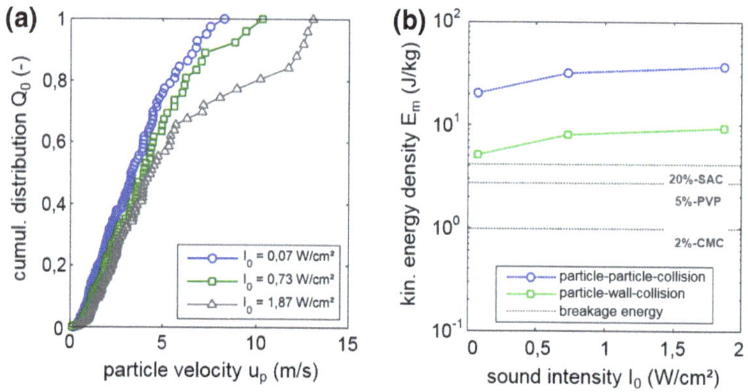

Fig. 25 Velocity distribution of the sound induced movement of glass particles as a function of the sound intensity I_0 (**a**) and the resulting kinetic energy density of a particle-particle or particle-wall collision (**b**)

Change of Particle Size Distribution

The findings of the previous chapter prove that there is also the potential to disperse agglomerates of higher strength by the application of ultrasound. Figure 26 shows that this process can also be carried out in experiments.

The images of the camera show the fluidized agglomerates and their fragments after 10 and 30 s of exposure to sound. While after 10 s some structures in the size range of the loaded agglomerates ($d_{ag} = 1000\ \mu m$) are still visible, significantly smaller fragments can be detected after 30 s. As a result, the progressive exposure to sound and the associated collision processes have led to a shift in particle size distribution towards smaller agglomerate sizes.

Figure 27 shows the density and cumulative distribution of an agglomerate collective which was irradiated over a period of 8 s. This shows a multimodal distribution, which in its basic form is exemplary for all recorded distributions. However, the values of the individual modal values differ, depending on the process conditions.

The modal values can be assigned to individual fracture forms. While the first three maximum values a, b and c represent a single primary particle, a two- and a three-particle configuration, the range d stands for larger fragments. The modal value at e shows the size of the loaded agglomerates.

The characteristic of the individual modal values can be seen as a measure of the degree of comminution of the agglomerates and is therefore dependent on the process parameters. Figure 28 illustrates the influence of different, relevant parameters on the cumulative distribution of sound exposed agglomerate collectives.

Figure 28a shows that an increase in sound intensity I_0 at constant test conditions, caused by a higher kinetic energy of the accelerated agglomerates, results in a shift of the curve to smaller particle sizes. Accordingly, a higher energy input leads to an increase in small and medium-sized fragments, while the proportion of large fragments or complete agglomerates is reduced. The dispersion effect is therefore correspondingly greater.

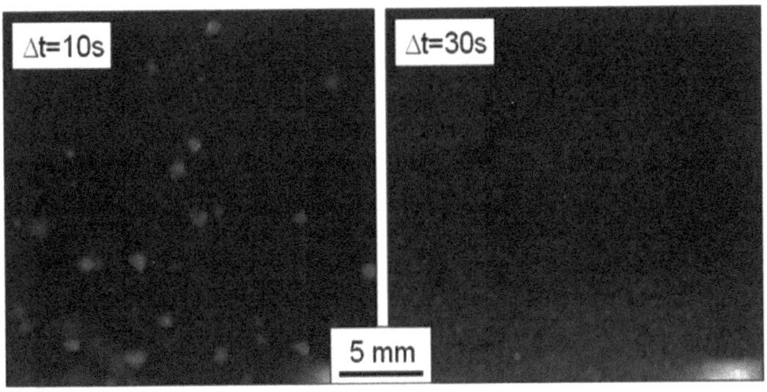

Fig. 26 Fluidized glass particle agglomerates in the ultrasonic standing wave field for sound exposure times of 10 s (left) and 30 s (right) at a sound intensity of $I_0 = 0.36\ W/cm^2$

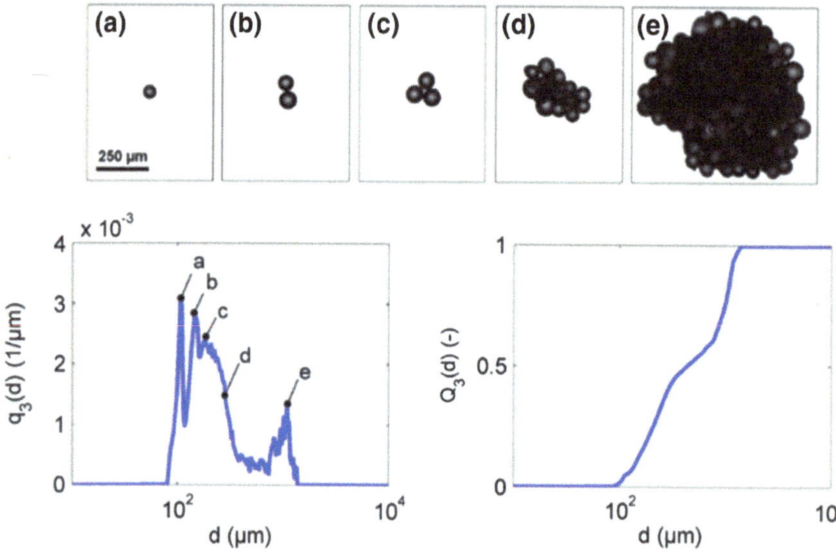

Fig. 27 Volume density and cumulative distribution of the size of PVP-bound agglomerates for a sound exposure duration of $\Delta t = 8$ s at a sound intensity of $I_0 = 0.36$ W/cm^2

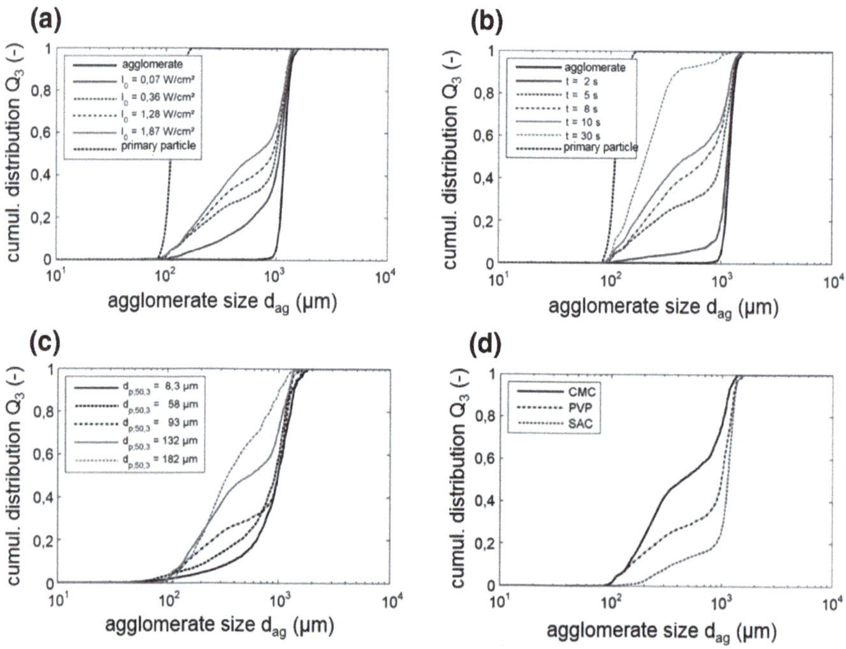

Fig. 28 Cumulative distributions of the fragment size of sound exposed agglomerate collectives by variation of the sound intensity I_0 (**a**), the exposure time Δt (**b**), the primary particle size $d_{50;3}$ (**c**) and the binder material (**d**)

The dispersion effect can also be increased by increasing the duration of application Δt (Fig. 28b). The longer a fluidized agglomerate collective is exposed to the sound field, the greater the number of collisions between the agglomerates or the resulting fragments. The number of fragments increases with each breakage, which in turn increases the number of collision events. For example, the experimental investigations after 30 s of exposure time show that almost no agglomerates can be found in the process chamber. The fragments are mostly small in size.

An increased agglomerate strength, as occurs when the primary particle size d_p is reduced, results in a reduced crushing effect, as can be seen in Fig. 28c. The higher the agglomerate strength, the more energy must be used to disperse agglomerates. Accordingly, a shift of the determined cumulative distributions to larger fragment diameters for agglomerates of smaller primary particles can be observed. The results of the binder variation shown in Fig. 28d also confirm that the SAC agglomerates with the highest strength have experienced the lowest degree of dispersion, while the dispersion of the CMC agglomerates with the lowest strength is the most advanced.

Dispersion of Single Agglomerates

In order to analyze the breakage behavior and breakage parameters of agglomerates during indirect dispersion in detail, experiments on individual agglomerates have been carried out in addition to dispersion experiments in the ultrasound field. The agglomerate collisions during indirect dispersion were represented as agglomerate-wall impact of individual agglomerates. The agglomerate has been accelerated by an air impulse from a nozzle. The impact velocity was regulated by changing the pressure and pulse duration of the air impulse. In order to keep the influence on the agglomerate as low as possible during launch, the agglomerate was levitated in a standing ultrasound wave in front of the nozzle. The flight and impact phases were recorded using a highspeed camera (see Fig. 29).

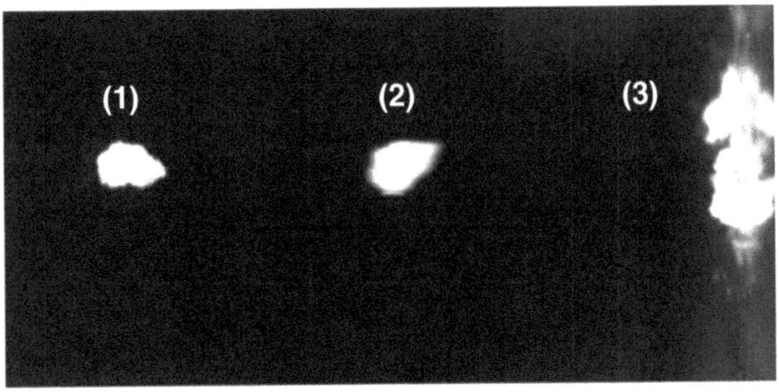

Fig. 29 High-speed camera image frequency of the agglomerate wall impact in a single experiment. At the beginning, in the levitated state (1), in the flight phase (2) and on impact (3)

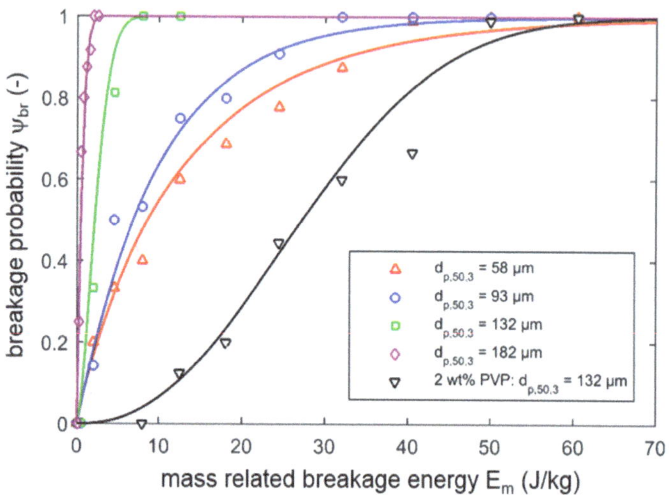

Fig. 30 Breakage probability of single liquid and solid bonded agglomerates at wall impact

Liquid bonded agglomerates with primary particle sizes of $d_{50.3} = 58–182$ µm at a relative humidity of $\phi = 50\%$ and solid bonded (PVP) agglomerates with a primary particle size of $d_{50.3} = 132$ µm were investigated in the test series. As for direct ultrasound dispersion, the images were evaluated with regard to the breakage probability of the agglomerates (see Fig. 30).

Due to the increasing agglomerate strength with decreasing primary particle sizes, an increase in the mass-related impact energy E_m required for breakage can be seen for liquid bonded agglomerates with decreasing primary particle size. The significantly increased strength of the solid-bonded agglomerate due to the PVP binder again shows a significant increase in impact energy in relation to the liquid bonded agglomerates. If these results are combined with the analysis of the collision velocities in section "Collision Velocity", conclusions can be made about the breakage behavior during indirect dispersion in the ultrasonic field. Figure 31 shows the median values of the kinetic energy density from the velocity analysis at different sound intensities and the median values of the breakage probability of the liquid and solid bonded agglomerates.

A comparison of the kinetic energy density calculated from the velocity median values shows that the energy level in the case of an agglomerate-wall impact for coarse primary particle sizes is significantly higher than the breakage energies of the moist agglomerates determined in the impact test. As particle sizes become smaller, the kinetic energy density outweighs the breakage energies. In the case of agglomerate-agglomerate impact, the kinetic energy density is significantly higher than the breakage energies of the liquid bonded agglomerates and is also slightly higher at higher sound intensities than the breakage energies of the solid bonded agglomerates. Thus, the kinetic energies achieved during indirect ultrasonic disper-

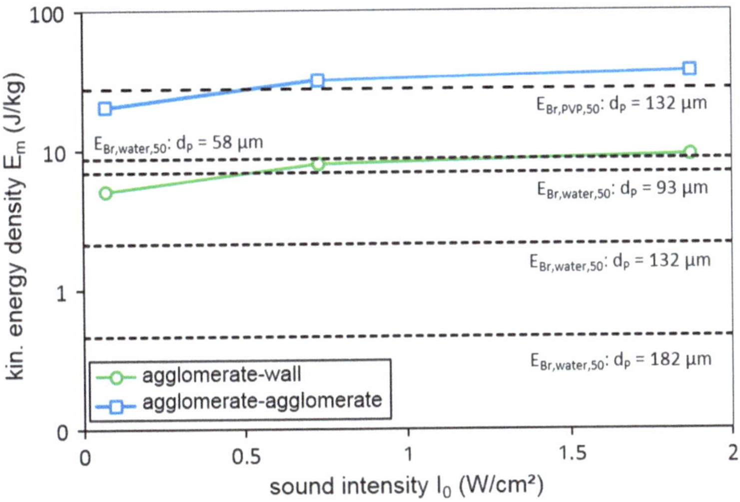

Fig. 31 Kinetic energy density of the agglomerate-wall or agglomerate-agglomerate collision and the median values of the fracture energies at different primary particle sizes

sion are high enough to reduce the agglomerate diameters not only by abrasion, but also by fracture events.

Numerical Investigation of the Breakage Behavior

Simulation Setup

The breakage behavior of agglomerates during indirect dispersion has been investigated numerically using DEM simulation. Spherical model agglomerates in three different sizes with diameters of approximately $d_{Agg} = 1.0–1.7$ mm were created. To take the particle size distribution into account, each agglomerate consists of 5 different primary particle size fractions which have been adapted to the experimentally measured distribution ($d_{P,50,3} = 132$ µm) of the primary particles in section "Particle System". Three agglomerates were randomly generated for each agglomerate diameter. The parameters of each agglomerate are shown in Table 5. To ensure comparability of the numerically generated agglomerates with the agglomerates used in the experiments in terms of porosity and particle density, the masses of the agglomerates were compared with each other (Fig. 32). The experimental agglomerates were prepared in a spray agglomeration as described in section "Particle System". The different size ranges of the experimentally investigated agglomerates compared to the numerically generated agglomerates can be explained by the fact that the mass of the smaller agglomerates could not be measured with sufficient accuracy due to the limitations of the balance. However, the trend line shows that the masses of the

Table 5 Agglomerate configuration for the investigation of the breakage behavior via DEM simulation (d_{agg}: agglomerate diameter; n_p: number of particles; c_n: coordination number; n_b: number of bonds)

Parameter	Agglomerate								
	Type I			Type II			Type III		
d_{Agg} (μm)	1087	1070	1070	1267	1280	1280	1693	1680	1683
n_P (–)	354	333	334	620	591	584	1457	1412	1397
c_N (–)	4.6	4.5	4.5	4.7	4.7	4.7	5.0	5.0	4.9
n_B (–)	906	832	863	1645	1558	1569	4064	3924	3912

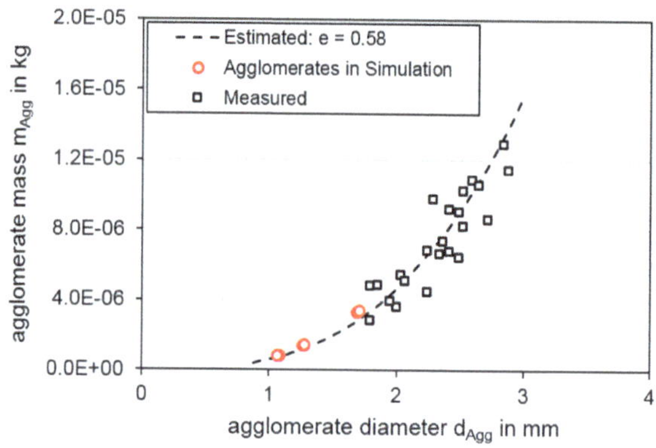

Fig. 32 Comparison of masses of the experimentally and numerically used agglomerates

numerically generated agglomerates are in accordance with the agglomerates used experimentally. The porosity of the agglomerates was calculated both in the experiment and in the simulation with $\varepsilon \approx 0.58$ and corresponds to the value estimated in the literature in section "Particle System".

Particle interaction on contact is described by using the Hertz-Mindlin model. SiO_2 was considered as material, as used in the experimental investigations for the particles. The material parameters set are listed in Table 6. Since only solid bonded agglomerates were investigated experimentally for indirect ultrasonic dispersion, a purely elastic stiff-bond model is used to represent the particle bonds, as used in the DEM simulations in section "Direct Agglomerate Dispersion". The bonds between the particles are assumed to be ideally cylindrical. Therefore, in accordance with [20] bond material parameters, such as tensile strength, cannot be used directly to reproduce real material behavior. Experimental comparisons are necessary in order to be able to set the individual parameters according to the breakage behavior. Therefore, the bond parameters were estimated with help of experimental comparisons. Only PVP was considered as a binder. The binding parameters are also shown in Table 6. The particle bonds were generated between directly adjacent particles. This corresponds to microscopic investigations of the experimentally used agglomerates. An example of a generated agglomerate and its binding structure is also shown in the table.

Table 6 Particle and bond parameters used in the DEM simulation (Reprinted with permission from [29].)

Particle parameters		Bond parameters	
Parameter	Value	Parameter	Value
Material	SiO_2	Bond model	Elastic stiff bonds
Contact model	Hertz-Mindlin	E (N/m^2)	1.2×10^9
ρ (kg/m^3)	2650	σ (N/m^2)	2.0×10^7
E (N/m^2)	6.9×10^{10}	τ (N/m^2)	2.0×10^7
e (–)	0.94	bond size factor r (–)	0.15
μ (–)	0.2	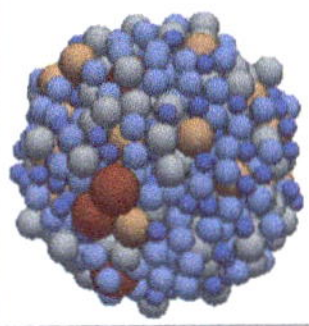	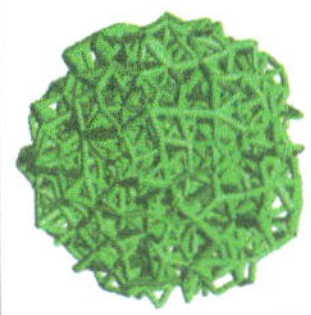
μ_r (–)	0.05		
ν (–)	0.17		

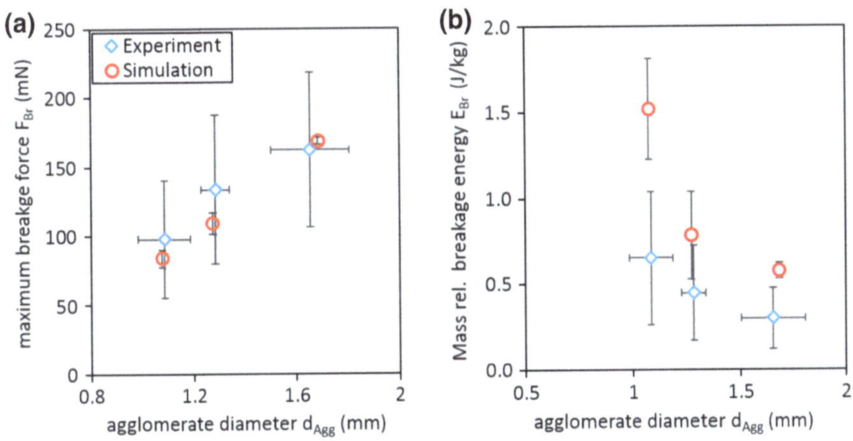

Fig. 33 Comparison of the breaking load (**a**) and the mass-related breaking energy (**b**) from the compression test of solid-bonded agglomerates in simulation and experiment

To validate the agglomerate and binding models, pressure tests were carried out both in the simulation and in the experiment. Experimentally, agglomerate sizes comparable with the simulation were used, which were agglomerated in a fluidized bed. The results are evaluated in terms of breaking load and breaking energy (Fig. 33).

The physical behavior of the agglomerates used experimentally can be reproduced in the considered agglomerate size range in the simulation. While the breaking load also agrees in its absolute values, the values of the breaking energy in the simulation are somewhat higher. In simulation and experiment, however, an exponentially decreasing course of the fracture energy can be determined. Images of the compression test and the viewing of the simulation data show that in both cases the agglomerate breaks along the pressure direction. Therefore, in experiments and simulations tensile forces are generated perpendicular to the compression direction, which ultimately leads to breakage. The fracture process itself is also based on identical mechanisms and alternative mechanisms such as shear can be excluded.

Numerical Investigation of the Impact Breakage

The model agglomerates validated by the compression test were subjected to an impact test (analogous to section "Dispersion of Single Agglomerates") to describe the breakage behavior during indirect dispersion. To evaluate the breakage, the results were evaluated with regard to the fragment size distribution in relation to the relative kinetic impact energy. For comparison, impact tests were also carried out in the experiment (section "Dispersion of Single Agglomerates") and the size distribution of the fragments was determined using an analytical microscope (Malvern Morphologi G3). Representative cumulative size distributions of the fragments from simulation and experiment are compared in Fig. 34.

The cumulative size distributions derived from the simulation and experiment exhibit similar behavior. With increasing impact speed the curves move upwards, which can be explained by the increasing proportion of smaller fragments. In addition, a jump between fine and coarse fragment sizes in the experiment as well as in the simulation at lower impact velocities is to be seen. Video analyses of the fracture process and the review of the simulation results show that this is still a very low degree of dispersion in which only small fragments flake off from the main agglomerate. At higher impact velocities, a significantly higher variation of fragment sizes

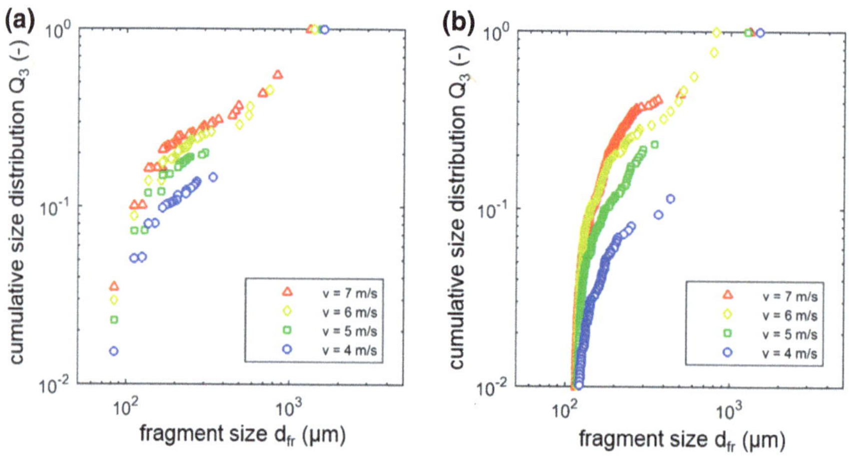

Fig. 34 Cumulative distribution of the fragments after agglomerate-wall impact in simulation (**a**) and experiment (**b**)

in the coarser fragment area can be observed, which indicates a stronger degree of dispersion of the agglomerate. In terms of quality, the results from simulation and experiment agree. To quantitatively evaluate the fragment size distributions from simulation and experiment, the results are described using the Gates-Gaudin-Schumann distribution:

$$P = \left(\frac{x}{x_0}\right)^m \tag{9}$$

The fragment fraction P of fragment size x is described using the size module x_0 and the distribution module m. The size module describes the maximum fragment size after fracture, the distribution module the width of the fragment distribution. The Gates-Gaudin-Schumann distribution requires a linear course of the fragment size distribution for double logarithmic application of the cumulative curve. The validity of this assumption for agglomerates after impact was simulatively confirmed by [21] in accordance with experimental results from [22]. Figure 34 shows this linear course for the coarser fragment size range and thus allows the application of the Gates-Gaudin-Schumann distribution. The bending of the curve in the fine fragment size range results from the fact that primary particles are already present there which no longer follow the typical distribution of the fragments. Therefore, only the coarse fragment size range was evaluated, in which primary particles have no influence on the distribution. Figure 35 shows the distribution module for solid-bonded agglomerates from experiment and simulation.

The distribution module decreases with increasing impact energy. This indicates a broader fragment size distribution resulting from the higher degree of fragmentation at higher impact velocities. The simulation values asymptotically approach a minimum value, which represents the fragment size distribution at the maximum degree of fragmentation. This asymptotic approach can only be seen to some extent in the experimental results. In general, however, the approximation to a minimum value can also be assumed here, since a maximum degree of fragmentation is also

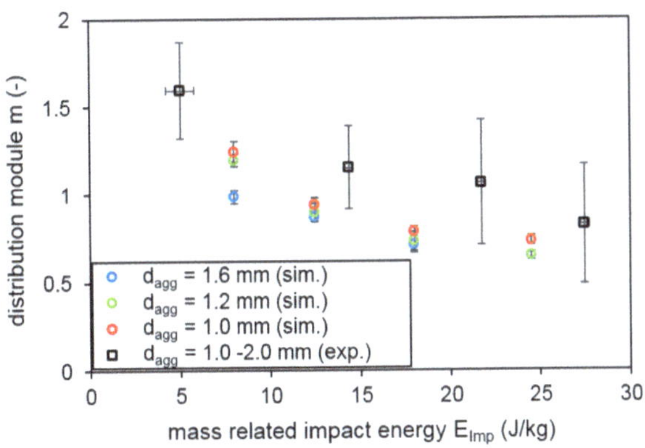

Fig. 35 Distribution module m of the fragment size distributions after agglomerate impact derived from simulation and experiment

achieved experimentally from a certain impact velocity. Furthermore, regarding the standard deviations of the mean values, the wide dispersion of the measurement results becomes apparent. This shows that a significantly higher number of experiments is presumably necessary in order to achieve an asymptotic approximation also in the experimental results. In absolute terms, the simulation results are just slightly below the values determined experimentally. In summary, the results from Figs. 34 and 35 show good agreement, so that the change in the degree of fragmentation of the experimental agglomerates can be described with the model agglomerates used in simulation.

Detachment of Wall Adhering Particles

Experimental Setup

The investigations for the acoustic detachment of solid particles adhering to walls have been carried out according to the procedure illustrated in Fig. 36. In a first step, a volume of 19 mm^3 of glass particles was dispersed onto a substrate using the Malvern Morphology G3 dispersion unit. This enables a defined quantity of particles to be homogeneously applied to a surface by pressure shock. Slides made of soda-lime glass whose material properties correspond approximately to those of the glass particles (72.2% SiO_2; 14.3% Na_2O; 6.4% CaO) were used as substrate.

The morphology G3 also has been utilized in the second step to measure the number and size distributions of the separated particles. The measured samples were subsequently subjected to ultrasound. For this purpose, the sound generation structure

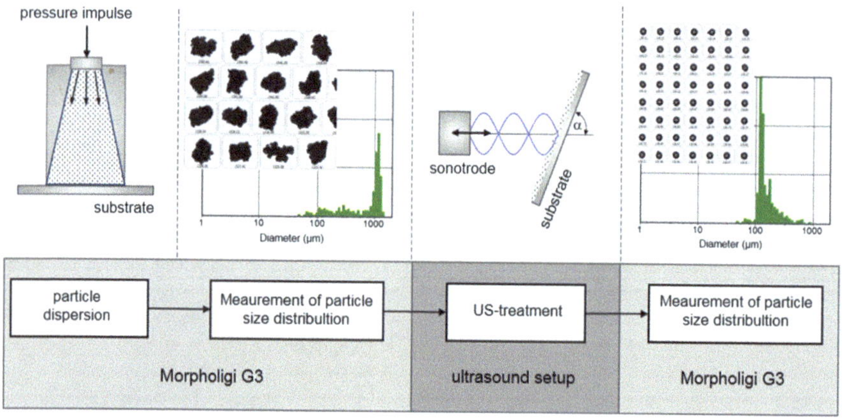

Fig. 36 Schematic representation of the experimental procedure for the investigation of ultrasound induced detachment of wall adhering particles [30]

Table 7 Parameter variation of the experimental investigations for sound-induced detachment of particles adhering to walls [30]

I_0 (W/cm^2)	f (kHz)	x_{us} (mm)	α (°)	t_{us} (s)	$d_{p;50,3}$ (μm)	φ (%)	Θ (°)
0.07–1.87	20; 30	25.76	0–90	2–30	58	35–85	14.5–103.4

shown in Fig. 4 was used in a horizontal arrangement, with the particle-laden substrate acting as a reflector. After the acoustic irradiation, the number and size distribution of the remaining particles on the surface were again recorded. The particle separation efficiency η_i of different size fractions i can be determined from the ratio of the particle number before $n_{i,0}$ and after acoustic irradiation n_i via the following correlation:

$$\eta_i = 1 - \frac{n_i}{n_{0,i}}. \tag{10}$$

The experimental investigations derive the detachment efficiency by varying the influencing parameters shown in Table 7.

The detachment efficiency of different size classes of the distributions determined with the Morphologi G3 allows a conclusion about the influence of the primary particle size d_p.

Representation of the Partially Developed Standing Wave Field

A variation of the angle of incidence α has a significant influence on the type of acoustic field generated, since a completely formed standing wave only occurs for an angle of incidence of 90°. Figure 37 shows the influence on the resulting acoustic field.

The left-hand graphics show the dimensionless distribution of sound radiation pressure and sound velocity for a continuous wave (sound incidence angle of 0°). Here, the input amplitudes are not amplified and the courses of pressure and velocity have the same phase position. In contrast, a sound incidence angle of 45° (middle) results in the presence of elements of a completely formed standing wave (right). Thus the characteristic nodes and anti-nodes exist in the courses. At the same time, the shape of the field is clearly deformed and the resonant amplification of the field sizes can only be seen in the area of the reflector. Here, the incident wave is amplified by a factor of 2.5. In the case of a fully developed standing wave, this value for the same boundary conditions is approximately 5.

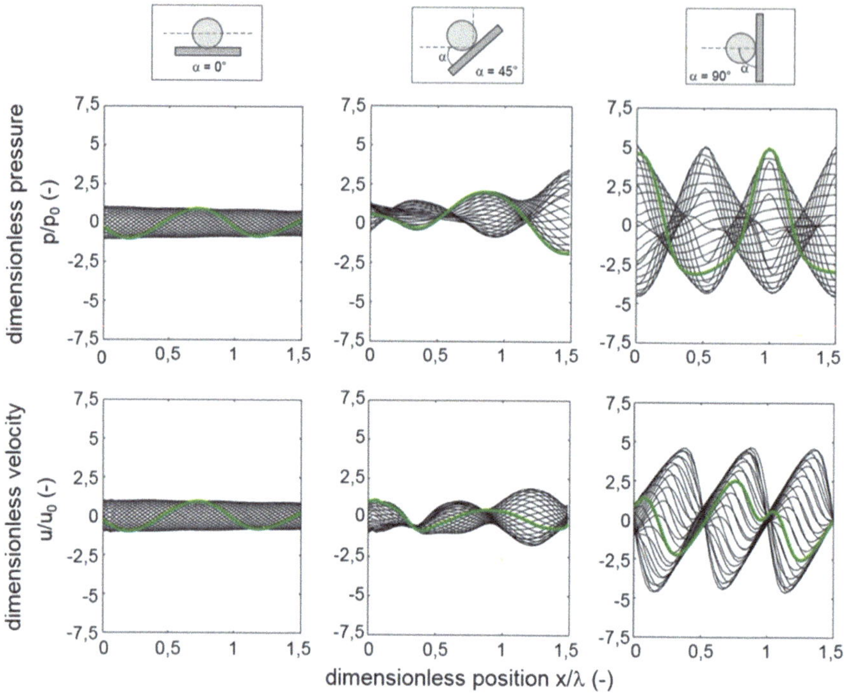

Fig. 37 Sound pressure and sound velocity distribution of a sound flow between the sonotrode and a wall (reflector) for a sound incidence angle of $\alpha = 0°$ (left), $\alpha = 45°$ (centre) and $\alpha = 90°$ (right) with $u_0 = 8$ m/s [30]

Detachment Efficiency

The wall adhering particles are located at a defined position on the surface of the substrate, cannot move freely in the sound field and therefore do not reach the areas of the highest energy densities, as is the case for the levitating agglomerates. In addition, the adhesive forces acting on a plate-sphere arrangement are higher than in the case of two spherical particles. For capillary and van der Waals bonds, for example, the forces can be assumed to be twice as strong. Figure 38 shows that it is still possible to overcome the adhesive forces.

The images of a particle-laden glass substrate before and after exposure to sound illustrate a significant reduction in the amount of particles due to the exposure. The intensity during the test was $I_0 = 1.87$ W/cm^2, the duration of sound $t = 5$ s and the angle of incidence $\alpha = 45°$. With this configuration acoustic drag forces were applied, which led to a separation of 90% of the particles.

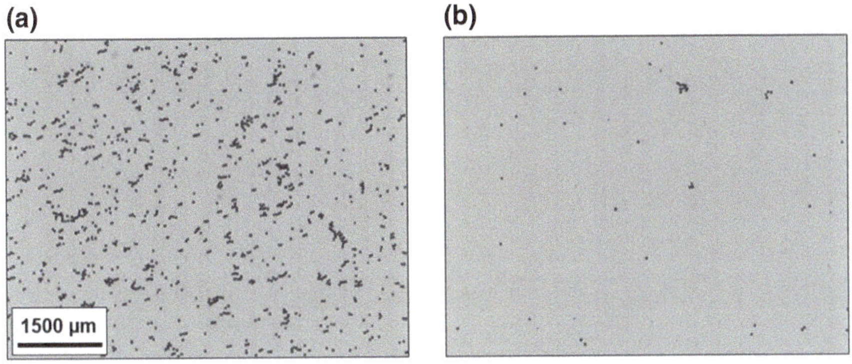

Fig. 38 Particle-laden surface before (**a**) and after (**b**) exposure to high-intensity ultrasound; f = 20 kHz; $I_0 = 1.87$ W/cm²; α = 45°

Variation of the Sound Incident Angle

The observation of the angle-dependent sound field (Fig. 37) allows the assumption that the angle of incidence has a clear influence on the effectiveness of particle separation. Accordingly, Fig. 39 shows the detachment efficiency η in varying the angle α.

The trend shows that the effectiveness is minimal for a standing wave (α = 90°) as well as for a parallel wall flow (α = 0°) and increases to a maximum value for angles between 30° and 60°. In the case of a frequency of f = 20 kHz at an intensity of $I_0 = 0.36$ W/cm², the efficiency of particle detachment can be increased from just over 30% to almost 70% by changing the angle. Figure 40 shows the curves of the maximum acoustic forces F_N and F_Q derived from the CFD simulation, which aim to remove a wall-sticking particle, depending on the angle of incidence.

Fig. 39 Detachment efficiency for different sound incident angles at f = 20 kHz and $I_0 = 0.36$ W/cm² [30]

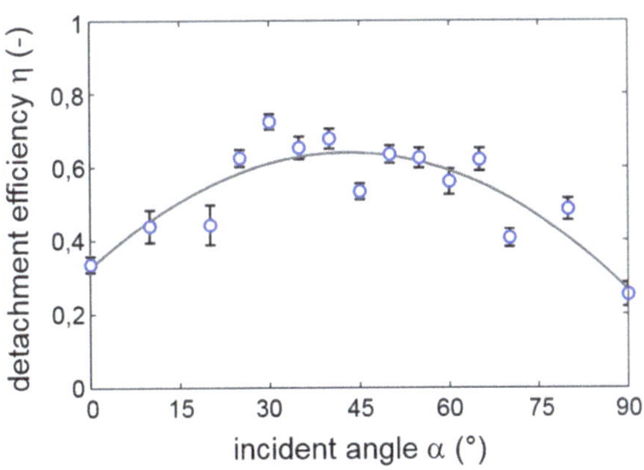

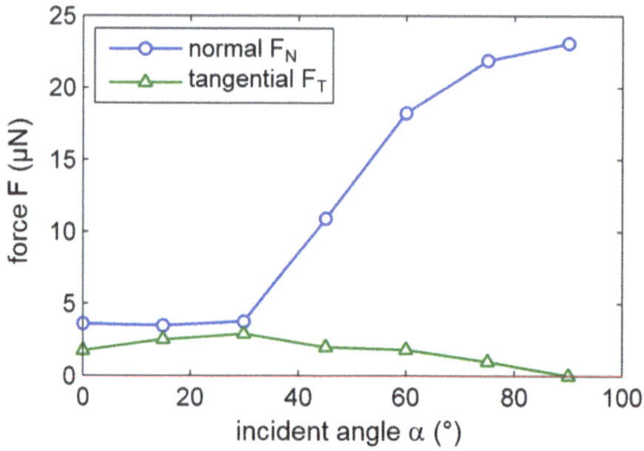

Fig. 40 Course of the amplitudes of normal force F_N and shear force F_N on a particle adhering to a wall ($d_p = 80\ \mu m$) in dependence of the sound incident angle α [30]

The tangential force reaches a maximum in the range in which most particles were detached in the experiment, at $\alpha = 45°$, while the normal force of $\alpha = 0°–90°$ increases steadily. The level of the normal force is significantly higher than the tangential force, which has no significant effect on the detachment efficiency. Accordingly, it can be assumed that the particle detachment is essentially influenced due to the effect of the tangential forces.

Variation of Sound Intensity

The development of the detachment efficiency at varying sound intensity I_0 is shown in Fig. 41 for sound incidence angles of 45° and 90°. In the left diagram the results are plotted for a frequency of 20 kHz and in the right diagram for a frequency of 30 kHz.

All curves show an increase in the detachment efficiency with increasing sound intensity. The level for a sound incidence angle of 45° is, as already observed in the previous chapter, above that of the curves for an angle of 90°. In addition, an increase in frequency leads to a significant reduction in particle removal. One reason for this is the lower amplitude, which can be transmitted at higher frequencies. On the other hand, the inertia of the particles increases with increasing frequency. If one considers an oscillation of constant amplitude, the maximum velocity gradients and thus also the inertia force increases with increased frequency. A wall adhering particle can thus follow the load changes of the oscillating loads with higher frequency ever more badly.

In contrast to the processes shown so far, which take into account the entire size spectrum of the separated particles, Fig. 42 shows the separation efficiency as a function of the sound intensity for different fractions of the distribution ($f = 20$ kHz and $\alpha = 45°$).

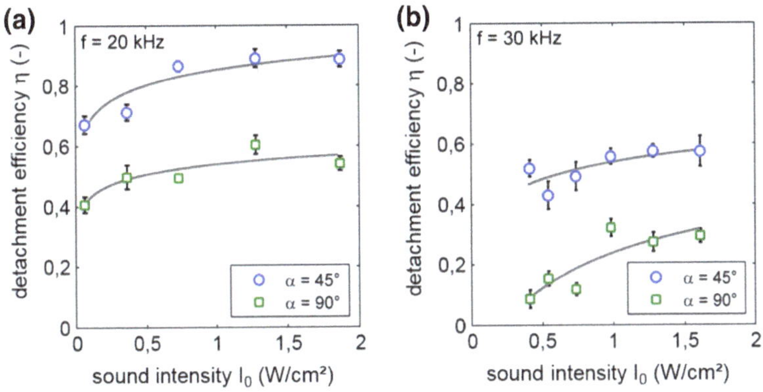

Fig. 41 Detachment efficiency η for different sound intensities I_0 at a frequency f of 20 kHz (**a**) and 30 kHz (**b**)

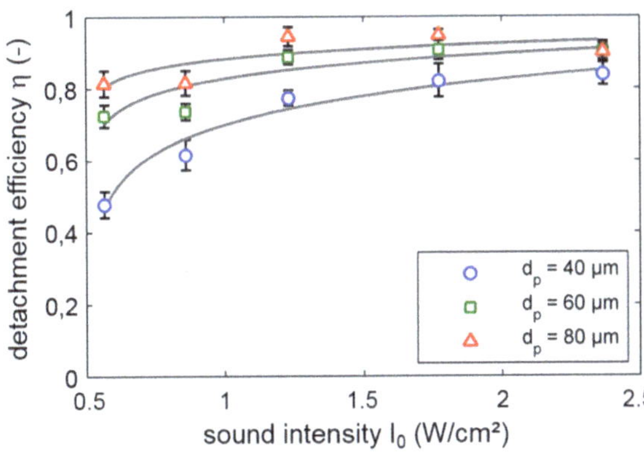

Fig. 42 Detachment efficiency η for different sound intensities I_0 at a frequency of f = 20 kHz[2], a contact angle Θ = 19.3° and an exposure time for t = 5 s

The size of the particles has a significant effect on the detachment. Detachment efficiencies of over 90% for particles with an average size of $d_p = 80\ \mu m$ are compared with values of about 80% for the fraction with a diameter of $d_p = 40\ \mu m$. This is due to the fact of increasing adhesive forces in relation to the weight force for smaller particle dimensions.

Variation of Relative Humidity

For a sound incidence angle of 90° and a sound intensity of 0.07 W/cm^2, the influence of the relative humidity ϕ on the detachment efficiency η is determined in Fig. 43. The frequency of the sound field is 20 kHz.

From the observations of the adhesive force (Fig. 12) the increase of the relative humidity leads to an increase of the adhesion force. Accordingly, the experimentally determined course shows a reduced number of detached particles with increasing

Fig. 43 Detachment efficiency η as a function of relative humidity ϕ at a frequency of f = 20 kHz, a contact angle Θ = 19.3° and an exposure time for t = 5 s [30]

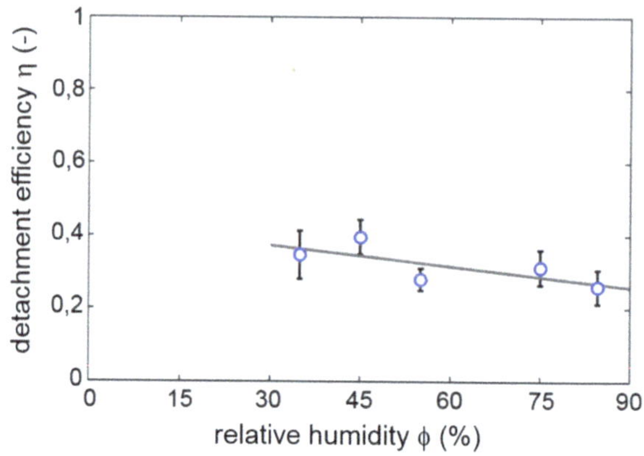

humidity and thus a lower detachment efficiency as a consequence of the increased adhesion. For an increase in humidity from 35 to 85%, η drops from around 40% to well below 30%.

Variation of the Contact Angle

The previous considerations of particle-particle contact behavior in agglomerates show a clear influence of the contact angle Θ on the dispersibility of the structures (section "Agglomerate Movement in the Standing Ultrasound Field"). Figure 44 shows the detachment efficiency of particle separation for different contact angles.

The larger the contact angle, i.e. the more hydrophobic the surfaces, the higher the number of detached particles. While the detachment efficiency reaches values of more than 80% for the hydrophobic particles, the efficiency in the case of hydrophilic particles is significantly below 60%. The reason is the change of the geometry of

Fig. 44 Detachment efficiency η for different contact angles Θ at a frequency of f = 20 kHz, a contact angle Θ = 19.3° and an exposure time for t = 5 s [30]

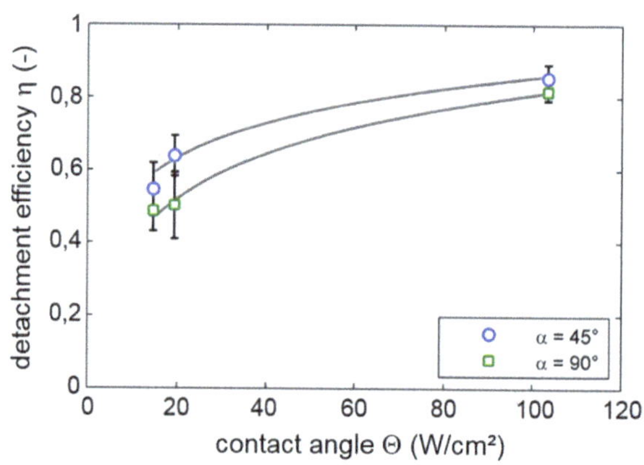

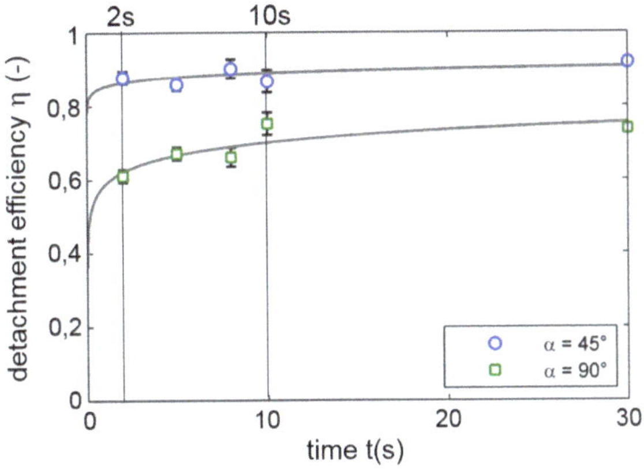

Fig. 45 Detachment efficiency h for different sound exposure times t at a frequency of f = 20 kHz and a contact angle Θ = 19.3°

the liquid bridge with change of the contact angle. The smaller the angle the more pronounced the curvature of the meniscus of the liquid surface. As a result, the negative pressure in the condensed liquid and thus also the adhesive force increases.

Variation of Sound Exposure Time

Removal of the particles is not an instantaneous process that is independent of the duration of the sound irradiation. Therefore, the influence of the duration of the sound effect t on the detachment efficiency has been derived as illustrated in Fig. 45.

The level of the curves changes only insignificantly over the sound exposure time between 10 s and 30 s. The detachment efficiency for an angle of incidence of 45° over the entire range is constant at values of $\eta = 90\%$ and also for an angle of 90° only minor changes (>10%) can be observed within the first 10 s. Accordingly, it can be assumed that the detachment is a very fast process. Due to the relatively high number of load changes due to the alternating stress, it is also assumed that particles whose adhesive forces exceed the acoustic resistance forces can also be detached. This can also explain the fact that even after a duration of several seconds, there are still minor changes in the removal efficiency.

Tensile Tests on Agglomerated Particles

Experimental Analysis of the Tensile Strength

The numerical investigation of ultrasonic dispersion has shown that, in addition to compressive forces, tensile forces also affect the particle bonds. Thus knowledge of the tensile strength of the agglomerates and their interparticulate bonds is required.

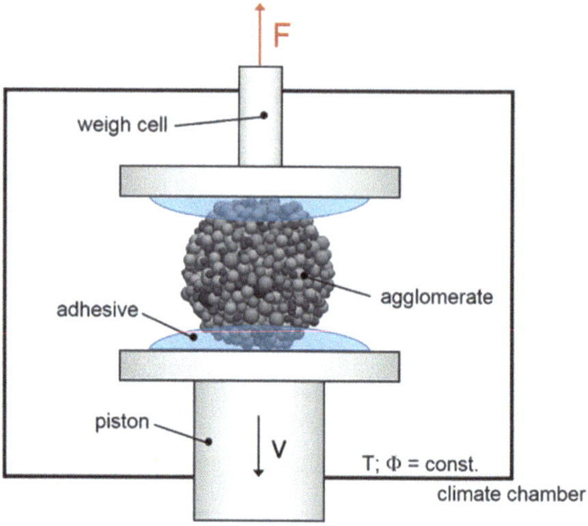

Fig. 46 Experimental setup for the measurement of the tensile strength of single agglomerates (Reprinted with permission from [29].)

Many studies have already attempted to determine the tensile strength of agglomerates indirectly by means of pressure tests. As discussed in [23], many experimental results have shown that no simple correlation between tensile and compressive strength is possible. Therefore, direct tensile tests were performed here to determine the tensile strength of individual agglomerates. A test setup based on the work from [24] was used. The structure consists of a weigh cell on which a frame is mounted, which places the plate below the load cell for weighing (Fig. 46).

If a downward pull comes to the plate, it can be measured as mass with the weigh cell. This mass can then be converted into a tensile force. Below the plate is another plate, which is mounted on a piston. The agglomerate was fixed to both plates with a fast-curing UV adhesive. Pulling forces have been applied onto the agglomerate by moving the piston downwards with a speed of 1 μm/s. Wet and solid bonded agglomerates with diameters of approx. $d_{agg} = 1–2$ mm were investigated. The fracture surface was determined using microscope and camera images. The breakage probability as a function of tensile strength and relative breakage energy is shown in Fig. 47.

Tensile strength as well as breakage energy exhibit trends towards smaller values with increasing fracture area because with increasing agglomerate size the probability of defects or larger pores increases, whereby the breaking force decreases relative to the fracture surface. Similar trends can also be seen in compressive tests with identical agglomerates. Figure 48a shows a direct comparison between the tensile strength from the tensile test and the compressive strength from the compression test.

The results in Fig. 48a illustrates on average higher values for the compression test. This is in accordance with [23]. With respect to the fracture surface, both compression and tensile tests show a similar decreasing trend with increasing fracture surface. In Fig. 48b–d correlations for the determination of tensile strength from compression

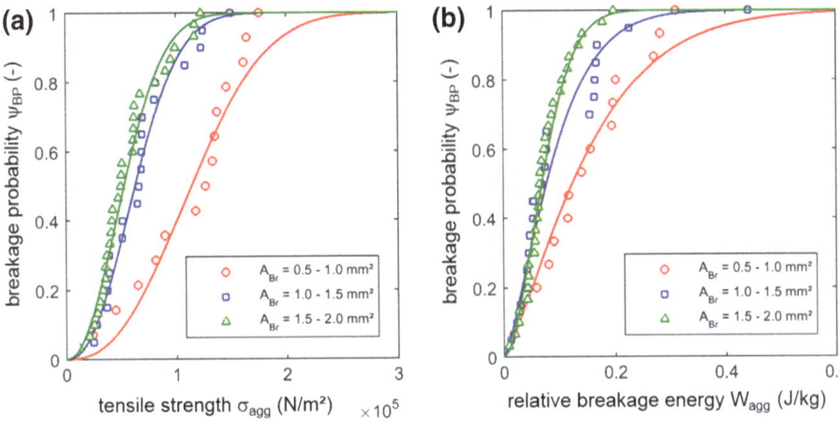

Fig. 47 Breakage probability of solid bonded agglomerates in the tensile test as a function of tensile strength (**a**) and relative breakage energy (**b**) (Reprinted with permission from [29].)

tests based on the fracture of spherical bodies were examined. The correlations are based on the approach from [26]:

$$\sigma_{Agg} = b\frac{F_{Compr}}{A_{Compr}}. \tag{11}$$

The compressive strength calculated from the compression test is multiplied by the factor b that ranges between 0.4 [25] and 0.7 [26], depending on the material and contact surface between plate and particle. The validity of these factors is discussed in [25]. The experimental agreement of Hiramatsu's calculations is based on the fact that the tensile stress at the surface can be seen almost constant for large areas of the contact surface to the plate and in many cases the fracture is initiated at the surface and not internally. Therefore the tensile strength at the surface is described more than the tensile strength within the agglomerate. According to Shipway, the factor of 0.4 should be closer to the actual tensile strength. However, it is also acknowledged here that this factor is only valid in certain cases and that the results of the indirect measurement of tensile strength must continue to be treated with caution. In (c) and (d) the compressive strength has therefore been subjected to the limits specified by Hiramatsu and Shipway. At a value of 0.4 in (c), the tensile strength determined from the compression test is significantly lower than the actual tensile strength from the tensile test. A value of 0.7 in (d) shows good agreement in the investigations. This is also confirmed by the determination of the actual value of b = 0.73 in (b). The value was determined by adjusting the trend lines, since the tensile strength of the agglomerates is dependent on the fracture surface and therefore simple mean values cannot be used. Furthermore an identical distribution of the agglomerate sizes investigated cannot be assumed in both investigations. The factor b must be considered here as a material parameter which varies depending on the type of agglomerate. The

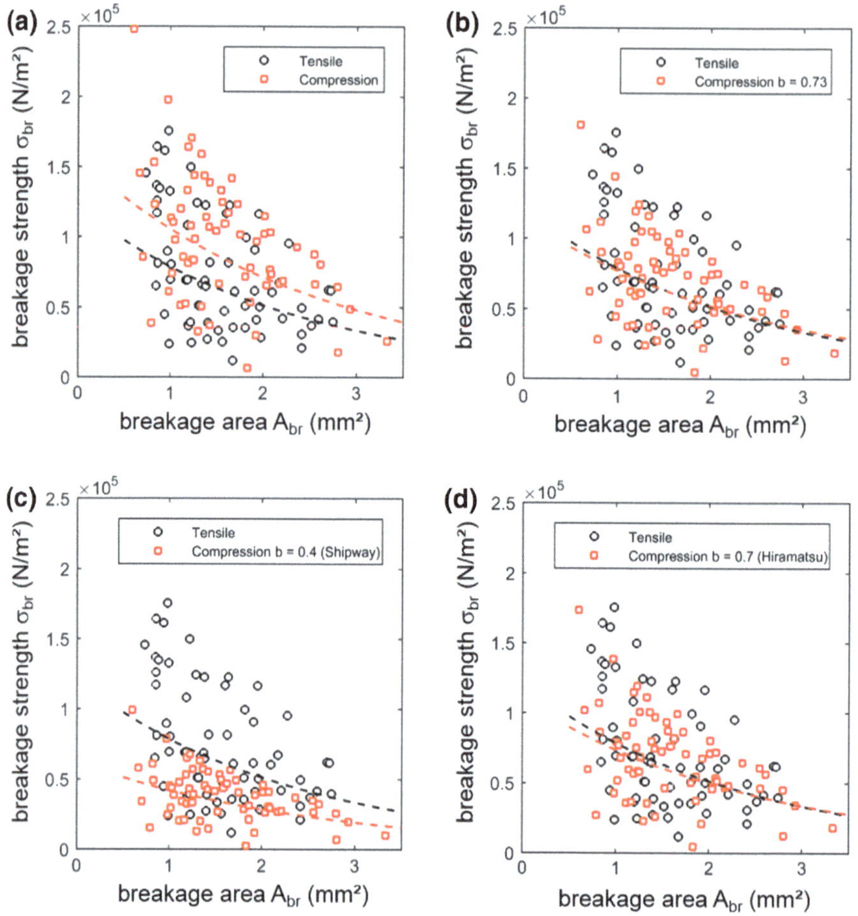

Fig. 48 Breakage strength from tensile tests and compression tests (**a**), with determined correction factor b = 0.73 (**b**), with b = 0.4 [25] (**c**) and b = 0.7 [26] (**d**) (Reprinted with permission from [29].)

agreement with the Hiramatsu data should therefore not be regarded as confirmation of the model for solid-bonded agglomerates.

Further comparisons between tensile test and compression test are made in Fig. 49. For this purpose, the relative breakage energy was applied above the maximum breakage force and fracture surface.

The comparison in Fig. 49a shows that in both tensile and compression tests the mass-related fracture energy decreases with increasing fracture area, i.e. increasing agglomerate size. This behavior is known from pressure experiments and can be explained, as in Fig. 48, by the increase in defects and larger pores. It is noticeable that the required breakage energy in the tensile test is significantly lower than in the compression test. In Fig. 49b the fracture energy is plotted over the fracture

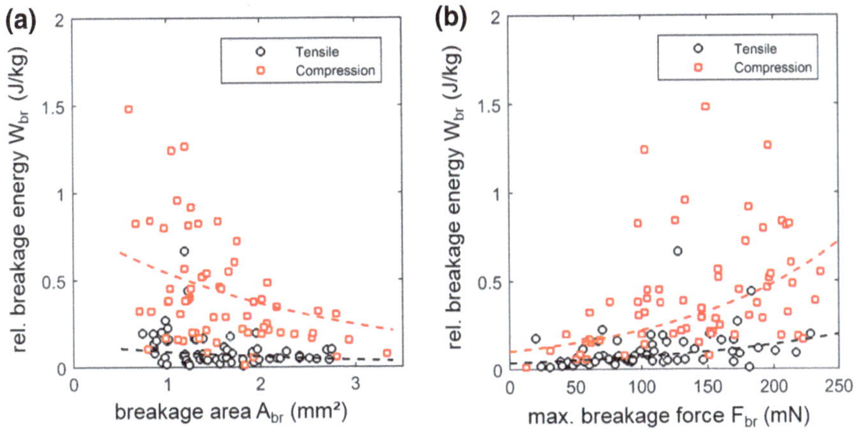

Fig. 49 Relative breakage energy of tensile and compressive tests as a function of breakage area (**a**) and maximum breakage force (**b**) (Reprinted with permission from [29].)

force. This makes it clear that higher relative breakage energy is required on average for the same breakage force in the compression test. The higher breakage energy can be attributed to the different breakage mechanisms. In the tensile test, tensile forces are applied directly parallel to the tensile direction. The breakage plane is perpendicular to the direction of tension. Since the particles are pulled apart directly, this process is only slightly influenced by rolling and sliding effects. During the compression test, the particles are pressed against each other. The proportion of rolling and constant movements is therefore larger. Stress analyses show that tensile forces act perpendicular to the compression direction, which leads to breakage along the compression direction [23]. Overall, higher compressive strengths and energies can be observed experimentally compared to the tensile strengths and energies.

Numerical Investigation of the Tensile Strength on Single Agglomerates

The tensile test on single agglomerates have been examined numerically using DEM simulation and compared with the experimental results. The same agglomerates and parameters were used for solid bridges as for the pressure and impact tests (see Tables 5 and 6). In order to guarantee a breakage of the agglomerate in the width and to ensure comparability with the experiment, the particles near the plate were provided with significantly higher bonding forces. The aim is to simulate the adhesive layer that fixes the agglomerate to the plate in the experiment. Figure 50 shows the initial situation of the tensile test in the simulation.

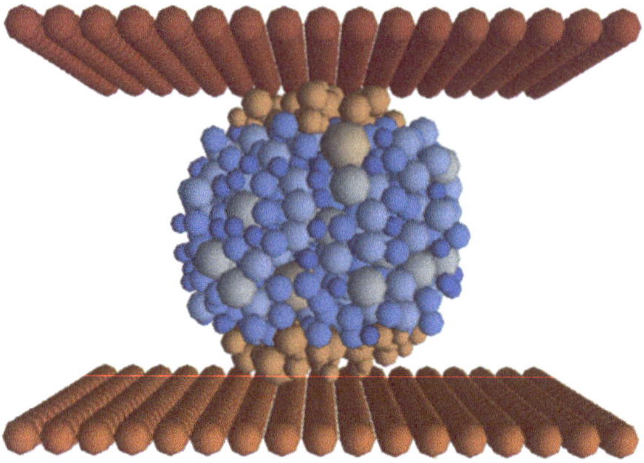

Fig. 50 Setup of the tensile test on single agglomerates in the DEM simulation. At the upper and lower end of the agglomerate, particles with higher bonding forces are visible for the simulation of the adhesive layer in the experiment. The blue particles show typical binding forces as they are present in the agglomerate

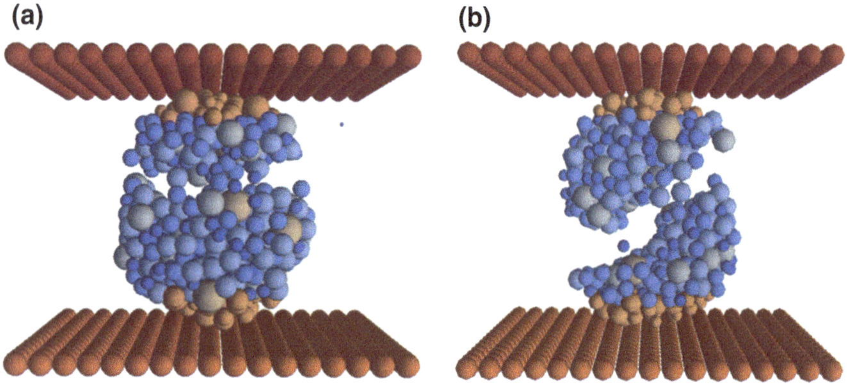

Fig. 51 Breakage characteristic of agglomerates after tensile test in DEM simulation (Reprinted with permission from [29].)

The particles representing the adhesive layer are connected to the two plates via the solid bridge model. To make this possible, both plates were generated from a monolayer of particles. In contrast to the experiment, the pulling velocity was set to $v_{pull} = 0.01$ m/s, as a simulation in real speed would be too time consuming. The same speed was used in the compression test. The result was qualitatively evaluated and compared with the fractures from the experiment. Figure 51 shows the fracture characteristics of two agglomerates as an example.

In the numerical investigation of the tensile test, two different breakage characteristics were observed for the agglomerates investigated. In both cases, the fractures were directed towards the center, i.e. not near the edge particles, with the smallest possible agglomerate cross-section. In Fig. 51a there is a break vertically to the

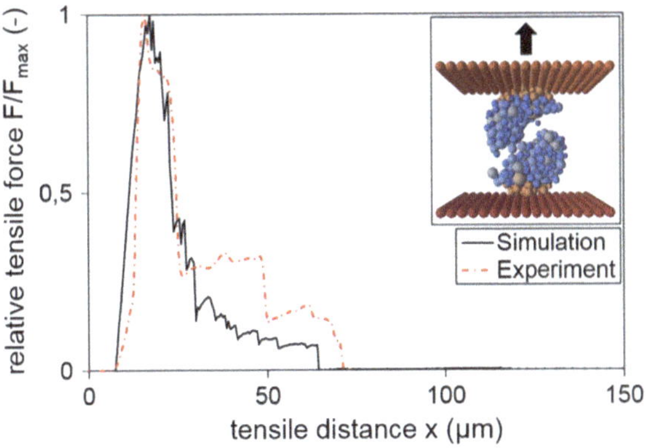

Fig. 52 Force-distance-curves (examples) from the tensile test of solid bonded agglomerates in comparison (Reprinted with permission from [29].)

direction of tensile, while in (b) there is a diagonal break. These fracture characteristics occurred in the agglomerates irrespective of their diameter. Similar breakages were observed in the experimental investigation. A comparison of the force-distance curves is shown in Fig. 52.

In the experiment, a steep increase in tensile force can be seen at the beginning, which reaches its maximum after just a few micrometers. This is followed by the primary fracture, during which the tensile force drops to a significantly lower level. This is followed by even smaller secondary fractures, which indicate that the fracture did not occur completely over the entire cross-section. Some individual particle bonds remained intact, which break as the tensile distance increases further. The simulation provides a comparable result. After a similarly rapid increase in tensile force, primary fracture also occurs here, followed by minor secondary fractures. With the help of the simulation data it can also be confirmed that the smaller secondary jumps are actually fractures and not the loosening of wedged particles when the fracture halves drift apart.

For a quantitative comparison, the mean tensile strength of the tensile tests carried out in the simulation was determined and compared with the mean tensile strength of agglomerates of the same size range $d_{agg} = 1.0–1.7$ mm. Figure 53 compares the tensile strengths. In comparison, the breakage strength of the compression test from simulation and experiment. The compressive test is also subjected to the determined conversion factor $b = 0.73$ from Fig. 48b.

The breaking strengths for the compression and the tensile tests are comparable. As already observed in the experiment, higher values are obtained in the compression test than in the tensile test. The conversion factor b for calculating the tensile strength from the compression test results is in agreement to the tensile strength from the direct tensile test from the experimental results. If this factor is used for the simulation results, the difference is slightly larger than in the experiment. However, if the two tests are compared, it becomes apparent that the difference between the tensile and compression tests in the experiment is larger (in percent) than in the simulation. The factor b, calculated from the simulation data, is thus somewhat higher at $b = 0.85$

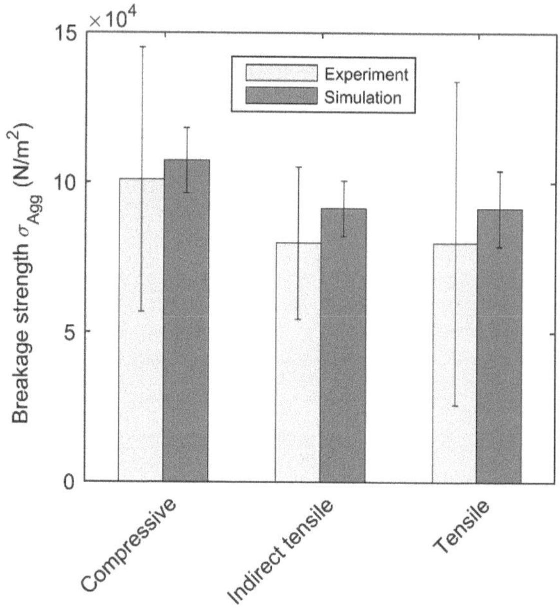

Fig. 53 Breakage strength of tensile and compressive test (simulation and experiment) (Reprinted with permission from [29].)

(experiment: b = 0.73). This may be attributed to inhomogeneities and defects in the experimental agglomerates which are not included in the numerically generated agglomerates. In addition, the number of runs is significantly lower than in the experiment, which means that a statistically reliable determination of b is not possible. Experiments would also require a significantly higher number of experiments.

Conclusions

The potential of high-intensity acoustic fields (ultrasonic flows) to overcome the cohesion of agglomerated and wall adhering solid particles in gaseous atmosphere has been demonstrated successfully. In principle, it is possible to overcome the adhesive forces by the acoustic resistance forces that arise as a result of the dynamic flow through the sound field. This could be proven by experimental and simulative investigations for relatively weak bonds (van der Waals and capillary bridges). The required dispersion intensity of the acoustic field has been derived as a function of primary particle size, ambient humidity and contact angle of the particle surface. A reduction in ambient humidity or an increase in the contact angle leads to an improvement in dispersibility.

A significantly greater dispersion effect can be achieved by the collision-related dispersion of agglomerates. Here, agglomerate-agglomerate or agglomerate-wall collisions lead to a comminution of the structures. The sound field is used for fluidization or acceleration of the agglomerates and is not the direct cause for the comminution. This indirect dispersion can also be used to disperse agglomerates with a significantly higher strength. In this way, it is possible to overcome even solid bonds of different stability. The relatively high kinetic energy of the accelerated agglomerates leads to a significant shift in the size distribution over the duration of the acoustic irradiation. For particle sizes below 100 μm, this displacement can extend as far as the distribution of the primary particles, depending on the process control. Also for particle sizes at the lower end of the micrometer scale, significant comminution by indirect dispersion has been demonstrated.

Separation of particles that adhere to surfaces by capillary and van der Waals forces is effectively possible by the use of ultrasound. As in the dispersion of agglomerates, larger particle sizes and higher contact angles have a positive influence on the detachment efficiency, while higher frequencies and ambient humidity negatively influence the efficiency. The duration of the acoustic irradiation, on the other hand, has no significant influence on the efficiency of the detachment.

Tensile and compression tests and simulations on individual agglomerates have shown that the tensile and compressive strength of the agglomerates differ in their absolute values. Correlation of both strengths via a general proportionality factor needs to be taken as a material dependent parameter and must therefore be determined for each type of agglomerate. Differences between tensile and compressive strength can also be observed with regard to the required breaking energy. In solid-bonded agglomerates this is significantly lower for tensile than for compression.

The comparison between tensile and compressive tests illustrates that the breakage strength of agglomerates must be differentiated according to the type of force applied and that the forces cannot be regarded as isotropic. A calculation of the tensile strength from compression tests by means of a correlation factor is only possible if this is determined experimentally beforehand for the corresponding agglomerate type from a comparison of direct tensile and compression tests.

References

1. Benes, E., Groschl, M., Nowotny, H., Trampler, F., Keijzer, T., Bohm, H., Radel, S., Gherardini, L., Hawkes, J. J., Konig, R., Delouvroy, C.: Ultrasonic separation of suspended particles. In: IEEE Ultrasonics Symposium, 7–10 October 2001, Atlanta, USA
2. Kühne, U., Fritsching, U.: Dosage of highly disperse powders by ultrasound agitated tube modules. Powder Technol. **155**(2), 117–124 (2001)
3. Hoffmann, T.L.: Environmental implication of acoustic aerosol agglomeration. Ultrasonics **38**(1–8), 353–357 (2000)
4. Noorpoor, A.R., Sadighzadeh, A., Habibnejad, H.: Influence of acoustic waves on deposition and coagulation of fine particles. Int. J. Environ. Res. **7**(1), 131–138 (2013)
5. Kusters, K.A., Pratsinis, S.E., Thoma, S.G., Smith, D.M.: Ultrasonic fragmentation of agglomerate powders. Chem. Eng. Sci. **48**(24), 4119–4127 (1993)
6. Visser, J.: Particle adhesion and removal: a review. Part. Sci. Technol. **13**(3–4), 169–196 (1995)
7. Fuhrmann, A., Marshall, J.S., Wu, J.: Effect of acoustic levitation force on aerodynamic particle removal from a surface. Appl. Acoust. **74**(4), 535–543 (2013)

8. Chen, D., Wu, J.: Dislodgement and removal of dust-particles from a surface by a technique combining acoustic standing wave and airflow. J. Acoust. Soc. Am. **127**(1), 45–50 (2010)
9. Rumpf, H.: Zur Theorie der Zugfestigkeit von Agglomeraten bei Kraftübertragung an Kontaktpunkten. Chem. Ing. Tech. **42**(8), 538–540 (1970)
10. Dadkhah, M., Peglow, M., Tsotsas, E.: Characterization of the internal morphology of agglomerates produced in a spray fluidized bed by X-ray tomography. Powder Technol. **228**, 349–358 (2012)
11. Butt, H.-J., Kappl, M.: Normal capillary forces. Adv. Colloid Interface Sci. **146**(1–2), 48–60 (2009)
12. Wan, K.-T., Smith, D., Lawn, B.: Fracture and contact adhesion energies of mica-mica, silica-silica, and mica-silica interfaces in dry and moist atmospheres. J. Am. Ceram. Soc. **75**(3), 667–676 (1992)
13. Kuttruff, H.: Physik und Technik des Ultraschalls, 1 Auflage. Hirzel Verlag (1988). ISBN 978-3777604275
14. Free Software Foundation Inc., OpenFOAM TheOpen Source CFD Toolbox—User Guide, Boston (2009)
15. Courant, R., Friedrichs, K., Lewy, H.: Über die partiellen Differentialgleichungen der mathematischen Physik. Math. Ann. **100**, 32–74 (1928)
16. Nyquist, H.: Certain topics in telegraph transmission theory. Trans. Am. Inst. Electr. Eng. **47**(2), 617–644 (1928)
17. Knoop, C., Fritsching, U.: Dynamic forces on agglomerated particles caused by high intensity ultrasound. Ultrasonics **54**(3), 763–769 (2014)
18. Šutilov, V.A.: Physik des Ultraschalls, Wien, 1 Auflage. Springer (1984). ISBN 978-3-7091-8750-0
19. Rabinovich, Y.I., Esayanur, M.S., Moudgil, B.M.: Capillary forces between two spheres with a fixed volume liquid bridge : theory and experiment. Langmuir **24**(21), 10992–10997 (2005)
20. Dosta, M., Dale, S., Antonyuk, S., Wassgren, C., Heinrich, S., Litster, J.D.: Numerical and experimental analysis of influence of granule microstructure on its compression breakage. Powder Technol. **299**, 87–97 (2016)
21. Kafui, K.D., Thornton, C.: Numerical simulations of impact breakage of a spherical crystalline agglomerate. Powder Technol. **109**(1–3), 113–132 (2000)
22. Arbiter, N., Harris, C.C., Stamboltzis, G.A.: Single fracture of brittle spheres. Trans. AIME **244**, 118–133 (1969)
23. Bika, D.G., Gentzler, M., Michaels, J.N.: Mechanical properties of agglomerates. Powder Technol. **117**(1–2), 98–112 (2001)
24. Schweiger, A., Zimmermann, I.: A new approach for the measurement of the tensile strength of powders. Powder Technol. **101**(1), 7–15 (1998)
25. Shipway, P.H., Hutchings, I.M.: Fracture of brittle spheres under compression and impact loading. II. Results for lead-glass and sapphire spheres. Philos. Mag. A **67**(6), 1405–1421 (1993)
26. Hiramatsu, Y., Oka, Y.: Determination of the tensile strength of rock by a compression test of an irregular test piece. Int. J. Rock Mech. Min. Sci. Geomech. Abstr. **3**(2), 89–90 (1966)
27. Lian, G., Thornton, C., Adams, M.J.: A theoretical study of the liquid bridge forces between two rigid bodies. J. Colloid Interface Sci. **161**(1), 138–147 (1993)
28. Knoop, C., Todorova, Z., Tomas, J., Fritsching, U.: Agglomerate fragmentation in high-intensity acoustic standing wave fields. Powder Technol. **291**, 214–222 (2016)
29. Wollborn, T., Schwed, M.F., Fritsching, U.: Direct tensile tests on particulate agglomerates for the determination of tensile strength and interparticle bond forces. Adv. Powder Technol. **28**(9), 2177–2185 (2017)
30. Knoop, C., Wollborn, T., Fritsching, U.: Detachment of solid particle systems from walls in an oscillating flow field. In: 9th International Conference on Conveying and Handling of Particulate Solids, 10–14 September 2018, London, Great Britain (2018)

Other Publications

31. Knoop, C., Fritsching, U.: Gas/particle interaction in ultrasound agitated gas flow. In: 20th International Congress of Chemical and Process Engineering CHISA, 25–29 August 2012, Prague, Czech Republic; Procedia Eng. **42**, 770–781
32. Knoop, C., Fritsching, U.: Numerical derivation of forces on particles and agglomerates in a resonant acoustic field. In: AIP Conference Proceedings, vol. 1558, no. 1, pp. 1115–1118 (2013)
33. Knoop, C., Fritsching, U.: Ultraschallangeregte Dispergierung agglomerierter Partikel in gasförmiger Atmosphäre. Chem. Ing. Tech. **86**(9), 1629 (2014)
34. Wollborn, T., Knoop, C., Fritsching, U.: Stability of brittle agglomerates in agitated gaseous environment. In: International Congress on Particle Technology PARTEC, 19–21 April 2016, Nuremberg, Germany, Book of Abstracts (Hrsg. VDI Verlag GmbH), Düsseldorf. ISBN: 978-3-18-092283-6
35. Wollborn, T., Knoop, C., Fritsching, U.: Particle and agglomerate dispersion in gas flow by collisions and flow induced stress. In: 9th International Conference on Multiphase Flow ICMF 2016, 22–27 May 2016, Firenze, Italy
36. Wollborn, T., Knoop, C., Fritsching, U.: Untersuchung der Dispergierung von Partikelagglomeraten unter Zug-und Druckbeanspruchung. Chem. Ing. Tech. **88**(9), 1359 (2016)
37. Knoop, C., Wollborn, T., Fritsching, U.: Detachment of solid particle systems from walls in an oscillating flow field. In: 9th International Conference on Conveying and Handling of Particulate Solids, 10-14 September 2018, London, Great Britain

Capillary Interaction in Wet Granular Assemblies: Part 1

Stephan Herminghaus, Ciro Semprebon and Martin Brinkmann

Introduction

Since the pivotal work of Haines and Fisher on capillary forces between spherical particles [1, 2], the mechanics of partially saturated granular beds has been addressed in various problems of geophysics ranging form crack formation and the role of moisture in the formation of sand dunes [3] to the mechanics of soils [4] and landslides. Cohesive forces induced by the surface tension of a wetting liquid are also relevant to process engineering and industrial production, for example wet granulation [5, 6] and molding of particulate materials. Fresh concrete, for example, consists of two granular materials, cement and sand, well mixed with water. Liquid binders need to be added to pharmaceutical substances in powder form before they can be molded into pills.

In the absence of external forces such as gravity or applied pressure gradients, the main driving force of liquid transport in a static assembly of particles is the surface tension of the liquid-air interface. This capillarity driven transport is controlled by many factors, such as local liquid saturation, surface energies of fluids and solids, as well as particle shape and roughness. Surface wettability, in particular, largely controls the morphology of the interstitial fluids [7] and is an important factor not only in capillarity driven transport at low saturation [8, 9] but also in forced imbibition into

S. Herminghaus
Max–Planck Institute of Dynamics and Self–Organization, Am Fassberg, 37077 Göttingen, Germany
e-mail: Stephan.Herminghaus@mpi-ds.mpg.de

C. Semprebon
Smart Materials & Surfaces Laboratory, Northumbria University, Newcastle upon Tyne NE1 8ST, UK
e-mail: ciro.semprebon@northumbria.ac.uk

M. Brinkmann (✉)
Experimental Physics, Universtät des Saarlands, 66123 Saarbrücken, Germany
e-mail: martin.brinkmann@pyhsik.uni-saarland.de

S. Antonyuk (ed.), *Particles in Contact*,
https://doi.org/10.1007/978-3-030-15899-6_8

a porous medium at high saturations [10]. However, in a fluidized state, the transport of liquid in a granular material also occurs by advection with the particles as they move. Again, the dynamics of capillary-driven liquid transport is an important factor in describing the migration of liquid in the fluidized granular assembly.

The mechanism of liquid transport in a granular bed is closely linked to the morphology of the interstitial fluids [7, 9]. Also, the magnitude of the cohesive forces is determined by the distribution of the liquid on the pore scale [11, 12]. A detailed description of the interplay of liquid morphology can be reached, for example, by fast confocal microscopy [13] and X-ray tomography [11]. A fundamental understanding of capillary cohesion and liquid transport could help to improve existing models of wet particulate materials for technological applications including wet granular aggregation [6] or particle coating [14].

Capillarity provides the theoretical framework to formulate models for the stability of liquid films wetting rough surfaces, the equilibrium shape of liquid clusters, and the cohesive forces these clusters exert on the particles. Wettability and surface energies are key factors in slow displacement of a fluid from a porous medium by a second fluid being immiscible to the first [10, 15]. The interaction of the advancing contact angle of the invading fluid with the local pore geometry of the medium selects the specific mode of interfacial advancement. The prevalence of certain modes of interface advancement is reflected in distinct patterns of the invading and defending fluids that develop during the displacement process [10, 12, 15]. Detailed information about the pore-scale morphology of interfaces between a wetting and a non-wetting fluid, for instance during drying or re-wetting in dense assemblies of particles, is highly relevant to understand the history dependent mechanics of wet soils.

This chapter starts with a brief overview of the different regimes of liquid saturations and time resolved measurements of liquid redistribution within a granular assembly in section "Liquid Saturation Regimes" followed by a short introduction to the fundamental concepts of capillarity in section "Theory of Capillarity". Numerical energy minimizations (NEM) as a method to compute mechanical equilibrium shapes of liquid-fluid interfaces in complex wetting geometries and capillary forces are outlined in section "Numerical Computation of Capillary Surfaces". The following section "Wetting of Rough Surfaces" summarizes an analytically amenable mean field theory of wetting and drying transitions on randomly rough surfaces in section "Some Fundamental Aspects of Rough Surface Wetting" while conditions for spontaneous wicking of a liquid film on rough model surfaces obtained by NEM are discussed in section "Film Spreading on Model Roughness". The results of NEM for liquid clusters in small assemblies of spherical particles in section "Capillary Forces Between Spherical Particles" illustrate the history dependence of multi-body capillary forces. Numerical investigations using a contact line friction model in section "Rolling Friction of Capillary Bridges Caused by Contact Angle Hysteresis" demonstrate how contact angle hysteresis creates a rolling friction of two spherical particles with a capillary bridge at the contact.

Liquid Saturation Regimes

Driven by the minimization of interfacial energy, a wetting liquid tends to maximize the area in contact to a solid while, at the same time, reduces the area of the free liquid interface. Consequently, a wetting liquid will first migrate into regions of a granular bed offering a high specific surface area. At low liquid saturations the liquid will be fully adsorbed by the roughness of the solid surfaces and forms a continuous film in the troughs and valleys. An example of a thin liquid film of water on the surface of glass beads is shown in Fig. 1a.

Excess liquid that cannot be accommodated in the surface roughness will be attracted to the contact points between neighboring particles where it forms capillary bridges. The saturation marking the transition between the liquid film regime and the regime of capillary bridges at the contacts is controlled by the surface area of the particles, as well as by the geometry and amplitudes of the surface roughness [7, 17]. Capillary bridges between spherical glass beads at low saturation display an irregular shape but become increasingly symmetric as the liquid saturation increases [7, 17], cf. the optical images in Fig. 1b, c. At even higher saturation, the capillary bridges coalesce and form large liquid clusters, as shown in the example Fig. 1d. To delineate the liquid clusters from single capillary bridges, we adopt the definition

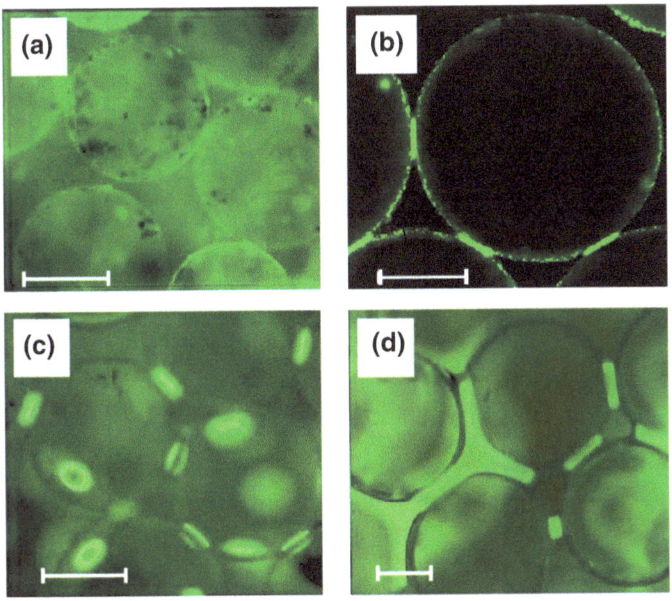

Fig. 1 Confocal microscopy images of wet glass beads in the liquid film regime (**a**) and in the asperity bridge regime (**b**). Fluorescence microscopy images of the pendular bridge regime (c) and the funicular regime (**d**). The scale bar represents 400 μm. Images taken by Mario Scheel (Soleil Synchrotron, France)

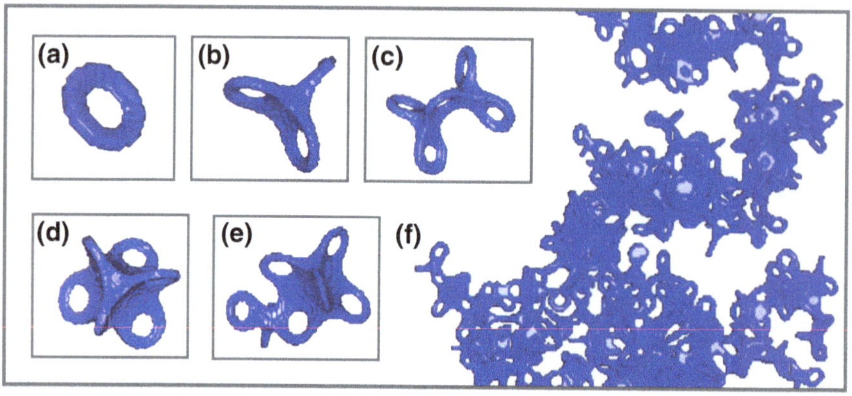

Fig. 2 Rendering of a funicular cluster of a wetting liquid in a dense bed of spherical grains obtained with X-ray micro-tomography. Central openings of the toroidal capillary bridges are artifacts caused by the finite resolution and the thin liquid film at the contacts of two spherical particles

that liquid clusters must simultaneously wet more than two particles. The transition between the 'pendular' liquid regime of isolated capillary bridges and the 'funicular' regime of clusters of coalesced bridges depends on the particle shape and packing density of the granular bed [7].

The liquid morphologies in the pendular and funicular regime, as well as the transition between both regimes have been studied by X-ray tomography [11, 18]. Examples illustrating the spectrum of liquid shapes in beds of spherical particles are displayed in Fig. 2. The most prominent liquid structures, even in the funicular regime, are axially symmetric capillary bridges forming at the contact between two spherical particles, cf. Fig. 2a. Once all particle contacts and small gaps between neighboring particles are saturated by capillary bridges, any additional liquid will be attracted to regions in the pore space that still offer a high specific surface area. Besides the region around the contact points, these regions are the narrow throats or 'nooks' [7] formed by three neighboring spherical particles in close contact. When filled by liquid each of these throats can connect three adjacent capillary bridges to form a bridge 'trimer', cf. the rendering in Fig. 2b. An angled 'pentamer' of five capillary bridges connected by two adjacent throats filled with liquid is shown in Fig. 2c.

Increasing the saturation further beyond the point where all available throats are saturated, the wetting liquid starts to invade the smallest pores which are typically bounded by a tetrahedral arrangement of four spherical particles. These particles are either in mechanical contact or display a small surface-to-surface separation. An example of a filled pore surrounded by six capillary bridges found in an X-ray microtomography image is depicted in Fig. 2d. The liquid cluster in Fig. 2e shows that capillary bridges of a such a 'hexamer' can be shared between different liquid clusters, here between a trimer and a hexamer of bridges.

It is apparent that highly ramified clusters of coalesced capillary bridges appear if the cluster of connected capillary bridges percolates the bed of spherical particles, cf. the example shown in Fig. 2f. In a bed of monodisperse spherical particles, this percolation occurs at a saturation of $S \approx 0.07$, here with respect to available interstitial space [11, 18]. Only for high saturations with a wetting liquid above $S \approx 0.25$, extended regions of fully saturated pore space appear in a bed of spherical particles [11, 18].

Liquid Transport in the Pendular Regime

Fluidization of a partially saturated bed of particles induced by shaking or shearing destroys the liquid structures shown in Fig. 2. Ideally, this input of energy furnishes the work to continuously break the network of capillary bridges and liquid clusters [19]. Constant rupture and re-formation of capillary bridges at particle contact favors the formation of a liquid film on the surface of the particles. Examples for this type of liquid film at low saturations are displayed in Fig. 1a, b. But once the external agitation is stopped, the particles settle into a stable configuration where they establish enduring mechanical contacts. At the same time, the wetting liquid is attracted to the contacts where it forms macroscopic capillary bridges. These liquid bridges grow over time until they eventually start to coalesce and aggregate into larger clusters, provided the saturation with the wetting liquid is sufficiently high.

Time-resolved x-ray micro-tomography measurements of liquid distribution in a bed of spherical particles [7] reveal that particle wettability and flow in thin liquid films is one the major factors that control the growth kinetics of capillary bridges and liquid clusters. While the volume distribution of water bridges in a bed of wettable glass beads shown in Fig. 3a becomes more peaked with time, the volume distribution in a bed of less wettable ruby beads shown in Fig. 3b does not display a significant change over the duration of the measurement. The typical equilibration time for the bridge volume in a pile of glass spheres is several minutes. This value is consistent with the assumption of a thin wetting film on the micro-rough glass surfaces. The advancing contact angle of the aqueous solution of KI in water on the glass spheres used in the X-ray micro-tomography experiments is $\theta_a \lesssim 15°$.

On ruby spheres, however, the significantly larger value of the advancing contact angle $\theta_a = (60 \pm 10)°$ is measured for the aqueous solution. Both glass and the ruby spheres in our experiments display a surface roughness with an RMS value below one μm pointing to the explanation that the large advancing contact angle on the ruby spheres is a consequence of the high material contact angle of water on an ideally plane ruby surface. The advancing contact angle on the ruby spheres is consistent with the value $(64 \pm 10)°$ reported in Ref. [20] for a water/water-vapor interface on an plane sapphire surface at $T = 30\,°\mathrm{C}$.[1]

---

[1]Ruby is chemically to sapphire (Al_2O_3) with traces of Cr^{3+} ions.

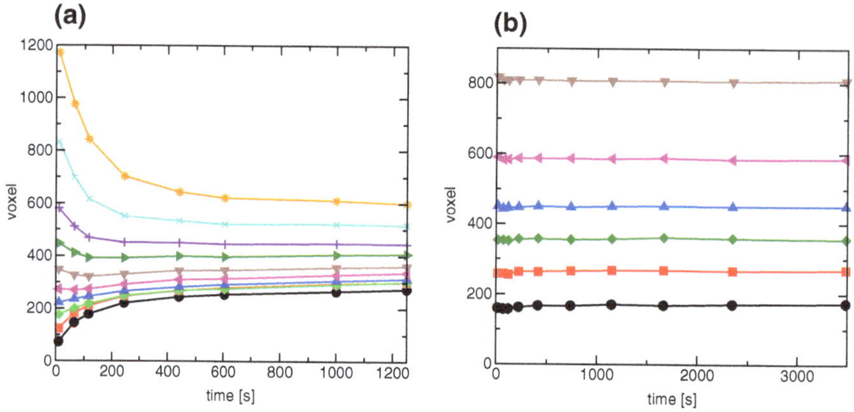

Fig. 3 Volume evolution of capillary bridges and liquid clusters observed in X-ray computer microtomography in beds of spherical particles. Liquid clusters and capillary bridges are first classified according their respective volume bin at the first measurement. Averages over all liquid objects in each class are plotted as a function of time. Shown are data for aqueous CsI solution in a bed of **a** spherical glass particles with a diameter of $D = (420 \pm 40)\ \mu\text{m}$ and **b** spherical ruby particles with $D = (600 \pm 5)\ \mu\text{m}$. Data reproduced from Ref. [16]

Theory of Capillarity

Enlarging the interfacial area between a liquid and a vapor phase or between two immiscible liquids requires work against surface tension. Liquids thus seek to minimize their interfacial area, leading to spherical droplet shapes in the absence of gravity or other external forces. Additional contributions to the interfacial energy have to be included to the total energy if the liquid is in contact to a solid. Similar to the energy of the free liquid interface, the free energy of the solid surface in contact to liquid or vapor must be taken into account.

Capillarity is the theory of the shape of liquid interfaces in, or close to, mechanical equilibrium. In a mechanical equilibrium, the interfacial configuration Σ_{lv} of a small non-volatile liquid droplet or thin liquid film in contact to a solid surface is a local minimum of the interfacial free energy:

$$\mathscr{E}\{\Sigma_{\text{lv}}\} = \gamma_{\text{lv}} A_{\text{lv}} + (\gamma_{\text{ls}} - \gamma_{\text{vs}}) A_{\text{ls}} \,. \tag{1}$$

The first term on the RHS of Eq. (1) represents the free energy of the liquid-vapor interface with area A_{lv} and interfacial tension γ_{lv}. The second term in Eq. (1) accounts for the free energy of the surface Σ_{ls} with area A_{ls} in contact to the liquid, and with the liquid-solid and vapor-solid surface tensions γ_{ls} and γ_{vs}, respectively. When dealing with non-volatile liquids, one has to consider local minima of the functional (1) under the subsidiary constraint of a fixed volume V of the liquid body. This situation is referred to as the 'volume controlled case'.

If the droplet in contact to the surface can exchange liquid with a reservoir that fixes the pressure difference $P \equiv P_\text{l} - P_\text{v}$ between the liquid (l) and vapor (v) phase and mechanically stable shapes of the liquid interface are local minima of the Grand interfacial free energy

$$\mathcal{G} = \mathcal{E} - P\,V\,. \tag{2}$$

The grand interfacial free energy takes into account the work received from or done at the volume reservoir, respectively. In contrast to the volume controlled case, this case is referred to as the 'pressure controlled case'.

If the vertical extensions Δz of the liquid droplet or film is small compared to the capillary length

$$L_c = \sqrt{\frac{\gamma_{lv}}{\Delta\rho g}}\,, \tag{3}$$

one can safely neglect the contribution of gravity or buoyancy to the total energy. The capillary length is typically on the order of millimeter or less. Contributions to the free energy that are described by interface potentials can be excluded if the typical distances between the liquid interface and the surface of the particles are larger than several tens to hundred nanometers [21, 22].

Extrema of the free energy $\mathcal{E}$ under the constraint of a fixed volume V necessarily satisfy two conditions. The first condition is expressed by the Young-Laplace equation:

$$P = 2H\,\gamma_\text{lv}\,, \tag{4}$$

that holds in every point of the interface Σ_{lv}. The mean curvature H is the sum of the two principal curvatures, or any pair of normal curvatures of the interface into orthogonal direction, see for instance Ref. [23]. In the absence of buoyancy, the capillary pressure $P = P_l - P_v$ does not depend on the position $\mathbf{r} \in \Sigma_\text{lv}$ on the interface and Σ_lv is a surface of constant mean curvature.

The second condition for an extremum of the free energy is the condition of Young and Dupré:

$$\gamma_\text{lv}\cos\theta_0 = \gamma_\text{sv} - \gamma_\text{sl}\,, \tag{5}$$

which expresses the mechanical equilibrium of three phase contact line. The equilibrium contact angle θ_0, also termed Young's or material contact angle, is solely determined by free energies per unit area of the surfaces in contact to the liquid or vapor phase.

Contact Angle Hysteresis

On any realistic surface that is partially wet by a liquid, the displacement of the three phase contact line is not fully reversible. In the limit of slow displacement the dissipated energy is proportional to the area swept out by the moving contact line.

The work per unit area, w_a, required to move the contact line into outward direction is not identical to the work per unit area, w_r, to move the three phase contact line inward. Even in the asymptotic limit of an infinitesimally small displacement velocity, the difference $w_a - w_r$ does not tend to zero.

Macroscopically, one observes a range of apparent contact angles of a static liquid-vapor interface on a solid surface. If the contact line moves slowly outward seen from the wet surface of the solid, the apparent contact angle equals the static advancing contact angle θ_a. If the contact line moves slowly inward one observes an apparent contact angle that is equal to the receding contact angle θ_r. However, if the three phase contact line is at rest, the local apparent contact angle may assume any value between the advancing and receding contact angle. The work per unit area w_a and w_r can be expressed through the advancing and receding angles by $w_a = -\gamma_{lv} \cos\theta_a$ and $w_r = -\gamma_{lv} \cos\theta_r$, respectively. Provided the motion of the three phase contact line is fully reversible, we have $w_a = w_r = \gamma_{ls} - \gamma_{lv}$ and $\theta_a = \theta_r = \theta_0$, i.e. vanishing contact angle hysteresis.

Stability of Interfacial Shapes

Any interfacial configuration that satisfies Eqs. (4) and (5) is a extremum of the interfacial energy $\mathscr{G}$ for the given capillary pressure P in Eq. (4). From this observation one can conclude that an extremum of the free energy Eq. (1) for a certain volume V is also an extremum of the Grand free energy Eq. (2) for a certain pressure P and vice versa. Hence, the set of interfacial equilibria in the volume controlled case and in the pressure controlled case are identical. The mechanical stability of the same state in the volume controlled case and in the pressure controlled case may differ. Apart from a constant, the free energies $\mathscr{G}$ and $\mathscr{E}$ are identical for all interfacial configurations enclosing the same volume V. Hence, any local minimum of $\mathscr{G}$ will be also a local minimum of $\mathscr{E}$, where the capillary pressure P in $\mathscr{G}$ can be regarded as a Lagrange multiplier to enforce the constraint of a constant liquid volume V. Mechanical stability in the pressure controlled case always implies mechanical stability in the volume controlled case. The converse statement, however, is not necessarily true as the liquid may exhibit more locally stable states in the volume controlled case in comparison to the pressure controlled case.

Numerical Computation of Capillary Surfaces

Due to a lack of high symmetries like, e.g. the rotational symmetry, the computation of the equilibrium shapes of a liquid body wetting three or more spherical particles is amenable only to numerical methods. In these situations, the equilibrium shape of a liquid-vapor interface in contact to the complex surface geometry of a solid can be computed by numerical minimization (NEM) of the interfacial energy. In all of the presented examples the freely available software surface evolver developed by K. Brakke [24, 25] is employed.

The central idea of the surface evolver is that shapes of liquid films or droplets are entirely fixed by the shape of their liquid-vapor interface. In the finite element approach the latter interface is represented by a mesh of small triangles spanning between N nodes. Hence, the contribution of the liquid-vapor interface to the free energy is the sum over the area of all triangles multiplied by the interfacial tension γ_{lv}. As a consequence the surface of the wet solid is not explicitly modeled and is accounted for by local constraints that are applied to the nodes of the three-phase contact line. During the minimization procedure the contribution of the wet solid surface to the total energy Eq. (1) is computed by integrating a suitable function over the three-phase contact line. In principle there is an infinite set of functions that satisfies the specific requirements for a particular surface geometry. In practice, however, the chosen form of the function should not be computationally costly and should lead to stable minimizations steps [24, 25].

During minimization of the interfacial energy, the condition of a constant liquid volume V_{l} is imposed through a global constraint on the shape Σ_{lv} of the liquid-vapor interface. Provided the liquid configuration is in a local minimum of the interfacial free energy (1), the Lagrange multiplier corresponding to the volume constraint that needs to be calculated in every minimization step is identical to the capillary pressure P. Alternative to the case of interfacial configurations enclosing a the same volume, one may consider the case where the volume of the liquid body is allowed to fluctuate but the capillary pressure is fixed by a volume reservoir.

A variety of gradient descent energy minimizations schemes are implemented in the surface evolver [24, 25]. A script language allows to extract geometrical quantities from the liquid-air interface during minimizations, e.g. the coordinates of the center of mass of the liquid body [26]. To maintain the quality of the triangulation, long and short edges of the triangulation are refined, respectively, removed from the mesh. This routines are applied after a number of minimization steps and keep the size distribution of the triangle sides in a desired range. Subsequent edge flipping allows to retain the quality of the mesh during large changes of the interfacial shape.

Besides the Laplace equation Eq. (4) on the free interface, the equation of Young-Dupré (5) are naturally met in a local minimum of the interfacial free energy. The contact angle θ_0 of the interface in contact to the solid is solely determined by surface energies on smooth parts of the solid surface. This is also the case for any droplet or liquid film that is modeled with the surface evolver. Effects of contact angle hysteresis are not amenable to numerical energy minimizations of the total interfacial energy.

A scheme to account for contact angle hysteresis with the surface evolver was proposed in Ref. [27]. At each instant in time, the interfacial energy is minimized while the position and shape of the three phase contact line is fixed. Once the conformation of the free interface is sufficiently close to a local minimum of the interfacial energy Eq. (1) for fixed volume, or (2) for fixed capillary pressure, the local contact angle is computed from the orientation of the small triangular surface elements that share a side with the three-phase contact line. The nodes of the contact line Γ_{slv} are then displaced into the direction of the local co-normal vector $\mathbf{n}_{\mathrm{s}} = \mathbf{N}_{\mathrm{s}} \times \mathbf{t}$ if the local contact angle $\theta(\mathbf{r})$ exceeds θ_a, where $\mathbf{N}_s$ is the local normal of the solid surface and $\mathbf{t}$ the tangent to Γ_{slv}. If $\theta(\mathbf{r}) < \theta_r$ the contact line is displaced into the direction of $-\mathbf{n}_s$. The procedure comprising a sufficient number of energy minimizations steps at fixed contact line Γ_{slv}, calculation of the local contact angle, and local displacement of the contact line $\mathbf{r} \in \Gamma_{slv}$ is repeated until $\theta(\mathbf{r})$ has sufficiently converged to a value within the interval $[\theta_r, \theta_a]$ for every point $\mathbf{r} \in \Gamma_{slv}$. Details of the numerical implementation and possible extensions of the model to include dynamic contact angle are outlined in Ref. [27].

Calculation of Capillary Forces

Capillary forces and torques acting on particle can be computed in two fundamentally different methods. The first method is based on the work injected or received during adiabatically slow reversible changes of the equilibrium conformations of the liquid. In the second method, the forces and torques are obtained by integrating the local forces acting on the particle surface.

Considering liquid in contact to two or more particles, the interfacial free energy E in a local minimum is a function of the particles coordinates $\bar{\mathbf{r}} = (\mathbf{r}_1, \mathbf{r}_2, ...\mathbf{r}_N)$ and liquid volume V where N is the number of particles. The coordinates may not only comprise the Cartesian coordinates of the particle's center of mass positions but also the relative orientation of the particles. Because the energy functional Eq. (1) may allow multiple local minima for a given liquid volume V and given positions $\bar{\mathbf{r}}$ of the particles, the function $E_\xi(\bar{\mathbf{r}}, V)$ can be multi-valued and forms several branches ξ. Analogous statements hold for the Grand free energy $\mathscr{G}$ and the corresponding energy landscape $G_\xi(\bar{\mathbf{r}}, P)$, where the capillary pressure P represents the accessible control parameter, instead of the liquid volume V. Equilibrium shapes of the liquid body corresponding to different branches ξ but for the same coordinates $\bar{\mathbf{r}}$ and volume V, can be distinguished by suitable order parameter related to the interfacial shape. For the sake of clarity, the index of the branches is suppressed in the following considerations.

If the interfacial energy $\tilde{E}(\bar{\mathbf{r}}, V)$ or the corresponding grand energy $\tilde{G}(\bar{\mathbf{r}}, P)$ as function of the particle coordinates $\bar{\mathbf{r}}$ is known, the capillary force $\mathbf{F}_i$ acting on particle i can be directly computed from partial derivatives of with respect to the Cartesian coordinates $r_{i\alpha}, \alpha \in \{x, y, z\}$:

$$F_{i\alpha} = -\partial_{i\alpha} E(\bar{\mathbf{r}}, V) \tag{6}$$

Despite its conceptual clarity, this approach is numerically costly because one needs to first scan a high dimensional energy landscape in terms of the all particle coordinates. Calculating the capillary forces directly from the configuration Σ_{lv} of the interface in a local minimum of $\mathscr{E}$ or $\mathscr{G}$, respectively, involves only integration over the three phase contact line [28].

To determine the total capillary force acting on a particle by a local integration of stresses, it is useful to consider a control surface at a small but finite distance to the particle surface. Since one is interested in the total force acting on a particular particle, this control surface must not intersect with any of the neighboring particles. The total capillary force on the particle is obtained by an integration of the normal component of the stress tensor over the control surface. Assuming mechanical equilibrium, this integral is independent on the particular choice of the control surface.

The total capillary force $\mathbf{F}$ acting on a particle can be split into a contribution of the isotropic pressure in the liquid and ambient fluid (superindex p) and a contribution is due to the interfacial tension (superindex t):

$$\mathbf{F} = \mathbf{F}^{(\mathrm{p})} + \mathbf{F}^{(\mathrm{t})} . \tag{7}$$

The first contribution $\mathbf{F}^{(\mathrm{p})}$ to the force is the capillary pressure $P \equiv P_{\mathrm{l}} - P_{\mathrm{v}}$ multiplied by the local surface normal $\mathbf{N}_{\mathrm{s}}$ of the solid, and integrated over the surface Σ_{ls} of the particle in contact to the liquid:

$$\mathbf{F}^{(\mathrm{p})} = \int_{\Sigma_{\mathrm{ls}}} \mathrm{d}A \; P \, \mathbf{N}_{\mathrm{s}} . \tag{8}$$

Provided that P is constant this contribution to the capillary force vanishes for particles that are completely immersed in one of the two immiscible fluids. Including the hydrostatic pressure contribution to P, this integral yields the force of buoyancy.

The second contribution $\mathbf{F}^{(\mathrm{t})}$ arises only in the presence of a three phase contact line Γ_{slv} on the particle, i.e. if the particle is partially wet. Locally, the interfacial tension γ_{lv} of the liquid-vapor interface Σ_{lv} pulls on the particle perpendicular to both the tangent vector $\mathbf{t}$ of the contact line Γ_{slv}, and the local normal vector $\mathbf{N}_{\mathrm{lv}}$ of Σ_{lv} in a point $\mathbf{r}$ on Γ_{slv}. The local interfacial force per unit length reads:

$$\mathbf{f} = \gamma_{\mathrm{lv}} (\mathbf{N}_{\mathrm{s}} \sin\theta - \mathbf{n}_{\mathrm{s}} \cos\theta) . \tag{9}$$

Figure 4 illustrates the definition of the two local orthonormal vector bases $\{\mathbf{t}, \mathbf{N}, \mathbf{n}\}$ of the liquid vapor interface Σ_{ls} and $\{\mathbf{t}, \mathbf{N}_{\mathrm{s}}, \mathbf{n}_{\mathrm{s}}\}$ on the surface Σ_{s} of the particle, as well as the local contact angle θ between Σ_{ls} and Σ_{s} in a point $\mathbf{r} \in \Gamma_{\mathrm{slv}}$. The surface normal of the free interface, $\mathbf{N}_{\mathrm{lv}}$, and the local normal of the particle surface, $\mathbf{N}_{\mathrm{s}}$, allows to express the local contact angle as

$$\cos\theta = \mathbf{N}_{\mathrm{lv}} \cdot \mathbf{N}_{\mathrm{s}} . \tag{10}$$

The vectorial line force Eq. (9) integrated over the contact line Γ_{slv} on the particle yields the total force that the liquid-vapor interface exerts on the particle:

$$\mathbf{F}^{(\mathrm{t})} = \int_{\Gamma_{\mathrm{slv}}} \mathrm{d}\ell \; \mathbf{f}(\mathbf{r}) . \tag{11}$$

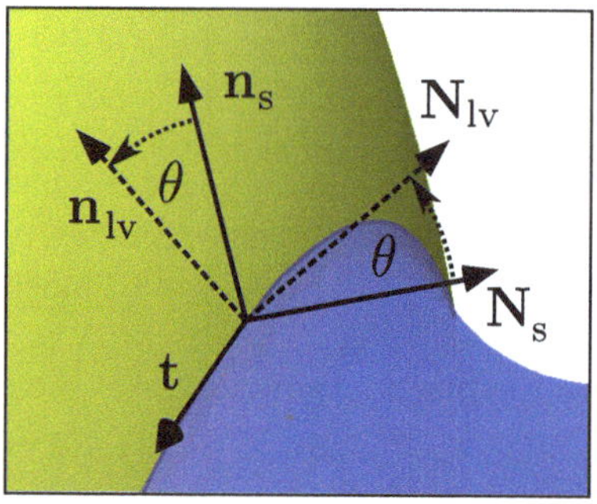

Fig. 4 Three phase contact line Γ_{slv} between the solid (yellow) and the free liquid interface Σ_{lv} (blue). Shown are the tangent $\mathbf{t}$ to Γ_{slv}, the local surface normal $\mathbf{N}_s$ to the solid surface, the local surface normal $\mathbf{N}_{lv}$ to the liquid-vapor interface Σ_{lv}, and the co-normal vectors $\mathbf{n}_s = \mathbf{t} \times \mathbf{N}_s$ and $\mathbf{n}_{lv} = \mathbf{t} \times \mathbf{N}_{lv}$. Reprinted with permission from [28]. Copyright (2016) by the American Physical Society

One must not forget that the decomposition of the total force into a surface tension part and a suction part depends on the particular choice of the control surface. This can be demonstrated for axially symmetric capillary bridge spanning between two spherical particles. Using the Delaunay curves, the surface of these bridges can be explicitly parameterized in terms of closed form expressions. The corresponding capillary forces of interfacial tension and suction can be computed, for instance, with respect to a central plane cutting the bridge perpendicular to the symmetry axis.

Another choice of the control surface would be a parallel surface placed at an infinitesimal distance to the surface of one of the two particles. While the sum of the two parts of the force is the same for either choices, the corresponding surface tension and suction forces are not identical. In our NEM, integrals of Eq. (9) over the contact line can be performed as summations over the corresponding expressions on the edges representing the contact line. Employing Stokes' theorem, one can re-express the components of the surface integral in Eq. (8) by a line integral over a suitable function [24, 25]. In contrast to the energy landscape approach, the integration of stresses is not restricted to the case of an ideal particle surfaces where the local contact angle is related to the surface energies by the Young-Dupré equation (5) [29, 30], and can be applied to the case of capillary bridges between spherical particles including the effect of contact angle hysteresis section "Rolling Friction of Capillary Bridges Caused by Contact Angle Hysteresis".

Wetting of Rough Surfaces

Some Fundamental Aspects of Rough Surface Wetting

Spontaneous spreading of a liquid film on the rough surface of a solid, as well as its retraction, is driven by interfacial energy. If the typical roughness amplitudes are on the scale of a micro-meter or lager, one can safely neglect contributions described by interface potentials. Even for surfaces with a roughness amplitude of only several tens of nanometers, the contribution of the effective interface potential can be considered as a small correction to the interfacial free energy. Consequently, the description of wetting phenomena on the basis of interfacial energies per unit area, as in the spirit of the previous section, can be assumed to be sufficiently accurate even for rough and chemically inhomogeneous (i.e., for real) surfaces. Aside from contact line pinning and contact angle hysteresis as discussed above, there are a number of fundamental aspects to wetting and spreading of partially wetting liquids on rough surfaces which are of major importance to wet granular piles and can be described in a rather concise way, as we will shortly discuss in what follows.

First of all, we should appreciate that the energy balances presented in the previous sections need to be reconsidered when the solid surface is not flat on small scales. In this case, completely covering a macroscopic (projected) area A with liquid leads to an energy difference of $(\gamma_{\mathrm{sl}} - \gamma_{\mathrm{sv}})\mathsf{r}A$, where r is the ratio of the total solid area divided by the projected area. The macroscopic contact angle, θ_m, the liquid will form under these circumstances is then given by

$$\cos\theta_m = \mathsf{r}\cos\theta_0. \tag{12}$$

as is seen directly from Eq. (5). It should be noted that for a granular material, the most convenient way of measuring θ_m is not by determining the macroscopic geometric angle of contact, but to measure the capillary pressure of imbibition of the liquid into a granular pile. These methodological aspects, however, shall not concern us here.

All of the discussion in the previous section remains valid, except for replacing θ_0 with θ_m. Since $\mathsf{r} \geq 1$, we generally have $\theta_m \leq \theta_0$ if the liquid wets the grains at least partially. Consequently, there is for any given roughness profile (and hence for given r) a certain finite microscopic contact angle at which θ_m vanishes, i.e. the surface appears completely wet phenomenologically, even if the substrate material is not. This angle is commonly called Wenzel's angle [31], θ_W, and is accordingly given by

$$\theta_W = \arccos\left(\frac{1}{\mathsf{r}}\right). \tag{13}$$

A typical random roughness profile of a worn surface, such as is usually present on the grains of any granular material, is shown schematically in Fig. 5a. At low liquid saturation of the granular pile, it is of central importance how much of the liquid resides within the troughs of the roughness on the grain surfaces. It is clear that the liquid surface must accommodate the equilibrium contact angle, θ_0, wherever it meets the solid surface. As this is on small scales, it appears well justified to assume that deviations from θ_0 due to contact angle hysteresis do not play a major role here. the second condition to be met is that the mean curvature of the liquid surface assumes its value H as prescribed by the experimental conditions. If the liquid is considered to be in thermal equilibrium with the surrounding vapor, its surface mean curvature is given by the Kelvin equation,

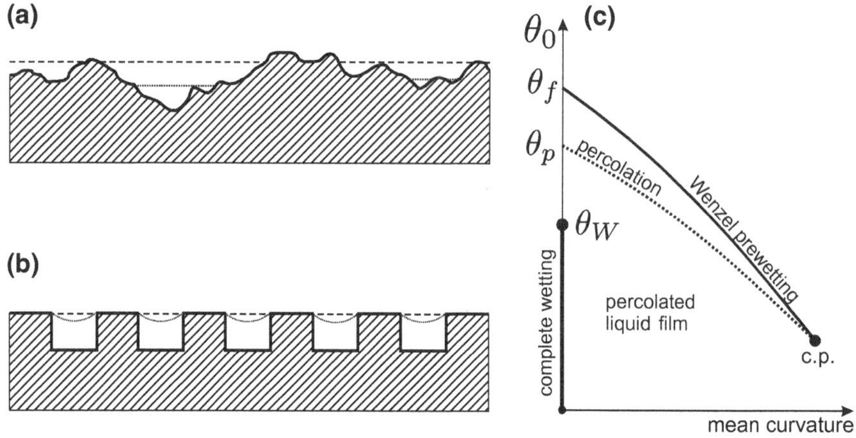

Fig. 5 Wetting of a rough surface: **a** random roughness as typically encountered on a real surface, **b** model roughness with rectangular profile, **c** phase diagram of wetting, in the plane spanned by the microscopic contact angle and the mean curvature of the liquid surface (distance from two-phase coexistence)

$$H = \frac{k_B T}{2\gamma_{\mathrm{lv}} v_m} \ln \frac{p_s}{p}, \tag{14}$$

where k_B is Boltzmann's constant, v_m is the molecular volume, and p and p_s are the vapor pressure in the gas phase and the vapor pressure at saturation, respectively.

If the roughness profile is not too steep, one can evaluate the effect of these two combined conditions for arbitrary random roughness [32, 33]. The resulting phase diagram is shown schematically in Fig. 5c, where the states of the system are presented in the plane spanned by the microscopic contact angle, θ_0, and the mean curvature of the liquid surface, H. For most practical situations in wet granular materials, the liquid is provided by mixing with the granular material, such that equilibration with the vapor phase does not play a significant role. Hence for the present discussion we will be dwelling close to the vertical axis.

For all microscopic contact angles θ_0 which are smaller than θ_W, the surface will appear completely wet by the liquid if the latter is at liquid-vapor coexistence. This is indicated as the bold line between the origin and the point at $(0; \theta_W)$. In the vicinity of this line (i.e., for small capillary pressure or if the liquid is close to coexistence), there will be some asperities sticking out of the liquid, but the troughs are filled deeply enough for the liquid forming a percolated film across the rough surface. This is important for the liquid to be able to spread along the grain surfaces without transport through the vapor phase to be necessary.

The salient feature in this phase diagram is the solid curve, which has been termed the Wenzel prewetting line [32–34]. As this line is crossed from below, the liquid cannot anymore accommodate the microscopic contact angle θ_0 wherever it meets the solid and at the same time retain a constant mean curvature H. As a result, the majority of the troughs of the roughness become empty, and film flow along the solid surface is no longer possible. In an idealized picture, the Wenzel prewetting line ends in a critical point (c.p.), where the size of the jumps in liquid coverage goes to zero. In practice, the line manifests itself as a strongly hysteretic region where abrupt changes in liquid coverage are encountered when either θ_0 or H are changed [34].

Another remarkable fact is that for Gaussian roughness, the extension of the prewetting line vanishes, such that the filling transition (between empty and filled troughs) occurs only in the θ_0 domain, at the filling angle θ_f [34]. However, since no roughness profile is precisely Gaussian, the extension of Wenzel prewetting into the range of nonzero H is always finite. While for Gaussian roughness the percolation transition is identical to the filling transition ($\theta_p \to \theta_f$), the percolation line lies slightly below the prewetting line for non-Gaussian roughness profiles [32].

Quantitatively, it turns out that for not too steep profiles, we have $\theta_p = \theta_f = 2\theta_W/\sqrt{\pi} \approx 1.13\theta_W$ [32]. Non-Gaussian surface roughness results in an increase of θ_f/θ_W [34], such that $\theta_p \geq 1.13\theta_W$. We can use this information to come up with a useful criterion for the presence of percolated liquid films within the granular pile. Let us assume we have measured θ_m and know θ through the interfacial energies of the involved materials. We have $\cos\theta_m = \cos\theta_0/\cos\theta_W$ (Eqs. (12) and (13)). As we have generally assumed shallow profiles, we can expand the cosine function as $\cos\theta \approx 1 - \frac{1}{2}\theta^2$ and find that for a percolated liquid film to be present along the grain surfaces, we should fulfill

$$\frac{\theta_m}{\theta} \leq 1 - \frac{\pi}{4} \approx 0.22 \tag{15}$$

Clearly, this all holds quantitatively only for sufficiently shallow roughness, but should persist also away from this condition qualitatively [33], i.e., for all typical random roughness encountered on the surfaces of real grains. Dynamical aspects can of course not be easily accessed on such random surfaces because of the enormous complexity they need for their numerical representation. It is thus reasonable, and quite common, to study dynamic effects instead on model roughness profiles, a typical representative of which is depicted in Fig. 5b. For such rectangular profiles, the filling transition takes place (at coexistence) at $\theta_f = \pi/2$ (dashed lines in Fig. 5b. For finite underpressure (i.e., away from coexistence), this happens at somewhat smaller contact angle (dotted lines in Fig. 5b. Furthermore, the question whether the liquid imbibing into the troughs forms a laterally percolated structure cannot be answered without knowledge of the three-dimensional geometry of the solid substrate. This said, we shall now turn to studying numerically the dynamics of wetting (and flow of liquid) in model geometries of surface roughness, which provide the substantial simplifications alluded to by Fig. 5 but can be studied with accessible amounts of computer power.

Film Spreading on Model Roughness

Plane surfaces decorated with posts of homogeneous cross-section provide a convenient starting point for numerical energy minimizations to investigate how the local geometry of the these model surfaces control the stability of a wetting film or spontaneous film formation. A useful example are regular arrays of posts of homogeneous cross-section that are arranged in a square unit cell, cf. also Fig. 6. Owing to the high symmetry of the array and post geometry it is not restrictive to compute only a small stripe of the interface close to the terminal contact line [35].

In agreement with wetting experiments on micro-fabricated topographies [35], the posts of the model surfaces are of uniform height and have a flat top face. If a wetting liquid spontaneously spreads from a reservoir at zero capillary pressure, the liquid forms a film of uniform thickness extending up to the height of the posts. On a naturally rough surface such a liquid reservoir could be a large feeding droplet or liquid bridges whose extensions are much larger than the typical length-scale of the surface roughness. Hence, to determine the factors that control spontaneous film

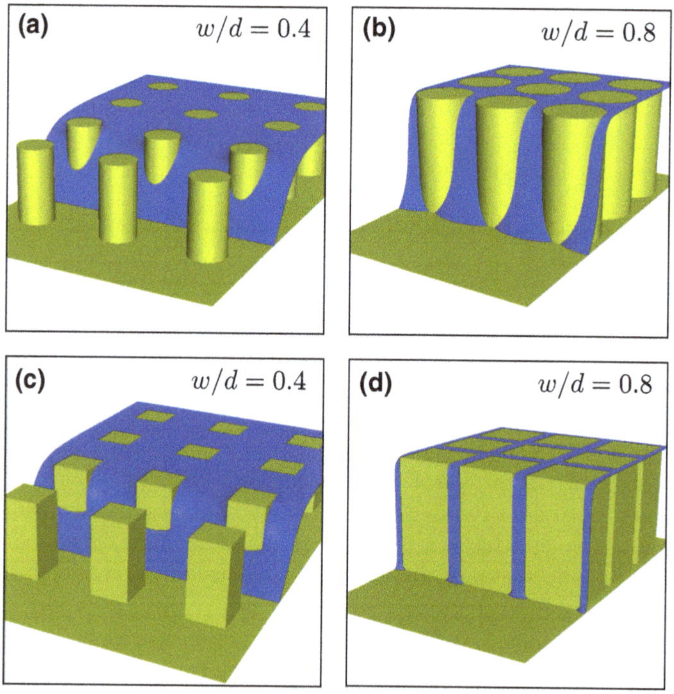

Fig. 6 Morphologies of the film edge obtained in numerical energy minimizations with the program surface evolver. The aspect ratio is $h/w = 2$ for all posts. Shown is a coalesced morphology in **a** and **c** show coalesced morphologies and the pinned morphology in **b** and **d**

spreading on the micro-patterned surface, it is sufficient to consider the terminal part of the liquid film in the numerical energy minimizations at zero capillary pressure.

A necessary condition for spontaneous wicking into a topographic micro-pattern can be derived from simple thermodynamic arguments. If the interfacial energy of a homogeneous film with a thickness given by the height of the posts is smaller than the free energy of the dry state, the liquid can be spontaneously drawn into the array of micro-posts. If the energy of the dry state is lower than the energy of the homogeneous film state, the liquid can spontaneously retract from the post array. To account for meta-stable states and the mechanical stability of the film edge, the thermodynamic arguments must be complemented with numerical considerations of the stability of the terminal interface of the film.

Two fundamental morphologies of the film edge close to the terminal contact line are found for posts with a circular and square cross-section [35]. In contrast to the pinned morphology shown in Fig. 6a, the interface of the coalesced morphology in Fig. 6b engulfs the base of the posts in the first row in contact with the liquid film. The mechanical stability of both interface morphologies is limited to a certain range of the ratio h/w of post height h to post width w, the ratio w/d of post width to the period d of the posts, and the material contact angle θ_0 of the substrate. The stability diagrams in Fig. 7a, b display regions of w/d and h/w allowing a mechanically stable film edge for posts with a circular and square cross-section, respectively, for $\theta_0 = 50°$.

Spontaneous spreading of the film is expected in the region of w/d and h/w where neither the coalesced nor the pinned morphology of the film edge exists and where the film state has a lower interfacial free energy than the dry state. The high sensitivity of

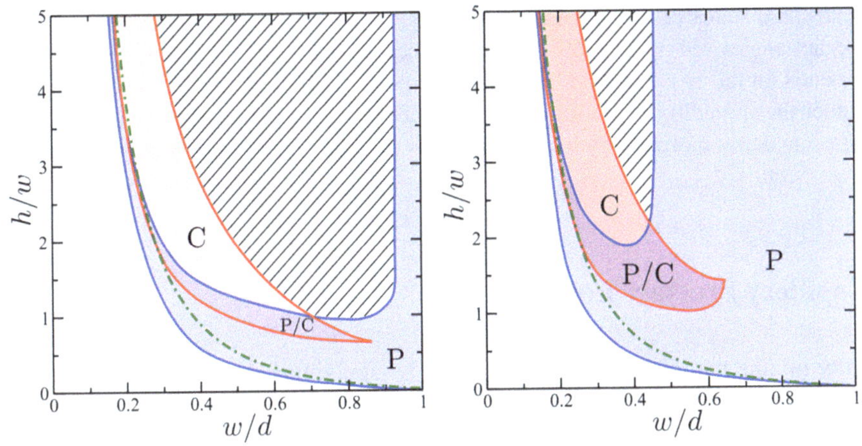

Fig. 7 Morphology diagram of the film edge in a square array of posts with **a** circular and **b** square cross section in terms of line fraction w/d and aspect ratio h/w. Indicated are regions where the pinned morphology (P) and the coalesced film morphology (C) are mechanically stable [35]. Spontaneous film spreading occurs in the hatched region. On the green dashed-dotted line, the dry state and the film state have equal free energies. The material contact angle of the substrate is $\theta_0 = 50°$

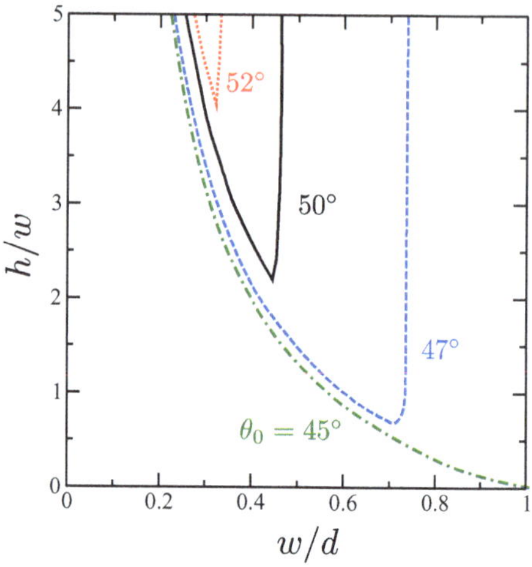

Fig. 8 Regions of line fraction w/d and aspect ratio h/w of square posts where spontaneous film spreading is predicted for a particular value of Young's contact angle θ_0, cf. also the hatched region in Fig. 7

spontaneous film spreading with respect to the material contact angle θ_0 is illustrated in Fig. 7c showing the region of control parameter where film spreading is predicted for various material contact angles θ_0 between 45° and 52° (Fig. 8).

For material contact angles $\theta_0 < 45°$ wetting liquid always spontaneously flows around the base of a post and encloses it, and the region of spontaneous film spreading coincides almost with the region predicted by the thermodynamic arguments [35]. For larger material contact angles $\theta_0 \gtrsim 60°$, however, spontaneous spreading of a film and coalescence at the base of a posts is fully suppressed by the presence of metastable states of the pinned morphology. In the intermediate range of material contact angles $45° < \theta_0 \lesssim 60°$, the mechanical stability of the pinned morphology depends on the cross-sectional shape of the posts. Sharps vertical edges effectively hinder the spreading of the liquid film around the base of square posts and stabilize the pinned film morphology in comparison to an array of posts with a smooth circular cross-section.

Capillary Pressure Equilibration

After mixing particles and a wetting liquid by applying a strong agitation, i.e. like fluidization by vertical shaking, one observes that the distribution of capillary bridge volumes becomes narrower with time [11, 18], cf. also the example for glass beads and an aqueous salt solution in Fig. 3. The convergence of volumes can be easily explained from two observations. The first observation is the existence of a thin liquid film on the surfaces of the particles. This film enables a transport of liquid after a fluidization, and allows a mutual exchange of liquid mass between bridges held by the same particle [36].The second essential feature is the spherical shape of the particle surfaces around a contact point, in combination with the small contact angle of the liquid interface at the solid surface. In this particular wetting geometry, and for bridges of sufficiently small liquid volume, the capillary pressure P is a monotonously growing function of the liquid volume V. Thanks to this feature, in contrast to freely suspended droplets, or droplets sitting on an ideal plane surface without contact angle hysteresis, capillary bridges are not subject to Ostwald ripening.

For capillary bridges spanning a gap with a finite surface-to-surface separation $s > 0$ and in a narrow range of V close to the smallest volume that permits the formation of a mechanically stable capillary bridge, the derivative $\partial_V P(s, V)$ may change sign. Negative derivatives $\partial_V P(s, V) < 0$ are also observed for capillary bridges of less wetting liquids with contact angles on the particles larger than 60°. In this case, the axial symmetry of the liquid bridge can be spontaneously broken [37, 38], leading to a 'popping-out' of the liquid body from the axis passing through the center of the two spherical particles.

Most of the works on capillary cohesion focus on the capillary bridge regime at low fluid saturations and assume axially symmetric interfaces forming a contact angle that is solely determined by the surface energies of the particles. For small particles with a diameter below a hundred nanometers, one may consider thin liquid films on the surfaces that is subject to a material specific disjoining pressure [39]. Despite the implications that arise from the assumption of a uniform material contact angle, it is realistic to account for surface roughness or heterogeneous chemical composition of the surface with a history dependent contact angle [30, 40].

Capillary Bridge Model

Contact angle hysteresis affects not only the evolution of cohesive forces of a capillary bridge during extension [41] but controls also the dynamics of liquid transport between capillary bridges in a static assembly of spherical particles [9]. An exchange of a non-volatile wetting liquids between capillary bridges on the same spherical particle through thin film flow [8, 9] is driven by gradients in capillary pressure between the three-phase contact lines of the bridges.

To account for contact angle hysteresis, one needs to first distinguish between capillary bridges with pinned or mobile contact lines. A suitable parameter to describe the shape of a capillary bridge with a pinned contact line is the opening angle β of the wet surface area, cf. also the sketch in Fig. 9. In this case, the capillary pressure of the bridge is a function $\tilde{P}_0(V, s, \beta)$ of liquid volume V, gap separation s, and opening angle β. However, for bridges with a freely sliding contact line and a prescribed contact angle θ, the capillary pressure is expressed by the function $P_0(V, s, \theta)$. In the numerical model, the capillary pressure P of a bridge with volume V is computed for various surface-to-surface separation s of spherical particles with radii R. The branches of the functions $\tilde{P}_0(V, s, \beta)$ and $P(V, s, \theta)$ are sketched in Fig. 10.

The interpolated capillary pressures P_0 and $\tilde{P}_0$ are complemented with the contact angle $\theta(V, s, \beta)$ for bridges with a pinned contact line and with the opening angle $\beta(V, s, \theta)$ for bridges sliding contact line. All functions, $\tilde{P}_0(V, s, \beta)$, $P_0(V, s, \theta)$,

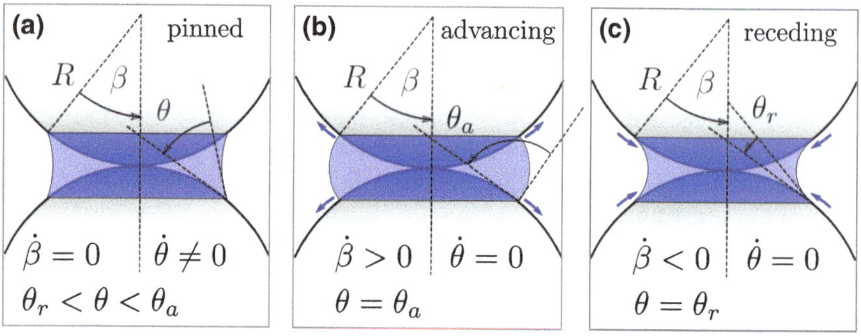

Fig. 9 Capillary bridge between two spherical grains of equal radii with a pinned (**a**), an advancing (**b**), and a receding (**c**) three-phase contact line. A capillary bridge is described by the opening angle β, the contact angle θ, the surface-to-surface separation of the spheres, and the pressure P, or volume V

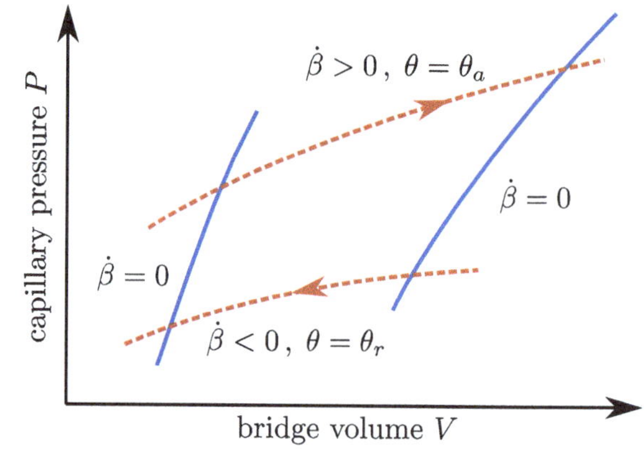

Fig. 10 Branches of capillary pressure characteristics $P(V, s, \theta)$ and $\tilde{P}(V, s, \beta)$ for volume V and surface-to-surface separation parameterized by contact angle θ and opening angle β of the capillary bridge, respectively. Reprinted with permission from [9]. Copyright (2015) by the American Physical Society

$\theta(V, s, \beta)$, and $\beta(V, s, \theta)$ are computed by NEM with the program surface evolver [25]. The restriction to axially symmetric bridges is justified for small contact angles $\theta < 30°$ and realistic volume $V/R^3 < 0.1$, and allows a reduction of the numerical complexity because the shape is fully described by the two-dimensional contour. The rupture distance of a bridge with fixed volume V is computed from the approximate expression

$$s_0^* = \left(1 + \frac{\theta}{2}\right) V^{1/3} \tag{16}$$

given by Willet et al. [42] and verified by our calculations. As shown in the subsequent work of Willet et al. Ref. [43], Eq. (16) remains a good approximation for the rupture separation of bridges with a pinned contact line if the opening angle β is not too small. In this case, the actual contact angle $\theta = \theta(V, s, \beta)$ of the bridge is introduced in Eq. (16) (Fig. 10).

The dependence of capillary pressure P on the liquid volume V and surface-to-surface separation s on the individual radii R_i of the two spherical particles $i = 1, 2$ is implemented by the scaling:

$$P(V, s, \theta) = \xi^{-1} P_0(\xi^{-3} V, \xi^{-1} s, \theta) \tag{17}$$

$$\tilde{P}(V, s, \beta) = \xi^{-1} \tilde{P}_0(\xi^{-3} V, \xi^{-1} s, \beta) \tag{18}$$

$$s^*(V, \theta) = \xi\, s_0^*(\xi^{-3} V, \theta) \tag{19}$$

where the dimensionless factor $\xi \equiv R_c/R_0$ accounts for the Derjaguin mean

$$R_c \equiv \frac{2R_1R_2}{R_1 + R_2} \tag{20}$$

of bead radii [42].

Network Model

The dynamics of fluid transport of many capillary bridges that mutually exchange liquid, like in the static assemblies of glass beads considered in X-ray tomography experiments [11, 18] can be studied in a network model. In such a coarse grained model each capillary bridge forms a node i of the network which are connected by bonds (i, j). Each bond represents the thin film on the surface of the particle between bridge i and j that allows a volume flux Q_{ij} being proportional to the difference of capillary pressure $P_i - P_j$, cf. also the sketch in Fig. 11.

Mass conservation implies that time derivative of the volume of bridge i is the sum

$$\dot{V}_i = - \sum_{j \in \mathcal{N}(i)} Q_{ij} = \sum_{j \in \mathcal{N}(i)} C_{ij} (P_j - P_i) \,, \tag{21}$$

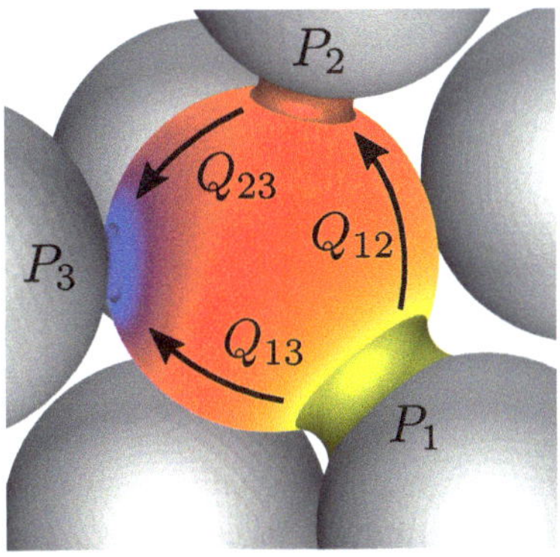

Fig. 11 Sketch illustrating the distribution of capillary pressure P_i and volume flux Q_{ij} in a wetting film around three capillary bridges i, j of different size

of volume fluxes to all further capillary bridges $j \in \mathcal{N}(i)$ on the same particle. The conductance coefficients C_{ij} are fully analogous to Maxwell's capacitance coefficients if the mobility of the liquid film is assumed to be constant. In this case, the capillary pressure in the film obeys a Laplace equation with Dirichlet boundary conditions on the contact line of the capillary bridges. Conformal mapping to the corresponding 2D electrostatic capacitance problem reveals that the dependence on the capillary bridge size and geometry is weak, which justifies setting $C_{ij} = 1$ throughout.

As the state of the capillary bridge in the presence of contact angle hysteresis is not only determined by its liquid volume V and separation s, one may consider the pair given by opening angle β and actual contact angle θ. Therefore, one needs to express the changes of the opening and contact angle through changes of the volume and separation of the particle surfaces. Employing the differentials

$$a \equiv \partial_V \beta(V_i, s_i, \theta_i)\, \dot{V}_i + \partial_s \beta(V_i, s_i, \theta_i)\, \dot{s}_i$$
$$b \equiv \partial_V \theta(V_i, s_i, \beta_i)\, \dot{V}_i + \partial_s \theta(V_i, s_i, \beta_i)\, \dot{s}_i$$

the time derivatives $\dot{\theta}$ and $\dot{\beta}$ can be expressed as:

$$\dot{\beta}_i = a\,, \quad \dot{\theta}_i = 0 \quad \text{for} \quad \begin{cases} a \geq 0 \text{ and } \theta_i = \theta_a \\ a \leq 0 \text{ and } \theta_i = \theta_r \end{cases} \tag{22}$$

$$\dot{\beta}_i = 0\,, \quad \dot{\theta}_i = b \quad \text{for} \quad \theta_r < \theta_i < \theta_a\,. \tag{23}$$

Variations of the gap separation s are relevant only if the particles move relative to each other, which simplifies the treatment of static assemblies of spherical particles.

It follows from Eq. (22) that the contact line of a bridge i can only slide outward ($\dot{\beta}_i > 0$) for a contact angle $\theta_i = \theta_a$. Capillary bridges with inward sliding contact lines ($\dot{\beta}_i < 0$), however, have to satisfy $\theta_i = \theta_r$. The set of equations (22) guarantees that the contact angle θ_i cannot become larger than θ_a or smaller than θ_r. The complementary case of an immobilized contact line is taken into account by Eq. (23): Any intermediate value $\theta_r < \theta_i < \theta_a$ of the contact angle θ_i conforms to a pinned contact line ($\dot{\beta}_i = 0$).

As a consequence of the contact angle hysteresis described by Eqs. (22) and (23), the state of a capillary bridge is not uniquely determined by the volume V and separation s. Depending on whether the contact line is advancing, pinned, or receding one needs to employ different expressions for the capillary pressure:

$$P_i = \begin{cases} P(V_i, s_i, \theta_a) \text{ for } \dot{\beta}_i > 0 \\ \tilde{P}(V_i, s_i, \beta_i) \text{ for } \dot{\beta} = 0\,. \\ P(V_i, s_i, \theta_r) \text{ for } \dot{\beta}_i < 0 \end{cases} \tag{24}$$

The temporal evolution of liquid volume $V_i(t)$, opening angle $\beta_i(t)$, and contact angle $\theta_i(t)$ for each individual bridge i is obtained from an integration of Eqs. (21–24).

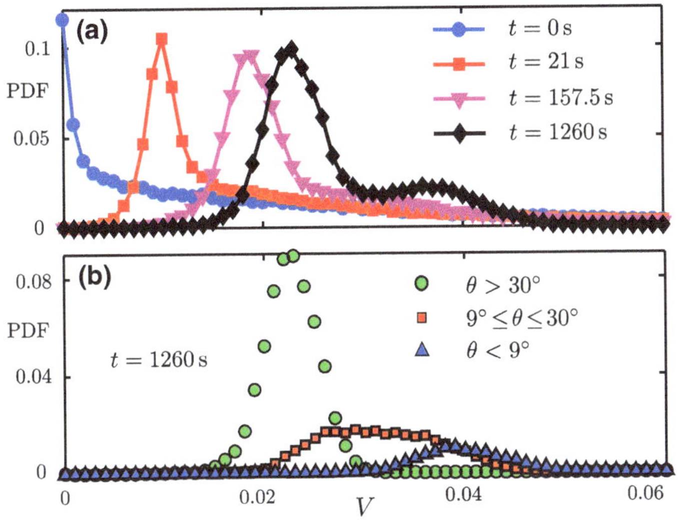

Fig. 12 **a** Probability density function (PDF) of capillary bridge volumes V obtained in the network model at different times after the shaking has been stopped at $t = 0$. **b** PDF of volumes for capillary bridges with advancing, pinned, and receding contact lines at a late stage of the equilibration process. Reprinted with permission from [9]. Copyright (2015) by the American Physical Society

This coarse grained model for the capillary pressure driven transport between capillary bridges is suited to follow the evolution of bridge volumes in a large assembly of capillary bridges and to extract statistically relevant quantities [9]. The probability distribution function shown in Fig. 12a reveals a first peak is formed at low volume already shortly after the preparation of intitial bridge volumes. With progressing time, this peak moves toward larger volumes where a second, shallow maximum is formed at later times. The nature of the first peak at small volumes and the shallow peak at larger volumes becomes clear from the probability distribution function for the volume of capillary bridges with pinned, advancing, or receding contact lines in Fig. 12b. The fast growing peak that moves to larger volumes corresponds to a bunch of growing capillary bridges whose three-phase contact line is sliding outward. The shallow maximum at large volumes, however, is due to a number of initially large capillary bridges with receding or pinned contact lines.

Large assemblies of communicating capillary bridges allow to study the effects of contact angle hysteresis not only on the evolution of bridge volumes but also on the driving force of the flow, i.e. the capillary pressure itself. The squared variance of the distribution of capillary pressures is shown in Fig. 13a in direct comparison to the evolution of the variance of bridge volume in Fig. 13b. While the variance of the pressure distribution tends to zero following the same power law for vanishing and for finite contact angle hysteresis, the spread of the volume distribution converges to zero only for a vanishing contact angle hysteresis. With contact angle hysteresis, the variance of the volume distribution saturates to a finite value. Evidently, a simple mean field model is already sufficient to explain the evolution of capillary bridge

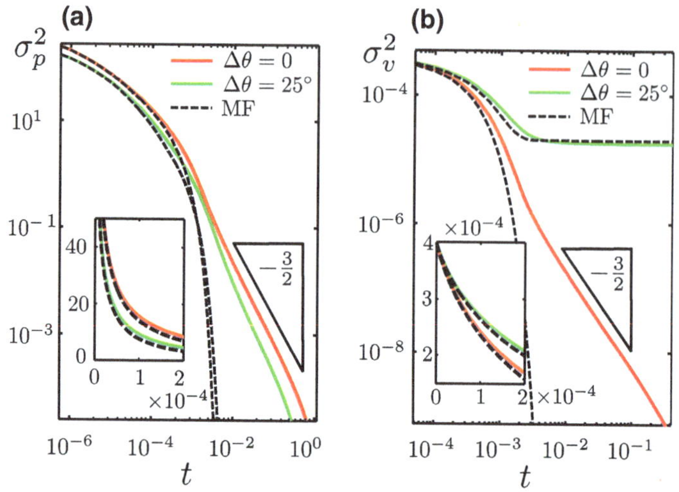

Fig. 13 **a** Square of the variance σ_P of the capillary pressure P of bridges as a function of time t. **b** Square of the variance σ_V of capillary bridge volumes V as a function of time t. The receding contact angle is $\theta_r = 7°$, the advancing contact angle either $\theta_a = 7°$ or to $\theta_a = 32°$. Note that all quantities are rescaled according average radius of the spherical particles, interfacial tension γ_{lv}, and the mobility of the liquid film. Reprinted with permission from [9]. Copyright (2015) by the American Physical Society

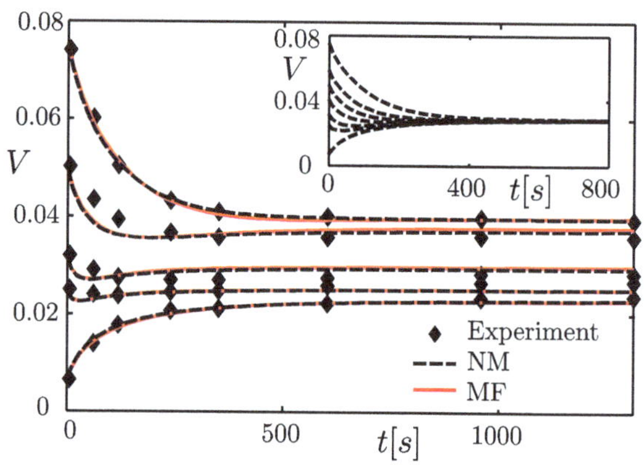

Fig. 14 Evolution of capillary bridge volumes predicted by the network model in comparison the X-tomography experiments [11, 18]. Inset: Network simulations for vanishing contact angle hysteresis. Reprinted with permission from [9]. Copyright (2015) by the American Physical Society

volumes for large contact angle hysteresis, cf. the dashed lines in Fig. 14a, b. Within this mean field model, the volume flux is driven by the difference of the local capillary pressure and the globally averaged pressure [9].

A comparison of the equilibration dynamics of large assembly of capillary bridges in the network and mean field model to the results of X-ray tomography imaging of wet glass beads [18] is shown in Fig. 14. In the simulated liquid equilibration, the capillary bridges are binned and labeled according to their volume at the time of preparation $t = 0$. Like in the plot shown in Fig. 3, "section Liquid Transport in the Pendular Regime", the plot in Fig. 14 displays the average volume of capillary bridges with the same label (initial volume bin) as a function of time. It can be clearly observed in the plot that, on average, capillary bridges with an initially small volume grow while bridges with a initially large volume shrink. As already mentioned for the data shown in Fig. 3, the curves do not tend to the same final value, and a certain spread of the volume distribution persists at asymptotically large times. However, for a vanishing contact angle hysteresis, the volume distribution of the capillary bridges in both network and mean field simulations for otherwise unchanged parameters tends to a very narrow peak, cf. the inset of Fig. 14.

Capillary Forces Between Spherical Particles

The majority of microscopic models of capillary cohesion in wet particulate materials are restricted to the pendular bridge regime at low liquid saturations. In many situations the assumption of disconnected capillary bridges on the particles is not justified and it is highly questionable to estimate cohesive forces in the funicular regime from force models of capillary bridges. Considering perfectly wetting liquids and dense beds of spherical particles, this pendular bridge regime is limited to a range of liquid contents $S < S^* \approx 4.2 \cdot 10^{-2}$ [11, 44]. For slightly higher liquid saturations $S > S^*$, the extensions of the capillary bridges on the spherical surfaces does not anymore permit the formation of isolated bridges on the particle [11]. Consequently, a certain fraction of the capillary bridges coalesces and transforms into funicular liquid clusters as shown in the tomography image in Fig. 2.

The magnitude of cohesion in the funicular regime is not only determined by the liquid saturation. The experimental data by Rumpf and Schubert in Ref. [45] shown in Fig. 15 reveal that the tensile strength of a packed bed of particles measured during drainage is much higher that the values measured at the same saturation during a re-invasion of the wetting liquid. A similar history dependence can be observed in

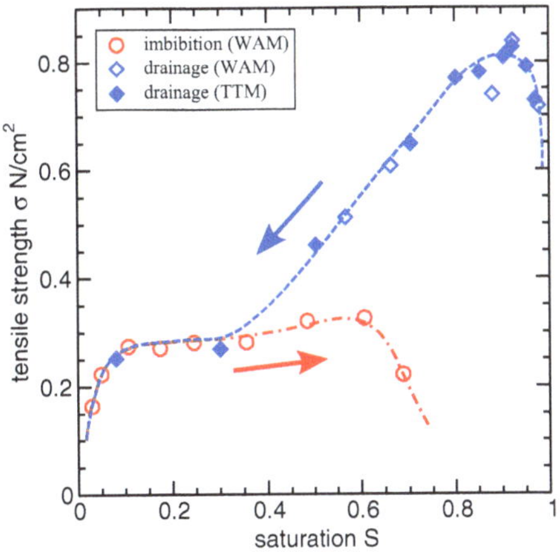

Fig. 15 Tensile strength σ of wet granular bed of lime stone particles with size as a function of the saturation S of the pore space with water for drainage (blue diamonds) and for imbibition (red circles). The lines serve as guide to the eyes. Data are reproduced from Ref. [45]

traces of the capillary pressure plotted against the saturation in a static bed of particles which is explained from different pore-scale distributions and morphologies of the fluids during drainage and re-invasion [46].

A hysteresis of cohesive forces in partially saturated particle beds and the underlying history dependence of the liquid morphology on the pore-scale can be demonstrated already for the most fundamental local structure, an arrangement of three spherical particles. In the absence of high symmetries, like the axial symmetry of an isolated capillary bridge, the liquid morphologies and the capillary multi-body interactions between three or more spherical particles can only be studied in numerical methods like NEM introduced in section "Numerical Computation of Capillary Surfaces".

Three Spherical Particles with One Gap

Capillary bridge trimers represent the smallest liquid cluster morphology in partially saturated beds of spherical particles and are the key to understand the cohesive mechanics in the funicular regime [11, 18, 28]. Trimers like the one shown in the example Fig. 16a are found at throats formed by three identical spherical particles where two of the three particles are in mechanical contact. Three dimensional imaging using X-ray micro-tomography show that configurations of three particles with two contacts and a single finite gap occur frequently in mechanically stable beds

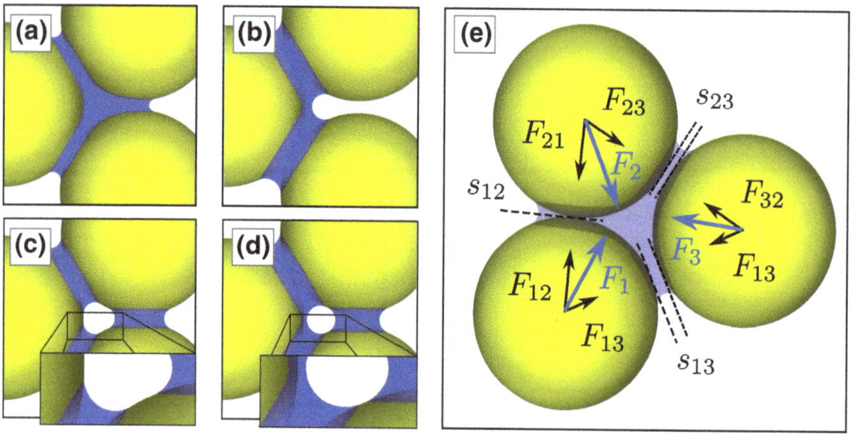

Fig. 16 All four possible morphologies of liquid in contact to three identical spherical particles with one finite gap and two contacts for a material contact angle of $\theta_0 = 5°$. Shown are **a** a trimer of three capillary bridges, **b** a dimer of two capillary bridges, **c** three capillary bridges and **d** a dimer and a capillary bridge. **e** Decomposition of the capillary forces $\mathbf{F}_i$ on three spherical particles $i = 1, 2, 3$ in a capillary bridge trimer into three central force pairs $\mathbf{F}_{ij} = -\mathbf{F}_{ji}$. Reprinted with permission from [28]. Copyright (2016) by the American Physical Society

of spherical particles [11]. Such a local configuration is uniquely specified by the surface-to-surface separation s_{12} between the two spherical particles 1 and 2 forming the gap. Since the surfaces of most particles are fully wet by the liquid, the contact angle of the liquid interface on the particles in the following results of NEM is set to the small value $\theta_0 = 5^\circ$.

In a certain range of the gap openings s_{12} and volume V or capillary pressure P, the wetting liquid may form a dimer of two capillary bridges leaving the gap dry, cf. also Fig. 16b. Dimers and trimers of capillary bridges belong to the same class regarding the topology of their liquid-vapor interface and can only be distinguished by the position of the meniscus in the gap [28]. Besides the possibility to find two or three isolated capillary bridges, as shown in Fig. 16c, the dimer morphology can coexist with a capillary bridge spanning the gap, cf. Fig. 16d. The latter morphology is found only in a very narrow range of the control parameters gap separation and liquid volume or capillary pressure.

In the volume controlled case, the regions of gap separation s_{12} and liquid volume V where capillary bridge trimers and dimers are found, is best illustrated in a morphology diagram Fig. 17a. Experimental studies of trimers of volatile liquids have shown that trimers may undergo two different types of interfacial instabilities during reductions of the volume [16, 28]. For small gap separations, a shrinking capillary bridge trimer may break up into three isolated capillary bridges when the two opposing liquid menisci in the center of the throat touch [16, 28], cf. (Fig. 17b). Alternatively to this 'burst' mode, one can observe a 'flip-in' instability where the interface of the capillary bridge at the gap is pulled to the center of the throat, cf.

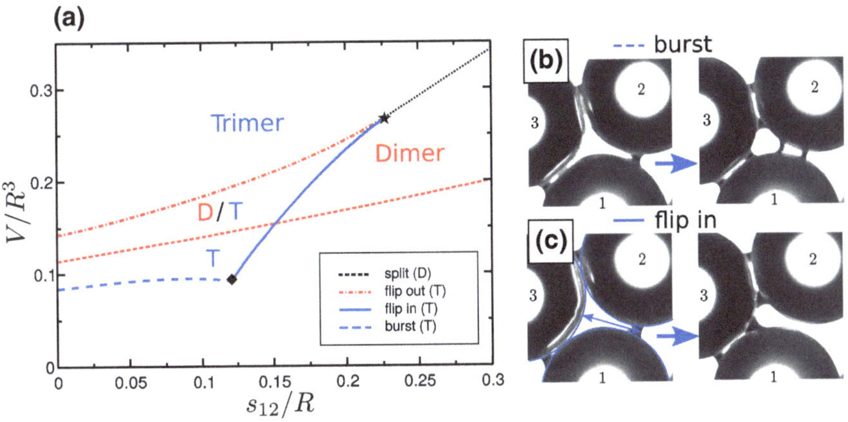

Fig. 17 **a** Morphology diagram of the trimer (T) and dimer (D) morphology between three identical spherical particles of radius R in the volume controlled case. The liquid volume is denoted by V, the surface-to-surface separation in the gap by s_{12}. The contact angle of the liquid on the surface of the spherical particles is $\theta = 5^\circ$. **b** Trimer decaying by a burst into three capillary bridges for narrow gap separation $s_{12}/R < 0.145$. **c** Discontinuous snap-in of a trimer and subsequent decay into two capillary bridges for a wide gap separation $s_{12}/R \gtrsim 0.12$

(Fig. 17c). For sufficiently small volumes this 'flip-in' transition (or the reverse flip-out) between a trimer and a dimer is discontinuous. A bistability of a capillary dimer and trimer can be found in a certain range of volumes and gap separations.

In order to quantify the capillary forces on the spherical particles in the presence of a certain liquid morphology, the attractive forces on the particles are computed directly from the suction force of the capillary pressure and the interfacial tension acting along the contact lines, as outlined in section "Calculation of Capillary Forces". For three spherical particles that are not in a co-linear alignment, and in the absence of contact angle hysteresis, it is always possible to decompose the total capillary force into force pairs acting on the sphere centers [28].[2] The possibility, however, of such a decomposition does not imply that the capillary forces in a trimer can always be represented as two-body forces. As the pair forces do not only depend on the gap opening of the respective gap but also on the opening of the remaining two gaps, the capillary forces between three spherical particles in the presence of a trimer are multi-body forces in general [28].

The map of the capillary force of a trimer and dimer Fig. 18 is defined only in regions of gap opening s_{12} and liquid volume V where either the trimer or the dimer state exist as a local minimum of the interfacial energy, cf. also the morphology diagram in Fig. 17a. Alternatively, we find regions of the control parameters s_{12} and V where both trimer and dimer states represent stable conformations of the liquid, which implies that the capillary forces F_{12} and F_{13} may be multi-valued functions for given control parameters (s_{12}, V) or (s_{12}, P).

The numerically computed capillary force map of the trimer in the volume controlled case Fig. 18a reveals that the force F_{12} across the gap is sensitive to the gap opening s_{12}, and falls off almost linearly with an increasing gap opening until the meniscus at the gap undergoes a flip-in. In contrast, the pair force F_{13} of a trimer acting at the two particle contacts depends only weakly on the gap openings s_{12}, cf. Fig. 18b. As expected from the distribution of the wetting liquid between the spherical particles, the force F_{12} across the gap in the presence of a dimer is small while the attractive force F_{13} at the particle contacts in the dimer state is almost as high as in the trimer state. Furthermore, the capillary forces of a dimer differ only slightly from those of two communicating capillary bridges comprising the same volume as the dimer. Analogous observations can be made for the forces of trimers and dimers in the capillary pressure controlled case shown in the force maps Fig. 18c, d. Since the capillary forces of a dimer are in terms of their magnitude very similar to those of two disconnected but communicating capillary bridges of the same total volume (or at the same capillary pressure), the cohesive forces of a dimer can be easily modeled by the forces of two ideal capillary bridges.

---

[2]This proof can be easily extended to four spherical particles whose centers are not co-planar.

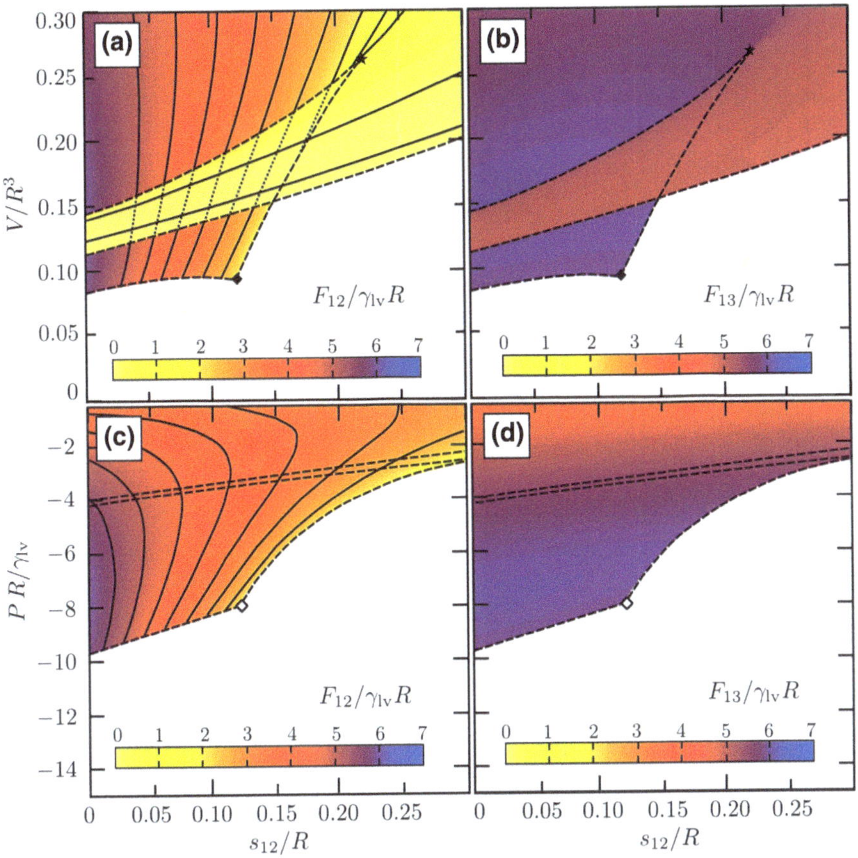

Fig. 18 Capillary forces F_{ij} of bridge trimer (dimer) between three identical spherical particles of radius R as a function of the surface-to-surface separation s_{12} for a fixed capillary pressure V in top row and for a fixed capillary pressure P in bottom row. The forces F_{12} across the gap are shown in (**a**) and (**c**) while the forces F_{13} at the contacts are displayed in (**b**) and (**d**). The contact angle is $\theta = 5°$. Reprinted with permission from [9]. Copyright (2015) by the American Physical Society

Trimer at Three Finite Gaps

The results of our NEM in section "Three Spherical Particles with One Gap" demonstrate that transformations of a capillary bridge trimer into a bridge dimer or vice versa are accompanied by large changes of the attractive forces acting across the gap. Based on the observation that the forces of a capillary bridge dimer can be well approximated by the forces of disconnected capillary bridges for the same total volume or capillary pressure, one can restrict the NEM to the proper trimer state where all three gaps are wet by liquid.

In accordance with the previously discussed case of one gap and two contacts in section "Three Spherical Particles with One Gap", the results of NEM for the

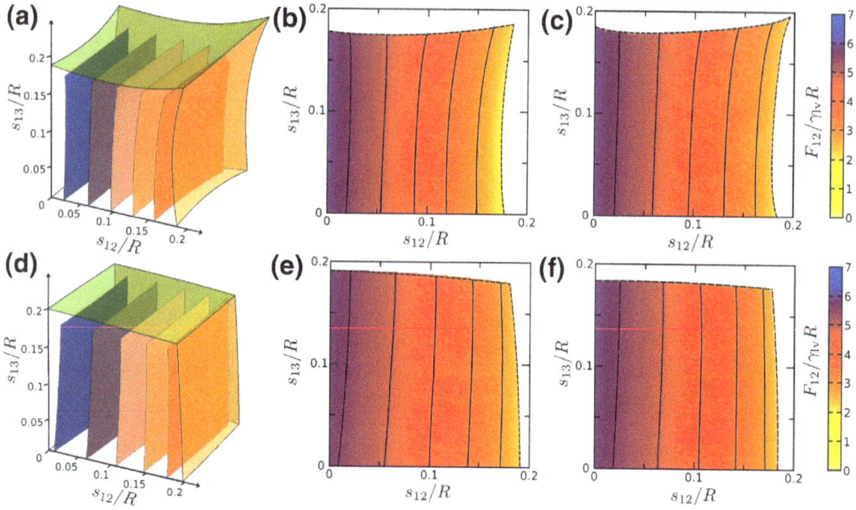

Fig. 19 **a** Isosurface plot of capillary forces F_{12} of a liquid bridge trimer as function of the gap openings s_{ij} at a fixed volume $V = 0.2$ for a $\theta_0 = 5°$. **b**, **c** Color density plots of F_{12} as a function of s_{12} and s_{13} for fixed gap openings $s_{23} = 0$ in (**b**) and $s_{23} = 0.17$ in (**c**). **d** Isosurface plot of F_{12} as a function of s_{ij} at a fixed capillary pressure $P = -4.5\ \gamma_{lv}/R$ for $\theta_0 = 5°$. **e**, **f** Color density plots of F_{12} as a function of s_{12} and s_{13} for fixed a gap opening $s_{23} = 0$ in (**e**) and $s_{23} = 0.17$ in (**f**)

general case of three finite gap openings demonstrate that the pair force F_{12} between particle 1 and 2 decreases almost linearly with the gap opening s_{12} until the trimer decays. The force F_{12} is hardly affected by the two remaining gap openings. This insensitivity can be observed in the three dimensional iso-surface plots of F_{12} shown in Fig. 19a where contours of different threshold values of F_{12} form an array of almost parallel planes stacked along the direction of the s_{12} axis. In addition, the region of the two remaining gap openings s_{13} and s_{23}, where the trimer state exists for a fixed volume V, assumes the shape of a slightly distorted cube being stretched along the diagonals, cf. Fig. 19a. Two-dimensional force maps Fig. 19b, c of F_{12} in terms of the gap openings s_{12} and s_{23} for two fixed values of s_{23} illustrate the aforementioned features.

In addition to the case of a prescribed trimer volume, analogous NEM were performed to determine the capillary forces of a trimer between spherical particles with three finite gaps for the case of a prescribed capillary pressure. Also here, the capillary forces at every of the three gaps depends almost linearly on the respective gap opening but only weakly on the remaining two gap openings, cf. Fig. 19d–f. The only qualitative difference between the pressure controlled case Fig. 19d–f and the volume controlled case Fig. 19a–c is the shape of the region in the space of the three gap openings (s_{12}, s_{13}, s_{23}), where mechanically stable capillary bridge trimers exist. In the capillary pressure controlled case, this region assumes the shape of a cube that has been shrunken along the space diagonals.

The numerical energy minimizations demonstrate that the capillary forces F_{ij} a of capillary bridge trimer depend mostly on the respective gap opening s_{ij} and may be considered as independent central force pairs, similar to cohesive force of an ensemble of isolated capillary bridges. Regarding the magnitude of the attractive forces, however, the cohesion of a capillary bridge trimer is larger than that of three isolated bridges in both, the volume and the capillary pressure controlled case.

Capillary Forces Between Four Spherical Particles

A wetting liquid in a bed of spherical will be found not only at the contacts and the throats formed by three adjacent spherical particles, but also in tight pores formed by four particles that are close to a tetrahedral arrangement. Such a 'hexamer' of six coalesced capillary bridges around a liquid filled tetrahedral pore is shown in the inset of Fig. 20. Capillary bridge hexamers in ideal tetrahedral pores with all particles in contact can be metastable for capillary pressures above $P \approx -12\,\gamma_{\mathrm{lv}}/R$, in contrast to the minimum pressure of $P \approx -10\,\gamma_{\mathrm{lv}}/R$ for a trimer of capillary bridges in an ideal throat.

The numerically computed attractive capillary forces at the gaps of a symmetric hexamer around a tetrahedral pore with opening s in Fig. 20 shows that the magnitude F of these pair forces in the pressure controlled ensemble is slightly higher than the forces of an ideal trimer or capillary bridge. These plots also demonstrate that the attractive forces at the gaps of a hexamer, trimer, and of single capillary bridges all decay as the gap opening s of the spherical grain increases. Hexamers not only create the largest capillary force in the pressure controlled case, but also exhibit the

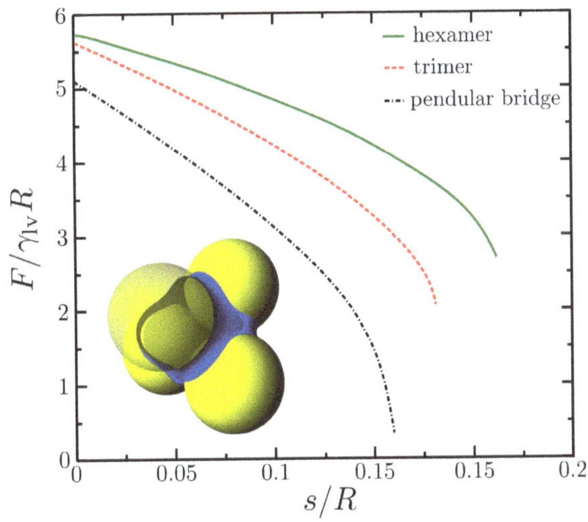

Fig. 20 Comparison of capillary forces F between spherical particles in a symmetric capillary bridge hexamer, a symmetric capillary bridge trimer, and of a capillary bridge as a function of the surface-to-surface separation s at a fixed capillary pressure $P = -4.5\gamma_{\mathrm{lv}}/R$. The maximum extension and the magnitude of the attractive force increases from the single bridge over the trimer to the hexamer

largest rupture distance. Hence, the formation of hexamers around a pore body will contribute to a further strengthening of the granular bed.

Rolling Friction of Capillary Bridges Caused by Contact Angle Hysteresis

Static contact angle hysteresis observed on virtually all solid surfaces is a consequence of the work dissipated during motion of the three phase contact line. To sustain the motion of the interface on a solid, a certain amount of mechanical work per time needs to be injected. The dissipated energy manifests itself not only as contact angle hysteresis but also as non-conservative forces or torques.

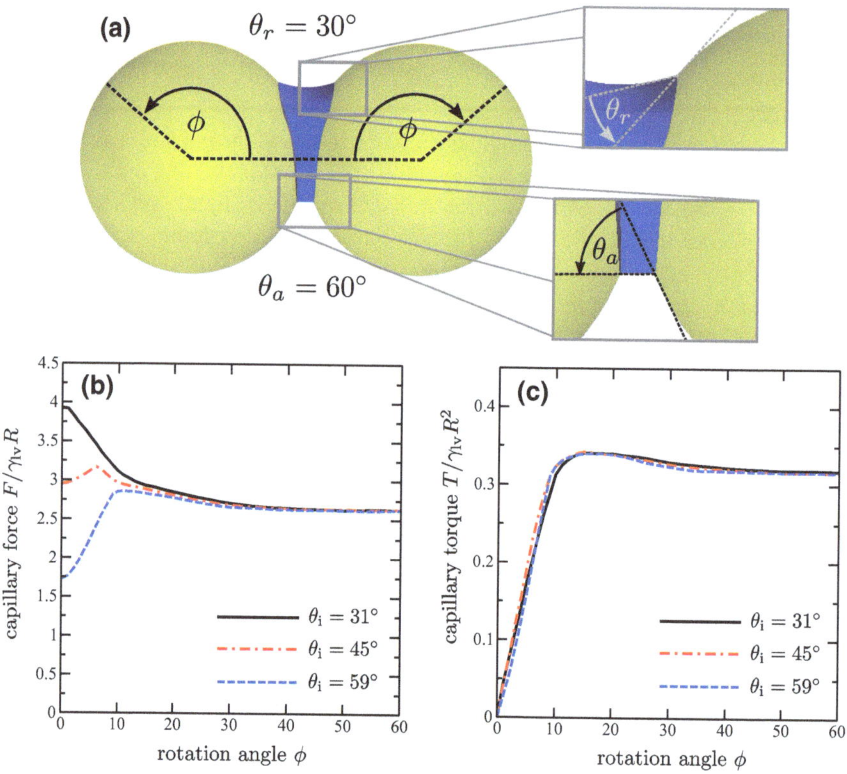

Fig. 21 **a** Capillary bridge with volume $V/R^3 = 0.1$ between two spherical particles rolling on each other with contact angle hysteresis $\theta_a = 60°$ and $\theta_r = 30°$. **b**, **c** Evolution of the capillary force and torque on the spherical particle as a function of the rotation angle ϕ for different initial contact angles θ_i

A relevant example illustrating how the dissipation of energy caused by contact angle hysteresis affects the mechanics of liquid bridges is given by two spherical particles that roll on each other without slip, while they hold a capillary bridge at their point of contact, cf. the renderings in Fig. 21a. This problem can be studied in a modified NEM algorithm where the contact line is displaced according to the local contact angle. An outline of this algorithm is found at the end of section "Numerical Computation of Capillary Surfaces".

The mechanics of capillary bridges that are deformed by a relative motion of the wetted surfaces is best explained by the simple example of an initially axially symmetric capillary bridge spanning between two identical spherical particles. The rolling motion of the solid spherical particles shall be sufficiently slow such that effects related to the bulk dynamics of the liquid can be neglected. Furthermore, the initial contact angle θ_i of the capillary bridge shall be different to both, the advancing contact angle θ_a and the receding contact angle θ_r.

Under these conditions, the deformation of the liquid interface is fully reversible for small rotation angles as long as the three phase contact line on the surface of the particles remains everywhere in the pinned state. As a consequence of the interface deformation, one finds a restoring torque that counteracts the rolling motion of the spherical particles, cf. the evolution of the torques in Fig. 21b. After a rotation by a certain finite angle, the local contact angle $\theta(\mathbf{x})$ along the contact lines of the deformed interface reaches the condition for de-pinning at some point. Either the contact angle has reached the advancing contact angle at a point in the rear of the capillary bridge or the contact angles has decreased so much that the receding contact angle is attained in a point of the contact line in the front of the capillary bridge.

Upon any further rotation into the initial direction, larger and larger parts of the contact line de-pin and move relative to the surface of the spherical particles. After de-pinning, the interface of the bridge converges to a stationary shape. The extension of the transition from the first phase where the contact line is partially or than $\phi \approx 45°$, cf. the example of the forces and torques shown in Fig. 21(b, c). Besides the quick build-up of the restoring torque, one observes also noticeable changes of the normal capillary force acting on the rolling contact. These changes are pronounced in the range of rotation angles $\phi \lesssim 45°$ but quickly cease with growing angle ϕ. Independent on the initial contact angle, the normal force and torque level off at a value that is independent on the initial contact angle of the capillary bridge.

The magnitude of the restoring torque T and the normal capillary forces F of a capillary bridge in the stationary interfacial shape between the rolling spherical particles are summarized in color density maps in Fig. 22 in terms of the advancing contact angle θ_a and receding contact angle θ_i. Points on the diagonal of the maps in Fig. 22 correspond to a vanishing contact angle hysteresis. Apparently, the normal force is the higher the lower the advancing contact angle. For a fixed advancing contact angle, the magnitude of F decreases as the receding contact angle is increased. Contrary to the normal capillary forces, the capillary torque attains its maximum for the largest contact angle hysteresis, i.e. the smallest receding contact angle and the largest advancing contact angle.

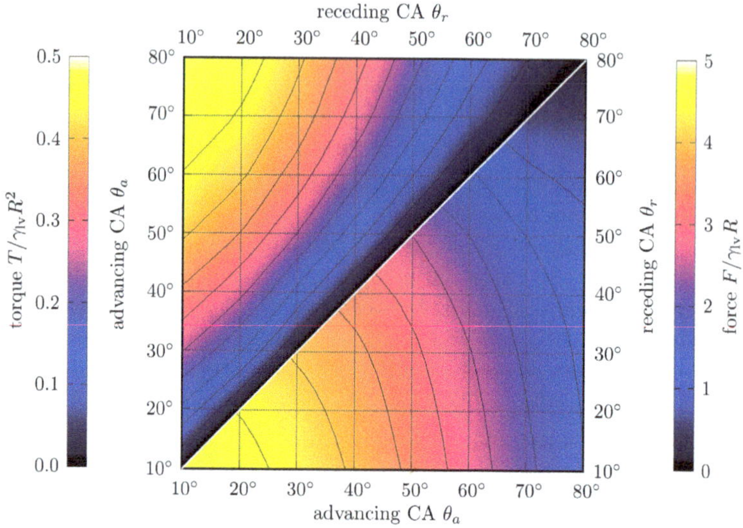

Fig. 22 Plot of the capillary forces and torques of two steadily rolling spherical particles induced by a capillary bridge with a volume $V/R^3 = 0.1$ as a function of advancing contact angle θ_a and receding contact angle θ_r

Appendix

Equivalence of Capillary Forces

By differentiation of the Grand interfacial free energy $\mathcal{G}$ Eq. (2) with respect to the coordinates $\bar{\mathbf{r}}$ one obtains the following identity:

$$\nabla_{\mathbf{r}_i} G(\bar{\mathbf{r}}, P) = \nabla_{\mathbf{r}_i} E(\bar{\mathbf{r}}, \tilde{V}(\bar{\mathbf{r}}, P)) - P\, \nabla_{\mathbf{r}_i} \tilde{V}(\bar{\mathbf{r}}, P)\,. \tag{25}$$

where

$$P = \tilde{P}(\bar{\mathbf{r}}, V) = \partial_V E(\bar{\mathbf{r}}, V)\,. \tag{26}$$

Employing now the chain rule of differentiation, the first term in Eq. 25) can be rewritten as:

$$\begin{aligned} \nabla_{\mathbf{r}_i} E(\bar{\mathbf{r}}, \tilde{V}(\bar{\mathbf{r}}, P)) &= \nabla_{\mathbf{r}_i} E(\bar{\mathbf{r}}, V)|_{V=\tilde{V}(\bar{\mathbf{r}},P)} \\ &+ \partial_V E(\bar{\mathbf{r}}, V)|_{V=\tilde{V}(\bar{\mathbf{r}},P)} \nabla_{\mathbf{r}_i} \tilde{V}(\bar{\mathbf{r}}, P)\,, \end{aligned} \tag{27}$$

and one arrives at:

$$\nabla_{\mathbf{r}_i} G(\bar{\mathbf{r}}, P) = \nabla_{\mathbf{r}_i} E(\bar{\mathbf{r}}, V)|_{V=\tilde{V}(\bar{\mathbf{r}},P)} \tag{28}$$

which shows the equivalence of capillary forces in the volume controlled and capillary pressure controlled case.

References

1. Haines, W.B.: Studies in the physical properties of soils: II. A note on the cohesion developed by capillary forces in an ideal soil. J. Agr. Sci. **15**(4): 529–535 (1925). https://doi.org/10.1017/S0021859600082460
2. Fisher, R.A.: On the capillary forces in an ideal soil; correction of formulae given by W. B. Haines. J. Agric. Sci. **16**(3), 492–505 (1926). https://doi.org/10.1017/S0021859600007838
3. Pye, K., Tsoar, H.: Aeolian Sand and Sand Dunes. Springer (2009)
4. Kim, T.H., Hwang, C.: Modeling of tensile strength on moist granular earth material at low water content. Eng. Geol. **69**(3), 233–244 (2003). https://doi.org/10.1016/S0013-7952(02)00284-3
5. Simons, S.J.R., Fairbrother, R.J.: Direct observations of liquid binder-particle interactions: the role of wetting behaviour in agglomerate growth. Powder Technol. **110**(1–2), 44–58 (2000). https://doi.org/10.1016/S0032-5910(99)00267-3
6. Iveson, S.M., Litster, J.D., Hapgood, K., Ennis, B.J.: Nucleation, growth and breakage phenomena in agitated wet granulation processes: a review. Powder Technol. **117**(1–2), 3–39 (2001). https://doi.org/10.1016/S0032-5910(01)00313-8
7. Herminghaus, S.: Wet granular matter. A truly complex fluid. In: Series in Soft Condensed Matter, vol. 6. World Scientific Publishing, Singapore (2013)
8. Lukyanov, A.V., Sushchikh, M.M., Baines, M.J., Theofanous, T.G.: Superfast nonlinear diffusion: capillary transport in particulate porous media. Phys. Rev. Lett. **109**(21), 214501 (2012). https://doi.org/10.1103/PhysRevLett.109.214501
9. Mani, R., Semprebon, C., Kadau, D., Herrmann, H.-J., Brinkmann, M., Herminghaus, S.: Role of contact-angle hysteresis for fluid transport in wet granular matter. Phys. Rev. E **91**, 042204 (2015). https://doi.org/10.1103/PhysRevE.91.042204
10. Singh, K., Scholl, H., Brinkmann, M., Di Michiel, M., Scheel, M., Herminghaus, S., Seemann, R.: The role of local instabilities in fluid invasion into permeable media. Sci. Rep. **7**, 444 (2017). https://doi.org/10.1038/s41598-017-00191-y
11. Scheel, M., Seemann, R., Brinkmann, M., Di Michiel, M., Sheppard, A., Breidenbach, B., Herminghaus, S.: Morphological clues to wet granular pile stability. Nat. Mater. **7**, 189–193 (2008). https://doi.org/10.1038/nmat2117
12. Melnikov, K., Mani, R., Wittel, F.K., Thielmann, M., Herrmann, H.-J.: Grain-scale modeling of arbitrary fluid saturation in random packings. Phys. Rev. E **92**(2), 022206 (2015). https://doi.org/10.1103/PhysRevE.92.022206
13. Butt, H.-J., Kappl, M.: Normal capillary forces. Adv. Colloid Interface Sci. **146**(1–2), 48–60 (2009). https://doi.org/10.1016/j.cis.2008.10.002
14. Turton, R.: Challenges in the modeling and prediction of coating of pharmaceutical dosage forms. Powder Technol. **181**(32), 186–194 (2008). https://doi.org/10.1016/j.powtec.2006.12.006
15. Blunt, M.J.: Multiphase Flow in Permeable Media: A Pore–Scale Perspective. Cambridge University Press (2017)
16. Scheel, M.: Experimental investigations of the mechanical properties of wet granular matter. Dissertation, University of Göttingen (2010). http://hdl.handle.net/11858/00-1735-0000-0006-B4BC-4
17. Halsey, T.C., Levine, A.J.: How sandcastles fall. Phys. Rev. Lett. **80**(14), 3141–3144 (1997). https://doi.org/10.1103/PhysRevLett.80.3141
18. Scheel, M., Seemann, R., Brinkmann, M., Di Michiel, M., Sheppard, A., Herminghaus, S.: Liquid distribution and cohesion in wet granular assemblies beyond the capillary bridge regime.

J. Phys.: Condens. Matter **20**(49), 494236 (2008). https://doi.org/10.1088/0953-8984/20/49/494236
19. Melnikov, K., Wittel, F.K., Herrmann, H.J.: Micro-mechanical failure analysis of wet granular matter. Acta Geotechnica **11**(3), 539–548 (2016). https://doi.org/10.1007/s11440-016-0465-5
20. Friedman, S.R., Khalil, M., Taborek, P.: Wetting transition in water. Phys. Rev. Lett. **111**(22), 226101 (2013). https://doi.org/10.1103/PhysRevLett.111.226101
21. Pompe, T., Herminghaus, S.: Three-phase contact line energetics from nanoscale liquid surface topographies. Phys. Rev. Lett. **85**(9), 1930–1933 (2000). https://doi.org/10.1103/PhysRevLett.85.1930
22. Hofmann, T., Tasinkevych, M., Checco, A., Dobisz, E., Dietrich, S., Ocko, B.M.: Wetting of nanopatterned grooved surfaces. Phys. Rev. Lett. **104**(10), 106102 (2010). https://doi.org/10.1103/PhysRevLett.104.106102
23. do Carmo, M.: Differential Geometry of Curves and Surfaces, Prentice Hall (1976)
24. Brakke, K.A.: The surface evolver. Exp. Math. **1**(2), 141–165 (1992). https://doi.org/10.1080/10586458.1992.10504253
25. Brakke, K.A., Klinowski, J., Mackay, A.L.: The surface evolver and the stability of liquid surfaces. Philos. Trans. R. Soc. Lond. A **354**, 2143–2157 (1996). https://doi.org/10.1098/rsta.1996.0095
26. Eral, H.B., de Ruiter, J., de Ruiter, R., Oh, J.M., Semprebon, C., Brinkmann, M., Mugele, F.: Drops on functional fibers: from barrels to clamshells and back. Soft Matter **7**, 5138–5143 (2011). https://doi.org/10.1039/C0SM01403F
27. Semprebon, C., Brinkmann, M.: On the onset of motion of sliding drops. Soft Matter **10**, 3325–3334 (2014). https://doi.org/10.1039/C3SM51959G
28. Semprebon, C., Scheel, M., Herminghaus, S., Seemann, R., Brinkmann, M.: Liquid morphologies and capillary forces between three spherical beads. Phys. Rev. E **94**(1), 012907 (2016). https://doi.org/10.1103/PhysRevE.94.012907
29. De Gennes, P.G.: Wetting: statics and dynamics. Rev. Mod. Phys. **57**(3), 827–863 (1985). https://doi.org/10.1103/RevModPhys.57.827
30. De Gennes, P.G., Brochard-Wyart, F., Quéré, D.: Capillarity and Wetting Phenomena: Drops, Bubbles, Pearls, Waves. Springer (2004)
31. Wenzel, R.N.: Resistance of solid surfaces to wetting by water. Ind. Eng. Chem. **28**(8), 988–994 (1936). https://doi.org/10.1021/ie50320a024
32. Herminghaus, S.: Wetting, spreading, and adsorption on randomly rough surfaces. Eur. Phys. J. E **35**, 43 (2012). https://doi.org/10.1140/epje/i2012-12043-8
33. Herminghaus, S.: Universal phase diagram for wetting on mesoscale roughness. Phys. Rev. Lett. **109**, 236102 (2012). https://doi.org/10.1103/PhysRevLett.109.236102
34. Dufour, R., Semprebon, C., Herminghaus, S.: Filling transitions on rough surfaces: Inadequacy of Gaussian surface models. Phys. Rev. E **93**, 032802 (2016). https://doi.org/10.1103/PhysRevE.93.032802
35. Semprebon, C., Forsberg, P., Priest, C., Brinkmann, M.: Pinning and wicking in regular pillar arrays. Soft Matter **10**, 5739–5748 (2014). https://doi.org/10.1039/C4SM00684D
36. Kohonen, M., Geromichalos, D., Scheel, M., Schier, C., Herminghaus, S.: On capillary bridges in wet granular materials. Physica A **339**, 7–15 (2004). https://doi.org/10.1016/j.physa.2004.03.047
37. Niven, R.K.: Force stability of pore-scale fluid bridges and ganglia in axisymmetric and non-axisymmetric configurations. J. Pet. Sci. Eng. **52**, 1 (2006). https://doi.org/10.1016/j.petrol.2006.03.015
38. Farmer, T.P., Bird, J.C.: Asymmetric capillary bridges between contacting spheres. J. Colloid Interface Sci. **454**, 192–199 (2015). https://doi.org/10.1016/j.jcis.2015.04.045
39. Dormann, M., Schmid, H.J.: Simulation of capillary bridges between nanoscale particles. Langmuir **30**, 1055–1062 (2014). https://doi.org/10.1021/la404409k
40. Eral, H.B., 't Mannetje, D.J.C.M., Oh, J.M.: Contact angle hysteresis: a review of fundamentals and applications. Colloid Polym. Sci. **291**(2), 247–260 (2013). https://doi.org/10.1007/s00396-012-2796-6

41. Pepin, X., Rossetti, D., Iveson, S.M., Simons, S.J.R.: Modeling the evolution and rupture of pendular liquid bridges in the presence of large wetting hysteresis. J. Colloid Interface Sci. **232**(2), 289–297 (2000). https://doi.org/10.1006/jcis.2000.7182
42. Willet, C.D., Adams, M.J., Johnson, S.A., Seville, J.P.K.: Capillary bridges between two spherical bodies. Langmuir **16**(24), 9396–9405 (2000). https://doi.org/10.1021/la000657y
43. Willet, C.D., Adams, M.J., Johnson, S.A., Seville, J.P.K.: Effects of wetting hysteresis on pendular liquid bridges between rigid spheres. Powder Technol. **130**, 63–69 (2003). https://doi.org/10.1016/S0032-5910(02)00235-8
44. Harris, C.C., Morrow, N.R.: Pendular moisture in packings of equal spheres. Nature **203**, 706–708 (1964). https://doi.org/10.1038/203706b0
45. Rumpf, H., Schubert, H.: The behavior of agglomerates under tensile strain. J. Cherm. Eng. Jpn. **7**(4), 294–298 (1974). https://doi.org/10.1252/jcej.7.294
46. Bear, J.: Modeling Phenomena of Flow and Transport in Porous Media. Springer (2018)
47. Richefeu, V., El Youssoufi, M.S., Radjaï, F.: Shear strength properties of wet granular materials. Phys. Rev. E **73**, 051304 (2006). https://doi.org/10.1103/PhysRevE.73.051304
48. Mani, R., Kadau, D., Or, D., Herrmann, H.-J.: Fluid depletion in shear bands. Phys. Rev. Lett. **109**, 248001 (2012). https://doi.org/10.1103/PhysRevLett.109.248001
49. de Ruiter, R., Semprebon, C., van Gorcum, M., Duits, M.H.G., Brinkmann, M., Mugele, F.: Stability limits of capillary bridges: how contact angle hysteresis affects morphology transitions of liquid microstructures. Phys. Rev. Lett. **114**, 234501 (2015). https://doi.org/10.1103/PhysRevLett.114.234501

Capillary Interaction in Wet Granular Assemblies: Part 2

Falk K. Wittel, Roman Mani, Konstantin Melnikov, Filippo Bianchi and Hans J. Herrmann

Nomenclature

α_{jk}	Opening angle
β	Opening angle
ε	Geometrical correction parameter
η	Fluid viscosity
η^{coh}	Cohesion number
γ	Shear deformation
γ	Surface tension
κ	Inverse flow resistance
λ	Drainage parameter
μ	Coulomb friction coefficient
μ_f	Internal friction coefficient
ω_{ij}	Conductance coefficient
Φ	Packing density
ϕ	Face-edge angle of tetrahedron
ψ	Filling angle of the meniscus
σ	Stress
τ	Shear strength
τ_i	Relaxation times
θ	Angle of internal friction
θ	Contact angle
ε	Strain
A	Area
a_i, b	Fitting parameters
C	Surface curvature

F. K. Wittel (✉) · R. Mani · K. Melnikov · F. Bianchi · H. J. Herrmann
Institute for Building Materials, ETH Zurich, Stefano-Franscini Platz 3, Zurich 8093, Switzerland
e-mail: fwittel@ethz.ch

S. Antonyuk (ed.), *Particles in Contact*,
https://doi.org/10.1007/978-3-030-15899-6_9

c	Cohesion
D	Macroscopic diffusion constant
F	Force
g_{ij}	Local conductance coefficient between two pores
H	Height
N	Number
P	Pressure
q	Differential stress
Q_{eff}	Net in- or outflow from a cell
q_{ij}	Liquid flux between neighboring cells i, j
R	Particle radius
$r_{1,2,men}$	Radii of curvature of bridges ($_{1,2}$) and menisci ($_{men}$)
$s_{ij/c}$	Particle distance/bridge rupture distance
t	Time
V	Volume
V_{ij}	Volume of liquid bridge between particle i and j
W_c	Liquid volume content
x, y, z	Cartesian coordinates
CD	Contact Dynamics
DEM	Discrete Element Method
NEM	Numerical Energy Minimization
pbc	Periodic boundary condition
v	Velocity

Introduction

Liquids are known to alter the mechanical behavior of granular systems dramatically what can be impressively experienced when creating sculptures such as the one shown in Fig. 1, that can only defy gravity as long as the sand is wet and partially saturated [25]. Cohesion by capillary forces, acting between neighboring grains, gives loose sand tensile strength. Increasing computer power allows to quantitatively study the nature of this phenomenon on the scale of single grains—also called pore scale—experimentally by computer tomography (part 1) and numerically on representative assemblies (part 2). These findings help to understand the processing of wet granulates, for example for granulation, molding or transport in the pharmaceutical industry, in materials or food processing.

For wet granular matter, one typically distinguishes four different states, characterized by increasing saturation: *pendular, funicular, capillary* and *slurry* [26, 45]. The *pendular state* is characterized by the sole existence of liquid bridges between grain pairs. The Laplace pressure of the liquid phase is negative resulting in attractive inter-granular forces. If more fluid is available, the *funicular state* is reached, where liquid bridges and fully saturated pores coexist, that still exert attractive forces. The

Fig. 1 ETH main building by ETH Prof. Gottfried Semper (1864) carved in sand (unknown artist from the 2017 AIV student sandcastle competition)

capillary state is reached when liquid fills almost all pores. Note that still concave menisci exist, resulting in a net cohesive force on the grain assembly. This condition distinguishes it from the *slurry state*, where the Laplace pressure in the liquid phase is equal to or larger than the air pressure. Without cohesion, the real material behavior is similar to the completely dry state [23]. On a geotechnical scale, landslides and debris flow are prominent examples of slurries, which are dangerous natural hazards occurring mostly in mountainous terrain when heavy rainfall mobilizes a large amount of debris. The mechanical properties of the granular soil are significantly altered before slope failure by the increasing presence of water. After cohesion reaches a maximum at lower saturation, capillary forces slowly decrease until the gravitational load exceeds cohesion and landslides can be triggered. This triggering effect can be studied by tracking the liquid distribution on the micro-scale. However, the complex nature of the problem requires larger assemblies of particles to be representative for the bulk. For larger liquid contents, also an astonishing richness of liquid body morphologies are revealed by computer tomography studies, which render direct computations based on surface minimization principles inapplicable. As we will show in the following, the findings from these simulations of Part 1 are valuable as direct input into grain scale models allowing to execute meaningful calculations on the microscopic level.

The organization of the chapter is as follows: Before focusing on models for wet granular assemblies, we demonstrate in section "Liquid in Multi-grain Assemblies" on two simple experimental setups the interplay of micro-scale forces, packing densities, liquid redistributions and resulting macro-behavior under tensile and shear loading. In section "Discrete Element Models for Granular Assemblies with Capillary Interactions" we describe model extensions to include the liquid into contact dynamics (CD) with spherical particles. After a brief description of CD, the essential phenomena and fundamental models for the pendular state are reviewed, and extensions to the funicular state are described. Section "Simulations for Wet

Particle Assemblies" starts with simulations of the pendular state where studies of fluid migration under shear are shown before the formation of liquid clusters and the loss of cohesion is studied for increasing liquid content.

Liquid in Multi-grain Assemblies

The mechanical behavior of granular packings is rather complicated. Liquids add complexity, as it introduces new timescales for fluid transport, depending on the transport mechanism. We demonstrate this for observations of force relaxation in a tensile split box setting and by liquid migration from shear bands in a shear cell.

Tensile Behavior in the Pendular State

The pendular state is characterized by the exclusive presence of liquid bridges. They form when two grains contact each other, start to act mechanically, once they overcome the effect of roughness [21], exchange fluid through the wetting film covering the particles [61] or the vapor phase [29] driven by Laplace pressure differences [56] and once excessively stretched, fracture [70]. Tensile failure of wet soils, for example, initiates landslides. Hence, it is important to understand the tensile stress relaxation in multi-grain assemblies as the cumulation of microscale forces and processes from capillary bridges, inter-particle friction, liquid flows and particle movements. We demonstrate the tensile relaxation behavior on a split cell box with rough inner surfaces (see Fig. 2a), filled by wet glass beads with mean radius of $\bar{R} = 61\ \mu\text{m}$, liquid volume content W_c below 2.4% and two distinct packing densities $\Phi \approx 0.41, 0.51$ by loosely filling and compacting via vibration (full description in Ref. [3]). One half of the split cell is moved at a constant velocity of $4 \cdot 10^{-6}$m/s, while the reaction force is measured on the second. When during the first extension the global strength is reached at a displacement of about 1.5–2.3$\bar{r}$, the position is kept, and the system relaxes for 30 min before the test is proceeded until complete separation (see a typical test in Fig. 2b). To reveal the effect of surface tension γ, one set of experiments was done with water ($\gamma = 72.8$mN/m) and another with added surfactant ($\gamma = 32.8$mN/m).

As one expects, the mean force of the first peak is about 2.5 times higher for the higher packing fraction. In dense packings ($\Phi \approx 0.51$), strains above which a crack becomes visible are about one order of magnitude lower than in loose packings and close to the peak force, while for $\Phi \approx 0.41$, particle rearrangements prohibit an early detection of the crack. Those observations are in agreement with earlier works on tensile strength [9, 14, 19, 33, 39, 49, 50, 64]. The higher strength has to do with a larger number of liquid bridges and smaller inter-particle distances resulting in increased resistance for spatial rearrangement, since each particle displacement involves the mobilization of a larger number of particles [22]. Smaller surface tension

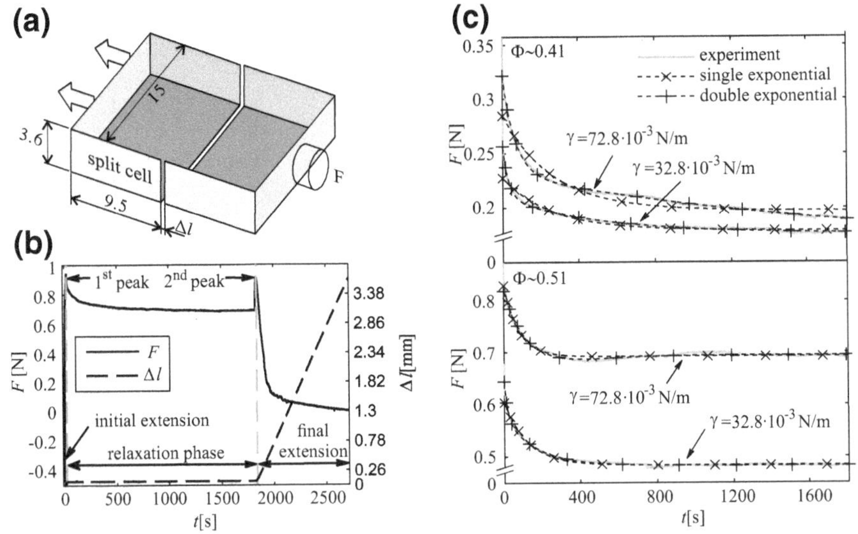

Fig. 2 **a** Split cell setup with dimensions in cm. **b** Prescribed displacement and reaction force for $\Phi \approx 0.51$ and $\gamma = 32.8$mN/m. **c** Force decay after stopping the initial extension for $\Phi = 0.41$ (top) and $\Phi = 0.51$ (bottom) (**b**, **c**). Reprint by permission from Springer Nature: Ref. [3]

results for both packing fractions in lower strength (see Fig. 2c). Using particle sizes, liquid content and assuming a connectivity of 6 bridges per particles, one can estimate with Eq. 5 a rupture distance $s_c \approx 24\,\mu$m. Note that displacement values at the peak force for $\Phi = 0.51$ are around 90 µm. The relaxation phase is characterized by a slowing down relaxation towards a steady state (see Fig. 2c) that can be fitted by exponential decay functions in the time domain:

$$F = \sum_{i=1}^{2} a_i e^{-t/\tau_i} + b \tag{1}$$

with elapsed time t, relaxation times τ_i and fitting parameters a_i, b. Fits are shown in Fig. 2c revealing, that low packing fractions Φ require at least $i = 2$, while high Φ can be described already with single exponential functions. The distinct relaxation behavior of loose and close packing is not affected by surface tension γ, while strength clearly depends on γ. During the relaxation phase one expects three phenomena: (i) liquid bridge ruptures, (ii) fluid redistribution and (iii) grain rearrangements (see Fig. 3b). Fluid redistribution between capillary bridges occurs through the thin wetting film surrounding rough particles, driven by differences in the Laplace pressure [61] and through the vapor phase [28]. Extension or dilation of the packing elongates bridges, resulting in increasing Laplace pressure, reduced capillary force (Eq. 4 in section "Models for Granular Assemblies in the Pendular State") and consequently fluid flux into bridges with lower Laplace pressure until they equilibrate or rupture [30–32]. The local fluid depletion results in locally reduced internal stress,

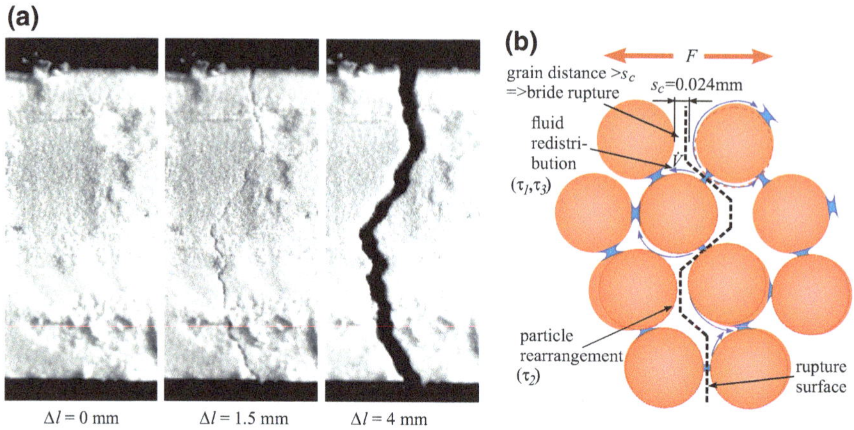

Fig. 3 **a** Fracture evolution $\Phi = 0.41$ and **b** sketch of relaxation phenomena. Reprint by permission from Springer Nature: Ref. [3]

internal load redistributions, bridge ruptures and consequently particle rearrangements. Rearrangements themselves modify the capillary bridge network, resulting in pressure variations between liquid bridges, driving fluid flux, and so forth. The timescale for water redistribution is about 300 s [23] while the one of the grain rearrangement is shorter and between 20 and 300 s, following [22]. The observed relaxation times for the double exponential fits of the loose packings are $\tau_1 = 50$ s and $\tau_2 = 1200$ s, indicating the presence of two different phenomena dominating during relaxation—the fast grain rearrangement and the slower fluid redistribution. In the highly packed samples, the fluid redistribution dominates, since one single exponential function sufficiently describes relaxation and particle mobility. Higher packing results in increased connectivity [16, 29], thus smaller bridge volumes and consequently faster fluid depletion and larger differences in Laplace pressure, hence shorter measured relaxation times ($\tau_1 \approx 145$ s).

Liquid Migration from Shear Bands

The process of fluid depletion in regions with strain localization, described in section "Tensile Behavior in the Pendular State" is relevant also in shear zones due to inherent dilation. To be able to observe fluid depletion, a stable shear band is realized in a circular split bottom cell [12], consisting of an outer L-shaped cylinder profile and an inner rotating cylinder with an attached disk, sketched in Fig. 4a (see also [36]). The shear cell is filled up to 15 mm with glass beads (radii $R = 0.85\pm$ 0.075 mm) and the homogeneously distributed liquid (1%) is fluorescent UV glue ($\gamma = 40$ mN/m; $\theta = 10°$; viscosity $\eta = 300$ mPa·s) and hardens quickly when irradiated with UV light. After shearing by rotating eight times with 10 min per rotation,

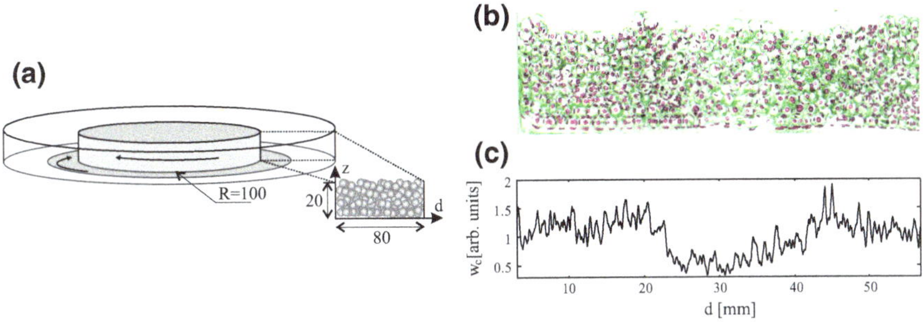

Fig. 4 **a** Circular split bottom shear cell with rotating core (gray) and units in mm. **b** 3 mm thick slice with liquid bridges (magenta) and particles (green), showing the fluid migration away from the shear band forming at about $d = 30$ mm. **c** Averaged spatial distribution of liquid content W_c in the cross section. Reprint by permission of APS: Ref. [36]

the glue is quickly hardened and thus stores the liquid configurations for subsequent evaluation. The entire pore space of the granulate is filled with colorless epoxy resin to obtain a solid block to be able to cut 3 mm thick cross sections that are suitable for fluorescent imaging under UV-light (see Fig. 4b).

It is evident from Fig. 4, that liquid is driven out of the shear band and accumulates towards the edges. Figure 4c shows the liquid content as a function of the distance d from the outer fixed cylinder wall averaged over heights $z > 3.4$ mm. The liquid content was evaluated by summing the pixels that belong to a liquid bridge and averaging over eight cross sections. Liquid depletion in the shear band could, in fact, be a cause for the shear softening behavior of wet granular systems. However, these observations are only valid in the pendular state up to a liquid content $W_c^{max} \leq 2.4\%$ where only liquid bridges form. To be able to understand and generalize such findings, suitable models for wet particle assemblies and simulations, described in the following are required.

Discrete Element Models for Granular Assemblies with Capillary Interactions

Simulations of the mechanical behavior of unsaturated granular systems are rather straightforward, as long as particles only experience inter-particle contact forces, Coulombian friction, and cohesive forces from liquid bridges in the pendular state. For increased liquid content, however, a richness of higher order liquid morphologies arises which continuously change due to deformation of the solid phase. In any case, to simulate the behavior of a granular system, Newton's equations of motion have to be integrated for the entire grain assembly to follow its time evolution [24]. The problem can be solved either explicitly, allowing particle overlap for contact force calculations [11], or implicitly using contact dynamics (CD) [46]. The first one is the method of choice when elastic deformations, represented via grain overlap, are relevant and contact interactions need to be resolved by tiny time steps. However, for perfectly rigid, frictional particle systems at high packing fractions but negligible particle elasticity, the contact dynamics method can be superior [27]. Implicit calculations of contact forces with the zero overlap constraint result in a high dimensional system of equations that require a solution at each time step. The substantially increased computational cost, however, is more than counteracted by several orders of magnitude larger time increments. In the following first CD is explained, before we describe extensions of the method to account for liquid bridges and finally arbitrary liquid bodies.

Contact Dynamics in a Nutshell

Contact Dynamics (CD), originally proposed by Moreau [46] is based on the assumption of perfectly rigid, frictional particles. In such systems, exclusively the rearrangements of particles determine the system's behavior [6]. Like other discrete element methods, CD calculates the trajectory of each particle and its rotation. A force F^p acts on each particle, that is composed of the contact forces F_i^c with all $i = 1, \ldots, n$ contacting particles and boundaries like walls, a gravitational force F^v, as well as forces from the liquid (see section "Geometric Representation of Liquid Bodies"). In CD a strict volume exclusion constraint is implemented, and the normal force component F^n can take arbitrary large positive values if the particle distance between surfaces of particles s_{ij} becomes zero, unless there is cohesion, to avoid surface interpenetration. Consequently, one must determine the position of the particle in an iterative process for which the F^n of each active contact is minimal such that there is no overlap of particles at the next time step. When $s_{ij} > 0$, the corresponding F^n vanishes, unless attractive cohesive forces F^{coh} act in the normal direction, e.g., in a range $0 < s_{ij} < s_c$. Sliding at particle contacts is prevented if the tangential component of the contact force F^t is smaller than the threshold determined by the Coulomb friction coefficient μ: $0 \leq F^t \leq \mu F^n$ (see Fig. 5), with equal values for the static and dynamic friction coefficients. F^t introduces a moment around the particle center resulting in particle rotation.

In dense packings, particles typically interact through a contact force network. Hence, a system of equations must be solved, e.g., by an iterative algorithm on all particles in a random order to obtain the global solution of contact forces [6]. The global contact force network converges iteratively toward a relaxed state providing the required forces for the integration of the equations of motion by an implicit Euler method.

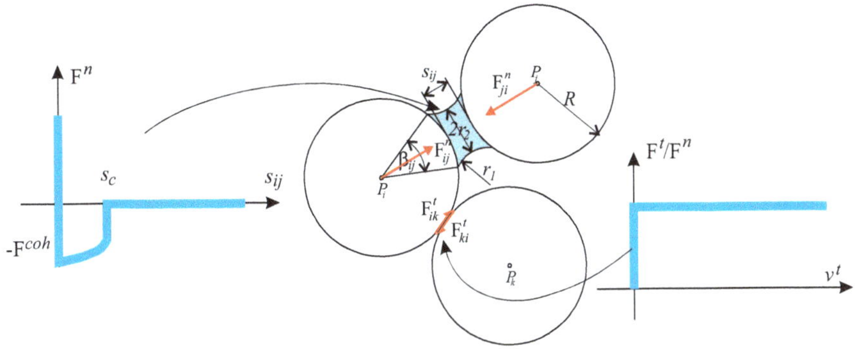

Fig. 5 Contact laws in CD with different particle configurations (center). (left) Repulsive normal force F^n with volume exclusion constraint for $s_{ij} = 0$ and attractive cohesive force F^{coh} with rupture distance s_c. (right) Sticking until tangential force $F^t < \mu F^n$. Reprint of central drawing by permission from Springer Nature: Ref. [43]

Models for Granular Assemblies in the Pendular State

Molecules of a liquid attract each other. However, at surfaces with a gaseous phase, the number of cohesive interaction partners of the same kind differs from inside the bulk, where molecules are equally pulled in all directions by their neighbors. Molecules at the surface experience a net force which is directed normal to the surface, resulting in surface tension γ or energy [69]. γ denotes the increase in free energy per unit surface area required to bring molecules from the bulk to the surface. Since interface molecules are in an energetically unfavorable state, the liquid tries to minimize its surface. As a consequence, a pressure jump ΔP occurs by passing across the interface, called Laplace pressure. For convex liquid surfaces like droplets, the pressure inside is larger (overpressure), while for concave interfaces, common for unsaturated porous media, it is negative. The Young–Laplace equation connects the geometry via surface curvature C and surface tension γ to the pressure difference ΔP:

$$\Delta P = \gamma C = \gamma(r_1^{-1} + r_2^{-1}), \qquad (2)$$

where $r_{1,2}$ are the principal radii of curvature. With increasing curvature C, ΔP increases linearly. A Euler-Lagrange formalism is used for the minimization of the surface under the condition of volume conservation [69] as demonstrated in Part 1 of this work.

When a liquid drop sits on a solid, the liquid-gas interface will touch the solid at a certain contact angle along the so-called wetting or triple line. The Young–Dupré equation describes the relation between the surface tensions of the three phases liquid (L), solid (S) and gas (G) in equilibrium: $\gamma_{SG} - \gamma_{SL} = \gamma_{LG} \cdot \cos\theta$ with the contact angle $\theta = 0°$ for complete wetting, $0° < \theta < 180°$ for partial wetting and $\theta = 180°$ for non-wetting situations [4, 15]. Note that on rough surfaces, measured

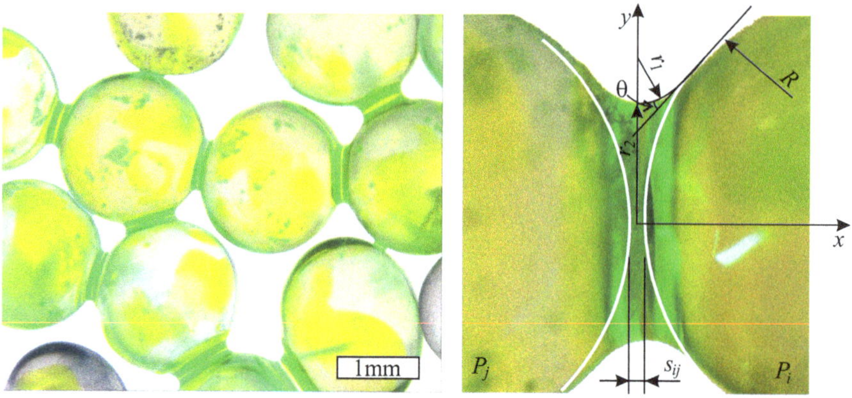

Fig. 6 Fluorescence microscope image for a partially saturated packing of glass beads in the pendular state and a single capillary bridge between two particles

contact angles can differ from the value predicted by the Young–Dupré equation. If the volume of a stable liquid body is increasing, the wetting line will advance once the advancing contact angle θ_A is reached. Vice-versa, volume reduction only results in a receding wetting line when reaching the receding contact angle θ_R [15]. The experimentally observed contact angles θ will be in the range $\theta_A \leq \theta \leq \theta_R$, depending on the history of the system and the difference $\theta_A - \theta_R$ is called contact angle hysteresis and has multiple practical implications in agitated granular packings at low saturation states [35, 38].

The presence of a liquid phase adds additional forces to particles connected by liquid bridges. These bridges form at contact points as soon as particles touch. No matter whether liquid is located in the wetting layer on a rough grain surface or in bridges, the total liquid volume must be conserved. At lower saturation, liquid bridges like the ones in Fig. 6 are the elementary units and dominantly observed basic liquid structure in the pendular state. Due to its relative simplicity, the pendular state received substantial attention by experimentalists and theoreticians alike, who developed approximate solutions. For small grains the effect of gravity on the shape of liquid bridges is negligible. The resulting rotational symmetry of the structure with respect to the line connecting the grain centers is an important simplification for the solution of the Young–Laplace equation (Eq. 2)

$$\Delta P = \gamma \left(\frac{\ddot{y}(x)}{(1+\dot{y}(x)^2)^{3/2}} - \frac{1}{y(x)(1+\dot{y}(x)^2)^{1/2}} \right), \tag{3}$$

where $y(x)$ describes the liquid-gas interface of the bridge in the $x - y$ plane shown in Fig. 6 [32]. Various suggestions were made for solving $y(x)$ with increasing accuracy. The toroidal approximation proposed by Ref. [14] describes $y(x)$ by a circular arc with constant radii $r_{1,2}$. This leads through Eq. 3 to an analytical relation of ΔP as a function of the separation s_{ij}, the contact angle θ and the bridge volume

V_{ij}. Unfortunately the assumption $r_1 = const$ is unrealistic (see for example Fig. 6 right) and one needs to solve Eq. 3 numerically with the boundary condition given by the contact angle θ as shown in Refs. [32, 58].

If ΔP is known, the attractive capillary force F_{ij}^{cap} can be calculated. It consists of two components: the forces $F_\gamma = 2\pi y(x)\gamma$ due to the surface tension acting on the profile $y(x)$ and $F_{\Delta P} = \pi y(x)^2 \Delta P$ due to the pressure drop ΔP [32]. Since F_{ij}^{cap} must be independent of the x-position one can evaluate it at the neck [45] and obtains $F_{ij}^{cap} = F_\gamma + F_{\Delta P} = 2\pi y(0)\gamma + \pi y(0)^2 \Delta P$. A closed form expression for the force acting between particles i and j of equal radius R was derived by [70]:

$$F_{ij}^{cap} = \frac{4\pi R\gamma \cos\theta}{2 + s_{ij}\sqrt{R/V_{ij}} + 5s_{ij}^2 R/V_{ij}}. \tag{4}$$

It is important to note that Eq. 4 is only valid for small capillary bridges up to $V \approx 0.03R^3$. It also gives a constant force, independent from V_{ij} at grain contact ($s_{ij} = 0$) contrary to real bridges. Due to particle roughness, the full capillary force F_{ij}^{cap} only acts above a threshold volume V_c. This can be considered using a Heaviside step function to switch on F_{ij}^{cap} when the volume reaches V_c [37]. From there on the respective capillary force F_{ij}^{cap} is calculated by the empirical relation Eq. 4 until the bridge ruptures by fulfilling the criterion described in Eq. 5. Additionally, real bridges can only be stretched to a volume dependent rupture distance s_c. Based on experimental observations, Willett et al. [70] proposed a refined expression with higher order terms to obtain the criterion for bridge rupture as

$$s_c \geq (1 + \theta/2)(V_{ij}^{1/3} + 0.1V_{ij}^{2/3}). \tag{5}$$

Fluid redistributes in wet particle systems in various ways. First, in the case of *bridge ruptures* the fluid that once was in the bridge is sucked back into the wetting layers of the two involved grains [61]. The larger one of the grains is with respect to the other, the larger fraction of water it will receive and the liquid portions can be estimated via a scheme described in Ref. [63]. Since the wetting film swells, its flow resistance drops and fluid gets quickly sucked into already existing bridges, because their Laplace pressures are low compared to the wetting layer. In the models, one can equally redistribute the fluid on the particle to all intact bridges of that particle [37], distribute it proportional to the pressure differences between individual Laplace pressures P_i and the film pressure $P_f = \gamma/R$ or inversely proportional to the distance of a bridge from the rupture point of the failed bridge L_i. Mani et al. [36] proposed a combination where the amount of liquid each bridge receives is $\Delta V_i = A(P_f - P_i)/L_i$ with the normalization factor A to assure volume conservation. Timescales for bridge failure and redistribution can be considered to be of the same order, and for slow loading, assuming instantaneous redistribution is sufficient. Due to liquid redistribution from bridge failures, locally the liquid content for the formation of higher order liquid structures, such as trimers, addressed in the next section, could be reached. To avoid this artificially, and to keep the simulation in the pendular state, one can define a constant upper bridge volume V_{max}, such as $V_{max} = 0.057R^3$ [56]. Whenever the volume exceeds V_{max}, the excess liquid can be temporarily stored in the liquid film of the particle to be available for possible bridge formations. The second redistribution process is *liquid flux between bridges* through the wetting layers, driven by pressure differences between bridges acting on the same particle. This equilibration is rather slow and can negligible if the lifetime of a capillary bridge is much smaller than the equilibration timescale, that is usually in the order of minutes [55]. The time scale for this equilibration decreases with the inverse flow resistance $\kappa \propto \eta^{-1}$ of the film with the fluid viscosity η. For $\kappa \to \infty$ pressure between all bridges is instantaneously equilibrated, and all liquid bridges have equal Laplace pressure. For $\kappa \to 0$ bridges never, and for $\kappa > 0$ the dynamics of fluid redistribution and pressure equilibration is fully controlled by η. *Equilibration through the vapor phase* takes even longer [28] and is therefore not considered in dynamic simulations.

Extension of Models for the Funicular State

Starting from the pendular state, an increase in liquid content ($W_c^{max} > 0.024$) leads to merging of liquid bridges, which generates structures of increased complexity called liquid clusters, comprising more than two particles. In the funicular state one observes distinct liquid clusters of various sizes, while in the capillary state all clusters merge to form one spanning cluster. Since most clusters require three dimensional particle assemblies, they are best studied by microtomography [2, 10, 56, 67]. Scheel et al. [55, 57] described the richness and variety of liquid cluster morphologies in a dense packing of glass beads at various saturation degrees. Just above the pendular regime liquid bridges start to overlap, merge and form closed three particle configurations with filled pore throats, called trimers (see Fig. 7). Larger clusters emerge by merging of smaller units when the Laplace pressure reaches a critical value estimated as $-4.46\gamma/R$ [56], resulting in the formation of rather open, branched structures. Two connected trimers are called pentamers (see Fig. 8b), three heptamers and so on (see Part 1). Four particles forming a tetrahedron with one trimer

Fig. 7 (left) Three grains with liquid bridges, (center) trimer with filled pore throat and (right) ruptured trimer with subsequent liquid redistribution

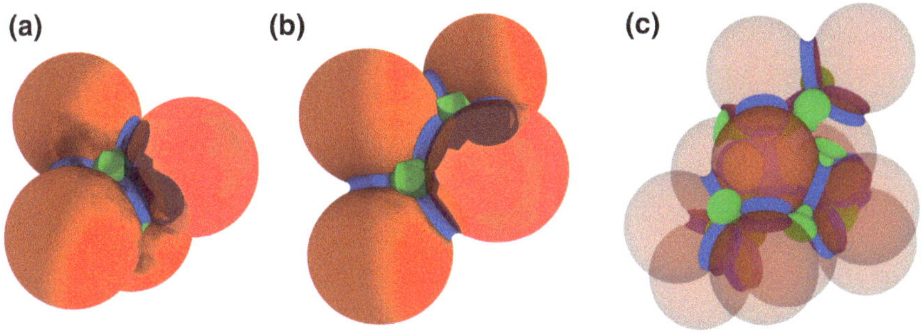

Fig. 8 **a** Tetrahedral cell **b** pentamer and **c** larger composite liquid cluster. Bridges are colored blue, menisci and pore throats green. Reprint of (a, b) by permission of APS: Ref. [42]

on each face are called tetrahedral cluster (see Fig. 8a). The pore inside can be filled or even contain a trapped gas bubble. The liquid clusters are formed when the fluid content increases, but can also decay when fluid drains from the system.

Contrary to single bridges, higher-order clusters lack rotational symmetry. Nevertheless, for trimers, numerical solutions could be found by Rynhart et al. [53] for equal particles for a given contact angle and constant mean curvature. Also solutions from numerical surface energy minimization like the Surface Evolver software [5] are available [62]. Such simulations give the shape of the liquid-vapor interface and the corresponding capillary forces. The analysis of Semprebon et al. [62] is limited to three particle configurations with a gap that varied. One observes a transition between fundamental morphological types like liquid bridge and trimer, showing slightly larger capillary forces in a trimer compared to the ones of three single bridges due to the filled pore throat (Fig. 7 left, center).

The distribution of liquid morphologies is highly history dependent on the particular wetting process. While mixing of grains with liquid and subsequent equilibration results in cluster morphologies observed by Scheel [55], other processes like spontaneous imbibition can lead to completely filled regions of the packing with a single wetting front [47].

Geometric Representation of Liquid Bodies

If one accepts reasonable downturns in numerical performance, the pore-space and liquid structures can be approximated rather close to their real shape. For a given granular packing, one calculates the pore-throat network via Delaunay triangulation [40]. Hence the system volume is subdivided into single tetrahedra, like the basic unit shown in Fig. 9b. The faces of the tetrahedron form the four pore throats, while the void space in each tetrahedron is called the pore body of the cell. Pore bodies can be empty, filled, or partially filled with liquid separated by menisci. The basic strategy consists in the decomposition and representation of higher geometrical configurations such as trimers, pentamers, tetrahedral clusters, and higher ones as a combination

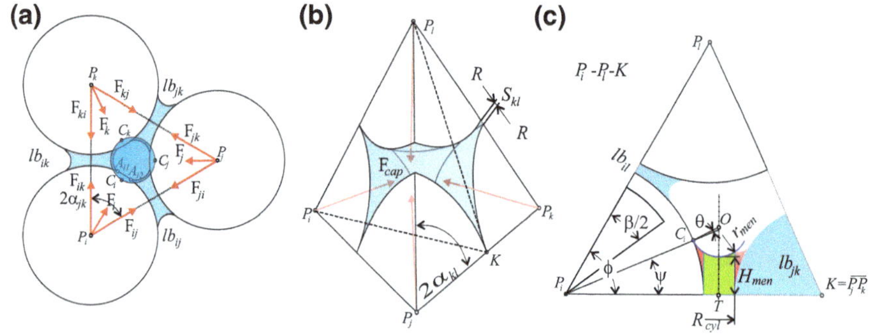

Fig. 9 **a** Trimers **b** tetrahedral cell and **c** cut through a meniscus (green) in the $P_i - P_l - K$ plane (see dashed line in (**b**)). Red regions highlight the volume that is not captured by the cylinder representation of menisci. Reprint by permission from Springer Nature: Ref. [43]

of elementary units such as liquid bridges, menisci, or filled pore bodies. Melnikov [41] describes this approach in detail.

We start with the liquid-gas interface between three grains—the menisci—that can exhibit rather complicated shapes since we already described bridge models in section "Models for Granular Assemblies in the Pendular State". Nevertheless, we simplify them to be spherical of radius r_{men} like Haines [20], what is in satisfying agreement with experimental observations [55, 56]. A cut through a pore throat is shown in Fig. 9c. The centers of the four possible menisci of a tetrahedral cell are located on the normal of each pore throat through the circumcenter of the respective face. With the contact angle θ and meniscus radius r_{men}, the exact position on this normal can be calculated following the method proposed in Ref. [17] (see Fig. 9c). Note that single menisci can only exist in combination with associated liquid bridges forming one single liquid body with its Laplace pressure determined by the curvature of the meniscus r_{men}^{-1} via the Young–Laplace equation (Eq. 2), as $\Delta P = \gamma 2 r_{men}^{-1}$.

Similar to the bridge, a capillary force $F_{i,j,k}$ from the meniscus will act on the involved particles i, j, k in addition to the forces from the liquid bridges. It acts in the direction given by the points $\overline{P_i T}$ (see Fig. 9c) with unit vector $\mathbf{e}_{iT}$ and is calculated as $F_{i,j,k} = \mathbf{e}_{iT} \cdot \Delta P A_{eff}$. The effective wetted area A_{eff} can be defined for example in a simplified fashion for particle i through the area of a triangle connecting the points $C_i A_{i1} A_{i2}$.

With the three basic units liquid bridge, meniscus, and filled pore body, in principle all liquid clusters can be composed. As an example, the smallest cluster is the trimer consisting of two menisci, located in neighboring tetrahedra on both sides of the common pore throat and their shared liquid bridges (see Figs. 7 and 9a). The volume of a cluster V_c with N_{imb} imbibed pore bodies and N_{men} menisci is given by

$$V_c(r_{men}) = \sum_{i=0}^{N_{imb}} V_{pore,i} + \sum_{j=0}^{N_{men}} V_{men,j}(r_{men}), \tag{6}$$

with the volume $V_{pore,i}$ of the imbibed pore body i and the volume $V_{men,j}$ of meniscus j including portions of the m associated liquid bridges with volumes V_i^{lb} of this meniscus:

$$V_{men,j}(r_{men}) = V_{cyl}(r_{men}) + (0.5 + \varepsilon) \sum_{i=0}^{m} V_i^{lb}(r_{men}). \tag{7}$$

One approximates the fluid volume of the pore throat V_{cyl} by a cylindrical body volume coaxial to the normal of the pore throat (Fig. 9c) with radius $\bar{P_i T} - R_i$ and its upper surface delimited by the meniscus. Note that the cylinder height H_{cyl} can be negative if in the neighboring tetrahedral cell the pore body is filled. In this case, the volume of the neighboring pore body needs to be reduced by the volume given by the intersection of the meniscus with the throat pore plane. The cylinder-sphere cap approximation systematically underestimates real volumes, what can be taken care of via a geometrical correction parameter ε in Eq. 7 that will be determined later on from experiments. Index m is between 0 and 3, depending on the number of connected liquid bridges. If a meniscus shares a liquid bridge with an adjacent meniscus, as for a pentamer in Fig. 8b, m for one of the menisci is reduced by 1 to avoid double counting of bridge volumes. Note that for every meniscus only half of each liquid bridge volume is considered since the other half is located in the opposed triangulation cell if its pore body is not saturated. For tetrahedral cells with filled pore body, the Laplace pressure pulls each of the four particles towards the center as shown in Fig. 9b. One represents the respective wetted area by the one of the liquid bridge with opening angle β_{ij} for cases with a liquid bridge between the corresponding particles. For simplicity, the force direction is not corrected and keeps pointing towards the cell center. For entirely immersed particles, forces from the pore body cancel each other out. Tetrahedral cells with four menisci may comprise an incompressible, trapped gas bubble.

To summarize, in a cluster, liquid bridges are always associated with one of its menisci creating a common liquid body with equal pressure defined by the meniscus radius r_{men}. One calculates the maximum liquid volume of the filled pore via the difference of the tetrahedral cell volume and the partial volumes of the four particles contained within the cell.

Evolution Laws of Liquid Bodies

Liquid clusters evolve when their volume changes due to sources, sinks, condensation, evaporation, or merging with other clusters as well as relative movements of involved grains. Stability criteria for drainage and imbibition with instantaneous jumps between stable configurations control the evolution of liquid clusters. These interface jumps are micro-mechanically associated with drainage or imbibition of pore bodies or throats or bridge rupture. A limited set of geometrical instability criteria for imbibition (***i1–i4***) and drainage (***d1–d3***) can be identified, similar to Refs. [17, 48], itemized below:

i1 When two liquid bridges with filling angles $\beta_{ij;ik}$ touch each other, they can form a new trimer by filling the pore throat, provided that sufficient liquid for a stable meniscus (drainage criterion ***d1***) is available. With the opening angle α_{jk} (see Fig. 9a, b) it reads $0.5(\beta_{ij} + \beta_{ik}) > 2\alpha_{jk}$.

i2 If a meniscus in a tetrahedral cell touches a single liquid bridge that is presently not part of the liquid body, the pore body is filled (Melrose criterion [44]). Hence, the respective meniscus with the filling angle of the meniscus ψ and the face-edge angle ϕ shown in Fig. 9c becomes unstable when $\psi + \beta/2 > \phi$.

i3 When centers of two menisci inside one pore body touch each other, they build a single spherical interface and can fill the pore body. However, in the case when all four menisci already exist, a gas bubble gets trapped.

i4 When a meniscus touches the opposite particle, the pore body is imbibed.

d1 When pore throats are reduced to a minimum size, given by the critical height of the menisci $H_{min}^{men} \geq \lambda R$ with the drainage parameter λ, they drain. Best agreement with experimental data is found for $\lambda \approx 0.15$ [42] (see Fig. 13).

d2 When the center of the meniscus of a neighboring cell touches the respective pore throat plane of an entirely saturated cell, the pore body becomes unstable. Instantaneously the liquid interface jumps to a new stable position, and the pore body is drained.

d3 Liquid bridge failure as described by Eq. 5.

Inside one liquid body instantaneously pressure equilibration and respective fluid flow are assumed. Transport in between spatially disconnected liquid clusters that share at least one grain, however, is assumed to occur via a liquid film on the particle surface, inspired by experimental observations by Refs. [34, 56]. The volume flux $\dot{V}_i$ into structure i is proportional to the local Laplace pressure differences of the involved liquid clusters. If N_i clusters connect to the same grain,

$$\dot{V}_i = \frac{R}{\gamma} \cdot \sum_{j=0}^{N_i} \omega_{ij}(P_j - P_i) \tag{8}$$

with the dimensionless conductance coefficient ω_{ij}, determining the equilibration timescale. ω_{ij} comprises details such as the distance between clusters or the number of clusters connected to the grain and others [35, 38]. However, for simplicity, typically a constant value for $\omega_{ij} = 0.01$ is chosen. With all the described model building blocks at hand, we can now turn to simulations.

Pressure-Controlled Simulation Variant

To this point, the liquid transport within the saturated region of the pore space was assumed to occur instantaneously. This hypothesis is a valid assumption for small liquid clusters. However, if larger filled regions of the pore space are concerned, the flow is limited by the permeability of the granular material and the viscosity of the fluid. For modeling of large liquid clusters connected to a reservoir, pressure-controlled models like classical imbibition models are a better choice. This can also be motivated by the calculation of Haines jumps in section "Granular Assemblies at Arbitrary Saturation" Fig. 16, where the pressure dependence on single imbibition events in the form of pressure jumps rapidly decreases with increasing cluster volume. Using the representation of the liquid-gas interface described above, it is also possible to introduce a pressure-controlled imbibition model which can simulate the unsteady liquid infiltration into the unsaturated granular material.

The pressure controlled simulation model is based on the assumption that pressure inside each pore is constant and pressure drops occur at pore throats (see Fig. 9b). The liquid flux q_{ij} occurs between neighbor cells i and j due to the local pressure gradient ΔP_{ij}. If all four neighbors $j = 0, \ldots, 4$ of a given saturated pore i are also completely filled with liquid, the continuity equation for this pore reads

$$\sum_{j=0}^{4} q_{ij} = 0, \tag{9}$$

meaning that the overall liquid volume in this pore is conserved. However, if one or more of the neighbor pores are empty like for all pores at the liquid front, the total flux into or out of the given cell i is not necessarily zero, hence

$$\sum_{j=0}^{4} q_{ij} = Q_{eff}. \tag{10}$$

Here, the net in- or outflow from the cell i is denoted as Q_{eff}. This volume change results in changes of the bounding menisci of this cell due to volume conservation since at least one neighbor cell is not saturated. The volume Q_{eff} is used to find the new pressure in the cell i and to recalculate its menisci. Due to in- or outflow the menisci can become unstable (criteria **i1–i4**, **d1–d3**) which results in imbibition of neighbor pores or in drainage.

The liquid flux occurs through pore throats which have different geometrical shapes and q_{ij} can be calculated based on the geometrical properties of the pore throats. Both, the Stokes and the Darcy equations assume a linear relationship between the pressure gradient and the flux is

$$q_{ij} = g_{ij}(p_i - p_j), \tag{11}$$

with g_{ij} being the local conductance coefficient between two pores with respective pressures p_i and p_j. g_{ij} takes into account the geometry of the pore throat calculated via an approach originally introduced by Bryant et al. [7]. In this approximation the flow through the pore throat is assumed to be a Hagen-Poiseuille flow through a cylinder of radius r and length l_{ij}:

$$g_{ij} = \frac{\pi r^4}{8\eta l_{ij}}, \tag{12}$$

where η denotes the viscosity of the fluid. Since the pore throat does not have a circular shape (see Fig. 9c), an approximation can be used for the radius $r = r_{eff}$. One of the possible definitions of the effective radius of the pore throat r_{eff} is given in Ref. [7] via $r_{eff} = 0.5 \cdot (r_c + r_e)$ where r_c denotes the radius of the inscribed circle inside the pore throat and r_e the radius of the circle with the same surface as the pore throat. One defines the effective flow path length l_{ij} as the distance between the centers of the cells i and j. The center of the cell is the point which is equidistant from the centers of all four particles. One uses the projection of this point onto the nearest cell face if this point lies outside of the cell body.

To run simulations, first the pore-throat network must be extracted, and entry pores with respective pressure must be defined. Then the instability criteria **i2–i4** are checked, resulting in pore imbibition. When one fills a new pore, one sets its pressure to the minimal possible value which allows the formation of stable menisci. The new pressure creates suction on the liquid phase, supporting the flow towards the liquid front. In the next step a system of linear equations based on Eqs. 9 and 10 is formulated and solved e.g. with the LAPACK sparse matrix solver provided in the *Armadillo* library [54]. The calculated flows are used to update the pressure in completely saturated pores. One uses the volume conservation constraint to update the pressure in the saturated boundary pores next to the liquid front.

Simulations for Wet Particle Assemblies

To construct the initial mono-disperse particle packing, a random sequential adsorption [13] of equally sized spheres inside the sample volume V_{sample} with periodic boundary conditions (PBC) is employed. One adjusts the solid fraction Φ_s by increasing particle radii R while particles can rearrange. An initial liquid volume is given to each particle, to keep the system in the pendular regime where only liquid bridges exist. The liquid content W_c is defined by the total volume of the liquid V_{liquid} normalized by the void volume V_{void} as $W_c = V_{liquid}/V_{void}$. V_{void} is calculated as the difference between V_{sample} and the total volume of all N_p grains R: $V_{grains} = 4\pi N_p R^3/3$. Consequently the porosity can be calculated with $\Phi = 1 - V_{grains}/V_{sample}$. We simulate different features of fluid transport and configurations in wet granular assemblies in the pendular (section "Fluid Motion in Sheared Unsaturated Granular Media") and funicular state (section "Granular Assemblies at Arbitrary Saturation"), as well as consequences on the rheological behavior (section "Fluid Induced Failure").

Fluid Motion in Sheared Unsaturated Granular Media

With the particle packing and initial homogeneous liquid content of $W_c = 1\%$ we simulate shear flow in the circular split bottom shear cell described in section "Liquid Migration from Shear Bands". One uses periodic boundary conditions for a small segment with arc length 0.0873 and hydrophobic cell walls. One considers gravity and calculates capillary forces via Eq. 4. Whenever particles contact, a liquid bridge is formed, while particle separation beyond s_c results in bridge rupture (criterion ***d3***) [23, 65]. One determines the fluid redistribution by the combined pressure-distance relation (section "Models for Granular Assemblies in the Pendular State"). The inverse flow resistance of the film κ (section "Models for Granular Assemblies in the Pendular State") controls the pressure equilibration.

When comparing the outcome of simulations in the steady state, reached after eight rotations, to the experimental data (section "Liquid Migration from Shear Bands") we see that for $\kappa = \infty$ (instantaneous pressure equilibration), the liquid content W_c is proportional to the number of contacts, but not to the measured liquid content (see Fig. 10b). For $\kappa = 0$, however, the steady-state liquid distribution shows good agreement with experiments and a clear depletion from the shear zone (Fig. 10a). Hence, in this system with glue as a liquid, equilibration times of liquid bridges are much larger than contact times of the beads in the shear band. Consequently, liquid redistribution is mainly taking place via bridge rupture events, leading to liquid fluxes away from the shear zone. One can generally conclude that in shear bands of wet particle packings, not only the coordination number but also the average bridge volume is reduced [36].

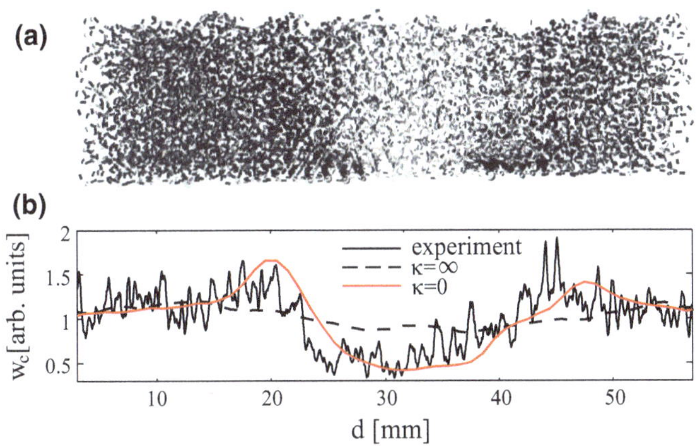

Fig. 10 **a** A slice perpendicular to the the shear band showing capillary bridges as lines of width and darkness proportional to their volume. **b** Liquid content as function of radial position after 8 rotations compared to the experiments (section "Liquid Migration from Shear Bands") for different values of κ. Reprint by permission of APS: Ref. [36]

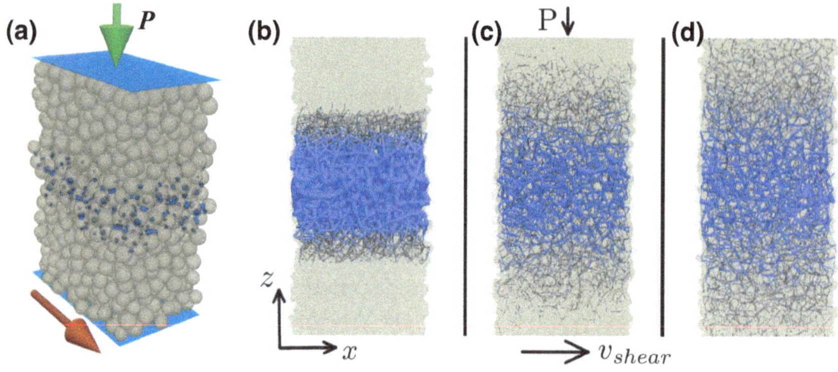

Fig. 11 **a** Plane shear cell with periodic boundaries and fluid initially located in the center. **b–d** Snapshots of the liquid configurations in a slice, located in the central part under shear at the beginning (**b**), at an intermediate stage (**c**) and at the end of the simulation (**d**). The bridge volume V_b is indicated by the width of the lines, connecting the sphere centers. Reprint (b–d) by permission of Springer Nature: Ref. [37]

In general, granular systems under shear can form quite different shear profiles, depending on the hydrostatic stress. High stresses exhibit alternating diffuse and localized shear that result after averaging in a rather linear shear velocity profile. For low confining pressure P (see Fig. 11a) capillary forces play an important role and stresses from the shear deformation are less capable of rupturing bridges. Consequently, the velocity profile depends on the fluid distribution, since wet grains stick together. In Mani et al. [37] the effect of the stress state on the transport behavior was studied by using a cohesion number $\eta^{coh} = F_c^{max}/PR^2$ giving the ratio of the largest cohesive force to the one exerted on a particle. For this purpose the particle packing was loaded with a plate at a variable pressure P and using periodic boundaries (in x, y directions) again, the system was sheared as depicted in Fig. 11a. Initially, liquid is only placed in bridges close to the center with a Gaussian distribution for the bridge volume $V_b(z) = A\exp(-(z - z_{wall}/2)^2/\sigma_0^2)$ with amplitude A and width σ_0.

In the snapshots of Fig. 11b–d one can easily see that the liquid is spreading toward the top and the bottom wall during the simulation. Quantitatively this is shown by the average bridge volume per particle Q in Fig. 12 for various cohesion numbers η^{coh} and at different deformations γ. For high confining pressure P ($\eta^{coh} = 0$) a nearly linear shear rate profile is measured, while for high η^{coh} particles in the wet zone adhere, forming a plateau in the shear rate profile that is more pronounced as η^{coh} increases and finally controls liquid distribution (Fig. 12). Due to the linear shear profile, the initially Gaussian liquid content profile is kept, pointing at diffusive transport [63]. The diffusion constant D is found to be a function of the initial moisture amplitude A of the type $D(A) = a/(1 + bA^{1/3})$ with $a = 0.633R^2$; $b = 2.936R^{-1}$. Qualitatively the decreasing diffusivity for increasing liquid content can be explained by larger rupture distances s_c of bridges and thus decreased rupture rates, since this is a main liquid redistribution mechanism, next to the diffusive motion of particles under shear

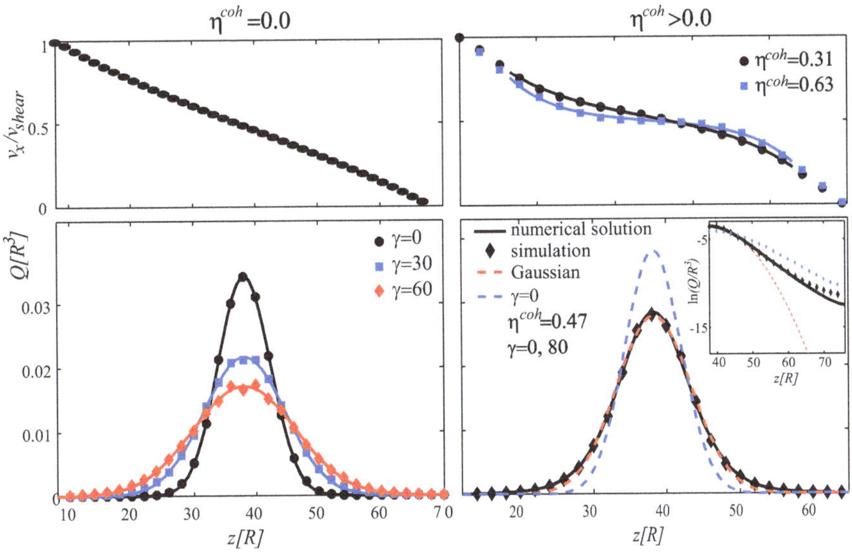

Fig. 12 Shear velocity profiles (top row) and average liquid contents as a function of the height z (bottom row) for large (left column) and low (right column) hydrostatic stresses. The inset (bottom right) highlights the deviation from the Gaussian distribution. Reprint by permission of Springer Nature: Ref. [37]

[8, 66]. Diffusion of particles and liquid bridge ruptures are also important processes for higher cohesion numbers. However, both depend on the reduced shear rate in the wet region. We observe the largest deviation from the diffusive Gaussian at the borders between the wet and dry region ($z \approx 50$) from the liquid profiles (Fig. 12). In practice, this results in the known difficulty of mixing dry and wet powder. On can improve the mixing process by decreasing η^{coh}, in other terms increasing the hydrostatic stress, since local shear rates determine the liquid migration [37].

Granular Assemblies at Arbitrary Saturation

The previously shown simulations (section "Fluid Motion in Sheared Unsaturated Granular Media") only consider the interaction between single particles with no need for the consideration of the pore space. In a first study, after the construction of the packing, all particles are immobilized for the remainder of the simulation to first focus on the evolution of liquid clusters as liquid is injected (removed) at a specific position or condensed (evaporated) at gas-liquid interfaces. The system is subdivided into tetrahedral cells by a Delaunay triangulation of particle centers, and liquid bridges are allowed to equilibrate pressure. Note that the Delaunay triangulation is the dual graph to a Voronoi diagram used in a comparable context in

Ref. [71]. In this simulation, liquid volume is the control variable. Consequently, volume changes, e.g., due to condensation into already existing bridges, result in an update of all liquid structures to make the Laplace pressure correspond to the changed volume. Changes in the liquid cluster morphology can trigger instabilities. We check the instability criteria (section "Evolution Laws of Liquid Bodies") for imbibition ***i1–i4*** and drainage ***d1–d3*** in ascending order, because first menisci must form before pore bodies can be filled. The entire cluster needs to be updated when one adds a trimer to a cluster due to criterion ***i1***. The same holds when one merges a cluster with a trimer. One eliminates identified instabilities through drainage or imbibition and performs a cluster update. Resulting pressure differences again drive liquid transport through liquid films (Eq. 8) that changes liquid clusters and so forth. One repeats this procedure until convergence before adding or removing any additional fluid. To ensure liquid volume conservation, one subsequently recalculates all liquid clusters. To speed up simulations, we interpolate from tabulated values for the liquid bridges [38]. Those originate from simulations with the numerical energy minimization method (NEM) of the software SURFACE EVOLVER as discussed in Part 1[5]. Note that the contact angle θ is kept constant at 5° in the following.

Calibration with trimer formation and decay: First we calibrate the model on experimental data from Ref. [55] by simulating three particles that form trimers when liquid is condensed at liquid-gas interfaces (criterion ***i1***) and decay, when liquid evaporates or opening angles change (criterion ***d1***). The liquid bridge pressure P is updated to correspond to the changing volume. During condensation, the bridges coalesce, and a new trimer is created at the critical dimensionless value P_{max}. Consequently, P drops, since the liquid needed to fill the pore throat which previously was part of the bridges that decreases the radius of the interface curvature. During evaporation, the trimer decays into three liquid bridges at the critical minimal dimensionless pressure P_{min}. Good agreement with experimental data is obtained with the geometrical correction factor $\varepsilon = 0.07$ (see Eq. 7) and the drainage parameter $\kappa = 0.15$ for the drainage criterion ***d1*** as shown in Fig. 13a. The critical minimal pressure P_{min} of a trimer strongly changes with the inter particle separation distance s_{ij} or the corresponding opening angles α_{ij} (definition see Fig. 9). When $s_{ij} \leq s_c$ (criterion ***d3***) a bridge and consequently the trimer can no longer exist. Figure 13b shows the pressure curves for a trimer when one of the opening angles α_{ij} is increased while the other two contacts remain closed. High P_{min} are associated with large values of α_{ij}. For imbibition this implies that trimers with non-contacting particles form at higher overall Laplace pressures compared to contacting ones. To prove the validity of this assumption, we compare the resulting dependence of P_{min} on α_{ij} with measurements (see Fig. 13b) from evaporation experiments. Our predictions are correct within experimental error bars and we observe two different regimes separated by an angle $\alpha_{ij} \approx 31.7°$. Below, the drainage criterion ***d1*** is fulfilled, while above bridge rupture (***d3***) occurs. Interestingly, the NEM results from Ref. [55] agree less with the experimental data than our simplified approach.

Liquid clusters in larger particle assemblies: In unsaturated granular packings, a multitude of different cluster morphologies and sizes coexist. Scheel evaluated them in his thesis for a random, monodisperse packing of spherical particles with

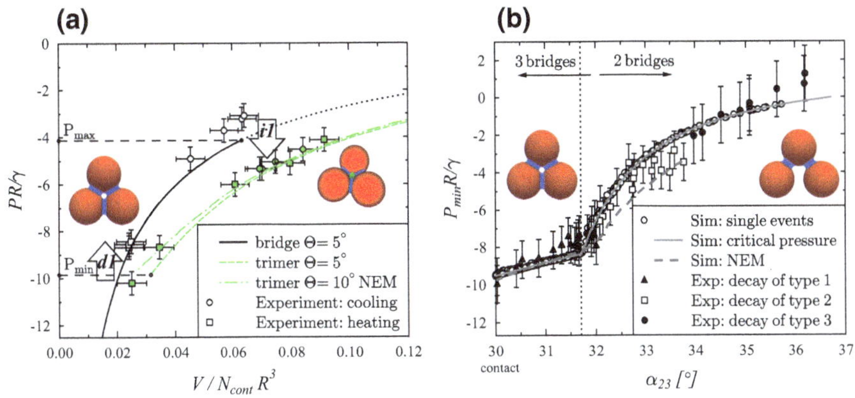

Fig. 13 **a** Comparison of the dimensionless Laplace pressure of a trimer and liquid bridges between three contacting particles as a function of their volumes V normalized by the number of contacts $N_{cont} = 3$ and the grain radius R^3. Trimers form at P_{max} and decay at P_{min} (for $\theta = 5°$). **b** Trimer decay: Dependence of P_{min} on the opening angle α_{23} for trimers. Three types of trimer decay are possible: (1) decay into two liquid bridges, (2) slow transition into two liquid bridges, and (3) decay into three liquid bridges. Only configurations with one gap opposite to the opening angle are presented. Experimental data and NEM results by Ref. [55]. Reprint by permission of APS: Ref. [42]

solid fraction $\Phi = 0.57 \pm 0.01$ at liquid contents in the range $0 < W_c < 0.2$ [55]. We simulate this case by incrementally condensing liquid into a particle packing consisting of 2000 particles with periodic boundary conditions in all three directions via the procedure described above. Figure 14 shows a snapshot of the liquid clusters present at the beginning of the funicular regime at $W_c = 0.03$. The distribution of morphologies shows a fast transition between the liquid regimes (see Fig. 15). At low W_c only bridges exist, while the regime above the pendular state shows a rapid increase in the number of clusters. It reaches a maximum when the merging into larger clusters starts to dominate. Finally, only a single spanning cluster remains. The simulations reproduce the general trends observed in the experiments. However, deviations due to finite sample sizes and experimental resolution are visible. A similar picture is found for the proportion of liquid stored in different morphologies as a function of W_c (see Fig. 15b) where one can see that at $W_c = 0.2$ more than 95% of the liquid volume is located in large clusters.

Evolution of single liquid clusters: When clusters grow or contract, pressure jumps originating from rapid imbibition of single pore bodies or their drainage are observed. These so-called Haines jumps have been observed experimentally by Berg et al. [2] and are due to volume conservation for the new configuration, resulting in increasing or decreasing curvature of menisci. As local instabilities drive the liquid interface propagation, we model it as a discontinuous process jumping from one stable configuration to the next one. We show the evolution of the pressure in one liquid body that forms by injecting liquid with a constant flux at a single point (Fig. 16). First, the dimensional cluster pressure increases, until due to trimer formation a sharp drop

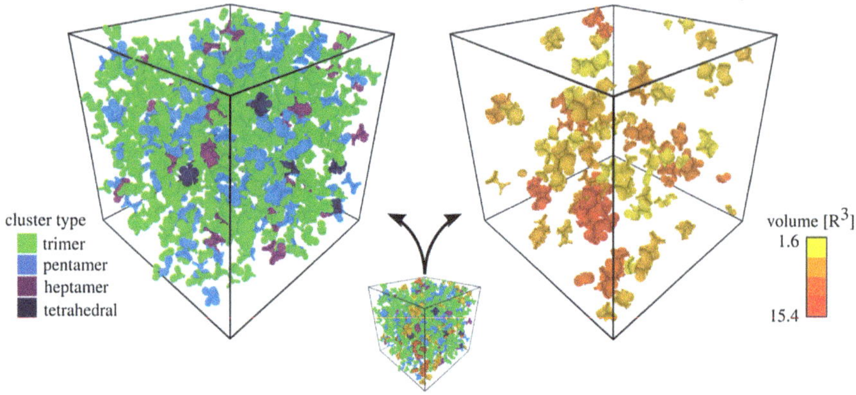

Fig. 14 Rendering of liquid clusters without bridges (center) at $W_c = 0.03$ split up into tow images for small (left) and large (right) clusters for clarity. Reprint by permission of APS: Ref. [42]

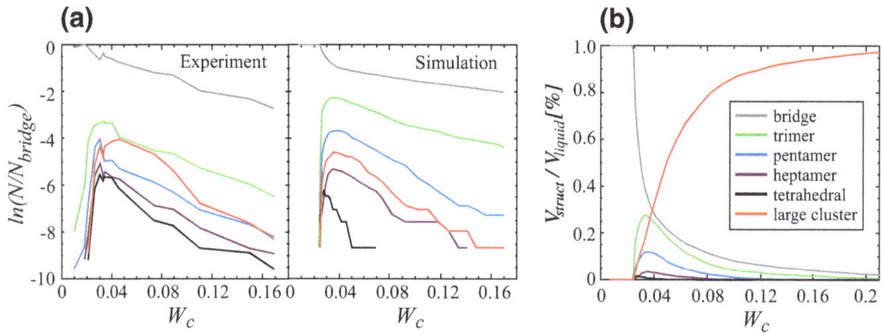

Fig. 15 **a** Comparison between simulation and the experiments of Ref. [55] of the number of different liquid cluster morphologies N scaled by the total number of existing liquid bridges N_{bridge} (at $W_c \approx 0.02$) for $\Phi_s = 0.57$ as function of the liquid content W_c. **b** Proportion of liquid V_{struct}/V_{liquid} stored in different morphologies as a function of W_c. V_{struct} is the summed volume of all morphologies. Liquid structures larger than heptamers are called large clusters. Reprint by permission of APS: Ref. [42]

occurs. Each drop is due to a local instability of filling a new pore body or a throat. The decreasing magnitudes of jumps with cluster size are due to the relation of the liquid volume needed for the filling of the next pore to the total cluster volume. With the conductance coefficient $\omega = 0.01$, the flux can be fast compared to the liquid transport through surface films. Jumps from small clusters can be compared to experimental data [56], while for larger clusters no data is available.

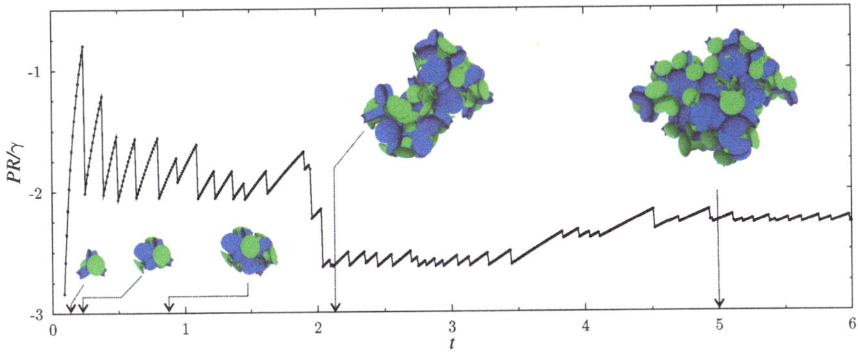

Fig. 16 Evolution of dimensionless Laplace pressure of a liquid cluster for liquid injection at a constant flow rate at one source point. Reprint by permission of APS: Ref. [42]

Pressure Controlled Imbibition Front Formation

In this simulation a discontinuous propagation of the liquid interface is studied in a dense packing with $N_p = 1750$ particles [41]. Unlike the simulations described up to now, the process is controlled by fluid that enters the system under a given pressure (compare section "Pressure-Controlled Simulation Variant"). We fill the entry cells in the upper boundary of an initially dry sample with liquid and set the pressure to a constant value of $PR/\gamma = -3$. Additionally, we consider the effect of gravity (hydrostatic pressure) by introducing a pressure gradient in the z-direction. The pressure on the upper boundary is maintained constant during the whole simulation. This boundary layer is thus assumed to be directly connected to a reservoir. Due to instabilities, we fill new pores as the interface propagates through the material. The filling of a new pore is only possible if enough liquid is available in the neighbor pores. Due to the filling of a new pore, the pressure of the liquid phase inside the pore is lower than in the neighboring pores since only a limited amount of liquid is available. Note, that in this model the sample has a finite permeability since fluid transport in now not instantaneous but exhibits viscous flow. Therefore, liquid first needs to be transported from the reservoir to the newly filled pores. After the filling of a new pore, a subsequent inflow through the neighbor pore throats leads to an increase in capillary pressure during the next time steps until reaching a steady state or a new instability. The filling of new pores creates a pressure gradient in the sample which leads to suction of liquid from the reservoir. Figure 17 shows snapshots of the liquid body at different time steps during the simulation, colored according to the pressure field. Figure 17a shows the filled entry pores at the initial time step ($PR/\gamma = -3$). During the next time steps, due to instabilities, new pores are filled Fig. 17b, c. One can see, that the pressure field varies throughout the liquid body. The pressure remains constant in the entry cells (upper layer). The newly filled cells at the interface have a significantly lower pressure of approx. $PR/\gamma = -10$. Finally, there are also cells with a higher pressure than in the entry cells due to gravity.

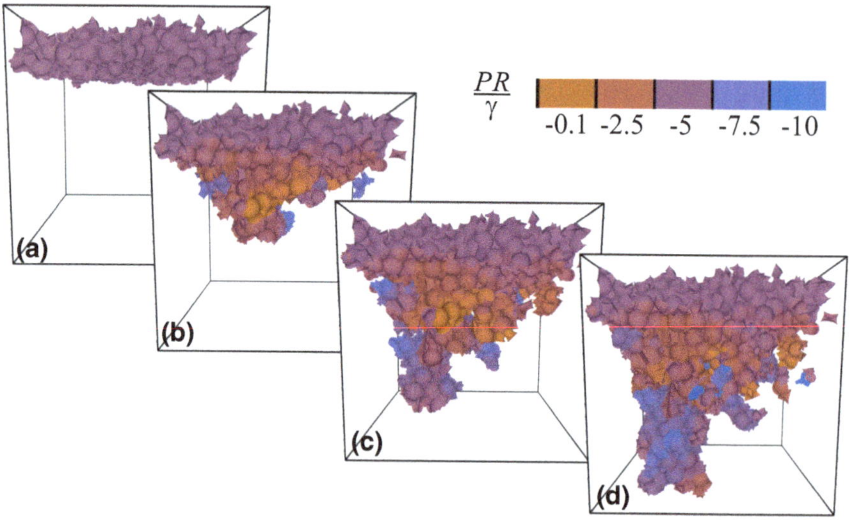

Fig. 17 Time evolution of an imbibition front in a sample with $N_p = 1750$ particles (not shown). The pressure in the upper particle layer is kept constant at $PR/\gamma = -3$. We show only the liquid phase consisting of filled pore bodies. Note that we use periodic boundary conditions in horizontal directions. **a** Initial front; **b** Front with finger formation and **c** front shortly after (**b**). **d** percolating cluster

Fluid Induced Failure

In a rigid porous body from sintered glass beads, one can omit the coupling between fluid and particles. In frictional granular systems, however, even at high hydrostatic stress states, a significant number of so-called spectator particles does not directly contribute to the contact force network. The coupling between fluid and particles changes particle configurations and liquid clusters add macroscopic cohesion via the Laplace pressure to the system. The fluid couples to the particles via the forces described for liquid bridges in Eq. 4 and pore throats and bodies in section "Geometric Representation of Liquid Bodies". Due to the complexity of higher order liquid bodies, described above, simulations are predominantly limited to the pendular state (e.g. [18, 35, 37, 51, 59, 60]). With simplified geometrical assumptions, described in section "Extension of Models for the Funicular State", however, this limitation can be overcome. In soil mechanics, the effect of the internal pressure field on the macroscopic behavior is of wide importance not just for sand sculptures, but for slope stability in general. Triaxial shear tests on cubic samples are simulated at different initial saturation states to capture the effect of hydrostatic stress on such systems [1, 43]. The cubic sample contains 5000 particles of mass density $\rho_p = 1$, radius $R = 1$ and resulting mass m_p. First, we assign random velocities to all particles. After randomization, the sample is confined by hydrostatic pressure on the box walls with suppressed contact friction $\mu = 0$ until a solid fraction of $\Phi_s = 0.6$

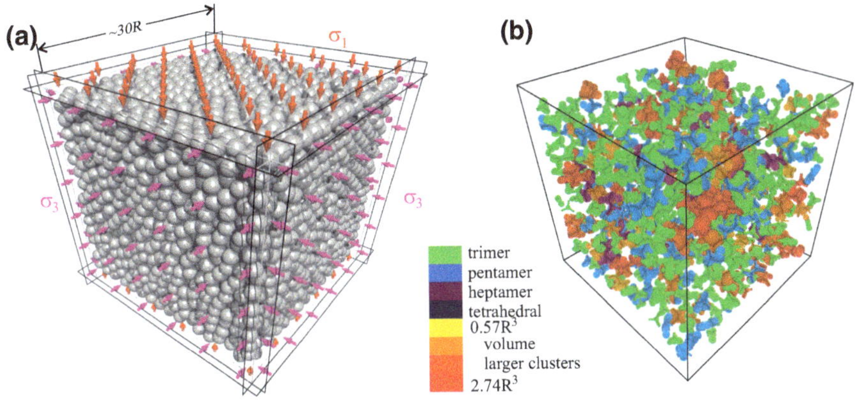

Fig. 18 **a** Setup of the triaxial shear test with confining walls and **b** liquid clusters emerging at liquid content $W_c = 3\%$ with colors representing cluster morphology and size. Reprint by permission from Springer Nature: Ref. [43]

is reached at a cube edge length L of about $32R$ (see Fig. 18). Triaxial tests require confining forces on each wall F_{conf}, that are calculated with the respective stress $\sigma_{1,2}$ and the true area of the wall A_{wall} involved. For the compression, μ is set to 0.3 and the surface tension $\gamma = 1$. The hydrostatic stress σ_3 is applied on all movable side walls, the bottom wall is fixed and the upper wall lowered at a constant strain rate $\dot{v} = v_w/L_z$ with the vertical displacement rate v_w that stays within the quasi-static limit. The differential stress is defined as $q = \sigma_1 - \sigma_3$ and axial ε_a, as well as volumetric strain $\varepsilon_V = \Delta V/V_0$ are calculated from the present and initial wall positions. The macroscopic behavior of the system was simulated with increasing initial liquid content W_c up to 40% and confining pressure $\sigma_3 = 20, 40, 60$ in units $u_p = m_p/(T^2 R)$ where T is the unit of time defined as $T = 1/\dot{v}$. To allow for a direct relation between the contact dynamical simulations with physical data, the ratio of inertial to confining forces can be used as proposed in Ref. [52]: $I = \dot{v}R\sqrt{\rho_p \sigma_3}$. Consequently for the three confining stresses this ratio is equal to $I_{20} \approx 2.2 \cdot 10^{-5}$, $I_{40} \approx 1.6 \cdot 10^{-5}$, $I_{60} \approx 1.3 \cdot 10^{-5}$.

Granular systems are typically dilating under load. The dependence on the confining pressure σ_3 is shown in Fig. 19a with the typical decrease in volumetric strain ε_V for increasing σ_3 similar to Ref. [59]. When comparing wet to dry cases, slightly higher values of ε_V are observed for $W_c = 1\%$ (see Fig. 19a). This observation could be due to attractive capillary forces that hinder spatial rearrangements. During the compression, $q = \sigma_1 - \sigma_3[u_p]$ is calculated as function of the axial strain ε_A. q rapidly increases for small ε_A, almost independently on the liquid content W_c what is in agreement with comparable experimental [68] and numerical observations [1, 59]. In the course of the compression, q reaches a constant, moisture dependent plateau. Note that the volume exclusion constraint of contact dynamics leads to a slightly exaggerated increase in q at the beginning of the loading, contrary to soft particle dynamics, where particles are allowed to overlap (see, e.g., Ref. [60]). Clearly one

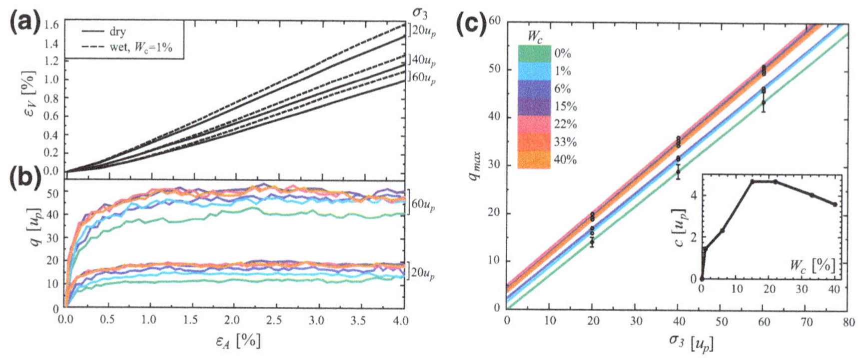

Fig. 19 **a** Volumetric strain ε_V and **b** differential stress q as a function of axial strain ε_A for different confining pressures σ_3. Liquid contents W_c have identical legend as **c**. **c** Mohr-Coulomb failure envelope with maximal values of differential stress q_{max} as a function of confining pressure σ_3. Inset: Dependence of cohesion c on W_c. Reprint by permission from Springer Nature: Ref. [43]

can observe an increase for the maximum differential stress q_{max} or shear strength for wet samples for identical σ_3 (see Fig. 19b, c). Shear strength τ versus confining pressure σ_3 can be approximated by a straight line passing through the origin for dry granulates as proposed by the Mohr-Coulomb failure criterion in the form $\tau = \mu_f \sigma + c$, where $\mu_f = \tan(\theta)$ is called internal friction coefficient, θ angle of internal friction and c cohesion. As seen in Fig. 19, similar behavior is found for wet cases ($W_c > 0$), only with $c > 0$ resulting from the sum of capillary forces of all liquid bridges and clusters. For the entirely dry and fully saturated state $c = 0$, a maximum in cohesion must exist, that is located in the funicular state in the interval $15\% \leq W_c \leq 22\%$, with a maximal increase in τ of about $5u_p$, independent on the confining pressure (see Fig. 19c inset). θ proves to be independent on W_c within our statistical error bars at $\theta \approx 35.9°$.

Since we resolve liquid structures, their evolution during compression is accessible. Exemplary the evolution of morphologies at one distinct liquid content $W_c = 6\%$ with confining pressure $\sigma_3 = 60u_p$ is shown in Fig. 20a. The values are related to the initial state to highlight the relative change during axial compression. As the system dilates, inter-particle distances increase and consequently single bridges rupture, and their liquid fuels the formation of new higher order clusters. Hence, mechanical deformation triggers multiple instabilities leading to trimer formation and consolidation of existing clusters into larger structures by coalescence. This evolution is exemplified by looking at a single compact liquid cluster, previously formed by liquid injection at a single position of an initial volume $V_0 \approx 100R^3$. As shown in Fig. 20b, the cluster volume is rather constant during compression, but significant changes occur in the number of individual units which constitute the cluster. One observes a spreading and loosening of the cluster orthogonal to the axial loading direction. The increasing number of menisci is due to the formation of new trimers or the merging of trimers with already existing ones in the cluster. Interestingly, the number of filled

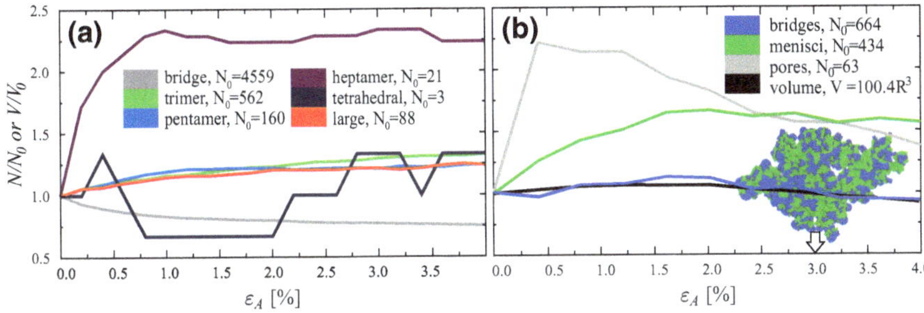

Fig. 20 **a** Evolution of liquid cluster morphologies relative to the initial configuration as function of ε_A for $W_c = 6\%$ and $\sigma_3 = 40u_p$. **b** Same as (**a**) but for the elemental parts of one large cluster. Reprint by permission from Springer Nature: Ref. [43]

pore bodies and liquid bridges decreases after reaching a local maximum, what is due to dilatation and thus increasing inter-particle distances. Note that the initial strong increase of menisci and filled pore bodies do not result in a higher cluster volume since Laplace pressure decreases when new pores are filled, resulting in a decrease of other menisci volumes as they move deeper into pore throats.

Conclusions and Outlook

With the fluid-particle models described in this chapter, one can study the shearing behavior of granular packings under different saturation levels, ranging from the dry material via the capillary bridge regime to higher saturation levels with percolating clusters, even up to the fully saturated state. The complexity of liquid morphologies is taken into account, implying the formation of isolated arbitrary-sized liquid or gas clusters with proper Laplace pressures. The core of our novel fluid-particle model is the discretization of the complex liquid morphologies by three basic liquid structures: bridges, trimers, and tetrahedral cells. Those are combined to form arbitrary-sized liquid clusters with proper Laplace pressures that evolve by fluid exchange via films on the grain surface. Depending on local conditions, changes of accessible liquid and the pore space morphogenesis, determined by the granular phase, liquid clusters can grow in size, shrink, merge and split. We represent the coupled solid phase by a discrete particle model based on Contact Dynamics, where capillary forces exerted from a liquid phase add to the motion of spherical particles. With the model, we study the failure of systems at various liquid contents under multi-axial stress states, the change in fluid distribution and morphologies during the loading due to the deformation of the pore space, drainage, and imbibition, as well as imbibition front instabilities like viscous fingering, using a pressure-driven model.

The usage of the models is demonstrated in studies on the fluid transport in sheared granular media, exhibiting the critical role of confining pressure on the diffusion mechanism. We show a study on the role of contact angle hysteresis in Part 1 of this work. Exact calculations of the liquid surface morphologies of the liquid bodies are impracticable for larger grain assemblies. However, we demonstrated how to use findings from NEM simulation (Part 1) in grain scale simulations with a large number of particles. It is likely that in the future, one extends the parameter space for NEM simulations concerning contact angles, surface tension, and variable particle radii, what would immediately allow for extensions of the described DEM models, e.g., to capture contact angle hysteresis, poly-disperse packings, and non-hydrophobic boundaries. In the current version of the model also a limitation to small strains up to 4–5% is inherent since we did not include updates and mappings of triangulation structures. For higher strains, the particle positions must be triangulated anew to provide a valid pore network. Another extension of the model could be the consideration of liquid transport through the gaseous phase to consider liquid with low vapor pressure or a low-pressure environment, as well as hydrostatic pressure, essential to understanding viscous fingering.

Acknowledgements We would like to thank Ciro Semprebon and Martin Brinkmann for providing tabulated values characterizing single liquid bridges, M. Rüggeberg for support in making Figs. 6 and 7, and the Deutsche Forschungsgemeinschaft (DFG) for financial support through grant no. HE 2732/11-1. The research leading to these results has also received partial funding from the People Programme (Marie Curie Actions) of the European Unions Seventh Framework Programme FP7 under the MUMOLADE ITN project (Multiscale Modelling of Landslides and Debris Flow) with REA Grant Agreement No. 289911, as well as from the European Research Council Advanced Grant No. 319968-FlowCCS.

References

1. Belheine, N., Plassiard, J.P., Donz, F.V., Darve, F., Seridi, A.: Numerical simulation of drained triaxial test using 3D discrete element modeling. Comput. Geotech. **36**(1), 320–331 (2009)
2. Berg, S., Ott, H., Klapp, S.A., Schwing, A., Neiteler, R., Brussee, N., Makurat, A., Leu, L., Enzmann, F., Schwarz, J.O., Kersten, M., Irvine, S., Stampanoni, M.: Real-time 3d imaging of Haines jumps in porous media flow. Proc. Natl. Acad. Sci. USA **110**(10), 3755–3759 (2013)
3. Bianchi, F., Thielmann, M., Mani, R.A., Or, D., Herrmann, H.J.: Tensile stress relaxation in unsaturated granular materials. Granul. Matter **18**(4), 75 (2016)
4. Blake, T.D., Ruschak, K.J.: Wetting: static and dynamic contact lines. In: Liquid Film Coating, pp. 63–97. Springer, Dordrecht (1997)
5. Brakke, K.A.: The surface evolver and the stability of liquid surfaces. Philos. Trans. R. Soc. Lond. A **354**, 2143–2157 (1996)
6. Brendel, L., Unger, T., Wolf, D.E.: Contact dynamics for beginners. In: Hinrichsen, H., Wolf, D.E. (eds.) The Physics of Granular Media, Wiley Online Books (2005)
7. Bryant, S., Blunt, M.: Prediction of relative permeability in simple porous media. Phys. Rev. A **46**(4), 2004–2011 (1992)
8. Campbell, C.S.: Self-diffusion in granular shear flows. J. Fluid Mech. **348**, 85–101 (1997)
9. Carr, J.F.: Tensile strength of granular materials. Nature **213**(5081), 1158–1159 (1967)
10. Coles, M.E., Hazlett, R.D., Spanne, P., Soll, W.E., Muegge, E.L., Jones, K.W.: Pore level imaging of fluid transport using synchrotron X-ray microtomography. J. Pet. Sci. Eng. **19**(1), 55–63 (1998)

11. Cundall, P.A., Strack, O.D.L.: A discrete numerical model for granular assemblies. Géotechnique **29**(18), 47–65 (1979)
12. Dijksman, J.A., van Hecke, M.: Granular flows in split-bottom geometries. Soft Matter **6**(13), 2901–2907 (2010)
13. Feder, J.: Random sequential adsorption. J. Theor. Biol. **87**(2), 237–254 (1980)
14. Fisher, R.A.: On the capillary forces in an ideal soil; correction of formulae given by W. B. Haines. J. Agric. Sci. **16**(3), 492–505 (1926)
15. de Gennes, P.G., Brochard-Wyart, F., Quere, D.: Capillarity and Wetting Phenomena: Drops, Bubbles, Pearls, Waves. Springer, New York (2004)
16. German, R.M.: Coordination number changes during powder densification. Powder Technol. **253**, 368–376 (2014)
17. Gladkikh, M.: A priori prediction of macroscopic properties of sedimentary rocks containing two immiscible fluids. Ph.D. thesis, University of Texas at Austin (2005)
18. Gröger, T., Tüzün, U., Heyes, D.M.: Modelling and measuring of cohesion in wet granular materials. Powder Technol. **133**(1), 203–215 (2003)
19. Haines, W.B.: Studies in the physical properties of soils: II. A note on the cohesion developed by capillary forces in an ideal soil. J. Agric. Sci. **15**(04), 529–535 (1925)
20. Haines, W.B.: Studies in the physical properties of soils: IV. A further contribution to the theory of capillary phenomena in soil. J. Agric. Sci. **17**(2), 264–290 (1927)
21. Halsey, T.C., Levine, A.J.: How sandcastles fall. Phys. Rev. Lett. **80**(14), 3141–3144 (1998)
22. Hartley, R.R., Behringer, R.P.: Logarithmic rate dependence of force networks in sheared granular materials. Nature **421**(6926), 928–931 (2003)
23. Herminghaus, S.: Dynamics of wet granular matter. Adv. Phys. **54**(3), 221–261 (2005)
24. Herrmann, H., Luding, S.: Modeling granular media on the computer. Contin. Mech. Thermodyn. **10**(4), 189–231 (1998)
25. Hornbaker, D.J., Albert, R., Albert, I., Barabási, A.L., Schiffer, P.: What keeps sandcastles standing? Nature **387**(6635), 765 (1997)
26. Iveson, S.M., Litster, J.D., Hapgood, K., Ennis, B.J.: Nucleation, growth and breakage phenomena in agitated wet granulation processes: a review. Powder Technol. **117**(1), 3–39 (2001)
27. Kadau, D., Schwesig, D., Theuerkauf, J., Wolf, D.: Influence of particle elasticity in shear testers. Granul. Matter **8**, 35–40 (2006)
28. Kohonen, M.M., Maeda, N., Christenson, H.K.: Kinetics of capillary condensation in a nanoscale pore. Phys. Rev. Lett. **82**, 4667–4670 (1999)
29. Kohonen, M.M., Geromichalos, D., Scheel, M., Schier, C., Herminghaus, S.: On capillary bridges in wet granular materials. Phys. A: Stat. Mech. Appl. **339**(1), 7–15 (2004)
30. Lambert, P., Chau, A., Delchambre, A., Régnier, S.: Comparison between two capillary forces models. Langmuir **24**(7), 3157–3163 (2008)
31. Lian, G., Seville, J.: The capillary bridge between two spheres: new closed-form equations in a two century old problem. Adv. Colloid Interface Sci. **227**, 53–62 (2016)
32. Lian, G., Thornton, C., Adams, M.J.: A theoretical study of the liquid bridge forces between two rigid spherical bodies. J. Colloid Interface Sci. **161**(1), 138–147 (1993)
33. Lu, N., Wu, B., Tan, C.P.: Tensile strength characteristics of unsaturated sands. J. Geotech. Geoenvironmental Eng. **133**(2), 144–154 (2007)
34. Lukyanov, A.V., Sushchikh, M.M., Baines, M.J., Theofanous, T.G.: Superfast nonlinear diffusion: capillary transport in particulate porous media. Phys. Rev. Lett. **109**(21), 214,501 (2012)
35. Mani, R.A.: Capillary interactions, shear thickening and liquid migration in wet granular media. Ph.D. thesis, ETH-Zürich (2014)
36. Mani, R.A., Kadau, D., Or, D., Herrmann, H.J.: Fluid depletion in shear bands. Phys. Rev. Lett. **109**(24), 248,001 (2012)
37. Mani, R.A., Kadau, D., Herrmann, H.J.: Liquid migration in sheared unsaturated granular media. Granul. Matter **15**(4), 447–454 (2013)
38. Mani, R.A., Semprebon, C., Kadau, D., Herrmann, H.J., Brinkmann, M., Herminghaus, S.: Role of contact-angle hysteresis for fluid transport in wet granular matter. Phys. Rev. E **91**(4), 042,204 (2015)

39. Mason, G., Clark, W.: Tensile strength of wet granular materials. Nature **219**(5150), 149–150 (1968)
40. Mellor, D.W.: Random close packing (RCP) of equal spheres: structure and implications for use as a model porous medium. Ph.D. thesis, Open University, Milton Keynes, U.K. (1989)
41. Melnikov, K.: Grain scale modelling of triggering mechanisms of rainfall induced slope failures. Ph.D. thesis, ETH Zürich (2016)
42. Melnikov, K., Mani, R.A., Wittel, F.K., Thielmann, M., Herrmann, H.J.: Grain-scale modeling of arbitrary fluid saturation in random packings. Phys. Rev. E **92**(2), 022,206 (2015)
43. Melnikov, K., Wittel, F.K., Herrmann, H.J.: Micro-mechanical failure analysis of wet granular matter. Acta Geotech. **11**(3), 539–548 (2016)
44. Melrose, J.C.: Wettability as related to capillary action in porous media. Soc. Pet. Eng. J. **5**(03), 259–271 (1965)
45. Mitarai, N., Nori, F.: Wet granular materials. Adv. Phys. **55**(1), 1–45 (2006)
46. Moreau, J.J.: Some numerical methods in multibody dynamics: application to granular materials. Eur. J. Mech.-Solids **13**, 93–114 (1994)
47. Morrow, N., Mason, G.: Recovery of oil by spontaneous imbibition. Curr. Opin. Colloid Interface Sci. **6**(4), 321–337 (2001)
48. Motealleh, S., Ashouripashaki, M., DiCarlo, D., Bryant, S.: Unified model of drainage and imbibition in 3d fractionally wet porous media. Transp. Porous Media **99**(3), 581–611 (2013)
49. Pierrat, P., Caram, H.S.: Tensile strength of wet granular materials. Powder Technol. **91**(2), 83–93 (1997)
50. Pierrat, P., Agrawal, D.K., Caram, H.S.: Effect of moisture on the yield locus of granular materials: theory of shift. Powder Technol. **99**(3), 220–227 (1998)
51. Richefeu, V., El Youssoufi, M.S., Radjaï, F.: Shear strength properties of wet granular materials. Phys. Rev. E **73**(5), 051,304 (2006)
52. Rognon, P.G., Roux, J.N., Wolf, D., Naaïm, M., Chevoir, F.: Rheophysics of cohesive granular materials. EPL (Europhys. Lett.) **74**(4), 644 (2006)
53. Rynhart, P., McLachlan, R., Jones, J., McKibbin, R.: Solution of the Young-Laplace equation for three particles. Res. Lett. Inf. Math. Sci. **5**(4), 119–127 (2003)
54. Sanders, C.: An open source C++ linear algebra library for fast prototyping and computationally intensive experiments. Technical report, NICTA (2010)
55. Scheel, M.: Experimental investigations of the mechanical properties of wet granular matter. Ph.D. thesis, Georg-August-Universität Goettingen (2009)
56. Scheel, J., Seemann, R., Brinkmann, M., Di Michiel, M., Sheppard, A., Breidenbach, B., Herminghaus, S.: Morphological clues to wet granular pile stability. Nat. Mater. **7**(3), 189–193 (2008)
57. Scheel, J., Seemann, R., Brinkmann, M., Di Michiel, M., Sheppard, A., Herminghaus, S.: Liquid distribution and cohesion in wet granular assemblies beyond the capillary bridge regime. J. Phys. Condens. Matter **20**(49), 494,236 (2008)
58. Scholtès, L.: Modélisation micromécanique des milieux granulaires partiellement saturés. Ph.D. thesis, Institut National Polytechnique de Grenoble - INPG (2008)
59. Scholtès, L., Chareyre, B., Nicot, F., Darve, F.: Micromechanics of granular materials with capillary effects. Int. J. Eng. Sci. **47**(1), 64–75 (2009)
60. Scholtès, L., Hicher, P.Y., Nicot, F., Chareyre, B., Darve, F.: On the capillary stress tensor in wet granular materials. Int. J. Numer. Anal. Methods Geomech. **33**(10), 1289–1313 (2009)
61. Seemann, R., Mönch, W., Herminghaus, S.: Liquid flow in wetting layers on rough substrates. EPL (Eur. Lett.) **55**(5), 698–704 (2001)
62. Semprebon, C., Scheel, M., Herminghaus, S., Seemann, R., Brinkmann, M.: Liquid morphologies and capillary forces between three spherical beads. Phys. Rev. E **94**, 012,907 (2016)
63. Shi, D., McCarthy, J.J.: Numerical simulation of liquid transfer between particles. Powder Technol. **184**(1), 64–75 (2008)
64. Tsunakawa, H., Aoki, R.: Tensile strength of wet granular materials. Chem. Eng. **36**(3), 281–286,a1 (1972)

65. Ulrich, S., Aspelmeier, T., Zippelius, A., Roeller, K., Fingerle, A., Herminghaus, S.: Dilute wet granular particles: nonequilibrium dynamics and structure formation. Phys. Rev. E **80**(3), 031,306 (2009)
66. Utter, B., Behringer, R.P.: Self-diffusion in dense granular shear flows. Phys. Rev. E **69**, 031,308 (2004)
67. Vogel, H.J., Roth, K.: Quantitative morphology and network representation of soil pore structure. Adv. Water Resour. **24**(3), 233–242 (2001)
68. Wang Qiong, Lade Poul V.: Shear banding in true triaxial tests and its effect on failure in sand. J. Eng. Mech. **127**(8), 754–761 (2001)
69. Widom, B.: Capillarity and wetting phenomena: drops, bubbles, pearls, waves. Phys. Today **57**(12), 66–67 (2004)
70. Willett, C.D., Adams, M.J., Johnson, S.A., Seville, J.P.K.: Capillary bridges between two spherical bodies. Langmuir **16**(24), 9396–9405 (2000)
71. Zhou, T., Ioannidou, K., Masoero, E., Mirzadeh, M., Pellenq, R.J.M., Bazant, M.Z.: Capillary stress and structural relaxation in moist granular materials. Langmuior **35**(12) 4397–4402 (2019)

Sintering—Pressure- and Temperature-Dependent Contact Models

T. Weinhart, R. Fuchs, T. Staedler, M. Kappl and S. Luding

Introduction

Granular materials—conglomerates of discrete particles—are of paramount importance in the chemical, pharmaceutical, agri-food, energy, high-tech, materials, manufacturing, mining and construction industries. Handling and processing of particulates is a huge market, with annual revenues of €660 billion [8], making such particulate materials the most widely manipulated substance after water [5]. Here, we focus on sintering, i.e., the process of compacting and transforming a free-flowing particulate mass into a solid object by applying heat and/or pressure, without melting it completely. Additive manufacturing, tabletting, and moulding are just a few applications in which sintering plays a role.

To better understand the sintering of granular materials, we take a bottom-up approach: First, the particle and contact properties of the particles that constitute the granular material are measured experimentally. This data is then used to construct contact laws and particle properties, paying special attention to pressure- and temperature-dependence. Those laws are then applied to model the granular

T. Weinhart (✉) · S. Luding
University of Twente, Enschede, The Netherlands
e-mail: t.weinhart@utwente.nl

S. Luding
e-mail: s.luding@utwente.nl

R. Fuchs
Evonik Technology & Infrastructure GmbH, Hanau, Germany
e-mail: regina.fuchs@evonik.com

T. Staedler
Universität Siegen, Siegen, Germany
e-mail: staedler@lot.mb.uni-siegen.de

M. Kappl
MPI for Polymer Research, Mainz, Germany
e-mail: kappl@mpip-mainz.mpg.de

S. Antonyuk (ed.), *Particles in Contact*,
https://doi.org/10.1007/978-3-030-15899-6_10

materials via Discrete Particle Simulations, which explicitly follow the motion of each individual particle by applying forces and torques that stem either from external (body) forces or from interactions with other particles or walls. Those simulations, properly calibrated, can then be used to predict the qualities of a sintered product (e.g., its tensile strength, elasticity, or brittleness) including the whole process chain: the cold, lose powder, the sintering process by increased pressure and heat, the cool-down, and the final testing phases.

This chapter summarises the advances made in modelling pressure-, temperature-, and time-dependent granular processes, sintering in particular, within the DFG SPP PiKo. The work has been published in several articles [11–13], in close collaboration by the groups of Luding, Staedler and Kappl. We focus here on the numerical models developed in this process; for details of the experimental work, see [11].

The discrete element method (DEM), here abbreviated as discrete particle method (DPM), is based on the contact forces and follows the motion of all particles in the system. Besides the repulsive/elastic-plastic and viscous forces, attractive forces due to cohesion still pose challenges. Of particular interest is dry cohesion due to van-der-Waals forces, which increases due to plastic (irreversible) deformation at the contacts [2, 3, 14, 15, 22, 25, 29, 31, 34, 35, 38, 41, 42, 44, 45, 50, 51]. Such forces are a source for agglomeration, sticking or caking, disturbing transport and storage of powders in many processes, and affecting the processes related to additive manufacturing. DPM has previously been shown to provide quantitatively accurate predictions [17, 30], if the microscopic contact laws are accurately calibrated. The understanding of particles in contact, by studying the mechanisms involved during contact and at separation (pull-off), was the main aim of most projects in PiKo. One result was to improve existing, or define new models or contact laws. Calibrating contact laws—either directly or indirectly, by contact or bulk measurements, involving experiments and simulations—was a central aim of various PiKo projects, e.g., for adhesive forces [19, 24]. However, taking experimental data for calibration is not trivial—not even for ideal spheres [11]; most powders are composed of micron-sized particles with irregular, sometimes fractal shapes [40]. An additional challenge for modelling sintering processes, is that micro-properties are not constant, but depend on the pressure and temperature applied to the particle bulk, and via force-chains to the single particle contacts. These dependencies are not well-known yet [11, 13, 29, 56].

We therefore approach the calibration process in two steps. In section “Contact Models”, we present a general DPM contact model that takes into account elasto-plastic, adhesive and frictional forces, as well as time-, pressure- and temperature dependent effects, as relevant e.g., during sintering or additive manufacturing. In section “Experimental Calibration of the Elastic-Adhesive, and Frictional Forces”, we show how to use existing and novel experimental methods to validate and calibrate this contact model; first, we calibrate the general properties, then we focus on the behaviour during sintering. The results of these experiments are presented and simulated in section “Initial Stage Sintering of Polymer Particles”.

Contact Models

In this section, we review notation, kinematics, and basic contact mechanics for a two-particle contact in section "Contact Mechanics". Based on this, a general contact is proposed, in section "Contact modelling" accounting for adhesive, elasto-plastic, dissipative and frictional forces, and complemented by the new model for the time-, pressure-, and temperature-dependent formation of sinter necks in section "Sintering Forces".

Contact Mechanics

Notation

We denote the mass of particle i as m_i, its centre-of-mass, or position, $\mathbf{r}_i$, its velocity $\mathbf{v}_i$ and the angular velocity around its centre-of-mass $\boldsymbol{\omega}_i$ (Fig. 1).

The definitions of the contact overlap δ_{ij} and the unit normal at the contact $\hat{\mathbf{n}}_{ij}$ between two particles i, j depend on the shape and size of the particles. For contacts between two spherical particles i, j with radii r_i, r_j, the overlap is $\delta_{ij} = r_i + r_j - |\mathbf{r}_i - \mathbf{r}_j|$ and the unit normal $\hat{\mathbf{n}}_{ij} = (\mathbf{r}_i - \mathbf{r}_j)/|\mathbf{r}_i - \mathbf{r}_j|$. For contacts between a spherical particle i and a planar surface w, with outward unit normal $\hat{\mathbf{n}}_w$ and crossing the point $\mathbf{r}_w$, we define the unit normal $\hat{\mathbf{n}}_{iw} = \hat{\mathbf{n}}_w$ and the overlap as $\delta_{iw} = r_i - (\mathbf{r}_w - \mathbf{r}_i) \cdot \hat{\mathbf{n}}_w$.

We further define the corrected radius $r_{ij} = r_i - \delta_{ij}/2$, the contact point $\mathbf{c}_{ij} = \mathbf{r}_i - r_{ij}\hat{\mathbf{n}}_{ij}$, the branch vector $\mathbf{r}_{ij} = \mathbf{r}_i - \mathbf{c}_{ij}$, the surface velocity of particle i at the contact point $\mathbf{v}_i^c = \mathbf{v}_i + \mathbf{r}_{ij} \times \boldsymbol{\omega}_i$, the relative surface velocity at the contact $\mathbf{v}_{ij} = \mathbf{v}_i^c - \mathbf{v}_j^c$, the relative angular velocity $\boldsymbol{\omega}_{ij} = \boldsymbol{\omega}_i - \boldsymbol{\omega}_j$, and the effective mass $m_{ij} = m_i m_j/(m_i + m_j)$.

For readability, we omit the ij-subscripts in the following sections.

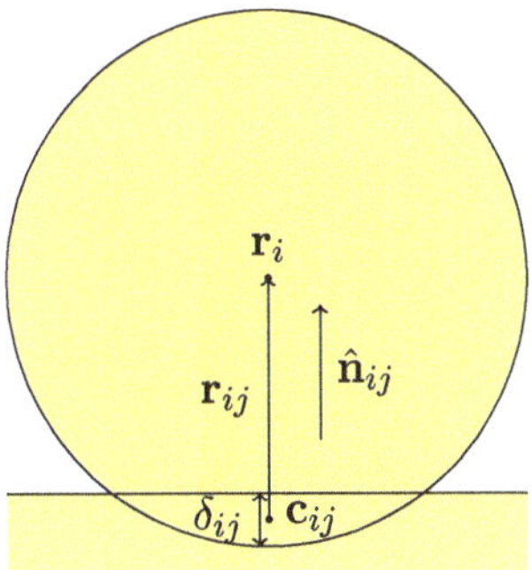

Fig. 1 Particle-wall contact

Decomposition of Contact Forces and Torques

Particles in DPM are assumed to be undeformable, but can slightly overlap. Thus, two particles interact via contact forces and torques acting at a single contact point, instead of a stress distribution over a contact area. These forces and torques, however, can be derived from integrating the stresses over the contact surface.

Integrating over the normal and tangential surface stresses yields the normal and tangential, lateral components of the contact force,

$$\mathbf{f} = f^{\mathrm{n}}\hat{\mathbf{n}} + \mathbf{f}^{\mathrm{l}}. \tag{1a}$$

Similarly, the torque around the particle's centre-of-mass can be decomposed into (*i*) a torque due to the tangential contact force, $\mathbf{r} \times \mathbf{f}^{\mathrm{l}}$, and (*ii*) remaining normal, $\mathbf{M}^{\mathrm{to}}$, and tangential, $\mathbf{M}^{\mathrm{ro}}$, components, yielding the total torque,

$$\mathbf{M} = \mathbf{r} \times \mathbf{f}^{\mathrm{l}} + \mathbf{M}^{\mathrm{ro}} + \mathbf{M}^{\mathrm{to}}. \tag{1b}$$

Objective Decomposition of Contact Velocities and Displacements

The decomposition by Luding [22], splits the relative translational and angular velocities and displacements into four components:
(a) The normal relative velocity,

$$v^{\mathrm{n}} = \mathbf{v} \cdot \hat{\mathbf{n}}, \tag{2}$$

(b) the lateral sliding velocity, due to tangential surface displacements at the contact,

$$\mathbf{v}^{\mathrm{l}} = \mathbf{v} - v^{\mathrm{n}}\hat{\mathbf{n}}, \tag{3a}$$

(c) the rolling "velocity" at which two surfaces roll over each other,

$$\mathbf{v}^{\mathrm{ro}} = r(\boldsymbol{\omega} \times \hat{\mathbf{n}}), \text{ and} \tag{3b}$$

(d) the torsion "velocity", measuring the normal relative angular velocity scaled by the effective radius,

$$\mathbf{v}^{\mathrm{to}} = r(\boldsymbol{\omega} \cdot \hat{\mathbf{n}})\hat{\mathbf{n}}. \tag{3c}$$

Each of these velocities is objective, meaning that the observer will measure the same velocities, even if (the frame of reference) is relocated by a finite translation and/or rotation, see Refs. [18, 22] for the general objectiveness of the rolling velocity. The normal relative velocity equals the time-derivative of the (normal) overlap,

$$\frac{d}{dt}\delta = \mathbf{v}^{\mathrm{n}}. \tag{4}$$

Similarly, the other velocities are given by their respective rates of change of the displacements,

$$\frac{d}{dt}\boldsymbol{\delta}^{\nu} = \mathbf{v}^{\nu}, \quad \nu = \mathrm{l, ro, to}, \tag{5}$$

where, however, (elastic) displacements are set to zero at the beginning of a contact and the displacements have to be obtained by integrating the velocities over time. We assumed here that the normal vector $\hat{\mathbf{n}}$ is constant over time, which is the case for particle-surface contacts; for particle-particle collisions, the displacements are calculated such that the directions of $\boldsymbol{\delta}^{\mathrm{l}}$, $\boldsymbol{\delta}^{\mathrm{ro}}$ stay perpendicular, while $\boldsymbol{\delta}^{\mathrm{to}}$ stays parallel to the normal vector, see [22] for details.

Contact Modelling

Next, we propose an objective contact model based on the above considerations. The normal force f^n is described in section "Normal Contact Force", the tangential forces and torques in section "Tangential Forces and Torques".

Normal Contact Force

We use the model by Luding [22], where the normal force f^{n} is decomposed into a repulsive, dissipative, and adhesive component,

$$f^{\mathrm{n}} = f_{\mathrm{ep}} - \gamma^{\mathrm{n}} v^{\mathrm{n}} - f_{\mathrm{a}}. \tag{6}$$

The repulsive (elasto-plastic) force due to mechanical contact interactions, $f_{\mathrm{ep}} = k^{\mathrm{n}}(\delta - \delta_p)$, is discussed in different variants below. Note that the stiffness k^{n}, the plastic deformation δ_p, the dissipation γ^{n}, and the adhesive force f_{a}, are not necessarily constants; different dependencies have been proposed for different materials and conditions. The dissipative force, $-\gamma^{\mathrm{n}} v^{\mathrm{n}}$, reduces the relative velocity between the particles and thus results in (dynamic) energy dissipation, while the elasto-plastic force also dissipates energy under quasi-static load-unload cycles. The adhesive force can represent both dry and wet situations, where dissipation is taking place in the latter case, i.e., due to cutting of a liquid bridge [36, 39].

The simplest elastic-dissipative contact model assumes constant stiffness k^{n} and dissipation γ^{n}, without plastic deformation $\delta_p = 0$. The result is a simple, analytically tractable, contact model with simple physical characteristics, namely a velocity-independent duration of collisions and restitution coefficient [22, 39]. This model is used in section "Initial Stage Sintering of Polymer Particles", for simplicity, as basis, since the elastic-dissipative properties are likely not dominating the more complex sintering mechanisms; this way, one can focus on the effects of other elements of the model.

In section "Experimental Calibration of the Elastic-Adhesive, and Frictional Forces", where we measure the contact properties for all the different degrees of freedom, we account for the variable stiffness by applying Hertz' model for purely elastic contacts between spherical particles [16], $k^{\mathrm{n},H} = \frac{4}{3}E^*a$, where the effective modulus $E^* = [(1-\nu_i)/(2G_i) + (1-\nu_j)/(2G_j)]^{-1}$ is a combination of the shear moduli G_i, G_j and Poisson's ratios ν_j, ν_j of the two materials, and the radius of the contact area is approximated by $a = \sqrt{\bar{r}\delta}$ for $\delta \ll \bar{r}$, with $\bar{r} = 2r_ir_j/(r_i + r_j)$ the harmonic mean of the particle radii. We further use a variable dissipation coefficient according to [47], $\gamma^{\mathrm{n}} = 0.0062\sqrt{k^{\mathrm{n},H}m}$, which results in a constant restitution coefficient [21].

For larger deformations, we need to account for plastic deformations at the contact. This is done by introducing a plastic overlap, $\delta_{\mathrm{p}} \geq 0$, the overlap at which the normal contact force vanishes. The stiffness increases with the plastic deformation, due to the increased contact area, with $k^{\mathrm{n}} = k_{\mathrm{l}}$ for zero plastic overlap, and during loading, interpolating up to the maximum of $k^{\mathrm{n}} = k_*$ at $\delta_{\mathrm{p}} = \delta_*$. This is expressed as

$$k^{\mathrm{n}} = k_{\mathrm{l}} + (k_* - k_{\mathrm{l}})(\delta_{\mathrm{p}}/\delta_*)^{\psi}, \tag{7}$$

in [22], where the maximal previous overlap was used instead of the plastic overlap. At first, the power $\psi = 1$ was used due to lack of calibration data [22, 23]. It was generalised to arbitrary ψ during the DFG SPP PiKo, in order to account for a more general non-linearity of the elastic-plastic contact law [39] and to allow calibration of the contact model with experimental measurements for a wider class of materials and under extreme conditions like during tabletting. It also includes the traditional Walton model [52] with constant k^{n} for $\psi = 0$. For spherical particles, $\delta_* \approx r/6$ ensures that the maximum stiffness is reached when the solid volume fraction approaches unity, i.e. when all void space between the particles is filled [22].

The plastic overlap increases if a new maximum overlap is achieved, up to the limit of δ_*, or drops if $\delta < \delta_p$,

$$\delta_{\mathrm{p}} = \begin{cases} \frac{k^{\mathrm{n}}-k_{\mathrm{l}}}{k^{\mathrm{n}}}\delta & \text{if } k^{\mathrm{n}}(\delta - \delta_{\mathrm{p}}) \leq k^l\delta \leq k_*(\delta - \delta_{\mathrm{p}}), \\ \delta & \text{if } \delta < \delta_{\mathrm{p}}. \end{cases} \tag{8}$$

This ensures that the repulsive force vanishes for $\delta \leq \delta_p$. The reduction in plastic overlap during unloading allows for the complex, hysteretic reloading behaviour sketched in Fig. 2.

The adhesive force at $\delta = 0$ can be measured as the pull-off force f_{pull} required to detach a particle from a surface. To avoid an abrupt change of adhesive force when a contact opens, the adhesive force is modelled as

$$f_{\mathrm{a}} = \begin{cases} f_{\mathrm{pull}} & \text{if } 0 \leq \delta, \\ k_c\delta + f_{\mathrm{pull}} & \text{if } -f_{\mathrm{pull}}/k_c \leq \delta < 0, \\ 0 & \text{if } \delta \leq -f_{\mathrm{pull}}/k_c. \end{cases} \tag{9}$$

For simplicity, we use $k_c = k_{\mathrm{l}}$.

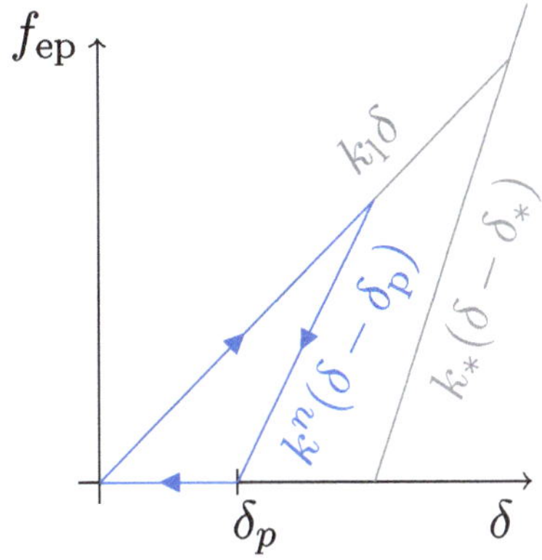

Fig. 2 Elasto-plastic repulsive force

Tangential Forces and Torques

The tangential sliding force and the rolling and torsion torques are assumed to resist the relative tangential and angular velocities, $\mathbf{v}^\nu$, and act against displacements, $\boldsymbol{\delta}^\nu$, with $\nu = \mathrm{l, ro, to}$. They can be modelled by coupling an elastic and a dissipative force,

$$\mathbf{f}^\nu = k^\nu \boldsymbol{\delta}^\nu - \gamma^\nu \mathbf{v}^\nu, \quad \nu = \mathrm{l, ro, to}. \tag{10a}$$

The rolling and torsion "forces" calculated here are not actual forces, but only used as intermediates to calculate the rolling and torsion torques,

$$\mathbf{M}^{\mathrm{ro}} = \frac{\lambda^{\mathrm{ro}}}{r} \mathbf{r} \times \mathbf{f}^{\mathrm{ro}}, \quad \mathbf{M}^{\mathrm{to}} = \lambda^{\mathrm{to}} \mathbf{f}^{\mathrm{to}}, \tag{10b}$$

where the length scales λ^{ro}, λ^{to} have to be specified.

Each of the three tangential forces/torques is assumed to have an *independent* yield criterion, where combined criteria are necessary for, e.g., the case of solid-bridges [4]. In the case of lateral slip this means: When the ratio of tangential to the nonadhesive normal force exceeds the static contact friction coefficient, $\mu^{\mathrm{l,s}}$, the contact surface yields until the ratio becomes smaller than or equal to the dynamic contact friction coefficient, μ^{l}. This is modeled by a yield criterion, truncating the magnitude of $\boldsymbol{\delta}^{\mathrm{l}}$ as necessary to satisfy

$$|\mathbf{f}^{\mathrm{l}}| \leq \mu^{\mathrm{l}}(f^{\mathrm{n}} + f_{\mathrm{a}}). \tag{11a}$$

The sliding friction coefficient μ^{l} is usually assumed to be constant (Coulomb type). While the static friction coefficient is generally larger than the dynamic friction coefficient, we assume for simplicity that $\mu^{\mathrm{l,s}} = \mu^{\mathrm{l}}$. Similarly, when the rolling or torsion torque-to-normal-force ratio becomes larger than the contact friction coefficient, μ^ν, the magnitude of $\boldsymbol{\delta}^\nu$ is cut as necessary to satisfy

$$|\mathbf{M}^{\nu}| \leq \mu^{\nu}\lambda^{\nu}(f^{\mathrm{n}} + f_{\mathrm{a}}), \;\; \nu = \mathrm{ro, to}. \tag{11b}$$

Here, we assume the friction coefficients μ^{ν} and adhesive forces f_a to be constant for the sake of a simple contact model.

Different models have been presented in the literature for the stiffness and friction coefficients. For Hertzian contacts, see "Normal Contact Force" the Mindlin [27] model assumes a non-linear sliding force with a spring stiffness depending on the contact radius (and thus on the normal load).

$$k^{\mathrm{l,M}} = 8G^{*}a, \tag{12}$$

where $G^{*} = [(2 - \nu_i^2)/G_i + (2 - \nu_j^2)/G_j]^{-1}$. This contact model is a simplification of the contact model of Mindlin–Deresiewicz [28], implemented in [47], where a varying spring stiffness $k^{\mathrm{l,MD}}$ is used that depends on the contact history. Note that if the spring stiffness k^{ν} varies during tangential displacement, the tangential force has to be computed incrementally [47].

Sintering Forces

It remains to specify how changes in pressure and temperature can lead to a permanent agglomeration of the particles. All parameters of the contact model—and the particle density—may vary with temperature [23], but we neglect this effect here for simplicity. Temperature-assisted sintering is described on the micro-scale by the formation of sinter necks at inter-particle contacts, caused by viscous flow of the particles' raw material at temperatures above the glass transition [26]. Pressure, on the other hand, causes elasto-plastic deformation of the contacts, which also leads to flow and possibly agglomeration. When, both pressure and temperature are applied, the two effects combine. Modelling this has been a particular challenge to existing sintering models, see e.g. [29], where sintering is modelled as a purely viscous, non-elastic process. Such models can be adapted to model temperature-assisted sintering under constant pressure [35], but do so by gradually transitioning between viscous and elastic contact models.

Here, we present a contact model, first published in [13], that allows for the simultaneous simulation of temperature- and pressure-assisted sintering. We extend the elastoplastic contact model of [22], which accounts for the deformation due to pressure, and apply a temperature-dependent viscous deformation-rate to the plastic overlap δ_{p},

$$\dot{\delta}_{\mathrm{p}} = \frac{f_{\mathrm{n}}\bar{r}}{n f_{\mathrm{a}}\tau}\left(\frac{\bar{r}}{x}\right)^{n-1}. \tag{13}$$

where $x = \sqrt{2}a$ denotes the sinter neck radius, see Fig. 3. For two particles without compression ($f_{\mathrm{n}} = f_a$) and of high stiffness (for which the contact duration

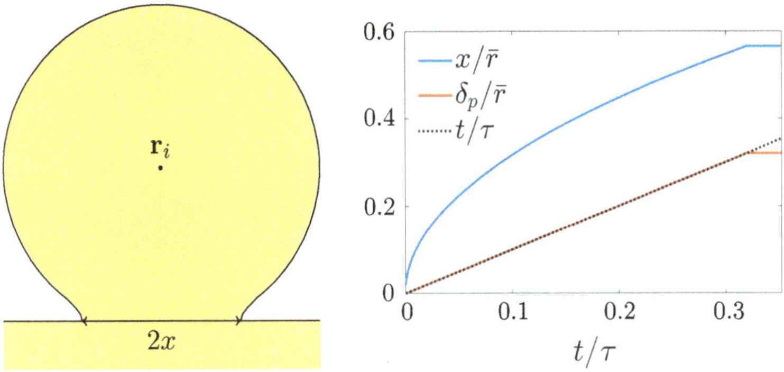

Fig. 3 (*left*) Sketch of sintered particle-wall contact; (*right*) contact radius x and plastic overlap for a particle-wall contact for $n = 1$, for the contact law used in section "Initial Stage Sintering of Polymer Particle". The neck radius plateaus when $\delta_p = \delta_*$ is reached

$\sqrt{m/k} \ll \tau$), one can assume that $\delta = \delta_p$, therefore the plastic overlap increases as $\delta_p/\bar{r} = \sqrt[n]{t/\tau}$. Therefore, τ denotes the *sintering time scale* and n the *sintering power law exponent*. Such a law can be rigorously derived from the force laws derived in [1] for surface sintering, which yields an exponent $n = 3$. For $n = 1$, we obtain the well-known Frenkel law [9] for viscous sintering, for which $x/\bar{r} \approx \sqrt{\delta/\bar{r}} \propto \sqrt{t}$, see Fig. 3; the remaining particle properties, see section "Initial Stage Sintering of Polymer Particles", do not significantly influence the result. If the increase of the particle radius due to sintering would be taken into account, we obtain the modified Frenkel law derived in [33]; for more details, see [12, 13].

Experimental Calibration of the Elastic-Adhesive, and Frictional Forces

We introduce and apply nanoindentation-based testing techniques for characterising the elastic, adhesive, and frictional forces of individual particle-surface contact, see Fig. 7. For brevity, we only provide a short description of the materials and experimental methods used. Further details can be found in [11] (Fig. 4).

Test Objects

Surfaces and Spherical Particles

The substrate material used was a single-crystalline Si(100) wafer, supplied by Siegert Wafer GmbH, Aachen. Surface topography changes of these samples were

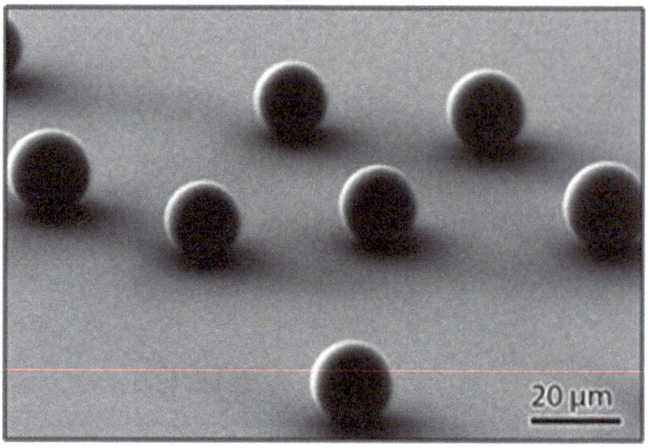

Fig. 4 Borosilicate glass spheres featuring a diameter of about 17 μm on a Si substrate

achieved through a slow etching process using H_2 plasma. Two different microwave powers (1600 and 1800 W) are used to sustain the plasma. The etching period was 20 min for all samples. This treatment leads to a variation in surface roughness due to the different etching efficiency of the H_2 plasma with different microwave power. Surface roughness was measured using atomic force microscopy, see Fig. 5 and Table 2.

A Focused-Ion beam (FIB) system (FEI Helios 600) was used to create rail-structure silicon samples (see Fig. 7), featuring a length of 100 μm and rail inclinations of 25°, 45° and 60°, respectively. Depth and width of the rails were chosen in such a way that the centre of mass of the particles was situated 1 μm below the wafer's surface level. This design effectively keeps the particles inside the rail during an experiment.

Regarding model particles, guided by the work of van Zwol et al. [49] we decided in favour of borosilicate glass beads (Duke Standards 9020).

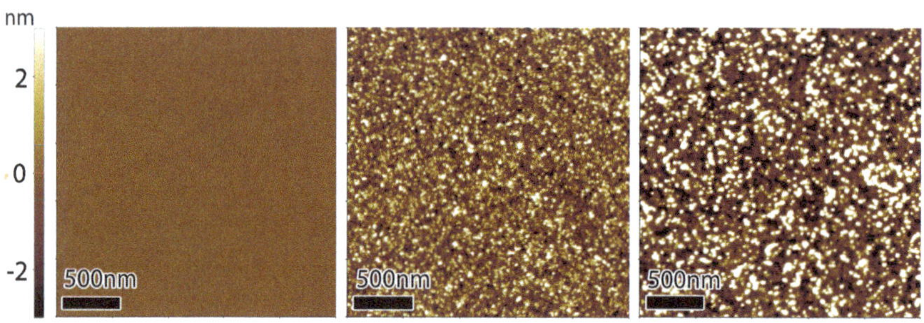

Fig. 5 AFM topography images of the Si surfaces clearly show the roughness change after MWCVD H_2 etching with 1600 W (middle) and 1800 W (right) power. The RMS variations of the surface heights are 0.3 ± 0.1 nm, 1.5 ± 0.2 nm and 2.7 ± 0.4 nm (upper panels, left to right)

Colloid and Other Probes

AFM colloidal probes were prepared by attaching the borosilicate glass spheres to Mikromasch NSC12 tip-less AlBS cantilevers with epoxy (Araldite 10 min, 2 components). The radii of the colloid probes and the quality (clean contact area) were determined by using scanning electron microscopy. In order to prepare colloid probes for the scanning nanoindenter (TriboIndenter, Hysitron Inc.) used in this work, the FIB was employed to create appropriate cavities into the apex of a commercially available diamond cube corner tip (Hysitron Inc.), see Fig. 7. Subsequently, the borosilicate spheres were fixed to this holder by means of a small amount of photosensitive acrylate-based adhesive glue (DIC Europe GmbH). A flat end diamond indenter (20 μm diameter, Hysitron Inc.) and non-glued spheres were utilised to sample rolling and torsional friction, as detailed in [10].

Normal Tests

Figure 6 shows a load-displacement curve obtained by using the nanoindenter colloid probe on a silica substrate. Elastic modulus and Poisson's ratio are assumed to be 71 GPa and 0.17 for both sphere and surface (Table 1), respectively. A simple Hertzian fit with a slightly larger radius of 9 μm was able to accurately reproduce the experimental data. The slightly stiffer response has to be attributed to a combination of surface roughness (about 0.7 ± 0.1 nm) and non-sphericity (i.e. local variation in actual radius).

According to Tabor [43], the yield stress can be approximated as one third of the hardness. For the silica surface used in this experiment, a measured hardness of 9 GPa results in a yield stress of 3 GPa, slightly above the maximum contact pressure of 2.8 GPa acting at the contact area with a flat substrate, based on the highest normal loads used in our sliding and rolling experiments, $f^{\mathrm{n}} \leq 3000\,\mu\mathrm{N}$. Therefore, with silica being the weakest material in this study, we have a predominantly elastic Hertzian contact in all experimental settings presented here.

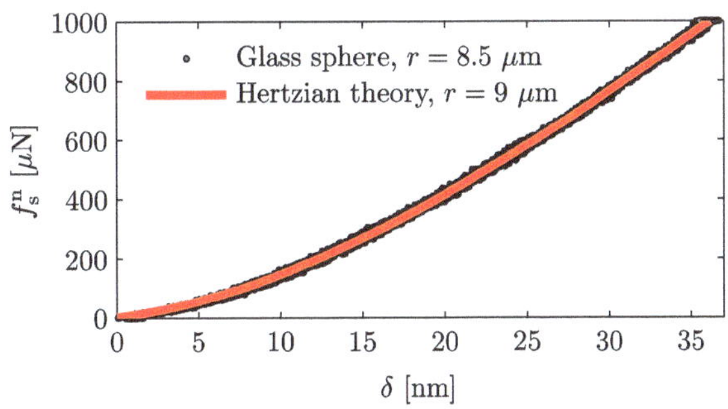

Fig. 6 Normal load f_s^{n} versus displacement curve δ. The red line shows a fit with Hertzian theory by assuming a radius of 9 μm, with materials from Table 1

Table 1 Material properties for substrates, particle and probes used for normal, sliding, rolling and torsion tests

	Substrate 1	Substrate 2	Indenter	Particle
Material	Si(100)	Fused quartz	Diamond	Borosilicate
Young's modulus	$E_s = 179\,\text{GPa}$	$E_s = 71\,\text{GPa}$	$E_i = 1140\,\text{GPa}$	$E = 71\,\text{GPa}$
Poisson's ratio	$\nu_s = 0.25$	$\nu_s = 0.17$	$\nu_i = 0.07$	$\nu = 0.17$
RMS roughness Δ (nm)	See Table 2		4.0 ± 1.0	0.7 ± 0.1
	(For sliding/rolling/rail)	(For normal testing)	(For rolling/rail)	$R = 8.5\,\mu\text{m}$

The adhesion measurements are summarised in Table 2. The measured pull-off forces, f_{pull}, decrease with increasing RMS roughness. As surface chemistry (amorphous SiO_2) and measurements conditions are kept constant, this finding is directly related to a diminished real contact area due to the roughness. The results are in good agreement with recent experimental results of Liu et al. [20], who studied the correlation between adhesion, tip radius and surface roughness in more detail.

Sliding Tests

In order to characterise the sliding friction of the particles in the predominant elastic contact regime, nanoindenter colloid probe tips have been used according to the schematic given in Fig. 7.

The sliding friction tests were performed in a nanoindentation-based setup using a TriboIndenter in combination with a Performech upgrade (Hysitron Inc.). The actual tests to characterise sliding friction were carried out at room temperature and $30 \pm 5\%$ relative humidity in a friction loop fashion, which is a common technique in the AFM community. In this case the probes have been scratched back and forth over a distance of 2 μm at a constant normal load and a speed of 1 μm/s under load control. The normal load was varied between 3 and 100 μN; five spheres were utilised probing each load five times at different surface spots for all Si surfaces, respectively. In turn, the corresponding absolute value of lateral force for each test was evaluated by a lateral displacement sensitive averaging of the difference in measured lateral force for forward and backward movement divided by two. In order to avoid artefacts originating from a change in the movement direction, only the central micron range of the friction loop was taken into account.

The results of the sliding friction tests are summarised in Fig. 8 left. A linear dependence between the measured lateral force f_s^1 and the applied normal load f_s^n can be seen. The linearity of these three curves indicates that the modified Coulomb friction law (11a) holds for the given experimental conditions. A linear fit of the data yields the sliding friction coefficient μ^1 along with an offset, which can be related to the adhesion f_a^n between the contact partners. The results from the linear regression are summarised in Table 2.

Table 2 Contact properties for sliding and rolling tests on different rough Si surfaces and effective friction values for rolling/torsion tests

Substrate 1 surfaces	Untreated	Etched (1600 W)	Etched (1800 W)	Section
RMS roughness Δ (nm)	0.3 ± 0.1	1.5 ± 0.2	2.7 ± 0.4	section "Test Objects"
Peak-to-valley (nm)	1.0 ± 0.1	10.5 ± 0.7	14.8 ± 1.0	
AFM pull-off force f_{pull} (μN)	3.2 ± 0.3	2.7 ± 0.3	1.9 ± 0.3	section "Sliding Tests"
Sliding friction $\mu_{\text{s}}^{\text{l}}$	0.23 ± 0.005	0.53 ± 0.005	0.65 ± 0.014	
Sliding adhesion f_{a}^{l} (μN)	12.00 ± 1.20	4.45 ± 0.55	7.76 ± 1.17	
Rolling fr. $\bar{\mu}^{\text{ro}}$ for $\bar{\lambda}^{\text{ro}} = \bar{r}$	$(7.9 \pm 0.3) \times 10^{-4}$	$(4.6 \pm 0.2) \times 10^{-4}$	$(3.5 \pm 0.3) \times 10^{-4}$	section "Rolling Tests"
Ro. adh. $\bar{f}_{\text{a}}^{\text{ro}}$ for $\bar{\lambda}^{\text{ro}} = \bar{r}$ (μN)	571.4 ± 56.1	1398.0 ± 82.7	2029.4 ± 149.2	
Rolling fr. $\bar{\mu}^{\text{ro}}$ for $\bar{\lambda}^{\text{ro}} = \bar{a}$	$(7.6 \pm 1.0) \times 10^{-3}$	$(1.8 \pm 0.7) \times 10^{-3}$	$(1.7 \pm 1.3) \times 10^{-3}$	
Ro. adh. $\bar{f}_{\text{a}}^{\text{ro}}$ for $\bar{\lambda}^{\text{ro}} = \bar{a}$ (mN)	2.6 ± 0.6	13.5 ± 5.6	12.6 ± 11.0	
Rail inclination (°)	25 ± 0.5	45 ± 0.5	65 ± 0.5	
RMS roughness Δ (nm)	2.5 ± 1.0	3.6 ± 1.4	4.8 ± 2.0	section "Test Objects"
Rail friction coeff. μ^{rail}	$(1.58 \pm 0.02) \times 10^{-3}$	$(2.42 \pm 0.09) \times 10^{-3}$	$(3.34 \pm 0.19) \times 10^{-3}$	section "Combined Rolling and Torsion Tests"
Rail adhesion $f_{\text{a}}^{\text{rail}}$ (μN)	173.4	57.8	191.3	
Torsion fr. $\mu_1^{\rightarrow}$ for $\lambda^{\rightarrow} = r$		$(5.6 \pm 0.5) \times 10^{-3}$		
Torsion fr. $\mu_1^{\rightarrow}$ for $\lambda^{\rightarrow} = a$		$(6.4 \pm 1.2) \times 10^{-2}$		

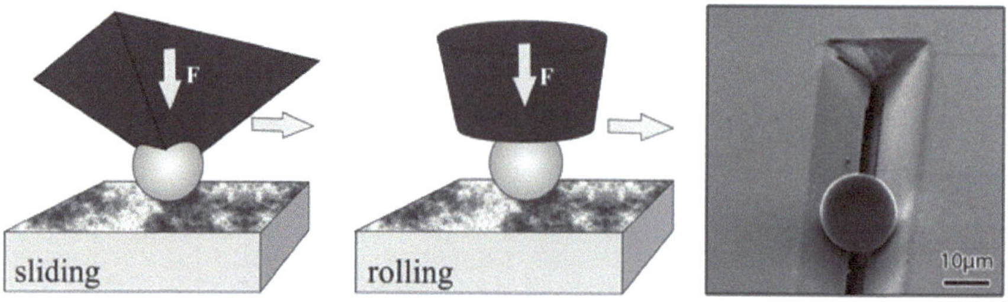

Fig. 7 Schematic of the test setups for sliding (left), rolling (middle) and rolling-torsion (right) measurements of individual fixed micron-sized borosilicate particles on a Si substrate

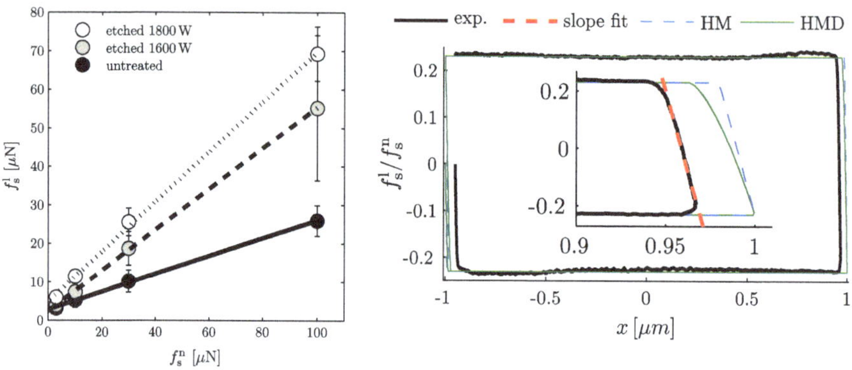

Fig. 8 (*left*) Measured lateral force f_s^l versus normal load f_s^n for sliding friction measurements. Lines represent simple linear fits, from which f_a^l, μ^l are extracted. (*right*) Shear force against displacement for the sliding experiment with the untreated substrate and $f_s^n = 100$ mN (thick black line). This is compared to DEM simulations using Hertz-Mindlin (blue dashed line) and Hertz-Mindlin–Deresiewicz (green line), with a fitted shear modulus of $G^{*,\mathrm{fit}} = 1.21$ GPa (thick red dashed line)

The larger values of sliding adhesion forces compared to the normal adhesion measurements potentially stem from the differences in the dynamics of the process that detaches micro-asperities from a surface and the process that initiates new contacts during shear of the sphere along the substrate.

Comparison to Model

For the sliding case, shown in Fig. 7 (left), the particle is attached to the indenter and thus moved with the same velocity as the indenter. The resulting shear force versus displacement curve is compared to an experimental friction loop in Fig. 8. The friction loop shows nearly linear elastic behaviour for short intervals after the reversal point (where the velocity switches directions). For the elastic behaviour, the force balance yields that the slope of the shear force versus displacement curve at the load reversal point equals the negative sliding spring stiffness, $d\, f_{\mathrm{el},i}^{\mathrm{l}}/dx = -k^{\mathrm{l}}$. For both the linear Mindlin and the nonlinear Mindlin–Deresiewicz [28] models, a stiffness of $k^{\mathrm{l}} = 8G^*a$ holds true at the reversal point.

The slope fitted to the experimental friction loop plotted in Fig. 8 yields $k^{\mathrm{l,fit}} = 2.219\,\mathrm{kN/m}$ for a normal load of $f_{\mathrm{s}}^{\mathrm{n}} = 100\,\mu\mathrm{N}$; this yields $G^{*,\mathrm{fit}} = k^{\mathrm{l,fit}}/(8a) = 1.36\,\mathrm{GPa}$, which is much smaller than the value calculated from the material parameters in Table 1 of a perfectly flat borosilicate sphere and a silicon substrate ($G^* = 11.87\,\mathrm{GPa}$). This result can be due to the nano-structure of the surfaces; however, it is difficult to interpret as the elasticity of the nanoindenter setup could possibly influence the slope observed in the friction loop.

After the sliding spring yields, the magnitude of the sliding force remains constant at $\mu_{\mathrm{s}}^{\mathrm{l}}|f_{\mathrm{s}}^{\mathrm{n}}|$. Fitting the experimental results for the untreated surfaces shows a constant friction coefficient $\mu_{\mathrm{s}}^{\mathrm{l}} \approx 0.23$. A linear fit of the lateral force against the contact area, $f^{\mathrm{l}} \propto \pi a^2 \propto (f^{\mathrm{n}} + f_{\mathrm{a}}^{\mathrm{n}})^{2/3}$, as suggested in [48], did not produce good, consistent results. For higher roughness, and a peak at the transition from the elastic to the yielding sliding force (not shown) suggests that the static sliding friction coefficient is higher than the dynamic one, however, this will not be discussed further here.

Rolling Tests

The nanoindentation test setup for the measurement of rolling friction of the microspheres is significantly easier compared to the sliding case. The spheres are first placed on the surface, then contacted with a flat end diamond indenter, carefully positioning the indenter tip onto the apex of the particle.

Preliminary test results with single scratch length of about $10\,\mu\mathrm{m}$ and a speed of $1\,\mu\mathrm{m/s}$ under load control, showed that a certain threshold [37] or rolling resistance moment [6, 32] was necessary to initiate particle rolling. In our case, the rolling of particles, i.e. a change in position detected by optical microscopy, was only observed for normal loads larger than $100\,\mu\mathrm{N}$.

Based on these results, a minimum normal load of $100\,\mu\mathrm{N}$ was selected for all rolling tests presented here to ensure rolling characteristics of the corresponding contact behavior. This $100\,\mu\mathrm{N}$ force to overcome the rolling resistant moment, was chosen independently of the load regime used in sliding tests. Again, in order to improve the reliability of the results, five beads were used for testing and each measurement, i.e. applied normal load, was taken five times for all Si surfaces, respectively.

Figure 9 shows the results of the rolling experiments. The mean rolling friction coefficient $\bar{\mu}^{\mathrm{ro}}$ was extracted from the friction loop data, as in the sliding experiments. The measured lateral force signals are 2–3 orders of magnitude smaller for rolling compared to sliding tests, which confirms that the particle is rolling over the surface. Surprisingly, in contrast to the sliding data, $\bar{\mu}^{\mathrm{ro}}$ decreases with increasing surface roughness, while the intercept with the abscissa shifts to the left, i.e., rolling adhesion $\bar{f}_{\mathrm{a}}^{\mathrm{ro}} \gg f_{\mathrm{a}}^{\mathrm{l}}$ increases with surface roughness. This suggests that a simple interpretation of this intercept as a measure of the normal adhesion force between the contacting partners is not straightforward in the case of a rolling contact.

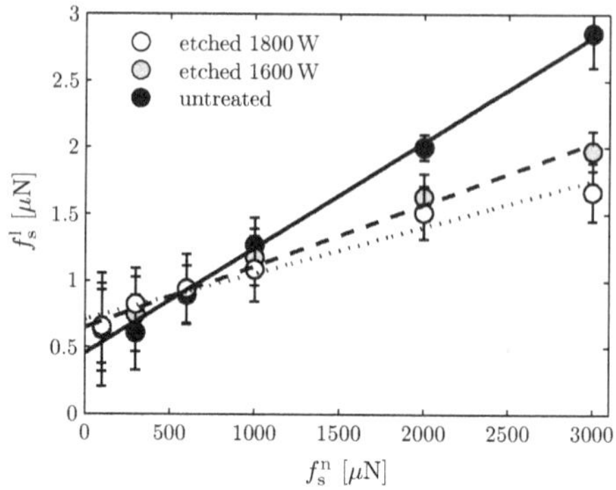

Fig. 9 Resulting lateral force f_s^l versus normal load f_s^n for rolling friction measurements. The lines represent simple linear fits to the corresponding data sets

Ideally, a non-adhering rolling sphere contacts with the underlying surface at only one point, which would result in a zero rolling friction since no torque can be created by a point contact. In our system, the finite contact surface, involving many single asperities on the sphere as well as on the substrate surface, leads to a non-zero rolling resistance due to various (three at least) dissipative phenomena. Rolling is a continuous generation of new contact area at the front and a continuous contact detachment at the rear of the contact (viewed in rolling-direction of the sphere). During this process, multiple surface mechanisms can dissipate energy [11]. Therefore, rolling friction appears to be attributable to various combinations of viscous, wet, and plastic effects at the individual asperities.

Analysis of the Rolling Measurements

An analysis of the forces relevant in the rolling experiment reveals that

$$f^{\mathrm{l}} = \frac{\bar{\mu}^{\mathrm{ro}}\bar{\lambda}^{\mathrm{ro}}}{\bar{r}}(f^{\mathrm{n}} + \bar{f_{\mathrm{a}}}^{\mathrm{ro}}), \tag{14}$$

where the bar denotes mean values for the two different surfaces.

Thus, the good approximation by a linear fit seems to suggest that the length scale λ^{ro} is proportional to the particle radius. The measured coefficients can be found in Table 2 for $\bar{\lambda}^{\mathrm{ro}}=\bar{r}$ and $\bar{\lambda}^{\mathrm{ro}}=\bar{a}$, where the contact radii been obtained using (2.2.1) and the data in Table 1.

In both cases, the rolling adhesion force, $f_{\mathrm{a}}^{\mathrm{ro}}$, is much higher than the pull-off force $f_{\mathrm{a}}^{\mathrm{n}}$, as the rolling friction appears to be large even for zero normal load. As discussed above, $\mu_{\mathrm{s}}^{\mathrm{l}}$ increases with higher surface roughness while $\bar{\mu}^{\mathrm{ro}}$ decreases. These results show that $\bar{\mu}^{\mathrm{ro}}$ is dominated by surface and interfacial forces such as *e.g.* water films while $\mu_{\mathrm{s}}^{\mathrm{l}}$ is dominated by mechanical frictional stick slip effects at asperities.

Combined Rolling and Torsion Tests

In combination with appropriate rail systems, the strategy to probe rolling friction, which has been introduced above, can also be exploited to study a combination of rolling and torsional friction. As far as the authors are aware, no scanning probe based technique has been proposed so far that would allow for the exclusive evaluation of torsion friction.

In order to access the combination of both mechanisms, we utilised Si-based rail structures featuring rail angles of 25°, 45° and 60°. Higher rail angles lead to an increased influence of the torsional contact mode.

Borosilicate spheres are placed inside the rails by AFM-based positioning (Fig. 7). In analogy to the rolling tests, the individual particles inside the rails are contacted and probed with a flat end diamond indenter. Careful positioning of the indenter tip with respect to the particle as well as alignment of the rail with regard to the scratch axis (movement direction) of the instrument are crucial.

The effect of scratch velocity can be considered negligible, since the results of the rolling test did not change while varying velocity from 1, 0.1 to 0.01 μm/s inside the rail system.

To estimate the rolling friction in the following rail experiments, both friction factor and rolling adhesion are fitted against the RMS roughness of the substrate, Δ, as an exponentially decaying value, $\bar{\mu}^{\mathrm{ro}} = b + c\exp(-\Delta/d)$, and $\bar{f}_{\mathrm{a}}^{\mathrm{ro}} = b' - c'\exp(-\Delta/d')$. with $b = 0.0002859$, $c = 0.0006475$, $d = 1.1548\,\mathrm{nm}$ and $b' = 3.046\,\mathrm{mN}$, $c' = 2.907\,\mathrm{mN}$, $d' = 1.5216\,\mathrm{nm}$, as shown in Fig. 10. The rolling friction factors in the rail experiments in section "Combined Rolling and Torsion Tests" are then assumed to satisfy the friction factors fitted to the rail roughness, see Table 2.

Experimentally, a controlled particle surface motion is possible by means of a Si based rail system featuring different inclinations. In the rail system, a controlled particle surface motion is possible, which results in an effective friction coefficient with contributions from both torsion and rolling resistance.

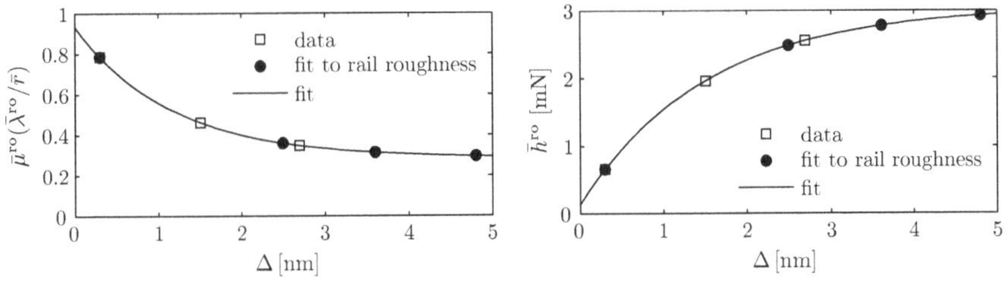

Fig. 10 Fit of the rolling friction coefficient $\bar{\mu}^{\mathrm{ro}}$ (top) and the rolling adhesion force $\bar{f}_{\mathrm{a}}^{\mathrm{ro}}$ (bottom) as functions of RMS roughness. This fit is used in section "Combined Rolling and Torsion Tests" to estimate the rolling friction in the rail experiments, based on the RMS roughnesses of the rail surface (dots)

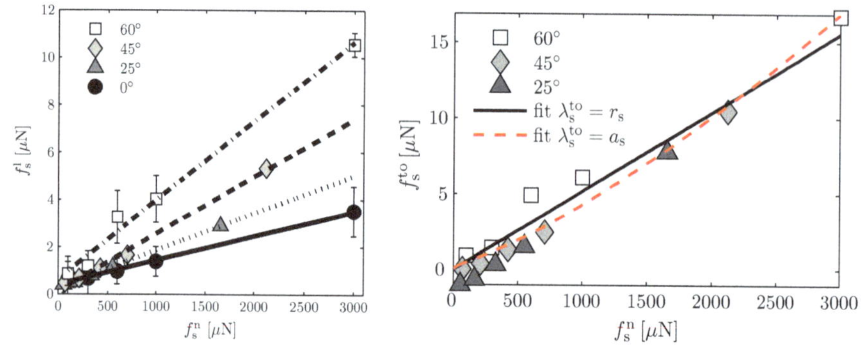

Fig. 11 (*left*) Resulting lateral force f^l versus effective normal load f_s^n for the rail system with different rail inclinations of 0°, 25°, 45° and 60°. All data sets are fitted with a linear relation. (*right*) Torsion contribution to the lateral force f_s^{to} versus effective normal load f_s^n. All data sets are fitted to Eq. (15); lines show fits of $\lambda_s^{to} = r_s$ (solid line) and $\lambda_s^{to} = a_s$ (dashed red line)

Fig. 11 shows the effective lateral forces, f^l, and normal loads, f^n, at the substrate contacts, which are calculated from the measured lateral force, f_i^l, and the applied normal load, f_i^n.

All results follow linear trends, confirming a modified Coulomb law, but the slopes, which are the effective friction coefficients μ^{rail}, show a clear correlation with increasing inclination of the rail, which will be further analysed in section "Analysis of Combined Rolling and Torsion".

The intercepts of the trend lines with the horizontal axis, taken as the values of rail adhesion, h^{rail}, are lower than for pure rolling. This suggests that adhesion plays a less dominant role during the combines rolling/torsion motion than for pure rolling on a plane surface.

Even though the experimental results presented up to now provide evidence of an increased effect of torsion with increasing inclination, a deeper insight into appropriate contact models is necessary to understand the evolution of the effective friction coefficient, μ^{rail}, with respect to inclination. In the following section, focusing on contact models and simulation, among other considerations, we strive to quantify the impact of rolling and torsional friction seen here.

Analysis of Combined Rolling and Torsion

An analysis of the forces relevant in the rail experiments [11] allows us to predict the rolling contribution to the lateral force measurement, and thus to extract the torsion component. This is plotted in Fig. 11.

The torsion friction shows an almost constant slope, which can be fitted by

$$\mu_s^{to} = 0.0056 \text{ for } \lambda^{to} = r, \tag{15a}$$

and

$$\mu_s^{to} = 0.064 \text{ for } \lambda^{to} = a. \tag{15b}$$

Both fits, plotted in Fig. 11, agree well with the data. Thus, the measurements are not able to distinguish if the torsion friction factor is independent of the normal force or scales with the (normal-force dependent) contact radius. The measured coefficient is about seven times smaller than the predicted value of $\mu_s^{to} = (2/3)\mu^1$ for $\lambda_\alpha^{to} = a_\alpha$, unless a smaller effective radius is used. A more detailed study with different particle radii is necessary to find correlations with either particle or effective contact radii.

Initial Stage Sintering of Polymer Particles

Here, we summarize experimental results on a thin bed of particles, heat-sintered but not fully melted [13]; its elastic properties are measured using a nanoindenter, which allows the calibration of the plastic deformation model for the forming and evolution of the sinter necks during time.

Preparation of Particles and Samples

Polystyrene spheres featuring nominal particle radii of 0.25, 0.5, 0.75 and 2 μm, molecular weights (M_w) of 111–312 kg/mol, and a glass transition temperature T_g of 94–99 °C were synthesised by dispersion polymerisation as reported in Zhang et al. [55] and stored in aqueous solution. For each particle radius, multi-layer films (<30 particle diameters) were realised by placing 10 ml of a particle suspension (1:1 volume ratio, PS particle:ethanol) on an oxygen-plasma hydrophilised glass substrate and dried for at least 24 h in air. Sintering was carried out in air on a heating stage at temperatures of 90, 95, 100, 105 and 110 °C. The rate of temperature change was set to be 15 °C/min during heating. Holding periods of 20, 40, 60, 90, 120 and 180 s were set at each temperature for each particle radius, respectively. After sintering, the samples were quickly cooled down to room temperature. The resulting average layer thickness reduced by up to 50% as the sintering temperature increases from room temperature to 110 °C.

Characterization of Microstructure by Slicing

The sintered PS particle layers were coated with a 100 nm Pt layer. A focused-ion beam combined with a scanning electron microscope (FIB/ SEM, Nova 600 Nanolab, FEI, USA), was then used for sectioning and imaging sequential 2D cross-sectional surface images, see Fig.12. A protective Pt layer of 1 μm thickness and $12 \times 8\,\mu m^2$ area was deposited on the sectioned sample. A volume of $12 \times 8 \times 6\,\mu m^3$ was milled slice-by-slice with 20 nm distance between two consecutive images. The actual 3D structures of the sintered films were reconstructed based on the as-recorded stacks

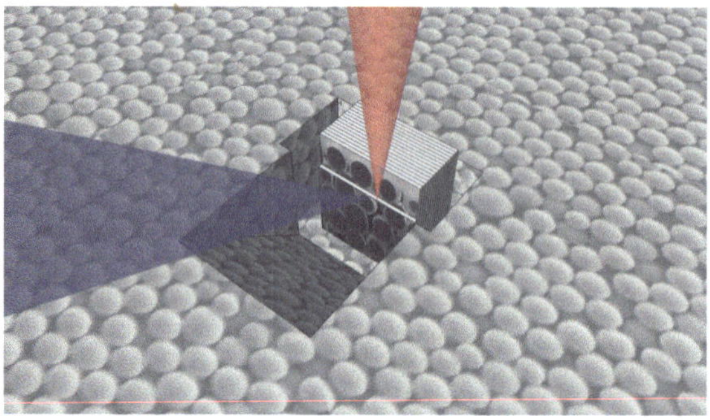

Fig. 12 Sinter bed sectioned via FIB/SEM for 3D reconstruction

of images using Amira 4.1 (Visage Imaging, San Diego, USA). The reconstruction process involved (i) alignment of the images, (ii) re-sampling and (iii) segmentation. The radius of at least 12 particles as well as the mean neck radius x was determined for each sintered sample.

Nanoindentation and Confocal Microscopy

Nanoindentation measurements were performed with an MFP NanoIndenter (Asylum Research, Santa Barbara, CA, spring constant k = 2390 N/m) equipped with a spherical ruby indenter of diameter d = 127 μm. Indentations were performed in load-controlled mode. The applied load varied between 1 and 4 mN with loading rates between 200 and 800 μN/s. Each sample was tested at least at 12 individual positions. The reduced elastic modulus (E_{red}) for each sample was obtained from the unloading portion of the load-displacement curve using the Oliver and Pharr method with a spherical area function [7].

The nanoindenter was then placed on the sample stage of a custom-built laser scanning confocal microscope (LSCM), which has the capability to measure in situ the real-time deformation within the sintered particle layers during nanoindentation. The ruby sphere was indented into the sintered fluorescently labelled PS particle film in displacement-controlled mode with a maximum displacement of 5 μm, while the structure was imaged by LSCM. For the LSCM imaging, the sintered films needed to be submerged in a liquid that matches the refractive index of polystyrene (Cargille Laboratories, USA).

Mechanical Properties

After sintering of the PS particles samples under varying temperature-, size- and time-conditions, different densifications of the powder layer are obtained. It was found that the reduced elastic modulus (E_{red}) of each sintered film increased by a

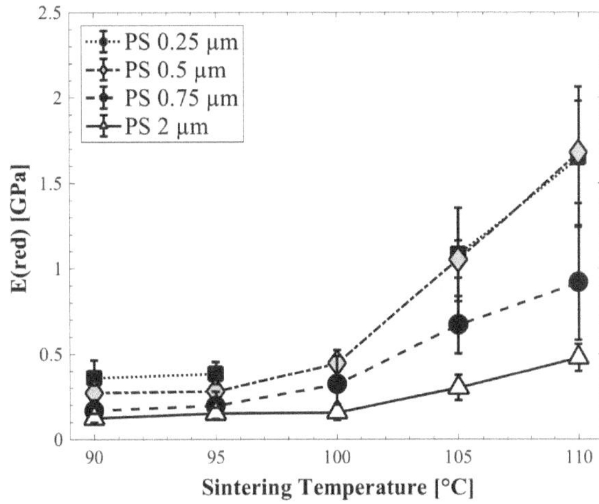

Fig. 13 Mechanical properties of PS multilayers with particle radii of 0.25–2 μm sintered for 60 s. E_{red} increases with the densification of the porous solids ($T > T_g$) which increases with decreasing particle size

factor of max. 6 for sintering temperatures above T_g as the sintering time increased from 20 to 180 s [12], while E_{red} also correlates strongly with the porosity of the film.

According to Mazur et al. [26], particles smaller than a certain limiting radius, r_{crit}, are predicted to sinter to uniform density regardless of the Newtonian viscosity. Consequently, the contact area initially grows much faster for particles with radius $r < r_{crit}$ than predicted by the classical sintering models [9, 33], which neglect the contribution of surface forces as well as the resultant plastic and elastic contact deformation in the early stages of sintering. Assuming an initial packing fraction of 58%, a critical particle radius of $r_{crit} = 1.055$ μm for sintering of PS particles is obtained.

Figure 13 shows E_{red} plotted against sintering temperature for layers made from PS particles with varying radii from 0.25 to 2 μm. Indeed, particles with $r \leq 0.5$ μm show a higher E_{red} compared to the larger particles, even for temperatures below T_g. This hints at an additional contribution of surface force and contact deformation, which leads to faster sintering in the initial stage. The variation of E_{red} (decrease) with particle size is a strong indication that there is a transition from sintering dominated by contact deformation for $r < r_{crit}$ to sintering dominated by viscous flow for $r > r_{crit}$. Consequently, larger particles require higher sintering temperatures and times to show the same mechanical properties as smaller particles.

In addition, in-situ real-time deformation within the sintered particle layers during nanoindentation was studied with the help of confocal microscopy. Cross-sectional xz-plane LSCM images for (a) not sintered films and films sintered at (b) 90 °C, (c) 100 °C and (d) 110 °C, all for 60 s, at maximum indentation depth, are shown in Fig. 14.

While the displacement of 5 μm is reached without exceeding the maximum force of the nanoindenter in case of Fig. 14a, b, the samples above T_g (Fig. 14c, d) show a smaller indentation depth at maximum load, due to a higher E_{red} of the porous sintered

(a)
(b)
(c)
(d)

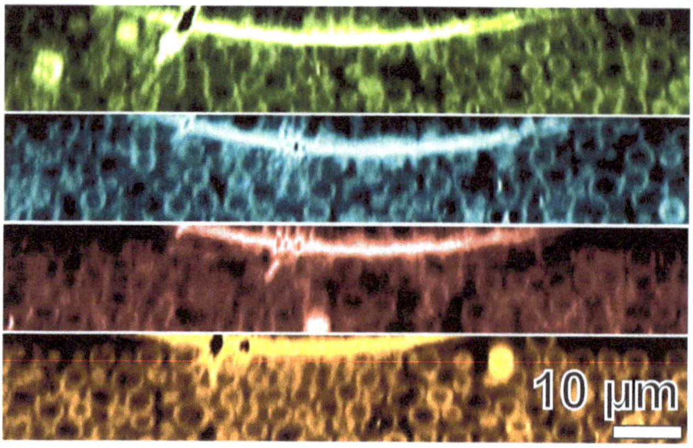

Fig. 14 LSCM image (xz plane) during indentation with a spherical tip for **a** not sintered PS multilayer film, **b** sintered at 90 °C for 60 s, **c** sintered at 100 °C for 60 s, and **d** sintered at 110 °C for 60 s. The maximum indentation decreases from (**a**) to (**d**) with increasing stiffness of the sample

layer. A plastically deformed region is observed after indentation for samples in Fig. 14a–c. The sample sintered at a temperature of 110 °C has a nearly fully elastic, reversible response. Such behavior can be attributed to a strong interconnection of the particles by sufficiently strong sintering necks.

Simulation Setup

Finally, we look at the analysis of the FIB/SEM data. A representative plot of the time-dependent increase of the sintering neck radius x at 110 °C for PS particles with radius of 1.5 µm is shown in Fig. 16, showing the increasing mean sinter neck radius as a function of time. We use this data to calibrate our sintering contact model, presented in section "Sintering Forces".

We model the sintering and indentation process with discrete particle simulations, using a small cubic domain of 60 µm, with a flat base wall and periodic boundary conditions in both width and length. We introduce 2178 particles of radii $1.35\,\mu\text{m} < r < 1.65\,\mu\text{m}$, producing a particle packing about 5 particle diameters thick (see Fig. 15). To simulate sintering in a reasonable amount of time steps, the collision time is scaled up to $t_c = 5\,\text{ms}$, which is still several orders of magnitude smaller than the time scales of gravity, sintering, and indentation. Therefore, using such artificially soft particles has little effect on the results. The packing is sintered using a sintering time scale of $\tau = 666\,\text{s}$ and $n = 1$ (thus assuming viscous sintering) to match the experimental results. Adhesive van-der-Waals forces were modelled as $f_a = (1.2\,\mu\text{N/m})\bar{r}$, see [9]. The restitution coefficient was chosen as 0.5 to determine the stiffness and dissipation parameters with the goal to quickly dissipate velocity fluctuations; for the plastic regime, the maximum unloading stiffness was set to $10k_1$.

Fig. 15 Vertical cut through the centre of the simulated layer during indentation

Comparing the simulated forces on the nanoindenter with the experimental results displays a rich phenomenology, qualitatively simular in experiments and simulations. However, a detailed discussion is beyond the scope of this paper, and the quantitative calibration of the model still requires further research. Nevertheless, next we extract the typical sintering neck radius from experiments and compare it with different models.

Sinter Kinetics

We plot the simulation results for various models together with the experimental results in Fig. 16, as obtained with 3D reconstruction. The experiments show a higher growth rate in the early stages of sintering compared to the prediction of the classical sinter models of Frenkel [9] and the modified Frenkel model [33]. The simulation of viscous sintering ($n = 1$, $\tau = 666$ s) underpredict the experimental data while agreeing well with the Frenkel and modified Frenkel model, up to the point where the majority of contacts reach the maximum plastic overlap ($t > 120$ s). Simulations

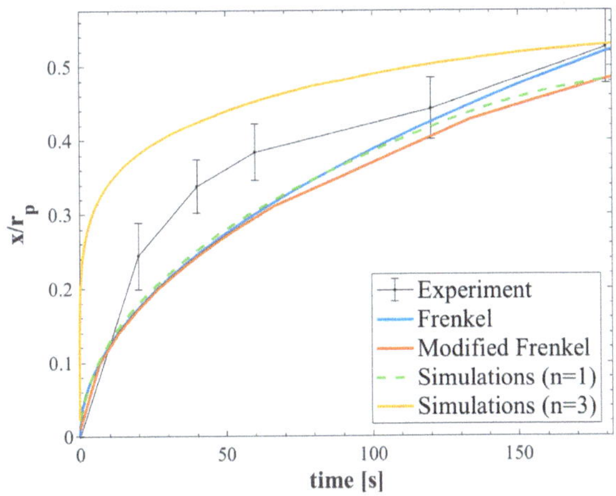

Fig. 16 Sintering kinetic of PS particle multilayer ($\mathbf{r}_p = 1.5\,\mu m$) at 110 °C compared with Frenkel model (blue line), modified Frenkel model (red line), and simulations of viscous sintering ($n = 1$, green line) and surface sintering ($n = 3$, orange line)

using the surface sintering model ($n = 3$, $\tau = 110\,\text{h}$) systematically over-predict the early-stage sintering rate up to about $t > 180\,\text{s}$. This indicates that viscous sintering is aided by an additional contribution of surface forces to the sintering process, but is not as strong as predicted by the extreme model with $n = 3$ which assumes ideal surface sintering.

Summary and Conclusions

This chapter begins with a short literature overview before the basics of contact models that involve pressure-, rate/time- and temperature-dependence are given. The model has some parameters that can be either directly or indirectly measured or calibrated experimentally. The former was carried out first, but only for well-defined spherical particles. Secondly, the temperature-driven growth of sinter necks was measured and compared to various models to indirectly determine the model and its parameters.

The direct measurements of the **pressure-dependence of contacts**—for all degrees of freedom in the relative motion of very small particles, such as normal adhesion and elasticity, sliding, rolling and torsion friction—was made possible by a relatively simple, but novel, nanoindenter setup. Compared with other techniques used in literature, a nanoindenter features the option to measure not only normal loads but also lateral forces and torques, both for sliding and rolling. Even though the interpretation/evaluation of the results still leaves various open questions, a big step was made towards understanding particle contacts and interactions and to directly obtain contact model parameters.

Normal force: Experimentally, the repulsive normal force between micron-sized borosilicate spheres and Si surfaces is well described by a Hertzian law, but the adhesion force decays with increasing surface roughness (on the scale of 0.3–3 nm). While the effect of the considerable surface roughness on the normal force is surprisingly small, it should lead to notable plastic deformations of the contacting micro-asperities.

Sliding, Rolling and Torsion: The influence of surface roughness and adhesion on pure sliding and pure rolling measurements was studied for the same particles on Si surfaces. In addition, novel rail experiments feature a combination of torsion and rolling motion. Due to the set-up, sliding and torsion occur only at the contact with the substrate; in contrast, rolling torques occur at both substrate and indenter contacts, i.e., the resulting rolling friction coefficient is an averaged quantity from the two contacts, which can involve different materials or surfaces. All motion types (pure sliding, pure rolling, mixed torsion and rolling) can be represented as relations between the measured lateral force and the applied normal load via an effective friction coefficient μ, and an effective adhesion, $f^{\nu} = \mu(f^{\mathrm{n}} + f_{\mathrm{a}}^{\nu})$, where ν stands for the motion type. The smallest normal forces typically display a mix of different motions, but the results are reliable for normal loads between 100 and $3000\,\mu\text{N}$.

All data can be fitted by either a linear or a non-linear model. The models assume independence between the adhesion and friction coefficients and the normal force. The former, linear model leads to constant friction factors for rolling and torsion (when scaled with the particle radius) that are much smaller than for sliding, but also to highly different predictions for the effective adhesion for the different motions, being highest for pure rolling. The latter, non-linear model takes into account the dependence of the rolling and torsion friction factors on the (predicted) contact radius by introducing it as a length scale into the yield criteria for the rolling and torsion torques. Unfortunately, the present set of data does not allow to decide which of the two models is the true one, so that both interpretations were presented in Ref. [11].

Sinter kinetics involve **temperature-, time- and size-dependent contact mechanisms** that result in an interesting sintering kinetics and mechanical properties during initial stage sintering. To indirectly calibrate the contact models, thin PS particle layers were investigated, utilising 3D tomography (FIB/SEM), nanoindentation and confocal microscopy. The experimental results indicate that the sinter kinetics and mechanical properties of particles with $r < 1\ \mu m$ are more caused by the contact deformation due to surface forces, whereas sintering of larger particles is characterised by viscous flow as the dominant mechanism. Consequently, larger particles require higher sintering temperatures and times to reach the same extent of sintering.

For a temperature- and pressure-dependent sintering model that includes the contribution of surface forces, we start from the elasto-plastic model of Luding [22, 23], and generalize it by introducing a rate of change for the permanent, plastic deformation at high temperatures. The contact model can simulate both contact sintering as well as compression, i.e., elastic repulsion, allowing the simulation of the free-flowing (cold) particles, their sintering, the cool-down and eventually the indentation tests in one single simulation framework (with one contact model, but parameters varying with pressure, temperature and time). Preliminary simulation results for either pure viscous or pure surface sintering (as based on theoretical predictions) respectively under- and over-predict the experimental results. A model taking into account both mechanisms (surface forces and viscous sintering), as well as the increase in particle radius due to sintering, may be able to fully explain the experimental observations.

Acknowledgements This work was carried out within the framework of the Key Research Program (SPP 1486 PiKo "Particles in Contact") grants LU 450/10, STA 1021/1 and KA 1724/1. The authors would like to thank the German Research Foundation (DFG) for their financial support during the SPP. The numerical simulations in this paper were carried out using the open-source code MercuryDPM [46, 53, 54], available via http://mercurydpm.org. This SPP and our research would not have been possible without Professor Tomas, who initiated the SPP, but sadly died and thus could not accompany us to the successful ending of the projects; we miss him and his inspirations about particles in contact very much. Special thanks also to Professor Antonyuk who led the SPP successfully to the end, as reflected by this book.

References

1. Besler, R., Rossetti da Silva, M., Rosario, J.J., Dosta, M., Heinrich, S., Janssen, R.: Sintering simulation of periodic macro porous alumina. J. Am. Ceram. Soc. **98**(11), 3496–3502 (2015)
2. Bierwisch, C., Polfer, P.: Scaling laws for implicit viscosities in smoothed particle hydrodynamics. In: EPJ Web of Conferences, vol. 140, p. 15008 (2017)
3. Breinlinger, T., Schubert, R., Hashibon, A., Kraft, T.: Modelling die filling for powders with complex rheology: a new DEM contact-model. In: Proceedings of 19th International Plansee Seminar. Plansee Group Service GmbH, Reutte, Austria (2017)
4. Brendel, L., Török, J., Kirsch, R., Bröckel, U.: A contact model for the yielding of caked granular materials. Granul. Matter **13**(6), 777–786 (2011)
5. de Gennes, P.G.: From rice to snow. In: Nishina Memorial Lectures. Volume 746 of Lecture Notes in Physics, pp. 297–318. Springer, Berlin/Heidelberg (2008)
6. Ding, W., Howard, A.J., Murthy Peri, M.D., Cetinkaya, C.: Rolling resistance moment of microspheres on surfaces: contact measurements. Philos. Mag., **87**(36), 5685–5696 (2007)
7. Fischer-Cripps, A.C.: A review of analysis methods for sub-micron indentation testing. Vacuum **58**(4), 569–585 (2000)
8. Freedonia: World Material Handling Products. Freedonia Group Inc. (2012)
9. Frenkel, J.A.: Viscous flow of crystalline bodies. J. Phys. **9**(5), 385–391 (1945)
10. Fuchs, R., Meyer, J., Staedler, T., Jiang, X.: Sliding and rolling of individual micrometre sized glass particles on rough silicon surfaces. Tribol. Mater. Surf. Interfaces **7**(2), 103–107 (2013)
11. Fuchs, R., Weinhart, T., Meyer, J., Zhuang, H., Staedler, T., Jiang, X., Luding, S.: Rolling, sliding and torsion of micron-sized silica particles: experimental, numerical and theoretical analysis. Granul. Matter **16**(3), 281–297 (2014)
12. Fuchs, R., Ye, M., Weinhart, T., Luding, S., Butt, H.-J., Kappl, M.: Sintering of polymer particle-experiments and modelling of temperature- and time-dependent contacts. In: Proceedings of the International Congress on Particle Technology, PARTEC (2016)
13. Fuchs, R., Weinhart, T., Ye, M., Luding, S., Butt, H.-J., Kappl, M.: Initial stage sintering of polymer particles. In: EPJ Web of Conferences, vol. 140, p. 13012. EDP Sciences (2017)
14. Harshe, Y.M., Lattuada, M.: Breakup of clusters in simple shear flow. J. Phys. Chem. B **120**(29), 7244–7252 (2008)
15. Hashibon, A.: A DEM model for history dependent powder flows. Comput. Part. Mech. **3**(4), 437 (2016)
16. Hertz, H.: Über die Berührung fester elastischer Körper. J. für die reine u. angew. Math. **92** (1882)
17. Imole, O.I., Kumar, N., Magnanimo, V., Luding, S.: Hydrostatic and shear behavior of frictionless granular assemblies under different deformation conditions. KONA Powder Part. J. **30**, 84–108 (2013)
18. Kuhn, M.R., Bagi, K.: Contact rolling and deformation in granular media. Int. J. Solids Struct. **41**(21), 5793–5820 (2004)
19. Laube, J., Drmann, M., Schmid, H.-J., Mdler, L., Colombi Ciacchi, L.: Dependencies of the adhesion forces between TiO_2 nanoparticles on size and ambient humidity. J. Phys. Chem. C **121**(28), 15294–15303 (2017)
20. Liu, D.L., Martin, J., Burnham, N.A.: Optimal roughness for minimal adhesion. Appl. Phys. Lett. **91**(4) (2007)
21. Luding, S.: Collisions & contacts between two particles. NATO ASI Ser. E Appl. Sci. Adv. Study Inst. **350**, 285–304 (1998)
22. Luding, S.: Cohesive, frictional powders: contact models for tension. Granul. Matter **10**(4), 235–246 (2008)
23. Luding, S., Manetsberger, K., Müllers, J.: A discrete model for long time sintering. J. Mech. Phys. Solids **53**(2), 455–491 (2005)
24. Mader-Arndt, K., Kutelova, Z., Fuchs, R., Meyer, J., Staedler, T., Hintz, W., Tomas, J.: Single particle contact versus particle packing behavior: model based analysis of chemically modified glass particles. Granul. Matter **16**(3), 359–375 (2014)

25. Marigo, M., Stitt, E.H.: DEM for industrial applications. KONA Powder Part. J. **32**, 236–252 (2015)
26. Mazur, S., Beckerbauer, R., Buckholz, J.: Particle size limits for sintering polymer colloids without viscous flow. Langmuir **13**(16), 4287–4294 (1997)
27. Mindlin, R.D.: Compliance of elastic bodies in contact. J. Appl. Mech. **16** (1949)
28. Mindlin, R.D., Deresiewicz, H.: Elastic spheres in contact under varying oblique forces. J. Appl. Mech. **20** (1953)
29. Nosewicz, S., Rojek, J., Pietrzak, K., Chmielewski, M.: Viscoelastic discrete element model of powder sintering. Powder Technol. **246**, 157–168 (2013)
30. Ormel, T., Magnanimo, V., Luding, S.: Modeling of asphalt and experiments with a discrete particles method. In: Conference Proceedings MAIREPAV7 2012 (2012)
31. Pasha, M., Dogbe, S., Hare, S., Hassanpour, A., Ghadiri, M.: A linear model of elasto-plastic and adhesive contact deformation. Granul. Matter **16**(1), 151–162 (2014)
32. Peri, M.D.M., Cetinkaya, C.: Adhesion characterization based on rolling resistance of individual microspheres on substrates: review of recent experimental progress. J. Adhes. Sci. Technol. **22**(5–6), 507–528 (2008)
33. Pokluda, O., Bellehumeur, C.T., Vlachopoulos, J.: Modification of Frenkel's model for sintering. AICHE J. **43**(12), 3253–3256 (1997)
34. Rojek, J., Labra, C., Su, O., Onate, E.: Comparative study of different discrete element models and evaluation of equivalent micromechanical parameters. Int. J. Solids Struct. **49**(13), 1497–1517 (2012)
35. Rojek, J., Nosewicz, S., Jurczak, K., Chmielewski, M., Bochenek, K., Pietrzak, K.: Discrete element simulation of powder compaction in cold uniaxial pressing with low pressure. Comput. Part. Mech. **3**(4), 513–524 (2016)
36. Roy, S., Luding, S., Weinhart, T.: Macroscopic bulk cohesion and torque for wet granular materials. In Conference for Conveying and Handling of Particulate Solids (2015)
37. Saito, S., Miyazaki, H.T., Sato, T., Takahashi, K.: Kinematics of mechanical and adhesional micromanipulation under a scanning electron microscope. J. Appl. Phys. **92**(9), 5140–5149 (2002)
38. Schilde, C., Burmeister, C.F., Kwade, A.: Measurement and simulation of micromechanical properties of nanostructured aggregates via nanoindentation and dem-simulation. Powder Technol. **259**, 1–13 (2014)
39. Singh, A., Magnanimo, V., Luding, S.: Mesoscale contact models for sticky particles. Powder Technol. (submitted)
40. Skorupski, K., Hellmers, J., Feng, W., Mroczka, J., Wriedt, T., Mädler, L.: Influence of sintering necks on the spectral behaviour of ITO clusters using the discrete dipole approximation. J. Quant. Spectrosc. Radiat. Transf. **159**, 11–18 (2015)
41. Steuben, J.C., Iliopoulos, A.P., Michopoulos, K.G.: Discrete element modeling of particle-based additive manufacturing processes. Comput. Methods **305**, 537–561 (2016)
42. Strege, S., Weuster, A., Zetzener, H., Kwade, A.: Approach to structural anisotropy in compacted cohesive powder. Granul. Matter **16**(3), 401–409 (2014)
43. Tabor, D.: Indentation hardness: Fifty years on—a personal view. Philos. Mag. A Phys. Condens. Matter Struct. Defects Mech. Prop. **74**(5), 1207–1212 (1996)
44. Thakur, S.C., Morrissey, J.P., Sun, J., Chen, J.F., Ooi, J.Y.: Micromechanical analysis of cohesive granular materials using the discrete element method with an adhesive elasto-plastic contact model. Granul. Matter **16**(3), 383–400 (2014)
45. Thakur, S.C., Ooi, J.Y., Ahmadian, H.: Scaling of discrete element model parameters for cohesionless and cohesive solids. Powder Technol. **293**, 130 (2016)
46. Thornton, A.R., Krijgsman, D.R., Fransen, R.H.A., Gonzalez Briones, S., Tunuguntla, D.R., te Voortwis, A., Luding, S., Bokhove, O., Weinhart, T.: Mercury-DPM: fast particle simulations in complex geometries. EnginSoft Newsl. Simul. Based Eng. Sci. **10**(1), 48–53 (2013)
47. Thornton, C., Cummins, S.J., Cleary, P.W.: An investigation of the comparative behaviour of alternative contact force models during inelastic collisions. Powder Technol. **233**, 30–46 (2013)

48. Tykhoniuk, R., Tomas, J., Luding, S., Kappl, M., Heim, L., Butt, H.-J.: Ultrafine cohesive powders: from interparticle contacts to continuum behaviour. Chem. Eng. Sci. **62**(11), 2843–2864 (2007)
49. van Zwol, P.J., Palasantzas, G., van de Schootbrugge, M., de Hosson, J.T.M., Craig, V.S.J.: Roughness of microspheres for force measurements. Langmuir **24**(14), 7528–7531 (2008)
50. Vastola, G., Zhang, G., Pei, Q.X., Zhang, Y.W.: Microstructure evolution during additive manufacturing. JOM **68**, 5 (2016)
51. Wahl, M., Brckel, U., Brendel, L., Feise, H.J., Weigl, B., Rck, M., Schwedes, J.: Understanding powder caking: predicting caking strength from individual particle contacts. Powder Technol. **188**(2), 147–152 (2008)
52. Walton, O.R., Braun, R.L.: Viscosity, granular temperature, and stress calculations for shearing assemblies of inelastic, frictional disks. J. Rheol. **30**(949) (1986)
53. Weinhart, T., et al.: MercuryDPM: fast, flexible particle simulations in complex geometries Part II: Applications. In: 5th International Conference on Particle-Based Methods—Fundamentals and Applications, PARTICLES 2017, pp. 123–134. International Center for Numerical Methods in Engineering (2017)
54. Weinhart, T., Tunuguntla, D.R., van Schrojenstein-Lantman, M.P., van Der Horn, A.J., Denissen, I.F.C., Windows-Yule, C.R., de Jong, A.C., Thornton, A.R.: MercuryDPM: a fast and flexible particle solver Part A: Technical advances. In: Springer Proceedings in Physics, vol. 188 (2017)
55. Zhang, L., DAcunzi, M., Kappl, M., Auernhammer, G.K., Vollmer, D., van Kats, C.M., van Blaaderen, A.: Hollow silica spheres: synthesis and mechanical properties. Langmuir, **25**(5), 2711–2717 (2009)
56. Zohdi, T.I.: Additive particle deposition and selective laser processing—a computational manufacturing framework. Comput. Mech. **54**(1), 171–191 (2014)

A Contact Model for the Discrete Element Simulations of Aggregated Nanoparticle Films

Valentin Baric, Jens Laube, Samir Salameh, Lucio Colombi Ciacchi and Lutz Mädler

Introduction

Nanoparticle films from the gas phase are applied in many fields, including battery materials [1, 2], gas sensing [3–5] or catalysis [3, 6]. Structural parameters such as a high porosity combined with a controlled pore size distribution and a percolating particle structure make them highly attractive. The film structure directly results from the particle film synthesis process. This synthesis includes particle formation in the gas phase at very high temperatures (flame synthesis), where primary particles form by nucleation and growth. Particle collision leads to incomplete coalescence of tens to hundreds primary particles and the formation of aggregates (Fig. 1). The size of the aggregates is determined by the temperature-time history provided by the flame. When the aerosol stream leaving the flame cools down sufficiently, the particles interact via van der Waals forces and, in the presence of humidity, solvation

V. Baric · J. Laube · S. Salameh · L. Colombi Ciacchi · L. Mädler (✉)
Faculty of Production Engineering, University of Bremen, Bremen, Germany
e-mail: lmaedler@iwt.uni-bremen.de

V. Baric
e-mail: v.baric@iwt.uni-bremen.de

J. Laube
e-mail: j.laube@hmi.uni-bremen.de

S. Salameh
e-mail: s.salameh@tudelft.nl

L. Colombi Ciacchi
e-mail: colombi@hmi.uni-bremen.de

V. Baric · S. Salameh · L. Mädler
Leibniz Institute for Materials Engineering IWT, Bremen, Germany

J. Laube · L. Colombi Ciacchi
Hybrid Materials Interfaces Group, Bremen, Germany

S. Antonyuk (ed.), *Particles in Contact*,
https://doi.org/10.1007/978-3-030-15899-6_11

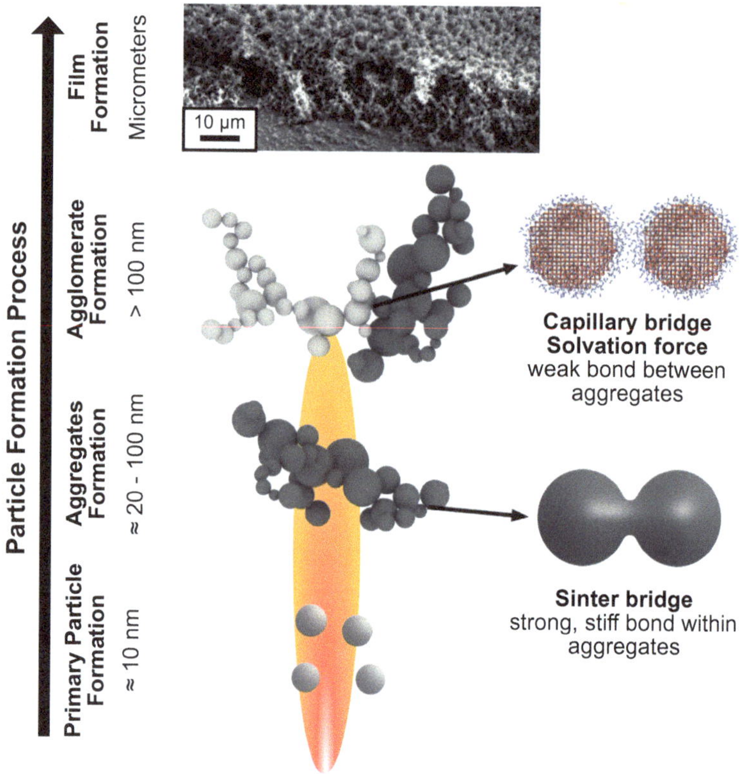

Fig. 1 Nanoparticle films from the gas phase are formed from primary particles of a few nanometers in diameter. In the early formation stages, the collision of primary particles leads to sinter bridge formation and stiff nanoparticle aggregates involving tens to hundreds of primary particles. In the film (or in agglomerates) in humid air, adhesive liquid bridges form between the aggregates. The combination of both binding interactions leads to (mechanically) stable nanoparticle films

and capillary forces [7]. Typical primary particles diameters are about 10 nm and the mobility diameter of aggregates is in the range of 20–100 nm [7].

Particle films formed from these aggregates usually have a porosity greater than 80% [8]. This high porosity leads to a poor resistance against mechanical stress, limiting their applicability. We could show with colloidal probe technique (CPT) that their compaction at pressures up to 2.6 MPa could mechanically stabilize these films. A significant increase in the films Young's Modulus, combined with a shift to smaller pore sizes, was observed.

An experimental investigation of the rearrangement of the particles during this compaction, however, is limited due to poor 3D-accessibility of single particles or aggregates at the nanoscale.

Simulation of such films can provide details of the particle-particle interactions within the films. For example, the discrete element method (DEM) is capable of

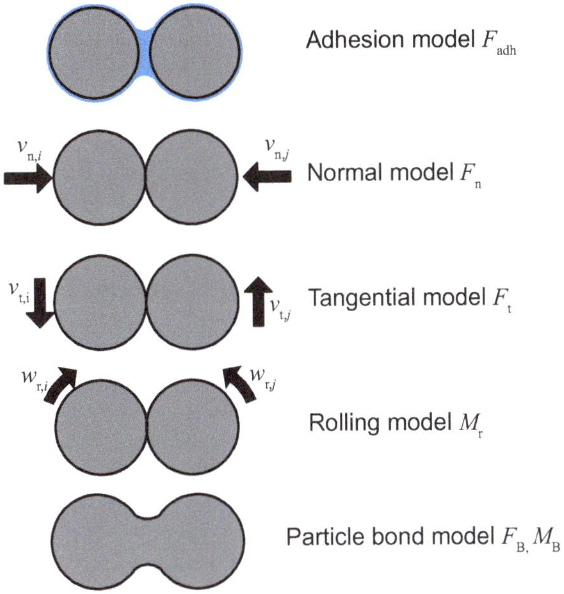

Fig. 2 The description of mechanical interactions between nanoparticles includes 5 model interactions. Weakly bound primary particles (aggregate-aggregate interactions) are modelled using an adhesion model (liquid bridges F_{adh}), a repulsive normal force (F_n), as well as tangential force (F_t) and rolling torque (M_r) models. Particles in aggregates are bound via stiff sinter bridges (F_B, M_B), which require a particle bond model

simulating millions of particles. Required contact models are widely available and established for particles with micrometer diameters (and larger) [9–11]. Body forces such as gravity and inertia determine the interactions of such large particles. Nanoparticles, however, have a significantly higher surface to volume ratio and interact mostly via adhesive forces, including capillary forces and van der Waals forces. Additionally, aggregates require the representation of sinter bridges between particles. The number of contact models for such nanosized particles is limited.

Here we present a contact model for DEM to represent aggregate and agglomerate interactions for nanoparticles smaller than 20 nm in diameter. Agglomerates are modelled considering adhesive forces between primary particles (F_{adh}), normal and tangential forces (F_n and F_t) and rolling torque (M_r). A particle bond model describes the sinter bridges in aggregates (F_B, M_B, Fig. 2).

Development of the Contact Model

All particle-particle interactions are determined by the crystal configuration of the primary particles and their environment. Quantum dynamic simulations based on the density functional theory (DFT) allow for the investigation of electronic structures of atoms in molecules and crystals. These calculations are based on first principles, and, therefore, can predict crystal structures under consideration of material composition and the environment. The simulation time, however, scales with the third power of the number of involved electrons. Commonly, DFT simulations calculate the structure of

Table 1 Comparison of simulation techniques. The density functional theory (DFT) simulates electrons to derive force fields. These force fields are utilized in molecular dynamic (MD) simulations to model particles based on atoms. From these, the contact models for DEM were derived. This allows to simulate millions of nanoparticles

	Density functional theory	Molecular dynamics	Discrete element method
Simulated entity	Electrons	Atoms	Particles
Simulation time per particle for 1 ns (1 CPU)	Years	Hours	Milliseconds–seconds
Reasonable number of particles	<1	<10	Millions
Simulation input	Exchange-correlation functionals and atomic pseudopotentials	Force fields	Contact models

crystal facets in a specific environment (e.g. considering humidity). Therefore, DFT simulations were utilized to derive force fields required for atomistic simulations (Molecular Dynamics, MD) [12–14].

Classical MD simulations resolve particles and water molecules on atomic level. The necessary input includes force fields that describe the interactions between neighboring atoms. These force fields are mainly governed by the crystal structure, material composition and the environment of the particle, as provided by DFT simulations.

Particle-particle interactions at ambient conditions were investigated based on MD simulations [15–17] which eventually resulted in the presented non-covalent DEM contact terms [18]. Typical simulation characteristics of DFT, MD and DEM simulations are given in Table 1.

Adhesion Model

Adhesion Between Nanoparticles

The most important difference between contact models at the nanometer and micrometer scales is the implementation of adhesion forces. These forces become more important due to the low mass of nanoparticles, which decreases the influence of body forces, such as inertia and gravity.

Van der Waals forces between two spherical particles with radius R_p are commonly described by Eq. 1 [19, 20].

$$F_{\mathrm{vdW}} = -\frac{A_{\mathrm{H}} R_p}{12\,\delta^2} \tag{1}$$

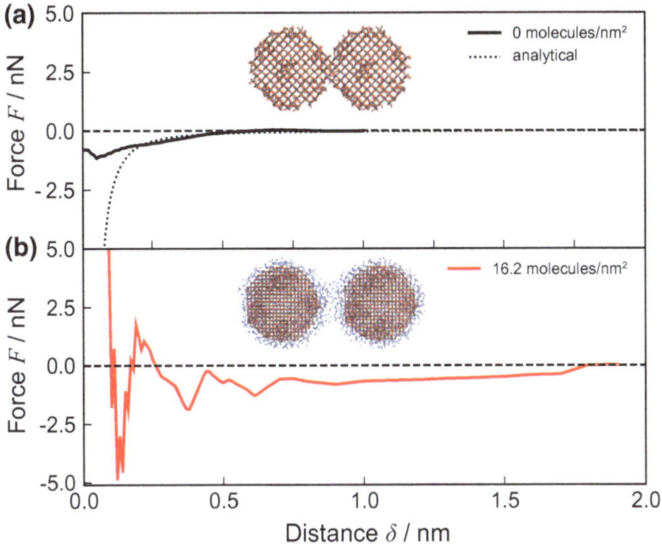

Fig. 3 Force-distance curves for two 4 nm TiO_2 rutile particles. **a** In vacuum and **b** with 16.2 water molecules/nm^2 particle surface area. In vacuum, van der Waals force attracts the particles. At standard conditions, capillary and solvation forces dominate the adhesion. Reprinted/adapted by permission from [16]

Depending on the Hamaker constant A_H (16.7×10^{-20} J for TiO_2 in vacuum [19]), the van der Waals force increases with decreasing distance δ of the two particles. This adhesion force is taken into account in MD simulations of two approaching spherical TiO_2 rutile particles as shown in Fig. 3a. The surface of the particles was terminated with 3 OH groups per nm^2 (up to 0.25 nm distance from the surface). At a surface-surface distance lower than about 0.5 nm, van der Waals forces attract the two particles. The MD simulations confirm the analytical solution of Eq. 1, considering that at 0.2 nm distance the OH groups at the surface become close enough to repel each other.

Typically, nanoparticles are processed and applied at ambient conditions. This leads to water vapour condensation at the particle surface, with formation of structured molecular layers around the particles [16] (Fig. 3b). Water molecules chemisorb at the surface with a density of 5.2 molecules/nm^2 (0.25–0.50 nm distance from surface). This is followed by weakly bound water molecules which are defined as 'physisorbed' water. The combined thickness of these water layers, h, is largely independent of the particle size. It depends on the water coverage and hence, the ambient humidity (vide infra, Fig. 5a). The computed adhesion force for a pair of particles at ambient conditions (relative humidity about 70%) is shown in Fig. 3b. The range and the shape of the attractive force totally differs from the force distance curve of dry particles. Even though the particles are separated by about 0.1 nm, the water layers already intersect, which results in a strong repulsive force. With increasing separation distance, the force becomes attractive and then starts to oscillate. The initially large amplitude of this oscillation (at 0.2 nm the interaction force is even repulsive) decays with increasing distance. This is the so-called solvation force [19, 21, 22]. Water molecules close to the particle surface are structured in layers. During particle approach these structures are disturbed and layers of water molecules are pushed out of the meniscus (or are pulled into the meniscus during particle retraction). This results in an oscillating pressure within the meniscus and hence, an oscillating adhesion force, whose amplitude depends highly on the local interface geometry between the particles. From about 0.9 nm distance, the solvation forces become negligible and an almost linearly decreasing force attracts the particles until it decays at about 1.8 nm separation distance. This constantly decreasing force can be attributed to a capillary force resulting from the liquid meniscus between the particles. Both forces (solvation force and capillary force) are present for a relative humidity greater than 0. The magnitude of the adhesion force as well as the rupture distance of the capillary bridge depend on the humidity, while the share of the solvation force is independent from the humidity [16].

At ambient conditions, these two force contributions dominate the adhesive interactions. While van der Waals forces are still considered in atomistic simulations, they can be neglected in the formulation of a DEM model term representing the non-covalent adhesion force between primary particles of particle aggregates at ambient conditions.

Capillary bridges are commonly described using the circular approximation (Eq. 2) and the geometrical parameters r_m, l_m and θ (Eqs. 3, 4) shown in Fig. 4b, as well as the molecular volume V_M, the gas constant R, temperature T, surface tension γ_L and relative humidity P/P_0.

$$F_{cap} = \pi \gamma_L R^* \sin(\beta)\left[2\sin(\Theta+\beta) + R^* \sin(\beta)\left(\frac{1}{r_m} - \frac{1}{l_m}\right)\right] \quad (2)$$

$$r_m = \frac{2R^*(1-\cos(\beta)) + D}{2\cos(\Theta+\beta)} \quad (3)$$

$$l_m = R^* \sin(\beta) - r_m[1 - \sin(\Theta+\beta)] \quad (4)$$

$$\frac{P}{P_0} = \exp\left[-\frac{\gamma_L V_M}{RT}\left(\frac{1}{r_m} - \frac{1}{l_m}\right)\right] \quad (5)$$

The application of the circular approximation under consideration of a surface tension $\gamma_L = 52.3$ mJ/m^2 (surface tension of bulk TIP3P water used in the simulations), a contact angle $\theta = 10°$ (hydrophilic TiO_2) and the fitted geometrical parameters β, r_m and l_m, however, failed to represent the computed capillary force. We found three necessary assumptions to apply circular approximation on particles with a diameter smaller than 20 nm. (1) The particle radius R^* needs to consider the chemisorbed water layer h around the particles and hence, represents $(R_p + h)^*$ (Fig. 4b). (2) The contact angle θ equals 0° due to the surface coverage of water. The adsorbed molecules prevent the formation of the macroscopic contact angle at these particle

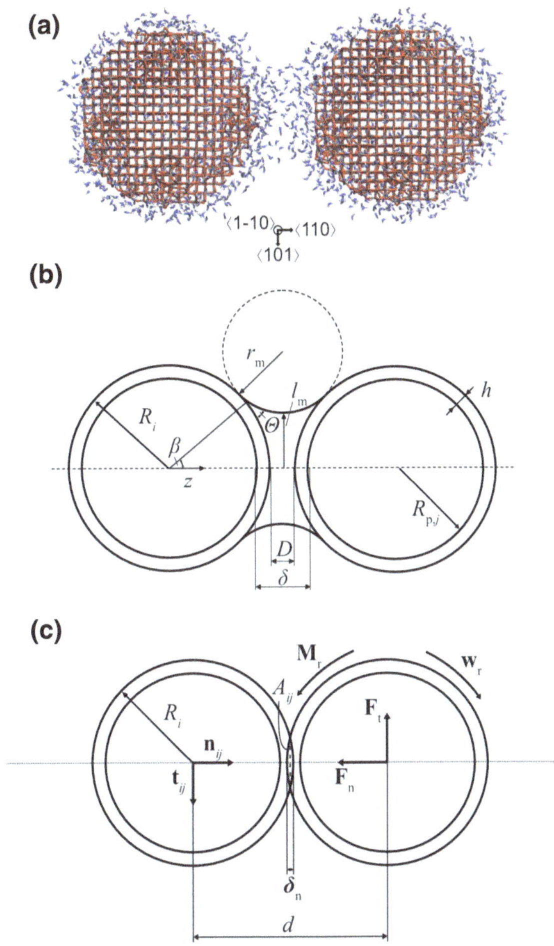

Fig. 4 **a** Molecular dynamic simulation setup with two 4 nm TiO_2 particles (oxygen = red, titanium = grey) and surrounding water molecules (oxygen = blue, hydrogen = white). **b** Relevant parameters to describe the capillary and **c** the representation in DEM. During contact the water layers (thickness h) around the primary particles intersect (intersection area A_{ij}) and determine the particle-particle interactions. The increase of the effective particle radius is considered in the interaction model. Reprinted/adapted by permission from [16, 18]

sizes. (3) The surface tension of water around particles with a diameter smaller than 20 nm is increased (compared to bulk water) due to particle curvature and water confinement in thin films.

The increase in surface tension is most pronounced for small particles and low humidity. This was shown from surface tension calculations of thin water films on flat TiO_2 slabs. When two slabs approach, a capillary bridge forms at a certain distance depending on the number of water molecules on the slabs. The surface tension calculated from the work of adhesion of this capillary bridge is shown in Fig. 5b (grey area). The capillary bridge forms or breaks at a different distance if the slabs are approached or retracted, leading to a hysteresis of the calculated surface tension value especially for large humidities. The surface tension of water around particles with a diameter greater than 10 nm converges to the surface tension obtained from the TiO_2 slab simulations. This indicates that the effect of particle curvature is

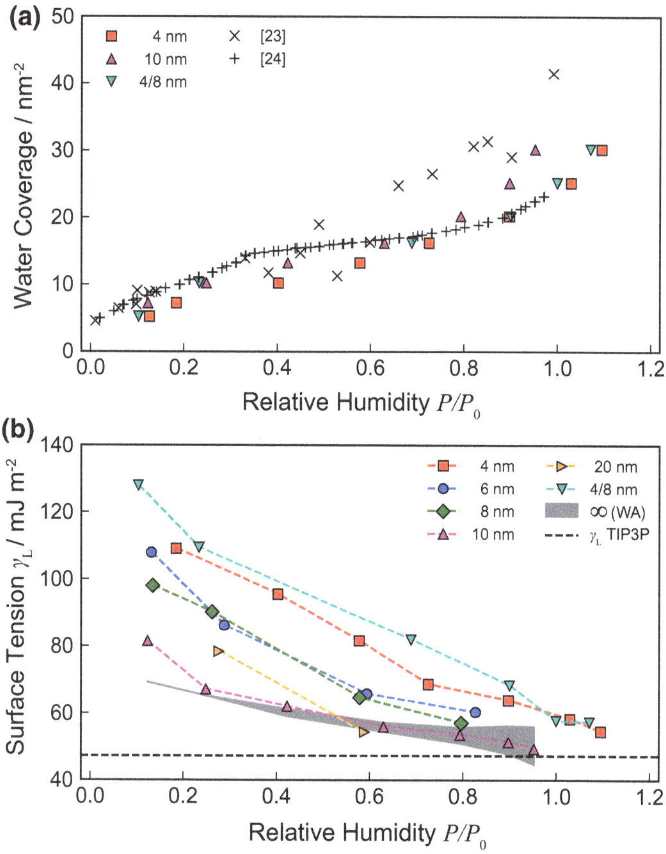

Fig. 5 Water coverage (**a**) and surface tension (**b**) depending on the relative humidity and particle size. The water coverage obtained from capillary fitting of the MD simulations was compared to literature [23, 24]. Small particles show an elevated surface tension, which converges to bulk values at a particle diameter of about 10 nm. The surface tension from the work of adhesion between two slabs is shown (WA, grey area shows hysteresis between approaching and retracting surfaces). The black dashed line displays the surface tension of the water model in the simulations (TIP3P, $\gamma_L = 52.3$ mJ/m^2). Reprinted/adapted by permission from [15]

only relevant for particles smaller than 10 nm in diameter. The water confinement in thin adsorbate films, however, needs to be considered for all particles within the relevant size range (3–20 nm diameter).

Modelling of the Solvation Force

The oscillating solvation force can be approximated using Eq. 7. This equation was fitted to the simulated forces to derive the decay length $\boldsymbol{\sigma}_{\mathbf{solv}}$ (0.21 nm [18]) and the amplitude $\boldsymbol{F}^{\mathbf{0}}_{\mathbf{solv}}$ (Eq. 8). The decay length describes the decreasing force amplitude $\boldsymbol{F}^{\mathbf{0}}_{\mathbf{solv}}$ with increasing particle distance D and is independent of particle size. The force amplitude $\boldsymbol{F}^{\mathbf{0}}_{\mathbf{solv}}$ increases with increasing particle size. The solvation force amplitude can be computed using the effective radius of the particles R^* (Eq. 6).

$$R^* = 2\frac{R_i \cdot R_j}{R_i + R_j} \tag{6}$$

$$F_{\text{solv}} = F^0_{\text{solv}} \cos\left(\frac{2\pi D}{\sigma_{\text{solv}}}\right) e^{-\frac{D}{\sigma_{\text{solv}}}} \tag{7}$$

$$F^0_{\text{solv}} = -11\,\text{nN} - R^* \cdot 2.4\,\text{nN/m} \tag{8}$$

Modelling of the Capillary Force

The capillary force can be represented using the circular approximation as described above. The geometrical parameters β, r_{m} and l_{m}, however, depend on the particle-particle distance and humidity. This needs to be considered when computing the capillary forces.

The following procedure allows to calculate the capillary force at distance D for a specific ambient humidity (exemplary values used in our work are given in brackets):

1. Calculate the liquid film height for the specific humidity, based on the data provided in Fig. 5a.
2. Calculate the surface tension for the specific particle size combination and humidity, based on the data provided in Fig. 5b.
3. Start with an initial angle β (45° was found to be a good initial value).
4. Calculate the geometric parameters r_{m} (Eq. 3) and l_{m} (Eq. 4) for the specific distance D using $\Theta = 0°$.
5. Calculate the relative humidity resulting from the geometrical parameters (Eq. 5).
6. If the deviation of the calculated and targeted humidity is above a defined threshold (0.0001), slightly increase β by a defined interval (1°) and repeat step 3 and 4. If the deviation increases with increasing β, decrease β instead and reduce the interval size of β.
7. If the deviation of calculated and targeted humidity is below the defined threshold, calculate the capillary force using Eq. 2.

Implementation of the Adhesion Model

The adhesion force is the superposition of the solvation force and the capillary force, therefore the adhesion model follows Eq. 9.

$$F_{\text{adh}} = F_{\text{solv}} + F_{\text{cap}} \tag{9}$$

The capillary model involves extensive computation of the correct forces. The shape of the meniscus depends on the particle-particle distance. Hence, Eqs. 2–9 must be solved for all particles at all timesteps. Therefore, we recommend the implementation via a tabulated potential. Many DEM software packages, such as LIGGGHTS, allow to import listed force-distance pairs and the interpolation between these points.[1]

Normal Contact Model

The normal contact of particles in DEM commonly includes a repulsive force $F_{n,r}$ in combination with a dissipating viscoelastic damping term $F_{n,d}$ (Eq. 10).

$$F_n = F_{n,r} + F_{n,d} \tag{10}$$

The repulsive component for the normal force was analyzed simulating the penetration of a TiO_2 particle into a flat TiO_2 surface in the presence of water adsorbate layers. The penetration distance with respect to the applied penetration force F_n was averaged over several locations on the slab. It was found that the Hertz function can represent the force-distance relationship for all relevant penetration distances [9]. The Hertz model is well established in DEM and follows a 3/2 order dependency of force and penetration distance (Eq. 11).

$$F_{n,r} = k_n \delta_n^{3/2} \sqrt{(0.5R^*)} \tag{11}$$

The model considers a contact stiffness k_n and the effective radius R^*, which depend on the radii of the interacting particles. During approach, the water molecules between the particles are compressed. As a response, the water layers start to generate a repulsive force at a compressed thickness h_n of approximately 0.36 nm [18].

The size-dependent contact stiffness k_n is calculated from the equilibrium distance ($F_{n,r}(0) = -F_{adh}(0)$). Therefore, the calculation of the stiffness requires previous computation of the adhesion forces. The physisorbed water layers increase the effective particle radius, which leads to a repulsive force already at greater distance R^*. The equilibrium distance in DEM, however, is chosen to be the particle radius R_p (without water layer). The values of k_n obtained for TiO_2 particles with diameters from 3–23 nm increase linearly from 20 to 45 GPa [18]. This is between the values for TiO_2 (210 GPa) and water or ice (2 GPa). A tabulated list for diameters in the range of 3–23 nm is given in [18].

$$k_n = \frac{-F_{adh}(0)}{\delta_n(0)^{\frac{3}{2}} \sqrt{(0.5R^*)}} \tag{12}$$

[1]This function was removed in LIGGGHTS 3.2 and needs to be implemented from previous versions.

The viscoelastic damping of the normal contact was estimated from two simulated TiO_2 particles. Starting at an initial distance greater than the equilibrium distance, the particles attracted each other due to capillary forces. The following fluctuation around the equilibrium distance [18] can be calculated using Eq. 13.

$$\mathbf{F}_{\mathrm{n,d}} = -\gamma_{\mathrm{n}} \mathbf{v}_{\mathrm{n}} \tag{13}$$

The resulting damping coefficient, γ_{n}, depends quadratically on the effective particle radius, which differs from Hertz-Theory: $\gamma_{\mathrm{n}} = 0.5 \times 10^6\ \mathrm{Ns/m}^3 \cdot R^{*2}$.

Tangential Force Model

The tangential force model was also derived from simulations of particles in contact with a TiO_2 surface slab [18]. A particle under load F_{n} was dragged over the surface and the rolling of the particle was suppressed. The resulting tangential force F_{t} followed a saw tooth-like stick slip behavior (Fig. 6b). The peak value $F^T_{t,max}$ linearly depends on the contact area A_{ij} (Eq. 15, Fig. 4c). This dependency on the area is fundamentally different from the linear relationship between F_{n} and F_{t} in macroscopic friction (Amontons' law).

$$F^T_{t,max} = \tau_T \cdot A_{ij} \tag{14}$$

$$A_{ij} = \pi \left[R_i^2 - \frac{\left(R_i^2 - R_j^2 + d^2 \right)}{4d^2} \right] \tag{15}$$

The period of the saw tooth λ_{t} is independent of the applied normal load, with a value of 0.5 nm. Therefore, the tangential force model is expressed as a spring-slider model represented in Eq. 16, in which δ_{t} denotes the tangential displacement along $\mathbf{t}_{ij}$ and $\lfloor x \rfloor$ is the floor function of x.

$$\mathbf{F}_{\mathrm{t}} = -\tau_{\mathrm{T}} A_{ij} \left(\frac{\delta_{\mathrm{t}}}{\lambda_{\mathrm{t}}} - \lfloor \frac{\delta_{\mathrm{t}}}{\lambda_{\mathrm{t}}} \rfloor \right) \mathbf{t}_{ij} \tag{16}$$

The tangential stiffness τ_{t} for TiO_2 (rutile) surfaces obtained from MD simulations is 0.65 GPa.

Rolling Torque Model

The rolling torque model was derived similarly to the tangential force model [18]. The TiO_2 particle under load F_{n}, however, could roll while it was dragged over the TiO_2 surface. The resulting force peaks were significantly lower than for the tangential forces. The rolling was characterized by rather constant forces and is therefore modelled using a constant friction force $F_{t,max}^R$ (Fig. 6c). This force also scales with the contact area A_{ij} (Eq. 15). The resulting torque follows Eq. 18.

$$F_{t,max}^R = \tau_R \cdot A_{ij} \tag{17}$$

$$\mathbf{M}_r = 0.5R^* \tau_R A_{ij} \tag{18}$$

The rolling friction stiffness τ_R for TiO_2 derived from MD simulations is 0.25 GPa.

Particle Bond Model

Sinter bridges in aggregates are known to be stiff and hardly breakable [25]. Parameterization and modelling of their mechanical properties and behavior, however, are missing. Based on the observation that sinter bridges can be approximated using cylinders, beam mechanics can be applied. TiO_2 nanoscopic beams were strained in

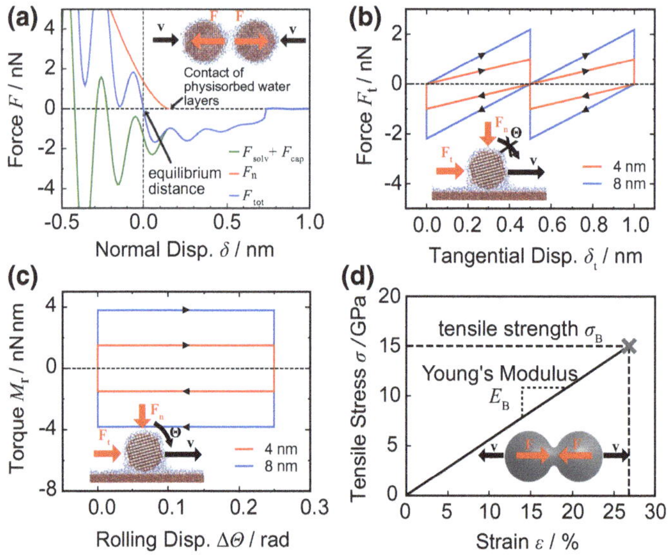

Fig. 6 Force-displacement curves for all implemented interaction models. **a** Adhesion force combined with the normal forces for two primary particles with 4 nm diameter. **b** The tangential force and **c** the rolling torque for particle with 4 and 8 nm diameter on a slab. **d** The sinter bridge model for normal strain follows a linear increase and a brittle rupture. The insets of (**b**–**d**) display the respective MD-simulation setup. Reprinted/adapted by permission from [18]

an AFM tensile test in Amin et al. [26]. The constant elongation of the beam resulted in a linear increase of the opposing force, followed from a brittle rupture. Therefore, we assume an elastic deformation and a brittle rupture for nanoparticle sinter bridges. Models, capable of representing such behavior have been implemented in DEM for a variety of applications, including polyvinylpyrrolidon (PVP) binders [27], maltodextrin agglomerates [28] and cementation of particles [29]. Most of them are based on (or similar to) the particle bond model introduced by Potyondy and Cundall [30]. The model covers particle displacement in normal and tangential direction as well as normal and tangential torque (torsion and bending, respectively). Equations 19–24 describe the particle bond model that was implemented into DEM.

$$\mathbf{F}_{\mathrm{B,n}} = \mathbf{L}^{\mathrm{n}} \frac{E_{\mathrm{B}}}{L_0} A_{\mathrm{B}} \tag{19}$$

$$\mathbf{F}_{\mathrm{B,t}} = \mathbf{L}^{\mathrm{T}} \frac{E_{\mathrm{B}}}{2L_0(1+\nu_{\mathrm{B}})} A_{\mathrm{B}} \tag{20}$$

$$\mathbf{M}_{\mathrm{B,n}} = \boldsymbol{\delta}^{\mathrm{n}} \frac{E_{\mathrm{B}}}{2L_0(1+\nu_{\mathrm{B}})} J \tag{21}$$

$$\mathbf{M}_{\mathrm{B,t}} = \boldsymbol{\delta}^{\mathrm{T}} \frac{E_{\mathrm{B}}}{L_0} I \tag{22}$$

$$\text{Normal breakage criterion:} \quad -\frac{\mathbf{F}_{\mathrm{B,n}}}{A_B} + \frac{|\mathbf{M}_{\mathrm{B,t}}|}{I} R_{\mathrm{B}} \geq \sigma_{\mathrm{B}} \tag{23}$$

$$\text{Tangential breakage criterion:} \quad \frac{|\mathbf{F}_{\mathrm{B,t}}|}{A_B} + \frac{|\mathbf{M}_{\mathrm{B,n}}|}{J} R_{\mathrm{B}} \geq \tau_{\mathrm{B}} \tag{24}$$

In the model, a sinter bridge is represented as a cylindrical bond with initial length L_0, radius R_{B} and cross-sectional area A_{B}. The forces and torques resulting from the sinter bridge are modelled using a Young's Modulus E_{B} and the poisson's ratio ν_{B}, as well as the moment of inertia I and the polar moment of inertia J. Forces and torques are calculated depending on the translational and rotational displacement L and δ, respectively. Subsequently, they are compared to a tensile strength σ_{B} (Eq. 23) and shear strength τ_{B} (Eq. 24), respectively. If one of the criteria is fulfilled, the sinter bridge breaks and irreversibly is removed from the simulation.

Summary and Implementation

Exemplary force-distance curves for pairs of particles are given in Fig. 6. The adhesion and normal contact model terms are combined for clarity. The combined force-distance curves show, that the initial 0.15 nm of contact (contact of physisorbed water layer until equilibrium distance) are still attractive (blue curve negative, Fig. 6a).

A graphical representation of the model terms and the most important equations are given in Table 2. Using these equations, the model can be implemented with low effort into a simulation package. An exemplary implementation into the open-source DEM software LIGGGHTS is provided on github.[2] This includes a program to compute and write the adhesion forces into a LIGGGHTS compatible table file and an example simulation setup. All terms are parameterized using the parameters described above and in the respective literature [18, 31].

Validation

The validation of the presented contact model requires suitable experiments to detect particle-particle interactions. As mentioned previously, the experimental access to single particles or aggregates is very challenging. Force-distance curves obtained from Atomic Force Microscopy (AFM), however, allow to retract small agglomerates or aggregates from a film of nanoparticles. While it remains challenging to characterize the exact structure of the particles retracted from the film, the force-distance curves show highly reproducible self-similar saw tooth-like profiles (Fig. 7 bottom right). Single particle interactions, such as rolling, sliding and particle-particle breakup determine these force-distance curves [17]. Therefore, these experiments were simulated to validate the contact models.

Nanoparticle films were computed using a primary particle and aggregate size distribution as found in [7]. The film generation procedure followed the model described in [32]. These films can represent thermophoretically deposited nanoparticle films with a high porosity up to 98%. The cantilever of the AFM experiment was modelled using a pyramid penetrating the film (Fig. 7). During penetration, the particles attached to the pyramid and subsequently were pulled out of the film during pyramid retraction.

Exemplary force-distance curves of such simulations (left) and experiments (right) are given in Fig. 7. The overall shape of the curves emphasizes the validity of the presented model. Both are characterized by maximum retraction lengths before final rupture lying between 1.5 and 2.5 μm. The overestimation of the simulated maximum retraction forces can be ascribed to the periodical boundary conditions on one side and to the assumption of rigid and unbreakable sinter bridges in these simulations on the other side (different from the presented particle bond model). Despite this simplified assumption, however, the simulations could represent the rearrangements

[2] https://github.com/UniHB-IWT/LIGGGHTS-Nano.

Table 2 Schematic representation of the implemented model terms and the respective equations

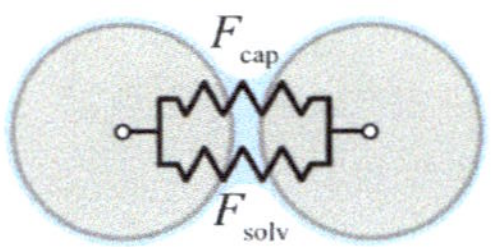	Adhesion model [15, 16, 18]	$R^* = 2\frac{R_i * R_j}{R_i + R_j}$ $F_{\text{solv}} = F^0_{\text{solv}} \cos\left(\frac{2\pi D}{\sigma_{\text{solv}}}\right) e^{-\frac{D}{\sigma_{\text{solv}}}}$ $F_{cap} = \pi \gamma_L R^* \sin(\beta)\left[2\sin(\Theta + \beta) + R^* \sin(\beta)\left(\frac{1}{r_m} - \frac{1}{l_m}\right)\right]$
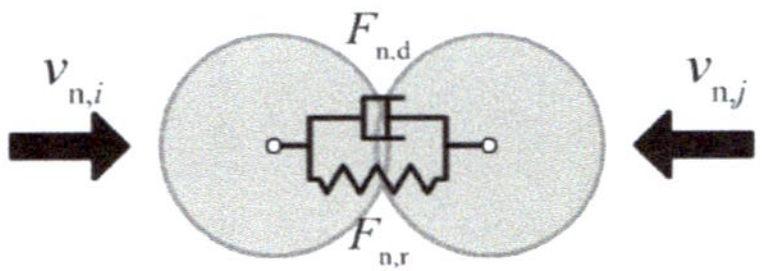	Normal model [18]	$F_{\text{n,r}} = k_{\text{n}} \delta_{\text{n}}^{3/2} \sqrt{(0.5R^*)}$ $\mathbf{F}_{\mathbf{n,d}} = -\boldsymbol{\gamma}_{\mathbf{n}} \mathbf{v}_{\mathbf{n}}$
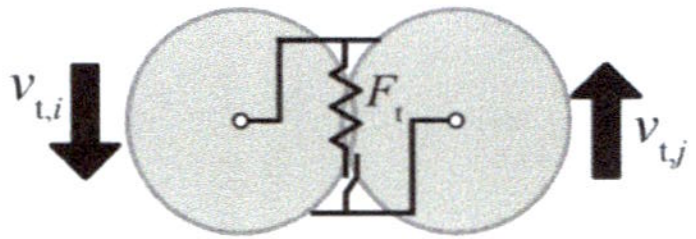	Tangential model [18]	$\mathbf{F}_{\text{t}} = -\tau_{\text{T}} A_{ij} \left(\frac{\delta_{\text{t}}}{\lambda_{\text{t}}} - \lfloor \frac{\delta_{\text{t}}}{\lambda_{\text{t}}} \rfloor\right) \mathbf{t}_{ij}$ $A_{ij} = \pi \left[R_i^2 - \frac{\left(R_i^2 - R_j^2 + d^2\right)}{4d^2} \right]$
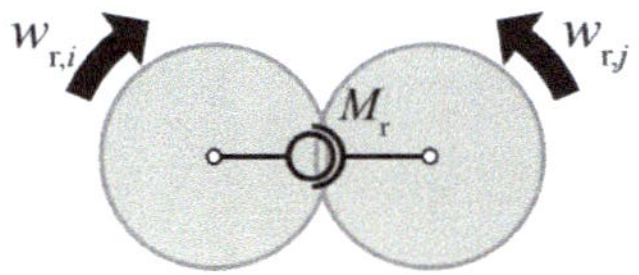	Rolling model [18]	$M_r = 0.5 R^* \tau_R A_{ij}$
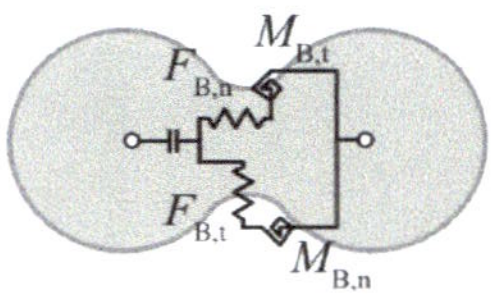	Particle bond model [30, 31]	$F_{\text{n}} = L^{\text{n}} \frac{E_{\text{B}}}{L_0} A_{\text{B}},\ F_{\text{t}} = L^{\text{t}} \frac{E_{\text{B}}}{2L_0(1+\nu_{\text{B}})} A_{\text{B}}$ $M_n = \delta^n \frac{E_B}{2L_0(1+\nu_B)} J,\ M_t = \delta^t \frac{E_B}{L_0} I$

during the retraction of aggregates or small agglomerates and attributed them to the adhesion, normal, tangential and rolling models.

Both in the simulations and the experiments it is visible that the forces during retraction (red curves) are significantly larger than during approach (blue curve). During approach, multiple capillary bridges form between the particles. This leads to a chain of aggregates pulled from the film during retraction. These capillary bridges are sufficient to cause restructuring of the aggregates, rather than particle-particle breakups. This restructuring, characterized by sliding and rolling events (tangential force and rolling torque models), leads to an increase in the attractive force on the pyramid during retraction. Only if further restructuring cannot maintain all particle-particle contacts, a particle-particle breakup causes a rapid decrease in attractive force, resulting in the characteristic saw tooth pattern. Both simulations and experiments are characterized by a final particle-particle breakup force of 1–3 nN.

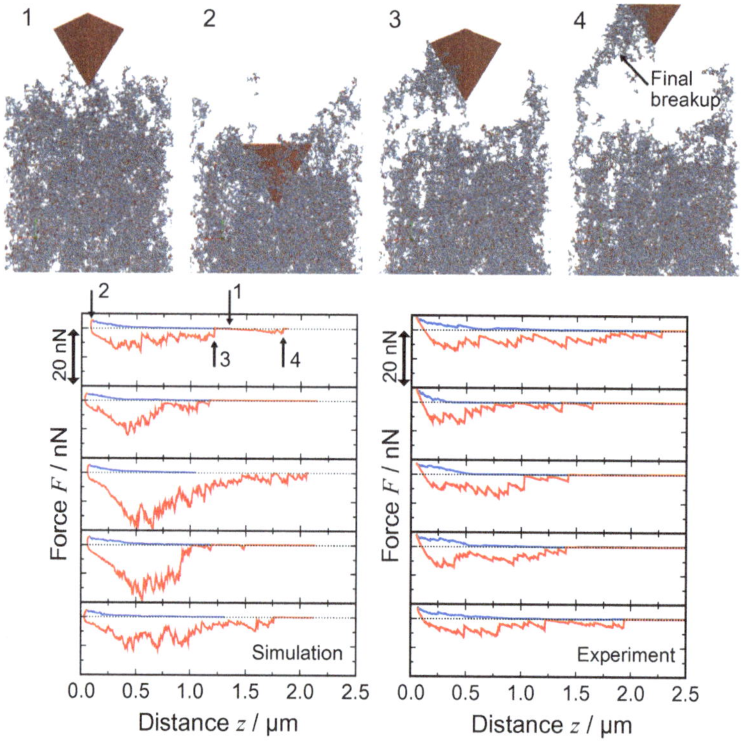

Fig. 7 Force-distance curves of DEM simulations (left) and AFM experiments (right). The snapshots 1–4 correspond to the marked positions of the trajectory (top). The particles are colored according to their diameter from blue (2 nm) to red (23 nm). The location of the final particle-particle breakup is highlighted in snapshot 4. Aggregates are assumed to be rigid in the simulations. Reprinted/adapted by permission from [18]

Application

Particle films with such high porosity (up to 98%) have a rather poor mechanical stability. Therefore, the films need to be stabilized to maintain their structural properties during processing and application. Thermal stabilization processes such as sintering result in particle coalescence, and therefore, in a decrease of the specific surface area. We could show that the low-pressure compaction at pressures up to 2.6 MPa results in a significant stabilization of the films [8]. This stabilization was accompanied by a decrease in porosity and a shift of the pore size distribution (PSD) towards smaller pores. The experimental characterization of the restructuring during this compaction, however, is challenging. Characterization methods, such as porosity and PSD from nitrogen adsorption isotherms or electrical film conductivity, result in integral film properties. As mentioned above, the film properties result from single-particle interactions. The complexity of such films, however, makes it difficult to attribute film properties to the contributions of particle interactions within aggregates and between aggregates. Especially the sinter bridges in aggregates play an important role in the compaction of films. These bridges are commonly assumed to be too stable to break [25]. Any information about their elastic behavior, however, remains missing. Simulations considering the presented particle bond model allow the effect of elastic sinter bridges on film restructuring to be isolated.

To this purpose, particle films with an initial porosity of 98% [32] were compacted in DEM simulations using the adhesion, normal, tangential, rolling and bond model terms [31]. Additionally, two alternative sinter-bridge terms were applied as reference tests: either (1) infinitely rigid, non-breakable aggregates, or (2) single particles without any particle bond model. The films were compacted using two walls. One wall was approaching the other with a constant velocity and the compaction pressure acting on the wall was calculated. It was found that the applied particle bond model significantly determines the film porosity at a given compaction pressure (Fig. 8b). Rigid aggregates resulted in the largest porosities at a given compaction pressure, overestimating experimental porosities obtained from nitrogen adsorption isotherms. Single particles underestimated these porosities, resulting in the lowest porosity with respect to compaction pressure. Porosities from simulations using the presented particle bond model (tensile/shear strength σ_{B}, τ_{B} of 15 GPa and Young's modulus E_{B} of 56 GPa), however, matched the experiments at pressures greater 2 MPa very well. Although none of the sinter bridges broke in the simulations, the capability of elastic deformation seems to be crucial for the compaction of aggregated nanoparticle films. This is further confirmed with a very good agreement of the PSD obtained from simulations and experiments (Fig. 8c). The inner 50% of all simulated pores deviate from the experiments by only a few nanometers. Solely very large pores found in the experiment are missing in the simulations. The probability for these pores is very low and they were already missing in the initial simulation domain of only 2 μm $\times$ 2 μm $\times$ 6 μm.

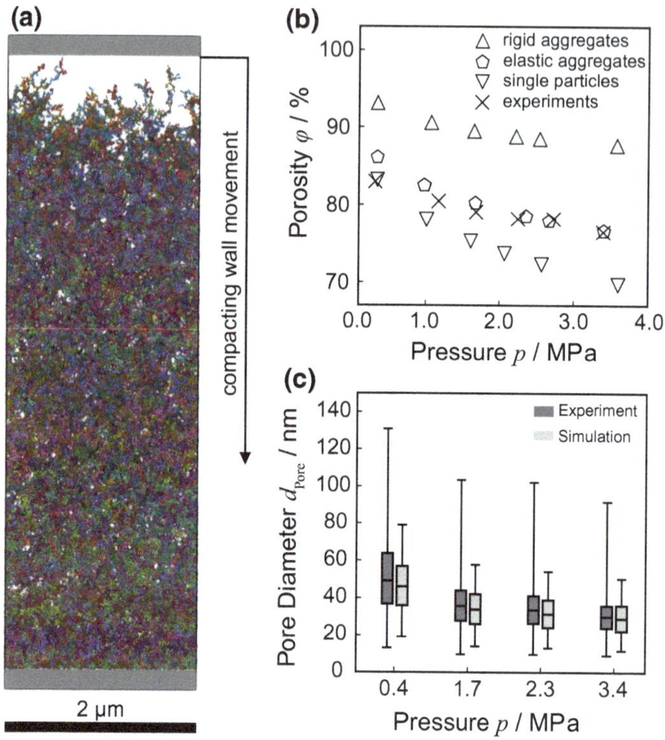

Fig. 8 Compaction of nanoparticle films using DEM. The film is continuously compacted using a moving wall (**a**), up to a specific compaction pressure is reached. The porosity (**b**) and pore size distribution (**c**) of the films as a function of the compaction pressure are in striking agreement with the experiments. Additionally, simulated porosities for rigid aggregates (inflexible sinter bridges) and films with single particles (no sinter bridges) were simulated. Reprinted/adapted by permission from [31]

Summary

Nanostructured surfaces and coatings play an important role in many processes that lead to applications of catalysts, batteries, sensors, composites. With the presented model, fundamental details of such coatings can be assessed. This leads to a better understanding of dynamic changes of these coatings during synthesis and processing such as mechanical compaction.

To this purpose, DEM model components were developed and implemented in the open-source DEM software LIGGGHTS that include adhesion forces (capillary and solvation forces), normal and tangential forces, rolling torque and particle bonds (sinter bridges) for polydisperse nanoparticle aggregates and agglomerates.

The DEM model was validated with data obtained in AFM experiments. Our simulations showed equivalent saw tooth patterns while we could attribute specific characteristics of these patterns to particle rearrangement events including rolling, sliding and particle-particle breakups.

The application of the presented models allowed us to analyze the influence of elastic aggregate deformation during nanoparticle film compaction. This is essential for evaluating porosities at specific compaction pressures. The application of an elastic particle bond model resulted in porosities and pore size distributions exactly matching the experimentally obtained data.

The presented model allows for versatile and detailed analyses of nanoparticle film properties such as elasticity, electrical conductivity and pore size distribution. Simulations based on the models can help relate integral film properties such as film elasticity and electrical film conductivity to local film properties including primary particle connectivity and aggregate size distributions.

References

1. Gockeln, M., Glenneberg, J., Busse, M., Pokhrel, S., Mädler, L., Kun, R.: Flame aerosol deposited $Li_4Ti_5O_{12}$ layers for flexible, thin film all-solid-state Li-ion batteries. Nano Energy **49**, 564–573 (2018)
2. Gockeln, M., Pokhrel, S., Meierhofer, F., Glenneberg, J., Schowalter, M., Rosenauer, A., Fritsching, U., Busse, M., Mädler, L., Kun, R.: Fabrication and performance of $Li_4Ti_5O_{12}$/C Li-ion battery electrodes using combined double flame spray pyrolysis and pressure-based lamination technique. J. Power Sources **374**(Suppl C), 97–106 (2018)
3. Dreyer, J.A.H., Riefler, N., Pesch, G.R., Karamehmedovic, M., Fritsching, U., Teoh, W.Y., Mädler, L.: Simulation of gas diffusion in highly porous nanostructures by direct simulation Monte Carlo. Chem. Eng. Sci. **105**, 69–76 (2014)
4. Kemmler, J., Schopf, S.O., Mädler, L., Barsan, N., Weimar, U.: New process technologies for the deposition of semiconducting metal oxide nanoparticles for sensing. Procedia Eng. **87**, 24–27 (2014)
5. Mädler, L., Roessler, A., Pratsinis, S.E., Sahm, T., Gurlo, A., Barsan, N., Weimar, U.: Direct formation of highly porous gas-sensing films by in situ thermophoretic deposition of flame-made Pt/SnO_2 nanoparticles. Sens. Actuators B: Chem. **114**(1), 283–295 (2006)
6. Pesch, G.R., Riefler, N., Fritsching, U., Colombi Ciacchi, L., Mädler, L.: Gas-solid catalytic reactions with an extended DSMC model. AIChE J. **61**(7), 2092–2103 (2015)
7. Salameh, S., Scholz, R., Seo, J.W., Mädler, L.: Contact behavior of size fractionated TiO_2 nanoparticle agglomerates and aggregates. Powder Technol. **256**, 345–351 (2014)
8. Schopf, S.O., Salameh, S., Mädler, L.: Transfer of highly porous nanoparticle layers to various substrates through mechanical compression. Nanoscale **5**(9), 3764–3772 (2013)
9. Di Renzo, A., Di Maio, F.P.: Comparison of contact-force models for the simulation of collisions in DEM-based granular flow codes. Chem. Eng. Sci. **59**(3), 525–541 (2004)
10. Luding, S.: Cohesive, frictional powders: contact models for tension. Granul. Matter **10**(4), 235 (2008)
11. Pasha, M., Dogbe, S., Hare, C., Hassanpour, A., Ghadiri, M.: A linear model of elasto-plastic and adhesive contact deformation. Granul. Matter **16**(1), 151–162 (2014)
12. Matsui, M., Akaogi, M.: Molecular dynamics simulation of the structural and physical properties of the four polymorphs of TiO_2. Mol. Simul. **6**(4–6), 239–244 (1991)
13. Schneider, J., Colombi Ciacchi, L.: A classical potential to model the adsorption of biological molecules on oxidized titanium surfaces. J. Chem. Theory Comput. **7**(2), 473–484 (2010)
14. Jorgensen, W.L., Chandrasekhar, J., Madura, J.D., Impey, R.W., Klein, M.L.: Comparison of simple potential functions for simulating liquid water. J. Chem. Phys. **79**(2), 926–935 (1983)

15. Laube, J., Dörmann, M., Schmid, H.J., Mädler, L., Colombi Ciacchi, L.: Dependencies of the adhesion forces between TiO_2 nanoparticles on size and ambient humidity. J. Phys. Chem. C **121**(28), 15294–15303 (2017)
16. Laube, J., Salameh, S., Kappl, M., Mädler, L., Colombi Ciacchi, L.: Contact forces between TiO_2 nanoparticles governed by an interplay of adsorbed water layers and roughness. Langmuir **31**(41), 11288–11295 (2015)
17. Salameh, S., Schneider, J., Laube, J., Alessandrini, A., Facci, P., Seo, J.W., Colombi Ciacchi, L., Mädler, L.: Adhesion mechanisms of the contact interface of TiO_2 nanoparticles in films and aggregates. Langmuir **28**(31), 11457–11464 (2012)
18. Laube, J., Baric, V., Salameh, S., Mädler, L., Colombi Ciacchi, L.: A new contact model for the discrete element method simulation of TiO_2 nanoparticles films under mechanical load. Granul. Matter **20**(28), 1–16 (2018)
19. Israelachvili, J.N.: Intermolecular and Surface Forces, 3rd edn. Elsevier, Oxford, USA (2011)
20. Hamaker, H.C.: The London—van der Waals attraction between spherical particles. Physica **4**(10), 1058–1072 (1937)
21. Israelachvili, J.N., Pashley, R.M.: Molecular layering of water at surfaces and origin of repulsive hydration forces. Nature **306**(5940), 249 (1983)
22. Israelachvili, J.N.: Solvation forces and liquid structure, as probed by direct force measurements. Acc. Chem. Res. **20**(11), 415–421 (1987)
23. Ketteler, G., Yamamoto, S., Bluhm, H., Andersson, K., Starr, D.E., Ogletree, D.F., Ogasawara, H., Nilsson, A., Salmeron, M.: The nature of water nucleation sites on TiO_2 (110) surfaces revealed by ambient pressure X-ray photoelectron spectroscopy. J. Phys. Chem. C **111**(23), 8278–8282 (2007)
24. Mamontov, E., Vlcek, L., Wesolowski, D.J., Cummings, P.T., Wang, W., Anovitz, L.M., Rosenqvist, J., Brown, C.M., Garcia Sakai, V.: Dynamics and structure of hydration water on rutile and cassiterite nanopowders studied by quasielastic neutron scattering and molecular dynamics simulations. J. Phys. Chem. C **111**(11), 4328–4341 (2007)
25. Seipenbusch, M., Rothenbacher, S., Kirchhoff, M., Schmid, H.J., Kasper, G., Weber, A.P.: Interparticle forces in silica nanoparticle agglomerates. J. Nanopart. Res. **12**(6), 2037–2044 (2010)
26. Amin, S.S., Li, S.Y., Wu, X., Ding, W., Xu, T.T.: Facile synthesis and tensile behavior of TiO_2 one-dimensional nanostructures. Nanoscale Res. Lett. **5**(2), 338–343 (2009)
27. Dosta, M., Dale, S., Antonyuk, S., Wassgren, C., Heinrich, S., Litster, J.D.: Numerical and experimental analysis of influence of granule microstructure on its compression breakage. Powder Technol. **299**, 87–97 (2016)
28. Spettl, A., Bachstein, S., Dosta, M., Goslinska, M., Heinrich, S., Schmidt, V.: Bonded-particle extraction and stochastic modeling of internal agglomerate structures. Adv. Powder Technol. **27**(4), 1761–1774 (2016)
29. Brendel, L., Török, J., Kirsch, R., Bröckel, U.: A contact model for the yielding of caked granular materials. Granul. Matter **13**(6), 777–786 (2011)
30. Potyondy, D.O., Cundall, P.A.: A bonded-particle model for rock. Int. J. Rock Mech. Min. Sci. **41**(8), 1329–1364 (2004)
31. Baric, V., Colombi Ciacchi, L., Mädler, L.: Compaction Induced Restructuring of Aggregated Nanoparticle Films Using the Discrete Element Method. Powder Technol. **342**, 773–779 (2019)
32. Mädler, L., Lall, A.A., Friedlander, S.K.: One-step aerosol synthesis of nanoparticle agglomerate films: simulation of film porosity and thickness. Nanotechnology **17**(19), 4783–4795 (2006)

Determination of the Adhesion Forces of Magnetic Composite Particles

Johannes Knoll, Frank Rhein and Hermann Nirschl

Nomenclature

A	Hamaker constant (J)
a_0	Particle surface distance (m)
B	Magnetic flux (T)
c	Particle number concentration (mol m^{-3})
d	Diameter (m)
E	Energy (J)
e_0	Electron charge (C)
F	Force (N)
H	Magnetic field vector (A m^{-1})
k_B	Boltzmann constant (J K^{-1})
M	Magnetization (A m^{-1})
n	Rotational speed (s^{-1})
N_A	Avogadro constant (mol^{-1})
R	Radius (m)
r	Vector between two magnetic dipoles (m, m, m)
rms	Root mean square roughness (m)
R_q	Roughness (m)
T	Temperature (K)
V	Volume (m^3)
z	Ion charge (–)
ϵ	Dielectric constant (C V^{-1} m^{-1})
ζ	Zeta potential (V)
κ	Debye-Hückel parameter (m^{-1})
μ_0	Vacuum permeability (N A^{-2})

J. Knoll · F. Rhein · H. Nirschl (✉)
Universität KIT, Karlsruhe, Germany
e-mail: hermann.nirschl@kit.edu

S. Antonyuk (ed.), *Particles in Contact*,
https://doi.org/10.1007/978-3-030-15899-6_12

ρ Density (kg m^{-3})

Indices

ad	Adhesive
c	Centrifugal
el	Electrostatic
l	Liquid
m	Magnetic
p	Particle
v.d.W	Van der Waals

Introduction

Over the last years magnetic particle systems received a higher importance in the pharmaceutical and medical field. Several possibilities of application of magnetic particles are passing clinical studies. A promising application is the injection of magnetic particles into the human body where the particles are drawn to a desired place, mostly cancer tissue by a magnetic field. Once in place, an alternating magnetic field is applied which leads to an oscillating motion with increasing temperature. As a result, the cancer cells are killed at higher temperatures. Another useful application is the fixation of drugs at the surface of the magnetic particles. Afterwards, the particles are again injected and positioned at the desired location by the magnetic field [1, 2].

In life science, functionalized magnetic particles are already used for bioproduct purification, the concentration of the bioproduct can be as low as 1 mg/l but the final purity has to be extremely high. Therefore, a selective purification step is necessary. One approach are functionalized magnetic particles which are added to a biosuspension and the target product binds to the specific surface functionalization. Subsequently, the suspension has to pass a magnetic filter where the magnetic particles attach to a magnetic matrix and the impurities are washed out of the system. This technology is called high gradient magnetic separation (HGMS). So far, the process usually works in batch mode, but a continuous working apparatus is already available [3, 4].

The construction of the devices is based on experimental investigations and the characterization of the process results from the information of the inlet and outlet. The reason is just the lack of information of the process inside the devices which prevents an optimal design.

The particles attach to the separation matrix due to magnetic forces, van der Waals forces and electrostatic interactions. Various investigations on both forces have been carried out. But so far no investigation of the interplay between them has been conducted. Hence the article wants to give an insight into the interaction between the magnetic, functionalized particles with a substrate but also in between the single particles. It is still challenging to directly measure the interaction forces which is solved here directly with an optical centrifuge where a magnetic field is integrated. This allows to predict directly the influence of a magnetic field on the interaction.

Theoretical Background of the Relevant Interaction Forces

The interaction inbetween the magnetic particles and the particle and the substrate is determined by the three main interaction forces, the van der Waals force, the magnetic force and the elctrostatic interaction. In the following the theoretical background and the governing equations for the approximation of the three contributions to the adhesion force are discussed.

Van der Waals Forces

The van Waals forces have to be taken into account for microscale particles. Important basic research was done by Israelachvili [5] and by Hamaker [6]. In their studies, neither deformations in the contact area surface nor roughness were considered. Derjaguin–Muller–Toporov and Johnson–Kendall–Robert considered the influence of particle deformation on adhesion [5, 6]. Rumpf and Rabinovich implemented a roughness correction term based on the work of Hamaker. In the case of two spheres with the radius R and the Hamaker constant A_H, the van der Waals force $F_{v.d.W.}$ can be calculated as

$$F_{v.d.W.,sph} = \frac{A_H R}{12\, a_0^2}\left(\frac{1}{1 + \frac{R}{1.485\, rms}} + \frac{1}{\left(1 + \frac{1.485\, rms}{a_0}\right)^2}\right), \qquad (1)$$

with the minimal intermolecular distance a_0 and rms as the root mean square roughness of the spheres [7]. The approximation is limited to nanoscale roughnesses.

Due to the different influencing factors, such as deformation and roughness, it is difficult to predict the adhesion of particles regardless of all theoretical advances. Another difficulty is the accurate determination of the Hamaker constant, which depends on the materials in contact but also on the surrounding liquid [6]. Therefore, diverse experimental setups were developed to measure the adhesion forces. The most common methods are atomic force microscopy [8–10] (AFM), the centrifuge technique [11, 12], measurements in a stream channel [13, 14], and, recently, the vibration method [15]. With all these techniques, adhesion forces between particles and plane substrates can be measured. However, direct measurement of the adhesion between particles is difficult. Usually, the AFM is applied in that case. As in particle wall measurements, AFM considers only one particle contact at the same time. In order to obtain adhesion information of a particle collective, a modification of the centrifuge method was realized.

Magnetic Force

Magnetism can be induced by two sources: Moving electric charges within a conductor or fluctuations of the electron position on an atomic scale, causing a permanent magnetic character of the body. In the given setup, the magnetic field is created by permanent magnets. The resulting magnetic field H is conservative (no work is done on a closed loop)

$$\nabla \times \boldsymbol{H} = 0, \tag{2}$$

and the magnetic flux B is free of divergence (the field runs in closed loops)

$$\nabla \cdot \boldsymbol{B} = 0. \tag{3}$$

In a non-magnetic medium, both fields are correlated by

$$\boldsymbol{B} = \mu_0(\boldsymbol{H} + \boldsymbol{M}), \tag{4}$$

with μ_0 being the permeability of vacuum and M the magnetization of the medium [16]. The magnetic force F_M on a particle with the volume V_P and the magnetization M_P within a magnetic field is given by

$$F_M = \mu_0 V_P M_P \nabla \boldsymbol{H}. \tag{5}$$

In the case of two dipoles μ_i and μ_j, the resulting force between them is

$$F = \nabla(\mu_j \boldsymbol{B}_i) = \frac{3\mu_0}{4\pi|r|^5}\left((\mu_i \boldsymbol{r})\mu_j + (\mu_j \boldsymbol{r})\mu_i + (\mu_i\mu_j)\boldsymbol{r} - \frac{5(\mu_i \boldsymbol{r})(\mu_j \boldsymbol{r})}{|r|^2}\boldsymbol{r}\right), \tag{6}$$

with r being the vector pointing from i to j [17]. Equation (6) only applies to permanent dipoles. For induced dipoles, which influence each other, the problem is more complex. In the case of superparamagnetic particles, each particle constitutes an induced dipole and no analytical solution is available. It can only be approximated by simulation.

Electrostatics

Electrostatic forces occur between two charged bodies. If both bodies carry the same charge type, they repel, if the charge is opposite, they attract each other. Solids usually carry negative charges when they are dissolved in water. As a result, a layer with positive charged ions forms around the solids, which a diffuse ion layer follows. For two spherical particles i and j with diameters d_i and d_j at the distance r, the electrostatic potential E_{el} can be described by the Derjaguin approximation. The electrostatic force F_{el} is then given by its gradient.

$$F_{el} = \nabla E_{el} = \frac{128\pi\, N_A c_{ion} k_B T}{\kappa^2} \frac{d_i d_j}{4|r|} \gamma_i \gamma_j \exp\left(-\kappa\left(|r| - \frac{d_i + d_j}{2}\right)\right)\left(\frac{1}{|r|} + \kappa\right), \tag{7}$$

where N_A is Avogadro's constant, c_{ion} is ion concentration, k_B is Boltzmann's constant, T is temperature and κ is the Debye–Hückel parameter. The Debye–Hückel parameter is given by

$$\kappa = \left(\frac{\epsilon\, k_B T}{\sum_i \left(c_{ion} z_i^2\right) N_A e_0^2}\right)^{-\frac{1}{2}}, \tag{8}$$

with ϵ being the permittivity, z the charge of the ion i, and e_0 the electron charge [18]. The parameter γ depends on the zeta potential ζ and is described by

$$\gamma = \tanh\left(\frac{z e_0 \zeta}{4 k_B T}\right). \tag{9}$$

Materials and Methods

Centrifuge Method

The centrifuge method has been applied in a series of adhesion force measurements. For the presented work a special set-up was used based on a laboratory centrifuge with an integrated optic. With the set-up two different configurations are possible. The first is used to measure the particle wall interactions. With the second, the particle particle interactions are addressed. Figure 1 shows a picture of the sample and the sample holder with a scheme of both configurations. The set-up consists of three main components: the glass cuvette, the steel cuvette holder and the polymer centrifuge adaptor.

The measurement of the particle wall interactions is carried out in three steps. First, the particles are placed on a plane surface, the so-called substrate. Second,

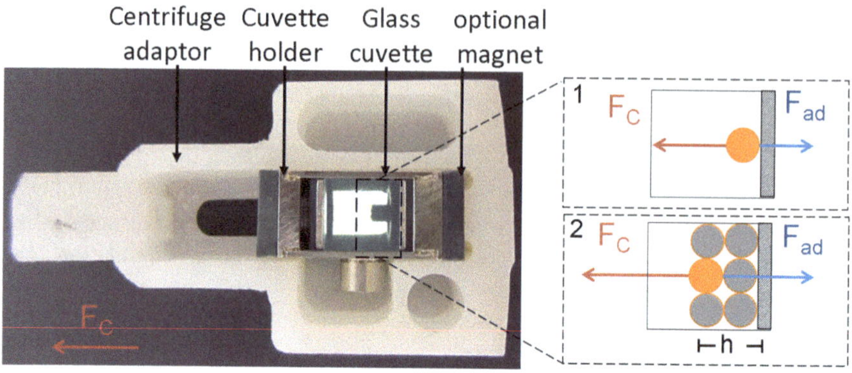

Fig. 1 Image of the sample holder and the scheme of the force balance of the centrifugal force F_C and the adhesion force F_{ad} during the measurement for a single particle (1) and a particle fixed bed with the fixed bed hight h (2)

the centrifugal force is applied and third, an image evaluation of pictures of the substrate before and after the centrifugation is performed. By gradually increasing the rotational speed, the particle adhesion rate depending on the rotational speed is obtained. For the measurement of the particle-particle interactions the disintegration of a particle fixed bed in the centrifugal field is considered. The used lab centrifuge has an integrated optic (type LumiSizer, LUM GmbH, Berlin, Germany) for transmission measurement, which makes it possible to measure the fixed bed height.

In the end, the adhesion force F_{ad} is calculated by a force balance with the centrifugal force F_C

$$F_{ad} = F_C = (\rho_P - \rho_l)V_p 4\pi^2 n^2 r, \tag{10}$$

including the particle volume V_P, the density difference $(\rho_P - \rho_l)$ of the particles and the liquid, the distance r of the particle from the axis of rotation and the rotational speed n.

The detailed set up and proceeding are reported in [18].

Image Acquisition and Analysis

The light microscope (type BZ-8000, Keyence Deutschland GmbH, Neu-Isenburg, Germany) is available for recording the particles on the substrate surface, for which transparent substrates are required. The microscope has an electronic controllable table in two directions and an integrated camera. This allows to capture an image section with several individual images and thus to enlarge it to a higher magnification. A 4× objective with a numerical aperture (NA) of 0.2 and a working distance (WD) of 20 mm and a 10× objective with an NA of 0.3 and a WD of 16 mm (both Nikon GmbH, Düsseldorf, Germany) are available for the system.

The images are evaluated for all configurations with the free available and open source software ImageJ. The individual evaluation parameters must be adapted for the respective objective, sample structure, particle size, etc. During image evaluation of the experiments in radial direction, coordinates are stored in addition to the number of particles. This makes it possible to track a particle movement as a result of centrifugation. In the orthogonal direction, however, this is not necessary because the dissolved particles move away from the surface.

Figure 2 summarizes the most important stages of image evaluation using the example of an experiment in orthogonal direction. Subfigure (a) shows the original capture. The inner bright circle is the substrate surface on which the particles are located. This is followed by a second, somehow darker circle. This is part of the PMMA cuvette that is illuminated. Subsequently, the sealing ring blocks the light. For image evaluation, the area of interest is selected for each image and the remaining image is discarded. Minor substrate and lighting irregularities can be corrected by

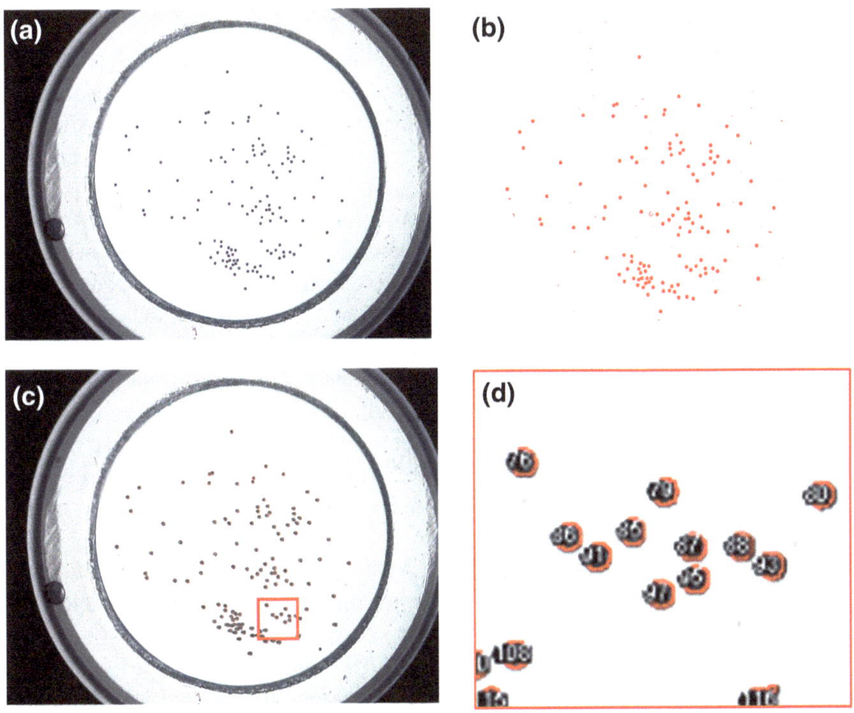

Fig. 2 Exemplary evaluation process for the configuration in normal direction for a PVC substrate with PS_{41} particles. **a** Initial image; **b** after the processing steps cutting out of working surface, removing the background (rolling = 50 light), threshold (0–231); **c** superposition of the result mask of the particle count and the original image; **d** enlarged image section of (**c**), a total of 121 particles

an adjustment of the background. The image is then converted to an 8-bit grayscale image. In this picture the color values are between 255 (white) and 0 (black). However, a black and white binary image is required for counting the particles. To be able to convert the grayscale image, a threshold is set. Pixels whose values are smaller than the threshold are assigned the value 1 and are therefore white, all pixels whose value is equal to or greater than the threshold are assigned the value 0 and are therefore black. In subfigure (b) the areas with a value greater than the threshold of 235 are marked in red for better visibility. Although all images are taken with the same settings, individual threshold settings are still required for many images.

For the centrifugation step, the sample is placed in the centrifuge with the centrifugation adapter. It is to be ensured that the substrate with the particles is closer to the axis of rotation than the detector plate so that the particles can be thrown off. The centrifugation time is 90 s including the acceleration time. The maximum acceleration time is 5 s.

Adhesive Force Measurement for Particle-Particle Contact

Measuring the adhesive force between two small particles ($d < 50\ \mu m$) is associated with certain difficulties. On the one hand the adhesive forces are smaller than in the particle-substrate contact, on the other hand the particles have to be positioned exactly. Such measurements are possible with an AFM, however, the measurements are very time-consuming.

In the course of this work, a different approach for measuring the adhesive force was developed. This is based on the observation of the adhesive forces within a particle fixed bed. For this purpose, a particle fixed bed is built up in the cuvette. In the LumiSizer, the cuvette with the particle fixed bed is installed in such a way that the centrifugal force dissolves the particle cluster. At a certain point there is a balance between the separating centrifugal force and the adhesive force of the particles in the aggregate. The position of the equilibrium height depends, among other things, on the rotational speed. The higher the rotational speed, the closer the boundary of the fixed bed approaches the wall of the cuvette. With the aid of the integrated optics, the position of the particle fixed bed boundary and thus the holding force of the uppermost particle layer can be determined.

The left side of Fig. 3 shows the cuvette with two variants of the substrate plate is shown. The glass cells used are the same as those used for radial particle-wall force measurements. The wires of the lower substrate plate are made of strongly magnetizable steel (1.4016) with a diameter of 0.6 mm and a height of 3.5 mm. On the right side of Fig. 3, the centrifuge adapter with cuvette holder, cuvette and substrates can be seen in the assembled state but without particles. The red rectangles mark the areas in which a magnet can be attached. Due to geometric limitations, two different magnet geometries are used. Cylindrical magnets can be inserted into the bottom of the cuvette, so that the magnetic field acts parallel to the centrifugal force.

Fig. 3 (Left) Cuvette; upper substrate plate is parallel to the centrifugal force for magnetic force tests, lower substrate plate with metal wires is orthogonal to the centrifugal force for magnetic tests; (right) centrifuge adapter with cuvette holder, the red rectangles mark the positions where a magnet can be attached during the tests

Fig. 4 Schematic diagram of the experimental arrangement; (left) magnetic force parallel to centrifugal force; (right) magnetic force orthogonal to centrifugal force

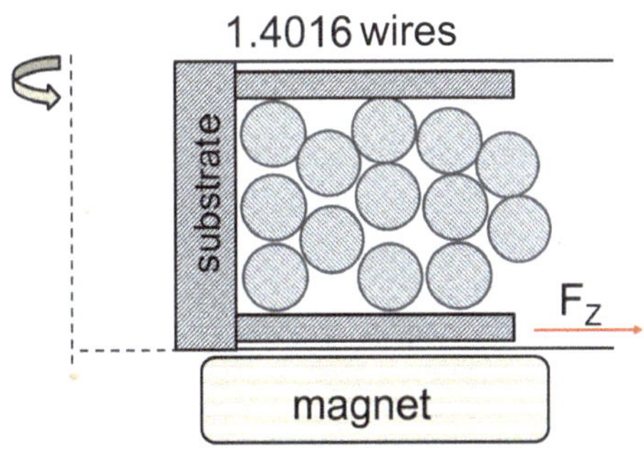

Block-shaped magnets can also be inserted into the side of the centrifuge adapter. Thus the direction of the magnetic force is orthogonal to the centrifugal force.

Figure 4 shows a schematic diagram of the experimental arrangements where the magnetic force is orthogonal to the centrifugal force.

Particle Preparation

Prior to the experiment, the particles have to be transferred from the mother solution into the salt solution. The new solutions are prepared in 2 ml safe-lock tubes (Eppendorf AG, Hamburg, Germany). The exact procedure depends on the application.

For the measurements of particle wall interactions, a few particles are necessary and 0.5 µl of the original particle solution are pipetted into 500 µl of the corresponding buffer. For the measurement of particle-particle interactions 300 µl are pipetted into 1500 µl of the corresponding buffer. The solution is then agitated for 10 s. A magnet is placed below the safe-lock tube. The magnet pulls the particles to the bottom and the supernatant is removed and replaced with new 500 μl or 1500, respectively, of the buffer. The washing step is repeated twice. Finally, the safe-lock tube is filled up to 500 μl.

Particles and Substrates

Commercial composite particles consisting of a polystyrene matrix with embedded magnetite colloids are used for the experiments. Due to their special structure, the particles are superparamagnetic, i.e. they are magnetized in an external magnetic field and are thus attracted by a magnet, but without a magnetic field they lose their magnetization again. The most important properties of the examined particles are summarized in Table 1. The particles marked with "1" are not surface modified. The particles marked with "2" have a chemical functionalization or a covalently bound protein type on the surface. All eight particle systems are based on the same basic particle system. The proteins are avidin (AV), streptavidin (SA), bovine serum albumin (BSA) and protein A (Prot A) with a load of 200 μg per gram particle.

In addition, particles with an anion exchange group (trimethylammoniumpro-pyl) (Merck, Darmstadt, Germany) were used. These particles, sold by the name MagPrep, have less defined properties than the model particles. Their primary particle size is 0.1 μm. However, they tend to agglomerate.

In the experiments, substrate refers to the adhesion partner for the particles during the investigation of the particle-substrate interactions. Polyvinyl chloride (PVC)

Table 1 Overview of the most important properties of the model particles

Particle system	DM (μm)	Density (kg/m^3)	Roughness R_q (nm)	M_S (Am^2/kg)	M_R (Am^2/kg)	ζ-Pot.* (mV)	pI (–)
[a]$PS_{19\mu m}$	18.8	1610	98	8.3	0.007		
[b]AV	10	1200	41	1.5	0.292	−35	3.9
[b]SA	10	1160	41	1.5	0.307	−45	3.3
[b]Prot A	10	1170	41	1.5	0.300	−45	3.4
[b]BSA	10	1180	41	1.5	0.290	−40	3.7
[b]COOH	10	1100	41	2.1	0.340	−45	2.7
[b]PEG-COOH	10	1160	41	2.1	0.340	−55	4.4
[b]NH_2	10	1080	41	2.1	0.340	−55	4.8
[b]PEG-NH_2	10	1080	41	2.1	0.340	−45	4.4

[a]Manufacturer microParticles GmbH, Berlin, Germany
[b]Manufacturer micromod Partikeltechnologie GmbH, Rostock, Germany
DM diameter, M_s saturation magnetization, M_R magnetic retention, *ζ-*Pot* zeta-potential at pH 7, *pI* isoelectric point

(hard, KTK Kunststofftechnik Vertriebs GmbH, Germany) and glass (soda-lime glass type, Gerhard Menzel GmbH, Brunswick, Germany) were used.

The surface roughness of the PVC substrates was modified by sandblasting and grinding. The sandblasting was carried out by glass beads with a diameter between 200 and 400 μm (apparatus type Strahlboss 140, Osu Günter Hessler GmbH & Co. KG, Bochum, Germany). An additional variation of the PVC substrate surface resulted from processing with sandpaper (grain size 100). While the surfaces of the unprocessed substrates and the substrates after sandblasting showed a very uniform surface structure, the surface after grinding was irregular and interspersed with grooves.

Results and Discussion

During the separation of proteins by means of magnetic separation, a fixed bed is formed on the magnetic matrix, which must be removed from the matrix after the impurities have been separated. Figure 5 shows a SEM image of a particle fixed bed on a wire, as it can be used in a HGMS unit.

As can be seen, the aggregate represents an extremely complex contact state for the individual particles. The object of the work was to investigate the interaction of the individual adhesive force components and their proportion of the total adhesive force both between the particles and between particles and a wall. The process particles are irregularly shaped, charged particles that are difficult to characterize. For this reason, commercial model particles, that were monodisperse and uncharged, were

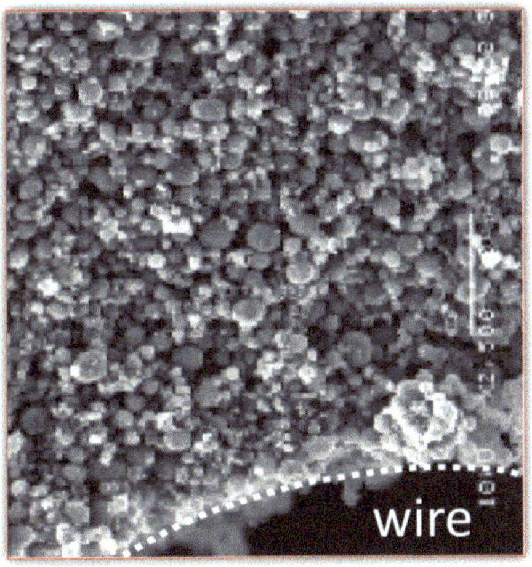

Fig. 5 SEM image of a fixed bed of magnetic particles on a magnetized wire. The diameter of the wire was about 1 mm

used in the first phase of the project. In the second step, investigations were carried out to determine the influence of bound proteins on the particle surface. This was also done with model particles on which, unlike the process particles, the proteins are covalently bound to the surface, which also allows the use of different buffer systems and salts. In the last step, the adhesion forces of process particles with different proteins were investigated. In the case of process particles, protein binding takes place via ion bonds, which limits the experiments to the use of a specific buffer system.

Influence of Surface Roughness

The adhesive force of particles is greatly reduced by surface roughness. The reason for this phenomenon is the short range of the operating van der Waals forces. Rabinovich extended the Hamaker model for the calculation of the adhesion force by a correction term accounting for roughness effects [7]. When regarding HGMS, an additional magnetic force acts in addition to the adhesive force. It had to be examined whether the magnetic force had an influence on the adhesive force. Figure 6 shows the adhesion forces of $PS_{19\mu m}$ particles in water on PVC of various roughness without and with a neodymium permanent magnet type N52, the strongest type of permanent magnet commercially available. The magnet generates a magnetic field of 0.5 T on the magnetic surface. At the distance at which the particles are located during the holding force measurement (1 mm), the magnet exerts a force of 3.3 nN on an individual particle. The lines show the adhesion force curve without magnet. They represent the reference values of the system. The symbols are the measured adhesion forces with magnet in place from which the magnetic force was subtracted. The values are thus corrected by the additional force and reflect the adhesive force [19].

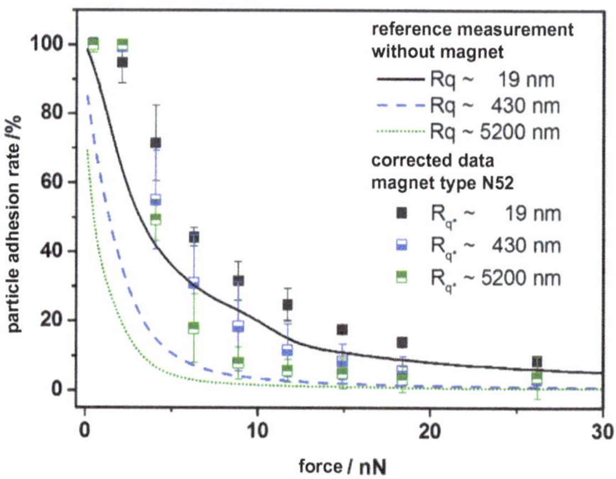

Fig. 6 Adhesion curves of $PS_{19\mu m}$ particles in water on PVC of various roughness; lines: adhesive forces without magnet; symbols: adhesive forces with magnet type N52 from which magnetic force is subtracted

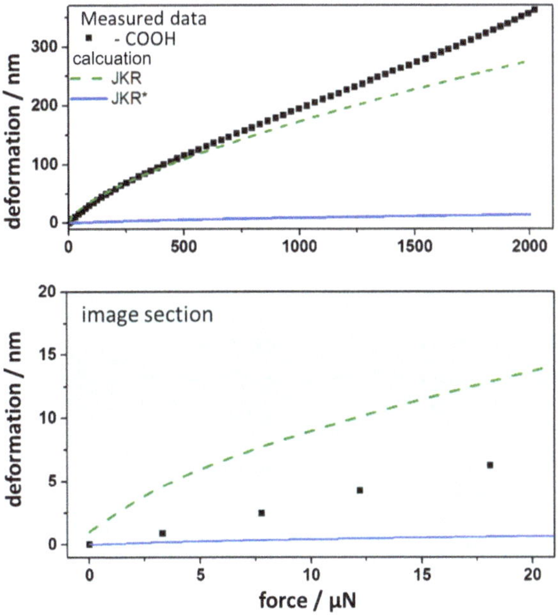

Fig. 7 Deformation curve of COOH particles d_{50} 9.8 µm on PVC $R_q = 19$ nm

The (corrected) adhesion values with magnet in place are above the reference values. The reason for this are deformations in the contact area. Nano indentation measurements could be carried out on selected particle systems. Figure 7 shows the measurement data for a COOH particle. Furthermore, the calculated deformation values based on the JKR and a modified JKR* model are plotted. In the lower part of the figure, the enlarged section of the coordinate origin is shown. Up to 600 µN, the JKR model almost completely matches the measured data. In contrast, the JKR* model predicts far too little deformation. In this force range, the overall geometry of the particle dominates and the roughness plays only a subordinate role. However, at very low deformation forces, the JKR* calculation approaches the measured date while the JKR model overestimates the deformation. In this area roughness is no longer negligible. The application of the JKR* model in the deformation calculation based on the magnetic forces is therefore justified. For the JKR* model, the radius correction shown by Rabinovich was used in the JKR model. The adhesive force F_{JKR*} thus results in

$$F_{JKR^*} = 3\gamma\pi R\left(\frac{1}{1 + \frac{R}{1.485\,R_q}} + \frac{1}{\left(1 + \frac{1.485\,R_q}{\mathrm{x}}\right)^2}\right), \tag{11}$$

with the surface energy γ, the particle radius R, the roughness R_q and the separation distance x.

Influence of Protein Loading

The particle systems with bound protein represent an approximation of the actual process conditions in which the type of protein and loading state of the particles vary. Figure 8 shows the adhesive force curves of the eight functionalized particles on glass in ultrapure water and in a 0.1 M citric acid solution whose pH was adjusted to the isoelectric point (pI) of the respective particle system. Although the particle systems are based on the same material, a wide distribution of the adhesive forces occurs. While the adhesive forces of SA, BSA and Prot A are above the measuring range in ultrapure water, those of the remaining particles are below. In the buffer solution, the particles are at the isoelectric point and thus have no net charge. The adhesive forces are therefore increased in the buffer solution. Electrostatic interactions have

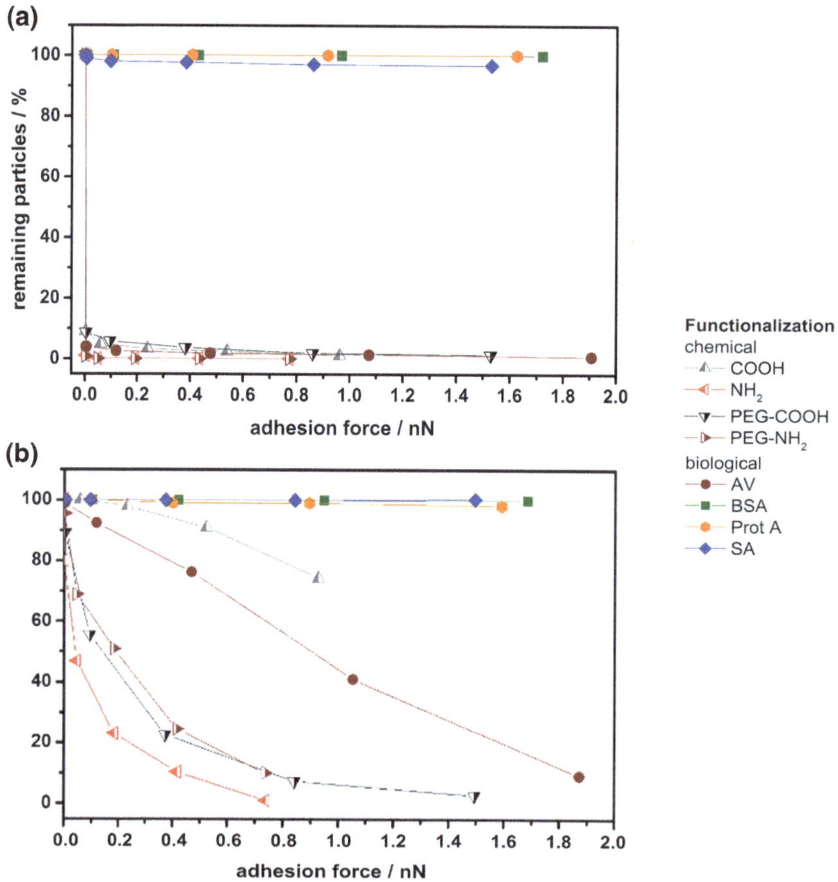

Fig. 8 Adhesion curves of the functionalized particles on glass in: **a** ultrapure water; **b** 0.1 M citric acid solution at the pI

an influence on the functionalized particles in ultrapure water. However, there is no direct correlation to the surface charges. All particles have a zeta potential of −35 to −55 mV.

Not only the zeta potential but also the hydrophobicity of the surface of functionalized particles is due to change. This applies in particular to particles with bound protein. Proteins are composed of hydrophobic and hydrophilic amino acids. The majority of the hydrophobic amino acids are accumulated on the inside of the protein and form a compact core [2, 3]. However, not all hydrophobic areas are located in the protein core but partly also on the protein surface [4]. In addition to the van der Waals force and electrostatic repulsion, hydrophobic interactions can also have an influence on the adhesive force. Salts can strengthen hydrophobic interactions (kosmotropic) or weaken (chaotropic). For AV and BSA the influence of the salt type was tested using the salts sodium dihydrogen phosphate (NaH_2SO_4; kosmotropic), sodium chloride (NaCl, neutral) and sodium thiocyanate (NaSCN, chaotropic) [5].

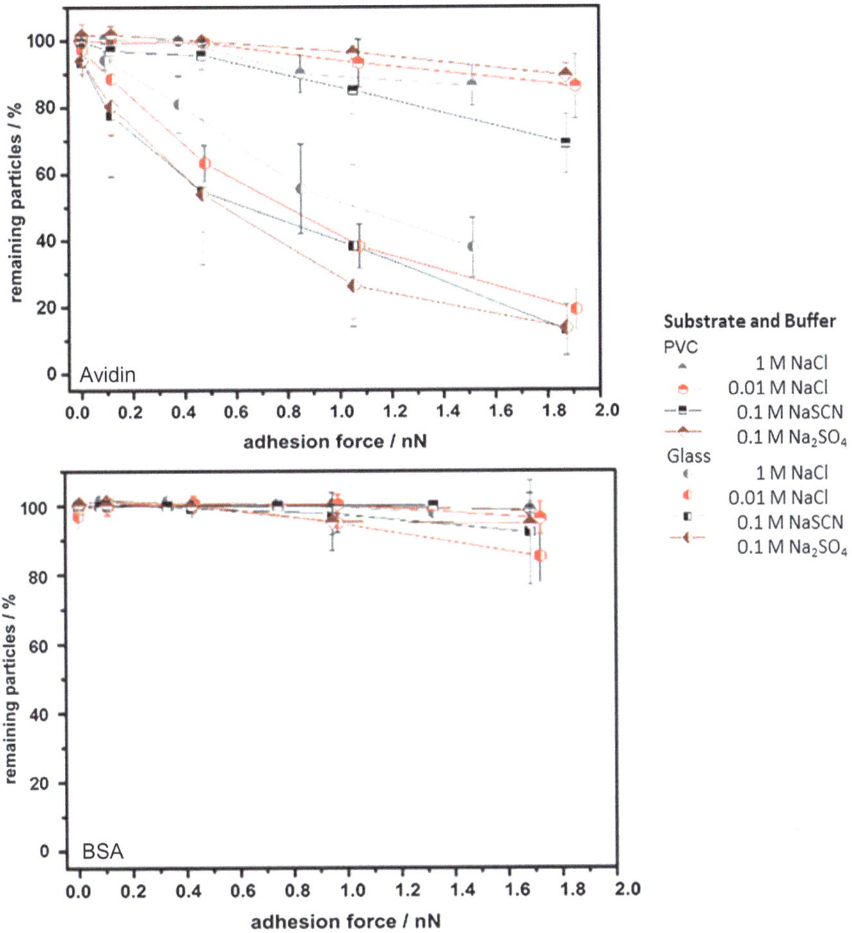

Fig. 9 Adhesion curves of particles on PVC and glass in different media: top—AV; bottom—BSA

Figure 9 shows the adhesive force of AV (top) and BSA (bottom) on PVC and on glass in the various salt solutions.

The influence of the used substrate and type of protein dominates over the influence of the salt type, as shown by the adhesive forces of AV. The maximum salt concentration of 1 mole per litre for NaCl and 0.1 mol per litre for NaH_2SO_4 and NaSCN must be taken into account. A higher salt concentration may have revealed a more distinct influence. However, the selected salt concentrations correspond to the range as it occurs in fermentation broths and are therefore process-relevant. The data also shows that just 0.01 mol per liter of NaCl are sufficient to prevent electrostatic repulsion, so that it is negligible in "real" processes.

In the second experimental series, the surfaces of the glass substrates were modified. The glass substrates were hydrophobized with 1H,1H,2H,2H,2H-

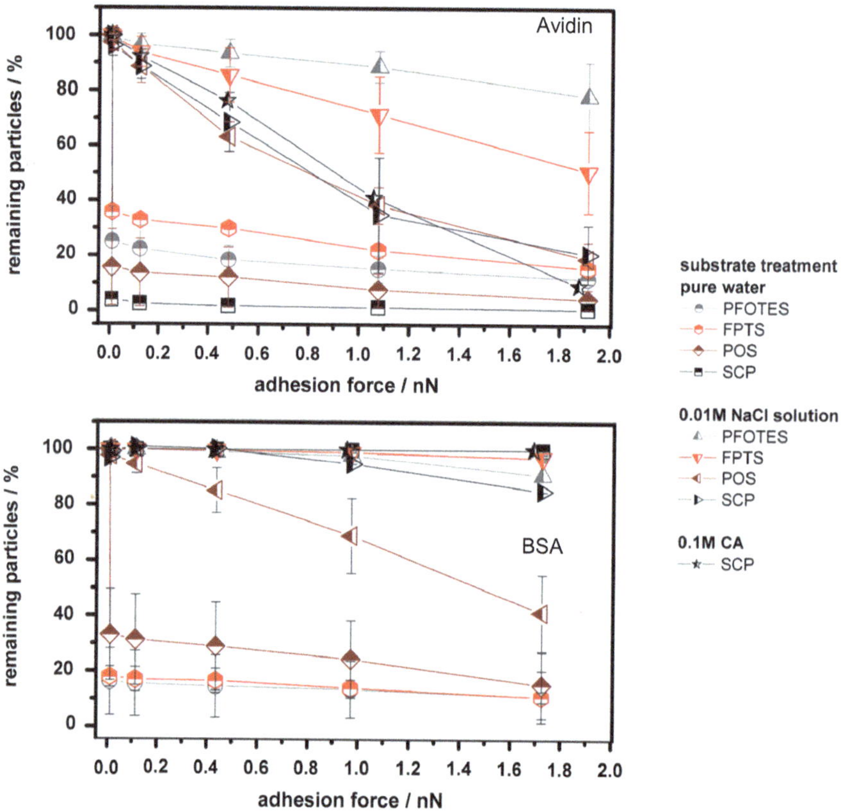

Fig. 10 Adhesion curves of functionalized particles on different glass substrates and in different media. Nomenclature of glass processing is as follows: PFOTES, FPTS—chemical surface functionalization with 1H,1H,2H,2H-perfluorooctyltriethoxysilanes or trichloro(3,3,3-trifluoropropyl)silanes; purified—non-functionalized glass surface purified with peroxomonosulphuric acid; untreated—ethanol purified glass surface; top—AV; bottom—BSA

perfluorooctyltriethoxysilanes (PFOTES, α = 109°) and trichloro(3,3,3-trifluoropropyl)silanes (FPTS, α = 92°) and purified with peroxomonosulphuric acid (POS, α = 15°), which increased the hydrophobicity compared to the untreated (with ethanol purified) glass (UBG, α = 54°). Figure 10, top, shows the adhesive force curves for AV on the four glass types in a 0.01 NaCl solution and in ultrapure water. For comparison, the adhesion curve is also plotted for the experiments in 0.1 M citric acid buffer (CA).

Both in ultrapure water and in the NaCl solution, the particles on the chemically functionalized—and thus hydrophobic—glasses exhibit higher adhesive forces than on the non-functionalized glasses. In NaCl solution, PFOFTES particles have the highest and FPTS particles the second highest adhesive forces. In ultrapure water, the order is reversed. Electrostatic effects may play a role, which are suppressed by the addition of NaCl.

Figure 10, bottom, shows the same data for particles functionalized with BSA. The BSA particles already had an adhesive force above the measuring range on the untreated glass in NaCl, so that any increase due to chemical functionalization cannot be measured. On the etched glass, the adhesive force of the particles decreased.

In a magnetic filter, however, not only the interactions between the particles and the magnetic matrix, but also between the particles themselves are crucial. The characterization of the interactions was carried out by determining the fixed bed height as a function of the centrifugal force. Figure 11 shows the fixed bed disintegration curves of PEG-COOH. The curves shown are those whose initial height matches best. Symbols of the same shape and color each represent the same test conditions.

Without a magnet, the adhesive forces between the particles are too low to be measured with the implemented method. This applies regardless of the used buffer

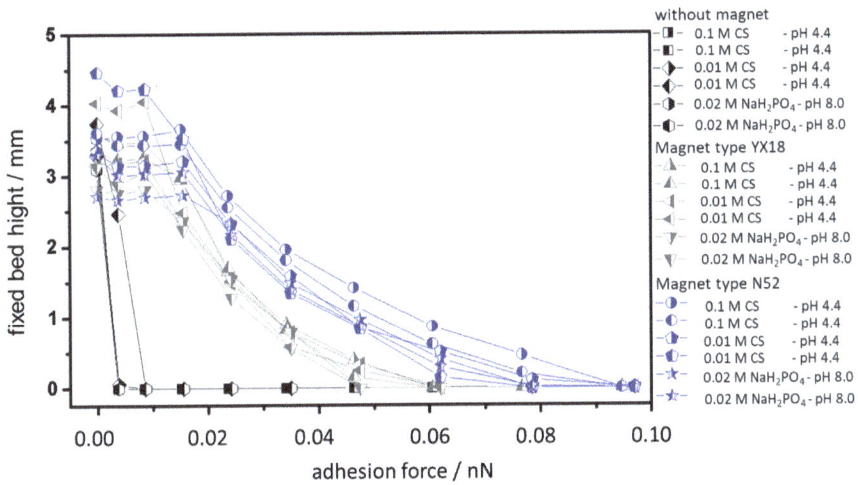

Fig. 11 Fixed bed disintegration curves of PEG-COOH in different buffers without magnet and with magnets YX18 and N52

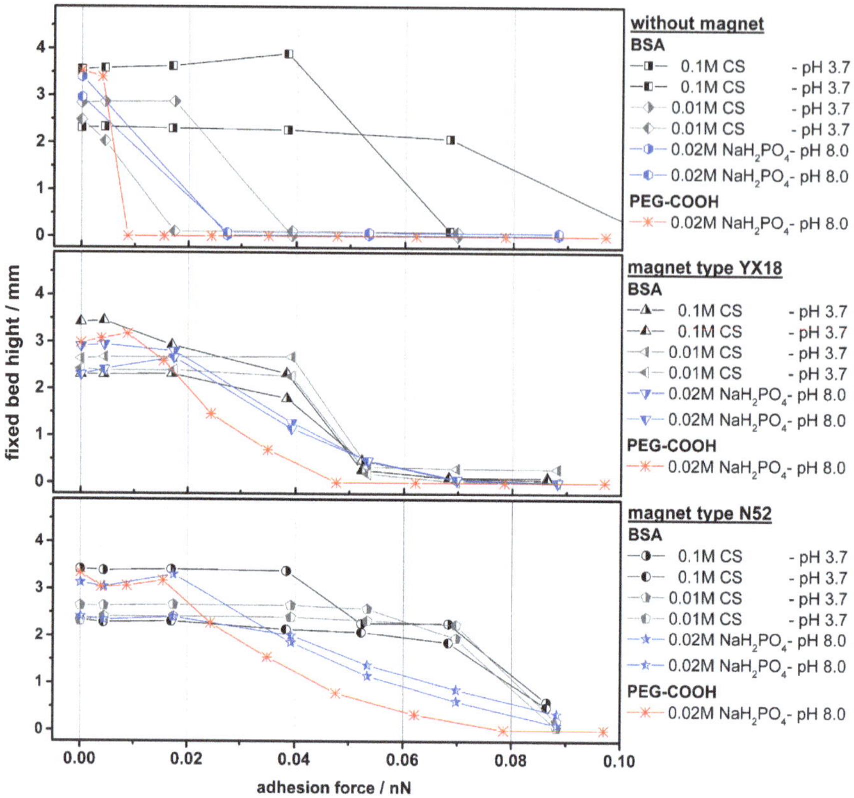

Fig. 12 Fixed bed disintegration for BSA in different buffers (CS = citric acid) without magnet and with magnets YX18 and N52; for comparison a curve of the unloaded base particles PEG-COOH is shown

system. Generally, the adhesive force increases with increasing magnetic strength. For the unloaded particles, the figure also shows that the fixed bed height is reproducible and independent of the buffer composition. In this case, the magnetic force plays the dominant role.

The fixed bed disintegration curves for BSA in different buffers are shown together with a reference curve of PEG-COOG in Fig. 12. For visualization, the measured curves are separated according to magnetic field strength and plotted one below the other. Compared to Fig. 11 the different visualization already shows that the particles behave differently. In contrast to the unloaded particles, the adhesive forces of the particles loaded with BSA were higher in the citric acid buffer than in the sodium dihydrogen phosphate buffer. In addition, the measurements with and without magnets cannot be so distinguished from each other as is the case without BSA. This can be seen particularly clearly in the measurements with the 0.1 M citric acid buffer. There the particles reach adhesive forces, as they also occurred during the measurements with the strong N52 magnets.

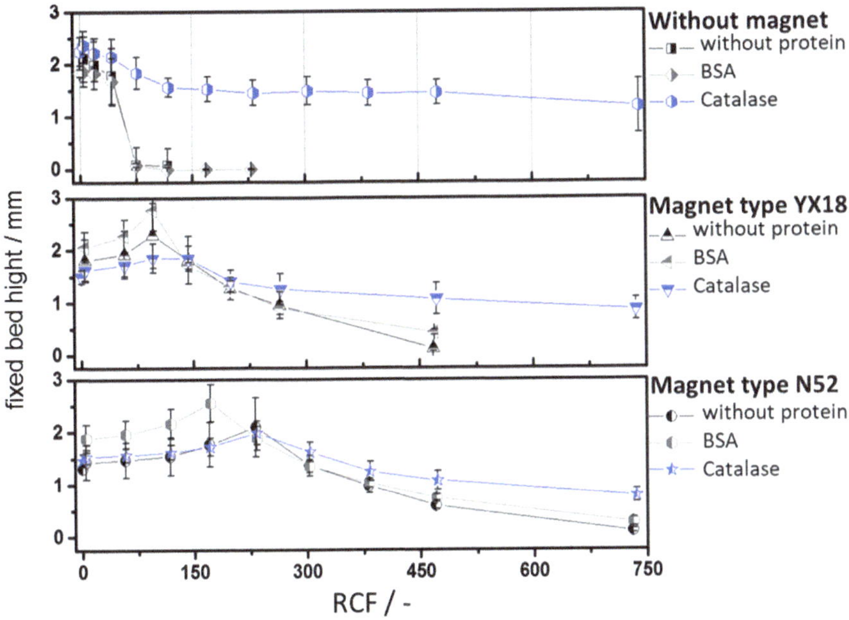

Fig. 13 Fixed bed disintegration curves of MagPrep particles without bound proteins, with 0.328 $g_{BSA}/g_{particles}$ and with 0.235 $g_{catalase}/g_{particles}$ in a 0.02 M Na_2PO_4 buffer each measured without magnet and with the magnet types YX18 and N52

Overall, the adhesive forces between the PS_{BSA} particles are higher than the forces between the initial particles without BSA. One reason for the increased adhesive force could lie in the hydrophobic character of BSA.

In the last step, MagPrep particles were used. These have an anion exchanger group that bind to the BSA and catalase in a 0.02 M NaH_2PO_4 buffer at pH 8. The particle load was 0.328 $g_{BSA}/g_{particles}$ and 0.235 $g_{catalase}/g_{particles}$. Figure 13 shows the disintegration curves of MagPrep particles without adsorbed proteins and with BSA. Furthermore, the results are shown for the cases without magnet and with the magnet types YX18 and N52. In this case, the fixed bed height data is plotted over the relative centrifugal acceleration. The data required for calculating the centrifugal force, such as particle size and particle density, cannot be clearly determined for MagPrep. Due to measurements dependent on the RCF (relative centrifugal force) no direct differences in the adhesive force of the particles can be generated, however the graph shows a change in the separation behavior for the same centrifugal accelerations or speeds and is therefore process-relevant. When looking at the figure, the different behavior of the fixed bed disintegration curve of Mag-$Prep_{catalase}$ without magnet is immediately noticeable. While the fixed beds of the other two particle systems already completely dissolved at RCF = 75, the fixed bed height of the $MagPrep_{catalase}$ decreases to 1.5 mm till RCF = 100 and remains constant up to RCF = 750. Already during handling, i.e. during sample preparation and filling of the cuvettes, the particles proved to be

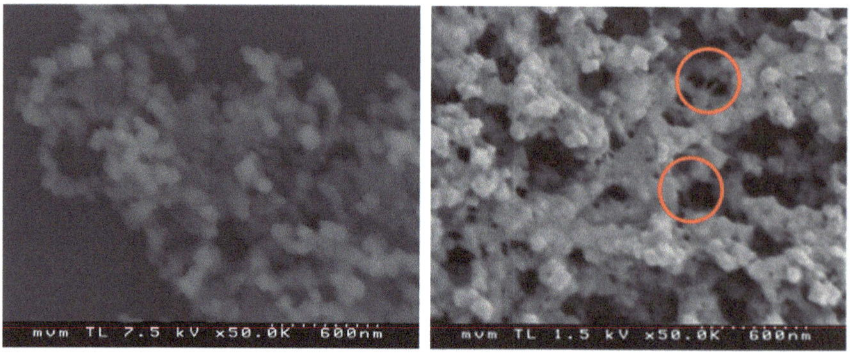

Fig. 14 SEM image of MagPrep particles without (left) and with catalase (right)

"sticky" and difficult to process. The curves of $\text{MagPrep}_{\text{Ca-talase}}$ particles are also higher than those of MagPrep and $\text{MagPrep}_{\text{BSA}}$ when magnets are used, but the differences are smaller. This behavior may be due to the formation of a network between the catalase molecules. In order to create as constant initial conditions as possible, the fixed beds are compacted with RCF = 280 before the experiment, whereby the upper particle layers experience a smaller pressure than the lower layers. The higher pressure causes the lower layers to come into stronger contact, so that stronger interactions can develop. However, the curves also show that the final height of the $\text{MagPrep}_{\text{catalase}}$ particle fixed bed decreases with increasing magnetic strength. This indicates that the particles are trying to align themselves along the magnetic field, partially breaking up the network structure.

Figure 14 shows SEM images of the unloaded particles and those loaded with catalase. The surface of $\text{MagPrep}_{\text{catalase}}$ particles looks "clogged" in comparison. Also, occasionally long thread-like structures are to be recognized, which cannot be seen in the case of unloaded particles. However, due to the small image section, these images can only be evaluated as an indication. The drying step necessary for preparing the SEM probes is particularly critical, during which the observed structures could also develop.

Integration of a Microscope into a Centrifuge for Direct Adhesion Force Measurements

In the above described investigations the evaluation of the adhesion force was based on image analysis methods. This setup does not allow a direct observation of the sample during the centrifugation process because the samples can just be analyzed before and after the experiment. Thus, no information of the particle motion is available during the centrifugation. For this reason a centrifuge with an integrated microscope was designed and built. The goal is to observe directly the adhesion of single particles for

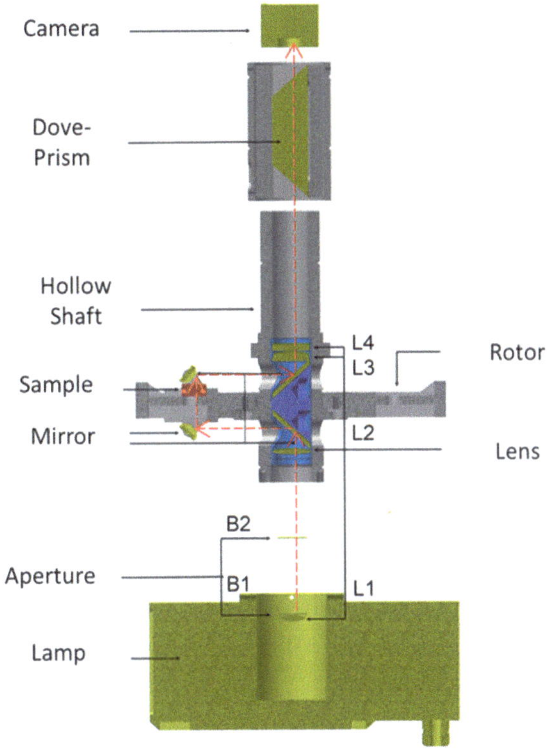

Fig. 15 Half section of the simplified draft of the optical centrifuge. L1–L4 are lenses, B1 und B2 are apertures, dashed line gives the optical path

a radial load and the behaviour of fixed beds in the centrifugal field. The requirement for the new set-up was to build a combination of a centrifuge and a microscope. The idea was to obtain a continuous image of a rotating sample with a fixed camera. In addition, the image should have a static orientation, in other words the image of the sample should appear fixed.

In Fig. 15 a scheme of the hollow shaft with the optical components and the rotor is shown. The set-up works if the optical path coincides with the axis of rotation. Therefore, the lamp and the camera had to be placed centrally to the hollow shaft. Four mirrors form a kind of double-periscope. They direct the light in the hollow shaft, through the sample and back into the hollow shaft. In order to achieve and enlarge the image, an objective lens was integrated. For details of the optical components and the design see [20].

The unique feature of the set-up is the derotation of the image. There are two possibilities to obtain a stationary image. Firstly, the camera should rotate at the same rotational speed. This could damage the camera. The second possibility is a derotation of the image with a prism. A so called 'dove' prism meets the optical and mechanical requirements. The dove prism derotates the image, if its rate of rotation is half the rate of rotation of the sample. This condition can be fulfilled mechanically.

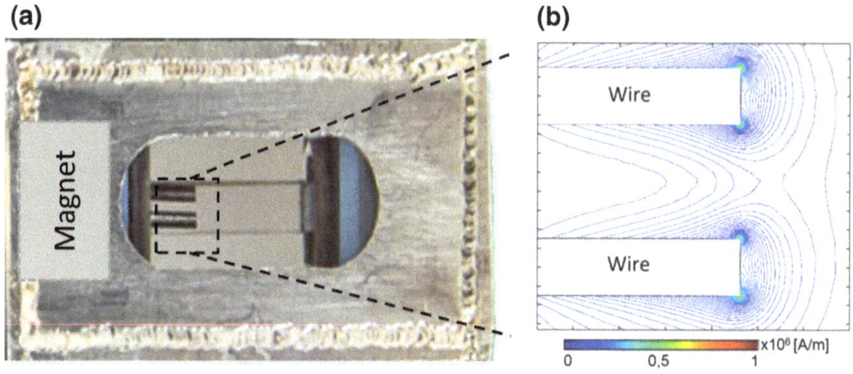

Fig. 16 **a** Image of the sample holder. Inside the cuvette magnetizable steel wires are located. **b** Cutout of the FEM simulation of the magnetic field; the simulation shows magnetic field lines for the field strength H around the wires

The framework condition for the design was to achieve a relative centrifugal speed of 2000, a field of view of three times two square millimeters and a sufficient resolution to observe particles with a diameter smaller than hundred micrometers. To prevent oscillations a massive concrete foundation including steel columns as fix points for the centrifuge was designed and connected to an electric drive.

Figure 16a shows an image of the sample holder. Inside the glass cuvette there are two steel wires which are used for a particle bulk characterization. In Fig. 16b the results of a FEM simulation of the magnetic field around the wires is sketched. The isolines show the magnetic field strength distribution with the highest density around the tips of the wires. Here the highest magnetic forces are expected.

Figure 17 gives two grayscale pictures of a glass substrate with polystyrene particles for a diameter of 41.1 μm in a NaCl solution. The black horizontal extending lines are the edges of the substrate. The centrifugal force acts from the left to the right. The comparison of the two pictures shows, that it is possible to track the movement of particles during centrifugation. Exemplarily, the position of two particles that moved tangentially to the substrate are marked with black circles. After some image corrections to depress the jittering of the video it was possible to improve the optical resolution compared to the requirements of the mechanical stability and an absolute maximum rotational speed of 1500 rpm was reached.

Figure 18 shows the adhesion curves of polystyrene particles with two different diameters on a glass substrate. Here, the centrifugal force acts perpendicular to the adhesion force. The percentage of remaining particles on the substrate surface is plotted against the centrifugal force and compared with the laboratory centrifuge (Lumisizer, LUM GmbH, Berlin). For the higher diameter the minimal rotation speed of the laboratory centrifuge is already too high and almost all particles detach. The data for the small particle sizes show a good agreement for both centrifuges. In the case of the optical centrifuge, the data are obtained with one experiment and lower separation forces could be analyzed.

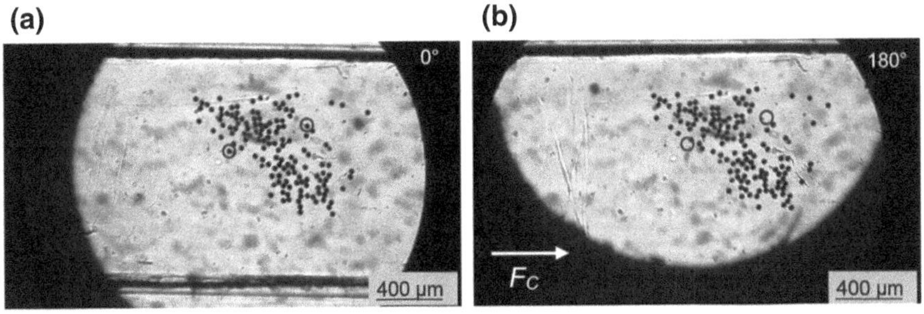

Fig. 17 Pictures of particles on a substrate in the optical centrifuge. **a** Rotor arm in initial position at an angle of 0°. **b** Snapshot during centrifugation at 100 rpm with rotor arm turned around 180°. Black circles mark moved particles for a tangential detachment

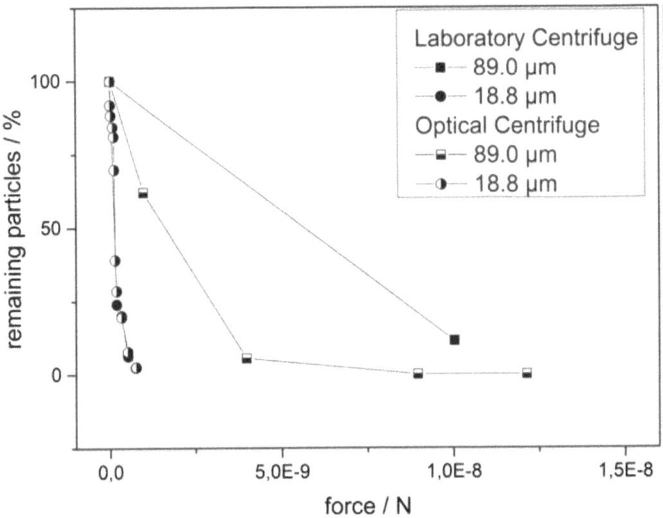

Fig. 18 Separation curve of polystyrene particles with two different diameters on glass for a tangential detachment direction; data comparison for the optical centrifuge with a laboratory centrifuge

For continuously working high gradient magnetic separation devices, the interactions between the particles are of interest [3]. The optical centrifuge provides a possibility to consider these interactions which are normally only accessible by simulations. Figure 19 describes an image sequence of a fixed bed disintegration of magnetizable polystrene particles in a NaCl solution. In order to mimic the process, two magnetizable wires are located in the sample volume. A magnet is mounted below the sample (see Fig. 16). With the given geometry, the magnetic field and its gradients are strongest at the end of the wires. This is the reason why the particles accumulate and form a bridge-like structure between the wire tips. At low centrifugation forces the bridge stays stable and no particles move away. At a RCF of 24 first particles start to detach at the center of the bridge. As a result the bridge tapers

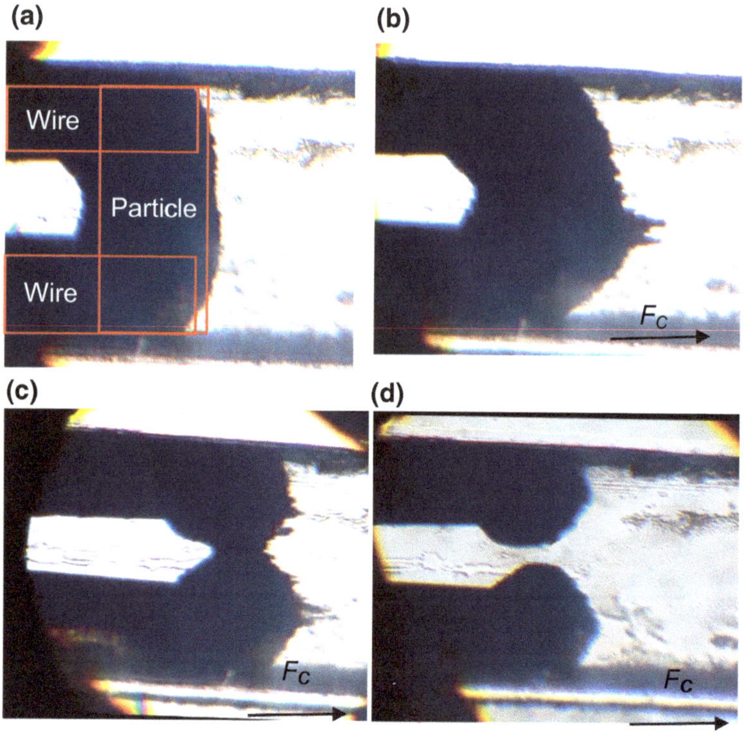

Fig. 19 Image sequence of a fixed bed disintegration with polystyrene particles ($d = 18.8\ \mu m$) in 0.01 M NaCl; **a** start of the experiment, formation of a bridge between the two wire tips; **b** RCF = 24, deformation of the bridge and particle separation in the middle of the bridge; **c** RCF = 95, the bridge tapers in the center and particles move from the bridge center to the wire tips where the separate; **d** RCF = 95, the force balance is reached at the steady state; in the images (**b**–**d**) the centrifugal force acts from left to the right

in the center with the area of the highest magnetic forces lies in the front of the bridge center. Particles that detach in the middle of the bridge are attracted by the end of the wires and it can be observed that the particles move over the surface of the bridge towards the end of the wires. When a critical thickness of the particle layer is reached, the centrifugal force exceeds the magnetic force again and the particles detach. Figure 19d shows the final stable state. The optical centrifuge delivers a tool for the investigation of particle interactions in a fixed bed. The only condition for the measurement is, that the interaction between the particles exceeds the gravitational force.

Conclusions

Adhesion force measurements have been performed for superparamagnetic particles depending on magnetic field strength. It was studied the material influence regarding particles with different sizes and substrates concerning different roughness and surface modification. In the case where the magnetic force is in the same magnitude as the van der Waals force it was shown that the magnetic forces have an enhancing effect on the van der Waals forces, according to the deformation in the contact area. The functionalization of the particle surface with proteins has a significant effect on the adhesion force. The influence of the functionalization and the substrate is more pregnant than the concentration of salt in the surrounding suspension.

Bulk adhesion of functionalized particles used for protein purification was measured. Particle sediments were built up and the disintegration process of the sediment was observed. The magnetic force was significantly stronger than the other intrinsic interparticle forces, such as van der Waals forces and electrostatic interactions. Depending on the particle magnetization, disintegration took place in one step for small magnetization and in two steps, when the sediment expanded before particles were removed. For with proteins functionalized magnetic particles, the necessary separation force was found to increase in the presence of protein on the particle surface.

A centrifuge with an integrated optic was designed. The arrangement of the optical components corresponds to a microscope configuration. The use of a mechanical image derotation device allows the application of a non-rotating camera. This gives the ability to observe the enlarged image of the sample during the centrifugation continuously and insitu.

References

1. Raj, K., Moskowitz, R.: Commercial applications of ferrofluids. J. Magn. Magn. Mater. **85**, 233–245 (1990)
2. Lubbe, A.S., Bergemann, C., Riess, H., Schriever, F., Reichhardt, P., Possinger, K., Matthias, M., Dorken, B., Herrmann, F., Gurtler, R., Hohenberger, P., Haas, N., Sohr, R., Sander, B., Lemke, A.J., Ohlendorf, D., Huhnt, W., Huhn, D.: Clinical experiences with magnetic drug targeting: a phase I stud with 4′-epidoxorubicin in 14 patients with advanced solid tumors. Cancer Res. **56**, 4686–4693 (1996)
3. Lindner, J., Wagner, K., Eichholz, C., Nirschl, H.: Efficiency optimization and prediction in high-gradient magnetic centrifugation. Chem. Eng. Technol. **33**, 1315–1320 (2010)
4. Lindner, J., Menzel, K., Nirschl, H.: Simulation of magnetic suspensions for HGMS using CFD, FEM and DEM modeling. Comput. Chem. Eng. **54**, 111–121 (2013)
5. Isrealachvili, J.N.: Intermolecular and Surface Forces, 3 edn. Academic Press (2011)
6. Hamaker, H.C.: The London-van der Waals attraction between spherical particles. Phys. A: Stat. Mech. Appl. **4**, 1058–1072 (1937)
7. Rabinovich, Y.I., Adler, J.J., Ata, A., Singh, R.K., Moudgil, B.M.: Adhesion between nanoscale rough surfaces: I. Role of asperity geometry. J. Colloid Interface Sci. **232**, 10–16 (2000)
8. Dos Santos Ferreira, O., Gelinck, E., de Graaf, D., Firscher, H.: Adhesion experiments using an AFM-parameters of influence. Appl. Surf. Sci. **257**, 48–55 (2010)

9. Jones, R.: Inter-particle forces in cohesive powders studied by AFM: effects of relative humidity, particle size and wall adhesion. Powder Technol. **132**, 196–210 (2003)
10. Vorburger, T.V., Dagata, J.A., Wilkening, G., Lizuka, K., Thwaite, E.G., Lonardo, P.: Industrial uses of STM and AFM. CIRP Ann. Manuf. Technol. **46**, 597–620 (1997)
11. Felicetti, M.A., Coury, J.R., Aguiar, M.L.: Centrifuge study on influence of particle size, compression and substratum on particle-surface adhesion force. In: Materials Science Forum, vol. 591–593, pp. 347–351 (2008)
12. Salazar-Banda, G.R., Felicetti, M.A., Goncalves, J.A.S., Coury, J.R., Aguiar, M.I.: Determination of the adhesion force between particles and a flat surface, using centrifuge technque. Powder Technol. **173**, 107–117 (2007)
13. Mercier-Bonin, M., Dehouche, A., Morchain, J., Schmitz, P.: Orientation and detachment dynamics of Bacillus spores from stainless steel under controlled shear flow: modelling of the adhesion force. Int. J. Food Microbiol. **146**, 182–191 (2011)
14. Dong, C., Lei, X.X.: Biomechanics of cell rolling: shear flow, cell surface adhesion, and cell deformability. J. Biomech. **33**, 35–43 (2005)
15. Ripperger, S., Hein, K.: Measurement of adhesion forces in air with the vibration method. China Particuol. **3**, 3–9 (2005)
16. Coey, J.M.D.: Magnetism and Magnetic Material. Cambridge University Press 2011
17. Lindner, J.: On Continuous Megnetically Enhanced Centrifugation in Large Scale Downstream Processing of Bioproducts. Dr. Hut Verlag (2014)
18. Knoll, J., Knott, S., Nirschl, H.: Characterization of the adhesion force between magnetic microscale particles and the influence of surface-bound protein. Powder Technol. **283**, 163–170 (2015)
19. Knoll, J., Nirschl, H.: Influence of the magnetic force on the van der Waals force of superparamagnetic composite particles. Powder Technol. **259**, 30–36 (2014)
20. Knoll, J., Knott, S., Nirschl, H.: Integration of a microscope into a centrifuge for adhesion force measurement of particles. Powder Technol. **305**, 147–155 (2017)

Deformation and Friction at the Microscale—From Model Experiments to Process Characterization

Wolfgang Peukert and Stefan Romeis

One of the grand challenges in Particle Technology is how to characterize the contact mechanisms and contact forces between particles. The mechanics of particles across all length scales down to the nanoscale are highly relevant for many applications including adhesion, friction, powder flow, comminution and tribology. Particular interesting new fields are additive manufacturing and 3D printing which "promise" a revolution in industrial manufacturing based on powder technologies. Mechanical particle properties are also the key parameters in discrete element models (DEM) which have been developed during the last decades and gradually reach "predictive power". In order to make DEM truly predictive, intrinsic material properties such as Young's modulus, hardness, plastic deformation or fracture toughness of the particles must be known quantitatively. However, measured mechanical particle properties (and their distributions) are only known in a few rare cases.

Elastic and inelastic particle properties have been measured systematically in a novel single particle compression device installed within a scanning electron microscope (SEM). Force-displacement information and electron images of deformed and broken particles can be obtained simultaneously. Quantitative data of intrinsic particle properties and their respective distributions are derived with high statistical relevance from force-displacement curves of single particles. At high strain, frictional forces become relevant which have been measured by frictional atomic force microscopy (AFM) and have been included in finite element (FE) models for the quantitative description of force-displacement curves. Particular challenging aspects of friction are the molecular effects between two surfaces in contact. Nonlinear optical spectroscopy was employed to shed light on the deformation of self-assembled monolayers during friction.

W. Peukert (✉) · S. Romeis (✉)
Institute of Particle Technology, Friedrich-Alexander-Universität Erlangen-Nürnberg, Erlangen, Germany
e-mail: wolfgang.peukert@fau.de

S. Romeis
e-mail: stefan.romeis@fau.de

S. Antonyuk (ed.), *Particles in Contact*,
https://doi.org/10.1007/978-3-030-15899-6_13

A quantitative model to describe the force-displacement of well-defined spherical particles and their shape evolution at different strains can be used to obtain so far unknown distributions of stress energy and stress number distributions in mills and other solids processing devices. Spherical ductile particles are stressed in the mill. Their altered shape allows deriving the absorbed energy distribution whereas the stress number distribution is obtained from a kinetic model which takes the proportion of the stressed particles into account. This way, fundamental studies of the mechanical behavior of single particles are directly connected to the analysis of the stressing conditions in technical mills.

Introduction

Developing physical models for systems where particles are in contact is a grand challenge of particle technology. A prerequisite for designing, modeling and understanding particulate processes is the availability of predictive models of individual unit operations or whole systems. Predicting energy consumption, throughput and the dynamic evolution of particle properties is of paramount importance. In the last years great progress has been made in all fields of particle technology. A systematic determination and evaluation of particle properties beyond size is, however, just beginning to develop. Especially information on surface, structural and mechanical properties of complex particle systems is still rare.

Classical unit operations of powder processing like mills, tableting machines, conveying systems, or fluidized beds exert forces on the processed particulate systems. From a unifying perspective stresses can be categorized according to the particular stress type, the stress energy and stress number and the related distributions [1]. In general, unit operation functions can be defined as the functional dependency between the operational parameters and the resulting stresses applied to the particles. Particles can be stressed by impact, compression, shear and their combinations. These complex functions might be determined by a proper modeling of the underlying two-phase flows, e.g. by means of computational fluid dynamics (CFD) and particle trajectory simulation coupled to DEM modeling. For the aforesaid simulations mechanical and tribological particle properties must be known. For microparticles, systematic studies of these properties are, however, widely absent. Furthermore, there is no comprehensive understanding of the contact mechanics, especially for small inter-particle distances at the lower nanoscale.

Particles react to applied stresses according to their material properties. The corresponding functional relationship is called material function. The material function depends on intrinsic material parameters and particle properties such as size and shape.

In this contribution we focus on the compression of spherical particles for small normal compression rates. In particular, we present studies on size dependent deformation of oxide particles and measure, for the first time, the distributions of the intrinsic materials parameters for statistically meaningful ensembles. The influence of contact friction on the deformation behavior of metal particles is studied. Novel techniques to study tribology and stress induced structural changes of ultrathin self-assembled adsorbate layers between two solid bodies are presented. Moreover, we present a novel technique to characterize the stressing conditions inside technical apparatuses: Our innovative approach is based on the deformation of mechanically well-characterized ductile particles.

Deformation of Single Particles Under Normal Loads

The mechanical properties of particles are of great practical importance. Our current understanding is, however, hampered by difficulties associated with testing of a sufficient number of individual particles: The particles have to be localized and stressed in a well-defined reproducible manner. At the same time small forces and deformations have to be recorded. By electron microscopy supported testing rigs, the aforesaid challenges can be overcome. So far, however, only a few in situ studies deal with the mechanical testing of individual (sub-) micron particles during uniaxial compression. Studies of other geometries (mostly pillars) are far more abundant and have highlighted the intriguing mechanical properties at the micro- and the nanoscale.

SEM Supported Microparticle Compression

For particle compression testing down to the submicron size range a custom-made scanning electron microscopy (SEM) supported particle tester has been developed [2]. A schematic drawing (a) and an image (b) are shown in Fig. 1. Briefly, the device consists of an AFM tripod scanner which is attached to a coarse positioning unit and force sensing springs which support the samples. The device can be attached to the standard sample holder of the used SEM (Gemini Ultra 55, Carl Zeiss, Germany). For better imaging the device is tilted by 4° with respect to the electron beam. A working distance as low as 7 mm was realized by the compact design.

The x- and z-ranges of the device account for 12.7 mm. The step width of the coarse positioning unit is 30 nm. The AFM scanner (fine positioning and measuring) has a nominal lateral range of $80 \times 80\ \mu m^2$. The vertical range is 15 μm. The piezo-stacks are equipped with strain gauges. Two interlocked steel foil springs are used to measure normal forces: Acting normal forces can be determined from Hook's law. The springs are fixed to the casing and a titanium rod which supports the sample. A linear variable differential transformer (LVDT) is used to measure the deflection of the springs. Read out of the LVDT is performed by lock-in technology. The spring constant (range of 50 N/m–50 kN/m) is determined from the eigenfrequency. Nominal resolutions in force and deformation account for 10 μN and 5 nm. Diamond flat punch indenters are used as probes to compress particles. Nominal diameters of 2 μm, 10 μm, 50 μm or 90 μm are typically used. The wide positioning range of the

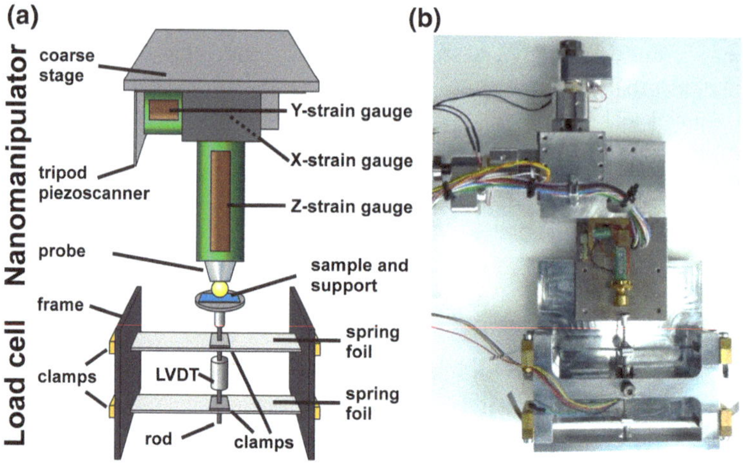

Fig. 1 Custom-made SEM supported manipulation device. The device consists of an upper part termed "Nanomanipulator" and a lower part called "Load cell". The Nanomanipulator performs all movements of the attached diamond tip. The Load cell acts as a force sensing sample support. **a** Schematic drawing of the device and **b** corresponding photographic image

device allows for testing of about 100 particles on a standard working day. Further details are given elsewhere [2].

Deformation and Fracture of Oxide Microparticles Under Normal Load

Size- and structure-dependent deformation and breakage has been investigated for sol-gel derived silica [3–5] and titania spheres [6]. In both cases amorphous and almost monodisperse particles can be precipitated over a wide size range. The structure of these sol-gel particles can be varied systematically by heat treatments at different temperatures: In the silica system cross-linking is enhanced. The amorphous titania particles can be crystallized.

The overall force-displacement (F-D) responses of the investigated silica and titania particles is comparable: Initially, the deformation is fully elastic. This regime is followed by linear elastic-plastic deformation and discontinuities which indicate the formation of cracks or fragmentation. Analytical models to describe particle deformation have been extensively reviewed by Paul et al. [7].

For sol-gel SiO_2 the heat treatments lead to enhanced cross-linking by condensation of internal hydroxyl groups. Figure 2a shows F-D data for SiO_2 particles (nominal mean diameter 500 nm) treated for 12 h at different temperatures. Figure 2b summarizes mean values for the reduced Young's modulus, the hardness from the CEB model and the yield pressure. Figure 2c, d exemplarily display the distributions of

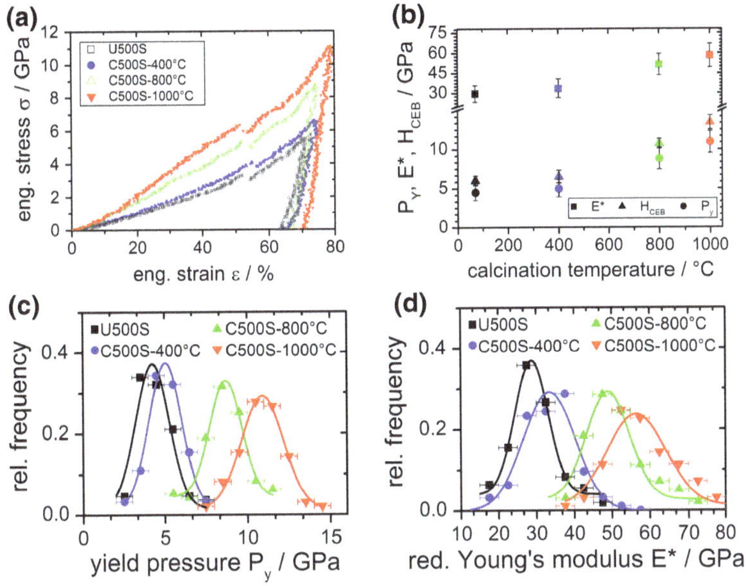

Fig. 2 **a** Force-displacement data for sol-gel SiO_2 particles heat-treated for 12 h at different temperatures. **b** Obtained mean values of the Young's modulus E*, the CEB hardness H_{CEB} and the yield pressure P_Y. The corresponding density distributions of the values are given exemplarily for the yield pressure P_Y (**c**) and the reduced Young's modulus E* (**d**)

the measured values for E* and P_y. It is noteworthy, that the mechanical properties follow normal distributions although the size distributions of the studied particles were quite narrow. It becomes obvious that mechanical properties reflect the chemical and structural changes: Starting significantly below the properties of fused silica the values of Young's modulus and CEB hardness are increasing over the course of the thermal treatments. After the treatment at 1000 °C the Young's modulus approaches the bulk value. At the same time, the yield strength and the hardness exceed typically reported bulk values. All particles show a significant amount of plasticity and the mechanism of deformation is not affected by the thermal treatments. For vitreous silica the observed plasticity could be partially attributed to an increase of material density directly below the contacts [5]. The increase in mechanical strength compared to bulk material is caused by the small size of the particles and the resulting low number of defects.

The influence of particle size was studied in the size range of 100 nm to 1.5 μm for heat-treated silica particles [4]. All particles showed an elastic-plastic deformation behavior. For the F-D data the normalization to the initial particle cross-section and diameter removed the influence of size. Particle failure, however, showed a strong size dependency: The results in Fig. 3 show that for particles below a critical size of 400 nm radial cracks are formed (indicated in green) and no separate pieces were observed. For larger particles, however, fragmentation into individual pieces occurred: Two

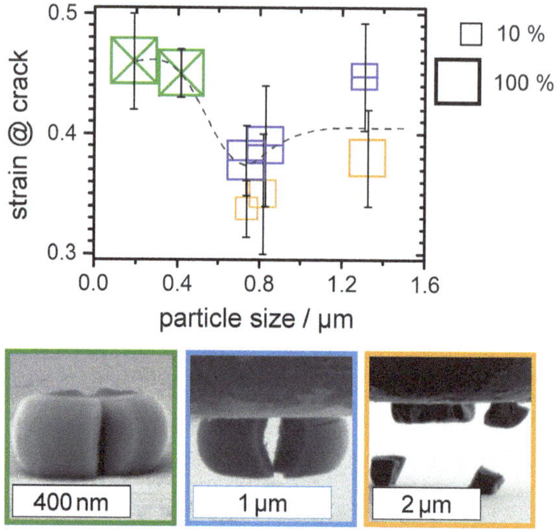

Fig. 3 Brittle-to-ductile-transition for vitreous SiO_2 expressed in terms of the maximum sustainable strain at the onset of cracking or fracturing; the box sizes indicate the relative probability of the observed failure modes. SEM images (bottom) of the observed fracture mechanisms from left to right: ductile crack formation without fragmentation (green crossed boxes), fracture into two separate pieces (blue dashed boxes), catastrophic failure and fracture into multiple fine pieces (orange boxes). Adapted from Romeis S., Schmidt, J. and Peukert W.: Mechanochemical aspects in wet stirred media milling, 2016 with kind permission of Elsevier

mechanisms, namely the separation into exactly two (indicated in blue) or into more than two fragments (indicated in orange), could be distinguished. The aforesaid size dependent cracking mechanism might be responsible for the observed formation of platelets during wet comminution of silica glasses [8–10].

Figure 4a shows a statistical analysis according to the Weibull descriptive statistics of the fracture stress values obtained for submicron sol-gel TiO_2 (blue circles), nanocrystalline (nc) anatase (black squares) and rutile (red crosses) particles: As expected, the normalized fracture force data falls on almost straight lines. From the Weibull distributions (maximum likelihood estimations indicated by the solid lines) the characteristic Weibull parameters, denoted as m, were deduced. The parameter m is a measure for the width of the Weibull distribution. An increasing value indicates a narrowing of the distribution. For anatase particles the narrowest distribution is found. The as-synthesized and the rutile particles exhibit significantly wider distributions, i.e. m is reduced. Commonly, Weibull statistics is discussed in terms of defect distribution in the investigated specimens: A broad Weibull distribution is caused by a wide distribution of defects. For nc anatase (grain size ~6 nm) and the polycrystalline rutile (grain size ~50 nm) particles the difference of the Weibull moduli is in excellent agreement with observed structural features: Generally speaking, small Weibull moduli are connected to large grain sizes, i.e. grain boundaries are defect sites where crack formation can be initiated. Furthermore, the grain size dis-

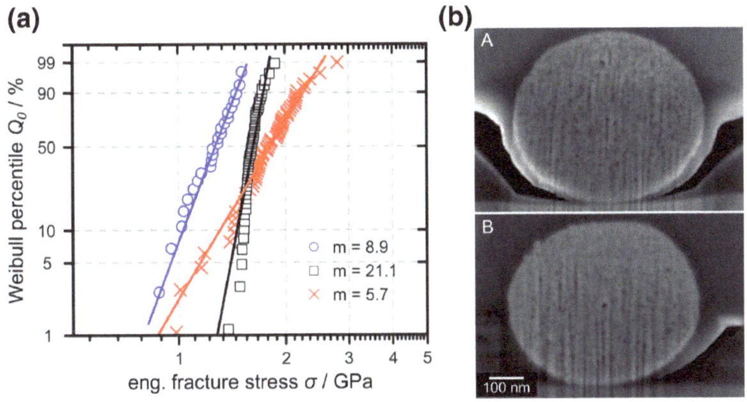

Fig. 4 **a** Weibull probability plots for cracked/fragmented as-prepared sol-gel (blue circles), nc anatase (black squares) and polycrystalline rutile (red crosses) particles. Straight lines are Weibull distribution functions with the corresponding shape parameters m denoted in the diagram. **b** SEM images of cross-sections prepared by FIB of an uncompressed (top) and compressed (bottom) anatase particle. Adapted from Herre P., Romeis S., Mackovic, M., Przybilla, T., Paul, J., Schwenger, J., Torun, B., Grundmeier, G., Spiecker, E., and Peukert W.: Deformation behavior of nanocrystalline titania particles accessed by complementary in situ electron microscopy techniques, 2017 with kind permission of John Wiley and Sons

tribution for the nc anatase particles is extremely narrow. Similar to the sol-gel silica particles the synthesized amorphous titania particles are porous and weakly cross-linked. The Weibull moduli for sol-gel silica and titania are in excellent agreement: For silica a modulus of 13.3 is found. Furthermore, characteristic strength values for a given survival probability of 1/e can be obtained from the Weibull plots. For the different particle ensembles, these so-called estimated scale parameters are 1.32 GPa (as-prepared), 1.65 GPa (nc anatase) and 1.94 GPa (rutile). In summary, it can be stated that the mechanical properties of the anatase particles clearly show their superior performance in terms of strength and reliability in comparison to conventional ceramics.

The deformation mechanisms for the anatase and rutile particles were obtained from HR TEM (high resolution transmission electron microscopy) and focused ion beam (FIB)-SEM experiments. For the crystalline particles the rotation and displacement of single crystallites in conical regions below the contacts was observed. SEM images of FIB cross sections for an uncompressed (top) and a compressed (bottom) nc titania particle are shown in Fig. 4b. For the uncompressed particle a homogeneous distribution of internal pores is visible (3–10 nm). The compressed sphere shows an inhomogeneity in the pore distribution, i.e. a conical volume below the upper contact region is densified. Due to the inhomogeneous contact situation no densification of the lower contact is observed [6]. The densification and the observed grain movements show that porosity is indeed a key factor for the plasticity of the anatase spheres. These observations are compatible with findings from pressure compaction: The initial densification is commonly attributed to grain movements and particle rearrangements. Shear stresses facilitate the rearrangement. The maximum shear stress acts at the contact edge at an angle of 45°. This leads to the formation of the aforementioned densified cones below the contacts. TEM analysis of the fractured nc anatase particles indicate that no defects are present within the grains. For the coarser grained rutile spheres deformation leads to the nucleation of dislocations within the single grains in direct contact with the diamond tip, i.e. the stress concentration in the asperity (grain) leads to local plasticity. The contact stresses are minimized and do not affect intergranular particle failure. Post mortem TEM performed for the particle fragments shows mostly defect-free grains. It is noteworthy that dislocation activity in rutile is commonly reported to be very weak and the contribution of grain plasticity is not yet fully understood and must be studied in more detailed.

Plastic Particle Deformation and the Role of Friction

The influence of contact friction was studied for the plastic deformation of polycrystalline gold particles (size 1 μm). Compression of the gold particles between the diamond flat punch and a silicon substrate was comprehensively studied by experimentally-informed frictional finite element (FE) simulations. To match FE simulations and experimental data the friction coefficients were determined by colloidal probe (CP) atomic force microscopy (AFM) measurements and the material parameters of the studied gold sphere were systematically varied [11].

For the best fit the yield strength for gold was set to 630 MPa. This high value is caused by grain boundary strengthening due to the fine grain size of the studied particles (20 nm ± 5 nm, by Scherrer evaluation). Our value is in good agreement with studies by Conrad and Jung [12]. CP AFM friction measurements on silicon and diamond substrates showed a linear relation of friction force and load. The friction coefficients were determined to $\mu_{dia} = 0.18 \pm 0.01$ and $\mu_{Si} = 0.30 \pm 0.01$. The CP cantilevers were prepared by using the custom-made SEM supported micromanipulation device: Gold particles were glued to the tip of AFM cantilevers by an electron-beam curable adhesive.

FE analysis was performed using a commercial FEM package (Ansys Inc., USA). Due to symmetry the system can be reduced to a 2-dimensional simulation, i.e. a half circle is compressed between two squares. The geometric setup of the simulation is shown in Fig. 5. The nodes on the axis of symmetry were restricted to parallel deflections. The lower square (silicon substrate) was fixed and the displacement was assigned to the top edge of the upper square (diamond tip). Geometries of the sphere and the boxes were set in agreement to the experiment. The squares were set to be ideal-elastic and the half circle was set to be elastic-perfectly plastic. Materials data for silicon and diamond were taken from literature.

Figure 6 shows the comparison between measured and calculated FEM data. It can be seen that friction needs to be taken into account to describe the measured data accurately at high strains above 0.4. The yield strength of the gold particles was found to be 630 MPa. Grain boundary strengthening might be responsible for this

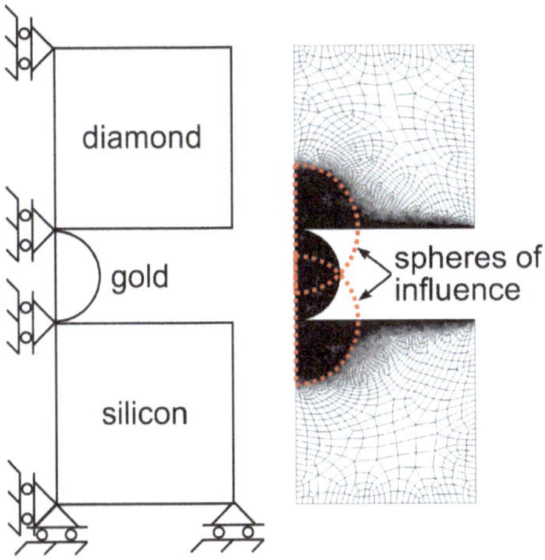

Fig. 5 Geometric setup of the used 2-D FE model. The axis of symmetry was fixed in horizontal direction and the bottom edge of the silicon flat was completely fixed. The mesh of the FE model is exemplarily shown for large strain simulations. Adapted from Paul J., Romeis S., Herre P., Peukert W.: Deformation behavior of micro-sized polycrystalline gold particles studied by in situ compression and frictional finite element simulations, 2015 with kind permission of Elsevier

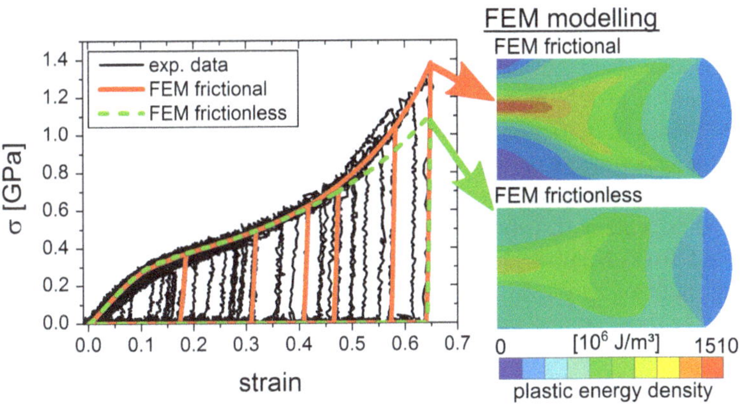

Fig. 6 Influence of contact friction on the uniaxial compression behavior of gold microparticles studied by in situ compression experiments (black lines) and experimentally informed (frictional) finite element calculations (red and green dashed lines). Adapted from Paul J., Romeis S., Herre P., Peukert W.: Deformation behavior of micro-sized polycrystalline gold particles studied by in situ compression and frictional finite element simulations, 2015 with kind permission of Elsevier

remarkably high yield strength. From literature a yield strength of 700 MPa has been reported for the grain size of 20 nm as measured by X-ray analysis [12].

The onset of plasticity was found at small strains of about 0.037% at the diamond-gold contact. Yielding might be facilitated by the rough surfaces of the particles. At high strains adhesion was observed prior to detachment of the diamond tip. Compared

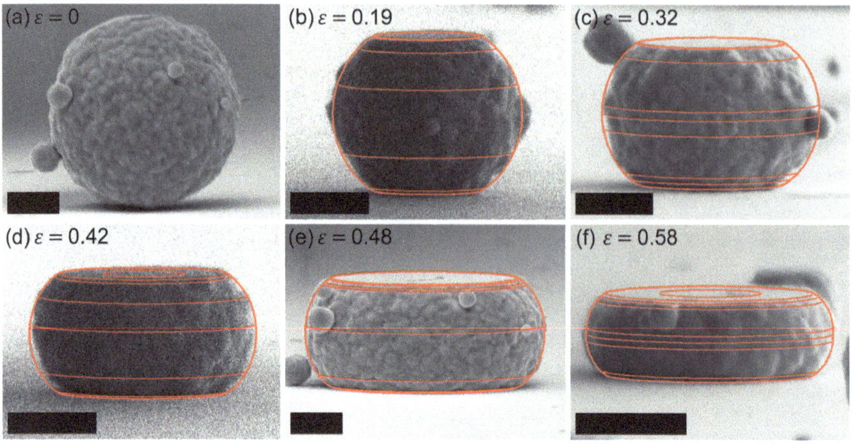

Fig. 7 SEM images of compressed gold microparticles and corresponding geometrical shapes obtained from frictional FEM (red solid lines). Scale bars correspond to 0.5 μm. Reproduced from Paul J., Romeis S., Herre P., Peukert W.: Deformation behavior of micro-sized polycrystalline gold particles studied by in situ compression and frictional finite element simulations, 2015 with kind permission of Elsevier

to the applied loads, however, the adhesion forces are small (near the resolution limit of our device) and are thus neglected.

A comparison of deformed gold particles at different strains and the corresponding predicted shapes from frictional FE analysis is given in Fig. 7. Clearly, the simulation matches the experiments perfectly. This remarkable result allows the determination of energy absorption from particle shape analysis and will be further exploited to measure stress energy distributions in mills, as will be discussed in section "Stress Energy and Stress Number Distributions in Mills".

The normalized hardness (the mean pressure) is shown in Fig. 8a as function of the strain. At small strains a maximum hardness of 2.5 Y is found. Tabor predicted an increase of up to 2.8 Y for metals [13]. For increasing strains, hardness decreases due to changes of the geometry. In the frictionless case, the normalized hardness approaches unity. This has also been reported for large metal particles [14, 15]. Friction leads to a further increase in hardness for strains exceeding 0.4. This effect can be understood by analyzing the differences of the individual contributions to the total expended energy density ΔW_{tot} (shown in Fig. 8b): For the frictional case less energy is spent at small strains while significantly more energy is expended at large strains.

While frictionless contacts are sliding freely, sliding is suppressed for frictional contacts at small strains. The tangential forces do not exceed the friction forces resulting in an altered stress state inside the particle. Sliding starts as soon as the tangential forces exceed the friction force. Since sliding results in energy dissipation, the particles harden at large strains: Only for large strains an appreciable portion of the total energy (~4%) is dissipated by friction. The contact area can be separated

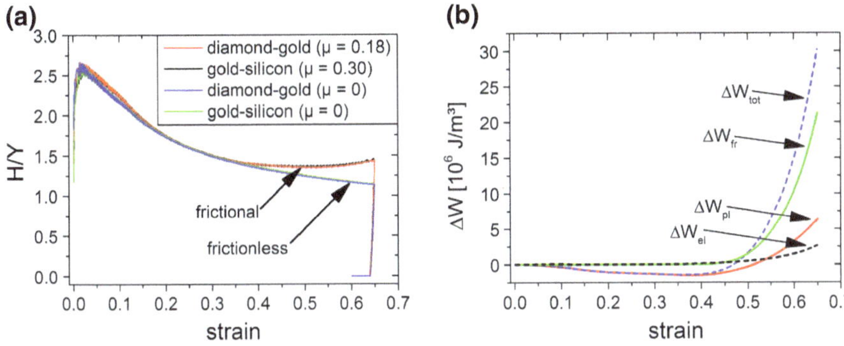

Fig. 8 **a** Normalized hardness (Y = 630 MPa) at the diamond-gold and gold-silicon contacts. **b** Difference of elastic-, plastic-, frictional- and total energy density between the frictional and frictionless case. Reproduced from Paul J., Romeis S., Herre P., Peukert W.: Deformation behavior of micro-sized polycrystalline gold particles studied by in situ compression and frictional finite element simulations, 2015 with kind permission of Elsevier

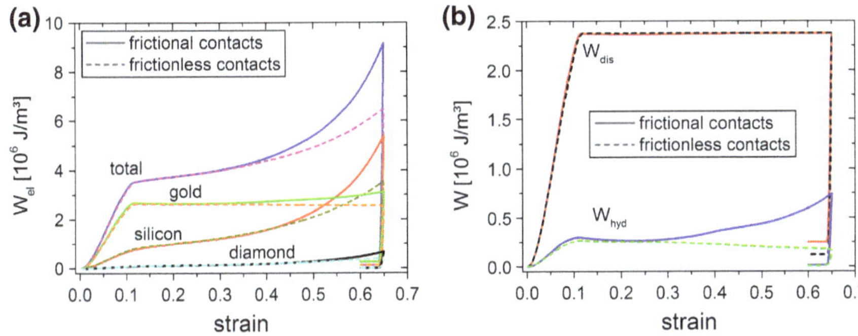

Fig. 9 **a** Total elastic energy density from FEM and its components for frictional and frictionless contacts. **b** Contributions to the elastic energy density of the sphere for frictional and frictionless contacts. Reproduced from Paul J., Romeis S., Herre P., Peukert W.: Deformation behavior of micro-sized polycrystalline gold particles studied by in situ compression and frictional finite element simulations, 2015 with kind permission of Elsevier

into a sticking and a sliding zone. At small strains the complete contact is sticking and only a tiny annulus around the contact edge slides. With increasing strain the proportion of the sliding area increases and only a small sticking region remains in the center of the contacts.

The total energy density W_{tot} in the system can be split into an elastic W_{el}, a plastic W_{Pl} and a frictional component W_{fr} (see Fig. 8b). In the frictional case the decrease of the total energy density for strains up 0.4 is caused by a reduced dissipation of plastic energy. At larger strains, however, the situation is reversed. Figure 9a shows the contributions to the total elastic energy density of the system, namely the deformation of the two plates and the sphere. Contributions from the diamond plate are negligible. The silicon substrate is responsible for the increase in total elastic energy density at large strains. The elastic energy density of the sphere can be further separated. The contributions of distortion energy density W_{dis} and hydrostatic energy density W_{hyd} are shown in Fig. 9b. The distortion energy is almost independent of the strain and approaches the theoretical maximum value: The maximum value of W_{dis} is reached in the entire particle. Small exceptions are only found at the centers of the contact areas and at a small zone near the curved outer shell of the sphere.

Since the maximum distortion energy density is approached in the whole particle, the yield strength can be estimated from the elastic energy released during unloading [11]:

$$Y = \sqrt{2 \cdot U_{el}/V \cdot E_{sphere}} \tag{1}$$

For a proper estimation the Young's modulus of the sphere must be known. Furthermore, the compression plates must be as stiff as possible. From FEM the applicability of Eq. 1 was confirmed for the range $10^{-3} < Y/E_{Sphere} < 10^{-1}$. For the gold particles a yield strength of 694 ± 116 MPa ($0.12 < \varepsilon < 0.30$) was calculated.

In summary, the influence of friction on the plastic deformation of the gold microparticles can be described as follows: Friction hinders plasticity near the contact areas. At the same time, however, plasticity is promoted at the center of mass of the sphere. At small strains, more energy is saved at the contact areas then spent additionally at the center. Hence, the particle appears softer. At large strains, however, the situation reverses and the particles become harder. At large strains the resistance to deformation is additionally increased by an increase in elastic energy and the frictional dissipation: At small strains, a conical region pushes into the sphere and causes plastic deformation close to the contact zones. The whole contact area is sticking and frictional energy dissipation plays a minor role. At larger strains the conical regions from the contacts converge at the center. The lateral displacement increases at the contact interfaces and tangential stresses increase at the contact interfaces. This results in the growth of the sliding areas and a rapid increase in energy dissipation due to friction.

Structure and Tribological Behavior of Self-assembled Monolayers Between Solid Bodies

Surface modifications with self-assembled monolayers (SAMs) have gained considerable technological importance. Their ability to include a wide range of chemical functionalities into extremely thin layers allows for tailored surface properties. Applications for SAMs are e.g. corrosion protection, sensors, lubrication, and tuning of wettability or the production of organic field effect transistors [16–19]. For lubrication SAMs are especially suitable when ultrathin coatings are required or the formation of menisci or adhesive bridges has to be avoided. The macroscopic functionality of SAMs is governed by their molecular structure, the order and especially the defects in the monolayer [20, 21]. SAMs comprising short alkyl chains exhibit an overall lowered order which results in a decreased friction force. To understand the frictional properties of SAMs it is paramount to investigate defect formation.

Reviews on tribology of SAMs studied by atomic force microscopy (AFM) have been given by Cheng et al. and Bhushan [21, 22]. Mainly alkanethiols on gold and alkylsilanes on silica have been studied. Conventional AFM studies represent the behavior between nanoscale bodies; a direct transferability to the technical relevant situation of microparticles in contact is not given. In this size range, however, the frictional behavior of SAMs has seen only little attention: Ecke and Butt studied the friction between silica spheres and a native or silanated SiO_2 by colloidal probe AFM [23]. For both surfaces a nonlinear friction-load behavior was observed and the friction forces were significantly lowered after coating. Ruths et al. [24] and Berman et al. [25] study silanated SiO_2 surfaces and alkyl phosphonic acid coated alumina surfaces and found reduced friction and high wear resistance.

Nonlinear optical spectroscopy techniques such as sum frequency generation (SFG) are surface sensitive and provide information on the local order. SFG is well-suited to study the order in SAMs even at embedded surfaces, i.e. while being subjected to mechanical stresses. Information on the structure and in particular the degree of order can be obtained. Conventional AFM is not capable to provide structural information during or directly after mechanical stressing. So far, there is only very limited information on changes induced during or after release of mechanical stresses [21, 26–29].

In particular, we study the defect formation and the tribology of octadecylphosphonic acid (ODPA) SAMs on Al_2O_3. The formation and self-healing of pressure induced defects is investigated by advanced nonlinear spectroscopy. The results are than brought together with CP AFM studies which shed light on the friction of particles on ultrathin layers.

Indentation and Self-healing of Self-assembled Monolayers

ODPA SAMs subjected to a normal load were studied by a custom-built manipulation device which is integrated into a self-made broadband SFG spectrometer (frequency range 2750–3050 cm^{-1}, spatial resolution 100 μm in y- and 200 μm in x-direction) [30]. Figure 10 shows a computer aided design model of the manipulation device (a), a schematic drawing of the contact situation (b) and a photographic image of the manipulation device installed into the SFG spectrometer (c).

In the manipulator a lens (BK7, radius 125 mm) is used to apply a pressure distribution onto the SAM. The SAM was prepared on an atomically smooth α-Al_2O_3 (1000) prism (rms roughness below 0.5 nm) which is transparent for the used laser wavelengths. The layer thickness exceeds the surface roughness, i.e. an intimate contact can be expected over the whole contact area [31, 32]. For details please refer to Meltzer et al. [30] and Paul et al. [33]. To model the contact pressure distribution Hertzian contact theory was used.

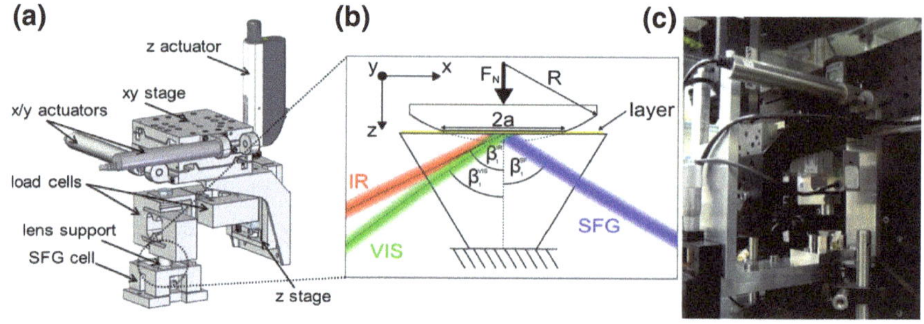

Fig. 10 **a** 3D CAD-model of the custom-made SFG manipulator. **b** Schematic drawing of the contact situation during normal loading. **c** Photograph of the device in the SFG spectrometer. A BK7 lens is used to apply a parabolic pressure distribution on the studied SAM which is prepared on an atomically flat α-Al_2O_3 (1000) prism

SFG spectra are typically obtained by focusing a tunable IR and a fixed Vis laser pulse at the interface. Due to second order process of SFG only information from the interface can be obtained at sub-monolayer sensitivity [30, 33]. Figure 11a (top) shows a SFG spectrum recorded for the freshly prepared ODPA SAM. Most prominent vibrational bands are located at 2880 and 2945 cm^{-1}. The band at 2880 cm^{-1} has been attributed to symmetric CH_3 stretching vibrations (CH_3^{ss}) of terminal methyl groups. The band at 2492 cm^{-1} has been assigned to a superposition of methyl Fermi resonance (CH_3^F). Antisymmetric methyl stretching vibration occurs at 2965 cm^{-1} (CH_3^{ss}) [34, 35]. For perfectly aligned ODPA molecules the contributions from the methylene groups are negligible due to local inverse symmetry. Band fitting of the symmetric methylene stretching vibrations in the spectrum yields an amplitude ratio of CH_2^{ss} (at 2852 cm^{-1}) to CH_3^{ss} (at 2880 cm^{-1}) of 0.1 [35, 36]. To study pressure-induced changes the vibrational modes CH_3^{ss} and CH_2^{ss} were used. A SFG spectrum for the SAM subjected to a load of 174 MPa is given in Fig. 11a (bottom). The presented data was corrected for intensity changes as caused by the contact situation. The comparison of the unperturbed and the compressed SAMs reveals distinct stress-induced changes [30]: The intensity of the CH_3^{ss} band shifts to 2870 cm^{-1} and its intensity decreases by ~99.3%. A third band appears at 2852 cm^{-1} (CH_2^{ss}). This band is accompanied by a strong increase in intensity around 2907 cm^{-1}. A comprehensive discussion is given by Meltzer et al. [30].

The increase of the CH_2^{ss}/CH_3^{ss} ratio and the relative amplitudes of CH_2 and CH_3 vibrations are excellent measures for stress-induced generation of *gauche*-defects [37–39]. The increase of the intensities of the CH_2 bands is caused by local breaks in inversion symmetry, i.e. by changes of the dihedral angles of CH_2-groups due to defect formation. This also affects the CH_3 groups in the vicinity [35].

The CH_2^{ss}/CH_3^{ss} for a loading–unloading cycle is given in Fig. 11b. All measurements were obtained at the center of the contact area. The applied pressure leads to a sharp increase of the CH_2^{ss}/CH_3^{ss} ratio. A maximum is reached for a pressure of 188 MPa and a further load increase leads to a decrease of the amplitude ratio. The

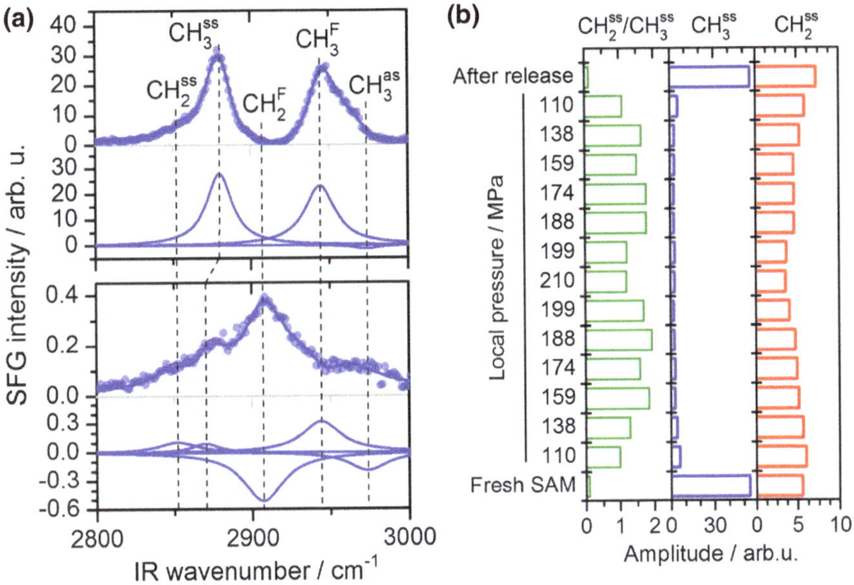

Fig. 11 **a** Top: SFG spectra of ODPA on α-Al_2O_3 (0001). The ODPA layer thickness was 2.2 nm. The molecular tilt angle accounted for about 30°. Bottom: ODPA SAM at a local pressure of 174 MPa of. **b** Changes of the CH_2^{ss}/CH_3^{ss} amplitude ratio and the CH_2^{ss} and CH_3^{ss} amplitudes at the center of the contact. Adapted with permission from Meltzer, C., Paul, J., Dietrich, H., Jäger, C.M., Clark, T., Zahn, D., Braunschweig, B., Peukert, W.: Indentation and Self-Healing Mechanisms of a Self-Assembled Monolayer—A Combined Experimental and Modeling Study. Copyright 2015 American Chemical Society

reversed behavior is observed for unloading and almost the original CH_2^{ss}/CH_3^{ss} ratio is restored. The majority of induced defects are thus reversed within the experimental time scale (<1 s). For a clear interpretation the CH_2^{ss} and CH_3^{ss} amplitudes are discussed separately: Once the pressure exceeds 155 MPa the influence of the CH_3^{ss} amplitudes is negligible. The CH_3^{ss} amplitudes depend on the average z-component of methylene groups, i.e. saturation indicates a constant density. The CH_2^{ss} amplitudes decrease due to a reduction of the average z-component of the methylene groups.

For further insights atomistic molecular dynamic (MD) simulations were performed [30, 40, 41]. ODPA surface densities of 3.8 and 5.07 molecules/nm^2 were studied. These densities correspond to the upper and lower end of the experimentally [42, 43] and theoretically [44] suggested surface coverages for alkyl-phosphonic acids on α-Al_2O_3. The pristine SAMs showed a high degree of order. By compressing the layers with a harmonic potential of the compressing sphere the molecular order in the contact area becomes perturbed: The average z-components of the CH_2 and CH_3 bond vectors increase and decrease, respectively. The pressurized molecules are kinked and twisted as can be seen in Fig. 12. For a comparison to the SFG experiments a geometric analysis of the CH_2 and CH_3 bond vectors has been performed: The pristine systems showed very narrow vibration vector distributions. For

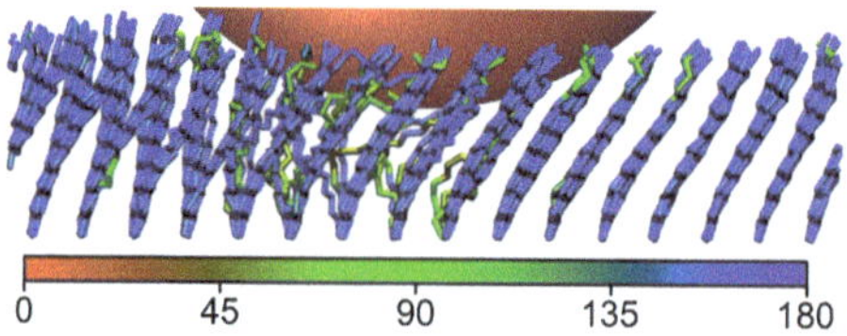

Fig. 12 Side view of a MD simulation of a ODPA layer (molecular density 3.8/nm^2) at 200 MPa. Only carbon chains are shown; the dihedral angle is color coded. The harmonic potential sphere is illustrated in red. Adapted with permission from Meltzer, C., Paul, J., Dietrich, H., Jäger, C.M., Clark, T., Zahn, D., Braunschweig, B., Peukert, W.: Indentation and Self-Healing Mechanisms of a Self-Assembled Monolayer—A Combined Experimental and Modeling Study. Copyright 2015 American Chemical Society

the compressed systems, however, highly spread distributions were obtained. For the pristine samples the CH_2 groups lie in the xy-plane. Hence, the dipole moment of symmetrical CH_2 vibrations perpendicular to the surface is small. The compressed systems have a stronger average z-dipole due to *gauche*-defects and the increased tilting of the molecules. In the pristine layer the CH_3 groups face upwards, i.e. a strong dipole moment results. In the systems under pressure the terminal CH_3 groups are pressed flat and the z-dipole is reduced. In the contact area an increase of the average z-component of the CH_2 bond vectors of up to 11 (5.07/nm^2, 500 MPa) and a decrease of the average z-component of the CH_3 vectors of up to 85% (3.80/nm^2, 200 MPa) was found, i.e. the CH_2/CH_3 ratio rises by a factor of ~35. The experimentally observed self-healing is also corroborated by MD studies: Several ps after pressure release the CH_3 signal intensity and the *gauche*-defect density return to their original state. Relaxation of the *gauche*-defects starts at the rim of the contact area and inverts the compression process. For the generation and relaxation of the *gauche*-defects two mechanisms are conceivable in the simulations, namely the rotation of the entire chain and the concerted generation of defects separated by one bond. Approximately 10–20% of the defects are produced and healed via the second mechanism. Single defect generation/healing occurs for the head groups and the double generation/healing mechanism dominates the anchor region. Furthermore, it is found that the densely packed system relaxes faster. This is caused by the positions of *gauche*-defects in the molecules. Only in the low density system the chains can tilt at the rim of the contact region. Kinks near the anchor groups result. These defects have to travel through the chain during self-healing. For the denser systems the molecules cannot tilt and shallow *gauche*-defects located in the upper part of the alkyl chain are formed [30].

The lateral extent of the contact region and the induced changes were studied by co-localized SFG spectroscopy. The laterally resolved CH_3^{ss} intensities at an applied normal force of 140 N (maximum pressure: 210 MPa) are exemplarily given in Fig. 13A (top). The Hertzian contact radius is indicated by the white circle. The CH_3^{ss} intensity shows a strong decrease toward the center of the contact area. Starting from the outer edge of the contact area, there is an up to two-fold increase in CH_3^{ss}

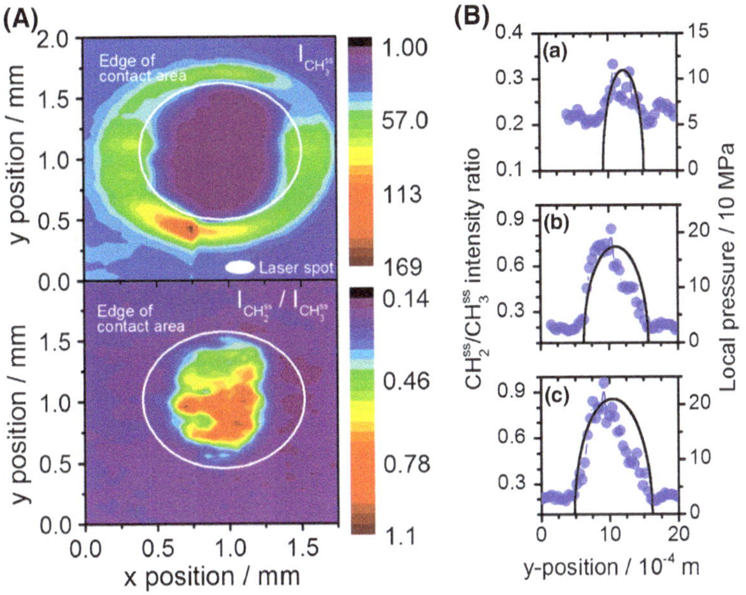

Fig. 13 **A** Co-localized SFG spectra of the contact area on an ODPA SAM at a normal force of 140 N. The Hertzian contact area is indicated by the white circles. A pressure of 210 MPa results at the center. **B** Comparison of CH_2^{ss}/CH_3^{ss} line scans through the contact centers for (a) 20 N, (b) 80 N and (c) 140 N. The calculated pressure distributions are indicated. Adapted with permission from Meltzer, C., Paul, J., Dietrich, H., Jäger, C.M., Clark, T., Zahn, D., Braunschweig, B., Peukert, W.: Indentation and Self-Healing Mechanisms of a Self-Assembled Monolayer—A Combined Experimental and Modeling Study. Copyright 2015 American Chemical Society

intensity. This effect is attributed to changes of the Fresnel coefficients once a close contact between the lens and the substrate is established but the molecular structure is not changed [30]. A comparison between the calculated and experimental contact area size is possible by the CH_2^{ss}/CH_3^{ss} intensity ratio (Fig. 13A (bottom)): Deviations from the value of a freshly prepared SAM are limited to the contact area. Figure 13B shows a comparison of the Hertzian pressure distribution in y-direction through the centers of the formed contacts for 120 N (top), 80 N (middle) and 20 N (bottom) and the corresponding spatially resolved CH_2^{ss}/CH_3^{ss} ratio (resolution 40 μm). For 20 N (110 MPa) the CH_2^{ss}/CH_3^{ss} the applied pressure distribution does not generate a significant number of defects. For 80 N (174 MPa) and 140 N (210 MPa) CH_2^{ss}/CH_3^{ss} intensity ratios of 0.7 and 0.8 are found close to the center. The gradual increase indicates that the defect density rises with pressure.

Friction and Lubrication Behavior of Self-assembled Monolayers

The friction and lubrication of the ODPA SAM on α-Al_2O_3 surfaces was studied by CP AFM [33]. Borosilicate glass spheres with a nominal diameter of 10 μm were used as colloidal probes. In comparison to the native α-Al_2O_3 surface significantly lower friction forces were observed on the ODPA coated sample. In comparison to earlier studies reported in literature [25, 45–47], which were exclusively performed with sharp AFM tips, a linear dependence with increasing load was observed only at high loads. At low normal loads a nonlinear dependence was observed. We attribute this behavior to the contact geometry. For the smooth transition between the aforesaid regimes two potential mechanisms, namely molecular plowing or the acting adhesion forces, are discussed.

For the measurements of friction-load (FL) curves areas 250 × 250 nm^2 were scanned (resolution 128 × 128 px) at normal loads of up to 4 μN. Experimental details and the calibration procedures are described by Paul et al. [33]. To obtain the average friction force (F_f) for a given normal load, so called friction loops, i.e. the deflection signal for trace and retrace for each line scan, were recorded. The difference between the signals gives the double of the mean friction force. The overall mean friction force was calculated by averaging all scan lines. FL curves for α-Al_2O_3- and ODPA-samples are shown In Fig. 14a. An enlarged view for the ODPA-samples is given in Fig. 14b. As can be seen the friction forces are significantly higher on the α-Al_2O_3-sample. A strictly nonlinear behavior is observed for the Al_2O_3-samples (Fig. 14a). For the ODPA sample (Fig. 14b) a clear transition from a nonlinear to

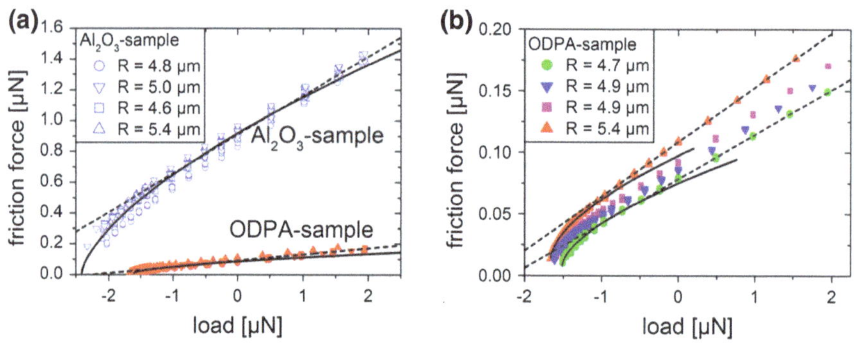

Fig. 14 **a** Friction-load curves measured on Al_2O_3- (open symbols) and ODPA-samples (solid symbols). The size of the colloidal probes is indicated in the legend. Nonlinear fit curves (solid lines) to the averaged data were obtained from the COS procedure [48] Linear fits (dashed lines) are given for comparison. In **b** the FL data measured on ODPA-samples are magnified. COS (solid lines) and linear (dashed lines) fits to the largest and smallest particle are given. Adapted with permission from Paul, J., Meltzer, C., Braunschweig, B., Peukert, W.: Lubrication of Individual Microcontacts by a Self-Assembled Alkyl Phosphonic Acid Monolayer on α-Al_2O_3 (0001). Copyright 2016 American Chemical Society

Table 1 Parameter α, shear strength τ, work of adhesion W, normalized adhesion force F_a/R, and coefficient of friction μ

	α	τ (MPa)	W (mJ/m²)	F_a/R (N/m²)	μ
Al_2O_3	0.03 ± 0.05	107.6 ± 8.3	78 ± 5	0.49 ± 0.04	0.253 ± 0.026
ODPA	0.25 ± 0.004	9.6 ± 0.5	57 ± 2	0.32 ± 0.01	0.039 ± 0.004

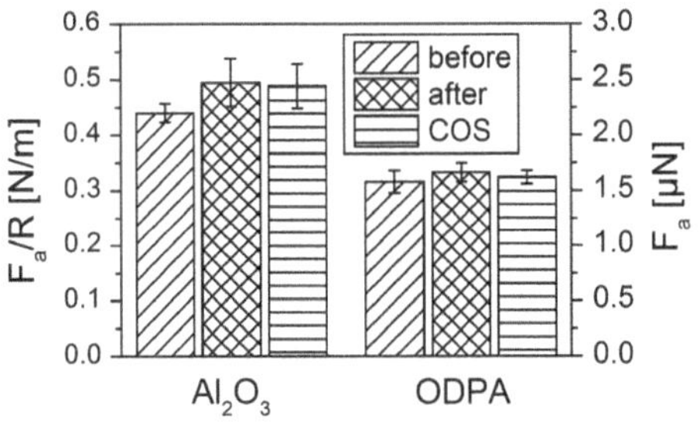

Fig. 15 Absolute and normalized adhesion forces measured before and after friction measurements. The values calculated from COS are given. Adapted with permission from Paul, J., Meltzer, C., Braunschweig, B., Peukert, W.: Lubrication of Individual Microcontacts by a Self-Assembled Alkyl Phosphonic Acid Monolayer on α-Al_2O_3 (0001). Copyright 2016 American Chemical Society

linear FL behavior occurs. For the Al_2O_3-sample the FL data is independent of the CP size. For the ODPA-samples, however, an increase in friction force with particle radius is observed.

Friction coefficients from the linear parts of the recorded FL data were calculated by Deraguin's modification of Amontons' law (direct proportionality between normal load and friction force). The Bowden-Tabor relationship (proportionality of F_F and real contact area) was used to extract the shear strength τ from the nonlinear parts of the FL data. To determine the shear strength from the FL data an empirical fitting procedure proposed by Carpick, Ogletree, and Salmeron (COS) [48], which is based on the Maugis–Dugdale model [49] of adhesive contacts, was used. Table 1 lists the obtained relevant parameters.

Al_2O_3-samples are well described by the DMT model (α close to zero). For the ODPA sample the contact behavior is shifted towards a softer adhesive contact (JKR model [50]) as indicated by the increase of α. Compared to the alumina surface the shear strength τ on the ODPA-sample is significantly reduced. The estimated values for the work of adhesion W are in good agreement with literature reports. To estimate the friction coefficients the linear regimes of the measured FL data at loads exceeding 1.5 μN were used. The friction coefficient of the Al_2O_3 sample exceeds the value of the ODPA-sample by more than a factor of 6.

Adhesion forces were measured before and after the friction measurements by using the colloidal probes. For each sample 100 individual force-distance were averaged. The results are summarized in Fig. 15.

Compared to the bare Al_2O_3 sample the adhesion force on the ODPA-samples is decreased. This is consistent with literature reports [45, 47] and can be attributed to the reduced work of adhesion. The adhesion forces before and after the friction measurements are similar, i.e. no persistent changes of the contact areas occur during the friction measurement.

Nonlinear FL behavior is often attributed to single asperities [23, 48, 51, 52]. From contact theories an approximate dependence of the contact area A on the load of $A \sim FL^{2/3}$ results [49, 50, 53]. The friction force is directly proportional to the real contact area A, i.e. $F_r \sim F_L^{2/3}$. This has been verified experimentally for several pairings [23, 47, 54]. The linear FL behavior, which is found on the ODPA samples (see Fig. 14b) for larger normal forces, is commonly attributed to several asperities which are in contact simultaneously: With increasing load the number of asperities in contact increases while the contact areas of the individual asperities remains constant. A transition from a linear to a nonlinear FL dependency expected as the asperities get deformed and the contact area increases. A transition in the opposite direction, however, cannot be explained by the aforesaid theories. Brukman et al. [47] reported a strictly linear FL on ODPA layers probed with standard cantilever tips and related this behavior to molecular plowing. We attribute the nonlinear FL behavior at low loads to the large radius of the colloidal probe particles. The relative contribution of molecular plowing, however, increases with the applied load due to indentation of the ODPA SAM and the resulting steeper cone angle at the edge of the contact area. Thus, the reason for the observed FL transition might be that friction behavior at low loads is governed by single asperity mechanics, while the effect of molecular plowing becomes dominant at higher loads.

Both adhesion and friction forces are significantly reduced on the ODPA covered samples (see Fig. 15). The relative reduction of the friction force is, however, smaller than the corresponding reduction of the adhesion force. According to Bhushan and Liu [55] compressed SAMs behave like carpets of molecular springs, i.e. individual springs can be compressed and oriented which reduces the shear and friction forces. The degree of induced changes depends on the molecular spring constant and the overall packing density of the monolayer. This model does not describe the effect of alkyl chain length satisfactory [46, 56]. Foster et al. studied the effect of chain length for alkyl phosphonic acid SAMs on alumina [46] and reported a substantial decrease of the friction coefficient for chain lengths of up to 8 methylene groups. For longer chain lengths saturation occurred. A similar dependence on the molecular order studied for different packing density has been reported by Perry et al. [20].

SFG spectroscopy (see section "Indentation and Self-healing of Self-assembled Monolayers") revealed a high degree order for the studied ODPA SAM. Due to their length the ODPA molecules fall in the saturation regime reported by Foster et al. [46]. Hence, the reported friction coefficients and shear strengths values should be close to the minimum attainable values for alkyl phosphonic acid SAMs subjected to the studied contact configuration. In the presented experiments changes of the *gauche*-defect density can be excluded: *Gauche*-defects form only under normal loads and are not affected by shear forces [57]. The saturation pressure for ODPA SAMs on α-Al_2O_3 (0001) accounts for 155 MPa [30], i.e. a normal load of 100 nN in the studied contact situation. Hence, the presented data lies almost exclusively in the saturation regime.

Stress Energy and Stress Number Distributions in Mills

Comminution as a part of interconnected solid processes is a widely used unit operation in process engineering and particle technology [58–62]. Various designs exist and thousands of mills are in operation. All mills are designed to stress particles, i.e. to transfer energy from the apparatus to the product. Fracture occurs when the released elastically stored energy is sufficiently high to induce crack opening [63]. Dependent on the design of the mill, stressing can occur predominantly by impact (one-sided) or compression (two-sided).

All mills can be characterized by the type of stressing, the transferred stress energy distribution (SE) and the stress number distribution (SN) [1]. These distributions describe how much energy is transferred and how often individual product particles are stressed. For stirred media mills convincing concepts, which link maximum SE and SN values to process parameters, were developed by Kwade, Schwedes and coworkers [59, 62, 64]: The maximum SE and SN values are derived from geometric considerations and the kinetic energy of the grinding beads. In a similar manner a measure for the cumulative stress energy can be derived for hammer mills [65] and for fluidized bed opposed jet mills. Recently, SE and SN distributions could be derived by coupled computational fluid dynamics—discrete element modelling [66–69].

Experimental studies on SE and SN distributions are rare. Recently, we have presented a methodology to measure SE and SN distributions in mills by using well-characterized monodisperse micro-particles. An experimental approach was also proposed by Iwasaki et al. [70, 71]: The authors assessed the transferred energy during comminution but did not determine SN or SE distributions. Our methodology is based on plasticity and the surprising observation that particles which have been deformed by a stirred media mill or by uniaxial single particle compression show a striking resemblance [72, 73], see Fig. 16. In both cases, spherical contacts form as pairs on opposed sides of the surfaces.

A continuous plastic deformation behavior of the particles makes it possible to determination the introduced deformation energy from geometry changes. So far, we have successfully used copper [72], aluminum [73] and silica sub-microparticle as single particle probes [8, 72]. For the latter, the particle size should be smaller than 500 nm [4].

An overview of our novel methodology is shown in Fig. 17: Stress numbers are obtained by counting contact pairs. Stress energy distributions are calculated by linking the size of the contact pairs to the corresponding volume specific energy measured by uniaxial particle compression.

To retrieve the deformation behavior of the single particle probes compression experiments were performed. The particle probes were compressed by a 10 μm flat punch on polished diamond substrates. Variations of the measured force displacement data was not observed within the range tip displacement rate of 20–2000 nm/s. Normalization to the initial cross-section and diameter of the particles was performed and no size dependencies were observed. An important aspect in this context is rate-

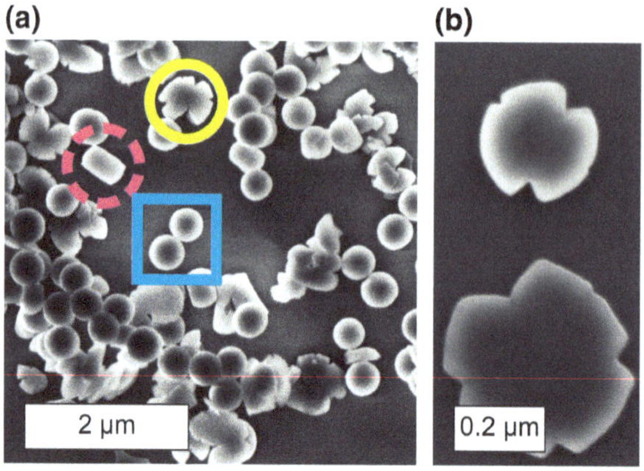

Fig. 16 **a** Silica spheres comminuted under controlled conditions. Yellow circle: Deformed particle with radial cracks. Pink dashed circle: Deformed particle showing parallel contacts and blue box: non-deformed particles. **b** Top-view images of silica compressed by uniaxial compression in the SEM. Adapted from Strobel, A., Schwenger, J., Wittpahl, S., Schmidt, J., Romeis, S., Peukert, W.: Assessing the influence of viscosity and milling bead size on the stressing conditions in a stirred media mill by single particle probes, 2018 and Strobel, A., Romeis, S., Wittpahl, S., Herre, P., Schmidt, J., Peukert, W.: Characterization of stressing conditions in mills—A comprehensive research strategy based on well-characterized model particles, 2017 with kind permission from Elsevier

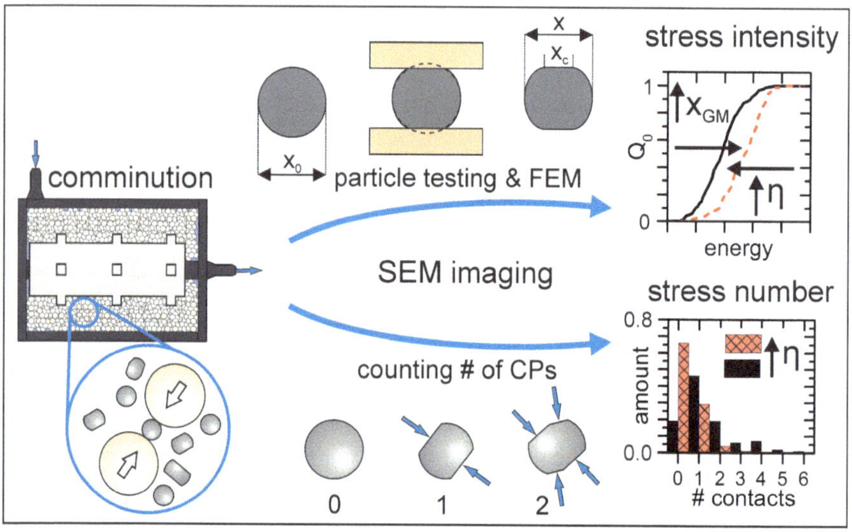

Fig. 17 Schematic representation of our innovative methodology to determine the stress intensity and stress number. Adapted from Strobel, A., Schwenger, J., Wittpahl, S., Schmidt, J., Romeis, S., Peukert, W.: Assessing the influence of viscosity and milling bead size on the stressing conditions in a stirred media mill by single particle probes, 2018 with kind permission from Elsevier

dependency. For the deformation rates which are accessible by our instrumentation we did not observe any influences.

For finite element analysis a 2-dimensional axisymmetric model was used [11]. As discussed above a hemicycle was compressed between two squares which represent the diamond substrate and the tip. Nodes on the axis of symmetry were limited to movements along the axis and the bottom edge of the substrate was fixed. For the compression plates an isotropic linear elastic material law was used. The hemicycle was set to be linear-elastic plastic. Experimental data was fitted by adjusting the yield strength and the work hardening coefficient. Neither from the experimental data nor from the simulations size dependencies were found in the size range from 1 to 100 μm. In the FE model no difference was observed for diamond plates or zirconia plates, which represent two approaching grinding beads. Contact friction was neglected.

Figure 18a shows measured normalized force (stress) versus normalized deformation (strain) data for copper particles. Consecutive loading-unloading cycles (black lines) are shown for one copper particle. The typical plastic deformation behavior of metal is observed, i.e. yielding occurs at a small deformation ($\varepsilon \ll 1\%$) and only little elasticity is recovered by unloading. The results from the fitted FE model are given by the red dashed line. In Fig. 18b the total volume specific energy E_V (obtained by integrating the measured F-D data) for different deformations is given. Open symbols denote data from single-step compression experiments. Data for the consecutive experiment shown in Fig. 18a is given by the filled stars. Again, the FE calculations (red dashed line) are in excellent agreement with the measured data. From FE calculations the link between changes of the particle geometry and the introduced deformation energy can be obtained: In Fig. 18c characteristic dimensions (see Fig. 18d) of the stressed particles (extracted from the FE model) are plotted for different maximum strains ε_{max}.

From SEM images of the single particle probes the number of contact pairs per particle and the areas of the formed contacts can be obtained. A minimum of 100 particles was evaluated for each sample. The relative number of particles with a given number of contact pairs $CP_i = N_i/N$ (N_i number of particles with i contact pairs, N total number of evaluated particles) can be obtained. From CP_i the effective stress number SN_{eff} (Eq. 2) and the relative stress number $SN_{\%}$ (Eq. 3) were calculated. The ratio of SN_{eff} to $SN_{\%}$ yields SN_{stress}, i.e. the average number of contact pairs per stressed particle (Eq. 4).

$$SN_{eff} = \sum_i CP_i \cdot i \tag{2}$$

$$SN_{\%} = 1 - CP_0 \tag{3}$$

$$SN_{stress} = \frac{SN_{eff}}{SN_{\%}} = \frac{\sum_i CP_i \cdot i}{1 - CP_0} \tag{4}$$

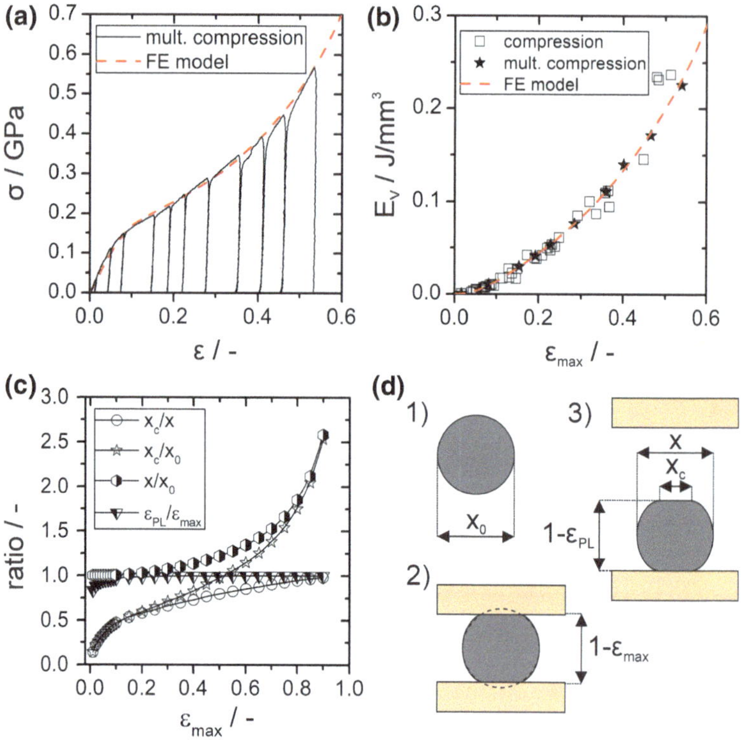

Fig. 18 **a** Experimental (solid) and FEM (dashed) data for the uniaxial compression of a copper particle probe. **b** Volume specific energy E_V given as a function of the particle deformation. **c** Changes of the geometry of an aluminum particle due to uniaxial compression. **d** Schematic representation of the compression: (1) prior loading, (2) maximum load and (3) after unloading. Adapted from Strobel, A., Schwenger, J., Wittpahl, S., Schmidt, J., Romeis, S., Peukert, W.: Assessing the influence of viscosity and milling bead size on the stressing conditions in a stirred media mill by single particle probes, 2018 and Strobel, A., Romeis, S., Wittpahl, S., Herre, P., Schmidt, J., Peukert, W.: Characterization of stressing conditions in mills—a comprehensive research strategy based on well-characterized model particles, 2017 with kind permission from Elsevier

Figure 19a shows the effective stress number SN_{eff} measured for batch comminution in a lab-scale wet stirred media mill. SEM images which correspond to processing times of 1.3 min (1), 7.7 min (2) and 130.0 min (3) are given below. In region I SN_{eff} increases linear with the processing time. In region II SN_{eff} exhibits a maximum whereby an average of 2.5 individual contact pairs were found per particle. In region III SN_{eff} decrease and approaches 1: The corresponding SEM image shows that the particles are flat and only one large contact pair remains. With processing time the probability to repeatedly stress a contact pair increases and flattening occurs. For short processing times (region I) two contact pairs are formed on average. It can be assumed that for this low number the contact pairs form independently. The probes thus can be classified with respect to the number of contact pairs (c.f. states CP_1 …

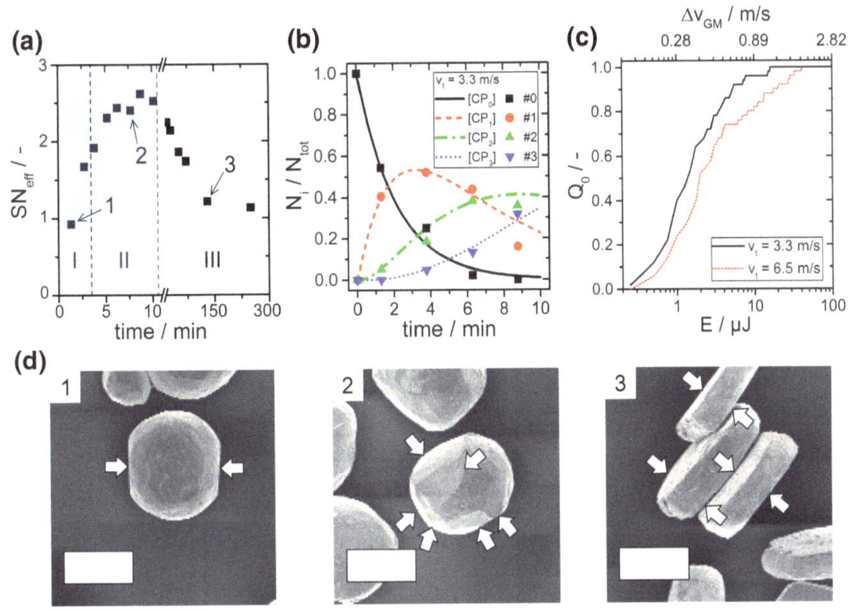

Fig. 19 Characterization of batch milling experiments by 3 wt% Cu particles ($X_{50,3}$ ~ 60 µm) in water-sucrose mixtures (210 mPas, Newtonian flow behavior). **a** Evolution of SN_{eff} for a fixed stirrer tip speed of 6.5 m/s. **b** Temporal evolution of the number of contacts. Fits from a model are given by the lines. **c** Measured stress energy distributions. **d** SEM images of samples taken after 1.3 min (1), 7.7 min (2) and 130.0 min (3). Scale bars are 50 µm. The arrows indicate contact areas. Adopted from Strobel, A., Romeis, S., Wittpahl, S., Herre, P., Schmidt, J., Peukert, W.: Characterization of stressing conditions in mills—A comprehensive research strategy based on well-characterized model particles, 2017 with kind permission from Elsevier

CP_6). The temporal evolution of the number of contacts can be described by a set of consecutive first order reactions (see Eq. 5). The fitting results are given by the solid lines in Fig. 19b: The agreement with the experimental data (symbols) is excellent. For stirrer tip speeds of 6.5 m/s and 3.3 m/s the rate constants k_1 are $(1.07 \pm 0.05)\ s^{-1}$ and $(0.49 \pm 0.02)\ s^{-1}$, respectively.

$$CP_0 \xrightarrow{k_1} CP_1 \xrightarrow{k_2} CP_2 \xrightarrow{k_3} CP_3 \xrightarrow{k_4} CP_4 \xrightarrow{k_5} CP_5 \xrightarrow{k_6} CP_6 \quad (5)$$

The stress energy distributions obtained for short processing times are given in Fig. 19c. As expected SE increases with the tip speed. Moreover, SE distributions span over wide energy ranges, i.e. 0.3–16.25 µJ (tip speed 3.3 m/s) or 0.2–41.25 µJ (tip speed 6.5 m/s). SE energy does not scale with the second power of the stirrer tip speed.

Figure 20 summarizes the key findings from characterizing a horizontal lab-scaled stirred media mill operated in passage mode [73]. The influences of grinding media size, stirrer tip speed and fluid viscosity are shown. For the first time, the relative

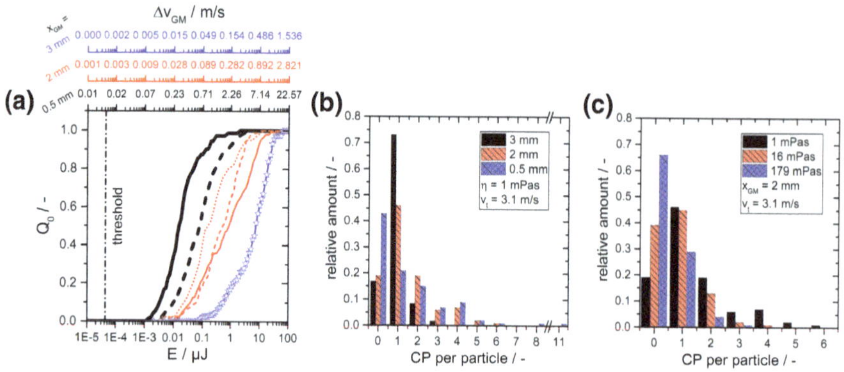

Fig. 20 Particle probe characterization of open-circuit milling experiments in a horizontal stirred media mill by 3 wt% Al particles ($X_{50,3}$ ~ 55 μm). **a** Stress energy distributions obtained for different milling bead sizes 3 mm (blue, open stars), 2 mm (red lines) and 0.5 mm (thick black lines). The depicted parameter are $x_{GM} = 0.5$ mm, $v_t = 3.1$ m/s, $\eta = 1$ mPas (thick black solid); $x_{GM} = 0.5$ mm, $v_t = 6.2$ m/s, $\eta = 1$ mPas (thick black dashed); $x_{GM} = 2$ mm, $v_t = 3.1$ m/s, $\eta = 1$ mPas (red solid); $x_{GM} = 2$ mm, $v_t = 3.1$ m/s, $\eta = 16$ mPas (red dashed); $x_{GM} = 2$ mm, $v_t = 3.1$ m/s, $\eta = 179$ mPas (red dotted); $x_{GM} = 3$ mm, $v_t = 3.1$ m/s, $\eta = 1$ mPas (open stars). **b** Relative number of deformed spheres for different milling beads. **c** Evolution of the relative number of deformed spheres for different viscosities. Adapted from Strobel, A., Schwenger, J., Wittpahl, S., Schmidt, J., Romeis, S., Peukert, W.: Assessing the influence of viscosity and milling bead size on the stressing conditions in a stirred media mill by single particle probes, 2018 with kind permission of Elsevier

influences of the milling bead dampening and particle inertia can be separated. For high fluid viscosities stressing efficiency is deteriorated due to the tendency of the probes to follow squeeze flows emerging during the approach of grinding beads. Bead dampening plays only a minor role. Furthermore, influences of bead segregation or short-circuit flows could be shown.

Figure 20a displays SE distributions for various process parameters. The minimum stressing energy accounts for 1.2 nJ. For different bead sizes at a fixed viscosity the mean stressing intensities E_{50} ($Q_0(E_{50}) = 0.5$) were 15.5 nJ (0.5 mm), 0.84 μJ (2 mm) and 7.70 μJ (3 mm). The overall increase of the SE with the milling bead size is in good agreement with the Kwade model. The absolute values, however, are significantly smaller than the calculated maximum values. The influence of tip speed was investigated for a fixed milling bead size (0.5 mm): Doubling the stirrer tip speed leads to a shift of the SE distribution. The mean SE increases by factor of 5. Increasing the suspension viscosity (by addition of sucrose) leads to a decrease of SE: E_{50} values are shifted to 0.52 μJ (16 mPas) and 0.15 μJ (179 mPas).

Further insights are obtained from the CP distributions and stress numbers: Fig. 20b shows the relative number of probes with a certain number of contacts as function of the milling bead size. Mainly one contact pair is found for 3 mm milling beads. The distribution for the 2 mm beads has the same center but is flattened and reaches to higher numbers. The highest amount of unstressed particles occurs for the grinding bead size of 0.5 mm. The stressed particles, however, show a high number of contacts, i.e. the stressed probe particles experience several stress events. We interpret these findings by zones of different capturing and stressing probabilities inside the mill [73]: Particles which flow through zones of low stressing probability remain unaffected. Probes which are entrained into zones of high stressing probability experience several events. The different zones might arise from an inhomogeneous distribution of the milling beads. The influence of solvent viscosity (1 mPas, 16 mPas and 179 mPas) is shown in Fig. 20c: Increasing the viscosity leads to a shifts towards lower contact numbers and a narrowing of the distribution. While the dependency of SE on viscosity can be explained by viscous dampening of the milling beads, the increasing amount of unstressed particles is related to changes of the capturing probability [74]: When two grinding beads approach each other, the fluid in-between escapes and a flow directed away from the line of approach is formed. The inertia parameter [75] characterizes the probes tendency to follow these flows. For the viscosities of 16 and 179 mPas the tendency to follow the fluid is remarkable. Hence, it can be concluded that the experimentally decreased milling efficiency with increasing viscosity might rather be attributed to a decreasing capturing efficiency of product between colliding beads then to milling bead dampening.

Conclusions

In many fields of Particle Technology, it is helpful to distinguish between process and materials functions. The process functions describe individual unit operations such as mills or classifiers in dependency of external parameters such as energy input, rotor revolution and machine geometry. As a result, the particles in the process or in the machine (such as mills, mixers or granulators) are stressed by these external forces and react according to their material properties. The particles react to the acting stresses primarily by elastic and inelastic deformation which can induce fracture in mills or particle degradation in other unit operations, e.g. during pneumatic conveying or mixing, or adhesion in granulators.

We addressed the material functions and measured distributed mechanical properties of well-defined single particles in a specifically designed manipulator in a scanning electron microscope. Elastic and inelastic material properties are derived from the measurement of a statistically relevant number of force-displacement curves. The experimental data can be described quantitatively by FEM simulations for spherical metal particles which also include information on friction at high strain. Remarkably, also shape evolution can be predicted quantitatively. This forms the basis for the model-based evaluation of stress energy and stress number distributions in stirred media mills and opposed jet fluidized jet mills. For the first time, the energy absorption in mills can be evaluated directly from stressed particles.

The developed methodologies for the systematic measurement of distributed mechanical particle properties and their impact on in-depth studies of particle stressing in mills opens new horizons for the understanding, quantitative modelling and optimization of mills. Also other unit operations, where particles are stressed or abraded will benefit from the presented approaches. The systematic measurement of mechanical property distributions of particles is also of high relevance for DEM models where measured particle properties are needed as input parameters for the predictive description of all processes where aspects of particles in contact are relevant.

References

1. Peukert, W., Segets, D., Pflug, L., Leugering, G.: Unified design strategies for particulate products. In: Advances in Chemical Engineering, Mesoscale Modeling in Chemical Engineering Part I, pp. 1–81. Elsevier (2015)
2. Romeis, S., Paul, J., Ziener, M., Peukert, W.: A novel apparatus for in situ compression of submicron structures and particles in a high resolution SEM. Rev. Sci. Instrum. **83**, 95105 (2012)
3. Romeis, S., Paul, J., Hanisch, M., Marthala, V.R.R., Hartmann, M., Taylor, R.N.Klupp, Schmidt, J., Peukert, W.: Correlation of enhanced strength and internal structure for heat-treated submicron Stöber silica particles. Part. Part. Syst. Char. **31**, 664–674 (2014)
4. Paul, J., Romeis, S., Mačković, M., Marthala, V., Herre, P., Przybilla, T., Hartmann, M., Spiecker, E., Schmidt, J., Peukert, W.: In situ cracking of silica beads in the SEM and TEM—effect of particle size on structure-property correlations. Powder Technol. **270**, 337–347 (2015)
5. Romeis, S., Paul, J., Herre, P., de Ligny, D., Schmidt, J., Peukert, W., de Ligny, D., Peukert, W.: Local densification of a single micron sized silica sphere by uniaxial compression. Scr. Mater. **108**, 84–87 (2015)
6. Herre, P., Romeis, S., Mačković, M., Przybilla, T., Paul, J., Schwenger, J., Torun, B., Grundmeier, G., Spiecker, E., Peukert, W., Przybilla, T.: Deformation behavior of nanocrystalline titania particles accessed by complementary in situ electron microscopy techniques. J. Am. Ceram. Soc. **330/100**, 556 (2017)
7. Paul, J., Romeis, S., Tomas, J., Peukert, W.: A review of models for single particle compression and their application to silica microspheres. Adv. Powder Technol. **25**, 136–153 (2014)
8. Romeis, S., Schmidt, J., Peukert, W.: Mechanochemical aspects in wet stirred media milling. Int. J. Miner. Process. **156**, 24–31 (2016)
9. Dörfler, A., Detsch, R., Romeis, S., Schmidt, J., Eisermann, C., Peukert, W., Boccaccini, A.R., Dörfler, A., Detsch, R., Schmidt, J., Eisermann, C., Peukert, W., Boccaccini, A.R.: Biocompatibility of submicron Bioglass® powders obtained by a top-down approach. J. Biomed. Mater. Res. **102**, 952–961 (2014)
10. Romeis, S., Hoppe, A., Eisermann, C., Schneider, N., Boccaccini, A.R., Schmidt, J., Peukert, W., Hoppe, A., Eisermann, C., Schneider, N., Boccaccini, A.R., Schmidt, J., Peukert, W., Bose, S.: Enhancing in vitro bioactivity of melt-derived 45S5 Bioglass® by comminution in a stirred media mill. J. Am. Ceram. Soc. **97**, 150–156 (2014)
11. Paul, J., Romeis, S., Herre, P., Peukert, W.: Deformation behavior of micron-sized polycrystalline gold particles studied by in situ compression experiments and frictional finite element simulation. Powder Technol. **286**, 706–715 (2015)
12. Conrad, H., Jung, K.: Effect of grain size from mm to nm on the flow stress and plastic deformation kinetics of Au at low homologous temperatures. Mater. Sci. Eng. A Struct. **406**, 78–85 (2005)
13. Tabor, D.: The Hardness of Metals. Oxford University Press, Oxford (2000)
14. Gane, N., Cox, J.M.: The micro-hardness of metals at very low loads. Philos. Mag. **22**, 881–891 (1970)
15. Chaudhri, M.M., Hutchings, I.M., Makin, P.L.: Plastic compression of spheres. Philos. Mag. A 493–503 (1984)

16. Bain, C.D., Whitesides, G.M.: Depth sensitivity of wetting: monolayers of omega-mercapto ethers on gold. J. Am. Chem. Soc. **110**, 5897–5898 (1988)
17. Rumpel, A., Novak, M., Walter, J., Braunschweig, B., Halik, M., Peukert, W.: Tuning the molecular order of C60 functionalized phosphonic acid monolayers. Langmuir **27**, 15016–15023 (2011)
18. Lin, Y.-Y., Gundlach, D.J., Nelson, S.F., Jackson, T.N.: Stacked pentacene layer organic thin-film transistors with improved characteristics. IEEE Electron. Device Lett. **18**, 606–608 (1997)
19. Chaki, N.K., Vijayamohanan, K.: Self-assembled monolayers as a tunable platform for biosensor applications. Biosens. Bioelectron. **17**, 1–12 (2002)
20. Perry, S.S., Lee, S., Shon, Y.-S., Colorado, J.R., Lee, T.R.: The relationships between interfacial friction and the conformational order of organic thin films. Tribol. Lett. **10**, 81–87 (2001)
21. Cheng, H., Hu, Y.: Influence of chain ordering on frictional properties of self-assembled monolayers (SAMs) in nano-lubrication. Adv. Colloid Interface Sci. **171–172**, 53–65 (2012)
22. Bhushan, B.: Nanotribology and Nanomechanics: An Introduction, 2nd edn. Springer, Berlin, Heidelberg (2008)
23. Ecke, S., Butt, H.-J.: Friction between Individual Microcontacts. J. Colloid Interface Sci. **244**, 432–435 (2001)
24. Ruths, M., Alcantar, N.A., Israelachvili, J.N.: Boundary friction of aromatic silane self-assembled monolayers measured with the surface forces apparatus and friction force microscopy. J. Phys. Chem. B **107**, 11149–11157 (2003)
25. Berman, A., Steinberg, S., Campbell, S., Ulman, A., Israelachvili, J.: Controlled microtribology of a metal oxide surface. Tribol. Lett. **4**, 43–48 (1998)
26. Yu, J., Banquy, X., Greene, G.W., Lowrey, D.D., Israelachvili, J.N.: The boundary lubrication of chemically grafted and cross-linked hyaluronic acid in phosphate buffered saline and lipid solutions measured by the surface forces apparatus. Langmuir **28**, 2244–2250 (2012)
27. Bhushan, B., Israelachvili, J.N., Landman, U.: Nanotribology: friction, wear and lubrication at the atomic scale. Nature **374**, 607–616 (1995)
28. Qian, L., Charlot, M., Perez, E., Luengo, G., Potter, A., Cazeneuve, C.: Dynamic friction by polymer/surfactant mixtures adsorbed on surfaces. J. Phys. Chem. B **108**, 18608–18614 (2004)
29. Cui, B., Zhang, L., Zhou, H., Zhang, J., Chen, J.: Preparation and tribological properties of self-assembled poly(amide amine)-Cu nanofilm on silicon. Surf. Interface Anal. **43**, 1377–1381 (2011)
30. Meltzer, C., Paul, J., Dietrich, H., Jäger, C.M., Clark, T., Zahn, D., Braunschweig, B., Peukert, W.: Indentation and self-healing mechanisms of a self-assembled monolayer—a combined experimental and modeling study. J. Am. Chem. Soc. **136**, 10718–10727 (2014)
31. Du, Q., Xiao, X.-D., Charych, D., Wolf, F., Frantz, P., Shen, Y.R., Salmeron, M.: Nonlinear optical studies of monomolecular films under pressure. Phys. Rev. B **51**, 7456–7463 (1995)
32. Beattie, D.A., Fraenkel, R., Winget, S.A., Petersen, A., Bain, C.D.: Sum-frequency spectroscopy of a monolayer of zinc arachidate at the solid-solid interface. J. Phys. Chem. B **110**, 2278–2292 (2006)
33. Paul, J., Meltzer, C., Braunschweig, B., Peukert, W.: Lubrication of individual microcontacts by a self-assembled alkyl phosphonic acid monolayer on α-Al_2O_3(0001). Langmuir **32**, 8298–8306 (2016)
34. Nishi, N., Hobara, D., Yamamoto, M., Kakiuchi, T.: Chain-length-dependent change in the structure of self-assembled monolayers of n-alkanethiols on Au(111) probed by broad-bandwidth sum frequency generation spectroscopy. J. Chem. Phys. **118**, 1904–1911 (2003)
35. Lambert, A.G., Davies, P.B., Neivandt, D.J.: Implementing the theory of sum frequency generation vibrational spectroscopy: a tutorial review. Appl. Spectrosc. Rev. **40**, 103–145 (2005)
36. Novak, M., Schmaltz, T., Faber, H., Halik, M.: Influence of self-assembled monolayer dielectrics on the morphology and performance of α, ω-dihexylquaterthiophene in thin film transistors. Appl. Phys. Lett. **98**, 93302 (2011)
37. Ward, R.N., Duffy, D.C., Davies, P.B., Bain, C.D.: Sum-frequency spectroscopy of surfactants adsorbed at a flat hydrophobic surface. J. Phys. Chem. **98**, 8536–8542 (1994)

38. Tyrode, E., Hedberg, J.: A comparative study of the CD and CH stretching spectral regions of typical surfactants systems using VSFS: orientation analysis of the terminal CH3 and CD3 groups. J. Phys. Chem. C **116**, 1080–1091 (2011)
39. Guyot-Sionnest, P., Hunt, J.H., Shen, Y.R.: Sum-frequency vibrational spectroscopy of a Langmuir film: study of molecular orientation of a two-dimensional system. Phys. Rev. Lett. **59**, 1597–1600 (1987)
40. Bauer, T., Schmaltz, T., Lenz, T., Halik, M., Meyer, B., Clark, T.: Phosphonate- and carboxylate-based self-assembled monolayers for organic devices: a theoretical study of surface binding on aluminum oxide with experimental support. ACS Appl. Mater. Interfaces **5**, 6073–6080 (2013)
41. Jäger, C.M., Schmaltz, T., Novak, M., Khassanov, A., Vorobiev, A., Hennemann, M., Krause, A., Dietrich, H., Zahn, D., Hirsch, A., Halik, M., Clark, T.: Improving the charge transport in self-assembled monolayer field-effect transistors: from theory to devices. J. Am. Chem. Soc. **135**, 4893–4900 (2013)
42. Liakos, I.L., Newman, R.C., McAlpine, E., Alexander, M.R.: Comparative study of self-assembly of a range of monofunctional aliphatic molecules on magnetron-sputtered aluminium. Surf. Interface Anal. **36**, 347–354 (2004)
43. Hauffman, T., Blajiev, O., Snauwaert, J., van Haesendonck, C., Hubin, A., Terryn, H.: Study of the self-assembling of n-octylphosphonic acid layers on aluminum oxide. Langmuir **24**, 13450–13456 (2008)
44. Luschtinetz, R., Oliveira, A.F., Duarte, H.A., Seifert, G.: Self-assembled monolayers of alkylphosphonic acids on aluminum oxide surfaces—a theoretical study. Z. Anorg. Allg. Chem. **636**, 1506–1512 (2010)
45. Hoque, E., Derose, J.A., Hoffmann, P., Mathieu, H.J., Bhushan, B., Cichomski, M.: Phosphonate self-assembled monolayers on aluminum surfaces. J. Chem. Phys. **124**, 174710 (2006)
46. Foster, T.T., Alexander, M.R., Leggett, G.J., McAlpine, E.: Friction force microscopy of alkylphosphonic acid and carboxylic acids adsorbed on the native oxide of aluminum. Langmuir **22**, 9254–9259 (2006)
47. Brukman, M.J., Oncins Marco, G., Dunbar, T.D., Boardman, L.D., Carpick, R.W.: Nanotribological properties of alkanephosphonic acid self-assembled monolayers on aluminum oxide: effects of fluorination and substrate crystallinity. Langmuir **22**, 3988–3998 (2006)
48. Carpick, R.W., Ogletree, D.F., Salmeron, M.: A general equation for fitting contact area and friction vs load measurements. J. Colloid Interface Sci. **211**, 395–400 (1999)
49. Maugis, D.: Adhesion of spheres: the JKR-DMT transition using a dugdale model. J. Colloid Interface Sci. **150**, 243–269 (1992)
50. Johnson, K.L., Kendall, K., Roberts, A.D.: Surface energy and the contact of elastic solids. Proc. R. Soc. A **324**, 301–313 (1971)
51. Jones, R.: From single particle AFM studies of adhesion and friction to bulk flow: forging the links. Granul. Matter **4**, 191–204 (2003)
52. Biggs, S., Scales, P.J., Leong, Y.-K., Healy, T.W.: Effects of citrate adsorption on the interactions between zirconia surfaces. Faraday Trans. **91**, 2921 (1995)
53. Derjaguin, B., Muller, V., Toporov, Y.: Effect of contact deformations on the adhesion of particles. J. Colloid Interface Sci. **53**, 314–326 (1975)
54. Berman, A., Drummond, C., Israelachvili, J.: Amontons' law at the molecular level. Tribol. Lett. **4**, 95–101 (1998)
55. Bhushan, B., Liu, H.: Nanotribological properties and mechanisms of alkylthiol and biphenyl thiol self-assembled monolayers studied by AFM. Phys. Rev. B **63**, 607 (2001)
56. Mino, N., Ogawa, K., Minoda, T., Takatsuka, M., Sha, S., Moriizumi, T.: Thin film characteristics of fluorine-substituted monolayers prepared by chemical adsorption from solution. Thin Solid Films **230**, 209–216 (1993)
57. Tutein, A.B., Stuart, S.J., Harrison, J.A.: Role of defects in compression and friction of anchored hydrocarbon chains on diamond. Langmuir **16**, 291–296 (2000)
58. Müller, F., Polke, R.F.: From the product and process requirements to the milling facility. Powder Technol. **105**, 2–13 (1999)

59. Stadler, R., Polke, R., Schwedes, J., Vock, F.: Naßmahlung in Rührwerksmühlen. Chem. Ing. Tech. **62**, 907–915 (1990)
60. Orumwense, O.A., Forssberg, E.: Superfine and ultrafine grinding—a literature survey. Miner. Process. Extr. Metall. Rev. **11**, 107–127 (1992)
61. Knieke, C., Steinborn, C., Romeis, S., Peukert, W., Breitung-Faes, S., Kwade, A.: Nanoparticle production with stirred-media mills: opportunities and limits. Chem. Eng. Technol. **33**, 1401–1411 (2010)
62. Kwade, A., Schwedes, J.: Wet comminution in stirred media mills. KONA **15**, 91–102 (1997)
63. Peukert, W.: Material properties in fine grinding. Int. J. Miner. Process. **74**, S3–S17 (2004)
64. Kwade, A.: A stressing model for the description and optimization of grinding processes. Chem. Eng. Technol. **26**, 199–205 (2003)
65. Juhnke, M.: Evaluation and scale-up of miniaturized rotor–stator impact mills. Powder Technol. **287**, 70–76 (2016)
66. Schilde, C., Beinert, S., Kwade, A.: Comparison of the micromechanical aggregate properties of nanostructured aggregates with the stress conditions during stirred media milling. Chem. Eng. Sci. **66**, 4943–4952 (2011)
67. Beinert, S., Schilde, C., Kwade, A.: Simulation of stress energy and grinding media movement within a wet-operated annular-gap mill using the discrete-element method. Chem. Eng. Technol. **35**, 1911–1921 (2012)
68. Beinert, S., Schilde, C., Gronau, G., Beinert, S., Schilde, C., Gronau, G., Kwade, A.: CFD-discrete element method simulations combined with compression experiments to characterize stirred-media mills. Chem. Eng. Technol. **37**, 770–778 (2014)
69. Beinert, S., Fragnière, G., Schilde, C., Kwade, A.: Analysis and modelling of bead contacts in wet-operating stirred media and planetary ball mills with CFD–DEM simulations. Chem. Eng. Sci. **134**, 648–662 (2015)
70. Iwasaki, T., Satoh, M., Takahashi, T.: Estimation of net energy applied to powders in a newly designed bead mill. Powder Technol. **119**, 95–101 (2001)
71. Iwasaki, T., Kim, J.Hwan, Satoh, M.: Characterization of media mills based on mechanical energy applied to particles. Chem. Eng. Sci. **61**, 1065–1073 (2006)
72. Strobel, A., Romeis, S., Wittpahl, S., Herre, P., Schmidt, J., Peukert, W.: Characterization of stressing conditions in mills—a comprehensive research strategy based on well-characterized model particles. Powder Technol. **305**, 652–661 (2017)
73. Strobel, A., Schwenger, J., Wittpahl, S., Schmidt, J., Romeis, S., Peukert, W.: Assessing the influence of viscosity and milling bead size on the stressing conditions in a stirred media mill by single particle probes. Chem. Eng. Res. Des. (2018)
74. Beinert, S., Kwade, A., Schilde, C.: Strategies for multi-scale simulation of fine grinding and dispersing processes: drag coefficient and fracture of fractal aggregates. Adv. Powder Technol. **29**, 707–718 (2018)
75. Löffler, F., Muhr, W.: Die Abscheidung von Feststoffteilchen und Tropfen an Kreiszylindern infolge von Trägheitskräften. Chem. Ing. Tech. **44**, 510–514 (1972)

Measurement of the Adhesion Moment of a Particle on a Wall in a Gaseous Environment and Comparison to Simulated Data

Alexander Haarmann and Eberhard Schmidt

Nomenclature

A	Hamaker constant (J)
a	Distance between atom and wall (m)
a_0	Contact distance (m)
d_p	Particle diameter (m)
d_w	Distance between the actual layer and the wall (m)
F_a	Adhesion force (N)
F_e	External force (N)
F_{el}	Electrostatic force (N)
F_{sp}	Spring force (N)
f	Flattening factor
k	Spring constant (N m^{-1})
Δl	Elongation (m)
M_a	Adhesion moment (Nm)
M_e	External moment (Nm)
$Q*$	Heat flow (J s^{-1})
r_{pe}	Penetration depth (m)
s_a	Lever arm "adhesion" (m)
s_e	Lever arm "external" (m)
ΔT	Temperature increase (K)
U	Contact potential (V)
ε	Relative dielectric constant
ε_0	Electric field constant (N V^{-2})
λ	Thermal conductivity (J s^{-1} m^{-1} K^{-1})
ρ	Atomic density of the material (m^{-3})

A. Haarmann · E. Schmidt (✉)
Institute of Particle Technology, Bergische Universität Wuppertal, Wuppertal, Germany
e-mail: eberhard@uni-wuppertal.de

S. Antonyuk (ed.), *Particles in Contact*,
https://doi.org/10.1007/978-3-030-15899-6_14

Introduction

The behaviour of particle collectives plays an important role for the industry. The question of how best to plan and realize processes that are part of the manufacturing, handling, portioning and packaging of solid bulk materials always arises; here, the best way is often the one with the least loss and effort. The macroscopic behaviour of a particle collective is dictated by the behaviour of its small components in interaction with each other and with their surroundings.

An experimental setup was realized to measure the adhesion moment of a particle on a wall in a gaseous environment. The experiments took place in the specimen chamber of an environmental scanning electron microscope (ESEM). For the measurement of the adhesion moment, a spring table system and a micromanipulator were installed to stress the particles. The experiments were carried out to compare the calculated results for both the adhesion force and moment from a previously developed simulation [1–5]. In the aforementioned publications, more information and literature can be found.

The simulation calculates the adhesion moment based on the adhesion force caused by the van der Waals interactions between the atoms contained in the particle volume and the wall. It does not take forces based on electrostatics or liquid bridges into account. The usage of an ESEM for the experiments allows the liquid bridges to be disregarded, as the humidity in the chamber is controllable. On the other hand, when one observes an experiment with the aid of an electron microscope, there will always be a certain amount of electrostatics in the chamber. An estimation of the influence of the electrostatic forces was made. With the right detectors and microscope settings, one can at least try to keep the charging low.

Adhesion Force

Calculation of the Adhesion Force

In this section, the focus is on the calculation of the adhesion force as it is done by the simulation. As mentioned above, a simulation was developed at the institute of particle technology (IPT) to calculate the adhesion moment. To do so, the first step is to calculate the adhesion force. The simulation was written in C++ and uses a desktop PC as platform. It is theoretically capable of calculating every particle size, although this takes a very long processing time as the particle size increases. The calculations were made mainly for four different particle diameters—103, 154, 206, and 411 nm—which were chosen because of the possibility to compare the calculated values with results of a former project at the IPT [6]. Beside these diameters, particles with diameters of 17 and 41 μm were simulated for comparison with adhesion forces measured at other universities. The small particles can be simulated in a few minutes; the 41-μm particle took several hours. This is because the particles are simulated as single atoms. As the number of atoms grows very fast with increasing particle size, it takes much longer time to calculate larger particles.

To gain further understanding of the adhesion force calculation, one should have a look at how the simulation works. First, a three-dimensional space is laid around the particle. The length of each side of the cubic space equals the diameter of the particle. Although it would be possible to speed up the simulation by reducing the empty space to be scanned, the decision was made to keep the cubic shape because thus it will be possible to place particles of any shape in this space, as long as the longest part of the particle equals the side length. After the space is set, the simulation starts to scan the given space in steps of one atomic diameter, starting at the origin of the coordinate system the particle is located in. There are three different ways to scan the space, which have been implemented by now. A cubic primitive, a cubic face-centred, and a cubic body-centred pattern can be scanned. These patterns reproduce the atomic lattices of materials. The atomic diameter of the material and the scan pattern are user-specified values and can be changed. The atom density in the material increases from the cubic primitive to the body-centred to the face-centred pattern.

In Fig. 1, the lowest two planes of a simulated particle are pictured. The perspective is from the wall towards the bottom of the particle. The atoms are magnified and pictured as small spheres with one atomic diameter. The pictures are made with the ray tracing software "POVRay". In Fig. 1a, the atoms are rectangular to each other both inside of one plane and in relation to the atoms of the second plane, which represents the cubic primitive pattern. In the cubic body-centred pattern in Fig. 1b, the atoms are still rectangular to each other in one plane, but the planes are shifted against each other, so that an atom always lies in a hollow between four other atoms of an above- or underlying plane. Finally, the cubic face-centred pattern is pictured in Fig. 1c, where not only the planes but also the atoms are shifted against each other and a higher package density is obtained. The pattern is created by modification of the steps during the scanning process. Other patterns would be possible, but every specialization makes the simulation less flexible, so the decision was made to stay with the implemented three patterns.

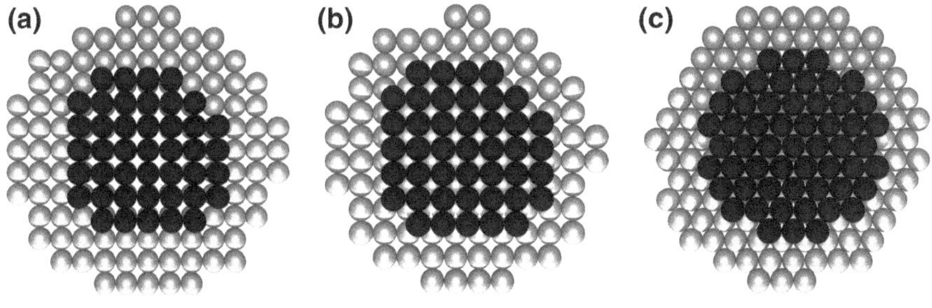

Fig. 1 Examples of the three scanning principles: **a** cubic primitive, **b** cubic body-centred, and **c** cubic face-centred pattern

There are two reasons why a particle in contact with a wall could be non-spherical. First, the particle would deform under the stress built up by the adhesion forces [7, 8]; second, since particles are stressed during manufacturing and procession, it is almost impossible that they should stay spherical. To pay attention to this fact, a flattening of the particle was made possible in the simulation. The flattening is a user-specified value. It is the product of a flattening factor f and the particle radius r_p.

$$Flattening = f \cdot r_p \tag{1}$$

Although the flattening factor is a user-specified value, the simulation will automatically run through different flattening factors. This can be changed easily, if desired. If one has a calculated or measured flattening factor or contact radius at hand, the simulation can use it right away. It was decided not to implement the calculation of the flattening, as it depends on various parameters and, even if it is calculated correctly, there could still be a deformation on real particles caused by forces acting in the process of handling that could not be considered by the calculation. Therefore, the calculation instead runs through a series of pre-set flattening factors, and it can be observed how this influences the adhesion force and moment (Fig. 2). With increasing flattening, an increase in the adhesion force of a particle-wall contact can be observed. This result was obtained with the simulation as well.

In Fig. 2, the adhesion forces for four simulated particles with different diameters are plotted against an increasing flattening factor f. The diameters of the particles are 103, 154, 206, and 411 nm and the flattening factor starts at a value of 2×10^{-4} and stops at a value of 0.5 with different steps in between, resulting in contact radii ranging from 2 to 356 nm. Taking a closer look at the curves, it can be seen that the diameter of the particles influences the adhesion force in a linear way. This can be seen by taking a specific value for the flattening factor and drawing a straight line rectangular to the x-axis over the four curves.

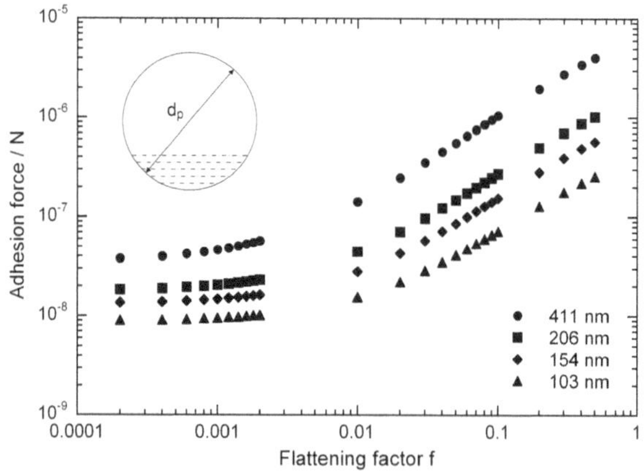

Fig. 2 Adhesion forces of four simulated particles with different diameters plotted against an increasing flattening factor f

The flattening process itself has a more complex influence, which can be seen by following the values for a single diameter. For very small flattening factors, the adhesion force increases slowly; then, the increase reaches a maximum, and at the high flattening factor values, the curve gradient becomes shallower. If the flattening of the particle was continued, this shallowing would become even more significant. The reason for this behaviour can be found in the way the particle is flattened in the simulation. In reality, the particle is deformed by some kind of stress and the material it consists of would perform an evasive movement; it could fill gaps in the particle structure or form some kind of bead at the contact point. Despite this fact, in the simulation, the material is simply cut off, causing an error as the adhesion force calculated is too small compared to the adhesion force of a real particle. This error is not that big for small deformations, but increases with the flattening. The benefit achieved from this simple point of view is a fast and simple simulation that concentrates on other effects influencing the adhesion force in a more significant way. It has to be said here that the main goal of the simulation is not to picture all the phenomena of contact mechanics but to estimate the expected adhesion moment of a particle-wall contact. The calculated adhesion forces are bigger than the already experimentally measured forces [1, 2], despite the fact that other interactions than van der Waals forces were not involved, and therefore the cut-off method error was neglected.

The adhesion force calculated by the simulation is based on the van der Waals attraction, as this interaction force is said to be the strongest of the short-distance interactions despite the liquid bridges [9], according to Eq. (2) [10]:

$$F_{atom-wall} = A/(2 \cdot \pi \cdot \rho \cdot a^4) \tag{2}$$

where A is the Hamaker constant for the particle material in interaction with the wall material through a medium, ρ is the atomic density of the wall material, and a is the distance between the examined atom and the wall. According to Israelachvili [10], Eq. (2) is valid even if intermolecular or interatomic forces are present, as long as the calculation of the Hamaker constant pays attention to these circumstances. Following Eq. (2), the van der Waals interaction decreases dramatically with increasing distance between the contact partners. So the question is how much influence the atoms in the upper part of the particle have on the adhesion force. Compared to the atoms close to the wall, this influence is insignificantly low. Because of this fact, repeated simulation runs were performed to figure out at which distance from the wall the van der Waals interactions between an examined atom and the wall become negligible. To achieve this, the adhesion force of a whole particle was calculated as usual. Then, the adhesion force was calculated again, and after each atom, the actual distance from the wall and the percentage of the total particle adhesion force reached was captured. It was found that there is an active volume above the wall with a thickness of less than 30 nm. The atoms in this volume are responsible for the van der Waals interaction between the particle and the wall (Fig. 3).

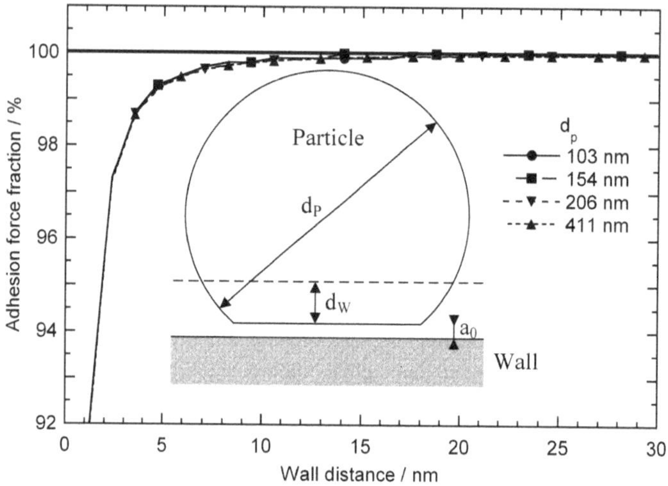

Fig. 3 Development of the adhesion force fraction based on the van der Waals interaction with increasing distance from the wall, d_P stands for the particle diameter, d_W for the wall distance, and a_0 for the contact distance of 4×10^{-10} m

To avoid any further errors in the force calculation, the decision was made to simulate the particle at up to 40 nm above the wall where 100% of the particle adhesion force should definitely be reached. This significantly decreased the simulation time and it became possible to simulate larger particles in a reasonable amount of time.

One problem that occurred during the comparison of the calculated and measured adhesion forces was that the simulation overestimated the adhesion force (see below). This was not expected as the simulation features only the van der Waals attraction while in reality there could also be adhesive forces caused by humidity or electrostatics. It is claimed that the overestimation is caused by one major difference between the simulated and the real particles. The simulated particle is built in a perfect way. There are no flaws or irregularities in the atomic lattice so that it would comply with the lattice of a monocrystalline material. To picture the flaws of a real material in the simulation, porosity of the particle was devised. A random number is assigned to each atom during the scanning process, and if that number does not equal a user specified threshold (which is defined by the porosity in percent), the atom is eliminated in the adhesion force calculation.

Finally, as the project does not only concentrate on smooth particles, a roughness was implemented in the simulation. This is not an overall surface roughness of the material but it means that one spherical structure is placed on the surface of the smooth particle. It was devised to examine the effect of such a structure on the adhesion moment, depending on the impact angle of the lateral force stressing the contact. By now, the roughness consists of the same material as the main particle, which means that the atomic diameter and lattice are the same and its diameter equals 1/10th of the main particle diameter. The roughness has only a small effect on the

Fig. 4 Image of a particle with a diameter of 40 nm, located on the wall. The particle is built in the cubic primitive lattice; it features a roughness, a flattening factor of 0.2, and a porosity of 10%

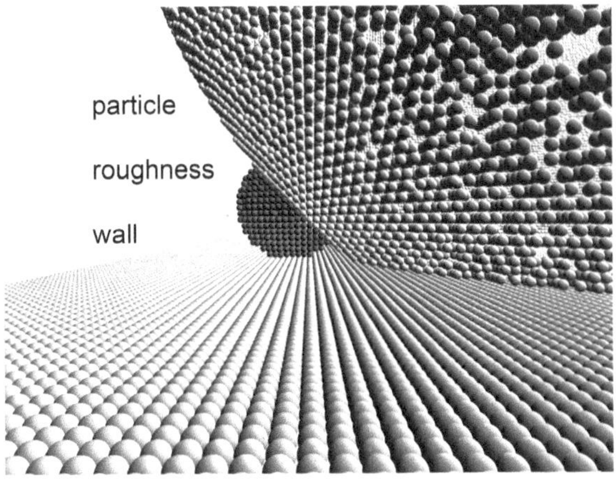

adhesion force since it does not increase the number of interacting atoms, but it has a significant effect on the adhesion moment where it influences the lever arm according to its location in the system (see below).

A simulated particle featuring flattening, roughness, and porosity is illustrated in Fig. 4. The depicted particle is hollow due to the limitations of the imaging software. Although the inner atoms are not shown, they are considered for the calculations. The particle has a diameter of 40 nm and a flattening factor of 0.2; it is built in a cubic primitive lattice and has a porosity of 10%. The first layer of wall atoms is also depicted.

Comparison to Experimental Results

As this project was part of a DFG priority program, there were related projects run at other universities. At the University of Magdeburg the adhesion forces of real particles were measured in atomic force microscope experiments [11]. These measured adhesion force values were compared to the calculated results of the simulation. This comparison can be seen in Fig. 5 where the adhesion force is plotted against an increasing flattening factor f. The experimental values, although there are only two of them, are the average values of many measurements. There was no accompanying statement regarding the flattening of the particles, but they were not artificially stressed. As can be seen, the unmodified particles build up less adhesion force than the cleaned ones. According to the considerations made, this could mean that the surface of the cleaned particles is smoother than that of the uncleaned ones. Residues and other kinds of roughness on the surface reduce the number of atoms located in the active volume.

The main result of the comparison in Fig. 5 is that the simulation matches the measured values of the adhesion force. This happens where the curves of the calculated

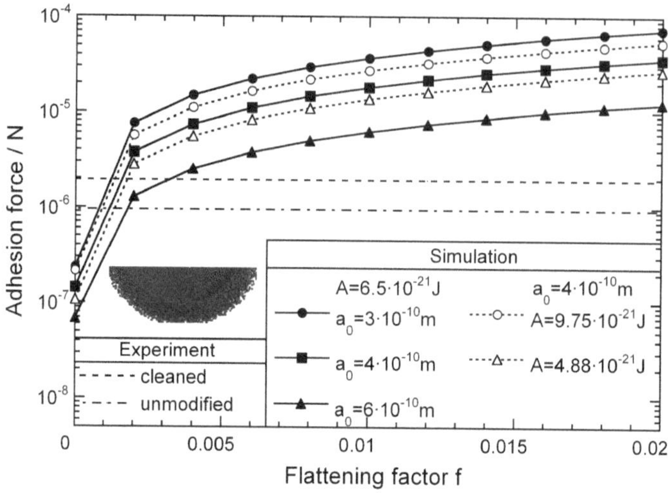

Fig. 5 Calculated and measured adhesion forces plotted against an increasing flattening factor f. The particle diameter is 17.3 mm. The experiments were made with unmodified and permonosulfuric acid-treated (cleaned) glass particles. In the simulation, the contact distance a_0 and the Hamaker constant A were varied

force cross the lines representing the measured force. There is only a very slight flattening there, but the flattening factor does not equal zero, which would be exactly the predicted behaviour for a particle in contact with a wall. According to the material stiffness, a small contact area will form due to the adhesion force, which is represented by the simulation. The simulation was run without any porosity in this case and the scan was performed in a cubic primitive pattern. Porosity would decrease the total force; a change in the pattern would increase it.

The Hamaker constant and the contact distance were varied in the simulation runs in order to examine their influence on the calculated adhesion force that led to the curves in Fig. 5. The curve with the filled-square markers represents $A = 6.5 \times 10^{-21}$ J and $a_0 = 4 \times 10^{-10}$ m, which is assumed as normal condition. An increase (filled triangles) of the contact distance decreases the adhesive force, and vice versa (filled circles). The inverse effect can be observed when the Hamaker constant is varied. An increase (hollow circles) of the value increases the adhesive force and a decrease (hollow triangles) makes the force smaller. The variation of the contact distance has the more significant influence. Variations of these values could be used to find explanations why the calculated adhesion force might be too high or too low compared to the measurements. Despite this, the influence of the surface of both the particle and the wall is important because surface structures define the number of atoms located in the active volume.

Adhesion Moment

Calculation of the Adhesion Moment

Smooth Particle with Flattening

If a particle adheres to a flat surface, it deforms. The contact models of Johnson, Kendall and Roberts (JKR) [7] and Derjaguin, Muller and Toporov (DMT) [8] predict this behaviour. If this particle gets stressed sideways by an external force, a resisting moment, called the adhesion moment (Fig. 6), must be overcome before the particle starts to roll.

The adhesion moment is influenced by the adhesion force acting between the particle and the wall and the lever arm, which is set by the radius of the contact zone. Depending on the stiffness of the particle material, the radius of the contact zone can vary. Thus, for two materials with a comparable density, the softer material would build up the higher adhesion moment. Two effects cause this phenomenon: The softer material produces a larger contact zone and therefore larger lever arm through the greater deformation. There is also more particle material in the significant volume above the wall where the van der Waals forces have a significant influence on the adhesion force. At this point, the significant volume should be explained.

The van der Waals interaction makes a significant contribution to the total adhesion force between two bodies if they are in extremely close proximity. If one prevents the existence of liquid bridges (due to lack of humidity in the specimen chamber) and keeps electrostatic forces low, the van der Waals attraction dominates the adhesion force if the distance between the contact partners is low [9]. In the aforementioned simulation, the force between an atom and a wall based on the van der Waals interaction is calculated by Eq. 2.

In the calculation, A is the Hamaker constant for the combination of the particle and wall material interacting with each other through the atmosphere of the ESEM specimen chamber. ρ is the atomic density of the wall material and a is the distance between the contact partners.

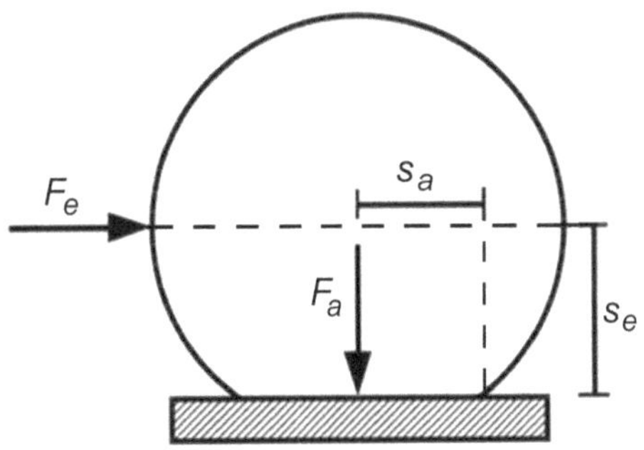

Fig. 6 Figure of the flattened particle adhering to a wall. The external force F_e stresses the particle sideways. It triggers a moment via the lever arm s_e that works against the adhesion moment caused by the adhesion force F_a via the lever arm s_a

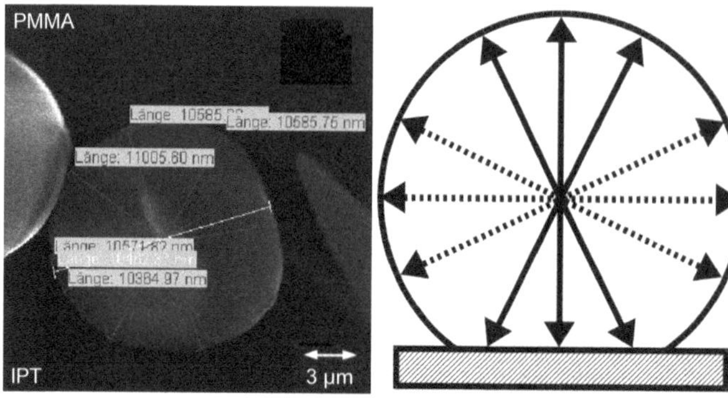

Fig. 7 Left: Measurement of the diameter of the particle and its height after the experiment. The particle was rotated by 90°. Both the diameter and the height are measured three times and the mean value is calculated and used. Right: Picture explaining how the three measurements of the diameter (dotted lines) and the height (solid lines) were made

As can be seen, the force based on the van der Waals interactions decreases by a power of four as the distance between the contact partners increases. This means that the material of the particle very close to the wall has the most influence on the total particle adhesion force. The volume of the particle above the wall where the material contributes significantly to the adhesion force is called the active volume.

To calculate the adhesion moment through the applied external force, the lever arms s_e and s_a (Fig. 6) must be known. In the experiments, the particle could be touched at the midpoint of its height, as mentioned before. This means that the lever arm s_e can be measured, since it is half of the measured diameter of the stressed particle, whereby the flattening of the particle must be subtracted from this value (Fig. 7).

Since the measurement of the diameter and height of the particle is also performed by means of picture analysis, a decision was made to measure three times and take the mean value. This is because it is not that easy to take precise measurements at this magnification, and although the experiments were carried out with care, it was not possible to be absolutely certain that the particle was really turned 90°. This is also hard to judge based on a 2-dimensional grey scale image with a lesser or greater amount of noise. With the measured flattening, it is possible to calculate the radius of the contact zone for the particle on the wall. The radius of the contact zone is the lever arm of the adhesion force.

Smooth Particle with Roughness

For a smooth particle with flattening, the examined system is symmetrical and it does not matter from which direction the external force attacks. The adhesion moment is defined through the adhesion force and the lever arm, which depends on the flattening.

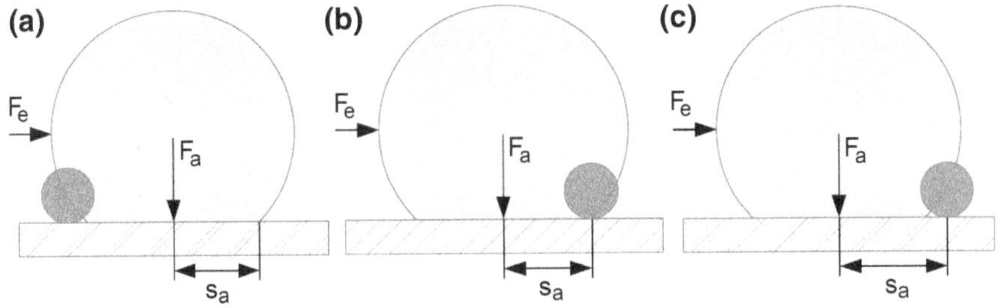

Fig. 8 Schematic pictures of the system external force F_e and the rough particle with the lever arm length. **a** Starting point of the simulation with the roughness in front of the particle, **b** the roughness is located at the edge of the contact zone (minimum adhesion moment), and **c** the roughness is located behind the particle (maximum adhesion moment)

But if a roughness is placed on the particle surface, this will change the behaviour of the adhesion moment as it might increase the lever arm, depending on the orientation of the external force to the roughness. To calculate the adhesion moment, the system of external force and point of contact of the roughness with the wall was rotated by 360° in 200 steps.

Three remarkable states were found, as illustrated in Fig. 8. In Fig. 8a, the starting point of the simulation of the adhesion moment is depicted for the rough particle. The point of contact of the roughness with the wall and the external force are in line. The roughness lies in front of the particle. The roughness itself produces a small adhesion force, which builds up an adhesive moment through the lever arm between its contact point with the wall and the edge of the main particle's contact zone. In addition to this adhesion moment, the main particle builds up an adhesion moment through the lever arm between its middle and the edge of the contact zone. Now the orientation of the external force and the contact point of the roughness start to rotate. The minimal adhesion moment for the rough particle is reached when the condition of the system equals the situation depicted in Fig. 8b. The contact point of the roughness is now in line with the edge of the contact zone. So, the roughness no longer has its own lever arm and the moment becomes zero. The moment of the main particle still exists and equals the moment of a smooth particle with the same flattening. As the system is rotated further, it reaches the condition pictured in Fig. 8c. The contact point of the roughness is again in line with the external force, but the roughness is now located behind the main particle, increasing the lever arm for the main particle's adhesion force. Although the roughness still does not build up its own adhesion moment, this lever arm extension leads to the maximum adhesion moment. After this point, the system completes the 360° rotation and regains the starting position.

In Fig. 9, the adhesion moment for the rough particle with two different flattening factors ($f = 0.0$ and $f = 0.5$) is plotted against the rotation angle. The left ordinate-axis scale corresponds to $f = 0.0$, and the right scale to $f = 0.5$. The three remarkable states illustrated in Fig. 8 can be found most obviously for $f = 0.0$, for the particle without flattening. The starting point is located on the left at zero degrees. The roughness

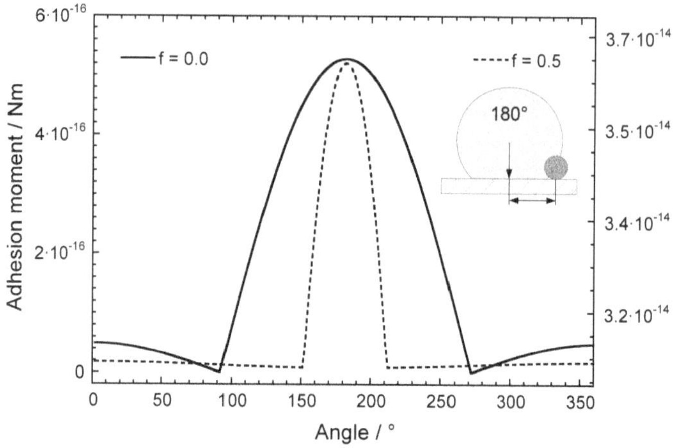

Fig. 9 Adhesion moment of the rough particle for two different flattening factors plotted against the rotation angle. The adhesion moment values on the left axis refer to a flattening factor of 0.0, the ones on the right axis, to 0.5

lies in front of the particle and in line with the external force. Although the main particle is not flattened, there is an adhesion moment caused by the adhesion force of the roughness and its lever arm between its own and the main particle's contact point with the wall.

As the system starts to rotate, the roughness moves around the main particle. The lever arm decreases as the roughness gets nearer to the main particle's contact point with the wall. At 90°, this point is reached and the roughness has no lever arm left. The adhesion moments of both the roughness and the main particle become zero due to the non-existing lever arms. As the rotation continues, the roughness moves behind the particle and, although it has no adhesion moment by itself, it creates a lever arm for the adhesion force of the main particle. As the system has rotated by 180°, the roughness lies right behind the particle and is again in line with the external force, and the maximum adhesion moment is reached. Further rotation repeats the values as the roughness moves back to the starting position along the other side of the particle. Looking at the curve for the flattened particle, it can be seen that the overall value for the adhesion moment is higher due to the always-present lever arm created by the contact zone. The influence of the roughness is almost negligible until it moves behind the edge of the contact zone.

Similar experiments were done by Ding et al. [12]. They measured a maximum resistance force of 128 nN for a particle of 23.2 μm in diameter. Assuming that the lateral force was initiated equatorially and the examined particle actually rolled, a value of about 1.5×10^{-12} Nm is obtained for the adhesion moment. A particle of this size was simulated in the cubic primitive lattice without porosity and with a small flattening factor of 0.02. The simulation calculates an adhesion moment of about 3.5×10^{-12} Nm.

As mentioned before, the moment depends on the adhesion force and the force is always overestimated without coarse adaption to the material. To compare the calculated values for the adhesion moment to more experimental data, experiments were started, which will be described below.

Materials and Methods

Particles

Spherical polymethyl methacrylate (PMMA) and bronze particles with a diameter of 40 μm were used for the experiments. Although it has been attempted to use other materials in the experiments, the best results were achieved with the PMMA particles. This is due to the ability of these particles to form a contact radius on the wall. This characteristic could not be observed with the bronze particles. Other used particles were insufficiently spherical, and a controlled manipulation was not possible. The movement direction was excessively dependent on the shape of the particles.

Since the electron beam of the ESEM can melt PMMA, the fact that the flattening at the contact point between the particle and the wall could be caused by a rise in the particle temperature above the melting point was checked. To estimate the temperature increase, the formula of Reimer and Pfefferkorn [13] was used:

$$\Delta T = Q^{*}/(2 \cdot \lambda \cdot r_{pe}) \tag{3}$$

The above formula can be used to estimate the temperature increase in the particle material irradiated by the electron beam. The increase is calculated from the heat flow caused by the electron beam divided by the thermal conductivity of the material λ and the penetration depth of the primary electron beam r_{pe}. Usage of the above formula with the experiment parameters gives an estimated temperature increase of 12 K. Assuming that the particle material is at the same temperature as the room in which it was stored, the starting temperature was about 293 K. The increase in temperature is not sufficient to reach the melting temperature of PMMA of about 433 K. The acting adhesive forces thus cause the deformation of the particle on the wall. The relatively soft PMMA particles are therefore more suitable for the experiments than particles consisting of more rigid materials such as bronze.

Another influence that should be considered besides the van der Waals forces and the adhesive characteristic of liquid bridges is that of electrostatic forces. Especially in the chamber of an ESEM, there must be electrostatic forces acting between the particle and the wall. The estimation of the magnitude and influence of electrostatic interactions was carried out in accordance with Stieß [14]:

$$F_{el} = \pi \cdot \varepsilon_0 \cdot \varepsilon \cdot U^2 \cdot d_p/(2 \cdot a_0) \tag{4}$$

In this formula, ε_0 is the electric field constant (8.85×10^{-12} N/V^2) and ε is the dielectric constant of air (1.00054). The values for U—the contact potential—are given by Stieß as 0.1–0.7 V. This gives a calculated electrostatic force in the range of 3.5×10^{-9} to 1.7×10^{-7} N. The simulated adhesion force for this particle is 6.7 $\times$ 10^{-5} N. The influence of the electrostatic force can be disregarded.

Measurement System

The experiments were carried out in the specimen chamber of a FEI Quanta FEG 450 Mk. II environmental scanning electron microscope (ESEM). The environment in the specimen chamber can be controlled so that liquid bridges can be avoided by keeping the moisture low. A spring table system with up to two micromanipulators can be installed in the specimen chamber (Fig. 10). Kleindiek Nanotechnik constructs the spring table and the micromanipulators.

The micromanipulator has two rotating axes and one linear axis. The rotating axes move the manipulator up and down and left to right with an accuracy of 5 nm and a range of 240°. The linear axis moves the tip in and out with an accuracy of 0.25 nm and a range of 12 mm. The manipulator is driven by piezoelectric nanomotors. To achieve large ranges, it moves like a ratchet. The stator is extended or contracted by electrical voltage. If it reaches one end of its movement range, it is moved to its starting position so fast that the manipulator ideally keeps its current position. The movement can then start again. Because of the extremely high accuracy of the linear axis, a decision was made to stress the particles in this direction. The problem that occurred was that it was not actually possible for any of the particles to be touched: They all jumped away. This phenomenon was first credited to the electrostatic charge, but the manufacturer stated that the ratchet movement of the linear axis occurs after every micrometre that this axis is moved. This means that the observed jumping might well be caused by the tip itself pushing the particles away while performing the ratchet movement. Afterwards, the rotating axis was used to stress the particle and the phenomenon disappeared.

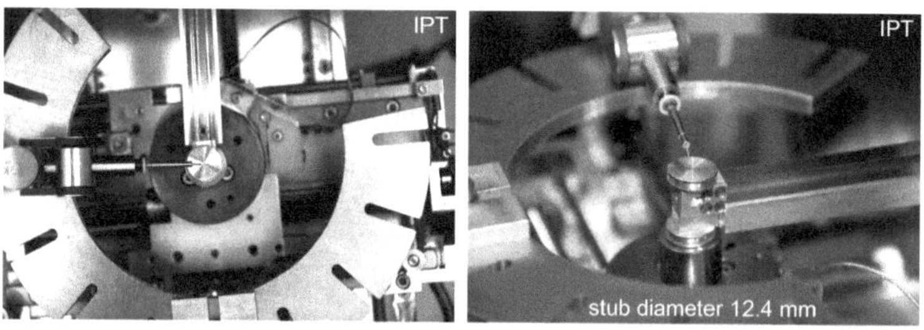

Fig. 10 The experimental setup in the specimen chamber of the ESEM (top view on the left and side view on the right). The micromanipulator with the tungsten tip and the stub mounted on the spring table can be seen

To measure the force, the spring table is connected tightly to the frame. The stub on top of the spring is inserted into a holder at the end of the spring. Depending on the length of the spring steel and its spring constant, different spring forces can be realized. In the experiments, a spring with a spring constant of 12.1 N/m was used. It is important to stress the particle at an angle of 90° relative to the spring. Thus, in Fig. 11, a pushing motion of the micromanipulator stresses the particle. As mentioned above, the stress was later applied with one of the rotational axes. Using the rotational axis for the left and right movement might cause a problem, since the force has to be applied to the particle at an angle of 90° relative to the spring, and the tip moves in a circular motion over the rotational axis of the micromanipulator. This problem is minor, since the tip is placed right in front of the particle at approximately 90° relative to the spring and it is not moved over large distances.

The spring table is one component of the measurement system; the other is the software that analyses the movement of the table and calculates the force needed for this movement. The measurement process can be described as follows: Several particles are placed randomly on the stub, which is then inserted into the specimen chamber with the installed spring table and the micromanipulator. The tip of the manipulator is placed in front of the particle but not in contact with it. Later in the experiment, the tip should touch the particle at the midpoint of its height. The control box for the software then has to be chosen. The software is able to track the pixels in the control box based on their grey scale value. It is therefore best to choose a place on the stub with a significant difference in brightness to track (Fig. 11).

The manipulator is then used to stress the particle sideways. As the manipulator tip continues to move, the table also moves, since the stressed particle adheres to its surface. With the increasing movement of the tip, the bending spring applies more and more force on the contact point between the particle and the stub. After some time, the force of the spring acting on the stub breaks the contact between the particle

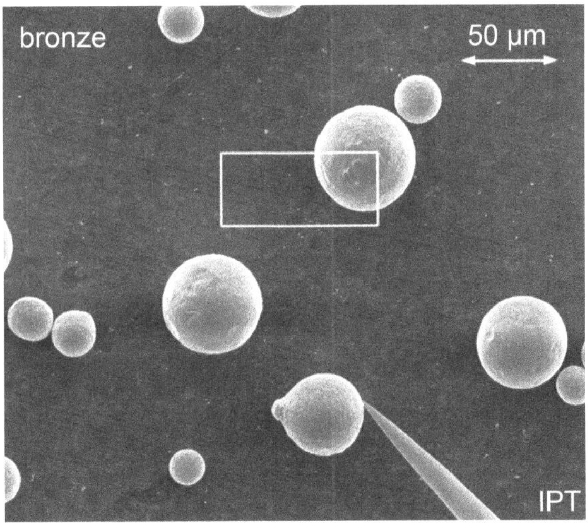

Fig. 11 Picture from the experiment. Bronze particles can be seen on a polished aluminium stub. The control box is placed over one of the non-stressed particles

and stub, and the particle starts to roll. A sliding movement of the particle would require more energy than a rolling movement. For this reason, as long as one assumes that the friction between the manipulator tip and the particle is small and the contact area between the particle and the wall is small as well, the particle should start to roll rather than to slide. The moment that must be overcome to make the particle roll is the adhesion moment. Only the force necessary to overcome the initial resistance is taken into account.

Pictures taken by the ESEM document the whole process. In the experiments, one picture per second was taken. The more quickly a picture is taken in an ESEM, the less scanning time is incurred. However, the longer the scanning time, the better the picture (noise reduction). Since the software performs an analysis based on pictures, one picture per second was a good compromise between quality and speed.

Following the experiment, the analysis software can be used to measure the force applied by the table. The software uses the tracking box to determine how far the table has moved. Based on this distance and with a known spring constant, a force can be calculated in accordance with Hooke's Law whereby F_{spring} is the force, k is the spring constant and Δl is the distance the table moves.

$$F_{sp} = k \cdot \Delta l \tag{5}$$

Figure 12 is shows the values that the software obtains through the picture analysis. The values are plotted against the experiment time in seconds. The movement of the manipulator tip occurs in the direction of the x-axis. Nevertheless, in this particular experiment, the y direction is also plotted. This is to show that the movement of the table in the y direction is negligible. For this reason, the movement in y direction will not be plotted or discussed further.

The movement in the x and y directions in pixels is what is actually measured by the software. The other values—the distance and force—are then calculated based on the set magnification and Hooke's law. Since the distance is used for the calculation of the force and is itself based on the magnification, the measurement becomes more accurate as the magnification increases. The downside is that the higher the magnification, the more image noise is produced with the same scanning time, and the entire pushing process can no longer be observed, since the view becomes narrower. Nevertheless, if an extremely small force or moment needs to be measured, a really high magnification might enable this.

Experimental Results

As the particle is stressed on the stub and moves, a force can be measured. Theoretically the behaviour of the system is clear. First, the adhesion moment must be overcome. This would result in a peak in the force curve. Then the particle rolls across the stub with a constant rolling resistance. The force to overcome this resistance

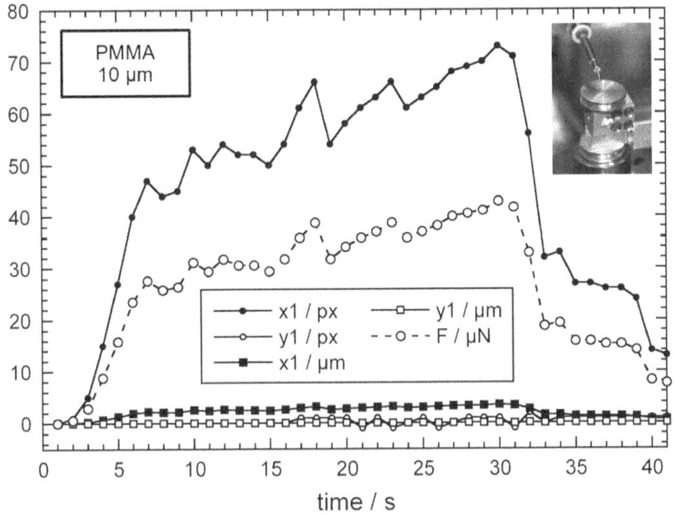

Fig. 12 Overview of the different values that can be measured and calculated by the software. The ordinate-axis is used for all values in their respective unit (px = pixel). The values are plotted against the experiment time. The deflection of the table can be seen in the x and y directions in px and as a distance in µm, as can the force F_{sp} built up by the spring table in µN

should be less than that required to overcome the initial moment. In reality, the particle surface is not perfect and the particle is not ideally spherical; furthermore, it moves across a stub that is not perfectly even. However, in this case, the stub surface was polished and quite smooth (root mean square roughness = 78 nm). There are therefore several peaks in the force curve, and most are even higher than the initial peak. Despite this, the first peak is considered as the force necessary to overcome the adhesion moment, since this is when the first rolling motion occurs. The other peaks might be the result of the surface roughness of both the particle and the stub or might be due to other effects. As the measurement experiments are still in the start phase, it has not yet been possible to collect much data. Some specific experiments will therefore be discussed. Figure 13 shows the force curve of one of these experiments.

The curve with the small solid circles is the more interesting one. There are several points to explain here. At second 3, there is a negative force. This might be caused by a jump-in-effect between the tip of the manipulator and the particle. As mentioned previously, the experiments start with the tip and the particle not in contact, so a jump-in-effect is possible. Subsequently, an increasing force is applied to the particle. It still adheres to the table and the contact is able to resist the increasing external force. At second 7, a first peak is reached, and afterwards the force decreases. The contact breaks and the particle starts to roll. At second 10 in this particular experiment, a second particle is touched with the manipulator tip, and the force rises steeply until second 18, when the tip rolls both particles (Fig. 14). As the second particle starts to

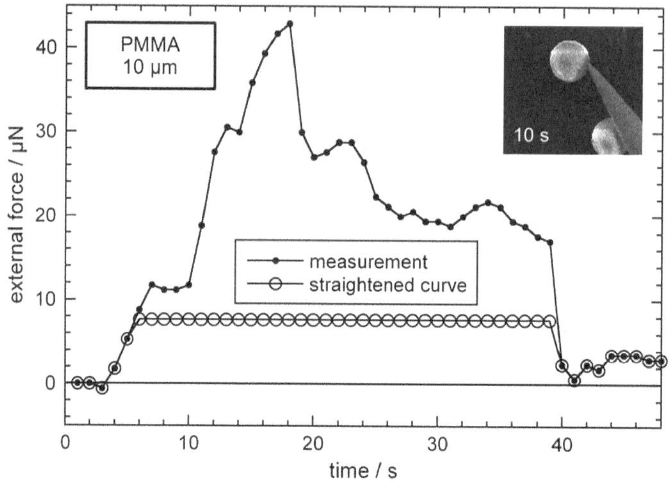

Fig. 13 The measured force plotted against the experimental time (solid circles) and the result of the same experiment with a curve straightened by the software (blank circles)

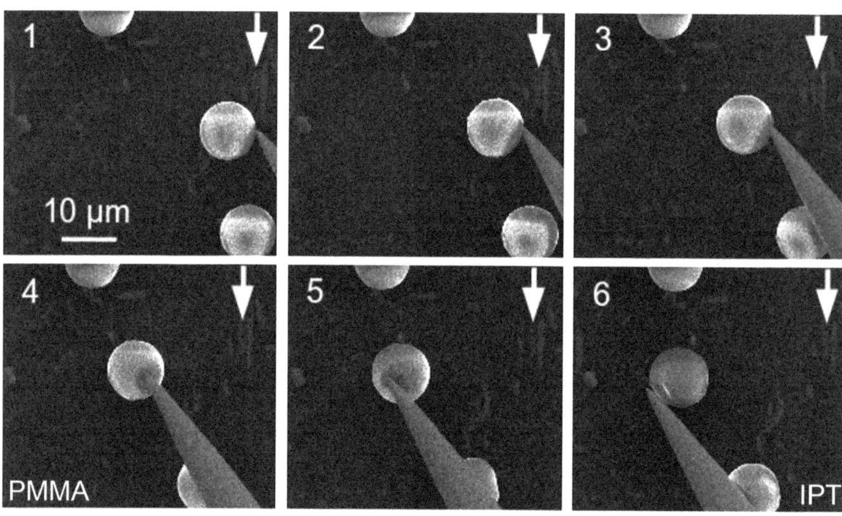

Fig. 14 Pictures from the experiment belonging to the force curves of Fig. 13. The experiment time for the pictures is as follows: 1 = 3 s; 2 = 6 s; 3 = 10 s; 4 = 20 s; 5 = 29 s; 6 = 44 s. The white pointer shows a striking pattern on the stub surface, indicating the table movement

roll too, the measured force decreases similarly to the behaviour of the first stressed particle.

In the curve of Fig. 13, another feature of the analysis software can also be seen. If the evaluation of the pictures is repeated, the software straightens the curve (blank circles). At first glance, the result is promising, since the particle rolls driven

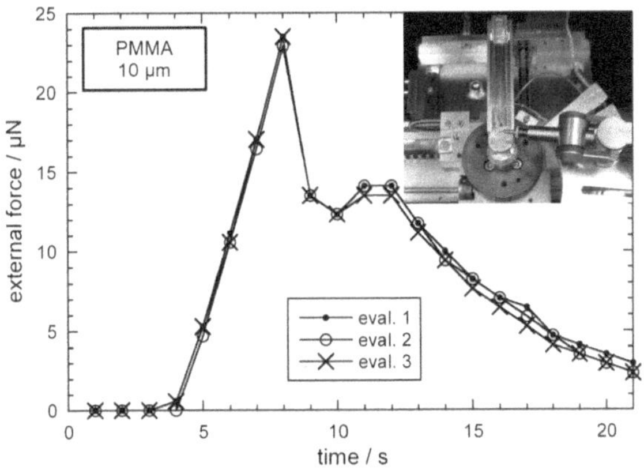

Fig. 15 Force curve for an almost perfect experiment. The stressed particle was rotated through 90°. No other peaks were measured. The measurement was evaluated three times with a different placement of the control box to confirm the curve

by a constant force after the adhesion moment is overcome. However, upon closer inspection of the pictures of the experiment, several events were taking place that cause a measurable force, viz. the second particle was touched at second 10 and both particles were stressed and moved across the stub. The straightening feature eliminates all these events. Following these results, straightening was avoided.

Figure 15 shows a curve for an almost perfect experiment. A particle lying completely alone on the stub could be stressed. It was rotated through 90° without any problems. This can be observed in the force curve. The adhesion moment is overcome at second 8, and after that the force decreases without any further peaks. It is worth noting here that the experiment was rather short and stopped after the rotation. The experiment was evaluated three times with the control box at different places in the picture. The result was confirmed through the evaluations. Once pictures have been taken, the pictures can be re-evaluated as often as desired.

The results were obtained through experiments where only PMMA particles were used. As mentioned previously, these particles are relatively soft, and might deform as a result of the heat applied to the material by the electron beam of the SEM. The deformable particles would be more likely to form a contact zone.

Without the contact zone, it is not possible for the adhesion moment to occur due to the missing lever arm. In experiments with harder bronze particles, an adhesion moment could not be measured with the specified measurement setup. The bronze particles were almost spherical and could be rolled without the application of any measurable force.

Another problem is a high dependency on the personal abilities of the operator of the system, since this makes the measurements less reproducible. It is important to bear in mind the fact that these experiments constitute only an initial step. Most of the process must be automated in order to make individual experiments more comparable. It would then also be possible to obtain larger data sets in a sensible amount of time.

References

1. Knoll, J., Nirschl, H., Haarmann, A., Schmidt, E: Adhesive behaviour of magnetizable particles in magnetic fields—simulation and verification. Chem. Eng. Technol. **2**(37), 357–363 (2014)
2. Haarmann, A., Mader-Arndt, K., Tomas, J., Schmidt, E.: Nanoskalige Betrachtung eines Partikel Wand Kontaktes – Vergleich zwischen Simulation und Experiment. Chem. Ing. Tech. **3**(86), 389–394 (2014)
3. Haarmann, A., Schmidt, E.: Simulation and measurement of the adhesion moment of a particle-wall contact on the atomic scale. Chem. Eng. Technol. **3**(39), 453–460 (2016)
4. Haarmann, A.: Haftmoment im Partikel-Wand Kontakt – Simulation und Messung des Haftmomentes in gasförmiger Umgebung, 1 Auflage. Shaker (2016). ISBN 978-3-8440-4880-3
5. Haarmann, A., Schmidt, E.: Adhesion moment of a particle on a wall in gaseous environment. Chem. Eng. Technol. **7**(41), 1484–1489 (2018)
6. Sonnenberg, J., Schmidt, E.: Numerical calculation of London-van der Waals adhesion force distributions for different superquadric shaped particles. Part. Part. Syst. Charact. (22), 45–51 (2005)
7. Johnson, K.L., Kendall, K., Roberts, A.D.: Surface energy and the contact of elastic solids. Proc. R. Soc. Lond. Ser. A **324**(1558), 301–313 (1971)
8. Derjaguin, B.V., Muller, V.M., Toporov, Y.P.: Effect of contact deformations on the adhesion of particles. J. Colloid Interface Sci. **2**(53), 314–326 (1975)
9. Rumpf, H.: Die Wissenschaft des Agglomerierens. Chem. Ing. Tech. **1**(46), 1–46 (1974)
10. Israelachvili, J.: Intermolecular & Surface Forces, 2 Auflage. Academic Press (1994). ISBN 0-12-375181-0
11. Mader-Arndt, K., Kutelova, Z., Tomas, J.: Microscopic particle contact adhesion models and macroscopic behavior of surface modified particles. In: Mittal, K.L., Jaisval, R. (eds.) Adhesion and Removal, pp. 105–156. Scrivener Publishing (2015). ISBN 978-1-118-83153-3
12. Ding, W., Howard, A.J., Murthy Peri, M.D., Cetinkaya, C.: Rolling resistance moment of microspheres on surfaces: contact measurements. Philos. Mag. **87**, 5685–5696 (2007)
13. Reimer, L., Pfefferkorn, G.: Raster-Elektronenmikroskopie, 2 Auflage. Springer (1977). ISBN 978-3-540-08154-8
14. Stieß, M.: Mechanische Verfahrenstechnik – Partikeltechnologie I, 3 Auflage. Springer (2009). ISBN 978-3-540-32551-2

Nanoindentation Based Colloid Probe Technique: A Unique Opportunity to Study the Mechanical Contact of Individual Micron Sized Particles

Thorsten Staedler, Katharina Diehl, Regina Fuchs, Jan Meyer, Aditya Kumar and Xin Jiang

Introduction and Motivation

A plethora of applications in pharmacy, cosmetics, food industry and other areas are directly linked to the research field of particle technology and contact mechanics in particular. The latter is a challenging topic as the characteristics of individual particle contacts and details of the contacting surfaces determine the behavior of whole ensembles. Consequently, in order to predict or model the behavior of such ensembles direct experimental access to the parameters describing the contact of individual particles is required [1, 2]. In the context of nanoparticles as well as particles featuring radii up to a couple of microns scanning probe microscopy represents today's most prominent method to sample the interaction of individual particles as well as particles and walls, i.e. the colloid probe technique [3, 4]. The corresponding strategies not only allow for the sampling of particle adhesion but also include the study of particle motions [5–7]. Unfortunately, this approach typically is limited with respect to load and particle size. Macroscopic techniques, on the other hand, fail to handle individual micron-sized particles with the required resolution in load and displacement. The advent of a new generation of high resolution depth sensing indentation instruments, which are typically called nanoindenters and now have access to a load regime previously reserved to AFM settings, however, paved the way for the option of transferring the colloid probe approach to a quantitative mechanical test platform, dramatically increasing load and particle size test regime.

As one of the key factors affecting the contacts of such particles in the low load regime certainly is given by the surface roughness of the contacting partners, the careful evaluation of the effect of surface roughness on normal, sliding, rolling and torsional contacts represents a central aim of this work. Here, an increasing roughness tends to increase the mean separation between two interacting bodies, which results

T. Staedler (✉) · K. Diehl · R. Fuchs · J. Meyer · A. Kumar · X. Jiang
Institute of Materials Engineering, Universität Siegen, Siegen, Germany
e-mail: thorsten.staedler@uni-siegen.de

S. Antonyuk (ed.), *Particles in Contact*,
https://doi.org/10.1007/978-3-030-15899-6_15

in a decrease of adhesion. Liu et al. [8] used the colloid probe technique to show the existence of an optimal amount of asperity yielding a minimum adhesion. This behavior was also reported by Rabinovich et al. [9] and Matope et al. [10]. The direct effect of surface roughness on particle motion was simulated by Karayem and Zakeri [11], who reported that the critical force for particle motion decreases when the particles are pushed on rough surfaces for rolling as well as sliding motion.

In the case of a sliding contact, investigations concerned with tribological phenomena have been carried out for atomically smooth surfaces [12, 13] as well as surfaces featuring different degrees of roughness [14]. Aside from a potentially strong adhesive contribution [15], in all cases Amontons' law is valid. However, a variation in friction coefficient can be measured and several models [16, 17] have been developed to correlate this variation with the corresponding variation in surface roughness. The most prominent theories have been proposed by Bowden et al. [16] and Greenwood et al. [18, 19] who suggested that the real contact area increase with applied load due to surface roughness. The results are in accordance with Amontons' law if plastic deformation of surface asperities was assumed.

Furthermore, qualitative results show that for particle rolling an external moment has to overcome a certain threshold [20], also known as critical moment or rolling resistance moment [21, 22] to initiate a rolling motion. These studies used a combination of lateral and normal load to initiate rolling, e.g., by using an AFM cantilever as a pushing tool. Rolling initialization with a controlled normal force was not realized until the work of Vilt et al. [23]. They presented a fundamental study of frictional properties of silica microsphere monolayers on silicon substrates by using ball-on-flat tribometer and reported that rolling resistance is highly dependent on load and sphere diameter. However, all of the above mentioned studies either require dedicated self-made set-ups or are limited with respect to the weight of the particle and/or the accessible load regime. With respect to these limits, the nanoindentation based approach presented here appears to be a suitable alternative.

Consequently, the opportunities of such an approach have been evaluated within the present study focusing on the contact mechanics of individual micron sized spherical particles. The goal was to use the nanoindentation based strategy to gain experimental access to particle adhesion as well as sliding, rolling and torsional friction of these particles in contact with Si surfaces as well as some selected additional wall materials such as diamond like carbon (DLC) and stainless steel. The majority of the results presented in the following paragraphs has already been published [24–28] and the reader is referred to these references for more details.

Choice of Contact Partners

The choice of particles as well as their contact partners (walls) within the framework of this study was in case of the particles strongly motivated by the availability of high quality commercial products or, in case of the wall material, a high safety in reproducibility of the synthesis/modification process. Consequently, SiO_2 micron-sized glass spheres were chosen as particles and mainly Si-wafers and their etched or DLC coated modifications were selected as walls. The following paragraphs provide some detailed information concerning these contact partners, which will be used in the experiments throughout this work.

SiO_2 Particles

Regarding model particles, guided by the work of van Zwol et al. [29], borosilicate glass beads provided by Duke Scientific were selected, which feature nominal particle radii of 10 μm (Duke Standards 9020, mean diameter 17.3 μm ± 1.0 μm, size distribution 2.0 μm std. dev.). The RMS roughness of these micro-spheres is 0.7 ± 0.1 nm [8]. Starting from these beads two types of particle probes were prepared. AFM based colloidal probes were produced by attaching (Araldite 10 min, 2 components, Epoxy) the micron-sized glass spheres to Mikromasch NSC12 tip-less AlBS cantilevers. In order to prepare colloid probes for the nanoindenter (TriboIndenter, Hysitron Inc.) used in this work, Focused-Ion beam system (FIB, FEI, Helios 600) was employed to create appropriate cavities into the apex of a commercially available diamond cube corner tip (Hysitron Inc.). Subsequently, the borosilicate beads were fixed to this holder by means of a small amount of photosensitive acrylate-based adhesive glue (DIC Europe GmbH), see Fig. 1a. The radii of the colloid probes and their quality (clean contact area) were determined by using scanning electron microscopy (SEM, ZEISS, Ultra 55). In case of the nanoindenter based probes, their mechanical integrity was subsequently tested by a series of indents with varying maximal load, ranging from 1 to 10 mN, into a smooth fused silica surface and sequential fitting of the resulting load-displacement curves by assuming a Hertzian contact [30]. The latter procedure was carried out to ascertained that all probes used in this study showed no hysteresis in load displacement curves monitored during normal loading into fused silica up to loads a hundred times larger than those used during the sliding experiments reported here, which was taken as strong indication that the glue employed to fix the beads does not cause any artifacts in the load signals gathered during normal and sliding testing.

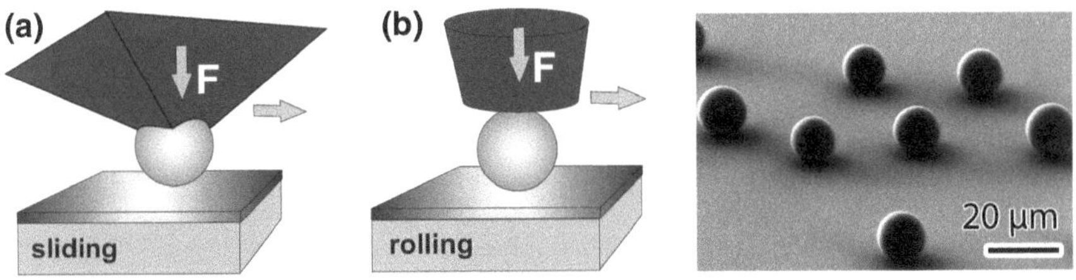

Fig. 1 Schematics of nanoindentation based colloid probe approaches in case of **a** a sliding and **b** a rolling/torsional contact (Reprinted with permission from [28].)

In context of the rolling and torsional experiments presented here, a flat end diamond indenter (20 μm diameter, Hysitron Inc.) and non-glued borosilicate beads were utilized to sample rolling and torsional friction (Fig. 1b), as detailed in [26]. Here, the misalignment of the indenter axis is below 1° and the RMS roughness of the central part of the punch was measured to be 4 ± 1 nm (on a lateral scale of 2.5 μm).

The actual experimental details concerning the test procedures of the different contact scenarios will be given in the corresponding section of this chapter, respectively.

Wall Material

The main substrate material used was a single-crystalline Si(100) wafer, supplied by Siegert Wafer GmbH, Aachen. Samples were cleaned by exposure to chloroform (CHCl3, 99%, Roth), afterward by rinsing with ethanol (EtOH, 97% with 1% petroleumether, Roth) and Milli-Q water, followed by drying in a stream of nitrogen. Surface topography changes of these samples were achieved through a slow etching process using H_2 plasma. The modification was carried out in an ASTeX A5000 microwave plasma chemical vapor deposition (MWCVD) reactor with an H_2 flow rate of 400 standard cubic centimeters per minute. Two different microwave powers (1600 W and 1800 W coupled with a gas pressure of 40 Torr and 45 Torr, respectively) are used to sustain the plasma. The etching period was 20 min for all samples. This treatment leads to a variation in surface roughness (Fig. 2) due to the different etching efficiency of the H_2 plasma with different microwave power. Afterwards, the modified samples were stored under ambient conditions for two weeks, allowing the formation of a thin natural oxide layer (~1 nm, see [31]). A FIB system was used to create rail-structure silicon samples (Fig. 3) featuring a length of 100 μm and rail inclinations of 25°, 45° and 60°, respectively. Depth and width of the rails were

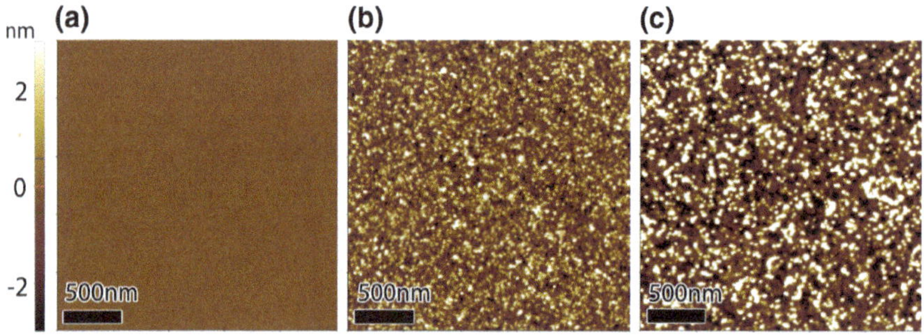

Fig. 2 AFM images of the surface of a silicon (100) wafer: **a** as received, **b** etched at 1600 W, and **c** etched at 1800 W, respectively (Reprinted with permission from [28].)

Fig. 3 Example of a rail system, which is used in the context of torsional testing, prepared by FIB (Reprinted with permission from [28].)

chosen in such a way that the center of mass of the particles was situated 1 μm below the wafer's surface level. This design effectively keeps the particles inside the rail during an experiment. SEM imaging was utilized to characterize the corresponding rail angles post preparation.

In Fig. 2, AFM topography images of the untreated as well as the etched surfaces are given (XE-100, PSIA). A clear change in topography of the Si samples with respect to increasing microwave power is visible from left to right. The AFM measurements were carried out with commercially available tips (ACTA, AppNano) featuring tip radii below 10 nm in non-contact mode. All samples were characterized at five different spots each. Scan-area and pixel resolution were chosen to be $2.5 \times 2.5\ \mu m^2$ and 512×512, respectively. The RMS roughness, at each of the five different spots, was calculated after a simple linear tilt correction in x- and y-direction. The mean RMS roughness is received as average of the RMS values of the five individual spots. Correspondingly, RMS values of 0.3 ± 0.1 nm, 1.5 ± 0.2 nm and 2.7 ± 0.4 nm are measured for microwave powers of 0 W, 1600 W and 1800 W, respectively. It turns out that the plasma etching process is capable of systematically changing the RMS roughness of the Si surfaces. As expected, a higher microwave power leads to an increase in surface damage and in turn to higher final roughness values. In this context, the distribution of surface heights features Gaussian characteristics for all three sample surfaces, indicating the random nature of this particular etching process.

The surface roughness of the rails varies with rail inclination between 2.5 ± 1 nm and 4.8 ± 2 nm, respectively. Larger inclinations show increased values due to the preparation procedure by FIB. The inclination itself was characterized by SEM based cross-sectional analysis, which was able to confirm the target inclination of 25°, 45° and 60°, respectively, with an error of 0.5°.

Normal Contact—Adhesion Measurements

As mentioned above, adhesion represents a significant parameter in the contact mechanics of micron-sized particles with each other as well as particle/wall contacts. Consequently, careful attention was payed to the adhesion forces acting on the particles in context of the experiments carried out within the framework if this study. In a preliminary study [25], it was possible to show the combined effect of surface roughness and particle size on the adhesion experienced by the corresponding AFM based particle probes. The results obtained agreed well with the ideas proposed by Rabinovich et al. [9] and the experimental findings reported by Liu et al. [8]. Subsequently, the adhesion between the borosilicate beads and the Si based surfaces presented in the previous section was measured in analogous manner.

AFM (Asylum Research MFP-3DTM AFM, Santa Barbara, CA) based adhesion measurements were carried out with the colloid probes. The colloid probe cantilever spring constant was evaluated by utilizing the thermal noise method [32]. Force-distance curves were measured using the force volume (force map) method, which automatically probes 20 × 20 points in a 50 × 50 μm^2 area. The adhesion forces were sampled using a z-rate of 1.0 μm/s. The loading force used was <5–10% of the adhesion forces. All experiments were conducted at room temperature and 33% relative humidity. The final adhesion force of a given sample was calculated as the mean value of 400 individual measurements.

The results of the colloid probe technique based adhesion measurements along with the RMS roughness values of the corresponding contact partner (wall) are summarized in Table 1. The measured adhesion forces decrease with increasing RMS value. As surface chemistry (amorphous SiO_2) and measurements conditions are kept constant, this finding is directly related to a diminished real contact area due to the roughness. As in the preliminary work, these results are in good agreement with the experimental results of Liu et al. [8], who studied the correlation between adhesion, tip radius and surface roughness in more detail.

The nanoindentation based colloid probes were also utilized to study adhesion. Here, as mentioned above, a series of normal loading experiments into fused silica preceded any further activities. This procedure assured the mechanical integrity of the tips and allowed for the identification of any unsuccessful preparation attempts. In the latter case loading and unloading segment of the indentation experiments would typically show some hysteresis hinting at an artifact caused by the glue, which is used to fix the bead to the apex of the diamond tip. For any successful probe preparation, however, the indents into fused silica would not show any hysteresis

Table 1 RMS values and corresponding adhesion force between the borosilicate beads and the Si based surfaces

	RMS (nm)	Adhesion (μN)
Si(100)	0.3 ± 0.1	3.2 ± 0.3
Etched (1600 W)	1.5 ± 0.2	2.7 ± 0.3
Etched (1800 W)	2.7 ± 0.4	1.9 ± 0.3

up to the maximum normal load of the setup (10 mN). In these cases, a simple Hertzian fit with a slightly larger particle radius than the one measured by SEM is able to accurately reproduce the experimental data. The slightly stiffer response has to be attributed to a combination of surface roughness (about 0.7 ± 0.1 nm for the borosilicate beads) and non-sphericity (i.e. local variation in actual radius). According to Tabor [33], the yield stress can be approximated as one third of the hardness. For the fused silica surface used in this experiment, a measured hardness of 9 GPa results in a yield stress of 3 GPa, which is slightly above the maximum contact pressure of 2.8 GPa acting at the contact area with a flat substrate, based on the highest normal loads used in all sliding and rolling experiments of this work (3 mN). Therefore, with fused silica being the weaker compared to Si(100), a predominantly elastic Hertzian contact in all deformation is expected.

The actual nanoindentation based colloid probe adhesion tests were conducted with a setup that included a linear piezo actuator. The latter was responsible for the positioning of the sample while the nanoindenter only tracked the normal contact load. This strategy allowed to decouple load and displacement of the nanoindenter enabling the measurement of small adhesion forces. It was observed that this setup, due to the combined noise floor of indenter and piezo actuator, was only able to quantitatively measure adhesion forces larger than 2.5 μN, a force that, in most cases, is already considered a high load for any commercially available AFM based colloidal probe technique. Therefore, this approach can be viewed as ideal expansion to the existing AFM based one and is well suited to handle larger contacts or strong adhering contacts, which feature adhesion forces beyond 2.5 μN. For this study, however, the results obtained by the AFM based approach, which are given in Table 1, will be employed.

Sliding Contact

All sliding experiments as well as the adhesion experiments mentioned above have been carried out at RT and 33 ± 5%RH. In general, the particle probes were mounted to a nanoindenter setup (Hysitron Inc.), which was utilized to carry out the measurements. Two sets of experimental data were acquired. In the first set, lateral force data was collected in a friction loop fashion, which is a common technique in the AFM community. The beads were slid over a surface length of 2 μm at a constant normal load and a speed of 1 μm/s. Normal loads of 3, 10, 30, and 100 μN were applied with the normal load being feedback controlled; five spheres were utilized probing each load five times at different surface spots for all Si surfaces respectively. The second data set was also performed in a friction loop manner. This time, however, ten consecutive loops at a constant normal load were executed. Scratch length was changed to 10 μm. Here, normal loads of 1, 3, 6, 10, 30, 60, and 100 μN were utilized, respectively. All other experimental parameters were chosen in analogy to the first data set.

The corresponding absolute value of lateral force for each test was evaluated by a lateral displacement sensitive averaging of the difference in measured lateral force for forward and backward movement divided by two. In order to avoid any artifact originating from a change in the movement direction only the central 50% of the friction loop was taken into account. On a side note, none of the tests, which have been carried out within the framework of the present study, showed any time dependent behavior that potentially could be attributed to the much softer, viscous layer of glue between particle and diamond tip.

The first lateral data set collected as described above was initially analyzed by simply assuming that the measured lateral force F is directly proportional to the applied normal load P modified by a potential adhesive load component P_{adh} with the proportionality factor being the friction coefficient μ, see Eq. (1). Such linear relationship is often described as modified Coulomb friction [12].

$$F = \mu(P + P_{adh}) \tag{1}$$

Analyzing the experimental data according to (1) yields two parameters, the friction coefficient μ as well as the adhesive load component P_{adh}, respectively. In this case, the adhesive load component represents the intercept with the abscissa for a fit of the whole normal load regime. Table 2 summarizes the findings for the three Si surfaces.

Two aspects of these results are somewhat surprising: (i) the etched Si surfaces show a significantly increased friction coefficient compared to the as received Si surface and (ii) P_{adh} does not agree with the measured AFM based adhesion forces, see Table 1. However, the resulting friction coefficient of the as received Si(100) wafer in contact with the spherical probe of 0.23 ± 0.05 is in good agreement with the work of Yu et al. [17], who probed the transition from stick to slip for the contact of micron-sized silica spheres on a Si(100) wafer depending on the relative humidity, as well as findings by Zhang et al. [34], who studied the sliding friction of silica colloidal probes on microsphere-patterned silicon surfaces. Further agreement is found with reports by Maharaj et al. [35] and Quintanilla et al. [36].

In order to address the inconsistency of the measured adhesion force (AFM) with respect to the one obtained from a linear fit to the lateral force data, it proved instructive to take a closer look at the fits of the first data set with respect to modified Coulomb friction (1) and the well-established contact theories of Hertz [30], Johnson, Kendall and Roberts (JKR) [37] as well as Derjaguin, Müller and Toporov (DMT) [38]—for the details here, please refer to [27]. While the modified Coulomb approach

Table 2 Summary of the results of a simply analysis of the sliding experiments (data set one) by assuming a modified Coulomb friction

	μ	P_{adh} (μN)
Si(100)	0.23 ± 0.05	12 ± 1.2
Etched (1600 W)	0.53 ± 0.05	4.5 ± 0.6
Etched (1800 W)	0.65 ± 0.14	7.8 ± 1.2

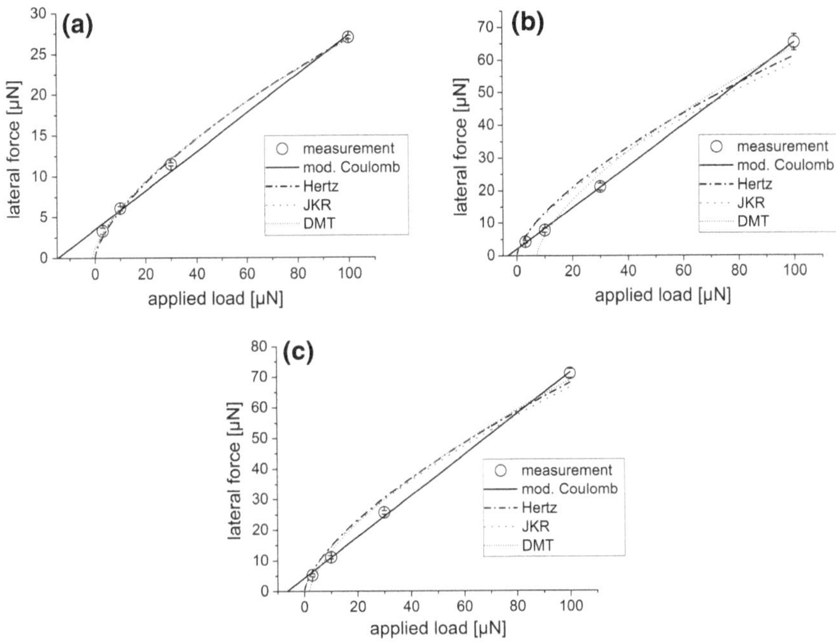

Fig. 4 Comparison between experimental data (set one) and a linear interrelationship (mod. Coulomb) as well as other contact models, for sliding contact between bead and surface; **a** as received, **b** etched at 1600 W, and **c** etched at 1800 W (Reprinted with permission from [27].)

is by far the best approximation in describing the correlation between lateral force and applied normal force for the etched Si surfaces; indicating a contact between particle and surface that can be viewed as a true multi-asperity contact resembling contact situations on a macroscopic level. The situation for the as received Si(100) surface differs. Here the Hertz contact theory as well as JKR and DMT are equally good and significantly better than the linear approach in describing the relationship between lateral and normal force for the full range of normal loads tested, see Fig. 4.

In order to resolve this issue, the same analysis was repeated with lateral forces obtained from the first friction loop of the second experimental data set. Comparing these results with the ones for the first experimental data set, a striking agreement between both series is found. However, in case of the as received Si surface in contact with the particle probe, the additional data points of the new data set are helpful to shed some light on the details of the sliding contact at low applied normal loads, see Fig. 5. The corresponding non-linear behavior with respect to the full normal load regime shows more resemblance to a single asperity contact.

By critically analyzing Fig. 5 one observes that neither the modified Coulomb relationship nor the other contact models, which were all fit to the whole data range up to 100 μN applied load, agree well with the measured data below 10 μN. In fact, a rather linear trend is observed for the measurement in the applied load regime

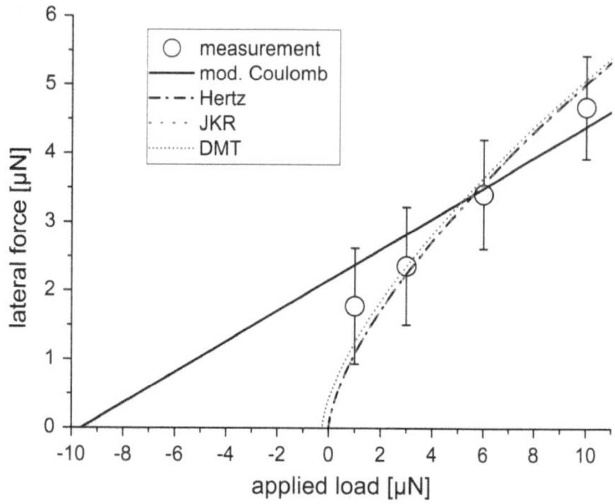

Fig. 5 Low load regime of the sliding contact between as received Si surface and glass bead along with corresponding fitting results (Reprinted with permission from [27].)

between 1 and 10 µN, which, however, features a steeper slope, i.e. higher friction coefficient, compared to the one obtained for linearly fitting the complete range of applied normal load. At the same time the intercept of the low load trend with the axes of applied load would lead to a better agreement with P_{adh} measured by AFM. Any remaining variations between the values received by AFM based adhesion measurements and the ones seen in context of the sliding experiments potentially stem from the differences in dynamics of the process, which detaches micro-asperities from a surface and one that initiates new contacts during shear of the sphere along the substrate. Unfortunately, in such a case, it is not possible to predict the appropriate load regime resulting in the adhesion between the contacting partners a priori. A true clarification on this issue, that ultimately also affects the contact of rougher partners, would, however, require access to measurements at negative applied loads. The corresponding modifications of the nanoindentation setup are work in progress and will be reported elsewhere.

After commenting on the question of the affective adhesive force component during sliding, it is still necessary to consider the issue of the relatively high friction coefficients in case of the etched Si surfaces. In [26, 28] it was reasoned that plastic deformation of surface asperities might be responsible for this phenomenon. A simple Hertzian contact with perfectly flat surfaces, even though predicting the normal force well, is not able to explain any onset of plasticity. Nevertheless, the pressure at individual small surface asperities will certainly be sufficient to initiate plastic deformation [39]. A quick check on this proposition by means of characterization of area wear marks generated at very low loads (<10 µN) confirmed the occurrence of plastic deformation. In these experiments the topography of the contacting surfaces showed a reduced roughness while no obvious wear debris was formed. Consequently, in order to gain detailed insight into the key mechanisms, the time dependency of lateral forces gathered during consecutive friction loops at a constant applied load over a fixed linear surface region was analyzed.

If the high friction coefficients obtained for etched surfaces are results of plasticity occurring on an asperity level during the sliding contact, a significant change in lateral forces after the first cycle of a multi-cycle sliding contact at constant applied load is expected. Indeed, such changes were observed for the second set of experimental data. Comparing the sets of lateral forces detected during the first as well as the tenths (last) cycle of sliding over as received and etched Si surfaces, respectively, again revealed a clear non-linear correlation between lateral load and applied load in case of the as received Si surface for the first as well as the tenth cycle. The rougher surfaces, on the other hand, displayed a linear interrelationship between lateral force and normal force in case of the first as well as the last cycle. The absolute values of detected lateral forces, however, behave significantly different for the two surfaces. While the lateral forces slightly increase for the as received surface, a dramatic drop in lateral forces is noted for the rougher surface between the first and tenths cycle. This change in detected lateral force with respect to test cycle for the two rough surfaces, however, is independent of applied normal load within the load regime studied here. The biggest drop in lateral forces is seen for the roughest sample tested. In this case, the lateral forces decrease to about 50% of their initial (first cycle) value during the course of ten cycles.

Finally, Fig. 6 displays a representative set of data collected at 60 μN constant applied load revealing details of the evolution of the lateral force during the ten consecutive cycles of sliding. This plot holds a wealth of information. It turns out, in case of the etched surfaces, that the results of the first cycle are clearly dominated by plastic deformation. After a few cycles the effect of any plastic events in the contact zone are significantly decreased. The etched surface featuring the lower RMS roughness reaches this kind of equilibrium surface state earlier compared to the rougher Si surface. The latter shows a slight final decreasing trend in lateral force that might very well continue beyond the tenths cycle. Even for the as received surface the occurrence of plastic events during the first cycle cannot be excluded. The small

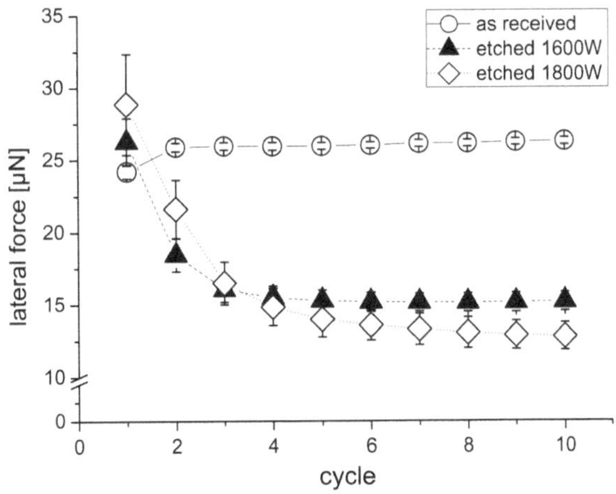

Fig. 6 Detailed evolution of lateral forces during the ten cycles of sliding at 60 μN constant applied normal load for all three surfaces tested in this study (Reprinted with permission from [27].)

increase in lateral force after the first cycle might reflect on a slightly lower contact roughness, which leads in turn to an increased real contact area and, therefore, an increased lateral force. Staying true to this line of thought, the fact that the absolute values of lateral forces for the rough surfaces after ten cycles are well below the one of the as received surface imply that the final surface states of these rougher surfaces feature a smaller real contact area compared to the as received one. The details of the evolution of the topography in the sliding track are a complex function based on the applied normal load, mechanical properties of the contact partners, number of cycles, and initial roughness as well as initial local curvature and local slope of surface asperities of both contact partners, respectively. In principle it is assumed that the topography of the contacting partners changes until the local contact pressure on an asperity level finally falls below the strength of the corresponding materials. Studying this evolution in more detail is quite a challenge as high resolution in surface characterization as well as high reproducibility of positioning of the glass bead is required in order to provide meaningful information. Dedicated efforts are in progress to derive appropriate strategies to succeed in this endeavor.

In order to check on the effect of any tribochemical reaction on the detected lateral forces, as reported by [17], the sliding experiments on the Si surface etched at 1600 W were repeated under a dry (2.4%RH) nitrogen atmosphere. A comparison between the previous results and the once under dry condition feature a very similar trend, indicating that with respect to the contact conditions studied here plastic deformation of individual surface asperities might play a more significant role compared to tribochemical reactions. It is, given the current set of experimental data, however, not possible to fully rule out the influence of tribochemistry on the results.

Rolling/Torsional Contact

The nanoindentation test setup for the measurement of rolling friction of the microspheres is significantly easier compared to the sliding case (Fig. 1b). Here, the borosilicate spheres featuring radii of approximately 10 μm are placed on the surface of interest prior to testing. In a second step individual beads are contacted with a flat end diamond indenter, carefully positioning the indenter tip with its axis aligned with respect to the apex of the spherical particle of choice. Preliminary test results with single scratch length of about 10 μm and a speed of 1 μm/s under load control, showed that a certain threshold [20] or rolling resistance moment [21, 22] was necessary to initiate particle rolling. In our case, the rolling of particles, i.e. a change in position detected by optical microscopy, was only observed for normal loads larger than 100 μN. Based on these results, a minimum normal load of 100 μN was selected for all rolling tests presented here to ensure rolling characteristics of the corresponding contact behavior. This 100 μN force to overcome the rolling resistant moment, was chosen independently of the load regime used in sliding tests. The final rolling tests were carried out under the same environmental conditions in analogy to the friction loop method presented above and with normal load ranging from 100 μN to

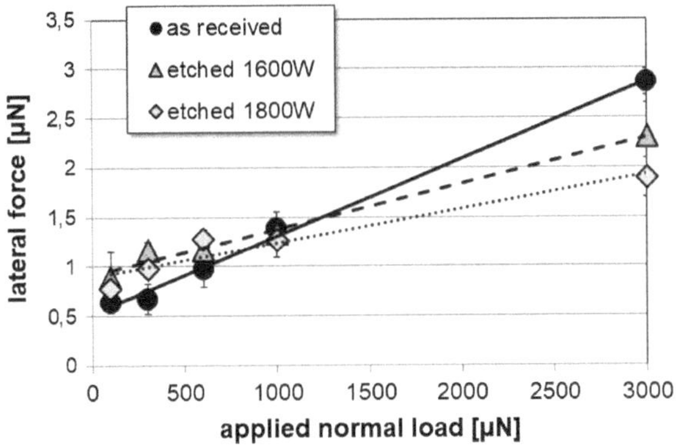

Fig. 7 Resulting lateral force values for rolling friction measurements. The lines represent simple linear fits to the corresponding data sets (Reprinted with permission from [26].)

3 mN. Again, in order to improve the reliability of the results, five beads were used for testing and each measurement, i.e. applied normal load, was taken five times for all Si surfaces, respectively. The absolute values of lateral forces for each test were calculated in analogy to the friction loop method presented for sliding friction tests.

Figure 7 shows the results of the rolling experiments. Again, a straight line fits the data although some additional scattering due to small variations of the radii of the different spheres used is observed. Processing the experimental data in an analogous manner to the sliding tests yields the mean rolling friction coefficient μ_{ro}. The measured lateral force signals are 2–3 orders of magnitude smaller for rolling compared to sliding tests, which confirms that the particle is rolling over the surface. Surprisingly, in contrast to the sliding data, μ_{ro} decreases with increasing surface roughness, while the intercept with the abscissa shifts to the left, i.e., rolling adhesion P_{adh_ro} increases with surface roughness. The latter finding suggests that a simple interpretation of this intercept as a measure of the normal adhesion force between the contacting partners is not valid in the case of a rolling contact. Ideally, a non-adhering rolling sphere contacts with the underlying surface at only one point, which would result in a zero rolling friction since no torque can be created by a point contact. In our system, the finite contact surface, involving many single asperities of the sphere as well as of the surface, leads to a non-zero rolling resistance. Our idea of rolling is a continuous generation of new contact area at the front and a continuous contact detachment at the rear of the contact area (viewed in rolling-direction of the sphere). As adhesive contact forces such as van der Waals interactions and water bridges from at the front and break at the backside of the sphere, energy is dissipated. This explains the decrease in rolling friction with higher surface roughness, as the AFM results show a decrease in adhesion with higher surface roughness. Thus, rolling friction appears to be dominated by adhesion, as supported by reports of other groups. One of the first overviews attributing rolling friction to adhesion was given by Briggs and Briscoe in 1976 [40]; a more recent study has been presented by Sumer and Sitti [7]. The corresponding results are summarized in Table 3.

In order to explore the relationship between lateral force and applied normal loads or rather applied pressure somewhat further, the rolling experiments shown

Table 3 Summary of the results of a simply analysis of the rolling experiments by assuming a simple linear interrelationship

	μ_{ro} (10^{-4})	P_{adh_ro} (μN)
Si(100)	7.9 ± 0.3	660 ± 30
Etched (1600 W)	4.6 ± 0.4	1960 ± 180
Etched (1800 W)	3.5 ± 0.6	2550 ± 550

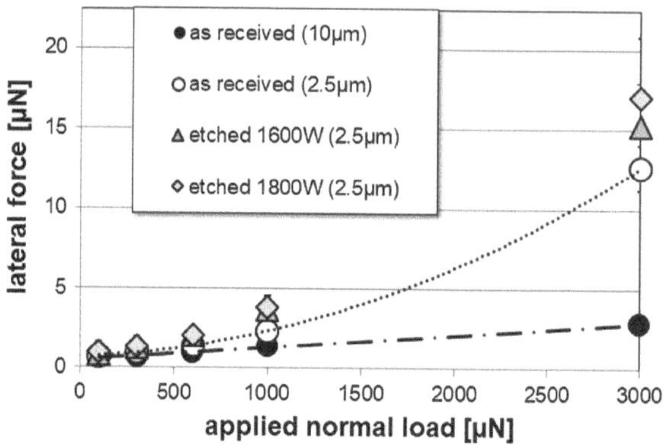

Fig. 8 Comparison of the results of rolling experiments with particles featuring 10 and 2.5 μm radii, respectively; for details refer to text (Reprinted with permission from [26].)

above have been repeated with borosilicate beads that feature nominal radii of only 2.5 μm. This is actually the only experiment within this report, which utilizes a particle size that differs from the 10 μm particle radius employed throughout the rest of the experimental work. The results are presented in Fig. 8. A significant change in the correlation between lateral force and applied normal load is observed. It clearly contains a non-linear term now. For example, the data set of the rolling experiments of the 2.5 μm beads on the as received silicon surface in Fig. 8 was fitted by a second order polynomial. The fit serves as guide to the eye but is well capable of reproducing the corresponding relationship. Such a behavior is likely to be attributed to irreversible processes during the rolling contact [2]. Assuming a yield strength between 3 and 4 GPa leads in turn to an onset of plastic deformation below an applied normal load of approximately 500 or 1000 μN, respectively. This corresponds very well with the experimental data, as this is obviously the load regime in which the behavior of the rolling beads that feature tip radii of about 2.5 μm starts to deviate from the simple linear trend of the rolling experiments carried out with the 10 μm beads. Additionally, it should be noted that in the plasticity dominated case of the experiments with small glass beads the rougher surfaces generally feature larger lateral forces at a given applied normal load. A trend which is quite the opposite compared to the adhesion dominated rolling in case of the larger beads. This observation is supporting the findings of Vilt et al. [23], who were studying the behavior of particle ensembles under rolling conditions.

In combination with appropriate rail systems (for example see Fig. 3), the strategy to probe rolling friction, which has been introduced above, can also be exploited to study a combination of rolling and torsional friction. As far as the authors are aware,

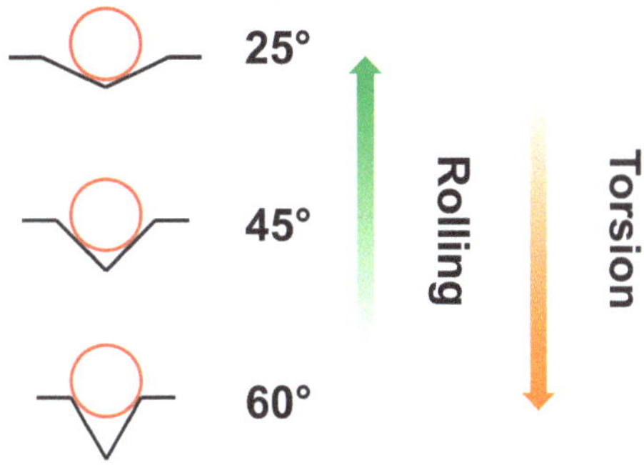

Fig. 9 Schematics of the rail based assessment of a combination of rolling and torsional friction

no scanning probe based technique has been proposed so far that would allow for the exclusive evaluation of torsion friction.

In order to access the combination of both mechanisms, Si based rail structures featuring rail angles of 25°, 45° and 60° were utilized, respectively. Higher rail angles lead to an increased influence of the torsional contact mode. Borosilicate spheres are placed inside the rails by AFM based positioning (Fig. 9). In analogy to rolling tests, the individual particles inside the rails are contacted and probed with a flat end diamond indenter. Careful positioning of the indenter tip with respect to the particle as well as alignment of the rail with regard to the scratch axis (movement direction) of the instrument are crucial. It should be mentioned, that the actual normal loads acting on the rail slopes are calculated by vector considerations. To ensure that no sliding of the particle takes place, an exemplary pure sliding test with a nanoindenter based colloid probe at a normal load of 100 μN in a 25° rail was carried out by fixing the indenter as in the main sliding experiments. The effect of scratch velocity on the rolling results was studied by varying velocity from 1, 0.1 to 0.01 μm/s inside the rail system.

From a mechanical point of view, torsion friction is due to a rotation of the sphere in normal direction and rolling friction is due to a rotation in tangential direction to the substrate. Experimentally, a controlled particle surface motion is possible by means of a Si based rail system featuring different inclinations. For a dedicated mathematical analysis of this contact scenario, please refer to the chapter of the group of Prof. Stefan Luding, Twente (NL). At this point here, the focus lies purely on the experimental considerations.

In the rail, these motions occur simultaneously which results in an effective friction coefficient with contributions from both torsion and rolling resistance. Sliding motion at the indenter occurs for normal loads below a minimum normal load of 100 μN, which is required to overcome the critical rolling resistance, this finding, as mentioned before, is based on optical images which show no change in position after the rolling test for loads smaller than this minimum. Below this threshold, sliding, rolling and torsion motion are presumably mixed, while above this threshold, pure rolling and torsion motion are active (no sliding). With higher inclination of the rail, the torsion contribution increases whereas the rolling part becomes comparatively smaller. The

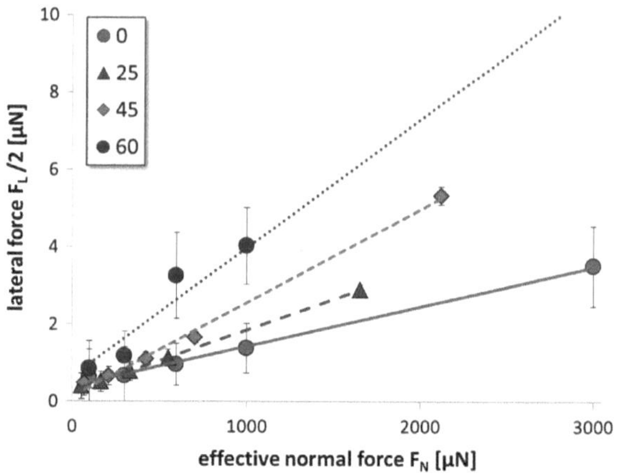

Fig. 10 Resulting lateral force versus effective normal load for the rail system with different rail inclination ranging from 0°, 25°, 45° and 60°. All data sets can be fitted with a simple linear relation (Reprinted with permission from [28].)

surface roughness of the rail slopes cannot be ignored and has a significant effect on the measurements. According to the correlation between rolling resistance and surface roughness, as discussed before, RMS values between 2 and 5 nm lead to measurably smaller lateral forces, see Tables 1 and 3. The difficulty lies in preparing comparable rail systems with a constant low surface roughness smaller than 1 nm RMS for all rail inclinations, which would allow to focus on the effect of normal loads. Until now, this is not realized and all presented values for the different rail systems have a systematic error due to variable surface roughness. In order to apply a given normal load at the substrate contact, a lower normal load has to be applied at the indenter. The measured lateral forces and effective normal loads, at the substrate contacts are shown in Fig. 10. All results follow linear trends, confirming a modified Coulomb law, but the slopes, which are the effective friction coefficients in the rail, show a clear correlation with increasing inclination of the rail.

In summary the partner group in Twente was able to derive information about the torsional friction, which in particular is an entity that can clearly be distinguished from sliding and rolling friction. For details, the reader again is directed towards the corresponding chapter of the group of Prof. Stefan Luding.

The independence from sliding can be experimentally shown with relative ease. To do so, a pure sliding test (nanoindenter colloid probe) on a 25° rail with a normal load of 100 µN was carried out. It results in an average lateral force of 26 µN ± 1.71 µN. This value is more than one order of magnitude higher compared to the measured lateral force with a single free rolling microsphere inside the 25° rail at a corresponding normal load. It is expected that for higher inclination and higher normal load the measured lateral force increases due to the larger torsion component. The intercepts of the trend lines with the horizontal axis, taken as the values of rail adhesion are lower than for pure rolling, which can be attributed to the higher surface roughness of the rails. This suggests that adhesion plays a less dominant role during these rolling/torsion experiments than for pure rolling on a plane surface. Even though the experimental results presented up to now provide evidence of an increased effect of torsion with increasing inclination, a deeper insight into appropriate contact models is necessary to understand the evolution of the effective friction coefficient with respect to inclination. Consequently, these aspects are handled elsewhere.

On a final note, no velocity dependence was observed for indenter velocities of 0.01, 0.1 and 1 μm/s, thus the rolling friction is not dominated by viscous effects [41] in this regime.

Summary and Conclusion

In this work the concept of the AFM based colloid probe technique was successfully transferred to a nanoindenter setup. By doing so it was possible to study motions such as sliding, rolling and torsion of individual very small particles. The nanoindenter was used as a platform where experiment and model/theory meet. A big step towards understanding particle interactions and to directly obtain contact model parameters was done. Compared with other techniques used in literature, sliding, rolling and torsion resistance of spherical micron-sized particles can be relatively easily measured with a nanoindenter featuring the option to measure not only normal loads but also lateral forces.

However, the interpretation/evaluation of the results still leaves various open questions. Experimentally, the core of all those issues lies in the unknown real contact area. Due to the actual size of the nominal contact area as well as the dimension of the asperities within this zone, any quantitative assessment of the true contacting area and its evolution over time for the contact scenarios presented here poses an immense challenge; which only can be addressed by further dedicated experimental work in combination with corresponding simulations.

Acknowledgements The authors would like to first and foremost thank the late Prof. Jürgen Tomas for assembling the Key Research Program (SPP 1486 PiKo "Particles in Contact"), which created a wonderful platform to meet and exchange ideas about contact mechanics of micron-sized particles and their impact on our everyday life. Further thanks go to Stefan Luding and Thomas Weinhart for many fruitful discussions as well as to Hao Zhuang for his great support in preparing rough silicon surfaces. This work is carried out within the framework of PiKo under grants STA 1021/1-1, STA 1021/1-2 and STA 1021/1-3.

References

1. Dominik, C., Tielens, A.G.G.M.: Resistance to rolling in the adhesive contact of two elastic spheres. Philos. Mag. A **72**, 783–803 (1995)
2. Tomas, J.: Adhesion of ultrafine particles—energy absorption at contact. Chem. Eng. Sci. **62**, 5925–5939 (2007)
3. Ducker, W.A., Senden, T.J., Pashley, R.M.: Direct measurement of colloidal forces using an atomic force microscope. Nature **353**, 239–241 (1991)

4. Butt, H.-J.: Measuring electrostatic, van der Waals, and hydration forces in electrolyte solutions with an atomic force microscope. Biophys. J. **60**, 1438–1444 (1991)
5. Sitti, M., Hashimoto, H.: Controlled pushing of nanoparticles: modeling and experiments. IEEE/ASME Trans. Mechatron. **5**, 199–211 (2000)
6. Sitti, M.: Atomic force microscope probe based controlled pushing for nanotribological characterization. IEEE/ASME Trans. Mechatron. **9**, 343–349 (2004)
7. Sümer, B., Sitti, M.: Rolling and spinning friction characterization of fine particles using lateral force microscopy based contact pushing. J. Adhes. Sci. Technol. **22**, 481–506 (2008)
8. Liu, D.-L., Martin, J., Burnham, N.A.: Optimal roughness for minimal adhesion. Appl. Phys. Lett. **91**, 43107 (2007)
9. Rabinovich, Y.I., Adler, J.J., Ata, A., Singh, R.K., Moudgil, B.M.: Adhesion between nanoscale rough surfaces. J. Colloid Interface Sci. **232**, 10–24 (2000)
10. Matope, S., Rabinovich, Y.I., van der Merwe, A.F.: Van der Waals interactions between silica spheres and metallic thin films created by e-beam evaporation. Colloids Surf. A **411**, 87–93 (2012)
11. Korayem, M.H., Zakeri, M.: Dynamic modeling of manipulation of micro/nanoparticles on rough surfaces. Appl. Surf. Sci. **257**, 6503–6513 (2011)
12. Gao, J., Luedtke, W.D., Gourdon, D., Ruths, M., Israelachvili, J.N., Landman, U.: Frictional forces and Amontons' law: from the molecular to the macroscopic scale. J. Phys. Chem. B **108**, 3410–3425 (2004)
13. Ruan, J.-A., Bhushan, B.: Atomic-scale and microscale friction studies of graphite and diamond using friction force microscopy. J. Appl. Phys. **76**, 5022–5035 (1994)
14. Koinkar, V.N., Bhushan, B.: Effect of scan size and surface roughness on microscale friction measurements. J. Appl. Phys. **81**, 2472–2479 (1997)
15. Otsuki, M., Matsukawa, H.: Systematic breakdown of Amontons' law of friction for an elastic object locally obeying Amontons' law. Sci. Rep. **3**, 1586 (2013)
16. Bowden, F.P., Tabor, D.: Friction, lubrication and wear: a survey of work during the last decade. Br. J. Appl. Phys. **17**, 1521–1544 (1966)
17. Yu, J., Chen, L., Qian, L., Song, D., Cai, Y.: Investigation of humidity-dependent nanotribology behaviors of Si(100)/SiO_2 pair moving from stick to slip. Appl. Surf. Sci. **265**, 192–200 (2013)
18. Greenwood, J.A., Williamson, J.B.P.: Contact of nominally flat surfaces. Proc. R. Soc. A: Math. Phys. Eng. Sci. **295**, 300–319 (1966)
19. Greenwood, J.A.: Constriction resistance and the real area of contact. Br. J. Appl. Phys. **17**, 1621 (1966)
20. Saito, S., Miyazaki, H.T., Sato, T., Takahashi, K.: Kinematics of mechanical and adhesional micromanipulation under a scanning electron microscope. J. Appl. Phys. **92**, 5140–5149 (2002)
21. Peri, M.D.M., Cetinkaya, C.: Adhesion characterization based on rolling resistance of individual microspheres on substrates: review of recent experimental progress. J. Adhes. Sci. Technol. **22**, 507–528 (2008)
22. Ding, W., Howard, A.J., Peri, M.M., Cetinkaya, C.: Rolling resistance moment of microspheres on surfaces: contact measurements. Philos. Mag. **87**, 5685–5696 (2007)
23. Vilt, S.G., Martin, N., McCabe, C., Kane Jennings, G.: Frictional performance of silica microspheres. Tribol. Int. **44**, 180–186 (2011)
24. Kumar, A., Staedler, T., Jiang, X.: Effect of normal load and roughness on the nanoscale friction coefficient in the elastic and plastic contact regime. Beilstein J. Nanotechnol. **4**, 66–71 (2013)
25. Kumar, A., Staedler, T., Jiang, X.: Role of relative size of asperities and adhering particles on the adhesion force. J. Colloid Interface Sci. **409**, 211–218 (2013)
26. Fuchs, R., Meyer, J., Staedler, T., Jiang, X.: Sliding and rolling of individual micrometre sized glass particles on rough silicon surfaces. Tribol. Mater. Surf. Interfaces **7**, 103–107 (2013)
27. Meyer, J., Fuchs, R., Staedler, T., Jiang, X.: Effect of surface roughness on sliding friction of micron-sized glass beads. Friction **2**, 255–263 (2014)
28. Fuchs, R., Weinhart, T., Meyer, J., Zhuang, H., Staedler, T., Jiang, X., Luding, S.: Rolling, sliding and torsion of micron-sized silica particles: experimental, numerical and theoretical analysis. Granul. Matter **16**, 281–297 (2014)

29. van Zwol, P.J., Palasantzas, G., van de Schootbrugge, M., de Hosson, J.T.M., Craig, V.S.J.: Roughness of microspheres for force measurements. Langmuir ACS J. Surf. Colloids **24**, 7528–7531 (2008)
30. Hertz, H.: Ueber die Berührung fester elastischer Körper. J. für die reine und angewandte Mathematik **92**, 156–171 (1882)
31. Morita, M., Ohmi, T., Hasegawa, E., Kawakami, M., Suma, K.: Control factor of native oxide growth on silicon in air or in ultrapure water. Appl. Phys. Lett. **55**, 562–564 (1989)
32. Butt, H.-J., Jaschke, M.: Calculation of thermal noise in atomic force microscopy. Nanotechnology **6**, 1 (1995)
33. Tabor, D.: Indentation hardness: fifty years on a personal view. Philos. Mag. A **74**, 1207–1212 (1996)
34. Zhang, X., Zhong, X., Meng, X., Yi, G., Jia, J.: Adhesion and friction studies of nano-textured surfaces produced by self-assembling Au nanoparticles on silicon wafers. Tribol. Lett. **46**, 65–73 (2012)
35. Maharaj, D., Bhushan, B.: Effect of spherical Au nanoparticles on nanofriction and wear reduction in dry and liquid environments. Beilstein J. Nanotechnol. **3**, 759–772 (2012)
36. Quintanilla, M.A.S., Goddard, D.T.: Lateral force microscopy with micrometer-sized particles: effect of wear on adhesion and friction. Wear **268**, 277–286 (2010)
37. Johnson, K.L., Kendall, K., Roberts, A.D.: Surface energy and the contact of elastic solids. Proc. R. Soc. Lond. Ser. A Math. Phys. Sci. **324**, 301–313 (1971)
38. Derjaguin, B.V., Muller, V.M., Toporov, Y.P.: Effect of contact deformations on the adhesion of particles. J. Colloid Interface Sci. **53**, 314–326 (1975)
39. Bhushan, B.: Contact mechanics of rough surfaces in tribology: multiple asperity contact. Tribol. Lett. **4**, 1–35 (1998)
40. Briggs, G.A.D., Briscoe, B.J.: Effect of surface roughness on rolling friction and adhesion between elastic solids. Nature **260**, 313–315 (1976)
41. Brilliantov, N.V., Pöschel, T.: Rolling friction of a viscous sphere on a hard plane. EPL (Europhys. Lett.) **42**, 511 (1998)

The Importance of Interactions Between Carrier and Drug Particles for the Application in Dry Powder Inhalers

Sarah Zellnitz, Niklas Renner, Yan Cui, Regina Scherließ, Martin Sommerfeld, Hartwig Steckel and Nora Urbanetz

Introduction

Diseases of the human respiratory tract such as asthma bronchial, chronic obstructive pulmonary disease (COPD) or local infections are preferentially treated with medicines administered via the pulmonary route. By doing so, the application of comparatively low doses is possible and an adequate drug distribution to its site of action is reached [1, 2]. At the same time, a reduced potential for adverse effects is observed compared to an oral or intravenous administration due to minimised systemic exposure of these drugs [3–5]. Besides the treatment of diseases which are locally limited to the airways, pulmonary drug delivery is also explored as an alternative to effectively administer small molecules or peptides to treat systemic diseases such as diabetes or osteoporosis [6, 7]. This is reasoned by a large surface area of the lungs, comparatively low local metabolic activity and the possibility to avoid first-pass metabolism [8]. In addition, the drug is rapidly distributed due to an extended blood flow [9].

For the administration of active pharmaceutical ingredient (API) particles to the lung, different types of inhalers including nebulizers, pressurized metered-dose inhalers (pMDIs), non-pressurized metered-dose inhalers (MDIs) and dry powder inhalers (DPIs) can be used. Formulations used in nebulizers, pMDIs and MDIs are liquid whereas solid formulations are used in DPIs.

S. Zellnitz · N. Urbanetz
Research Center Pharmaceutical Engineering GmbH, Graz, Austria
e-mail: sarah.zellnitz@rcpe.at

N. Renner · R. Scherließ · H. Steckel
Department of Pharmaceutics and Biopharmaceutics, Kiel University, Kiel, Germany
e-mail: rscherliess@pharmazie.uni-kiel.de

Y. Cui · M. Sommerfeld (✉)
Multiphase Flow Systems, Institute for Process Engineering, Otto-von-Guericke-University, Halle/Saale, Germany
e-mail: martin.sommerfeld@ovgu.de

S. Antonyuk (ed.), *Particles in Contact*,
https://doi.org/10.1007/978-3-030-15899-6_16

Advantages of the use of solid formulations and thus of DPIs are the avoidance of propellants, high patient compliance, high dose carrying capacity, high drug stability and reduced risk of microbial contamination compared to liquid formulations [1]. Moreover DPIs are small, cheap, more or less easy to use and do not require a coordination of actuation and inhalation. One weakness of some DPIs is that the amount of drug that can reach the lung varies and depends on the inspiration flow rate of the patient.

A considerable percentage of asthma and COPD medication is formulated as dry powders intended for inhalation. In order to reach the lower airways, the drug needs to have an aerodynamic particle size between 1 and 5 μm [10, 11]. Furthermore, an adequate flowability of the powder is mandatory to ensure processability and accurate dose-metering. These two requirements stand in direct contradiction to each other. For this particular size range, cohesive forces are predominant in relation to gravitational forces due to relatively low weight and large surface area of the particles [12]. To counteract poor flowability and secure dose uniformity as well as processability of the powder, two main formulation strategies have been developed, namely carrier-free and carrier-based systems. The latter approach comprises coarse and desirably inert particles in the size range of 50–200 μm with adequate flowability. During a blending step, drug agglomerates are broken up while these particles get attached to the carrier surface. As drug particles interact with the carrier via adhesive forces, these blends are also referred to as interactive mixtures.

During inhalation, the drug needs to be detached again from the carrier to allow delivery to the target site. The detachment is triggered by the patient´s inspiration. But at the same time, it is widely governed by inter-particle interactions between carrier and active pharmaceutical ingredient (API). The extent to which these forces occur is highly dependent on physio-chemical properties of carrier and API [13]. Therefore, an in-depth understanding of the impact of material characteristics on the performance of the dry powder inhalate is needed to overcome the challenges in the development of dry powder inhaler (DPI) formulations, e.g. insufficient drug detachment from the carrier during inhalation. Besides the formulation and patient's inspiratory flow, the inhalation device plays a key role for this process. Its conception and dimensioning of the different parts have to be evaluated carefully [14]. These inter-particle interactions determine the dispersion mechanism of the drug powder from the carrier particles.

For achieving high drug delivery efficiency during inhalation, the fine drug particles need to be detached from the carrier within the inhaler device through the stresses in the air flow, collisions between carrier particles or collisions with device walls. If detachment does not take place, the drug particles sticking on the carrier surface will impact in the mouth and the throat. The fluid dynamic detachment of the drug powder in the complex airflow of an inhaler is realised by acceleration/deceleration of the carrier particles (i.e. inertial force), flow shear gradients and turbulent stresses [15]. In order to deliver a high amount of detached drug particles a proper balancing between the adhesive forces between carrier and drug and the removal forces during the inhalation process is necessary. Thus, inter-particle interactions between API and carrier play a key role in this kind of formulations. On the one hand they have to be

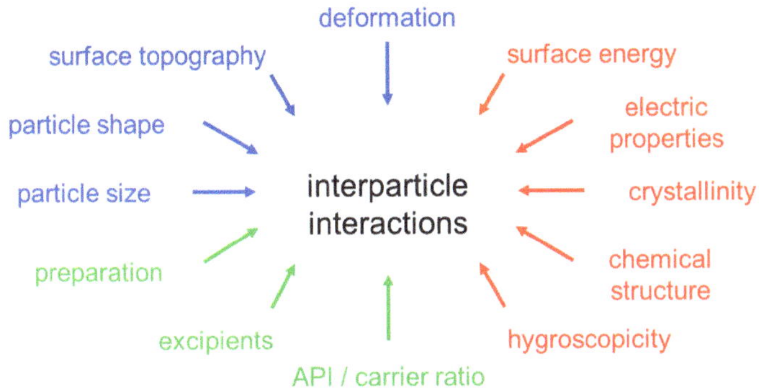

Fig. 1 Summary of inter-particle interactions in dry powder inhalers

high enough that the API sticks to the carrier surface, thus ensuring proper handling as well as uniform dosing, and on the other hand low enough that drug detachment from the carrier surface during inhalation is guaranteed.

Inter-particle interactions are affected by various factors like physicochemical properties. In Fig. 1 highlighted in red (surface energy, electric properties, crystallinity, chemical structure and hygroscopicity), factors that are related to either the carrier or the API particles, highlighted in blue (deformation, surface topography, particle shape and particle size) and factors that are related to the formulation itself, highlighted in green (preparation, excipients and API/carrier ratio). Furthermore altering one of those factors often affects and changes one or more other factors influencing inter-particle interactions.

According to literature one major factor influencing inter-particle interactions is the carrier surface topography [16, 17]. Therefore, different attempts to modify the surface topography of commonly used carrier particles like lactose or mannitol were performed with the main goal to increase drug detachment from the carrier particles. Such attempts include crystallization of lactose particles from different media and under different crystallization conditions [18–20] and particle smoothing by coating lactose particles with aqueous lactose solutions [21, 22]. Another attempt to smoothen particles is dispersing the carrier material in a dispersion media and subsequent elimination of the dispersion media [23–27]. Spray drying was described to be a suitable method to generate mannitol particles with modified surface roughness [28–30]. Other attempts implement the effect of mechanical stress on lactose surface properties e.g. wet smoothing in a high shear mixer [2, 31] particle smoothing with a high speed mixer in the presence of a small amount of wetting solvent [32], surface processing with a high speed elliptical-rotor type powder mixer [33] as well as milling lactose with various mill speeds to avoid batch to batch variability and to make the carrier particle surface homogenous [34]. However, in most of these cases beside the surface topography also other properties impacting inter-particle interactions, like particle size or shape of the carrier particles, were altered. For example, Maas reported carrier surface modification of mannitol by spray drying at different

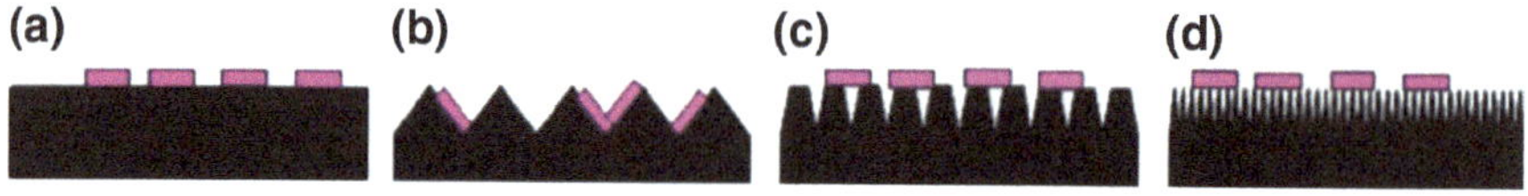

Fig. 2 Schematic of potential API carrier adhesion scenarios. Modified from [17]

conditions [35]. Spray drying changed the surface roughness, however, the particle shape changed as well.

When changing the surface topography or roughness of the carrier the contact area between carrier (black) and API (pink) can either be increased (Fig. 2a, b) or decreased (Fig. 2c, d) and thus inter-particle interactions can also either increase or decrease. Only when a surface roughness is introduced that is smaller with respect to the API particle size the acting adhesion forces are decreased and further drug detachment facilitated. By contrast, when a surface roughness is introduced that is larger compared to the API particle size, the API particles might be entrapped in these cavities and thus drug detachment is hindered. Therefore, the shade of surface roughness compared to the size and shape of the API largely impacts inter-particle interactions and thus drug detachment.

That is also why there are different and contradictory findings related to the correlation of particle roughness and the detachment of the API particles from the carrier surface during inhalation often expressed as the fine particle fraction (FPF). The FPF gives the percentage of API particles that are detached from the carrier during inhalation and that has an aerodynamic particle size smaller than 5 μm and thus is able to reach the deep lung. Therefore, the FPF may be defined as the main parameter describing the performance or efficiency of a DPI carrier system. For example Kawashima documented that the rougher the carrier particles the lower the FPF [36]. The surface roughness in that study resulted from crevices on the lactose surface. By contrast, Chan reported that rougher lactose carrier particles lead to a higher fine particle fraction [21]. In that study the surface roughness resulted from the presence of immobilized lactose fine particles on the surface that lead to microscopic undulations on the lactose surface rather than to crevices. These contradictory findings illustrate that it has to be differentiated between roughnesses and that in order to increase the FPF the right kind of surface roughness has to be introduced.

Besides the balancing of inter-particle interactions the inhaler design needs to ensure sufficiently high flow stresses on the carrier particles for yielding high detachment rates. Hence the detachment of drug particles from the carrier is a prerequisite for a high respirable fraction in vivo that is equivalent to a high fine particle fraction (FPF) in vitro. The FPF is determined by a standardised procedure using cascade impactors and represents the ratio of the fine particle dose (FPD) and the mass of API leaving the inhaler (emitted dose (ED)) in %. The FPD is the mass of drug particles below 5 μm. The FPF is a common parameter to compare the performance of different inhaler types and different formulations among each other.

The impactors used and the procedure of determining the FPF are described in the monograph "Aerodynamic assessment of fine particles for preparations for inhalation" of the European Pharmacopoeia. Impactor data are an important and principal part of marketing authorization applications for new dry powder inhaler formulations. Consequently, so far the design and optimisation of inhalers is mainly based on experimental analysis. Essential for the performance of an inhaler is a high amount of drug powder (i.e. fine particle fraction) that is able to leave the inhaler and reach the deep lung. As already has been mentioned the determination of the FPF is mainly done using different types of cascade impactors [37]. Numerous experimental studies have been conducted for analysing the influence of inhaler devices on the FPF [38–41]. However, these experimental approaches are rather cumbersome, time-consuming and may be associated with measurement errors. In the development of new inhaler formulations or new inhalers it would be beneficial to employ techniques where data acquisition is more rapid. Therefore, it would be advantageous developing a drug detachment model which could be used in connection with the application of computational fluid dynamics (CFD) for analysing inhaler performance and overcome the time consuming impactor experiments.

Computational fluid dynamics (CFD) has proven to be an excellent tool for analysing and evaluating flow structures in inhaler devices, carrier and drug particle motion, as well as other particle-scale phenomena being relevant for drug particle detachment and dispersion in different inhaler devices. CFD (computational fluid dynamics) has been applied to dry powder inhalers over about 15 years. Consequently, until today numerous publications on CFD applications to DPIs were published having different objectives as for example partly summarized by Wong et al. [42], Ruzycki et al. [43] and Sommerfeld and Schmalfuß [44]. The first very extensive numerical study on different phenomena in DPI was probably conducted by Coates et al. [45–47]. Initially, these numerical studies were performed for analysing the influence of the geometrical design of the inhalers on the developing flow structure at steady-state conditions. This concerns for example the mouthpiece geometry [47] and length [45], the effect of grid dimensions (normally mounted at the entrance to the mouthpiece) on flow and particle motion [45] and the influence of a rigid capsule on flow structure as examined by Coates et al. [48]. Some of these studies also consider the trajectories of carrier particles for visualisation purpose using the classical Lagrangian tracking approach (i.e. based on point-particle assumption). In the studies of Danovan et al. [49] carrier particle motion in different inhaler designs was analysed. Milenkovic et al. [50] studied particle deposition in inhalers based on a critical velocity for particle sticking. An overview on the application of CFD, combined with Lagrangian particle tracking, for inhaler predictions is given by Wong et al. [42] and Ruzycki et al. [43]. The usefulness of CFD combined with simple Lagrangian particle tracking (i.e. numerical visualisation of particle trajectories) for developing new inhaler designs was demonstrated by de Boer et al. [51] and Behara et al. [52]. Only recently, also the combination of CFD with DEM (discrete element method) was used for studying the destruction of fine particle agglomerates within inhaler devices (see Tong et al. [53]) and the flow detachment of fine agent particles from larger carrier particles (see e.g. Yang et al. [54]). In this respect it should be noted that DEM does not resolve the flow structure around the particle cluster, consisting of carrier particle and a large number of fine drug particles, since all particles are treated as point masses. A multi-scale approach was used by van Wachem et al. [55] for analysing drug particle detachment and also wall deposition in a special knee-type inhaler.

The inter-particle forces between carrier and API together with the dispersion forces generated by the inhaler during inhalation and the deposition forces in the human respiratory tract determine the fate of the administered formulation [56]. Consequently, with the support of new computational techniques together with particle engineering technologies [57], this is likely to result in a new generation of inhaler devices and formulations that allow improved and targeted drug deposition in the lung [56].

Objectives, Outline of This Research

The aim of the present project was two-fold; the experimental analysis of surface modifications of carrier particles and the resulting adhesion properties for different drug particles, as well as the numerical (CFD) investigation and modelling of flow-induced and wall-impact-generated detachment of the drug particles from the carriers using the experimentally obtained adhesion properties.

Understanding and controlling inter-particle forces is crucial for good and efficient carrier based DPI performance. One approach to tailor inter-particle-interactions between drug and carrier is the surface modification of the carrier particles [58]. Current carrier materials like lactose and or mannitol lack of the possibility to change surface topography without changing size and shape as well. Consequently, in the experimental part of the study, glass beads were selected as model carrier as well characterisable and uniform system to allow the modification of individual surface characteristics in terms of roughness without alteration of other properties. These spherical carriers should be matched with spherical API particles of varying characteristics (different hydrophobicity) to be able to model interactive blends with different extent of particle-particle interactions. The extent and effect of these interactions should be experimentally determined to allow implementation into simulation calculations. For this, both preparative activities to transfer different APIs to uniform, spherical particles and extensive characterisation methods had to be applied. Of special importance is the aerodynamic characterisation of such systems, which allows interpretation of the impact of particle-particle interactions within the interactive blends.

The experimental work followed a systematic step-wise approach. For the first set of experiments physical glass beads modification via ball milling and additionally plasma etching were explored. The impact of surface modifications of the glass beads on the aerosolization performance was then investigated with salbutamol sulphate, a model API representing hydrophilic APIs. Additionally adhesion forces were deter-

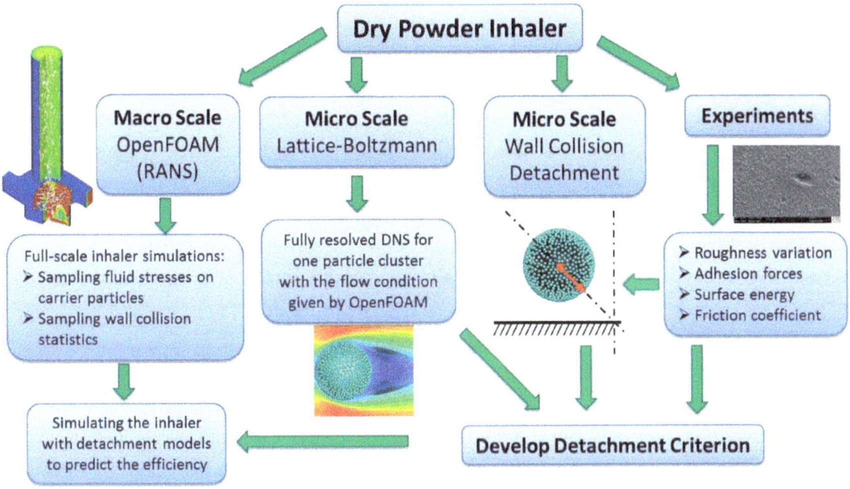

Fig. 3 Sketch of the multi scale approach adopted for analysing drug powder detachment from carrier in an inhaler due to flow stresses and wall impacts and connection between simulations and experiments

mined between carrier and API, to serve as input for simulations and to understand how adhesive forces can be tailored via surface modification.

In the second set of experiments surface modification via ball milling was expanded based on the results form first experiments. Further, a focus was set on spray drying of four additional APIs to generate spherical particles. Thereby a broad range of distinct APIs should be covered. From these four APIs, two were selected in order to evaluate the performance dependence of the modified carriers in relation to APIs with distinct properties. In a last step, budesonide was selected for determining the adhesion forces between carriers and API. Budesonide was chosen due to its distinct properties compared to salbutamol sulphate regarding hydrophobicity. This approach shall allow understanding the impact of a broad range of carrier surface topography in combination with distinct APIs on inter-particle interactions and thus DPI performance. Ideally, with the gained knowledge, identifying the optimal carrier for a given API (with certain properties) shall be possible. Further, with the help of simulation for a given carrier the ideal device geometry could be determined.

The developed Lagrangian detachment models will eventually allow the numerical prediction of inhaler performance and efficiency and hence constitute the basis for inhaler optimisation. For these objectives a multi-scale analysis is adopted here (Fig. 3). First the flow field through the entire inhaler device (here the Cyclohaler®) is calculated numerically using OpenFOAM® and the fluid dynamic stresses acting on different sized carrier particles are recorded and statistically analysed [59]. With these flow conditions micro-scale numerical simulations based on the Lattice-Boltzmann method (LBM) are conducted for a fixed carrier particle covered with hundreds of drug particles [59, 60]. From the simulated fluid stresses acting on the drug particles the possibility of detachment by pull-off, rolling and sliding was determined based on measured van der Waals forces and friction coefficients as well as other contact properties. The van der Waals forces acting between carrier particles and drug particles as well as the adhesion surface energy of untreated and physically modified glass beads was determined from the static contact angles of water and α-bromonaphthalene experimentally. Moreover, the effect of turbulence on drug particle detachment was analysed by direct numerical simulations by LBM, revealing the effectiveness of strong flow fluctuations on drug aerosolisation [61].

Furthermore, a carrier particle wall collision model was developed and tested (i.e. micro-scale simulations) for describing the inertia-induced drug particle detachment when the carrier is rapidly decelerated upon a wall collision [62]. Drug detachment might occur through lift-off, sliding or rolling. Nearly half of the drug particles on the carrier surface (i.e. those located on the front hemisphere facing the wall) can be detached by lift-off. Based on these studies, the Lagrangian carrier particle tracking [44] was supplemented by the developed carrier-wall collision model and was used to predict drug detachment probability within an inhaler device for two stationary flow rates and different carrier particle sizes. It was found that the averaged wall collision detachment for 1024 tracked carrier particles in a Cyclohaler was around 95%, which is of course much higher than known from experience [63]. Therefore, also the deposition of drug particle on the inhaler walls needs to be accounted for and modelled for example according to the suggestions of Milenkovic et al. [64].

Experimental Studies on Carrier Surface Modification

The fact that by modifying the surface topography/roughness of lactose or mannitol particles, not only the surface topography, but also other factors influencing interparticle interactions are altered, makes lactose and mannitol unsuitable as carriers to study the influence of surface topography on the performance of DPIs. That is why glass beads are used as model carriers in the present study. Glass beads are available in different sizes, they have ideal geometry and, most importantly, they allow various prospects of surface modification, without affecting other properties like particle size and shape. This allows the attribution of changes of DPI performance solely to the surface topography of the differently modified glass beads used. Thereby glass beads should help to bring the development of DPIs on a scientific level, and to avoid trial and error due to better knowledge about the relation of surface topography, interparticle interactions and DPI performance.

However, it has to be mentioned that glass beads are neither typical carrier particles nor safe for inhalation from a medical point of view. In this study glass beads are used as model carriers to understand how the surface topography of carrier particles in DPIs should ideally be to get a high deposition of API particles in the lung. For the application of glass beads containing dry powder inhalates to humans, a grid has to be implemented in the inhaler device retaining the carrier particles in the inhaler device.

Modification

In order to generate glass beads with different shades of surface roughness, physical surface modification by friction and impaction in a ball mill (Ball Mill S2, Retsch, Haan, Germany) was performed. Glass Beads in the size range of 400–600 μm (x_{50} = 537.3 μm ± 7.1 μm) were kindly provided by SiLibeads® (SiLibeads® Glass Beads Type S, Sigmund Lindner GmbH, Warmensteinach, Germany) and cleaned with Piranha Solution (3:7 H_2O:H_2SO_4) followed by a standard clean (1:1:5 H_2O_2:NH_4OH:H_2O) prior to use. Acetic acid, ammonium hydroxide, methanol, sulphuric acid and hydrogen peroxide were purchased from Lactan Chemikalien und Laborgeraete Vertriebsgesellschaft m.b.H & Co. KG, Graz, Austria.

For the first set of experiments, untreated glass beads (GB_UT) were processed for 4 and 8 h at 424 rpm with two different grinding materials (quartz (Q) and tungsten carbide (TC)) with varying hardness. Quartz has a Mohs hardness of 7.0 and tungsten carbide of 9.5, compared to glass that shows a Mohs hardness of 6.5. Tungsten carbide was provided from Wolfram Bergbau und Huetten AG, St. Martin i.S., Austria and quartz was obtained from Quarzwerke Austria GmbH, Melk, Austria. Thereby, glass beads with different shades of roughness could be prepared. Additionally, physical surface modification was performed by plasma etching (TEGAL901). The gases used were CF_4/O_2 and the ratio was 120sccm/15sccm at 1800mTorr. The plasma power was 200 W and the exposure time 1 min.

In the second part of the study, GB_UT were also cleaned with Piranha solution followed by the standard clean. Afterwards, glass beads were incubated with hydrofluoric acid (HF) for 10 min and thoroughly rinsed with purified water for several times (HF 10 min). One batch each of these glass beads was then ground with TC in a Retsch PM 100 ball mill (Retsch GmbH, Haan, Germany) for 4 and 8 h, respectively (HF + TC 4h and HF + TC 8h). Additionally, untreated glass beads were treated again with TC for four and eight hours as well (TC 4h and TC 8h). TC remaining on glass surfaces was removed in an ultrasonic bath by washing with purified water.

For preparing adhesive mixtures for inhalation and studying performance of the novel carriers and inter-particle interactions, APIs with distinct hydrophobicity and hydrophilicity were selected. The selected APIs were spray dried in order to generate spherically shaped particles. In combination with spherical glass beads as model carrier, this approach allows studying inter-particle interactions with minimal effect of particle shape and consequently particle orientation. For the first set of experiments salbutamol sulphate, a quite hydrophilic API was selected as model API and purchased from Selectchemie (Zuerich, Switzerland). Spray drying conditions were selected in order to generate inhalable sized particles based on work from [49]. Resulting particles showed a mean particle size (x_{50}) of 3.1 ± 0.4 μm and span of 1.9 and are shown in Fig. 4.

Fig. 4 SEM image of spray dried model API salbutamol sulphate

Table 1 Parameters for spray drying of budesonide (BUD), formoterol fumarate (FF) and tiotropium bromide (TB) and resulting product characteristics (n = 3, mean ± standard deviation)

	T_{inlet} (°C)	Spray feed rate (mL/min)	Solid content (% w/w)	Spray gas flow (L/h)	Drying gas flow (m^3/h)	X_{50} (µm)	Span	γ_{std} (mJ/m^2)
BUD	120	3.2	10	414	40	2.9 ± 0.1	2.4	48.5 ± 0.3
FF	120	2.1	4.5	357	40	2.7 ± 0.1	2.1	34.0 ± 0.6
TB	150	3.7	4	451	40	2.7 ± 0.1	2.0	48.1 ± 0.9

Further model APIs of varying hydrophilicity, namely budesonide (BUD), formoterol fumarate (FF) and tiotropium bromide (TB), were also transformed to spherical particles by spray drying (see Table 1 for parameters). For this, statistical experimental design was employed to select optimal parameters to obtain spherical particles in the inhalable range (Table 1 and Fig. 5). All trials were carried out with a B290 Mini Spray Dryer (Buechi Labortechnik AG, Switzerland) equipped with a high performance cyclone. Aqueous solutions were processed in open cycle mode, where heated air led to solvent evaporation. Due to safety reasons, the B290 was connected to a B295 Inert Loop (Buechi Labortechnik AG) and was operated in closed cycle mode when working with organic solvents. Here, nitrogen was used as drying gas to create inert conditions and prevent the danger of potential explosion. The final products were stored in a desiccator over silica gel (Sigma Aldrich Chemie GmbH, Munich, Germany) immediately until further use.

Characterization

The surface modified rough carrier particles were characterized with respect to particle size (laser diffraction and image analysis), shape (image analysis) and surface

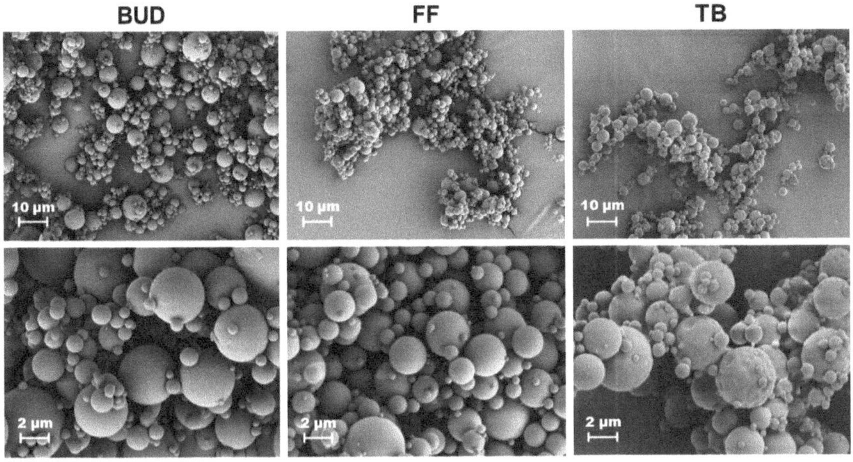

Fig. 5 SEM micrographs of spray dried APIs at respective magnifications of 1000× (top) and 5000× (bottom). From left to right: budesonide (BUD), formoterol fumarate (FF), tiotropium bromide (TB) (from [80])

topography (Scanning Electron Microscopy (SEM), and Atomic Force Microscopy (AFM)). Additionally, inter-particle forces between differently engineered carrier particles and selected API particles were determined (colloidal probe AFM).

Particle Size and Shape

When modifying the surface of lactose or mannitol carrier particles for example by spray drying, besides surface topography typically particle size and particle shape are influenced by the modifying procedure as well. As already mentioned, one reason for choosing glass beads as model carriers for the present work was that by modifying the surface topography particle size and particle shape are not changed by the surface modifying procedure necessarily. To confirm this, the particle size and the particle sphericity of untreated glass beads and the physically treated glass beads were analysed by Littringer et al. [65].

The particle size of the untreated as well as physically modified glass beads was investigated using a QICPIC (Sympatec GmbH, Clausthal-Zellerfeld/Germany) image analysis instrument. Samples were dispersed using a dry dispersing system (Rodos/L, Sympatec GmbH) and a vibrating chute (Vibri, Sympatec GmbH). The dispersion pressure applied was 0.2 bar. The measuring time was 6 s and 450 pictures/s were taken, so the total number of analysed particles per measurement was 400–7000. Data evaluation was done using the software Windox 5 (Sympatec GmbH). The mean particle size was described by the Feret mean diameter. The Feret diameter is derived from the distance of two tangents to the contour of the particle in a well-defined orientation. The Feret mean diameter is the mean value of the Feret diameters over all orientations (0–180°).

Besides particle size, the sphericity of untreated and physically modified glass beads was investigated also using the QUICPIC image analysis instrument using the same setup and same parameters as described above. Particle sphericity was calculated using the software Windox 5 (Sympatec GmbH). As a result values between 0 and 1 can be obtained. A value of 1 is given by a perfectly spherical particle. The smaller the value, the more irregular is the shape of the particle.

Figure 6 shows that physical surface modification does not substantially decrease the particle size of the glass beads that might have been occurred due to attrition or breakage. Further, sphericity was not changed via surface modification. This allows later on to attribute changes in adhesion forces and API detachment from the carrier solely to the surface topography neglecting the impact of size and shape of the carriers.

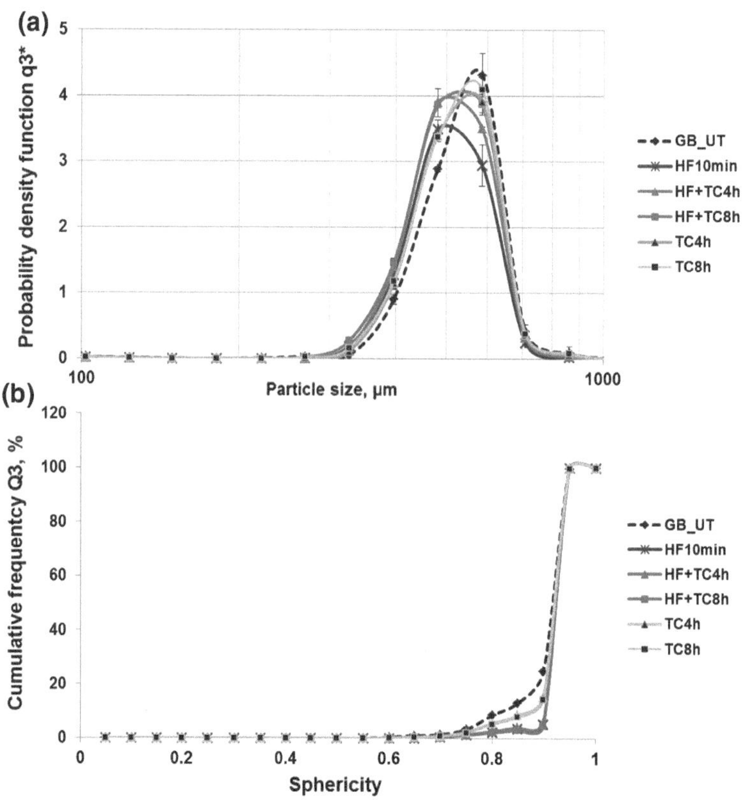

Fig. 6 **a** Particle size of untreated glass beads and modified batches (average of n = 3 measurements ± standard deviation), **b** sphericity of untreated glass beads and modified batches (from [80])

Particle Surface Topography

The surface of the modified glass beads and the quality of the adhesive mixtures was examined using a scanning electron microscope (Zeiss Ultra 55, Zeiss, Oberkochen, Germany) operating at 5 kV. Samples were gold palladium sputtered prior to analysis.

Figure 7 shows SEM pictures of untreated glass beads and physically modified glass beads. No changes of the surface topography can be observed at an image width of 882.2 μm, the untreated as well as physically treated glass beads look the same. However, SEM images at higher magnifications (image width 228.7 μm) showed that grinding with materials that are harder and smaller than the glass beads leads to roughnesses only visible at large magnifications. It appears that the surface of the TC treated glass beads shows larger surface irregularities than the surface of Q treated glass beads. This was expected and is well in accordance with the Mohs hardness (quartz 7 and tungsten carbide 9.5) of the grinding materials. From the SEM images it can also be concluded that an 8 h grinding time leads to more pronounced surface changes than a 4 h grinding time. This is true for to quartz as well as tungsten carbide treatment. By contrast, plasma etching leads to a more pronounced and different kind of surface roughness compared to glass beads treated in a ball mill.

SEM images from the second set of treatment experiments are shown in Fig. 8. Basically, results show same trends. Roughness seems to increase with increasing

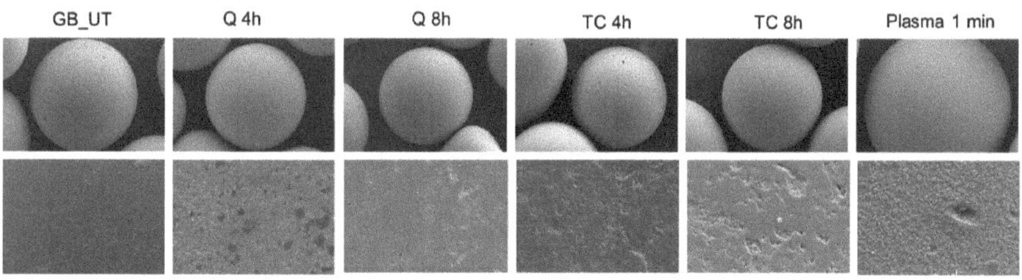

Fig. 7 SEM images of the first set of untreated and modified glass beads

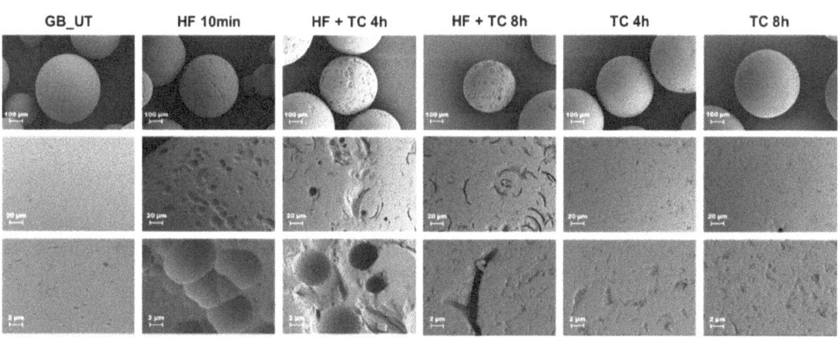

Fig. 8 Untreated glass beads and glass beads with micro-roughness (HF-treated ± TC-ground) and nano-roughness (TC-ground), 100-fold (upper), 500-fold (middle) and 2500-fold (bottom) magnification (from [67] with permission)

grinding time of TC. Whereas, HF treatment results in micro roughness as visible by the large indentions or break-offs, grinding with tungsten carbide only slightly changed surface morphology compared to untreated glass beads.

Particle Surface Roughness

The surface roughness of untreated and modified glass beads was determined via atomic force microscopy (VEECO, Dimension, Nanoscope IIIa, Bruker AXS GmbH, Karlsruhe, Germany). The glass beads were fixed on commercially available glass slides with a two component epoxy resin adhesive (UHU GmbH & Co. KG, Baden, Germany) prior to analysis. The surface roughness was measured with tapping-mode imaging in air using standard silicon tips OMCLAC160TS-WZ (Olympus, Tokyo, Japan). The data analysis program Gwyddion 2.25 (Gwyddion—Free SPM data analysis software under the terms of the GNU General Public License, Petr Klapetek, David Nečas, and Christopher Anderson, Czech Metrology Institute, http://gwyddion.net/) was used to calculate the root mean square roughness (R_{rms}). The R_{rms} value gives the root mean square of the measured height deviations from the mean height value of each scan line and is an amplitude parameter. Bead curvature was subtracted for each AFM image before roughness calculation in order to separate local surface roughness from the overall curvature of the beads. For each type of modified glass beads at least 4 different beads were analysed and on each glass bead three to four 10 μm $\times$ 10 μm squares were analysed. For each square a roughness value was calculated [58].

R_{rms} values in Fig. 9 show that the R_{rms} values for ball mill treated glass beads are in accordance with observations from SEM images. Glass beads treated with tungsten carbide for 8 h show the highest R_{rms} value (19.3 $\pm$ 4.7 nm) and thus exhibit the roughest surface. By comparison, grinding with quartz for 4 h leads to the smallest R_{rms} value (11.9 $\pm$ 2.5 nm) and so to the least rough surface. The tapping mode images presented in Fig. 10 further show that surface roughness looks different depending on the type of grinding material used. Processing with quartz (Fig. 9a) leads to a finer roughness compared to processing with tungsten carbide, where especially the latter leads to a higher depth of the roughness. This is well in accordance with the hardness of the grinding materials and the processing time. Compared to all samples processed in a ball mill, plasma etched glass beads show a distinct surface structure and a higher surface roughness (42.5 $\pm$ 12.45 nm). Untreated glass beads show comparably high RMS values with a high standard deviation. This could be explained by occasional appearing voids and clefts on the surface. As the R_{rms} value determined with AFM is a measure of the roughness amplitude, such irregularities, if present in the measuring area, result in high R_{rms} values and also a high standard deviation.

Glass beads being used for the second part of the study showed a low R_{rms} value of 21.2 $\pm$ 7.4 nm. This is in accordance with SEM images and represents the quite smooth surface of the glass beads without voids and clefts. This time, particular focus was set on selecting smooth parts on the glass beads surface excluding voids and clefts. In accordance with the first set of experiments, milling with TC for 4 h

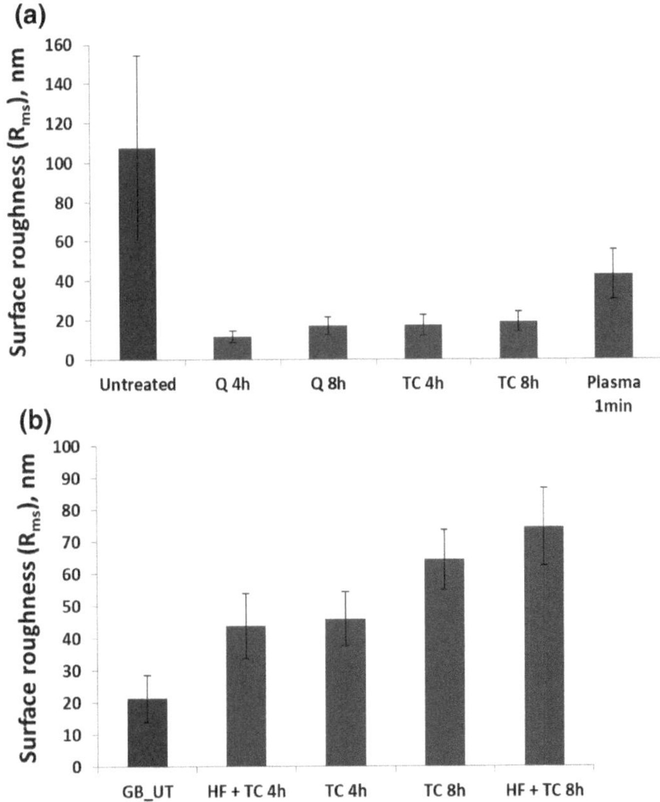

Fig. 9 Root mean square roughness of untreated glass beads and physically modified batches from the first part (**a**) and second part (**b**) of the study ((**b**) from [67] with permission)

increased the surface roughness to 46.0 ± 8.3 nm (TC 4h), while an 8 h treatment enhanced the R_{rms} value further to 64.2 ± 9.3 nm (TC 8h), which was a more pronounced change than in the first set of experiments. These differences could be related to distinct batches of glass beads used for the modification, the distinct type of ball mill or the measurement principle, where the selection of representative spots on the glass beads surface is challenging. It can be derived from Fig. 9b (AFM surface) that the density of surface irregularities was increased compared to TC 4h during the extended milling. Taking the micrographs from AFM into account, HF 10 min exhibited a drastically increased R_{rms} value of 69.7 ± 4.8 nm due to the respective frequency and depth of extended asperities located on the scan area. It needs to be emphasised that the R_{rms} does not reflect the frequency of occurrence of surface roughness. Instead, it provides information about the level of asperities. If AFM micrographs are additionally taken into account, the extended milling time protocol did only increase the occurrence of surface roughness area-wise, but also the depth of asperities. Due to the experimental setup, it was not possible to quantify

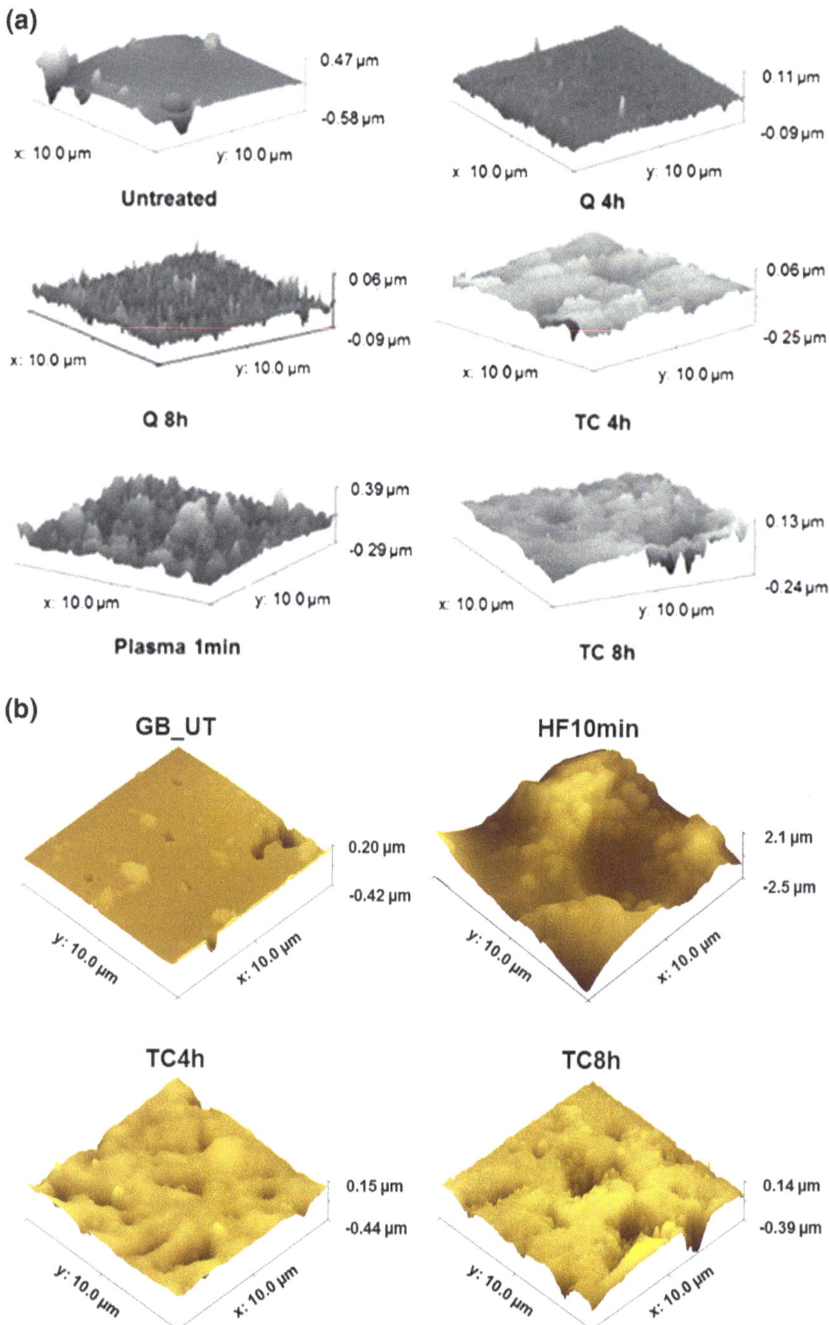

Fig. 10 Surface topography of **a** untreated glass beads (GB_UT) and beads being treated with Q, TC and Plasma and **b** untreated glass beads (GB_UT) and beads being treated with HF and TC, respectively ((**b**) modified from [67] with permission)

roughness of HF + TC 4h and HF + TC 8h across the whole surface as sensitivity of the cantilever could solely be adjusted to micron or submicron irregularities. This limitation of AFM is widely accepted and has been acknowledged in literature [66]. Therefore, it was decided to focus investigations on regions on the carrier surface where no evident alteration through HF had taken place. These findings correspond well to the R_{rms} values of TC 4h and TC 8h (Fig. 9b).

Summing up, the surface roughness of glass beads, intended for the use as carrier in dry powder inhalers, can be successfully modified via mechanically treatment in a ball mill and plasma etching. Overall, mechanical treatment in a ball mill leads to surfaces with roughnesses in the nanometer scale. Depending on the hardness of the selected grinding material and the grinding time the surface roughness can be tailored. Harder grinding materials and increased grinding times lead to more pronounced surface roughness. In our case, grinding with tungsten carbide leads to rougher surfaces than grinding with quartz. Treatment with hydrofluoric acid, in contrast, introduces microscale roughness. However, surface modification does not change the mean particle size and shape of the glass beads.

Adhesion Force Measurements

The adhesion force between glass beads and two spray dried model APIs (salbutamol sulphate and budesonide) was determined via atomic force microscopy (VEECO, Dimension, Nanoscope IIIa, Bruker AXS GmbH, Karlsruhe/Germany) operated in contact mode. The untreated and surface modified rough glass beads were glued on commercially available glass slides with a two component epoxy resin adhesive (UHU GmbH & Co. KG, Buehl/Baden/Germany). The colloidal probes were prepared by attaching the 3 μm sized API particles on the tip of a tipless AFM cantilever (NSC 12/tipless/no Al, NT-MDT Anfatec Instruments AG, Oelsnitz/Germany) with the help of an micro-manipulator (3 D oil-hydraulic manipulator MMO–203, Narishige Group, Tokyo, Japan). This was realized by spreading the API particles on a glass slide and attaching them to the cantilever tip by moving the cantilever towards a particle with the micro-manipulator and collecting it from the glass slide with the help of the adhesive that has been attached to the cantilever tip before [59].

The scan size was 5 μm × 5 μm and the imaging scan rate was 1 Hz. The scan area was divided into a grid of 25 equidistant points and on each point 50 force curves were taken, leading to a total of 1250 force curves acquired on each glass bead. For each glass bead/API combination a mean adhesion force was calculated out of the 1250 force curves. This procedure was repeated three times using a new pair of API and glass bead every time. Finally, a mean value was calculated out of the three means obtained for the adhesion force between the three different salbutamol sulphate particles and the corresponding glass beads. This was done in order to estimate trends and reduce the variation within the measured force results. Extraction of adhesion force data from the AFM force spectroscopy raw data was carried out with automated routines written in LabView [59].

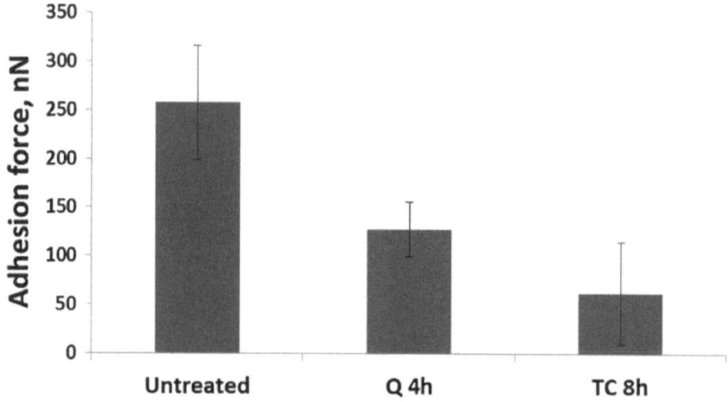

Fig. 11 Adhesion force (±standard deviation) between GB_UT or modified glass beads and spray dried SBS

For the first set of adhesion force measurement the two ball mill treated glass beads with lowest and highest surface roughness were chosen and salbutamol sulphate (SBS) as hydrophilic model API. Figure 11 shows that the adhesion force between untreated glass beads and SBS is higher compared to the adhesion force between surface modified glass beads and salbutamol sulphate. This can be explained by the rougher surfaces of the modified glass beads. By introducing surface roughness the contact area between the glass beads and the SBS particle is reduced and hence also the adhesion force between them. TC 8h treated glass beads show a higher surface roughness than Q 4h treated glass beads, and so is the adhesion force lower between glass beads modified with TC 8h and salbutamol sulphate.

Similarly, adhesion forces between HF and TC treated glass beads and budesonide as hydrophobic model API were assessed (Fig. 12). It is obvious, that the measured adhesion force between the glass bead and the hydrophobic model API is generally lower than observed for SBS being a hydrophilic model API. This can be explained by the difference in interaction forces, which mostly comprise van der Waals forces for BUD, whereas SBS will also interact via hydrogen bonds. Further it can be seen that the introduction of micro roughness by HF does not change the adhesion of spray dried budesonide to the glass bead, whereas a simultaneous micro- and nano-roughness (HF + TC batches) reduces the adhesion of BUD. However, there is no difference observed between glass beads treated with HF and then with TC, compared to TC treatment only.

Summing up, experimental work described above showed that the adhesion force between surface modified glass beads and salbutamol sulphate is remarkably lower than the adhesion force between untreated glass beads and salbutamol sulphate. This was related to the rougher surfaces of the modified glass beads and the thereby reduced contact area between modified glass beads and API particles [59]. The same is true for budesonide and untreated versus surface-modified glass beads with nano-roughness, but adhesion forces are generally lower for the rather hydrophobic budesonide. Adhesion force data of salbutamol sulphate were incorporated in

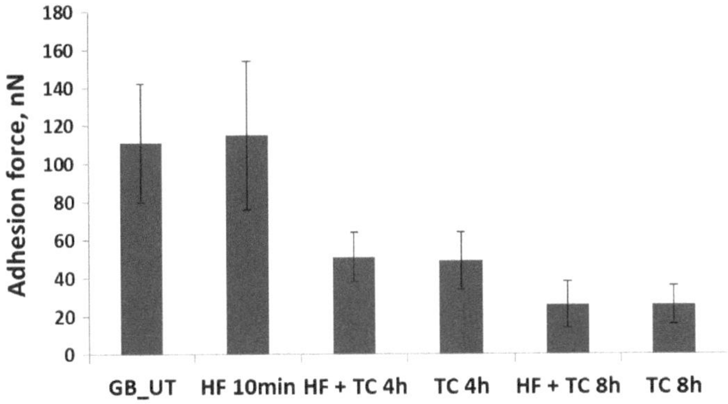

Fig. 12 Adhesion forces (±standard deviation) between GB_UT or modified glass beads and spray dried budesonide (from [80])

the simulations and used to calculate the detachment probability of API particles (section "Micro-scale CFD: API Flow Detachment from the Carrier").

Preparation of Blends

Besides the detachment of drug particles from the carrier surface the adhesion of drug particles onto the carrier surface is also influenced by the carrier surface properties. Thus, the same amount of API weighed in for the preparation of the adhesive mixture does not necessarily mean that the carrier surface is covered to the same extent with drug particles. In order to compare different carrier and model systems among each other the dependency of the true surface coverage (TSC) of surface modified and untreated glass beads after mixing and handling has been investigated depending on the amount of API that has been weighed in for the preparation of adhesive mixtures. Adhesive mixtures were prepared with 3 different calculated surface coverages (100, 50 and 25%). Calculations were done with the help of the projected surface area of API particles on the glass beads surface and based on the principle of the densest package of API particles on the glass bead surface, described in detail elsewhere [67]. The true surface coverage was obtained from the drug content analysis from 10 samples taken for the blend homogeneity determination. Knowing the calculated surface coverage and comparing it with the true surface coverage also a mass ratio of glass beads and salbutamol sulphate could be calculated by knowing the mass of adhesive mixture in each sample and the related mass of API obtained from HPLC analysis. By comparing the true mass ratio and the calculated theoretical mass ratio of glass beads and API, the true surface coverage in % could be calculated. As there was a quite good correlation between the true and calculated surface coverage, mixtures with same true surface coverage that can be compared among each other could be prepared.

The API/carrier blends were prepared for distinct calculated surface coverages (CSC) in a Turbula blender TC2 (Willy A. Bachofen Maschinenfabrik, Muttenz, Switzerland). 15 g of glass beads and the corresponding amount of API were weighed into a stainless steel mixing vessel via the double sandwich method. Mixing parameters were 45 min at 22 rpm for SBS, 30 min at 20 rpm for BUD containing blends and blends containing FF were blended for 35 min at 20 rpm followed by 5 min at 30 rpm.

Blend homogeneity was determined by taking a total of 10 samples of 150 mg from various locations in the mixing vessel. The samples were dissolved in 20.0 mL of the appropriate solvent and investigated via HPLC. Form the actual drug content, true surface coverage could be calculated.

Aerodynamic Assessment of DPI Performance

The aerodynamic assessment of blends containing salbutamol sulphate was performed according to the European Pharmacopoeia (preparations for inhalation: aerodynamic assessment of fine particles, Ph. Eur., 7.0) using Apparatus E (Next Generation Impactor (NGI), Copley Scientific, Nottingham, United Kingdom). Prior to each experiment the small cups of the impactor were coated with 2 ml, the large cups with 4 ml of a coating agent (solution of 5% of a mixture of glycerol and polyoxyethylene-20-cetyl ether (95:5) in isopropanol) in order to prevent re-entrainment of the particles (a detailed description of the method can be found at [3]). 150 mg (±20%) of the adhesive mixtures were filled into hard gelatine capsules of size 3 (G-CAPS®, GoCaps GmbH, Edlingen, Germany) manually. For each adhesive mixture two capsules were discharged into the impactor directly after each other. For adhesive mixtures with 25% calculated surface coverage an additional third capsule was discharged to increase the amount of salbutamol sulphate deposited in the impactor and to facilitate the HPLC quantification. The amount of drug in each cup, the pre-separator, the introduction port plus the mouthpiece and the inhaler was subsequently determined via HPLC analysis.

For BUD and FF containing blends, aerodynamic assessment was also performed utilising the Next Generation Pharmaceutical Impactor (a detailed description of the method can be found in Renner et al. [67]). In brief, each of the seven stages, the micro-orifice collector (MOC) and the impaction plate of the pre-separator were coated with stage coating consisting of ethanol (51%), glycerol (34%) and Brij® 35 (15%) and allowed to dry completely to prevent particle re-entrainment. 250 mg of the blend under investigation were filled into hard gelatine capsules size 3 with a spatula. Prior to impaction analysis, capsules and inhalation device were deionised with a Static Line LC discharging bar (HAUG GmbH & Co. KG, Leinfelden, Germany) to minimize electrostatic charging. When using interactive mixtures with CSCs of 100% or 50%, three capsules were used for each NGI run. Investigations concerning blends with a CSC of 25% made it necessary to release the content of five capsules to guarantee quantifiable amounts of drug across the whole impactor. Drug content of the individual samples was quantified via HPLC analysis.

According to the European Pharmacopoeia (preparations for inhalation, Ph. Eur., 7.0) the fine particle dose (FPD), the emitted dose (ED) and the fine particle fraction (FPF) were calculated. The ED represents the mass of active emitted from the inhaler and found in the whole impactor (mouthpiece adaptor, introduction port, pre-separator, impaction stages). The FPD gives the mass of API exhibiting an aerodynamic diameter of <5 μm and. FDP divided by ED gives the FPF indicating the amount of drug particles emitted from the inhaler and being small enough (<5 μm) in order to reach the deep lung. The FPF may be defined as the main parameter describing the performance of a DPI system.

The Cyclohaler® is a capsule based single dose device with a low resistance to airflow. For each inhalation manoeuvre, a capsule is placed in the compartment in the base of the inhaler and by squeezing two buttons inwards on the base of the inhaler the capsule is pierced so that the powder can be released during inhalation. It is a passive device where the patient has to generate the airflow dispersing the powder during inhalation.

Figure 13 displays the FPFs of mixtures containing SBS and untreated and surface modified glass beads with the same surface coverage (23.5 ± 5.5%). To avoid multilayer formation and cluster formation of the SBS particles the desired true surface coverage was set to 25%. Lower standard deviations for the FPFs of modified glass beads indicate more reproducible drug delivery compared to untreated glass beads. This can be attributed to the occasional appearing valleys present on the surface of untreated glass beads that result from the manufacturing process and/or transport and lead to quite inhomogeneous surfaces [58]. API particles might hide in these deep clefts and voids on the glass beads surface and thus might not be detached during inhalation. Surface treatment, either in a ball mill or by plasma etching, results in homogenous, rough surfaces that allow a more constant drug detachment and thus more reproducible FPFs.

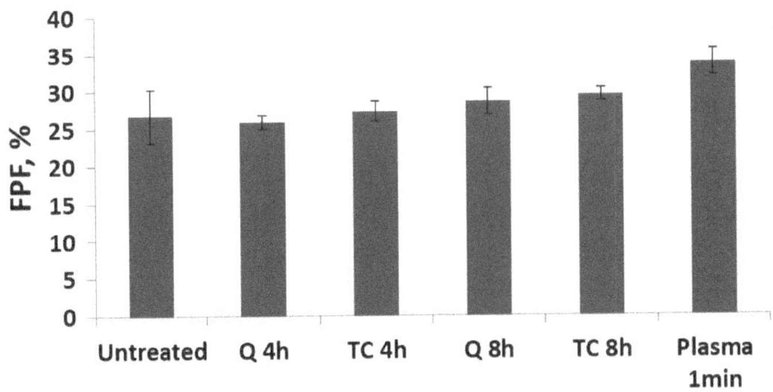

Fig. 13 Fine particle fractions from blends of glass beads and SBS with same surface coverage of (23.5 ± 5.5%)

Compared to untreated glass beads, the FPFs of mixtures containing glass beads treated with TC for 8 h and Plasma for 1 min are higher. No difference between untreated glass beads and Q 4h and TC 4h treated glass beads can be observed. The homogenously rough surfaces that have resulted from the latter seem to be as effective as the inhomogeneous surfaces of untreated glass beads, with their occasional appearing deep voids and clefts, when considering the amount of API particles that can be detached during inhalation. Moreover, when comparing the surface modified carriers among each other, Fig. 13 indicates that the FPF increases with increasing surface roughness of the glass beads.

To better understand the relationship between surface roughness and FPF, an attempt was made to correlate the surface roughness R_{rms} of the modified glass beads with the FPF. Untreated glass beads were excluded from the correlation as their surface is quite inhomogeneous due to the presence of few, but deep valleys on their surface.

SEM images already indicated and AFM analysis confirmed that glass bead etching with plasma for 1 min leads to the highest R_{rms} value and thus to the roughest surface. By comparison, grinding with Q for 4 h leads to the smallest R_{rms} value and so to the least rough surface. Grinding with Q for 8 h and TC for 4 h and for 8 h, however leads to surface roughnesses in between. This trend can also be observed when considering the FPF. The FPF tends to increase more or less linearly with the surface roughness, with a coefficient of determination R^2 of approximately 0.83. The correlation coefficient resulting from the correlation of the FPF and the surface roughness is 0.909 and indicates a positive correlation. This is well in accordance with our expectations and findings from Kawashima et al. that found that carrier particles with increased surface roughness and surface area lead to higher FPFs [36]. By introducing surface roughness to the glass bead surface, the contact area between the glass beads and the API particles is reduced and further also inter-particle interactions. Therefore, with increasing surface roughness and the accompanying reduced contact area, drug detachment is facilitated and as a consequence the FPF is increased [67].

In the second part of the study, the experimental data set was expanded to interactive blends with formoterol fumarate as an API of intermediate hydrophobicity and budesonide, which is a hydrophobic API. Again blends with different calculated surface coverages of 25, 50 and 100% were prepared. From the assessment of drug content and calculation of corresponding true surface coverage it was obvious that the degree of particle-particle interactions of these APIs with the unmodified and modified glass beads was different to SBS. Further it could be seen that true surface coverage and calculated surface coverage correlated, but were not the same (TSC was lower than calculated in any case) (Fig. 14). This was due a substantial proportion of API adhering to the inner walls of the mixing vessel after blending. The degree of drug recovery from the mixing vessel was determined to be dependent on the respective modification. For FF, the highest drug loading was reached on GB_UT. This was due to the physio-chemical properties of drug and carrier. For the latter, free silanol groups bring hydrophilicity to the surface. FF is also rather hydrophilic compared to BUD. In consequence, interaction between carrier and FF will be conveyed by

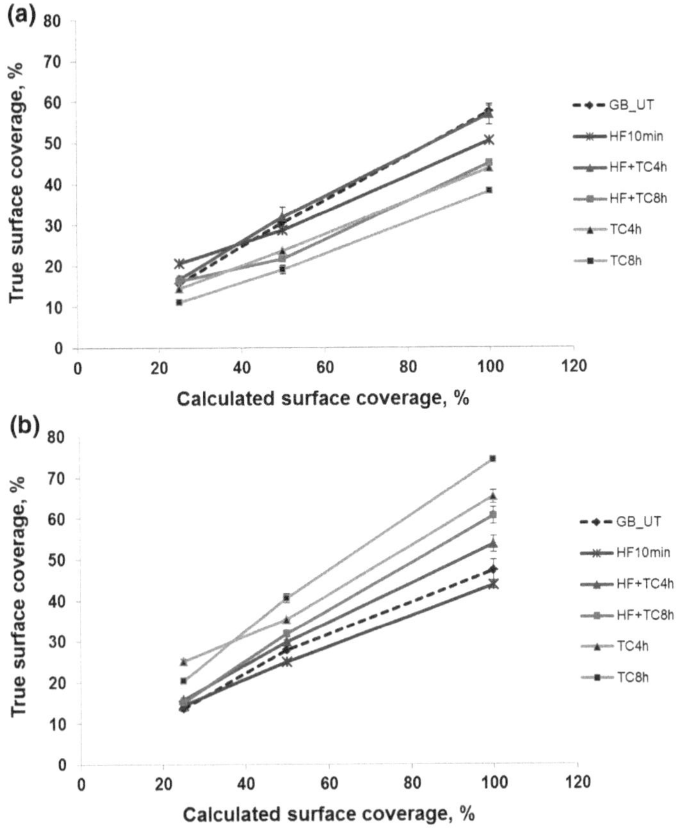

Fig. 14 Correlation between CSCs and the corresponding TSCs (+standard deviation) for interactive mixtures containing GBs and FF (**a**) and BUD (**b**) (n = 10) (modified from [67] with permission)

hydrogen bonds to a certain extent leading to comparably high adhesion forces for GB_UT and the API. Grinding with TC introduced submicron surface roughness as confirmed by AFM. Thus, a reduction of contact area between carrier and API is expected. Here, less contact area ultimately led to weaker binding of FF to the GB surface causing to the lowest TSCs for TC Q 4h and TC 8h. In theory, BUD should bind to the hydrophilic GB surface to a limited extent on account of a minor potential for interaction (mainly van der Waals forces) relative to FF. This was proven by lower TSCs for blends containing GB_UT and BUD compared to those with GB_UT and FF and additionally by comparing blends with BUD to those with HF 10 min and FF. In the case of BUD, roughness on nano-scale led to an enhanced effective drug loading. It is suggested that the hydrophobic drug particles are more likely to bind to the hydrophilic GB surface, if contact area is reduced.

To allow comparison of fine particle data regardless differences in true surface coverage, a normalisation of aerodynamic results was performed and FPF was nor-

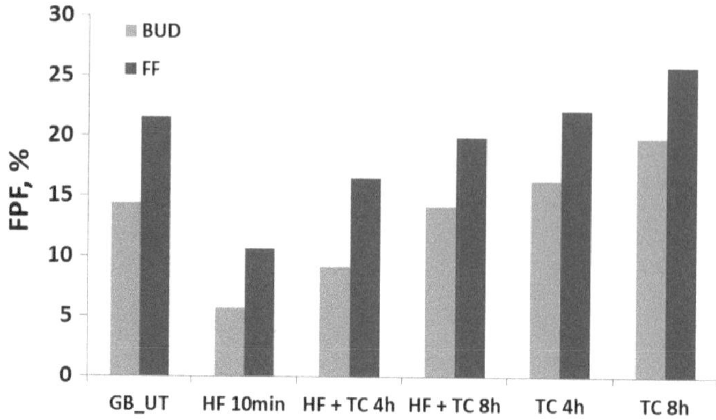

Fig. 15 Resulting FPFs at 30% TSC for untreated and modified GBs blended with FF (black bars) or BUD (grey bars) (modified from [67] with permission)

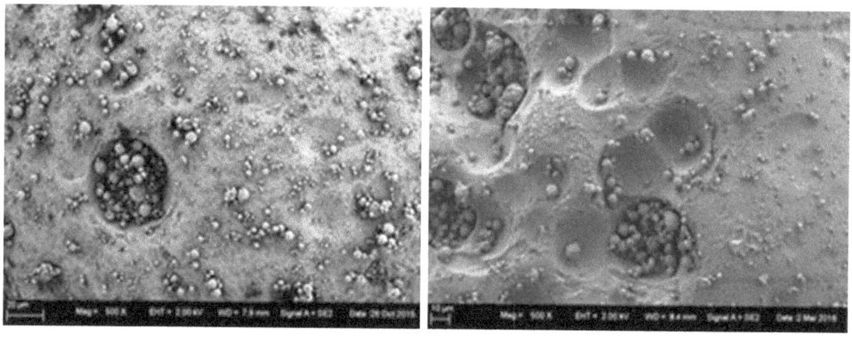

Fig. 16 SEM micrographs of the interactive mixture containing glass beads treated with HF 10 min and BUD at 50% CSC before (left) and after inhalation (right), (from [67] with permission)

malized to a fictive TSC of 30% for each type of GB to directly compare the different blends [67].

As shown in Fig. 15, it can be observed, that BUD blends in general gave a lower fine particle fraction than corresponding FF blends. This is in line with the aforementioned observation that BUD particles adhered better to the model carriers with nano-roughness than FF despite its predominantly hydrophobic interactions. A significant decrease of the respirable fraction compared to untreated glass beads was observed in the presence of deep cavities (μm range obtained from HF treatment) for both APIs and SEM pictures provide a conclusive explanation for this (Fig. 16 left). After blending, API particles had accumulated in those discontinuities. This observation has been ascribed to an enhanced potential for adhesive forces in those regions. In addition, API particles located in those active sites potentially find shelter from the airstream during inhalation, which causes a persisting attachment onto the carrier surface.

In case of the Cyclohaler, removal forces during inhalation are mainly based on drag and lift force, while inertial impaction only plays a minor role for this device. Figure 16 (right) shows indentations remaining occupied with drug particles. Even though a high flow rate of 100 L/min was applied, the kinetic energy of the air stream was evidently not high enough to detach the API causing a drastically low respirable fraction. The same trend was existent when comparing FPFs of blends containing HF + TC 4h or HF + TC 8h to those containing TC 4h or TC 8h, respectively. The presence of nano-scale roughness induced an increase in FPF, which is true for FF as well as BUD. Presumably, these submicron irregularities reduced the contact area between carrier and API and consequently diminished the potential for interaction between the two. This led to lower adhesive forces ultimately facilitating drug detachment to enhance the FPF. Again, the extended milling time showed an even more pronounced effect as superior FPFs were achieved when using TC 8h.

Compared to SBS, budesonide as well as formoterol fumarate showed lower FPFs, indicating better drug detachment from glass beads for more hydrophilic APIs. Lower FPFs for budesonide compared to salbutamol sulphate were already reported in literature and were attributed to the more cohesive nature of BUD. This is contrasting to the adhesion forces determined in the present study, which were higher for SBS compared to BUD. As adhesion forces were only determined for single particles, it can be concluded that other properties related to the bulk like press on forces during blending and/or cohesivity and agglomeration tendency of the API play a dominant role for the aerodynamic performance as well.

Concluding, the FPF can be tailored by the use of surface modified rough glass beads as carrier particles. The FPF increases with increasing surface nano-roughness. Moreover, drug detachment and thus the FPF from the surface modified carrier particles varies less compared to untreated glass beads. However, when micro-roughnesses are introduced, due to increased adhesion forces, and shelter of drug particles versus the airflow, API particles are likely to stick in these regions leading lower FPFs compared to untreated glass beads.

Influence of Inhaler Device on Aerodynamic Performance: The Unihaler

Up to this point, focus has been put on the formulation (carrier and API characteristics) to explain aerodynamic performance. However, the inhaler also plays a crucial role in overall performance of an inhalation product and influences of the individual devices can be shown experimentally as well as in simulation (see section "Inhaler Comparison with Carrier-Wall Collision Induced Drug Detachment"). In addition to the Cyclohaler another capsule-based inhaler, the Unihaler was used in the course of this study. The general concept of the inhaler is similar to the Cyclohaler, but geometry and thus, flow through the device is different. Further, device resistance is

Table 2 Dimensioning of the upper part, grid and bottom part of the inhaler in sizes small (S), medium (M) and large (L); CSA = cross sectional area

Dimensioning	S	M	L
Mouthpiece, CSA (mm^2)	7.06	21.30	35.54
Grid, free grid area (mm^2)	20.71	35.76	50.80
Grid, stanchions width (mm)	7.06	21.30	35.54
Air inlet, CSA (mm^2)	11.82	17.70	23.57
De-agglomeration chamber, height (mm)	7.00	8.00	9.00

higher compared to the Cyclohaler, thus the typical flow rate for in vitro assessment according to the Ph. Eur. is 60 L/min for the standard combination of geometries (in contrast to100 L/min for the Cyclohaler) and was maintained throughout the study. Adding a second device configuration for the simulations allows the verification of the model being developed for the Cyclohaler and testing its general applicability. The overall aim of the experimental part was the identification of inhaler parts with a significant impact on dispersion, de-agglomeration and consequently on drug delivery efficiency. This was implemented with DoE approaches utilising a modular inhaler system [68] of the Unihaler. Different dimensioning of inhaler parts (Table 2) was defined as respective factors within the experimental design. FPF, device retention (DR) and specific device resistance (SDR) were set as responses to be investigated. Blends were prepared according to [69] with a fixed drug content of 1.0% utilising micronized budesonide and GB_UT to allow comprehensive evaluation.

Figure 17 visualises the effect of the different inhaler parts on the FPF of BUD, whereby values in the range of 14.1 ± 1.3% to 31.4 ± 2.1% were obtained. The geometry of all three components showed a significant effect on the FPF.

Increasing the dimensioning of each inhaler part was unfavourable for the aerodynamic performance. Accordingly, highest FPF was exhibited by the device composed of a small upper and bottom part as well as a small grid (Modular_S), while lowest FPF was experienced for overall large dimensioning (Modular_L). These modifications all increase air velocity and turbulence in the inhaler. In an earlier study [69], an opposite effect was documented. This might seem contradictory at first sight, but is entirely coherent when experimental conditions are considered. The present study featured a consistent flow rate leading to varying pressure drops and thus different air stream velocities in the device for the different combinations of inhaler parts. The earlier study implemented the opposite approach, setting a constant pressure drop of 4 kPa, as defined by the Ph. Eur. as standard conditions. This resulted in varying flow rates and thus a constant air stream velocity. It had not been possible to apply this setup here due to the fundamental differences in characteristics of used carriers. Density and particle size are much lower for lactose compared to GBs, thus GBs needed certain airflow to get moved at all. By decreasing the mouthpiece diameter, inhaler resistance is increased. As a consequence, a lower flow rate would be needed to maintain the 4 kPa pressure drop which lowers the kinetic energy of the airstream and the level of turbulence. This did not allow movement of the dense glass beads at

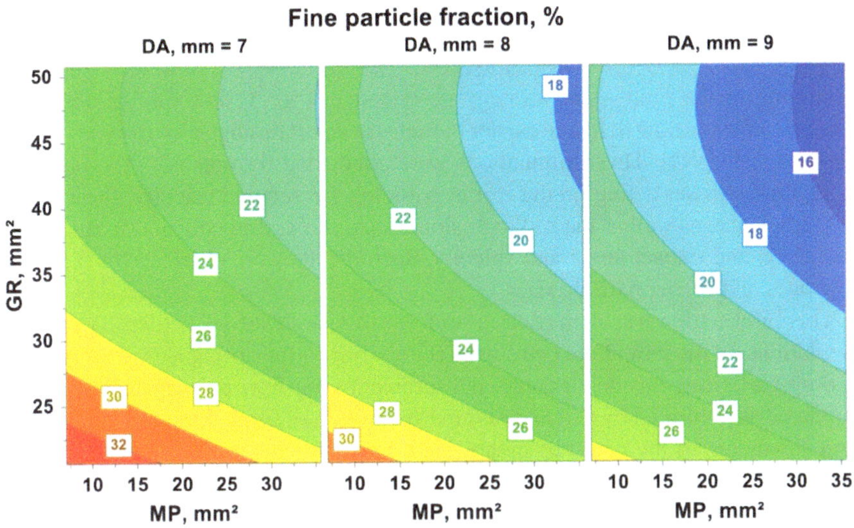

Fig. 17 Contour plot of the FPF for BUD as a response of the factors mouthpiece cross sectional area (MP) on the x-axis and grid (represented by the free grid area (GR)) on the y-axis. Results are given for the three different bottom parts represented by de-agglomeration chamber (DA) heights of 7, 8 and 9 mm (from [80])

all. A small grid was needed to achieve higher respirable fraction in the present study. It is assumed that a smaller free grid space decelerates emptying of the capsule chamber. By that, the kinetic energy of the airstream was applied to the interactive mixture for the whole duration of the inhalation manoeuvre. This enhanced de-agglomeration of the interactive mixtures to ultimately increase the respirable fraction. This corresponds to findings from the earlier study and indicates that a slow and controlled discharge of the de-agglomeration chamber is desirable regardless of carrier properties. For the hydrophobic BUD, the bottom part (air inlet and de-agglomeration chamber) significantly influenced the FPF. Tightening the air inlet and reducing the size of the de-agglomeration chamber enhanced the aerodynamic performance. On the one hand, this is also due to an increased air stream velocity at constant flow rate through a reduction of the cross sectional area of the air inlet. On the other hand, a smaller de-agglomeration chamber decrease the distance particles have to be lifted to leave the inhaler. The latter factor is certainly negligible for conventional carriers with a relatively low specific weight, but is relevant for high density GBs. It can be concluded that BUD needs to exit the inhaler persisting on the carrier and will undergo drug detachment and dispersion on its way through the mouthpiece. These findings are supported by the results of simulation studies (see section "Inhaler Comparison with Carrier-Wall Collision Induced Drug Detachment").

Numerical Studies and Modelling of API Detachment

According to the multi-scale strategy described in Fig. 3, first simulations were conducted for the flow field and carrier particle transport through a swirl-type inhaler the Cyclohaler [44]. These simulations were conducted for obtaining the statistics of the fluid stresses acting on the carrier particles, i.e. relative velocity, shear stress and turbulence, as well as the wall collision statistics, i.e. wall collision location and frequency, and impact angle and velocity. Such results were obtained for different flow rates and carrier particle sizes.

This information was then used to perform particle-resolved Lattice-Boltzmann simulations for the flow about one cluster (i.e. carrier particle occupied with hundreds of drug particles). Besides laminar and turbulent plug flow also shear flows were simulated as presented by Cui and Sommerfeld [60, 61]. From these simulations the fluid forces on all the resolved mono-sized drug particles with different diameter were evaluated. With measured adhesion properties [59] the detachment probability through lift-off, sliding and rolling by fluid stresses was evaluated in dependence of carrier Reynolds number, surface properties and degree of coverage with drug particles.

Finally, a carrier-wall collision model with drug particle detachment due to inertia was developed by Cui and Sommerfeld [62]. With this model again Lagrangian simulations for a Cyclohaler in comparison with the Unihaler, developed at the University of Kiel, were conducted for determining the total wall-collision induced drug detachment as briefly presented by Sommerfeld et al. [63]. These simulations revealed that the carrier-wall collision process is very effective and almost 100% of the drug powder was detached for both geometries.

With that, the multi-scale cycle is closed and numerical models are available to optimize inhaler geometries with regard to FPF. Missing is just the description of drug powder deposition at inhaler walls [43, 56], which can drastically reduce FPF. The development of a model for drug deposition during carrier-wall collision could be supported by DEM (discrete element method)-simulations [70].

Macro-scale CFD: Flow Stresses on Carrier Particles and Wall Collision Statistics

The numerical calculations of the stationary flow through an inhaler were conducted using the open-source-code OpenFOAM®. Thereby, the Reynolds-averaged conservation equations (abbreviation: RANS) were solved, using the k-ω-SST (shear stress transport) turbulence model for closure [44]. In the resulting final stationary flow field, after convergence was reached, the dispersed particle phase was simulated in the Lagrangian fashion by tracking a certain number of carrier particles accounting for a number of fluid forces being described below, as well as the influence of turbulence on particle motion. Two-way coupling, i.e. the influence of the particles on flow field and turbulence, was neglected in the present study, in order to more clearly emphasize the effects of different flow rates, particle sizes and fluid forces on carrier particle transport in the inhaler.

Particle tracking was conducted by considering various forces as summarized by Sommerfeld and Schmalfuß [44] in order to elucidate their contributions. The most important forces for the considered relatively large particles are the drag, gravity/buoyancy, and transverse lift through shear and rotation [71]. The pressure force and added mass which also were applied for completeness were of course in such a gas-particle flow of minor importance, which will be shown also below. As the particles were allowed to rotate (mainly induced by wall collisions) also the change of particle angular velocity through the viscous torque was calculated. Hence, the following set of partial differential equations for all the particles was solved:

$$\frac{d\overrightarrow{x_P}}{dt} = \overrightarrow{u_P} \tag{1}$$

$$m_P \frac{d\overrightarrow{u_P}}{dt} = \overrightarrow{F_D} + \overrightarrow{F_{AM}} + \overrightarrow{F_{PG}} + \overrightarrow{F_{LS}} + \overrightarrow{F_{LR}} + \vec{g} V_P(\rho_P - \rho_F) \tag{2}$$

$$I_P \frac{d\overrightarrow{\omega_P}}{dt} = \vec{T} \tag{3}$$

here $\overrightarrow{x_P}, \overrightarrow{u_P}, \overrightarrow{w_P}$ are the particle location, translational and rotational velocity vectors, ρ_p and ρ_F are the particle and fluid densities. The forces in Eq. 2 are the drag F_D, the added mass F_{AM}, the fluid inertia F_{PG} (i.e. part of the pressure force), the slip-shear lift force F_{LS}, the slip-rotational lift F_{LR} and the last term in Eq. 2 combines gravity and buoyancy. Note the gravity direction was assumed to be aligned perpendicular to the inhaler centre-line, i.e. horizontal arrangement of the inhaler. The change of particle angular velocity is governed by the fluid dynamic torque T acting on the particles. More details on the applied coefficients in the forces valid for spherical particles at higher particle Reynolds numbers are found in Sommerfeld et al. [71] and Sommerfeld and Schmalfuß [44].

Turbulence transport effects of the particle motion was modelled by using the instantaneous gas velocity in the equation of motion, which is the resolved cell-based mean fluid velocity interpolated to the particle position; plus a random generated fluctuation velocity component. This fluctuation value was generated from the local turbulent kinetic energy (i.e. assuming isotropic turbulence) using a single-step correlated Langevin approach [71]. The required local turbulent time and length scales were obtained from the values of k and ω, the turbulent kinetic energy and the specific dissipation rate.

The wall collision of the carrier particles was modelled based on solving the impulse equations in connection with Coulomb's law of friction [72, 73] and distinguishing between sliding and non-sliding collision. This model calculates the change of translational and rotational velocities of the particles upon wall impact with respect to impact angle and velocity. In the present simulation constant restitution and friction coefficients were assumed; i.e. independent of impact angle. The standard case

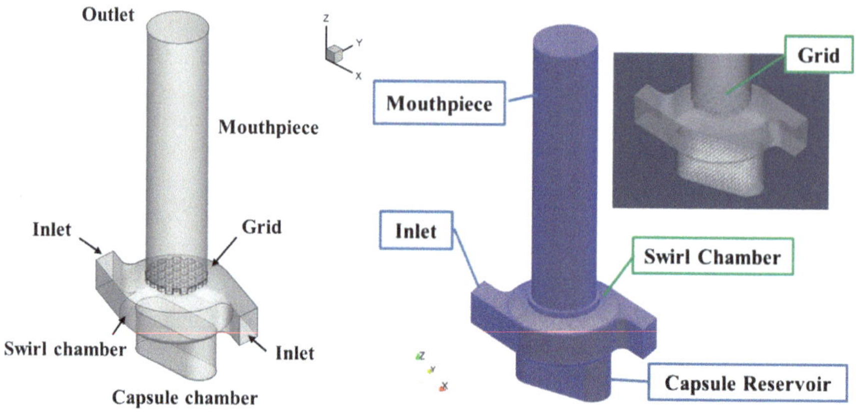

Fig. 18 Geometry of a typical swirl-flow inhaler device (left: Cyclohaler®) with definition of the coordinates system and numerical grid used for the inhaler discretisation with magnified swirl chamber and capsule reservoir, showing the initial carrier particle position (right)

Table 3 Summary of the two considered flow rates of 70 and 100 l/min, number of grid cells and from the simulations obtained maximum velocities in different regions of the swirl flow inhaler (Cyclohaler®)

Flow rate (l/min)	Number grid cells	Inlet velocity (m/s)	Outlet velocity (m/s)	Velocity near grid (m/s)	Velocity reservoir (m/s)	Swirl chamber (m/s)
70	300,000	25.0	13.5	70	48	43
100	400,000	36.0	19.3	80	75	64

was a restitution coefficient of e = 0.96 and a friction coefficient of $\mu = 0.01$. In case other values were used, this will be specified. The particles considered in this study were mono-sized glass beads (density $\rho_p = 2500$ kg/m^3) with diameters of 50, 110 and 500 μm.

In the present numerical studies a swirl type inhaler, namely a Cyclohaler®, is considered as illustrated in Fig. 18. Initially 1024 carrier particles were regularly distributed in the capsule chamber within a cuboidal volume covering the entire chamber (see Fig. 18 right). The initial particle translational and rotational velocities were set to zero. A capsule was not considered in the present calculations. In the following, particle phase results will be compared using all forces mentioned above with those only considering drag and gravity/buoyance, different wall collision parameters, different flow rates and different particle diameters.

The flow simulations were conducted for fixed and stationary flow rates of 70 and 100 l/min through the Cyclohaler® illustrated in Fig. 18 together with the numerical grid. The grid number was different for the two flow rates as summarized in Table 3. Here also typical velocity magnitudes within the Cyclohaler are provided. For the higher flow rate very high velocities between 64 and 80 m/s are found in the base of the inhaler (i.e. capsule reservoir, swirl chamber and grid). In the mouthpiece naturally the velocity magnitude is much lower, i.e. for 70 l/min: 13.5 m/s and for 100 l/min: 19.3 m/s. The results presented in this part of the report are partly taken from the review article of Sommerfeld et al. [63].

The two calculated flow fields (i.e. for 70 and 100 l/min) with velocity magnitude and velocity vectors as well as turbulent kinetic energy are shown in Fig. 19. Within the inhaler base the highest velocities are found, whereas the largest turbulence is found near the grid, which also extends into the swirl chamber for the high flow rate. The velocity vectors indicate the effectiveness of the solid grid at the mouthpiece entrance in reducing the swirl, whereby also the wall collision rate in the mouthpiece is reduced [45], as will be shown below. Several cross-sectional velocity fields of the tangential component (Fig. 19b) show the development through the mouthpiece. The

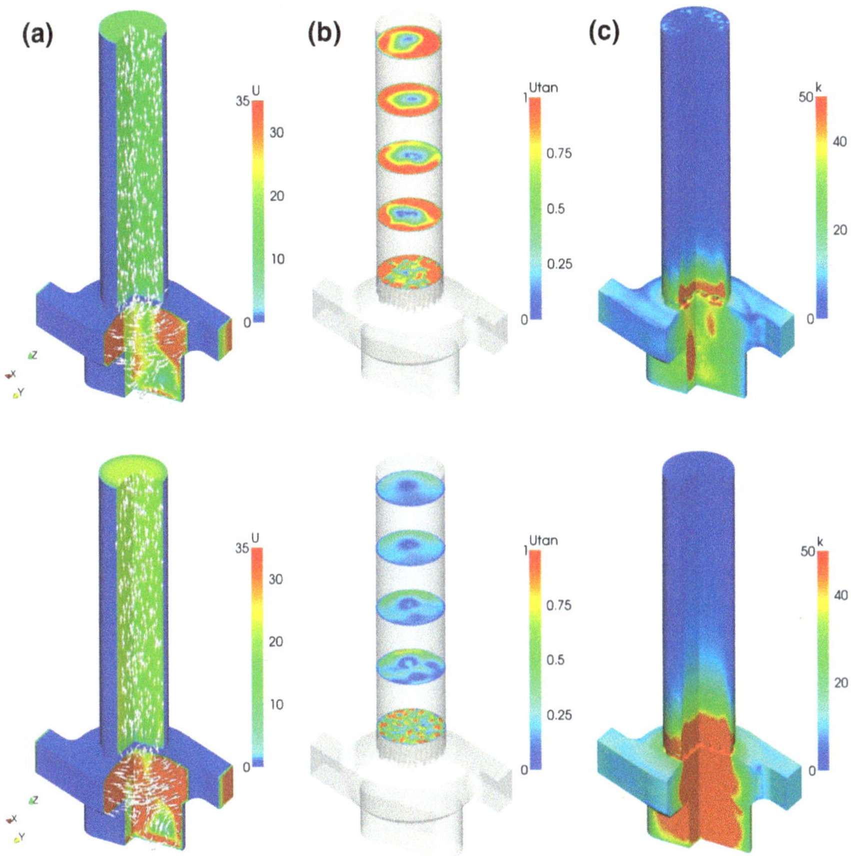

Fig. 19 Visualisation of calculated stationary air velocity field through the swirl inhaler (i.e. Cyclohaler®); **a** total velocity modulus and velocity vectors near the inhaler wall; **b** magnitude of tangential air velocity in various cross-sections of the mouthpiece (colour scales: magnitude of velocity in m/s); **c** turbulent kinetic energy (colour scale: m^2/s^2); (flow rates 70 l/min upper row and 100 l/min lower row) [63]

highest tangential velocities are found in the vicinity of the wall and a vortex core is also identified which possibly might exhibit a precession. With increasing flow rate the velocity magnitude in the swirl chamber becomes higher and the turbulence level is remarkably increased.

Typical particle trajectories within the inhaler for the three mono-sized particles considered are shown in Fig. 20. Even the smallest particles with 50 μm show rather straight paths, indicating that their response to turbulence is negligibly small. The Stokesian response time scale for such particles, $\tau_P = \rho_P \, D_P^2 / (18 \, \mu_F)$, is 0.019 s. With the bulk velocity of these particles (i.e. 13 m/s in the mouthpiece for 70 l/min, see Table 3), the length scale of particle motion is about 244 mm. The average integral time scale of turbulence was found to be about 0.1 ms in the swirl chamber and 0.3 ms in the mouthpiece, hence the turbulent Stokes number is very large. Consequently, also the smallest particles are basically bouncing from wall to wall (i.e. wall-collision dominated behaviour [71]) retaining however a swirling motion throughout the mouthpiece, which is most pronounced for the 500 μm particles. Here, many particles show a saltation-like motion along the mouthpiece wall (Fig. 20c). This is mainly due to the centrifuging effect, since the particles, despite of the solid

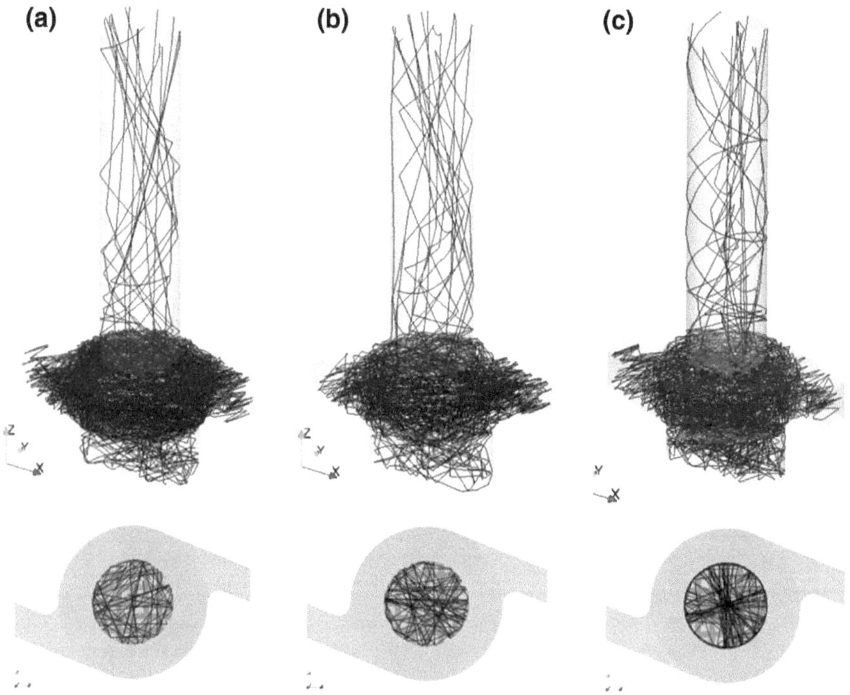

Fig. 20 Calculated representative particle trajectories for mono-sized particles through the inhaler; upper row: particle trajectories in the entire inhaler, swirl chamber and mouthpiece; lower row: cross-sectional view of particle trajectories only in the mouthpiece **a** 50 μm, **b** 110 μm, **c** 500 μm (100 l/min, all forces, e = 0.96 and μ = 0.01) [63]

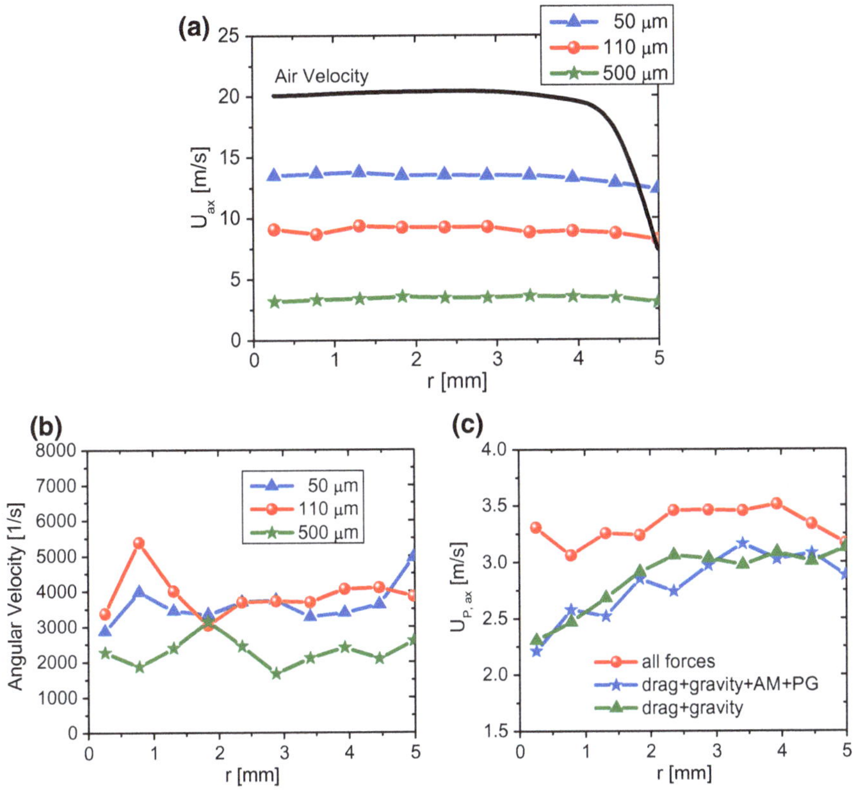

Fig. 21 Comparison of averaged particle velocity profiles at the mouthpiece outlet for a flow rate of 100 l/min, **a** axial outlet velocity profile for different sized particles in comparison with the air velocity, all forces considered, **b** angular particle velocity for 50, 110 and 500 μm-particles, all forces considered, **c** comparison of axial particle outlet velocity profiles for different forces considered, 500 μm-particles (see also Sommerfeld et al. [63])

grid; retain a small swirling velocity throughout the mouthpiece which will be shown below. It is also clear from these results (Fig. 20) that the number of particle-wall collisions increases with particle size and is extremely high within the swirl chamber.

First the velocity characteristics of the particles at the outlet of the mouthpiece are analysed (Fig. 21) by considering the different particle sizes and then the action of the different forces. For the particles the velocity statistics were derived from all 1024 particles leaving the inhaler and plotted versus the radius. Hence, the mean velocity profiles for the particles are not based on too many data. The axial outlet velocity for the air at a flow rate of 100 l/min is compared with the averaged axial particle velocity in Fig. 21a. The air velocity shows the expected profile, while the particle velocity profiles are almost flat as a result of their inertia yielding a bouncing from wall to wall. With increasing particle size, expectedly the particle axial velocities become lower and consequently slip velocities increase. As mentioned above, particles acquire considerable angular velocities due to wall impacts (Fig. 21b) which however will be damped by the viscous torque (see Eq. 3) when they move through the flow. Resulting, the 110 μm-particles show mean angular velocities of around 4000 1/s with an almost flat profile across the outlet. The 50 μm-particles have only slightly lower values, however, the large 500 μm-particles which have smaller wall impact velocities and hence acquire lower angular velocities during wall impact, have only mean values of about 2000 1/s. Therefrom; one may conclude that the slip-rotation lift will have a considerable influence on particle transport.

Now the influence of the forces is analysed for the 500 μm-particles (Fig. 21c). The particle velocity profiles with drag and gravity/buoyancy is very similar to that additionally considering added mass and pressure term (i.e. fluid inertia). Consequently, these additional forces could be neglected, which is of course expected for solid particles in gas flows [71]. Now we add also the transverse lift forces (i.e. due to shear and rotation), and as a result, the averaged particle velocity is much higher at the mouthpiece outlet. For the 500 μm-particles the difference is roughly 17%. In the case of the smaller 110 μm-particles the effect of the lift forces is even somewhat higher, probably be due to the larger angular velocity [44]. From these findings one may conclude that for the calculation of carrier particle motion in an inhaler with extremely high velocities and shear gradients as well as large wall collision frequencies, transverse lift forces are essential and need to be accounted for in any numerical calculation.

In the following part the fluid stresses acting on the carrier particles along their way through the inhaler for the 110 μm-particles and both flow rates are considered. For estimating the flow detachment of drugs through the instantaneous relative velocity between carrier and flow, the PDFs (probability density functions) are evaluated for the entire residence time of all injected particles. For both flow rates, two peaks in these PDFs are identified (Fig. 22a). The lower peak (around 10–12 m/s) is associated with particles moving through the mouthpiece as may be seen from Fig. 22b showing a scatter plot of all registered relative velocities along the height of the inhaler (note $z = 0$ is the upper wall of the swirl chamber, just underneath the grid). The second peak in the magnitude of the relative velocity is coming from particles moving within the base of the inhaler (i.e. swirl chamber and capsule reservoir). This maximum is about 28 m/s for the 70 l/min and about 42 m/s for the higher flow rate of 100 l/min. The consideration of all forces gives a slight shift of the PDF to the left to lower relative velocities (Fig. 22a). As shown above, accounting for all forces raises the particle velocity and hence reduces slip velocity (Fig. 21c). With the higher peak values of the relative velocity one obtains for the low flow rate a particle Reynolds number of $Re_p = 200$ and for 100 l/min a value of $Re_p = 300$.

The averaged PDFs of the shear rate seen by all the carrier particles are shown in Fig. 22c. The median value (at 50% of all counts) is around 14,000 1/s for the low flow rate and values up to 60,000 1/s are observed. Raising the flow rate gives a shift of the PDF to higher values. Actually, the particle resolved cluster simulations [59] showed that flow shear does not remarkably alter drug detachment. The gas velocity fluctuations seen by the particles were calculated from the turbulent kinetic energy as $\sigma = \sqrt{2/3k}$. The resulting values have a peak at around 3 m/s and reach

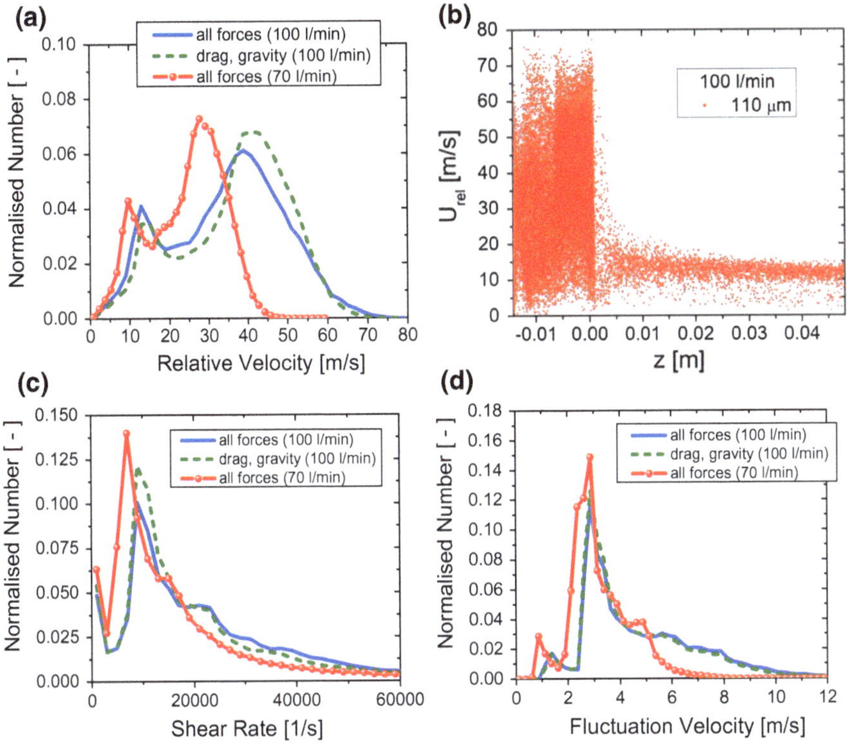

Fig. 22 Velocity statistics determined for all carrier particles moving through the entire inhaler for two flow rates (70 and 100 l/min) and different forces acting on the particle, 110 μm-particles; **a** PDF's of instantaneous relative velocity of all particles on their way through the inhaler; **b** correlation between instantaneous relative velocity and location registration within the inhaler, all forces 100 l/min; **c** PDF's of the shear rate of all particles on their way through the inhaler; **d** PDF's of the gas phase velocity fluctuation of all particles on their way through the inhaler [63]

up to 6 m/s for the low and up to 12 m/s for the high flow rate (Fig. 22d). Therefore, the randomness of turbulence with very high levels has a huge effect on the drug detachment as was demonstrated by Cui and Sommerfeld [61]. A consideration or neglection of transverse lift forces has little influence on the PDFs of shear rate and velocity fluctuation (Fig. 22c, d).

Carrier particle-wall collisions are immensely important for drug particle detachment [70]. From the particle trajectories in Fig. 20 it is obvious that with increasing particle size and hence inertia the wall collision frequency increases drastically. The averaged wall collision number related to one particle during the entire residence within the inhaler is, $N_{WC} = 64$, 100 and 300 for the 50, 110 and 500 μm particles, respectively. Within the base of the inhaler most of the wall collisions occur. As shown above (Fig. 21a) the transport velocity of the carriers decreases with particle size due to their inertia. This is also reflected in the PDFs (probability density

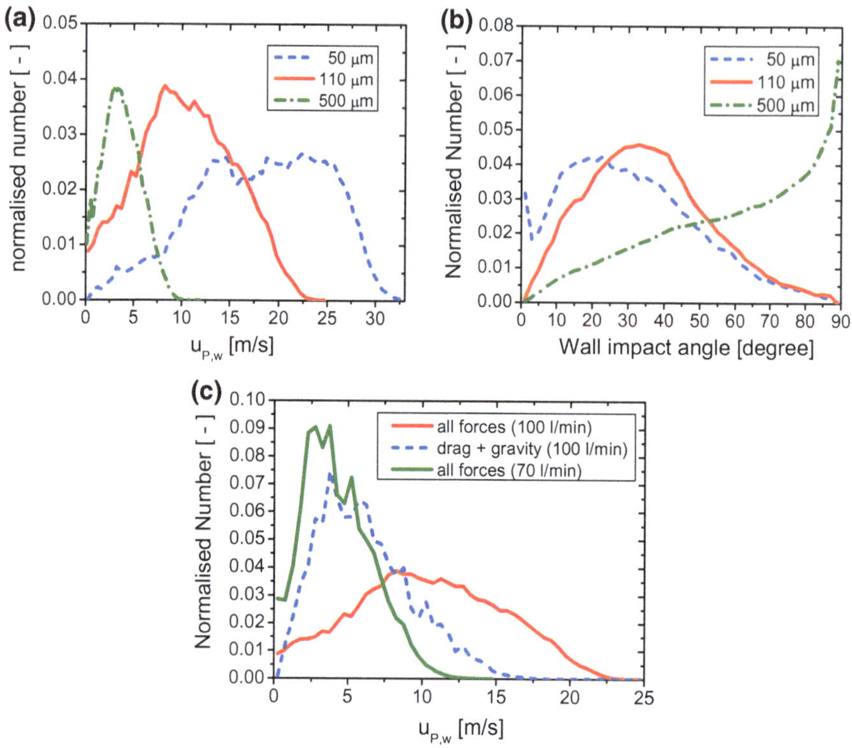

Fig. 23 Influence of particle size (mono-sized) on wall collision behaviour in a Cyclohaler; **a** PDFs of the modulus of particle-wall impact velocity (flow rate 100 l/min, all forces), **b** PDFs of particle-wall impact angle measured between wall and particle trajectory (flow rate 100 l/min, all forces); **c** PDFs of the magnitude of particle-wall impact velocity considering different forces and flow rates for 110 μm-particles ($e = 0.96$ and $\mu = 0.01$) [63]

functions) for the absolute value of wall impact velocity considering all forces at 100 l/min (Fig. 23a). The impact velocity for the 50 μm-particles ranges from zero to more than 30 m/s with a plateau between 13 and 27 m/s. On the other hand the very large particles (i.e. 500 μm) are much slower only up to 10 m/s with a maximum at about 3 m/s. Hence, the mean wall impact energy would be quite similar for the small and large particles. The wall impact velocities for the 110 μm-particles lie in between, ranging up to 22 m/s. Important for drug powder detachment during wall collisions is also the impact angle [70].

As a consequence of the relatively large particle inertia for all considered sizes (see Fig. 20) the impact angles range over the entire spectrum (i.e. 0°–90°) with a higher probability for smaller angles when the particles are not too large (Fig. 23b). Actually, the PDFs are log-normal distributed for the two smaller particle sizes (i.e. 50 and 110 μm). However, for the very large particles (500 μm) an impact angle close to 90° is most dominant due to the high probability of the particles bouncing from wall to

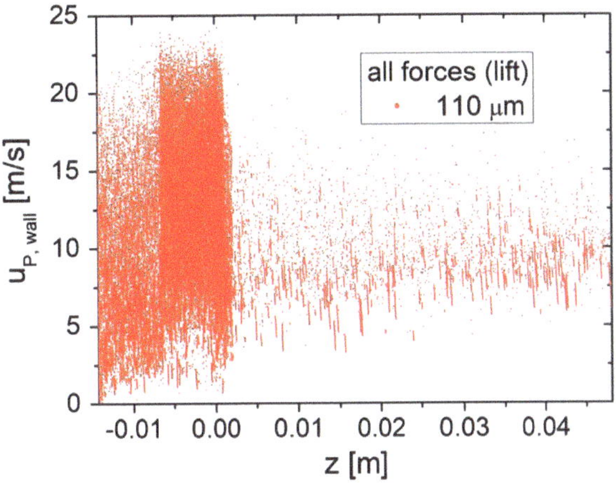

Fig. 24 Scatter plot of the impact velocity versus impact location (i.e. along the entire inhaler) for a flow rate of 100 l/min and 110 μm particles considering all forces ($e = 0.96$ and $\mu = 0.01$)

wall within the swirl chamber. These simulations were conducted accounting for all forces. In accordance with the observed lower particle velocity when only considering drag and gravity (Fig. 21c) also the PDF for the magnitude of the wall impact velocity is reduced when neglecting the important transverse lift forces (Fig. 23c). For this condition the wall impact velocity for the 110 μm-particles reaches only up to 15 m/s with a maximum around 5 m/s. As expected, a flow rate reduction to 70 l/min is associated with lower wall impact velocities ranging only up to about 10 m/s for the 110 μm-particles (Fig. 23c).

Subsequently, the location of wall collisions within the inhaler is evaluated (Fig. 24), considering scatter plots of the impact velocity magnitude versus the location of impact. As to be expected from the particle trajectories, most wall collisions occur within the inhaler base (i.e. the capsule chamber and the swirl chamber). Three typical regions of wall impact velocity, location and number are clearly identified in Fig. 24. The highest probability with the largest velocities is found in the swirl chamber. Within the capsule reservoir the wall collision number is lower with smaller impact velocities. The wall collision number within the mouthpiece is much lower with impact velocities of around 10 m/s, when considering all forces. This is in accordance with the results presented by Coates et al. [45] when considering the full grid.

Very important for the simulation of particle transported through an inhaler, where particle motion is essentially wall-collision dominated, is the wall collision model and the associated parameters [73], namely restitution ratio and friction coefficient. These parameters were varied and the result is shown in Fig. 25 for three different combinations of the wall collision parameters, i.e. reducing restitution and increasing wall friction.

The profiles of the particle axial velocity at the mouthpiece exit for the 500 μm-particles show a comparatively large difference for the 3 cases considered (Fig. 25a). A reduction in the restitution ratio e (defined as the normal rebound velocity after collision to that before) has only small effects on the axial particle outlet velocity,

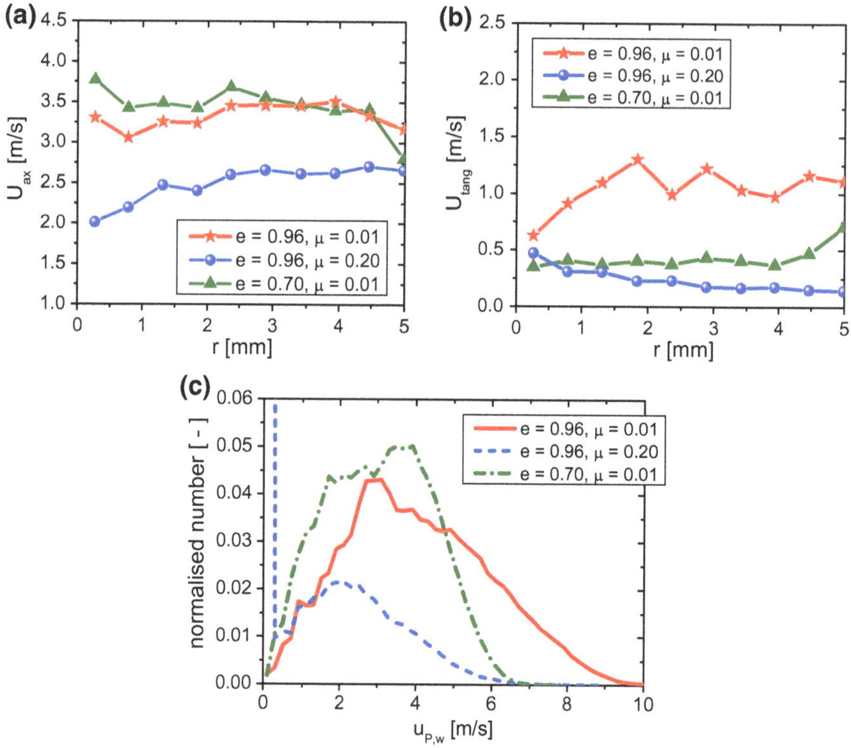

Fig. 25 Influence of the parameters for the wall collision model (i.e. Restitution coefficient e and friction coefficient μ) on the particle velocity characteristics within the inhaler (Cyclohaler®, flow rate 100 l/min, all forces, 500 μm-particles); **a** profiles of the axial particle velocity at the mouthpiece outlet; **b** profiles of the tangential particle velocity at the mouthpiece exit; **c** PDFs of the modulus of the particle wall impact velocity obtained from all wall collisions within the inhaler (see also Sommerfeld et al. [63])

however, the tangential outlet velocity is strongly reduced, more than by 50%, when reducing e from 0.96 to 0.7 (Fig. 25b). Very important is the selection of the wall friction coefficient (which describes the change of the wall-parallel particle velocity). Both velocity components are drastically reduced when increasing μ from 0.01 to 0.2 (Fig. 25a, b). The tangential velocity profiles indicate that the particles, despite the inserted grid, still perform a spiral motion through the mouthpiece (Fig. 25b). The magnitude of the tangential velocity is however very much lower than the axial component.

Of course also the magnitude of the particle wall impact velocity is affected strongly through the selection of the model parameters (Fig. 25c). The PDFs for the standard case (i.e. $e = 0.97$, $\mu = 0.01$) show a wide spectrum of impact velocities up to 10 m/s for the flow rate of 100 l/min. A reduction of the restitution coefficient yields a narrower PDF of the impact velocity modulus as expected from the simulated outlet velocities (Fig. 25c). Increasing the wall friction to $\mu = 0.2$ results in a very high probability for a velocity magnitude of below 0.4 m/s, stemming from those particles colliding with the grid (this was seen on a scatter plot of impact velocity versus location and is not shown here). The maximum wall impact velocity is also about 6 m/s, but the probability of the higher impact velocities is drastically reduced compared to the other two cases.

This section allows for very important conclusions to be respected when conducting Lagrangian calculations of carrier particle motion within inhalers. First of all, as the carrier particles are normally quite large, it is essential to take into account transverse lift forces due to shear flow and particle rotation. Rotational lift is induced by the high angular velocities acquired by the particles during wall collision. The lift forces mainly increase particle velocity and wall impact velocity [44]. Moreover, turbulence experienced by the carrier is very huge. Although particle motion is hardly affected, thereby drug particle detachment might be promoted. Finally, the parameters needed for modelling the wall collision process have a huge influence on the particle velocities within the inhaler and therefore need to be selected carefully. It is advised to obtain these values from an experiment.

Micro-scale CFD: API Flow Detachment from the Carrier

The analysis of a possible flow induced drug detachment from one carrier particle was performed on the basis of direct numerical simulations (DNS), resolving carrier and drug particles by the numerical grid. This allows determining the fluid dynamic forces acting on the drugs and comparing them with measured adhesion forces to judge detachment. For these simulations the Lattice-Boltzmann method (LBM) was used [59–61]. More details about this method are described in Dietzel and Sommerfeld [74]. The accuracy of these simulations and the required grid resolution are presented in Cui and Sommerfeld [60]. The majority of the results presented in this part of the report are taken from the recent review article of Sommerfeld et al. [63].

The numerical simulations were conducted for particle clusters consisting of one spherical carrier (in this case $D_{carrier} = 100\,\mu m$) and a mono-layer (no contact between neighbouring drug particles) of randomly distributed fine spherical mono-sized particles (i.e. sizes of $D_{fine} = 2.45$ and $5\,\mu m$ were considered). The spatial distribution of fine powder on the surface of the carrier was created through a random procedure with the degree of coverage as a target value. The coverage was calculated as the cross-section area of all fine particles divided by the surface area of a sphere with the diameter ($D_{carrier} + D_{fine}$). The coverage was varied between 10% and 50% in the present study, which results for the size ratio of $5\,\mu m/100\,\mu m$ in a total number of fine particles between 176 and 882. Simulating a cluster with the size ratio $2.45\,\mu m/100\,\mu m$ requires of course much more drug particles which are for the same range of coverage 699 and 3497 drug particles, respectively. In all cases the drug particles were resolved by 6 grid dimensions over the diameter, which was found to be sufficient [60].

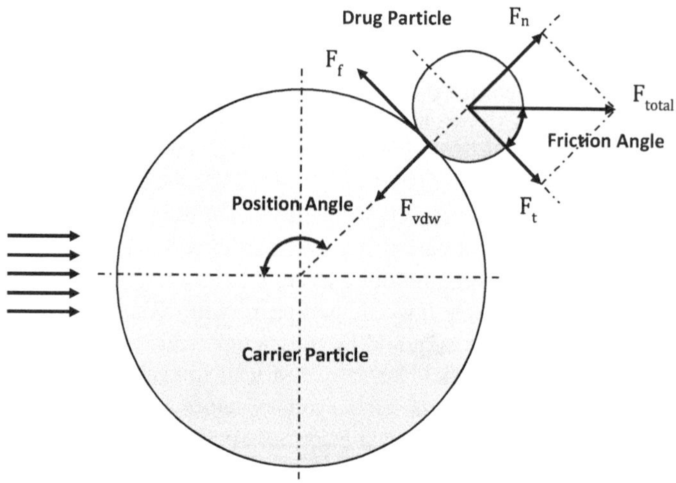

Fig. 26 Illustration of forces acting on the fine drug particles positioned on the large carrier particle with F_{total} the total fluid dynamic force, F_n the normal and F_t the tangential fluid force, and F_{vdw} and F_f the van der Waals force and the friction force (taken from [63])

For the flow simulations, the particle cluster was centrally fixed (i.e. also no rotation was considered) in a rectangular flow domain and exposed to different kinds of stationary flow situations with different inlet velocities. These were a laminar plug flow, a shear flow and a turbulent plug flow being specified at the inlet boundary. At the outlet a gradient-free condition is applied and at all side faces symmetry boundary conditions are being used. In these simulations first the carrier particle Reynolds number was varied being defined with the diameter of the carrier particle $D_{carrier}$ and the inflow velocity U_0 (in the case of shear flows this velocity is the value at the carrier particle centroid):

$$Re = \frac{\rho \, D_{carrier} \, U_0}{\mu} \tag{4}$$

For the other cases the shear rate and the turbulence intensity up to 9% were varied ($I = \sqrt{2/3k}/U_0$). For generating a turbulent plug flow a special inflow generator was developed by Cui and Sommerfeld [61]. The total number of grid cells and the number of refinement regions was selected according to the considered case and the carrier Reynolds number. From the simulations the forces acting on the fine drug particles in dependence of their location on the carrier surface (i.e. position angle) and the Reynolds number *Re* were analysed considering the different flow situations. The forces relevant to possibly detach the fine particles are illustrated in Fig. 26. The total fluid force can be separated in a normal and a tangential component. The adhesion force (i.e. van der Waals) is acting against the normal fluid force and the tangential force is balanced by the friction force. Both fluid forces are relevant for a

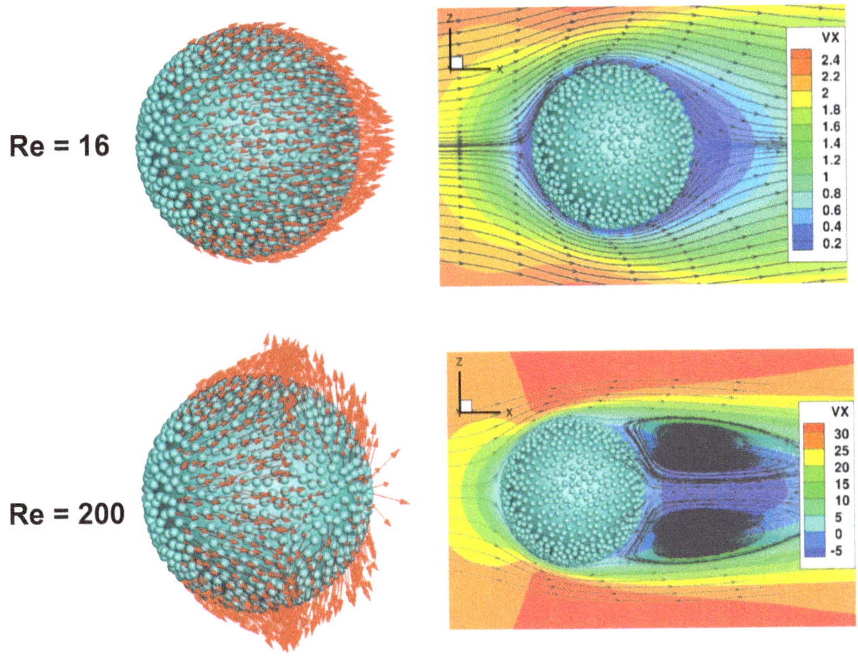

Fig. 27 Flow field around the particle cluster and resulting direction of total fluid forces on the fine particles for two Reynolds numbers (Re = 16 and 200), left: distribution of drug particles with vectors indicating the fluid force direction, right: colour contours for the stream-wise velocity (scale in m/s) and streamlines (degree of coverage 50%, $D_{fine}/D_{carrier} = 5/100$) (taken from [63])

potential detachment by lift-off, sliding or rolling [75]. The gravity force is about 8 orders of magnitude smaller than these forces and therefore was not considered here.

Exemplarily, all total force vectors on the fine particles are shown in Fig. 27 for two carrier particle Reynolds numbers together with the flow field around the cluster. It should be emphasised that the force vectors only indicate the direction and not the magnitude for clarity (i.e. otherwise many force vectors would not be visible). The directions of the forces are strongly correlated with the flow structure around the cluster. For the lower Reynolds numbers no wake separation occurs and consequently the force vectors in the wake region are mostly directed away from the carrier in the stream-wise direction. Around the equator region the force vectors are almost parallel to the carrier surface. For the highest Reynolds number the region of wake separation is clearly visible from the direction of the force vectors. Within the wake, with the backward flow along the cluster surface, the velocity vectors are directed radially outward with respect to the centre-line of the geometry.

The procedure of evaluating the fluid dynamic forces on the drug particles in dependence of position angle was already described in detail before [59, 60]. The results showed that lift-off by fluid dynamic forces is not possible even at the highest Reynolds number, since the normal force was much smaller than the measured van der

Table 4 Measured adhesion properties between glass spheres carrier particles (size ≈ 500 μm) and salbutamol sulphate drug particles (size range 1–5 μm) with mean value and standard deviation, namely adhesion forces (van der Waals force), wall friction angle and adhesion surface energy for untreated and modified glass beads (Cui et al. [59]); as mentioned previously the surface treatment was done in a ball mill with grinding material quartz (x_{50} = 32.17 μm) for 4 h (**Q 4h**) and with grinding material tungsten carbide (x_{50} = 23.58 μm) for 8 h (**TC 8h**)

Carrier particle	Adhesion force (nN)	Friction angle (°)	Friction coefficient (–)	Surface energy (mJ/m^2)
Untreated	257.48 ± 59.67	4.49 ± 0.19	0.079 ± 0.003	53.36 ± 7.94
Q 4h	127.83 ± 28.23	4.79 ± 0.34	0.084 ± 0.006	73.68 ± 5.61
TC 8h	62.48 ± 52.96	6.34 ± 0.36	0.111 ± 0.0063	65.76 ± 2.99

Waals forces (i.e. adhesion forces) for all considered surface modifications (Table 4). Only in the case of turbulent flow interacting with the cluster some drug particles might be detached by lift-off due to the strongly fluctuating normal forces at high turbulence levels [61].

The results presented in the following are focussing on the detachment probability due to sliding and rolling, determined for the different simulation conditions including the measured adhesion properties (i.e. adhesion force, friction coefficient and adhesion surface energy) for the three different surface textures of the carrier particles presented in Table 4 [59]. These two kinds of mechanically treated glass beads was a selection out of different tests considering also chemical treatments and a combination of mechanical and chemical conditioning. More details on the conducted experiments are provided in section "Modification". Moreover, the experimental procedures for obtaining adhesion forces the friction angle and the surface energy are presented in section "Adhesion Force Measurements" and Cui et al. [59].

The criteria for deciding on a possible detachment by sliding or rolling were presented by Cui and Sommerfeld [60] and are therefore not repeated here (see also [75]). For all the drug particles on one carrier surface these criteria were applied (i.e. considering the simulated fluid forces) and whenever detachment by sliding or rolling was possible this drug particle was counted. The total number of possibly detached particles was then divided by the total number of the drug particles initially attached to the carrier giving the value for the detachment probability. Since for each case or condition at least 4 simulations were conducted with different random distributions of the drugs, this calculation was repeatedly done for better statistical reliability.

The dependence of the detachment probability on the carrier particle Reynolds number is shown in Fig. 28, for the size ratio 5/100 (i.e. 5 μm drug and 100 μm carrier). The influence of surface treatment (i.e. measured adhesion properties) on sliding and rolling detachment is provided in Fig. 28a, b and Fig. 28c shows the effect of surface coverage by the drug particles. All the curves have a very typical curve which is known from surface detachment of fine particles through a shear flow [76].

For low *Re* the detachment probability is zero as the fluid dynamic forces are too small, then it begins to rise at a certain critical *Re*-number and approaches at higher *Re* some kind of limiting value. As to be expected, when increasing the

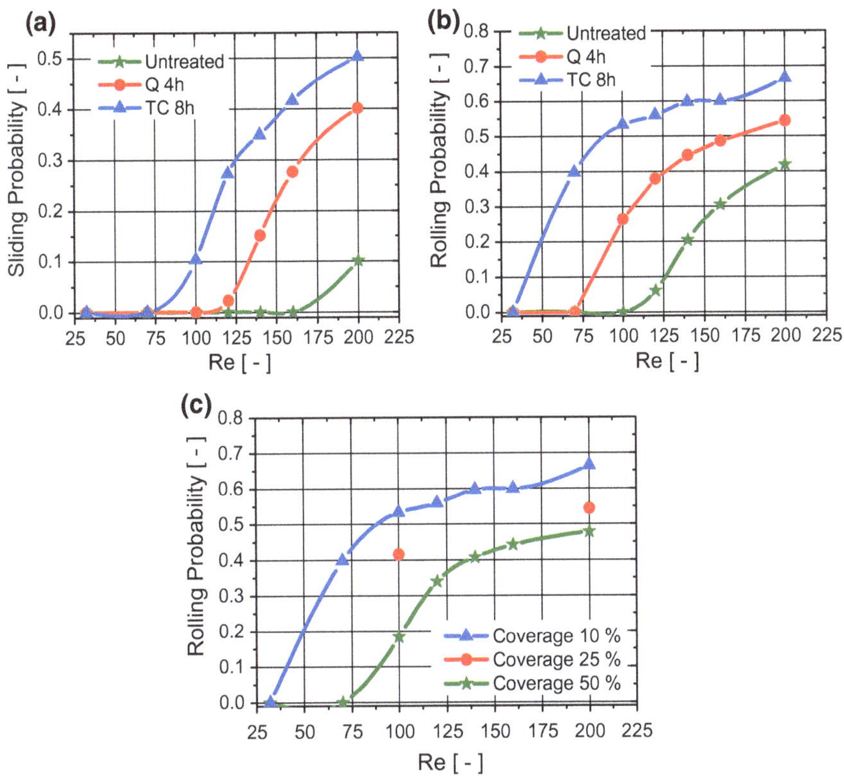

Fig. 28 Drug particle detachment probability in dependence of carrier particle Reynolds number ($D_{fine}/D_{carrier} = 5/100$); **a** sliding and **b** rolling detachment probability for different surface treatment (see Table 4, coverage degree 10%); **c** rolling detachment probability for different coverage degrees (surface treatment TC 8h, $D_{fine}/D_{carrier} = 5/100$) (taken from [63])

surface roughness the adhesion force is diminished and the detachment probability becomes larger for the same Reynolds number. Without surface treatment the sliding detachment probability is very small and begins only beyond Re = 150 whereas rolling begins already at Re = 100. For Q 4h the detachment probability is already significantly increased for both sliding and rolling. However, for the carrier particles treated for 8 h with tungsten carbide (TC 8h) the detachment probability through sliding reaches 50% and is larger than 60% for rolling at Re = 200; for this case the lowest adhesion force is obtained. Rolling detachment always occurs at lower critical Reynolds numbers than sliding. However, it was also shown that at a given Reynolds number rolling occurs always first, before sliding [60]. For the interpretation of these results it should be kept in mind that the detachment probability was calculated for one fixed orientation of the cluster with respect to the oncoming flow and one fixed flow velocity. In reality the instantaneous relative flow velocity will strongly vary along the trajectory of the cluster through the inhaler and also the cluster might rotate due to shear flow interaction and wall collisions.

Finally, also the influence of the degree of coverage on the rolling detachment probability was calculated. The detachment probability increases remarkably when decreasing the degree of coverage (Fig. 28c). This is the result of the higher fluid dynamic forces on the drug particles for a sparse coverage, since the flow goes more likely around the entire drug particles [60], whereas for high coverage the flow goes more or less over the drug particle layer.

It should be noted that so far rather large drug particles were considered. For smaller drugs the acting fluid dynamic force will be of course strongly reduced as these particles will be more embedded within the boundary layer developing around the cluster [60]. Moreover, the computational demands are increasing for the same grid resolution of the drug particles. Nevertheless, also 2.45 μm drug particles with a coverage degree of 23% were considered in order to be comparable with measured distributions of salbutamol sulphate drug particles. An additional issue is the measured van der Waals force, which also showed a distribution due to random variations of the surface structure, but additionally due to the size spectrum of the drug powder [77]. This randomness was considered in the calculation of the detachment probability for all drug particles on the carrier. For each drug particle a new van der Waals force was randomly drawn from the measured distribution (Fig. 29a), which is a subset of the TC 8h-results presented in Table 4, however with $F_{vdW} = 47.9 \pm 30.9$ nN (called SBS17) and all other parameters as provided in Table 4.

The result for the sliding and rolling detachment probabilities for this stochastic approach is compared in Fig. 29b, c together with those by using only a mean value of the van der Waals force, as it also was done before. Note that for these small drug particles the simulations were conducted up to higher carrier Reynolds numbers in order to obtain an appreciable detachment probability. Nevertheless, the values are considerably lower with the mean value of the F_{vdW} in the case of 2.45 μm-drug particles compared to the larger 5 μm-drug particles (compare with Fig. 11). The sliding probability with random variation of F_{vdW} is considerably larger than when only using a mean value (Fig. 29b). As a result, the sliding probability is only 18% at Re = 260. The rolling detachment probability is already quite high when using the mean of F_{vdW} and with random F_{vdW} it is becoming remarkably larger for the lower Re regime (Fig. 29c). At the highest Re the value of about 38% is reached for both cases.

Finally, the effect of turbulence on sliding and rolling drug detachment will be introduced (see also Cui and Sommerfeld [61]). The coverage degree for these simulations was 50% and the size ratio again $D_{drug}/D_{carrier} = 5/100$. The carrier particle Reynolds number was varied between 80 and 200 and the turbulence intensity ($I = \sqrt{2/3k}/U_0$) could be increased up to 9%. Turbulence in the incoming flow yields strong fluctuations in the fluid dynamic forces acting on each drug particle. The temporal variation of normal force on a drug particle located at a position angle of 85° (region of high normal force in the laminar case) is presented in Fig. 30. Although the details of the fluctuations cannot be identified in this figure, it is obvious that the peak values exceed 40 nN which is in the same order as the measured

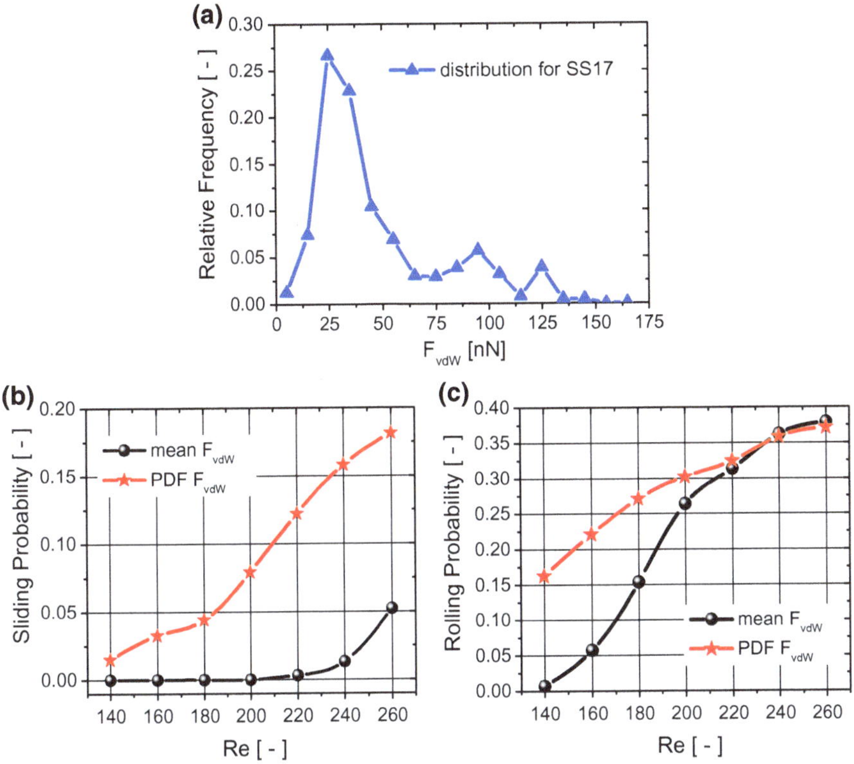

Fig. 29 Influence of the random variation of the van der Waals force on the drug particle detachment; **a** distribution of the measured van der Waals force between modified glass beads (TC 8h) and salbutamol sulphate particles (mean diameter 2.45 μm, SBS17, $F_{vdW} = 47.9 \pm 30.9$ nN, coverage degree 23%); **b** sliding detachment probability in dependence of carrier Reynolds number comparing mean and random F_{vdW}; **c** rolling detachment probability in dependence of carrier Re comparing mean and random F_{vdW} (taken from [63])

van der Waals force for glass carrier and salbutamol sulphate particles (see Fig. 29a). Consequently, also a good fraction of drug particles will be detached through lift-off.

Naturally, also sliding and rolling detachment will be possible which was obtained using the associated criteria and averaged over all drug particles and over a large number of time steps, i.e. 10,000 [61]. The influence of turbulence intensity on resulting sliding and rolling detachment probabilities is depicted in Fig. 31a for a carrier Reynolds number of 200. Again rolling has a higher probability than sliding. Starting from the laminar result both probabilities rise remarkably with turbulence intensity. Rolling reaches even a value of almost one and sliding reaches 0.75 at a turbulence intensity of 9% and is still further increasing. The Reynolds number dependence of the turbulent detachment probabilities for sliding and rolling is compared in Fig. 31b with the results obtained for laminar flow. For the turbulent case the standard deviations of the detachment probabilities are also indicated as vertical

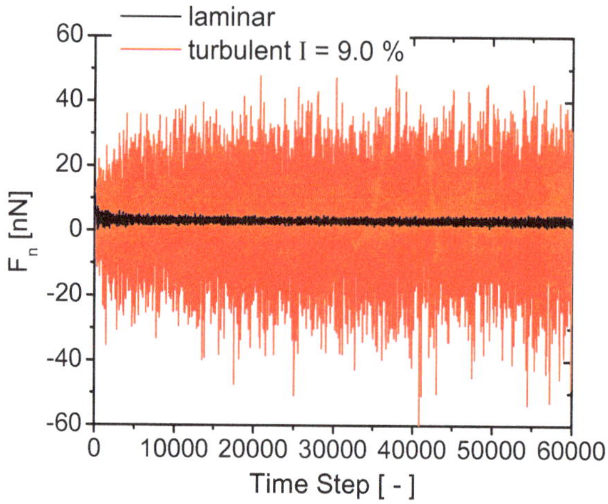

Fig. 30 Temporal variation of the normal fluid force on the drug particle which experiences the largest time-averaged normal force on the carrier surface (i.e. at position angle = 85°, see Cui et al. [59]) for both laminar and turbulent flows (Re = 200, U_0 = 30 m/s, coverage degree = 50%, $D_{drug}/D_{carrier}$ = 5/100, simulation period N = 60,000); mean normal force laminar: 3.15 nN, turbulent 3.19 nN, standard deviation of normal force laminar 0.68 nN, turbulent 14.56 nN [63]

bars. Consequently, turbulence remarkably increases detachment probability of the drug powder from the carrier particles. The macro-scale simulations of drug particle motion through an inhaler revealed that turbulence intensity can even reach much higher values up to about 20% (see Fig. 22d).

Inhaler Comparison with Carrier-Wall Collision Induced Drug Detachment

As shown above [44] the number of wall collisions for carrier particles moving through a swirl-type inhaler is quite large, which however is strongly related to the location of carrier injection as well as the wall collision parameters used. Hence, wall collisions should be mostly responsible for drug powder detachment [70]. In line with the discussion presented above on the applicability of the DPM (discrete particle method) a model was developed for describing the inertial drug detachment during a wall impact of carrier particles [62]. This model was implemented in the frame of the Euler/Lagrange approach using the code OpenFOAM [44].

In order to reduce the complexity of the calculation of the carrier-wall collision process, it is assumed that the carrier is an elastic body and the wall is a rigid body. The model assumes for simplification that between the spherical carrier and the plane smooth wall no drug particles are present. The collision between an elastic particle and a solid wall can be separated into two phases: the compression phase and the

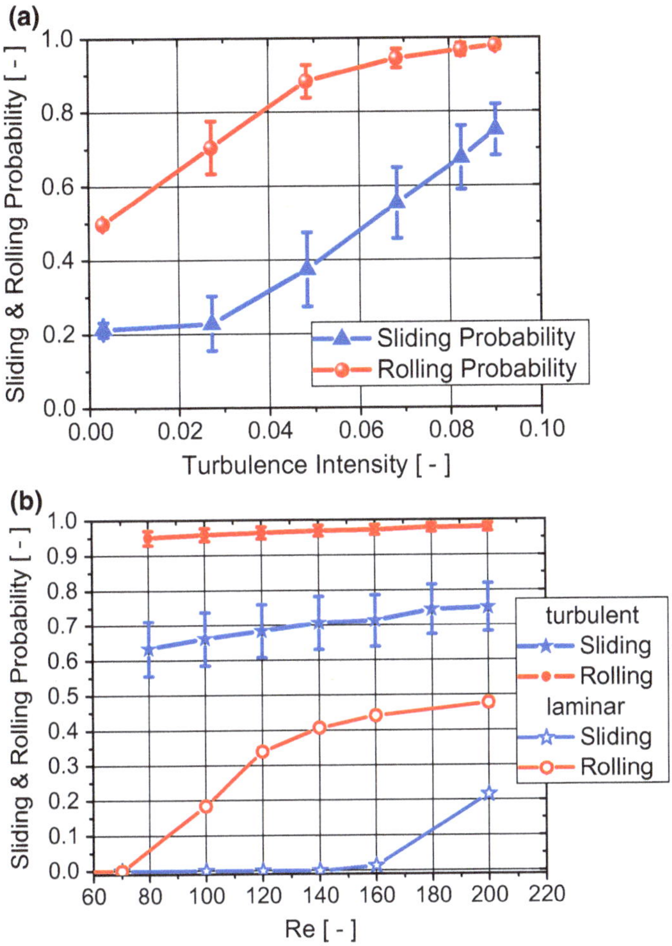

Fig. 31 Sliding and rolling drug detachment probability under turbulent flow conditions (surface treatment TC 8h, size ratio $D_{drug}/D_{carrier} = 5/100$, coverage degree 50%); **a** sliding and rolling detachment probability in dependence of turbulence intensity (Re = 200); **b** turbulent sliding and rolling detachment probability in dependence of carrier Reynolds number (I = 9%) and comparison with laminar results (taken from [63])

recovery phase. In the compression phase, part of the kinetic energy of the particle is converted into the elastic potential energy; in the recovery phase, the elastic potential energy is transformed back into the kinetic energy [78, 79]. Depending on the phase in which the particle slides along the wall, there are usually three cases: the particle stops sliding in the compression phase, the particle stops sliding in the recovery phase, or the particle continues to slide throughout the compression and recovery phases [79]. In the sliding period, the particle's angular velocity is increased by the friction force, while the tangential velocity is decreased. When the particle surface velocity through rotation is equal to the tangential velocity, the particle thereafter rolls along the wall without sliding. In this non-sliding period, the friction force disappears and the angular velocity of the particle is modified by the rolling resistance. For carrier particle-wall collisions, Cui and Sommerfeld [62] proposed a novel particle-wall collision model that can determine the temporal evolution of the particle velocities and the inertia forces acting on the carrier during the whole collision process. This model requires two coordinate transformations, in order to rotate the velocities of the carrier in the inertial frame of reference into the particle frame of reference [62]. The results presented in this section are taken from the review article of Sommerfeld et al. [63].

By using this model, it is possible to calculate the maximum inertia forces acting on the drug particles, fictitiously attached to the carrier, during the carrier-wall collision. The number of fictitious drug particles homogeneously distributed in a mono-layer depends on the assumed degree of coverage [59]. The particle size ratio between the drug and the carrier is 2.45 μm/100 μm, so that the tracked carrier particles have an effective diameter of 104.9 μm (particle density $\rho_p = 2.5$ g/cm^3). The degree of coverage was assumed to be 23% in this case, yielding 1585 fictitious drug particles of 2.45 μm. With the help of measured adhesion properties for carrier-drug it is then possible to estimates the detachment probability for all drug particles on the carrier. Here the surface treatment method TC 8h [59] was considered and the drug particles consisted of salbutamol sulphate using a mean van der Waals adhesion of $F_{vdW} =$ 62.48 nN and the other parameters specified in Table 4. The friction coefficient between carriers and inhaler walls was assumed to be $\mu = 0.1$.

In addition, we assume that the drug particles after a carrier-wall collision and the determined number of detached drug particles are again distributed evenly on the surface of the carrier. Therefore, the final detachment probability of the j_{th} carrier particle at the end of simulation is defined as:

$$\mathrm{FLOP_j} = \prod_{i=1}^{N_j} \mathrm{LOP_{j,i}} \tag{5}$$

where $LOP_{i,j}$ is the lift-off (detachment) probability on the j_{th} carrier at its i_{th} collision with the wall, N_j is the total collision count on the j_{th} carrier during its way through the inhaler.

Euler/Lagrange calculations were conducted using OpenFOAM by applying the k-ω-SST turbulence model and carrier particle tracking was performed by considering drag, gravity/buoyancy and transverse lift due to shear and particle rotation [44]. Two-way coupling, i.e. the influence of the particles on the flow, was neglected in this study. Turbulent dispersion of the carrier particles, although not very important (see Fig. 20), was accounted for through the single-step Langevin model [58]. Initially the carrier particles are placed within a cuboidal volume covering the entire capsule chamber (see Fig. 18). For the Cyclohaler 1024 carrier particles were released, while in the case of the Unihaler with a different design these were 1805 particles.

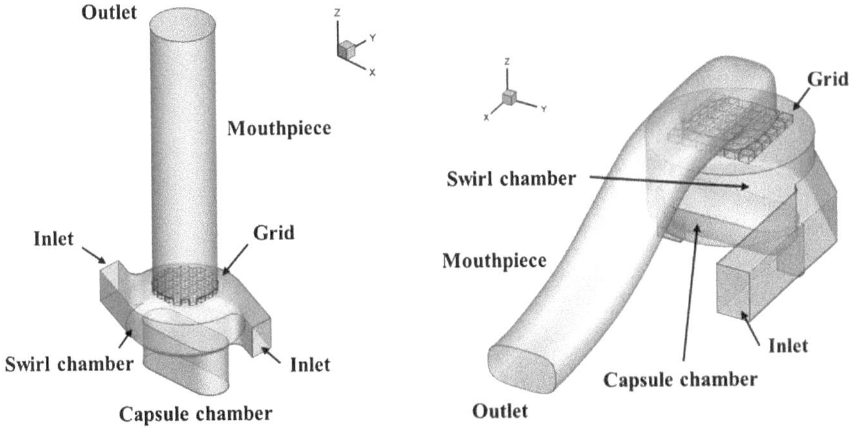

Fig. 32 Geometries of the two considered swirl-type inhalers with inlet, swirl and capsule chamber, grid and mouthpiece; (left) commercial Cyclohaler®; (right) modular Unihaler (University of Kiel, Friebel [69])

Drug detachment through carrier-wall collisions described by the new model was analysed for two different swirl-type DPIs. The first DPI is the commercial Cyclohaler®, and the second one the so-called Unihaler which was developed at University Kiel in the frame of a doctoral thesis [69]. Here, a geometry corresponding to the Modular Version L was simulated, which had shown best results for realistic powder blends during the development of the inhaler. Both swirl-flow inhalers (Fig. 32) have tangential air inlets entering into a swirl-chamber underneath which the capsule chamber is located. Then the inhalation-induced flow is passing a grid for reducing swirl, and finally leaves the inhaler through a mouthpiece. It should be noted that the dimensions of the two DPIs are very different so that for the same flow rate different velocities and a different pressure drop are resulting. The cross-section areas of the outlet are 86.59 mm^2 and 35.54 mm^2 for the Cyclohaler® and Unihaler, respectively. The symmetric double inlet of the Cyclohaler® has a total cross-section area of 46.3 mm^2, while the single inlet Unihaler has only 23.57 mm^2.

Because of the different geometries and flow fields, also different numerical grids were used for both inhalers as summarised in Table 5. In all simulations presented here, only stationary flows are considered.

The simulated relationship between flow rate and pressure drop for the two inhalers is shown in Fig. 33. It is clear that the Unihaler with the more complex geometry has an almost twice as large pressure drop as the Cyclohaler. For in vitro DPI testing, it is recommended to compare the emitted fine particle fraction (i.e. the percentage of the total mass of fine drugs that leave the inhaler in relation to the mass of drugs initially in the capsule) of drug particles of different DPIs at the same pressure drop. Therefore, here the two DPIs are compared at the same pressure drop of 4 kPa (horizontal dashed line in Fig. 33), which is the highest pressure drop that can be normally realised by patients. The corresponding flow rates of Cyclohaler® and Unihaler at the pressure

Table 5 Summary of the two simulated inhalers with flow rates, number of grid cells and from the simulations obtained maximum velocities in different regions of the inhalers

Inhaler, flow rate	Number grid cells	Inlet velocity (m/s)	Outlet velocity (m/s)	Velocity near grid (m/s)	Velocity reservoir (m/s)	Swirl chamber (m/s)
Cyclohaler: 90 l/min	400,000	32.4	17.3	58.6	54.1	46.2
Unihaler: 60 l/min	1,444,712	42.7	28.1	48.6	69.2	65.8

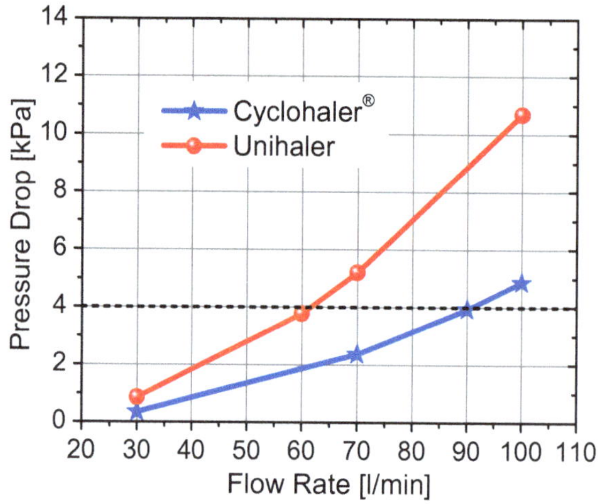

Fig. 33 Numerically simulated pressure drop in dependence of flow rates for the two dry powder inhalers Cyclohaler and the Unihaler [63]

drop of 4 kPa are 90 l/min and 60 l/min, respectively. The characteristic maximum velocities in certain regions of the inhalers are also listed in Table 5. It is obvious, that the velocities in the different regions of the inhaler are remarkably larger for the Unihaler, although the flow rate considered is lower. Only the flow velocity through the grid is smaller for the Unihaler resulting from the different grid structure.

The numerical results for the velocity and turbulent kinetic energy fields for the two DPIs at the pressure drop of 4 kPa are shown in Fig. 34. The highest magnitudes of velocities and turbulent kinetic energies are found in the swirl chamber and near the grid for both inhaler types. The magnitudes of velocities and turbulent kinetic energies of Unihaler are much higher (i.e. about two times larger) than the results for the Cyclohaler®. This is mainly because the Unihaler only has one inlet, which yields a more inhomogeneous air flow field within the swirl chamber. In addition, the grid dimension of the Unihaler is smaller than that of the Cyclohaler® so that turbulence generated by the grid is much stronger in the case of the Unihaler. The inhomogeneity of velocity and turbulent kinetic energy in the mouthpiece is reduced largely due to the inserted grid at the entrance of the mouthpiece. In particular, with the help of the design of the converging, curved mouthpieces of the Unihaler, the velocity and turbulent kinetic energy fields of the Unihaler are a bit more homogeneous in the

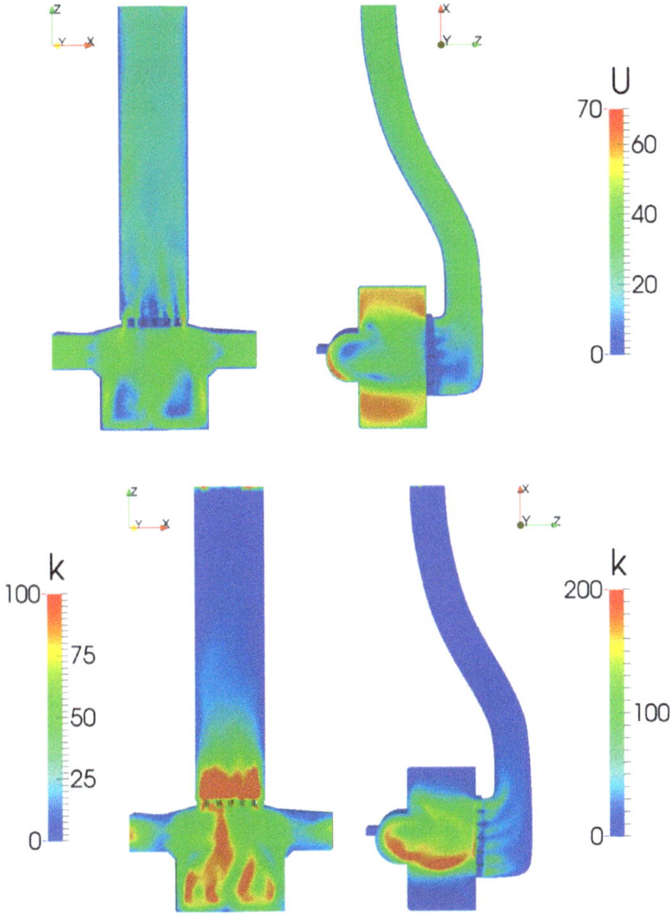

Fig. 34 Visualisation of the velocity magnitude (upper graphs, colour scale in m/s) and the turbulent kinetic energy (lower graphs, colour scale in m^2/s^2, please note the different scales of the turbulent kinetic energy) of calculated stationary flow field through the two swirl-type inhalers; for each row, left: Cyclohaler® with 90 l/min; right: Unihaler with 60 l/min (pressure drop of both DPIs: 4 kPa) (taken from [63])

mouthpiece than that of the Cyclohaler®. Previous studies [59, 61] prove that the high slip velocities (i.e. relative velocity between the fluid and the particle) and turbulent kinetic energies can help to improve the flow-induced detachment probability of drug particles. Therefore, the Unihaler is expected to have a better performance with regard to the emitted FPF of drug particles than Cyclohaler®.

The studies using the particle-wall collision model showed that the inertia force-induced detachment is much larger than the detachment induced by fluid dynamic stresses. During carrier-wall impact, the drug particles can be detached by lift-off, sliding or rolling. The lift-off probability for one carrier increases with increasing

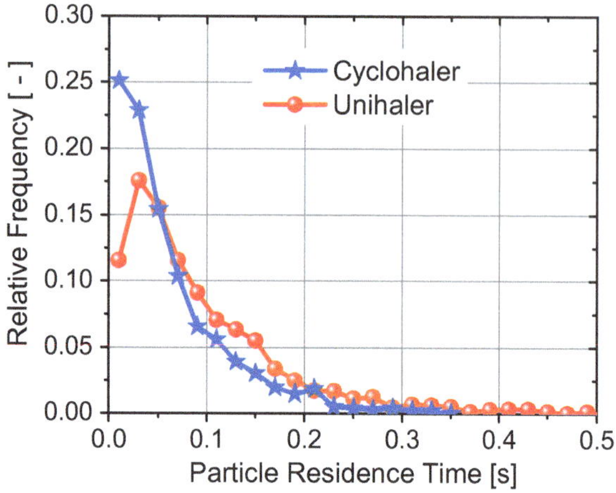

Fig. 35 PDFs of the particle residence time per tacked carrier particle; mean particle residence time in the Unihaler: 0.096 ± 0.091 s; mean particle residence time in the Cyclohaler®: 0.061 ± 0.057 (pressure drop of both DPIs: 4 kPa, effective carrier diameter: 104.9 μm, friction coefficient between the carrier and the wall: 0.1, surface treatment method: TC 8h) [63]

impact angle and has a maximum value of 50%, since only half of the drug particles experience a negative normal force which is able to lift-off the drug particle (i.e. on the carrier hemisphere facing the wall). The rest of the drug particles cannot be detached by lift-off but might detach later through a rolling detachment if the carrier surface is smooth. The influence of different flow rates, friction coefficients and drug particle sizes on the drug detachment probability have been studied by Cui and Sommerfeld [62] in detail. Therefore, in the present study, we focus on the comparison of two DPIs by using the same simulation setups considering only the lift-off probability and the resulting detached drug powder fraction due to all wall collisions.

The PDFs (probability density functions) of the carrier particle residence time obtained from all tracked particles is shown in Fig. 35. Due to the higher flow rate, the residence time distribution for the Cyclohaler is narrower and extends only up to about 0.3 s. On the other hand, the Unihaler has a more complex geometry resulting in a broader PDF ranging up to about 0.4 s. The mean particle residence time in the Unihaler and the Cyclohaler® are 0.096 s and 0.061 s, respectively.

The PDFs of the wall collision counts per carrier particle moving through the entire inhaler show remarkable differences for the two configurations (Fig. 36). The Unihaler shows a broad distribution with two maxima with around 80 and 230 wall collisions per carrier particle and even carriers with up to 600 wall collisions are found. The averaged collision counts per tracked carrier particle are for the Unihaler 212, being much larger than for the Cyclohaler® with only 55 wall collisions. The difference to the result presented above (i.e. 100 wall collisions per particle for a flow rate of 100 l/min and 110 μm-particles) is caused not only by the flow rate, but also by the modified wall collision model and the higher wall friction of $\mu = 0.1$. Naturally the wall collision number is first of all responsible for the drug powder detachment, and then it also depends on wall collision angle and impact velocity [62].

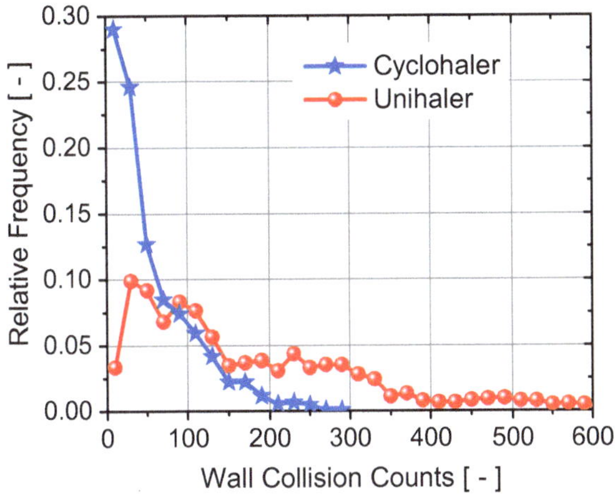

Fig. 36 PDFs of the collision counts per tacked carrier particle; collision counts for the Unihaler: 212.14 ± 320.28; collision counts for the Cyclohaler®: 54.70 ± 51.47 (pressure drop of both DPIs: 4 kPa, effective carrier diameter: 104.9 μm, friction coefficient between the carrier and the wall: 0.1, surface treatment method: TC 8h) [63]

In Fig. 37 the final lift-off probability (i.e. when the carrier particles leave the inhaler) is plotted versus the wall collision counts for all carrier particles. For the Unihaler with the higher wall impact velocities (see Table 5) close to 100% of the drug particles are detached through the wall collisions, almost independent of the number of wall collisions experienced by the carrier particle. This gives eventually a total mean detachment rate of 0.999. The Cyclohaler in contrast, shows clearly an increase of the detachment rate with growing number of wall collisions. When the carrier has only 2 wall collisions about 65% of the drug particles are detached and with 4 wall collisions this value increases to about 90%. The total mean detachment rate is consequently also very high, namely 0.989. This is of course contrary to experience where the fine particle fraction is much lower. The reason for that has been mentioned already above, namely drug particle deposition on the inhaler walls during carrier wall impact, as demonstrated by Ariane et al. [70]. It is expected that the Unihaler has less deposition due to the higher velocities, whereby the drug particles may rebound from the wall or deposited drug particles may be also re-entrained by the air flow.

Conclusion

For understanding and tailoring inter-particle interactions within DPIs, glass beads have shown to be an appropriate model carrier that allows tailoring the surface roughness without affecting particle size and shape. Using almost ideally spherical spray

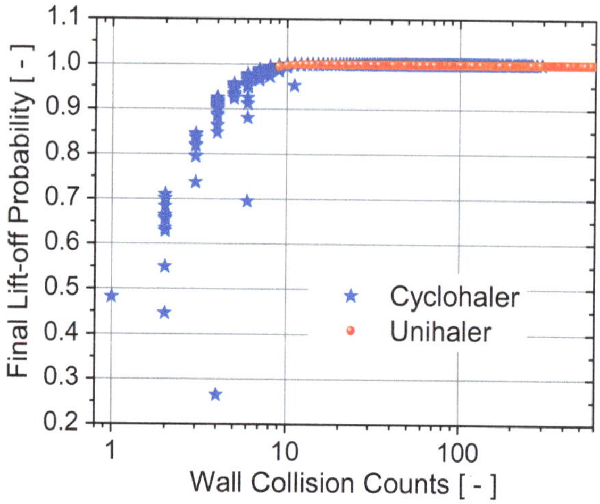

Fig. 37 Relation between final lift-off probability and wall collision counts for all tracked particles for both inhaler configurations; total final lift-off probability of Unihaler: 0.999 ± 0.0002; total final lift-off probability of Cyclohaler®: 0.989 ± 0.0523 (pressure drop of two DPIs: 4 kPa, effective carrier diameter: 104.9 μm, degree of coverage: 23%, friction coefficient between the carrier and the wall: 0.1, surface treatment method: TC 8h, 2.45 μm salbutamol sulphate (SBS17) drug particles), result taken from [63]

dried API particles with minimal differences of particle shape and thus particle orientation on the carrier surface the determination of adhesion forces was used to evaluate how inter-particle interactions were changed by the carrier surface roughness. Performance tests of selected carrier API combinations were correlated with the adhesion forces in order to reveal the impact of changing the adhesion forces on inhaler performance. Further experimental data were incorporated in the simulation work in order to calculate detachment probabilities and for validation.

More precisely, experimental work showed that the adhesion force between surface modified glass beads and salbutamol sulphate, representing a hydrophilic API, is remarkably lower than the adhesion force between untreated glass beads and salbutamol sulphate. This was related to the rougher surfaces of the modified glass beads and the thereby reduced contact area between modified glass beads and API particles [67]. The same is true for budesonide and untreated versus surface-modified glass beads with nano-roughness, but adhesion forces are generally lower for the rather hydrophobic budesonide [77]. However, the introduction of micro-roughness did not reduce the adhesion of spray dried budesonide to the glass beads surface. Instead, API particles are likely to accumulate in these regions impeding drug detachment.

Besides particle surface topography, the inhaler device plays a crucial role in drug dispersion from the carrier and overall performance of an inhalation product. Therefore, different inhaler geometries were studied experimentally and in parallel, their impact on the detachment probability evaluated via simulation. Via an experimental DoE approach and utilising a modular inhaler system, inhaler components with a significant impact on dispersion, de-agglomeration and consequently on drug delivery performance were identified. Here it became obvious that glass beads behave differently from conventional carriers like lactose and or mannitol. Density and particle size are much lower for lactose compared to GBs (glass beads), thus GBs need a higher airflow to get moved at all. This reversed the impact of inhaler geometry found for lactose-based blends and thus needs to be taken into account when working with this model system.

The numerical calculations (CFD) have provided more insight into particle transport and drug detachment in carrier-based formulations. A multi-scale approach was adopted to develop modelling approaches which are required for numerically predicting the efficiency of inhaler devices (i.e. the prediction of the Fine Particle Fraction (FPF)). This approach combines fully resolved simulations of a carrier covered with several hundreds of small drug particles for analysing flow detachment, experiments on surface modified carrier particles for providing the adhesion forces and other contact parameters as well as theoretical models to describe inertia-induced drug detachment upon carrier-wall collisions.

These studies showed that flow-induced drug detachment is only possible if the van der Waals force between carrier and drug is very weak (i.e. requiring carrier surface treatment with high roughness), even though complex inhaler geometries (swirl-type inhalers) with high relative velocities between carrier and flow were considered. High turbulence of the inhaler flow, generated within the swirl chamber or downstream the solid grid, can considerably support drug detachment since the fluid dynamic forces on the drug particles show strong random variations. The considered swirl-type inhalers (i.e. Cyclohaler and Unihaler) induce a large number of carrier-wall collisions, of course increasing with particle size. Hence, inertial-induced drug detachment during carrier-wall impact is very effective for drug detachment, so that both inhalers showed almost 100% drug aerosolisation. This is known not to occur in reality. The reason for this difference is seen in the possibility of drug deposition on inhaler walls during carrier-wall collisions or the transport of drug powder in the airflow; deposition effects were neglected so far. After such a model will be also implemented in the Euler/Lagrange approach developed, realistic fine particle fractions will be most likely predicted. A detailed review of the potential and constraints for the application of CFD combined with Lagrangian particle tracking to inhaler simulations was recently published by Sommerfeld et al. [63] along with suggestions for the method to be used for point-particle tracking and the models required for such an approach.

Bringing together the findings from experiments and numerical studies one may conclude the following:

- The surface modification of glass bead carriers, and hence the obtained adhesion forces between carrier and drug, should be tailored according to the type of inhaler considered.
- For inhaler geometries and flow structures yielding a low wall collision rate, quite weak adhesion forces are needed to ensure flow detachment. Turbulence in such inhaler types of course may remarkably enhance detachment.

- In the case of inhalers with high wall collision rates (e.g. swirl-type inhalers) drug detachment by inertia is very effective so that flow detachment plays a minor role and adhesion forces may be allowed to be higher.
- However, for high carrier-wall collision rates drug-wall deposition might deteriorate the fine particle fraction.
- Consequently, drug deposition on inhaler walls needs more attention in the future including the experimental analysis of adhesion behaviour.

This work presents the systematic investigation of the effect of carrier surface properties on the aerodynamic performance of inhaler devices and therefore contributes substantially to a deeper comprehension of inter-particle interactions between carrier and API. Additionally, present findings showed that efficient drug delivery is highly dependent on the interplay between device and formulation.

Acknowledgements The authors acknowledge the funding of this research project by the German Research Foundation (DFG) within the priority program SPP 1486 "Particles in Contact". The authors would also like to thank the research group of Dr. Michael Kappl and Dr. Regina Fuchs (Max Planck Institute for Polymer Research, Mainz) for providing technical support and equipment for AFM measurements. Further, the authors would thank Hartmuth Schroettner from FELMI-ZFE (Institute of Electron Microscopy and Nanoanalysis) for providing SEM images.

References

1. Rau, J.L.: The inhalation of drugs: advantages and problems. Respir. Care **50**, 367–382 (2005)
2. Pham, D.-D., Fattal, E., Tsapis, N.: Pulmonary drug delivery systems for tuberculosis treatment. Int. J. Pharm. **478**, 517–529 (2015)
3. Onoue, S., Yamada, S.: Pirfenidone in respirable powder form for the treatment of pulmonary fibrosis: a safer alternative to the current oral delivery system? Ther. Deliv. **4**, 887–889 (2013)
4. Thiringer, G., Svedmyr, N.: Comparison of infused and inhaled terbutaline in patients with asthma. Scand. J. Respir. Dis. **57**, 17–24 (1976)
5. Zhou, Q.T., Leung, S.S.Y., Tang, P., Parumasivam, T., Loh, Z.H., Chan, H.-K.: Inhaled formulations and pulmonary drug delivery systems for respiratory infections. Adv. Drug Deliv. Rev. **85**, 83–99 (2015)
6. Kwok, P.C.L., Chan, H.-K.: Pulmonary delivery of peptides and proteins. In: van der Walle, C. (ed.) Peptide and Protein Delivery, pp. 23–46. Elsevier Science, San Diego, CA, USA (2011)
7. Garcia-Contreras, L., Smyth, H.: Liquid-spray or dry-powder systems for inhaled delivery of peptides and proteins? Am. J. Drug Deliv. **3**, 29–45 (2005)
8. Wall, D.A.: Pulmonary absorption of peptides and proteins. Drug Deliv. **2**, 1–20 (2008)
9. Kwok, P.C.L., Chan, H.-K.: Pulmonary drug delivery. Ther. Deliv. **4**, 877–878 (2013)
10. Chrystyn, H.: Is total particle dose more important than particle distribution? Respir. Med. **91**, 17–19 (1997)
11. Ashurt, I., Malton, A., Prime, D., Sumby, B.: Latest advances in the development of dry powder inhalers. Pharm. Sci. Technol. Today **3**, 246–256 (2000)
12. Kawashima, Y., Imai, M., Takeuchi, H., Yamamoto, H., Kamiya, K., Hino, T.: Improved flowability and compactibility of spherically agglomerated crystals of ascorbic acid for direct tableting designed by spherical crystallization process. Powder Technol. **130**, 283–289 (2003)
13. Podczeck, F.: The relationship between physical properties of lactose monohydrate and the aerodynamic behaviour of adhered drug particles. Int. J. Pharm. **160**, 119–130 (1998)

14. Shur, J., Saluja, B., Lee, S., Tibbatts, J., Price, R.: Effect of device design and formulation on the in vitro comparability for multi-dose dry powder inhalers. AAPS J. **17**, 1105–1116 (2015)
15. Telko, M.J., Hickey, J.A.: Dry powder inhaler formulation. Respir. Care **50**, 1209–1227 (2005)
16. Lohrmann, M.: Adhäsionskräfte in interaktiven Mischungen für Pulverinhalatoren (2005)
17. Young, P.M., Roberts, D., Chiou, H., Rae, W., Chan, H.-K., Traini, D.: Composite carriers improve the aerosolisation efficiency of drugs for respiratory delivery. J. Aerosol Sci. **39**, 82–93 (2008)
18. Larhrib, H., Martin, G.P., Marriott, C., Prime, D.: The influence of carrier and drug morphology on drug delivery from dry powder formulations. Int. J. Pharm. **257**, 283–296 (2003)
19. Zeng, X.M., Martin, G.P., Marriott, C., Pritchard, J.: The influence of crystallization conditions on the morphology of lactose intended for use as a carrier for dry powder aerosols. J. Pharm. Pharmacol. **52**, 633–643 (2000)
20. Zeng, X.M., Martin, G.P., Marriott, C., Pritchard, J.: The use of lactose recrystallised from carbopol gels as a carrier for aerosolised salbutamol sulphate. Eur. J. Pharm. Biopharm. **51**, 55–62 (2001)
21. Chan, L.W., Lim, L.T., Heng, P.W.S.: Immobilization of fine particles on lactose carrier by precision coating and its effect on the performance of dry powder formulations. J. Pharm. Sci. **92**, 975–984 (2003)
22. Iida, K., Todo, H., Okamoto, H., Danjo, K., Leuenberger, H.: Preparation of dry powder inhalation with lactose carrier particles surface-coated using a Wurster fluidized bed. Chem. Pharm. Bull. **53**, 431–434 (2005)
23. Adi, H., Larson, I., Stewart, P.: Use of milling and wet sieving to produce narrow particle size distributions of lactose monohydrate in the sub-sieve range. Powder Technol. **179**, 95–99 (2007)
24. Dickhoff, B.H.J., de Boer A.H., Lambregts, D., Frijlink, H.W.: The effect of carrier surface and bulk properties on drug particle detachment from crystalline lactose carrier particles during inhalation, as function of carrier payload and mixing time. Eur. J. Pharm. Biopharm. **56**, 291–302 (2003)
25. El-Sabawi, D., Price, R., Edge, S., Young, P.M.: Novel temperature controlled surface dissolution of excipient particles for carrier based dry powder inhaler formulations. Drug Dev. Ind. Pharm. **32**, 243–251 (2006)
26. Iida, K., Hayakawa, Y., Okamoto, H., Danjo, K., Leuenberger, H.: Preparation of dry powder inhalation by surface treatment of lactose carrier particles. Chem. Pharm. Bull. **51**, 1–5 (2003)
27. Islam, N., Stewart, P., Larson, I., Hartley, P.: Lactose surface modification by decantation: are drug-fine lactose ratios the key to better dispersion of salmeterol xinafoate from lactose-interactive mixtures? Pharm. Res. **21**, 492–499 (2004)
28. Littringer, E.M., Mescher, A., Eckhard, S., Schröttner, H., Langes, C., Fries, M., et al.: Spray drying of mannitol as a drug carrier—the impact of process parameters on product properties. Dry. Technol. **30**, 114–124 (2012)
29. Maas, S.G., Schaldach, G., Littringer, E.M., Mescher, A., Griesser, U.J., Braun, D.E., et al.: The impact of spray drying outlet temperature on the particle morphology of mannitol. Powder Technol. **213**, 27–35 (2011)
30. Littringer, E.M., Mescher, A., Schroettner, H., Achelis, L., Walzel, P., Urbanetz, N.A.: Spray dried mannitol carrier particles with tailored surface properties—the influence of carrier surface roughness and shape. Eur. J. Pharm. Biopharm. **82**, 194–204 (2012)
31. Ferrari, F., Cocconi, D., Bettini, R., Giordano, F., Santi, P., Tobyn, M., et al.: The surface roughness of lactose particles can be modulated by wet-smoothing using a high-shear mixer. AAPS PharmSciTech **5**, 1–6 (2004)
32. Young, P.M., Cocconi, D., Colombo, P., Bettini, R., Price, R., Steele, D.F., et al.: Characterization of a surface modified dry powder inhalation carrier prepared by "particle smoothing". J. Pharm. Pharmacol. 1339–44 (2002)
33. Iida, K., Inagaki, Y., Todo, H., Okamoto, H., Danjo, K., Luenberger, H.: Effects of surface processing of lactose carrier particles on dry powder inhalation properties of salbutamol sulfate. Chem. Pharm. Bull. **52**, 938–942 (2004)

34. Steckel, H., Markefka, P., TeWierik, H., Kammelar, R.: Effect of milling and sieving on functionality of dry powder inhalation products. Int. J. Pharm. **309**, 51–59 (2006)
35. Maas, S.G.: Optimierung trägerbasierter Pulverinhalate durch Modifikation der Trägeroberfläche mittels Sprühtrocknung (2009)
36. Kawashima, Y., Serigano, T., Hino, T., Yamamoto, H., Takeuchi, H.: Effect of surface morphology of carrier lactose on dry powder inhalation property of pranlukast hydrate. Int. J. Pharm. **172**, 179–188 (1998)
37. Steckel, H., Mu, B.W.: In vitro evaluation of dry powder inhalers II: influence of carrier particle size and concentration on in vitro deposition. Int. J. Pharm. **154**, 31–37 (1997)
38. Srichana, T., Martin, G.P., Marriott, C.: Dry powder inhalers: the influence of device resistance and powder formulation on drug and lactose deposition in vitro. Eur. J. Pharm. Sci.: Off. J. Eur. Fed. Pharm. Sci. **7**, 73–80 (1998)
39. Telko, M.J., Dsc, A.J.H.: Dry powder inhaler formulation. Respir. Care **50**, 1209–1227 (2005)
40. Chew, N.Y.K., Chan, H.-K.: The role of particle properties in pharmaceutical powder inhalation formulations. J. Aerosol Med. **15**, 325–330 (2002)
41. Newman, S.P., Busse, W.W.: Evolution of dry powder inhaler design, formulation, and performance. Respir. Med. **96**, 293–304 (2002)
42. Wong, W., Fletcher, D.F., Traini, D., Chan, H.-K., Young, P.M.: The use of computational approaches in inhaler development. Adv. Drug Deliv. Rev. **64**, 312–322 (2012)
43. Ruzycki, C.A., Javaheri, E., Finlay, W.H.: The use of computational fluid dynamics in inhaler design. Expert. Opin. Drug Deliv. **10**, 307–323 (2013)
44. Sommerfeld, M., Schmalfuß, S.: Numerical analysis of carrier particle motion in a dry powder inhaler. J. Fluids Eng. **138**, 041308 (2015)
45. Coates, M.S., Fletcher, D.F., Chan, H.-K., Raper, J.A.: Effect of design on the performance of a dry powder inhaler using computational fluid dynamics. Part 1: Grid structure and mouthpiece length. J. Pharm. Sci. **93**, 2863–2876 (2004)
46. Coates, M.S., Chan, H.-K., Fletcher, D.F., Raper, J.A.: Effect of design on the performance of a dry powder inhaler using computational fluid dynamics. Part 2: Air inlet size. J. Pharm. Sci. **95**, 1382–1392 (2006)
47. Coates, M.S., Chan, H.-K., Fletcher, D.F., Chiou, H.: Influence of mouthpiece geometry on the aerosol delivery performance of a dry powder inhaler. Pharm. Res. **24**, 1450–1456 (2007)
48. Coates, M.S., Fletcher, D.F., Chan, H.-K., Raper, J.A.: The role of capsule on the performance of a dry powder inhaler using computational and experimental analyses. Pharm. Res. **22**, 923–931 (2005)
49. Donovan, M.J., Kim, S.H., Raman, V., Smyth, H.D.: Dry powder inhaler device influence on carrier particle performance. J. Pharm. Sci. **101**, 1097–1107 (2012)
50. Milenkovic, J., Alexopoulos, A.H., Kiparissides, C.: Flow and particle deposition in the Turbuhaler: a CFD simulation. Int. J. Pharm. **448**, 205–213 (2013)
51. de Boer, A.H., Hagedoorn, P., Woolhouse, R., Wynn, E.: Computational fluid dynamics (CFD) assisted performance evaluation of the Twincer^{TM} disposable high-dose dry powder inhaler. J. Pharm. Pharmacol. **64**, 1316–1325 (2012)
52. Behara, S.R.B., Longest, P.W., Farkas, D.R., Hindle, M.: Development and comparison of new high-efficiency dry powder inhalers for carrier-free formulations. J. Pharm. Sci. **103**, 465–477 (2014)
53. Tong, Z.B., Yang, R.Y., Chu, K.W., Yu, A.B., Adi, S., Chan, H.K.: Numerical study of the effects of particle size and polydispersity on the agglomerate dispersion in a cyclonic flow. Chem. Eng. J. **164**, 432–441 (2010)
54. Yang, J., Wu, C.-Y., Adams, M.: Three-dimensional DEM-CFD analysis of air-flow-induced detachment of API particles from carrier particles in dry powder inhalers. Acta Pharm. Sin. B **4**, 52–59 (2014)
55. van Wachem, B., Thalberg, K., Remmelgas, J., Niklasson-Björn, I.: Simulation of dry powder inhalers: combining micro-scale, meso-scale and macro-scale modeling. AIChE J. **63**, 501–516 (2017)

56. Hoppentocht, M., Hagedoorn, P., Frijlink, H.W., De Boer, A.H.: Technological and practical challenges of dry powder inhalers and formulations. Adv. Drug Deliv. Rev. **75**, 18–31 (2014)
57. Rahimpour, Y., Kouhsoltani, M., Hamishehkar, H.: Alternative carriers in dry powder inhaler formulations. Drug Discov. Today **19**, 618–626 (2014)
58. Zellnitz, S., Redlinger-Pohn, J.D., Kappl, M., Schroettner, H., Urbanetz, N.A.: Preparation and characterization of physically modified glass beads used as model carriers in dry powder inhalers. Int. J. Pharm. **447**, 132–138 (2013)
59. Cui, Y., Schmalfuß, S., Zellnitz, S., Sommerfeld, M., Urbanetz, N.A.: Towards the optimisation and adaptation of dry powder inhalers. Int. J. Pharm. **470**, 120–132 (2014)
60. Cui, Y., Sommerfeld, M.: Forces on micron-sized particles randomly distributed on the surface of larger particles and possibility of detachment. Int. J. Multiph. Flow **72**, 39–52 (2015)
61. Cui, Y., Sommerfeld, M.: Application of Lattice-Boltzmann method for analysing detachment of micron-sized particles from carrier particles in turbulent flows. Flow Turbul. Combust. **100**, 271–297 (2018)
62. Cui, Y., Sommerfeld, M.: The modelling of carrier-wall collision with drug particle detachment for dry powder inhaler applications. Powder Technol. **344**, 741–755 (2019). https://doi.org/10.1016/j-powtec.2018.12.067
63. Sommerfeld, M., Cui, Y., Schmalfuß, S.: Potentials and constraints for the application of CFD combined with Lagrangian particle tracking to dry powder inhalers. Eur. J. Pharm. Sci. **129**, 299–324 (2019)
64. Milenkovic, J., Alexopoulos, A.H., Kiparissides, C.: Deposition and fine particle production during dynamic flow in a dry powder inhaler: a CFD approach. Int. J. Pharm. **461**, 129–136 (2014)
65. Littringer, E.M., Zellnitz, S., Hammernik, K., Adamer, V., Friedl, H., Urbanetz, N.A.: Spray drying of aqueous salbutamol sulfate solutions using the Nano Spray Dryer B-90—the impact of process parameters on particle size. Dry. Technol. **31**, 1346–1353 (2013)
66. Flament, M.-P., Leterme, P., Gayot, A.: The influence of carrier roughness on adhesion, content uniformity and the in vitro deposition of terbutaline sulphate from dry powder inhalers. Int. J. Pharm. **275**, 201–209 (2004)
67. Renner, N., Steckel, H., Urbanetz, N., Scherließ, R.: Nano- and microstructured model carrier surfaces to alter dry powder inhaler performance. Int. J. Pharm. **518**, 20–28 (2017)
68. Friebel, C., Steckel, H., Müller, B.W.: Rational design of a dry powder inhaler: device design and optimisation. J. Pharm. Pharmacol. **64**, 1303–1315 (2012)
69. Friebel, C.: Rational development of an inhalation system. Ph.D. Thesis, Kiel University (2010)
70. Ariane, M., Sommerfeld, M., Alexiadis, A.: Wall collision and drug-carrier detachment in dry powder inhalers: using DEM to devise a sub-scale model for CFD calculations. Powder Technol. **334**, 65–75 (2018)
71. Sommerfeld, M., van Wachem, B., Oliemans, R.: Best Practice Guidelines for Computational Fluid Dynamics of Dispersed Multiphase Flows. ERCOFTAC (European Research Community on Flow, Turbulence and Combustion). ISBN 978-91-633-3564-8 (2008)
72. Sommerfeld, M.: Modelling of particle-wall collisions in confined gas-particle flows. Int. J. Multiph. Flow **18**, 905–926 (1992)
73. Sommerfeld, M., Huber, N.: Experimental analysis and modelling of particle-wall collisions. Int. J. Multiph. Flow **25**, 1457–1489 (1999)
74. Dietzel, M., Sommerfeld, M.: Numerical calculation of flow resistance for agglomerates with different morphology by the Lattice-Boltzmann method. Powder Technol. **250**, 122–137 (2013)
75. Ibrahim, A.H., Dunn, P.F., Brach, R.M.: Microparticle detachment from surfaces exposed to turbulent air flow: controlled experiments and modeling. J. Aerosol Sci. **34**, 765–782 (2003)
76. Goldasteh, I., Ahmadi, G., Ferro, A.R.: Monte Carlo simulation of micron size spherical particle removal and resuspension from substrate under fluid flows. J. Aerosol Sci. **66**, 62–71 (2013)
77. Zellnitz, S., Schroettner, H., Urbanetz, N.A.: Influence of surface characteristics of modified glass beads as model carriers in dry powder inhalers (DPIs) on the aerosolization performance. Drug Dev. Ind. Pharm. 1–8 (2015)

78. Crowe, C.T., Schwarzkopf, J.D., Sommerfeld, M., Tsuji, Y.: Multiphase Flows with Droplets and Particles, 2nd edn. CRC Press, Boca Raton, U.S.A (2012). ISBN 978-1-4398-4050-4
79. Tsuji, Y., Morikawa, Y., Tanaka, T., Nakatsukasa, N., Nakatani, M.: Numerical simulation of gas-solid two-phase flow in a two-dimensional horizontal channel. Int. J. Multiph. Flow **13**, 671–684 (1987)
80. Renner, N.: Understanding interparticle interactions in dry powder inhalation: glass beads as an innovative model carrier system, p. 181. Dissertation, Department of Pharmaceutics and Biopharmaceutics, Kiel University (2017)

Rapid Impact of Nanoparticles on Surfaces

Alfred Weber, Christian Schöner, Manuel Gensch, Alexander Werner and Thorsten Pöschel

Introduction

Collisions between particles or between particles and a wall play an important role in many process engineering applications such as deposition (e.g. adhesion on fibers in filters), dry dispersion and comminution. During the particle-particle and the particle-wall collisions a fraction of the kinetic energy of the relative motion is lost, for example as heat or plastic deformation. The portion of the regained energy after bouncing can be expressed by the coefficient of restitution ε_n which is the ratio of rebound velocity v_r to impact velocity v_i:

$$\varepsilon_n = \frac{v_r}{v_i}.$$

In more detail, there are two coefficients of restitution, i.e. ε_n and ε_t, which express the ratio of rebound and impact velocities for the normal respectively tangential components of the velocity relative to the contact plane. This differentiation is important since the contact phenomena due to relative velocity in normal and tangential

A. Weber (✉) · M. Gensch · A. Werner
Institute of Particle Technology, Technical University of Clausthal, Clausthal, Germany
e-mail: alfred.weber@tu-clausthal.de

M. Gensch
e-mail: manuel.gensch@tu-clausthal.de

A. Werner
e-mail: alexander.werner@tu-clausthal.de

C. Schöner · T. Pöschel
Institute for Multiscale Simulations, Friedrich-Alexander-Universität, Erlangen, Germany
e-mail: ch.schoener@web.de

T. Pöschel
e-mail: thorsten.poeschel@fau.de

S. Antonyuk (ed.), *Particles in Contact*,
https://doi.org/10.1007/978-3-030-15899-6_17

directions have completely different causes and influence the mechanics of the particle collision in different ways.

In the macroscopic case, i.e. when the particle can be described as macroscopic body, the normal coefficient suffices $0 \leq \varepsilon_n \leq 1$, which can be deduced from the dynamic deformation of the particle during the collision and therefore from its viscous material properties. This statement remains true for the case of adhesive interaction. The limits $\varepsilon_n = 1$ and $\varepsilon_n = 0$ correspond to the cases of perfect elastic collision and of complete loss of the relative velocity after collision, i.e. agglomeration, respectively. The normal coefficient of restitution results therefore from bulk material properties of the particle. In contrast, the tangential coefficient of restitution ε_t is a consequence of the viscous volume properties and the surface properties such as roughness. The roughness is induced by the microscopic surface structure and therefore the tangential coefficient of restitution is even for macroscopic particles a consequence of its microscopic properties. Thus, for given material properties, the normal coefficient can be deduced analytically from continuum mechanical (i.e. macroscopic) calculations, while for analogous calculations of the tangential coefficient, phenomenological models must be taken into account. The tangential coefficient satisfies $-1 \leq \varepsilon_t \leq 1$, exhibiting two elastic limits $\varepsilon_t \pm 1$, where $\varepsilon_t = 1$ corresponds to a perfectly smooth surface and $\varepsilon_t = -1$ to a complete reversal of the tangential impact velocity (perfect elastic gear-wheel). We would like to emphasize that $\varepsilon_t = 1$ and $\varepsilon_n \neq 1$ are not consistent, since a dissipative collision in normal direction always entails a dissipative collision in tangential direction. As a consequence of these features, for macroscopic particles, ε_n can be deduced from material properties while for ε_t further assumption have to be made. Both coefficients, ε_n and ε_t, are not constant but depend on the vectorial impact velocity.

In the case of nanoparticles, the situation is much more complex: Due to the noticeable discrete atomic structure, the particles can no longer be considered as homogeneous (bulk) bodies. Therefore, microscopic details of the geometric structure determine the collision properties and so ε_n and ε_t. In addition, the particles can no longer be considered macroscopic bodies, i.e. viscoelastic and plastic deformation of the particles during the collision have to be related to the rearrangement of the atomic structure. Finally, in contrast to macroscopic bodies, thermal motion of the atoms is no longer decoupled from particle motion. As a consequence, not only the transformation of kinetic energy into thermal energy is observable, but also the opposite with the result, that the condition $\varepsilon_n < 1$ holds true only on average, but not strictly [17, 28]. In analogy, also the condition $-1 \leq \varepsilon_t \leq 1$ is not strictly fulfilled for nanoparticles. Even the counterintuitive $\varepsilon_n < 0$ can be observed as a result of the fact that nanoparticles can exhibit very large rotation, see [21–23, 27], while for macroscopic particles, $\varepsilon_n \geq 0$ as good approximation [43].

During a normal impact of a spherical particle onto a plane wall, velocity, material properties of particle and substrate, but also their surface properties will define the transformation of kinetic and potential energy into elastic and plastic deformation, as well as into heat and surface waves. The situation with respect to the particle velocity is shown in Fig. 1 which is based on own calculations further outlined in section "MD Contact Models to Describe the Collision Process at Ultra Short Impact Loadings of

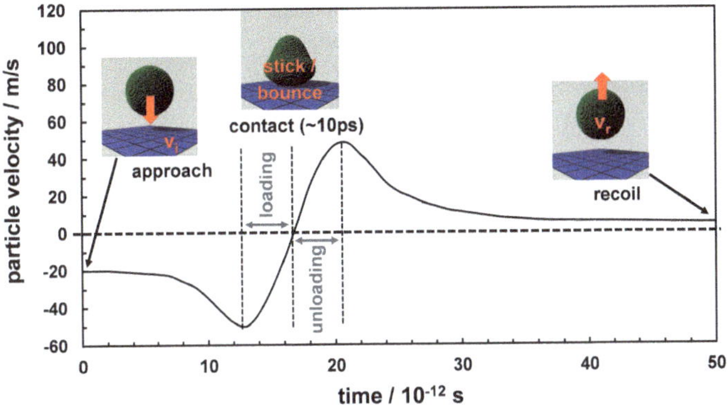

Fig. 1 Ag nanoparticle (5 nm) undergoing collision with a rigid wall: (i) approaching phase, (ii) contact phase (consisting of loading and unloading) and (iii) recoil phase

Nanoparticles". When a particle is approaching the wall, its kinetic energy may be increased by acceleration due to transformed adhesion energy. This effect can lead to impact velocities which are above the original approach velocity [50, 56]. When the regained elastically stored energy overcomes the adhesion energy, the particle will bounce back into the gas. For large particles and at high impact velocities the plastic deformation dominates the collision process and adhesion phenomena become less important. As a first approximation, the critical transition radius R_γ between adhesion and plastic dominated regime can be shown as:

$$R_\gamma = \frac{\gamma}{v_i}\sqrt{\frac{3}{2\rho_p Y}} \tag{1}$$

Hereby, γ stands for the surface tension, v_i for the impact velocity, ρ_p for the particle density and Y for the yield pressure for plastic deformation. For typical values of $\gamma = 0.5\,\mathrm{N\,m^{-1}}$ and $Y = 109\,\mathrm{Pa}$ the transition for metal particles with an impact velocity of $10\,\mathrm{m\,s^{-1}}$ is a diameter of about 40 nm.

At sufficiently high impact velocities, the yield pressure Y is surpassed in the center of the contact area of the impacting particle, thus plastic deformation starts to be observable asides elastic deformation. Wang and John used this model to describe the charge transfer to the particle during a collision process of microparticles [53]. The charge transfer happens during the formation of the contact, driven by the contact potential. The charge of the rebounding particle depends on the contact potential, the size of the contact area and on the duration of the collision process.

As mentioned above, rebounding of the particle starts when the impact velocity exceeds a certain limit, known as critical velocity v_{min}. According to [56], the critical velocity of a particle with radius R and density ρ_p can be related to material properties:

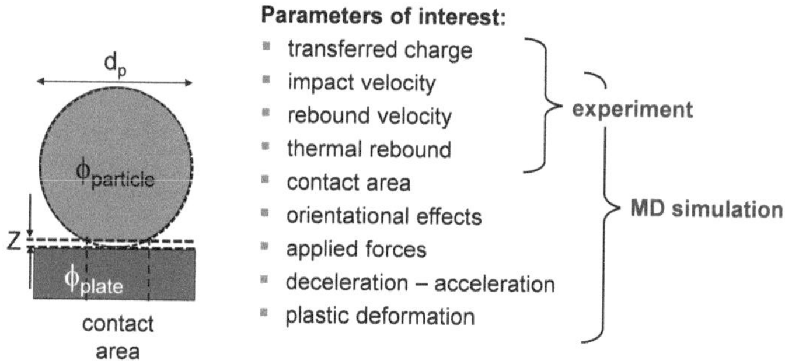

Fig. 2 Left: schematic diagram of the interesting parameters at maximum particle compression corresponding to $v = 0$ in Fig. 1. Right: ways to extract them

$$v_{min}^2 = \frac{4\gamma^2}{3^{1/3}5\rho_p Y R^2}$$

with the material parameter already used in Eq. 1. However, there is large hesitancy to what extent bulk material parameters may be applied for nanoparticles. For instance, the yield pressure Y of copper generally increases with decreasing particle size, known as the Hall–Petch effect. For very small particles, an inverse Hall–Petch effect is observed [46], say a decrease! Therefore, macroscopic rationales using bulk material parameters have to be reworked for nanoparticle-wall collisions. Since these collisions occur on picosecond time scale Fig. 1, the details of the collision process cannot be resolved with experimental methods but have to be examined by Molecular Dynamics (MD) simulations. On the other hand, certain parameters such as the transferred charge would involve a detailed modeling of electronic band structures and their deformation during collision which would require elaborate quantum mechanical calculations. The experiment can at most measure the charge state before and after the collision. Therefore it is assumed, that the charge does not change once the contact has ceased, which is reasonable as long as thermionic emission is avoided, say, for particles in a carrier gas not being at too high temperatures. From the aforementioned it is clear, that a combination of experiments and MD simulations is required to arrive at a good, nearly comprehensive picture of the phenomena occurring in the nanoparticle-wall collisions. The parameters of interest are outlined in Fig. 2 together with the route over which they are determined, i.e. MD simulation or experiment. Since some of the parameters are approachable in both ways, they will serve as mutual validation cases.

Considering the impact of nanoparticle agglomerates onto walls, things become even more interesting. In addition to material properties, agglomerate collision behavior is also influenced by their morphology which itself may undergo substantial structural changes starting from rearrangement of individual primary particles and branches all the way to fragmentation. While agglomerate-wall collisions comprising large particles have been thoroughly investigated in experiment and numerical studies [24, 47–49], the mechanical properties of nanoparticle agglomerates have moved into the focus of scientific research only within the last decade. The first impaction experiments with nanoparticle agglomerates were performed by Froeschke et al. using a low pressure impactor and TEM analysis to determine the degree of fragmentation [8]. Antony et al. performed Discrete Element Method (DEM) simulations with spherical agglomerates of 100 nm particles. They showed, that the degree of fragmentation depends on the ratio of kinetic energy to surface tension, the Weber number, which is influenced by the inter-particle bond strength [1]. Sator et al. used a MD model to investigate the fragmentation behavior of agglomerates [35]. They found that the total number of fragments related to the total number of primary particles is proportional to the fraction of broken bonds. Besides simple scaling laws for number and size distribution of the fragments, they also showed that the fragmentation curve scales logarithmically with the introduced energy which agrees with own results for metal and silica nanoparticle agglomerates [44, 45]. All these investigations are restricted to normal impaction.

Therefore, we give a comprehensive description of the impact phenomena occurring in single nanoparticle-wall and nanoagglomerate-wall collisions, respectively. For this purpose, the experimental setup to achieve controlled particle-wall collisions in normal and oblique impact will be outlined in section "Development of a Low Pressure Impactor for Normal and Oblique Impaction Experiments on Single Nanoparticles and Nanoparticle Agglomerates at Velocities up to 300 m s^{-1}". Then, a MD contact model will be presented, which is capable of recovering the mechanical parameters during the ultra short collision process and validating the experimental data in section "MD Contact Models to Describe the Collision Process at Ultra Short Impact Loadings of Nanoparticles". The combination of simulation and experiment allows for extraction of material parameters of single nanoparticles and to check to which extent classical continuum models may be applied to nanoparticle-wall collisions when modified parameters are used, see section "New Method to Obtain Material Values (Critical Velocity, Yield Pressure, Elastic Modulus) of Nanoparticles During Collision Processes". In section "Applicability of Macroscopic Theories to Describe the Mechanical Behavior of Nanoparticles in Particle-Wall Collisions", solely experimentally accessible results of the contact charging of single nanoparticle collisions with walls are discussed. Some of this data points towards changes in the atomic structure occurring in the nanoparticles during impaction. Therefore, in the following sections, "Introduction and Characterization of a Single Parameter Description of the Lattice Orientation of Nanoparticles" and "Impact Properties of Nanoparticles in Dependence of Their Lattice Orientation", the yet unmentioned anisotropy of nanoparticle-wall impacts is motivated and a one-parameter description of the particle lattice orientation is introduced. Utilizing this description, the influence of the orientation anisotropy on the collision parameters are investigated within MD simulations. Considering nanoparticle agglomerates, section "Expansion of the Systematic to Describe the Fragmentation of Nanoparticle Agglomerates by Means of Fragmentation Probability and Fragmentation Function" discusses our findings related to fragmentation behavior. A general approach is presented, describing the fragmentation behavior based on fragmentation probability and size distribution of the fragments. Bouncing of agglomerates and their fragments is investigated in section "Influence of Particle Size and Material as Well as Impact Velocity and Angle on the Bouncing and Fragmentation Behavior". In particular, the influence of particle size, impact velocity and impact angle on bouncing and fragmentation behavior will be treated. Concluding, optimal process parameters for the dry dispersion of nanoparticle agglomerates will be outlined as a principal application in section "Identification of Optimal Process Parameter for the Continuous Dry Dispersion of Nanopowders".

Development of a Low Pressure Impactor for Normal and Oblique Impaction Experiments on Single Nanoparticles and Nanoparticle Agglomerates at Velocities up to 300 m s^{-1}

Experimental Setup: Normal Impact

Due to their low inertia, the impact of nanoparticles with high velocities can only be realized at low pressure conditions where friction forces with the carrier gas are significantly reduced. The overall experimental setup is shown in Fig. 3. Nanoparticles were produced with a spark discharge generator (SDG) from electrodes of the material of interest. The nanoparticle agglomerates may be completely sintered in the subsequent tube furnace to obtain individual dense spherical particles. After charging the particles in a bipolar diffusion charger, a partial aerosol flow is classified in a Differential Mobility Analyzer (DMA) while the rest is discharged through an absolute particle filter. The concentration of the monomobile particles entering the Single Stage Low Pressure Impactor (SS-LPI) is measured with a Condensation Particle Counter (CPC).

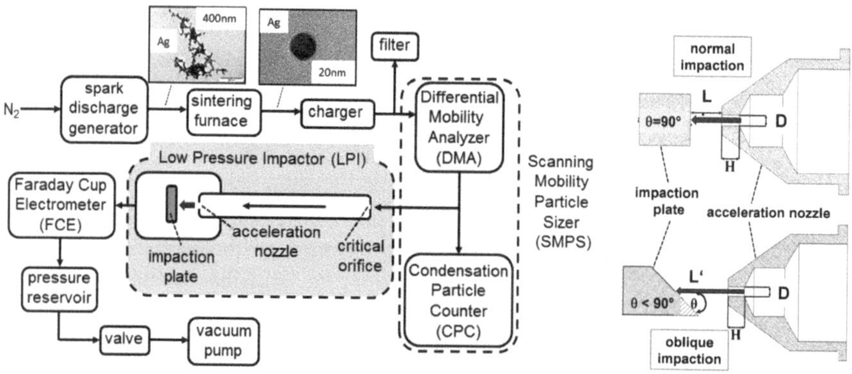

Fig. 3 Left: experimental setup for the impaction of spherical monodisperse Ag and Pt nanoparticles in a low pressure impactor, right: impactor geometries for normal and oblique impaction (adapted from [10])

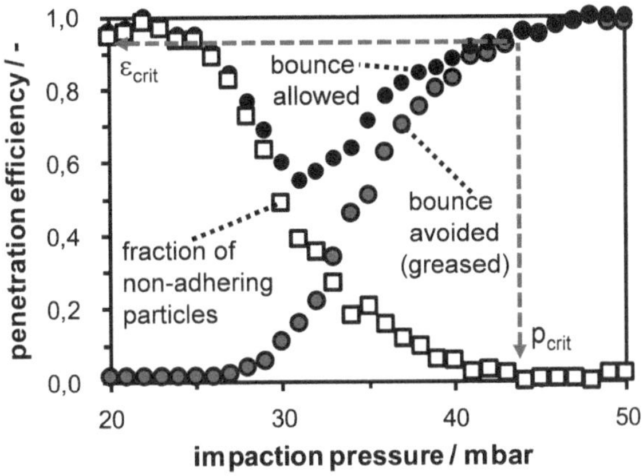

Fig. 4 Measured penetration curves of Ag nanoparticles for untreated (bounce allowed) and greased impaction plate and corresponding bouncing fraction (non-adhering): penetration ε_{crit} is related to pressure p_{crit} leading to the critical velocity v_{crit}, where bouncing starts

The flow rate into the SS-LPI is limited by a critical orifice to 0.3 l min^{-1}. To avoid the interference of the sonic flow regime at the critical orifice and the flow in the acceleration nozzle, an equilibration tube of 1 m length is introduced. In order to measure a separation curve, a pressure tank was evacuated with a vacuum pump, then the valve to the pump was closed and the connection to the SS-LPI was opened. Due to the aerosol flow through the critical orifice the pressure in the SS-LPI increased continuously (pressure scanning mode). The fraction of charged particles which were not deposited or which bounced from the impaction plate was monitored with a Faraday Cup Electrometer (FCE) behind the impactor. To obtain the deposition curve, the impaction plate was greased, suppressing bouncing cf. Fig. 4. From the measurement without grease the bouncing curve was obtained as difference of the two curves as further detailed in [38]. From the bouncing curve the critical velocity was determined as the initial velocity where the first particles started rebounding.

The determination of the coefficient of restitution, cf. section "Determination of the Coefficient of Restitution (CoR)" requires to know the incident and rebound velocities, v_i and v_r, respectively. Since the direct measurement of the impaction velocity is not possible, a method was developed to deduce it from the process conditions and the particle properties using the following three parameter model [31]:

$$v_i = v_{max}(p) \cdot \overline{v}_{imp}(Stk^*, {}^{L}/_{D}) \cdot \chi_{log}(Stk^*, {}^{L}/_{D}, {}^{H}/_{D}) \tag{2}$$

where v_{max} is the maximum gas velocity at the acceleration nozzle outlet, $\overline{v}_{imp}$ the non-dimensional impact velocity at a given Stokes number Stk^* and the correction function χ_{log} accounting for the lag of the particle motion behind the gas flow. L, H and D denote the geometrical dimensions of the SS-LPI as indicated in Fig. 3. The Stokes number used here is slightly modified compared to the classical definition and is given by:

$$Stk^* = \frac{2\rho_p d_p^2 C_C v_{max}}{9\mu D} \tag{3}$$

where ρ_p is the particle density, d_p the particle diameter, C_C denotes the slip correction factor (i.e. Cunningham correction) and μ is the viscosity of the surrounding gas. v_{max} is calculated assuming an ideal and incompressible gas as well as a parabolic velocity profile at the nozzle outlet. The non-dimensional impact velocity $\overline{v}_{imp}$ can be written as:

$$\overline{v}_{imp}(Stk^*, {}^{L}/_{D}) = \frac{-B({}^{L}/_{D})}{Stk^* + A({}^{L}/_{D})} + 1 \tag{4}$$

where the empirical constants amount to $A \equiv 0.328$ and $B \equiv 0.692$ for the present geometry [31]. Finally, the factor χ_{lag} accounts for insufficient acceleration of the particles given by:

$$\chi_{lag} = 1 - 0.58 \exp - \frac{\overline{S}/\overline{S}_{99}}{0.19} - 0.32 \exp - \frac{\overline{S}/\overline{S}_{99}}{0.022} \tag{5}$$

with the non-dimensional stopping distance and necessary acceleration length:

$$\overline{S}_{99} = \frac{4\tau v_{max}}{D} = 2Stk^* \qquad \overline{S} = \frac{H + L}{D} - 0.5 \tag{6}$$

Oblique Impact

The evaluation of the experiment to study the influence of the tangential component of the impact velocity on the particle-wall collision requires careful consideration regarding the impaction angle. This is due to the fact that the effective impaction angle of a particle on an inclined surface does not only depend on the particles' size but also on its velocity, both of which are part of the main parameter governing the particle impact characteristics, the Stokes number. As shown in Fig. 5 for two geometric angles of the impaction plate, the effective impaction angle approximates

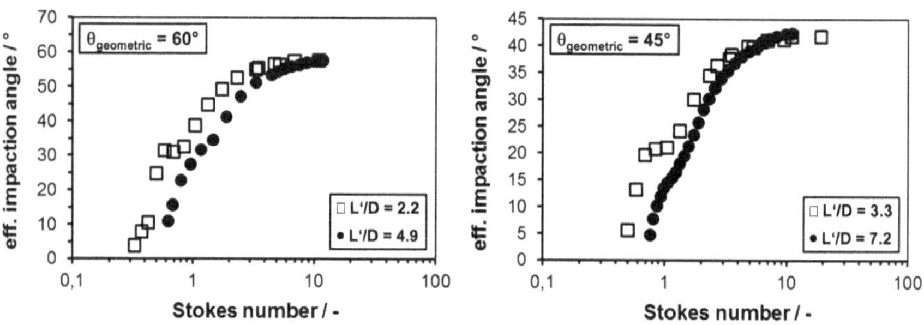

Fig. 5 Left: evolution of the effective impaction angle as a function of the Stokes number. Left: for $\theta_{geometric} = 60°$ and right for $\theta_{geometric} = 45°$ (adapted from [10])

the geometric one only for Stk approaching 10. CFD simulations and particle tracking analysis were performed to establish reliable correlations between particle properties and effective angle of impact.

Determination of the Coefficient of Restitution (CoR)

As outlined above, to calculate the coefficient of restitution also the rebound velocity is needed, which is determined by introducing a stagnation domain on the impaction plate in the form of a well cf. Fig. 6. The particle accelerated in the jet will not hit the extended impaction plate, but will enter the stagnation domain and be decelerated. It hits the bottom of the well with a reduced velocity and bounce, depending on the particle relaxation time and the rebound velocity, up to a certain height S (corresponding to a stopping distance). When $S < T$, the particle remains in the well and will eventually be deposited to the wall. When $S > T$, the particle can leave the well, reenter the main gas stream and finally reach the FCE to be counted. By varying the pressure p in the impactor, the inset of particle release can be measured from which the according rebound velocity can be deduced. The well depth T is adjusted exactly by attaching the bottom plate of the well to a micrometer screw. The assumptions and the details of the evaluation procedure are outlined in [30, 38]. Here it is important to emphasize, that the normal coefficient of restitution can be determined from the pressure where the first particles leave the well and the well depth T:

$$\varepsilon_n = \frac{v_r}{v_i} = \frac{T}{\tau(p)v_i(0) - T} \tag{7}$$

where $\tau(p)$ is the particle relaxation time at the pressure p.

Besides particle size and material also the material of the bottom plate of the well will influence the particle bouncing behavior, which in the present study was an Al plate with an oxygen layer on the surface. Therefore, the plate material was considered much harder than the particle material implying preferentially deformation of the particles at impact.

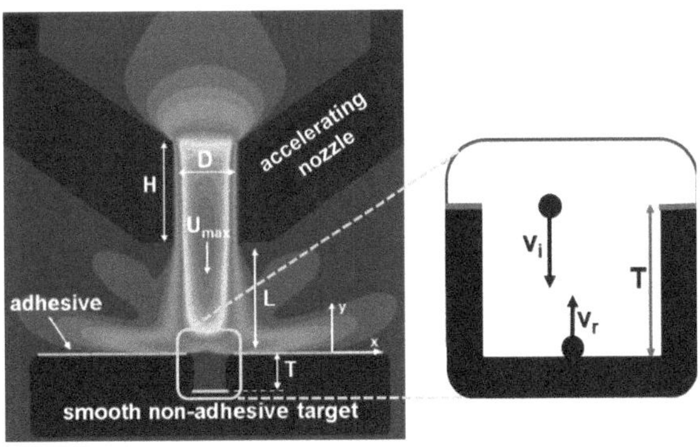

Fig. 6 Schematic diagram of the modified SS-LPI to determine the coefficient of restitution using a well of depth T and the principle to measure the rebound velocity v_r

Detection of Nanoparticles and Their Fragments

While deposition and bouncing of the charged nanoparticles on insulating walls was measured with a Faraday Cup Electrometer (FCE) cf. results Fig. 4, for agglomerates an additional characterization tool was employed. The nanoparticle agglomerates were impacted onto a TEM grid which was places on the impaction plate. Using image analysis, the projection area of the classified agglomerates was determined. Then changes of the projection area due to deformation all the way up to fragmentation could be detected, as long as the coverage of the TEM grid was kept low in order to avoid overlapping of individual agglomerates. In particular, this technique allows to obtain not only the incident of fragmentation as a function of the impact velocity, but also to evaluate the size distribution of the fragments which will be the basis to describe the fragmentation process completely section "Expansion of the Systematic to Describe the Fragmentation of Nanoparticle Agglomerates by Means of Fragmentation Probability and Fragmentation Function". The use of fragmentation probability and fragmentation function resembles the breaking probability and breaking function known from comminution.

For the impaction on conducting walls, the particle charge may change depending on the difference of the material work functions and the collision regime. In this case, particles may leave the impaction plate with a higher charge state or even with a reversed charge. The measurement of the particle charge with FCE after bouncing will provide new insights into the material parameters of nanoparticles undergoing ultrashort collisions with solid conducting walls section "New Method to Obtain Material Values (Critical Velocity, Yield Pressure, Elastic Modulus) of Nanoparticles During Collision Processes".

MD Contact Models to Describe the Collision Process at Ultra Short Impact Loadings of Nanoparticles

The impact simulations were performed as force-based MD calculations. For the interactions between the atoms of the particle the Embedded Atom Model (EAM) was used where the energy of the atom i is given by:

$$E_{i,EAM} = \mathrm{F}\left(\sum_{j \neq i} \rho(r_{ij})\right) + \frac{1}{2}\sum_{j \neq i} \varphi(r_{ij}) \tag{8}$$

F is the embedding energy functional, ρ the atomic electron density function and φ a pair potential interaction. This model reliably reproduces the main properties of bulk crystals [7].

The interaction between wall atoms and particle atoms was described using a cut-off and smoothed Lennard-Jones potential:

$$E_{i,LJ} = 4\varepsilon \sum_{j \neq i} \left(\left(\frac{\sigma}{r_{ij}} \right)^{12} - C \left(\frac{\sigma}{r_{ij}} \right)^{6} \right) + C_1 r_{ij} + C_2 \quad (9)$$

where ε is the depth of the energy well and σ the characteristic Lennard-Jones distance for which literature values for silver were chosen [12]. The constant C represents the strength of the attractive part of the energy. To simulate a weakly adhesive contact a value of $C = 0.35$ was chosen. Such a contact represents the situation when a covering oxide layer affects the interaction strength. The constants C_1 and C_2 serve to smooth out $E_{i,LJ}$ at the cutoff distance for which $r_c = 2.5\sigma$ was chosen. In contrast to the particles, the wall was modeled to be ideally stiff with no possibility for the atoms to move. More details about the wall structure and the preparation of the particles can be found in [38]. At the beginning of the simulation, the particle was placed outside of the Lennard-Jones interaction range, rotated randomly and released with the desired initial velocity. Particle sizes were 5, 10 and 15 nm and the initial velocity varied from 10 to 90 m s^{-1} in steps of 5 m s^{-1}. The results of 100 simulations with random rotation were averaged for each particle size and each initial velocity.

New Method to Obtain Material Values (Critical Velocity, Yield Pressure, Elastic Modulus) of Nanoparticles During Collision Processes

The critical velocity for bouncing is an important measure for the adhesion forces at work during the particle-wall contact. The higher the adhesion forces, the stronger the acceleration of the particle towards the wall cf. Fig. 1. In the elasto-plastic range this leads to an increase of the energy dissipation due to plastic deformation which in turn enhances the sticking probability. With the method outlined above cf. Fig. 4, the critical velocities have been measured for Ag and NaCl nanoparticles. In the double-logarithmic plot of Fig. 7, it becomes obvious that above a certain particle

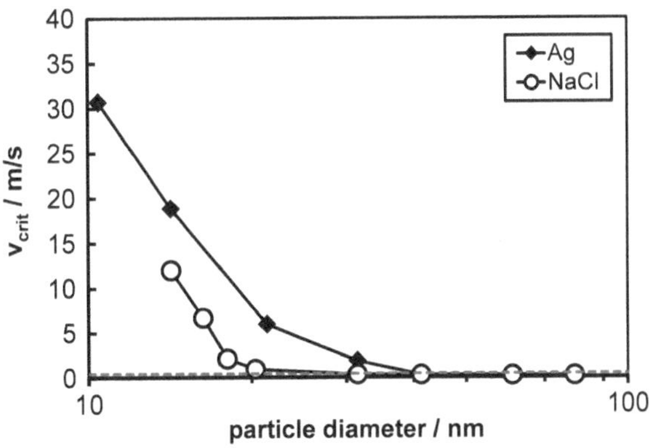

Fig. 7 Measured critical velocities for Ag and NaCl nanoparticles impacting on mica targets as a function of the particle size

size (40 nm for Ag and 20 nm for NaCl, respectively) the values of v_{crit} are small, typically below 0.5 m s^{-1}, indicating a low level of energy dissipation, i.e. mainly elastic deformation of the particles. For smaller particles, there is a rapid increasing with decreasing particle diameter which points towards enhanced energy dissipation, i.e. beginning of plastic deformation. However, the slope of the increase does not agree with the classical inverse relationship between critical velocity and particle diameter known for the micrometer size range cf. Eq. 1. As will be shown below, plastic deformation of 40 nm Ag particles does not set in below 40 m s^{-1}. Therefore, the steep slope needs to be explained in a different way. In fact, the impaction surface consisting of mica is commonly considered to be molecularly smooth. However, investigations by Ostendorf et al. [26] show that the remaining potassium ions on the surface after cleavage react in the presence of humidity with the carbon complexes to potassium carbonate particles with sizes in the range from 0.6 to 5 nm and with a surface number density of ca. 40000 μm^{-2}. This means that the average distance between two protuberances is about 6 nm. For a 40 nm Ag particle the contact circle is, even only under the influence of van der Waals adhesion forces, already larger than this distance. Therefore, larger particles experience the protuberances as roughness increasing the separation and reducing the adhesion energy. Smaller particles, however, experience the full contact to the wall and need higher energies for bouncing. From the MD simulations shown for instance in Fig. 1, impact and rebounding velocities can be calculated which determine the coefficient of restitution. As shown in Fig. 8, the MD results agree quite well with the values measured with the well method Fig. 6, although the smallest measured particles were only 18 nm. However, within experimental uncertainties, no clear size dependence of the coefficient of restitution could be observed. More importantly, Fig. 8 shows for the first time, that the MD simulations of the nanoparticle-wall collisions agree quantitatively with the first available experimental data of the normal collision processes. In particular,

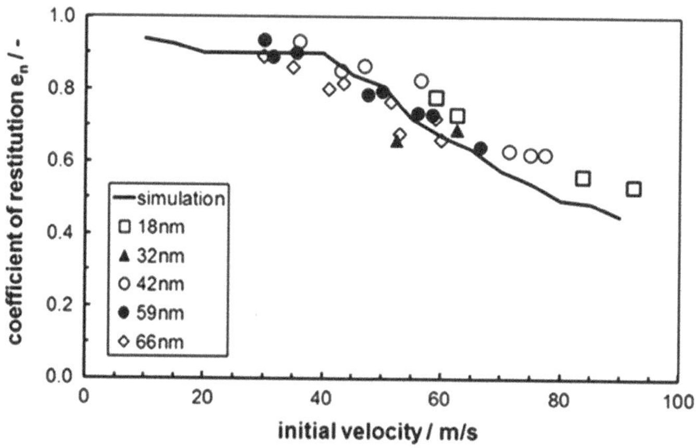

Fig. 8 Numerically calculated 95-percentiles (for 15 nm Ag particles) and measured values of the coefficient of restitution for various particle sizes

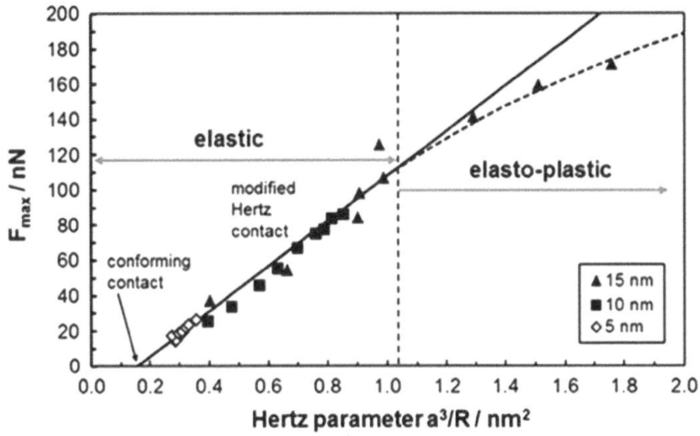

Fig. 9 Maximum applied force for Ag nanoparticles of different sizes as a function of the Hertz parameter: the results of the MD simulation deviate from the elastic behavior (indicated by the solid line) at higher Hertz parameters

the validation of the MD simulation opens the details of the extremely short collision as shown in Fig. 1, which cannot be accessed experimentally cf. Fig. 2. By means of MD simulation, temporal evolution of parameters such as applied force, resulting particle contact area and particle temperature can be determined: Using the quantitative results of the MD simulation, the relation between applied force and resulting contact area was evaluated for Ag nanoparticles impaction on a stiff wall. The results are presented in Fig. 9 in the form of a Hertz diagram. As assumed in the Hertz approach, the applied force should linearly scale with the Hertz parameter a^3/R, where a is the contact radius and R the particle radius and pass through the origin. From the slope of the fitting line, the Young modulus is deduced to be $E = 82$ GPa which is somewhat lower than the bulk value of silver $E_{bulk} = 104$ GPa. A good agreement of the elastic behavior of nanoparticles with their bulk counterparts was expected from the results of [2], where it was shown that at least down to about 30 nm, the Young modulus deviates only marginally for most metals. However, while the magnitude of the elastic particle behavior is close to the bulk conduct, the elastic regime seems to be largely extended for nanoparticle collisions compared to the bulk performance as shown below.

The observation that the line does not pass through the origin is an indication for a conforming contact, which means, that the particle experiences already a deformation in the contact area without external load. This is due to the weakly adhesive potential employed in the MD simulation. For higher Hertz parameters the simulation results start to deviate more and more from the elastic behavior entering the elasto-plastic regime. As discussed above only the largest nanoparticles with diameters of 15 nm reach this regime while smaller particles are not leaving the elastic regime not even for the highest velocities realized here, which were in the range up to $90\,\mathrm{m\,s^{-1}}$. The results shown in Fig. 9 lead to the surprising conclusion that the mechanical behavior of nanoparticles can be described by continuum mechanics approaches when

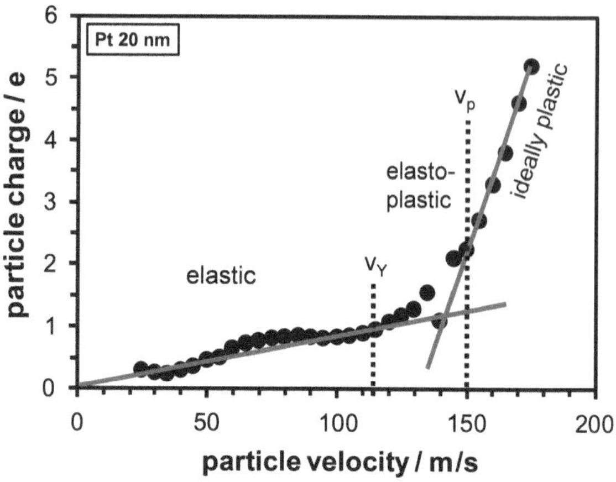

Fig. 10 Charge of the Pt particles rebounding from a hard Au impaction target: the transition from elastic via elasto-plastic to plastic behavior is indicated by different slopes. The three regimes are separated by the velocities v_Y and v_p

modifying the equations to use the real material parameters (e.g. E-modulus) and to include the adhesion which becomes significant for nanoparticle-wall collisions. The particle impaction characteristics can be described at least with statistical significance with macroscopic models down to a critical particle size. However, further investigations with deformable targets have to be performed to confirm this conclusion. To investigate the limit of the purely elastic behavior, impaction experiments with Pt nanoparticles on a hard Au target and with Ag nanoparticles on a hard Pt target, respectively, were performed at larger impaction velocities. Due to charge transfer between the conducting surfaces and the particle the influence of the particle velocity on the charge acquired by the particles could be measured. As shown in Fig. 10, a significant change of the slope indicates a change of the mechanical regime. While the transition from elastic to elasto-plastic behavior is considered to occur at at velocity v_Y when the applied pressure p_m in the contact zone exceeds the yield pressure Y by 10 %, $p_m > 1.1Y$, the transition to ideally plastic behavior is expected to appear at values of $p_m > 2.8Y$. According to the theory by Wang and John,

$$v_Y = \left(\frac{\pi^2 Y^5 (k_p + k_s)^4}{4\rho_p}\right)^{0.5} \tag{10}$$

where $k_i = \frac{1-\nu_i^2}{\pi E_i}$ with ν_i the Poisson ratio of species i and E_i as Young's modulus of species i, species being particle and substrate. Therefore, when assuming that Young's modulus and Poisson ratio are constant [50] and the substrate very hard as for $E_s \rightarrow \infty$, the yield stress Y can be calculated from the measured values of v_Y. As discussed in [3], the nanoparticles show an increased stiffness at high impaction velocities shifting the transition to plastic deformation to much higher velocities. For Ag and Pt nanoparticles the yield stress depends on the particle size as shown in Fig. 11. As expected the yield stress increases with decreasing particle size. However, the size dependence is weaker with a power law exponent equal to $m = -0.25$ than

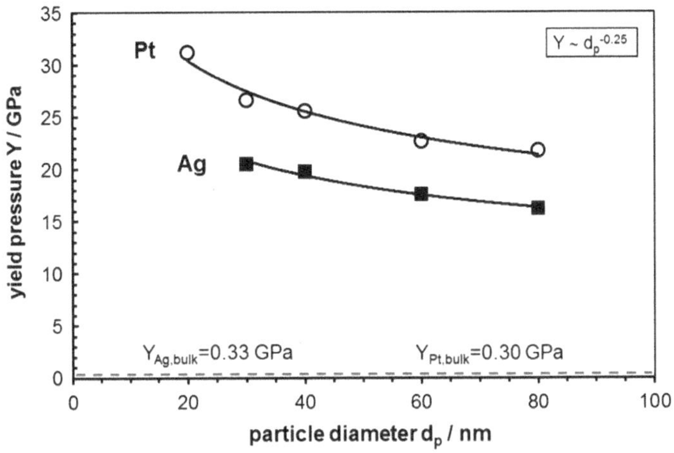

Fig. 11 Yield stress for Pt and Ag nanoparticles deduced from Eq. 9 using the velocity v_Y form the particle charge versus velocity curves which indicated the transition from elastic to elasto-plastic deformation (adapted from [10])

reported by other researchers. Kim and Greer studied the yield strength of gold nanopillars [15] and found an inverse proportionality $m = -1$ which was confirmed by Kiener and Minor for Cu pillars [14] and by Richter et al. for Cu nanowhiskers [34]. Nowak et al. measured the yield stress of silicon nanospheres [25] and found an inverse proportionality. However, these results are hardly comparable with the extremely short contact times of a few 10 ps which implies very high strain rates in the order of $10^9 s^{-1}$. In addition, the assumption of a rigid target may not be correct in evaluating Eq. 10. Consequently, as a result of the high particle hardness, the target material will yield first and absorb a part of the collision energy. Plastic deformation of the particle will not occur before the dynamic hardening of the substrate surface is high enough to reach the particle yield strength. The calculated yield strength is therefore likely to overshoot its real value. Further studies with very hard targets, such as finely sputtered sapphire or diamond substrates, will be necessary to explore the substrate influence on the resulting particle yield pressure.

Applicability of Macroscopic Theories to Describe the Mechanical Behavior of Nanoparticles in Particle-Wall Collisions

While the modified Hertz theory seems to be applicable to describe the elastic behavior of nanoparticles, the situation is less clear in the elasto-plastic regime. Several approaches have been undertaken to cover this regime with established theories. An extensive, but still manageable approach is the contact model of Tsai et al. [51].

This model assumes a normal impact of uncharged particles which are softer than the target and which exhibit uniform and evenly distributed surface roughness. In addition, the model, which considers elastic, elasto-plastic and ideally plastic deformation, assumes that the surface energy does not depend on the size of the contact area. The model starts from an energy balance for the particle [51]:

$$E_{kin1} = E_{kin2} + E_{adh} + E_{def} + E_{asp} \tag{11}$$

where E_{kin1} and E_{kin2} are the kinetic energies of the particle before (1) and after (2) the collision, respectively, E_{adh} is the adhesion energy, E_{def} is the energy for plastic deformation and E_{asp} is the energy needed to flatten the surface asperities. The energy losses during the collision are given by the last three terms in Eq. 11 so that the rebound velocity can be written as:

$$v_r = \sqrt{v_i^2 - \frac{2(E_{def} + E_{adh} + E_{asp})}{m}} \tag{12}$$

and the critical velocity for rebound, i.e. the minimum initial velocity for which bouncing occurs, as:

$$v_{cr} = \sqrt{\frac{2(E_{def} + E_{adh} + E_{asp})}{m}} \tag{13}$$

Finally, the coefficient of restitution ε_n is deduced from Eq. 12 to:

$$\varepsilon_n = \frac{v_r}{vi} = \sqrt{1 - \frac{2(E_{def} + E_{adh} + E_{asp})}{mv_i^2}} \tag{14}$$

For velocities v_i smaller than v_Y, the contact is merely elastic and the energy losses occur only due to adhesion [51]:

$$E_{adh} = \gamma\pi \left(\frac{a_{max}}{2^{2/3}}\right)^2 \tag{15}$$

where γ is the surface energy per unit area and the maximum radius of the contact circle a_{max} is obtained from the Johnson–Kendall–Roberts (JKR) theory [51]:

$$a_{max}^3 = \frac{R}{K}\left(F + 3\gamma\pi R + \sqrt{6\gamma\pi RF + (3\gamma\pi R)^2}\right) \tag{16}$$

where F is the external load, R is the particle radius and the material constant K is defined as:

$$K = \frac{4}{3}\left(\frac{1 - \nu_p^2}{E_p} + \frac{1 - \nu_s^2}{E_s}\right)^{-1} \tag{17}$$

with ν and E being the Poisson ratio and Young's modulus, respectively, for particle and surface. For a non-adhesive contact, say $\gamma = 0$, Eq. 16 reduces to the well know Hertz equation already used here to evaluate the data in Fig. 9. For $v_i > v_Y$ plastic deformation occurs. In this regime the elastically stored energy is given by:

$$E_{el} = \frac{8}{15}\frac{a_Y^5 E^*}{R^2} \tag{18}$$

where E^* is the reduced elastic modulus and a_Y the radius of the contact area, where plastic flow sets in:

$$a_Y = \frac{3}{4}\frac{1.1Y\pi R}{K} \tag{19}$$

Finally, the energy stored in the plastically deformed zone is given by [51]:

$$E_{def} = 0.55\pi Y \frac{a_Y^2}{2R}\left(a_{max}^2 - a_Y^2\right) \tag{20}$$

Equation 14 was evaluated for spherical Ag nanoparticles with ideally smooth surfaces impacting on a smooth target according $E_{asp} = 0$. Since the continuum mechanical approaches, such as the one of Tsai et al. [51] used here, do not account for effects due to particle orientation, the results can be directly compared to the experimental results which are related to the first particles that bounce. Using an oxidized aluminum surface for the well impaction experiments and adjusting the surface energy γ to $0.8\,\mathrm{J\,m^{-2}}$ and the yield pressure to 8.5 GPa, a good agreement between measurements and model calculation is obtained as shown in Fig. 12. While ε_n for particles larger than 60 nm converges towards a single curve, a flatter trend of the $\varepsilon_n = f(v_i)$ relationship is observed for smaller particles, which is due to the increasing

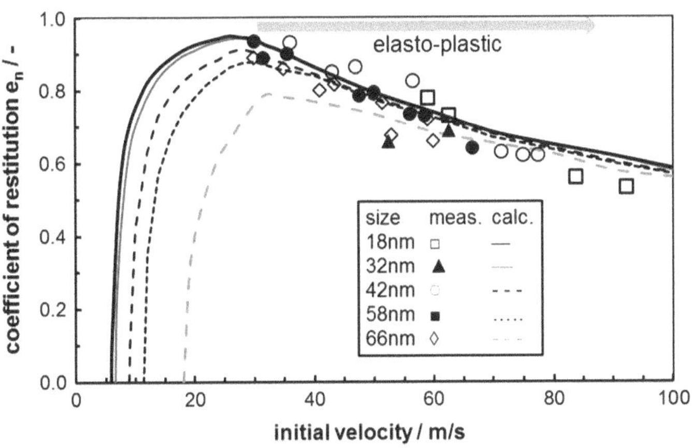

Fig. 12 Coefficient of restitution as function of initial particle velocity. Comparison of measured values with the model of Tsai et al. [51] using a surface energy of $\sigma = 0.8\,\mathrm{J\,m^{-2}}$

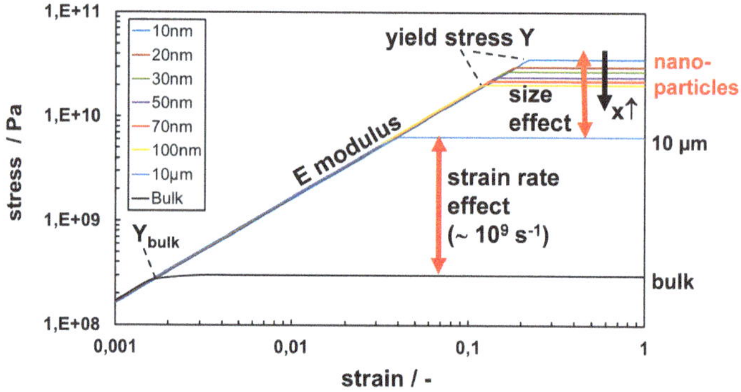

Fig. 13 Stress-strain correlation for nanoparticles: continuum parameters such as elastic modulus are still applicable for the impaction of nanoparticles but extend much further than for bulk materials due on the one hand ultra-fast collision kinetics (contact hardening by strain rate effect) and on the other hand size dependent material parameters such as yield pressure. This combination makes nanoparticles behave extremely elastic in particle-wall collisions

significance of the surface forces compared to the mass forces in the elastic regime. In the elasto-plastic regime the plastic deformation dominates against the adhesion energy and the curves approach each other. Consequently, for $v_i \gg v_Y$ the coefficient of restitution becomes independent of particle size and scales with $\varepsilon_n \sim v_i^{-0.5}$. This indicates that with the appropriate material parameters, the collision behavior of nanoparticles can also in the elasto-plastic regime be described with continuum mechanics approaches. However, certain refinements need to be introduced into the macroscopic models. For better results, the size-dependent yield pressure for instance needs to be considered, which was assumed in this first estimate as constant. However, as outlined in Fig. 11 for the yield pressure, such size-dependent parameters may, after further improvements of the technique, provide these refinements.

Based on the experimental and simulation results, a stress-strain curve for rapid collisions of nanoparticles with walls is presented in Fig. 13. It becomes clear that for not too small particles the elastic nanoparticle behavior corresponds to the continuum behavior but extends to much higher stresses before yielding is observed. A good part of this extension is due to hardening contact effects at extremely high shear rates, i.e. are not directly related to the particle size. However, there is also a size-dependent part of the yield pressure which increases with decreasing particle size cf. Fig. 11. It was shown that the collision of nanoparticles is not fundamentally different from the collision behavior of microparticles, at least for impact velocities below about 100 m s^{-1}. When the particles are larger than ca. 20 nm, their surface structure behaves as continuum and their collision behavior can reliably be described by means of continuum mechanics approaches if size dependent parameters such as the yield pressure are known. This information may either be extracted from MD simulations cf. Figs. 8, 9 or obtained from experiments cf. Fig. 10. For particles with sizes significantly smaller than 20 nm, where the collision results depend crucially on the initial particle orientation, neither continuum mechanics approaches nor experiments can recover the dependence of the coefficient of restitution, for example, on the initial particle orientation. The collision behavior of such small nanoparticles remains reserved to MD simulations as presented in the next chapters.

Introduction and Characterization of a Single Parameter Description of the Lattice Orientation of Nanoparticles

Initial Orientation of the Particle and the Parameter Ω

To investigate the orientation dependency, we determined 1,000 uniformly distributed random rotations of the particle with respect to the wall. Considering the embedding sphere of the cluster, the aim of this rotation is to assure that (a) each point of the sphere is located at the south pole with equal probability, that is, when the particle moves in negative z-direction, each point has the same probability to touch the target plane first, and (b) the angle between the orientations of the lattices of the target material and the particle is equally distributed. Such a transformation is achieved using the method by Miles [20], where a rotation axis is determined by the center of the sphere embedding the particle and a randomly chosen point on its surface [18]. The particle is then rotated around this axis by the random angle α with probability density

$$p(\alpha) = \frac{2}{\pi} \sin^2 \frac{\alpha}{2}; \qquad \alpha \in [0, \pi]. \tag{21}$$

In the following, the orientations are characterized by the orientation parameter Ω which arises from the coordinate transformation of the load direction from reference into crystal coordinate system, see Fig. 14:

$$\Omega(\gamma_1, \gamma_2, \gamma_3) \equiv \gamma_1^2\gamma_2^2 + \gamma_2^2\gamma_3^2 + \gamma_3^2\gamma_1^2 \tag{22}$$

$$= \left(\cos^2\phi \sin^2\phi - 1\right)\sin^4\theta + \sin^2\theta. \tag{23}$$

In Eq. 22, γ_i are the direction cosines from the coordinate transformation and ϕ, θ the axes of the spherical coordinate system, see Fig. 14.

Before describing the properties of Ω, let us discuss the symmetry of the problem: The projection of a cubic unit cell onto the the unit sphere delivers 48 spherical triangles as can be seen from Fig. 15 (left). These triangles are equivalent due to the symmetry of the fcc structure, see Fig. 15 (right).

Therefore, Ω is completely determined on the whole sphere by its values on the spherical triangle bound by the points $(0, 0, 1)$, $(1, 0, 1)/\sqrt{2}$, $(1, 1, 1)/\sqrt{3}$, which we call *critical triangle* in agreement with the literature, e.g. [36], see Fig. 15 (left).

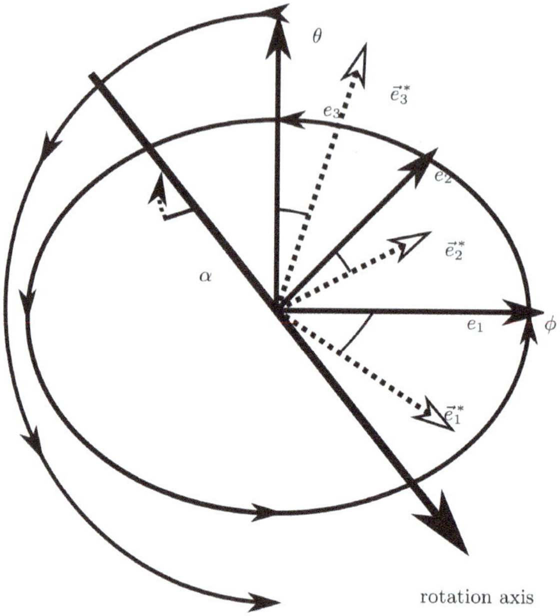

Fig. 14 Definition of the polar and azimuthal spherical coordinate system angles, θ and ϕ (curved line arrows), and the canonical standard coordinate system vectors, $\mathbf{e}_1 \equiv (1, 0, 0)$, $\mathbf{e}_2 \equiv (0, 1, 0)$, $\mathbf{e}_3 \equiv (0, 0, 1)$. By application of the orthogonal rotation given by the random rotation axis and the angle α, it is transformed into the coordinate system $\{\mathbf{e}_1^*, \mathbf{e}_2^*, \mathbf{e}_3^*\}$. The direction cosines used in Eq. 22 are given by $\gamma_i \equiv \mathbf{e}_i \cdot \mathbf{e}_i^*$

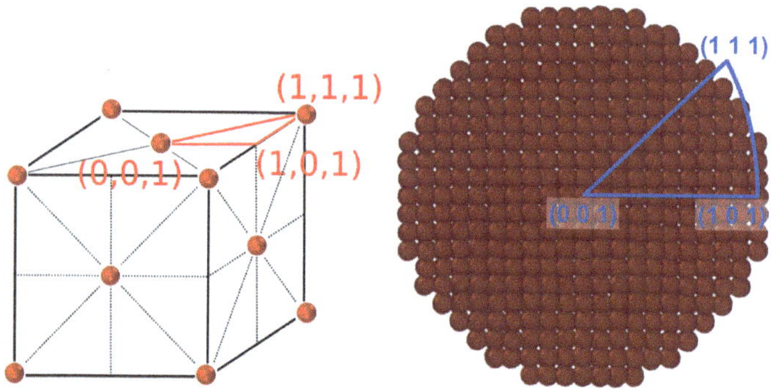

Fig. 15 Definition of the *critical triangle*. Left: the projection of the cubic unit cell onto the unit sphere delivers 48 spherical triangles. Right: layer of atoms located at the center of the particle in initial (non-rotated) position with the critical triangle superimposed (blue). Due to the symmetry of the fcc structure, all 48 triangles are equivalent

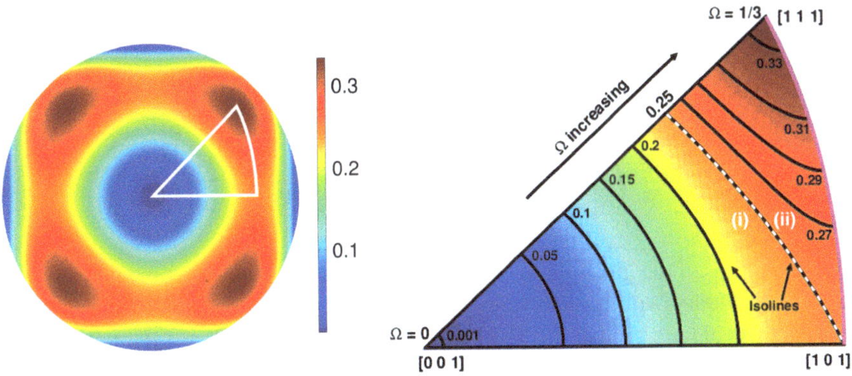

Fig. 16 Left: the function $\Omega\,(\phi,\theta)$ on the unit sphere. The position of the critical triangle is indicated. Right: on the critical triangle, Ω is a strictly monotonous function of θ and ϕ. For the computation of the probability distribution, $P_\Omega(\Omega)$, we divide the critical triangle into the areas (i) and (ii), separated by the isoline $\Omega = 1/4$ (dashed line). The pink line shows the big arc used for the computation of the distribution function for area (ii), Eq. 29, see the text for explanation

The Probability Measure P_Ω

To obtain the cumulative probability distribution, $\mathrm{P}_\Omega(\Omega)$, which will be used to analyze the properties of the impact with respect to the orientation of the particle, we calculate the integrals $\mathrm{P}_\Omega(\Omega < x)$, $0 \le x \le \frac{1}{3}$ exploiting the symmetry of the fcc structure, that is, by reducing the problem to the calculation of the distribution on the critical triangle.

On the critical triangle, Ω depends strictly monotonously on θ and ϕ with $\Omega\,((0,0,1)) = 0$, $\Omega\left((1,0,1)\,/\sqrt{2}\right) = 1/4$, and $\Omega\left((1,1,1)\,/\sqrt{3}\right) = 1/3$, see Fig. 16. For convenience of integration, we divide the critical triangle into the areas (i) and (ii) separated by the isoline $\Omega = 1/4$, see the dashed line in Fig. 16:

1. $0 \le \Omega \le \frac{1}{4}$
2. $\frac{1}{4} < \Omega \le \frac{1}{3}$.

The integration domain due to area (i) can characterized as limited by $0 \le \phi \le \pi/4$ and $0 \le \theta \le \tilde{\theta}$, where $\tilde{\theta}$ is a function of Ω and ϕ. This function is obtained by solving Eq. 23 for θ with respect to Ω and ϕ in this critical triangle:

$$\tilde{\theta}(\Omega,\phi) = \arcsin\sqrt{\frac{-1+\sqrt{1+4\Omega(\mathrm{cs}^2\phi-1)}}{2(\mathrm{cs}^2\phi-1)}}\,, \tag{24}$$

with $\mathrm{cs}^2\phi \equiv \cos^2\phi\,\sin^2\phi$. The tilde in Eq. 24 indicates that this solution is restricted to the critical triangle. Using the identity $\cos(\arcsin(x)) = \sqrt{1-x^2}$ and taking into account that the area of one triangle is $\frac{4\pi}{48}$, we obtain inside area (i)

$$
\begin{aligned}
P_\Omega\left(\Omega \leq \frac{1}{4}\right) &= \frac{48}{4\pi}\int\limits_0^{\frac{\pi}{4}}\int\limits_0^{\tilde{\theta}(\Omega,\phi)} \sin(\theta)\,\mathrm{d}\theta\,\mathrm{d}\phi \\
&= \frac{12}{\pi}\int\limits_0^{\frac{\pi}{4}}\left[1-\sqrt{1-\frac{-1+\sqrt{1+4\Omega(\mathrm{cs}^2\phi-1)}}{2(\mathrm{cs}^2\phi-1)}}\right]\mathrm{d}\phi\,.
\end{aligned}
\tag{25}
$$

The boundaries of the integrals corresponding to area (ii) are more complicated since here the boundary with respect to ϕ depends on Ω. This third side of the triangle is part of the big arc containing $(1, 0, 1)/\sqrt{2}$ and $(1, 1, 1)/\sqrt{3}$, see the pink line in Fig. 16 (right). Therefore, a normal vector to it is given by $(1, 0, -1)$, concluding $x = z$ on this side of the triangle. Now let θ^*, ϕ^* be the restrictions of the spherical coordinates to this boundary to area (ii). Then

$$\cos\left(\phi^*\right)\sin\left(\theta^*\right) = \cos\left(\theta^*\right), \tag{26}$$

$$\theta^*(\phi) = \operatorname{arccot}\left(\cos\left(\phi^*\right)\right). \tag{27}$$

Inserting Eq. 27 into Eq. 23 and rearranging delivers the boundary of area (ii) (pink line in Fig. 16 (right)) as a pure function of Ω,

$$\phi^*(\Omega) = \arccos\sqrt{\frac{1-\Omega+\sqrt{1-3\Omega}}{1+\Omega}}. \tag{28}$$

Consequently, the limits of area (ii) are given by $\phi^*(\Omega) \leq \phi \leq \frac{\pi}{4}$ and $\tilde{\theta}\left(\Omega = \frac{1}{4}, \phi\right) \leq \theta \leq \tilde{\theta}(\Omega, \phi)$. For the computation of the remaining part of the distribution function (area (ii)) we exploit the just denoted formula for $P\left(\Omega = \frac{1}{4}\right)$, and write

$$
\begin{aligned}
P_\Omega\left(\Omega > \frac{1}{4}\right) &= P_\Omega\left(\Omega = \frac{1}{4}\right) + \frac{12}{\pi}\int\limits_{\phi^*(\Omega)}^{\frac{\pi}{4}}\int\limits_{\tilde{\theta}(\frac{1}{4},\phi)}^{\tilde{\theta}(\Omega,\phi)} \sin(\theta)\,\mathrm{d}\theta\,\mathrm{d}\phi \\
&= P_\Omega\left(\Omega = \frac{1}{4}\right) + \frac{12}{\pi}\int\limits_{\phi^*}^{\frac{\pi}{4}}\left[\sqrt{\frac{2\mathrm{cs}(\phi)+1}{2\left(\mathrm{cs}(\phi)+1\right)}} - \sqrt{1-\frac{-1+\sqrt{1+4\Omega(\mathrm{cs}^2\phi-1)}}{2(\mathrm{cs}^2\phi-1)}}\right]\mathrm{d}\phi\,.
\end{aligned}
\tag{29}
$$

The derivatives of the integrals in Eqs. 25 and 29 can be calculated analytically to yield the probability density, $p_\Omega(\Omega)$. Since this expression is rather cumbersome, for convenient practical application we provide a fit to the ansatz

$$p_\Omega^{\mathrm{fit}}(\Omega) = a\ln\left|\Omega - \frac{1}{4}\right| + b\,\Omega + c\,, \tag{30}$$

where a, b, c are real numbers for both sides of the singularity at $\Omega = \frac{1}{4}$:

$$p_\Omega^{\text{fit}}(\Omega) = \begin{cases} -1.402 \ln\left(\frac{1}{4} - \Omega\right) - 2.493\,\Omega - 0.4388 & \text{for } \Omega < \frac{1}{4} \\ -1.328 \ln\left(\Omega - \frac{1}{4}\right) + 2.989\,\Omega - 1.287 & \text{for } \Omega > \frac{1}{4}\,. \end{cases} \tag{31}$$

Integrating $p_\Omega^{\text{fit}}(\Omega)$ delivers handy equations for $P_\Omega^{\text{fit}}(\Omega)$:

$$P_\Omega^{\text{fit}}(\Omega) = \begin{cases} \Omega\,(0.9632 - 1.2465\Omega) + \\ \quad +1.402\left(\frac{1}{4} - \Omega\right)\ln\left(\frac{1}{4} - \Omega\right) + 0.4859 & \text{for } \Omega \leq \frac{1}{4} \\ \left(\Omega - \frac{1}{4}\right)\left[1.4945\left(\Omega - \frac{1}{4}\right) - 1.328\ln\left(\Omega - \frac{1}{4}\right) + 0.78825\right] + \\ \quad +0.6489 & \text{for } \Omega > \frac{1}{4}\,. \end{cases} \tag{32}$$

We wish to point out that this fit is universal for the probability density of Ω for a randomly rotated particle impacting the plane. It is independent of any material properties but only restricted to the fcc lattice structure. The quality of the fit can be assessed in Fig. 17 (top) showing the analytical solution for the probability density $p_\Omega(\Omega)$, according to Eq. 29 together with the fit given in Eq. 31 and the results of a Monte Carlo sampling. The coefficient of determination (R^2-value) of the fit is $R^2 \geq 0.999$ for both parts with 10^7 uniformly distributed sampling points on $\left[\delta, \frac{1}{4} - \delta\right]$ and $\left[\frac{1}{4} + \delta, \frac{1}{3} - \delta\right]$. The value $\delta = 10^{-6}$ is needed to deal with the discontinuity of the density such that near the pole about 6×10^{-6} of the total range of Ω remains unsampled, which is good enough for all practical considerations. The curves are plotted together with the values obtained for the 1,000 random orientations

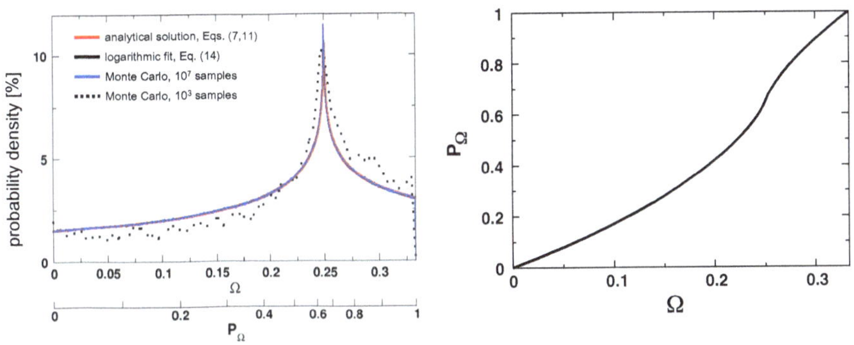

Fig. 17 Left: probability density, $p_\Omega(\Omega)$, of a randomly rotated particle. The figure shows the analytical solution, Eq. 29, the logarithmic fit, Eq. 31, and the results of a Monte Carlo sampling (in extension of the 1000 random orientations, a total of 10^7 random orientations were determined to check coincidence with the other curves). The simulation data set coincides up to good agreement with the other data sets. The analytical solution, Eq. 29, the fit, Eq. 32 and the MC sampling even up to line width. Right: corresponding cumulative probability, $P_\Omega(\Omega)$. The function $P_\Omega(\Omega)$ is used to draw the second horizontal axis in the top figure

of the particle shown in Fig. 17. The bottom panel of Fig. 17 shows the cumulative probability distribution, $P_{\Omega}(\Omega)$, according to Eqs. 25 and 29 which we will use in the subsequent text.

Impact Properties of Nanoparticles in Dependence of Their Lattice Orientation

Characteristics of Inelastic Interaction

We investigated the impact of a particle of random orientation and position as described above. In particular, we consider four characteristics of inelastic collisions, that is, dissipative interaction:

1. the amount of plastic deformation,
2. the maximal contact force, F_{max},
3. the coefficient of normal restitution, e_{n}, and
4. the sticking probability, p_{s}.

All of these characteristics of the crystalline particle are intimately related to plane gliding. Figure 18 shows the 4 slip planes and corresponding 3 directions to each slip plane, amounting to a total of 12 slip directions. The four colors belong to the different layers of slip planes. The sensitivity of a crystalline particle against sliding due to stress in a certain direction is characterized by the Schmid factor [36, 37]:

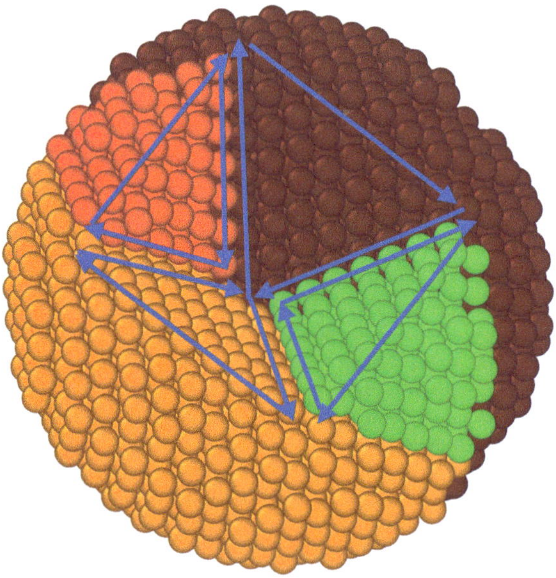

Fig. 18 Slip planes and slip directions of the fcc structure inside of a particle. The four colors distinguish the different stacks of fcc slip planes

according to Schmid's law, the critical resolved shear stress, τ, relates to total stress, σ, applied to a material in a certain direction via $\tau = \sigma m = \sigma \cos\varphi \cos\vartheta$ where φ is the angle of the stress, σ, with the glide plane and ϑ is the angle of the stress, σ, with the glide direction. The Schmid factors are then defined as $2\cos\varphi\cos\vartheta$ with the corresponding values of φ, ϑ. As the Schmid factors, especially the largest and second largest, are key parameters to characterize plastic deformation of crystalline materials under stress, in many places we will refer to these numbers. We will show, however, that for the description of the impact dynamics of nano-scale particles considered here, Ω is more significant that the largest Schmid factor.

Plastic Deformation

We quantify the plastic deformation of a particle due to an impact by the number of atoms which change their neighborhood relations. The neighborhood of an atom is defined by the set of other atoms located in a sphere of radius 1.5 next neighbor distances of the lattice and the neighborhood relations of an atom is called changed if the set of neighbors before the impact differs from the set after the impact. Because of the finite temperature of the impacting particle there is a certain thermal noise in the neighborhood, concerning in particular the atoms close to the surface whose total binding energy is low. The average amount of atoms which change their neighborhood due to thermal motion amounts to approximately 3 for the parameters used. Figure 19 shows typical examples of particles at the instant of maximal compression when the center of mass velocity changes its direction. Rows in Fig. 19 correspond to the same impact velocity, v_i, columns correspond to the same value of P_Ω characterizing the angular orientation. The degree of plastic deformation is coded by color.

Figure 20 shows the plastic deformation as a function of the impact velocity v_i and the orientation measure, P_Ω. The data points are sampled with increments of $\Delta v_\mathrm{i} = 10\,\mathrm{m/s}$ and $\Delta P_\Omega = 0.025$. For each data point, (v_i, P_Ω), we averaged over 1000 impacts at different orientations all characterized by the same values of v_i and P_Ω.

As expected, the degree of plastic deformation increases with increasing impact velocity. From the plot Fig. 20 (bottom), which is averaged with respect to velocity, we see that for $v_\mathrm{i} = 10\,\mathrm{m/s}$, the amount of plastic deformation is much higher for $\mathrm{P}_\Omega > 0.5$ as compared to $\mathrm{P}_\Omega < 0.5$. A band of high relative plastic deformation moves to lower values of P_Ω with increasing v_i until $v_\mathrm{i} \approx 100\,\mathrm{m/s}$. This can be understood from the fact that the particle surface is not perfectly spherical due to its crystalline structure: In the [1, 1, 1] direction, corresponding to $\mathrm{P}_\Omega = 1$ ($\Omega = 0.\overline{3}$) and the [0, 0, 1] direction, corresponding to $\mathrm{P}_\Omega = 0$ ($\Omega = 0$), see Fig. 16, the surface of the particle is terminated by very small portions of crystal planes. For $\mathrm{P}_\Omega = 1$, the three outermost layers contain 12, 61 and 102 atoms and are of maximal planar density. In contrast, for $\mathrm{P}_\Omega = 0$, the three outermost layers contain 32, 69 and 88 atoms, and the layers are of sub-maximal packing density. At low impact velocities, plastic deformation develops in form of irreversible plane gliding, that is, shearing of

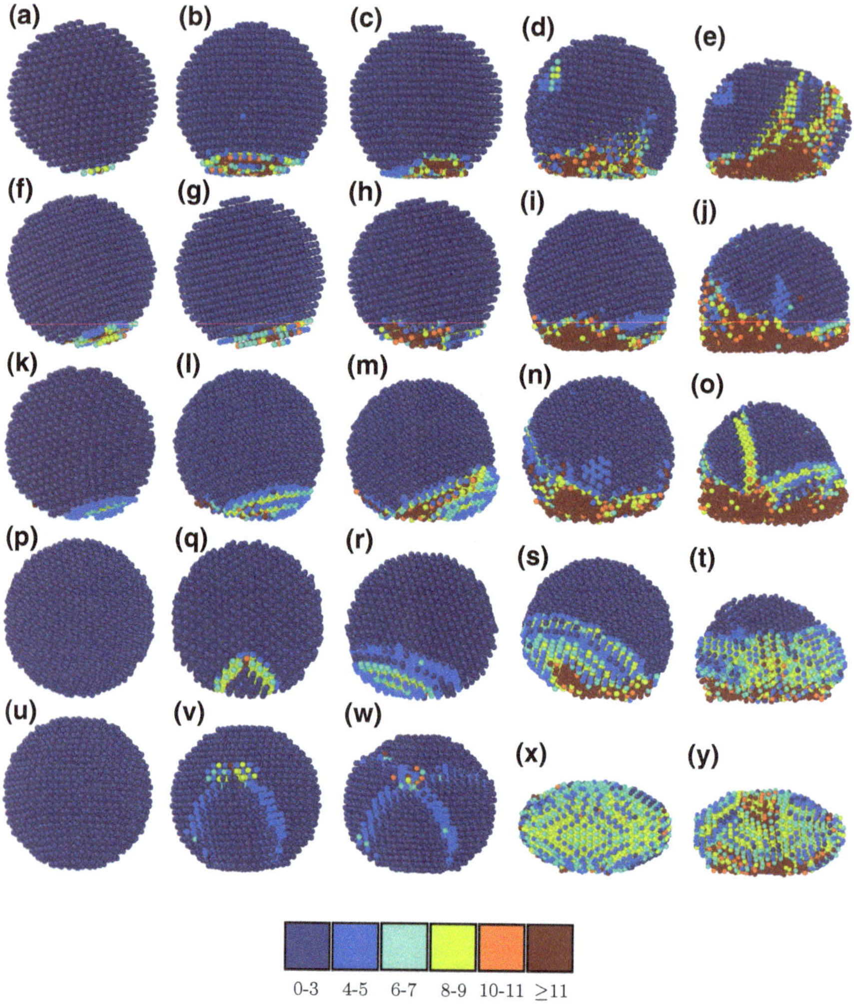

Fig. 19 The degree of plastic deformation of a particle impacting a plane in perpendicular direction depends on the impact velocity, v_i, and the relative orientation of the lattice structures of the particle and the plane. The figure shows examples of particles impacting the plane at $v_i = (40, 100, 150, 300, 400)$ m/s (columns from left to right) and at angular orientation characterized by $P_\Omega = 1, 0.75, 0.5, 0.25, 0$ (rows from top to bottom). The images show the particles at the instant of maximal compression when the center of mass velocity changes its direction. The number of changed neighbors of the atoms is coded by color. The labels (a–y) refer to the marks in Figs. 20, 22 and 24

the outermost layers, since atoms located in these layers have only one neighboring crystal layer. Since plane gliding happens only in planes of maximal planar density, the amount of plastic deformation is bigger for $P_\Omega > 0.5$ as compared with the

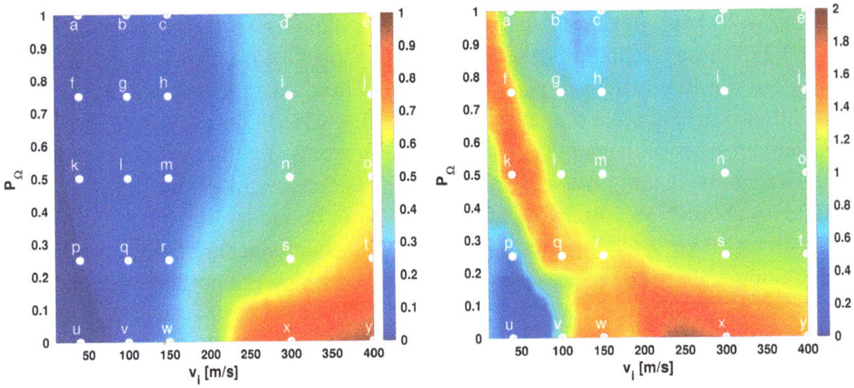

Fig. 20 Left: plastic deformation as a function of the impact velocity, v_i, and the orientation, P_Ω. Color codes for the fraction of atoms with changed neighborhood. Right: the same data but normalized for each velocity individually. The color indicates the plastic deformation (fraction of atoms with changed neighborhood) normalized by the plastic deformation at the given velocity but averaged over all orientations, Ω. The marks (a–y) refer to the labels in Fig. 19 showing a representative of an impact with the corresponding (v_i, Ω) combination. The labels (a–y) refer to the marks in Fig. 19

Fig. 21 Averaged values of the largest six Schmid factors S_i (P_Ω) with respect to P_Ω

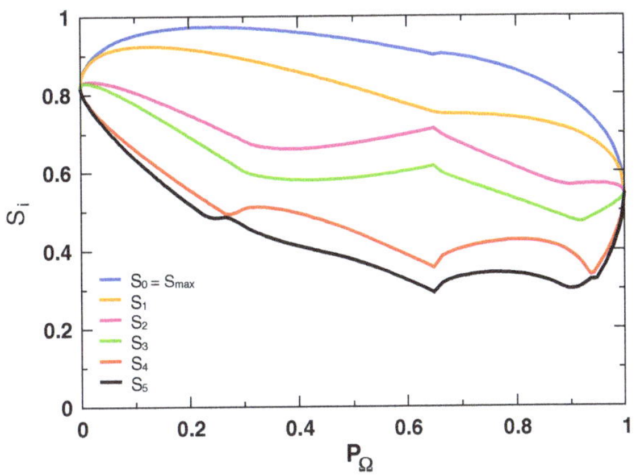

orientations $P_\Omega < 0.5$. Consequently, as can be seen from Fig. 21, the largest Schmid factor is always larger than 0.8, therefore, for the cases $P_\Omega < 0.5$, the plane gliding is reversible and happens more on the inside of the particle, see Fig. 19 bottom left images.

As velocity increases, the force due to the impact causes plastic deformation also for orientations corresponding to smaller P_Ω. The mentioned band structure comes from the fact that the shear angle for the outermost layer is equal to 0 for $P_\Omega = 1$ and increases for orientations with smaller values of P_Ω such that the average number of the dislocated atoms increases with v_i due to increased impact energy and, thus, the number of dislocated atoms with larger P_Ω decreases relative to the average.

For $v_i < 100$ m/s, the lowest values of P_Ω show almost no plastic deformation (see Fig. 20). For these orientations, the stress due to impact leads only to reversible plane gliding but not to plastic, say persistent deformation. At $v_i \approx 100$ m/s, we observe a transition of irreversible plane gliding due to increased impact energy, leading to persistent changes of the neighborhood for many atoms simultaneously. Essentially, two cases can be distinguished: Either a single dislocation travels through the entire particle on a certain slip direction, and hits the other boundary of the particle, or two dislocations hit each other to also generate a persistent stacking fault, see Fig. 20. At this point, surface or close-to-surface effects become unimportant regardless the orientation, since the number of atoms changing their neighbors due to irreversible plane gliding, be it shearing of the outermost layers or stacking faults in the inside or a mixture.

The behavior at larger impact velocity can be understood from the discussion of the Schmid factors characterizing the sensitivity of a crystalline particle against sliding due to stress in a certain direction, see section "Characteristics of Inelastic Interaction. For slow forcing and given orientation, the largest Schmid factor determines whether slip occurs, where a minimum of 45° between impact plane and crystal layer is required classically. Since Ω respectively its probability measure P_Ω describe the orientation of the crystalline structure of the particle with respect to the target, obviously, the maximal Schmid factor, S_{max} and Ω must be related, see Fig. 21. The relation between S_{max} and Ω is not a mathematical function since several orientations Ω and respectively P_Ω belong to the same value of S_{max} and vice versa. Such a relation exists only for the sum of all Schmid factors of the fcc lattice:

$$\Omega = \frac{1}{2} - \frac{3}{8}\sum_i S_i^2 \tag{33}$$

Nevertheless, Fig. 21 shows that the six largest Schmid factors grow from $P_\Omega = 1$ to $P_\Omega = 0$, except for some intervals where the S_i are nearly constant and some rather short intervals where they even decrease. Thus, as a rule of thumb, small values of Ω respectively P_Ω correspond to *good slip systems*, that is, only small deformation due to compression is required to activate a second, third or fourth slip plane. Therefore, for orientations corresponding to large values of P_Ω, stress is released by shear of the outermost layers, that exhibit quite weak slip systems, but only one layer is neighboring, weakening the cohesive attraction. This effect is of microscopic nature and cannot be observed for macroscopic impaction which is implied by weak adhesion and very high volume to surface ration. While the maximum Schmid factor characterizes slip for slow (quasi-static) deformation, it is not entirely adequate for stress due to impact at high velocity as the dynamics is due to shocks and other non-equilibrium effects. As a consequence not only the slip plane corresponding to the maximum Schmid factor is activated but also other slip planes related to other Schmid factors (in particular the second largest) become activated. Moreover, close to the contact zone, atoms leave their fcc lattice positions and are densified. This process consumes a lot of energy and thereby the total amount of atoms getting plastically deformed depends less on v_i as compared to smaller values of P_Ω.

Starting from $v_i \approx 200$ m/s, this effects becomes dominant for the lowest 15% of P_Ω, where many slip planes are activated, causing plastic shear deformation additionally to the irreversible plane gliding. Eventually at $v_i = 300$ m/s, almost all atoms are involved in plastic deformation for this part of the distribution. As velocity is further increased, the impact energy becomes so large that most of the fcc structure is converted upon impact. This deformation causes local transformations of the crystal structure leading to mostly bcc, corresponding to larger values of free energy and a more compact and, thus, pressure resistant unit cell.

Before discussing the main macroscopic characteristics of the impact, the maximal contact force, F_{max}, the coefficient of restitution, e_n, and the sticking probability, p_s, quantitatively in dependence on the orientation of the impact, we wish to remind the significance of Ω: Obviously, the unique description of the orientation of the particle needs two parameters, θ and ϕ, see Fig. 14. However, as we show here, certain combinations of θ and ϕ lead to the same macroscopic behavior of the impact, characterized by F_{max}, e_n, and p_s. It turns out that the two dimensional manifold (θ, ϕ), may be expressed by a one-dimensional manifold, Ω. That is, impacts characterized by the same value of Ω reveal the same characteristics, despite they correspond to different combinations (θ, ϕ). The reason for this mapping is the fact that not θ and ϕ directly, but the Schmid factors (in particular the two largest ones) are responsible for the impact behavior, supported by Eq. 33, expressing Ω in terms of the Schmid factors.

Maximal Contact Force, F_{max}

The maximal interaction force during a collision as a function of impact velocity, v_i, and orientation, P_Ω, is shown in Fig. 22. For $v_i \lesssim 100$ m/s, the orientation dependent details of the crystalline structure at the contact point have a significant influence on the interaction force. This can be understood from the following argument: For $P_\Omega = 0$, a crystal plane of sub-maximal planar density is parallel to the impact plane and the largest Schmid factor is $\frac{2}{3}\sqrt{2} \approx 0.82$. The two largest Schmid factors increase with P_Ω up to about $P_\Omega \approx 0.2$, see Fig. 21. Therefore, for decreasing values of $P_\Omega \lesssim 0.15$, stress transmission is getting weaker, explaining the island of high maximal contact force $F_{max}(P_\Omega \lesssim 0.15, v_i \lesssim 100\text{m/s})$. For $P_\Omega = 1$, the outer layer is of maximal crystal density and oriented parallel to the wall, where the two largest Schmid factors assume the value $\frac{\sqrt{6}}{9} \approx 0.54$. Since the shear angle is also close to 0 and consequently, the impact affects the entire impacting body, that is, the stress leads to a transfer of linear momentum into angular momentum leading to particle rotation, see Fig. 19 top left. This explanation emphasizes that the detailed shape of the surface of the particle due to its crystalline structure affects the impact behavior significantly for low impact rate, in particular for the orientations described

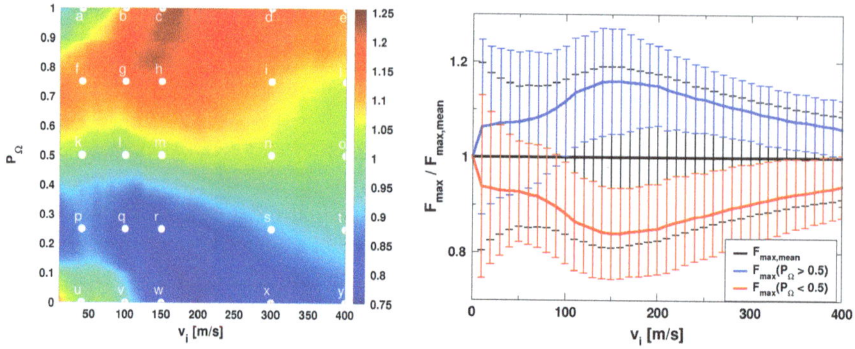

Fig. 22 Left: maximal contact force, $F_{\max}(v_{\mathrm{i}}, \mathrm{P}_{\Omega})$, normalized with respect to velocity. The labels (a–y) refer to the marks in Fig. 19. Right, blue curve: normalized maximal force, $\langle F_{\max}\rangle_{\mathrm{P}_{\Omega}\geq 0.5}(v_i)/\,\langle F_{\max}\rangle_{\mathrm{P}_{\Omega}}(v_i)$ where $\langle F_{\max}\rangle_{\mathrm{P}_{\Omega}\geq 0.5}(v_i)$ stands for the average over all impacts at v_i and orientations with $\mathrm{P}_{\Omega} \geq 0.5$ and $\langle F_{\max}\rangle_{\mathrm{P}_{\Omega}}(v_i)$ stands for the average over all impacts at v_i and all orientations, P_{Ω}. Red curve: same for $\mathrm{P}_{\Omega} < 0.5$. Vertical lines are error bars indicating the standard deviations. The color of the error bars correspond to the average data of the same color. Black curve and error bars correspond to all orientations. Here, the average is identical unity, of course

by $\mathrm{P}_{\Omega} \gtrsim 0.85$ and $\mathrm{P}_{\Omega} \lesssim 0.15$. This effect is less significant for larger impact rate and also for other values of P_{Ω}, as the shear angle increases and starting from the second sheared crystal layer, its already a form of irreversible plane gliding. From this argument, we can understand the large values of the maximal contact force, $F_{\max}\left(\mathrm{P}_{\Omega} \gtrsim 0.85,\, v_{\mathrm{i}} \lesssim 100\,\mathrm{m/s}\right)$, visible in Fig. 22.

Starting at $v_{\mathrm{i}} \approx 100\,\mathrm{m/s}$, plane gliding becomes dominant and, therefore, the properties of the slip system, characterized by the largest Schmid factors govern the impact behavior. Here, the pertinence of the here introduced parameter Ω characterizing the impact behavior becomes particularly obvious, since the description of the particle behavior upon impact via the single measure P_{Ω} becomes particularly simple as compared to the description by the Schmid factors, see Fig. 22 for $v_{\mathrm{i}} \gtrsim 100\,\mathrm{m/s}$. We will first explain the behavior of the system by means of the Schmid factor formalism and subsequently restate the argument in terms of Ω. For $v_{\mathrm{i}} \approx 120 - 200\,\mathrm{m/s}$, the relative values for the contact force span from $75 - 125\%$, the largest observed value.

Having in mind the mechanism of plastic deformation in dependence on P_{Ω}, we can easily understand the relative maximal contact force $F_{\max}^{\mathrm{rel}}(v_{\mathrm{i}}, \mathrm{P}_{\Omega})$, shown in Fig. 22 (top): A decreasing value of P_{Ω} describes the increasing activation of the slip systems of the fcc structure. Therefore, the particles' resistance against volume shear/plane gliding decreases from $\mathrm{P}_{\Omega} = 1$ to $\mathrm{P}_{\Omega} = 0.5$ and so does the maximal contact force. The same argument holds for values $\mathrm{P}_{\Omega} = 0$ to $\mathrm{P}_{\Omega} = 0.5$. The values of the largest and second largest Schmid factors are large for small values of P_{Ω}, where even the largest value becomes mediocre for values close to $\mathrm{P}_{\Omega} = 0.5$. For large shear stress, shear along a single plane corresponding to the largest Schmid factor is not sufficient to release all stress and, thus, other shear planes are activated corresponding to the second largest and further Schmid factors.

When the impact velocity is further increased, dislocation emission can be observed for all orientations, see Fig. 19 right columns. Additionally, we notice flattening of the contact area regardless of the orientation, due to very large impact energy. In this region, thus, we observe a combination of compression and plastic shear. For such impact parameters, the variation of the relative contact force decreases. The description of the orientation by P_Ω allows to subdivide the possible particle orientation into families of classes revealing similar behavior. For example, Fig. 22 (bottom) shows the maximal contact force as a function of the impact velocity, normalized by the average value for all orientations for impacts with the same velocity. Thus, the average normalized force assumes the value 1 for all velocities, by definition. If we plot the data separately for classes of orientations belonging to $P_\Omega < 0.5$ and $P_\Omega > 0.5$ (red and blue lines), we see that P_Ω indeed classifies the orientations in a meaningful way. This can be quantified by the standard deviations of the cases $P_\Omega < 0.5$ and $P_\Omega > 0.5$ which are much smaller than the standard deviation of the averaged data (black line).

After this detailed discussion, we can restate the arguments by means of the novel parameter Ω: *The orientation* $P_\Omega = 0$ *stands for good slip systems corresponding to small contact force at impact. With increasing* P_Ω *the particle behaves more and more rigid since slip in the volume of the particle becomes more and more unfeasible. Thus, maximum contact force is achieved for orientations corresponding to* $P_\Omega = 1$ *where plastic deformation dominates.* The relevance of the parameter P_Ω can also be seen in the examples shown in Fig. 19. In conclusion, the introduction of Ω allows for a convenient one-parameter description of the impact behavior.

Coefficient of Restitution, e_n, and Sticking Probability, p_s

The coefficient of normal restitution, e_n, defined as the ratio of the normal components of the rebound velocity and the impact velocity, and the sticking probability, p_s, at which the rebound velocity drops to zero, are important global characteristics of a particle impacting a plane. Here we discuss the dependence of these parameters on the impact velocity and in particular on the orientation of the particle prior to impact, shown in Fig. 23. For small impact velocity up to about $100 - 200$ m/s, the coefficient of restitution reveals large scatter indicated by large error bars (variance). This is again due to the crystalline structure of the particle and the variations of the slip properties in dependence on the orientation of the crystalline particle structure with respect to the target. Similar as this orientation characterized by the Schmid factors or Ω, respectively, has large effect on the interaction force discussed at length in section "Maximal Contact Force, F_max", it affects also the global properties, e_n and p_s. This coincidence appears quite natural as the coefficient of restitution is a direct consequence of the interaction force, that is, given the interaction force as a function of impact velocity, the coefficient of restitution can be derived by

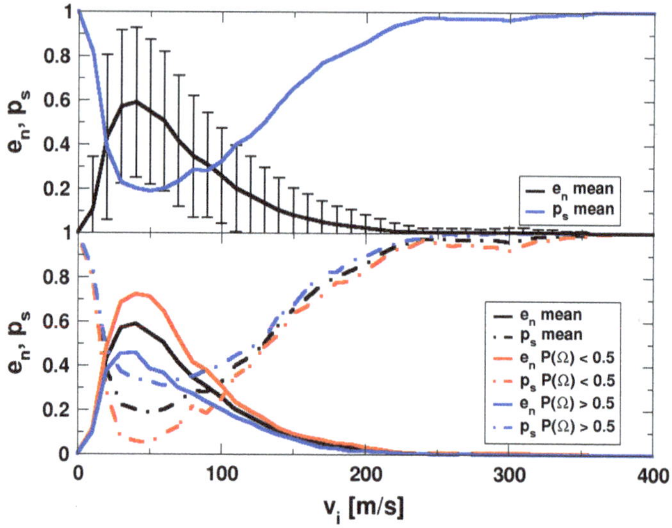

Fig. 23 Coefficient of normal restitution, e_n, and sticking probability, p_s, as functions of the impact velocity. Top: expectation values of e_n and p_s averaged over all orientations and corresponding error bars. Bottom: expectation values of e_n and p_s evaluated separately for arbitrary P_Ω (black lines), $P_\Omega \le 0.5$ (red lines) and $P_\Omega \ge 0.5$ (blue lines). See the text for discussion

integrating Newton's equation of motion. Examples for such analytical calculation for homogeneous (non-crystalline) materials have been done for viscoelastic spheres [29, 40, 42] and cylinders [39], simplified linear dashpot forces [41] and adhesive viscoelastic materials [4, 5].

The coefficient of restitution as a function of the impact velocity, averaged over all angular orientation of the impacting particle as drawn in upper part of Fig. 23 reveals large error bars indicating a variance of the same size as the data itself. Analyzing the data separately for $P_\Omega < 0.5$ and $P_\Omega > 0.5$ (bottom part), we obtain a clear separation of the data, that is, the average values for $P_\Omega < 0.5$ and for $P_\Omega > 0.5$ are well separated, indicating that P_Ω indeed groups families or angular orientations of similar behavior caused by the slip properties along different directions in the crystalline particle material. Figure 24 depicts this property of the Ω-description in more detail. In the top panel, we can clearly see very different behavior for the families of orientation, $P_\Omega < 0.5$ and $P_\Omega > 0.5$ which is particular prominent for small impact velocity, $v_i \lesssim 100\,\text{m/s}$, in agreement with the integral presentation shown in Fig. 23 bottom. Obviously, at large impact velocity, $v_i \gtrsim 150\,\text{m/s}$, a lot of irreversible, thus dissipative, shear takes place at any orientation, leading to small coefficient of restitution. In this velocity interval, the expectation values for $P_\Omega < 0.5$, $P_\Omega > 0.5$, and all values of P_Ω do not deviate much and the variance is small, see Fig. 23 bottom.

The situation is different for small impact velocity, $v_i \lesssim 100\,\text{m/s}$. Here the orientation of the particle prior to impact makes a significant difference. For small P_Ω, slip is active already for small stress (good slip system), thus, much energy

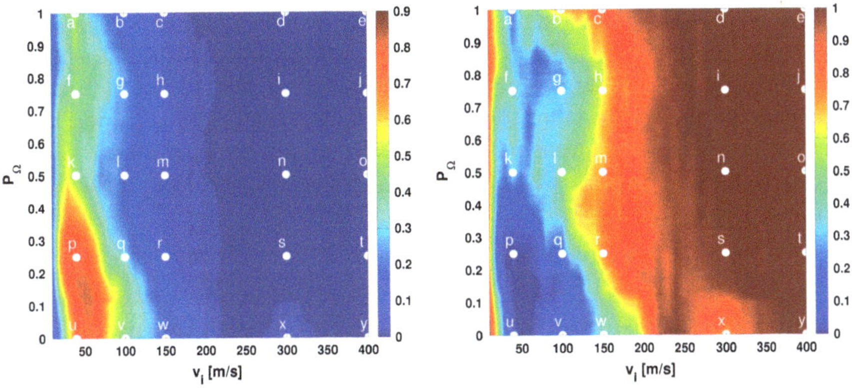

Fig. 24 Coefficient of normal restitution, e_n, and sticking probability, p_s, (both indicated by color) as functions of the impact velocity, v_i, and the angular orientation, P_Ω. The labels (a–y) refer to the marks in Fig. 19. For discussion see the text

can be stored in reversible slip deformation during the impact. When the contact terminates, the particle relaxes and the elastically stored energy is re-transformed into rebound velocity. Therefore, for small velocity and small P_Ω we find large values of the coefficient of restitution, see Fig. 24 (top). For large P_Ω we have a bad slip system, therefore, the particle cannot elastically store significant amount of energy. Instead, large part of the kinetic energy of the impacting particles is dissipated due to plastic (non-reversible) deformation. Consequently, for small v_i and large P_Ω we obtain small values for the coefficient of restitution. The coefficient of restitution assumes its largest possible value $e_n \lesssim 1$ at $P_\Omega \approx 0.14$, where the largest Schmid factor attains its maximal value 1.0, see Fig. 21.

The sticking probability, p_s, is closely related to the coefficient of restitution. It describes the situation that the impacting particle loses so much energy due to an impact that the post-collisional velocity is not sufficient to overcome the attractive adhesion force, see Eq. 9, such that the coefficient of restitution drops to zero. Figure 24 (bottom) shows the sticking probability as a function of the impact velocity, v_i and the orientation, P_Ω. According to the close relation of p_s and e_n its behavior can be understood using exactly the same arguments as for the discussion of $e_n(v_i, P_\Omega)$ above.

Both, p_s and e_n, become special at *very* low impact rate, $v_i \lesssim 15$ m/s. In this range, the attractive part of the interaction force, Eq. 9 causes the particle sticking to the surface, independently of the orientation. Even for perfectly elastic interaction corresponding to elastic slip, the energy of the post-collisional velocity would not be sufficient to overcome the attractive force. This effect can be seen in both panels of Fig. 24 in form of a small vertical strip of nearly homogeneous color at the very left side.

An interesting feature can be seen in Fig. 24 (bottom) for $P_\Omega \lesssim 0.15$ and $v_i \approx 300$ m/s, close to the label (x): Here the sticking probability reveals a sharp increase. This effect is due to a *pancake flattening* (compare Fig. 19 (x) against the neighboring sub-figures, (r), (s), (w)). At this value of Ω four slip planes are activated simultaneously and the energy of the impacting particle is, thus, sufficient to cause a large (pancake-like) deformation. The behavior of the crystalline nanoparticle in this parameter region deviates strongly from the mechanics of a homogeneous, viscoelastic particle impacting a plane where sudden changes in the macroscopic behavior are not found. For a more detailed discussion see reference [37].

Expansion of the Systematic to Describe the Fragmentation of Nanoparticle Agglomerates by Means of Fragmentation Probability and Fragmentation Function

While so far the collision behavior of single sphere-like nanoparticles with walls has been investigated, the impaction of nanoparticle agglomerates will be discussed in the following. In order to characterize the fragmentation on a more general basis, concepts from the comminution will be adopted regarding fragmentation probability and fragment size distribution. As outlined above, the structural changes of agglomerates due to impaction was evaluated by image analysis of TEM micrographs. In Fig. 25, the projected area of the agglomerates and their fragments is shown as a function on the impaction energy for three different geometrical impaction angles. Since the agglomerates were size-selected with a DMA before impaction the initial projection area is known and indicated in the figure. For normal impaction, an increase of the projection area is observed which was attributed to a flattening of the fragile agglomerates. This effect was less pronounced for thermally pretreated agglomerates with stronger inter-particle bonds (not shown here). However, with increasing impaction energy the projection area decreases due to fragmentation.

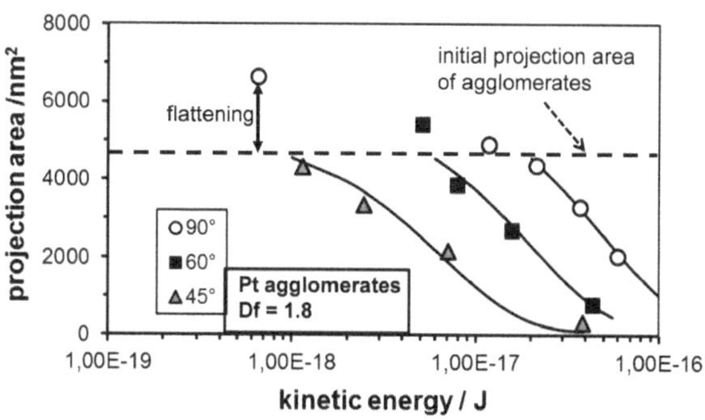

Fig. 25 Average projection area as a function of impact energy for three different impaction angles (90, 60, 45 °C). At low impaction velocities, the agglomerates show first an increase on the projection area which is due to flattening of the fragile Pt agglomerates. Then, at higher impaction velocities, fragmentation sets in significantly reducing the projection area (adapted from [10])

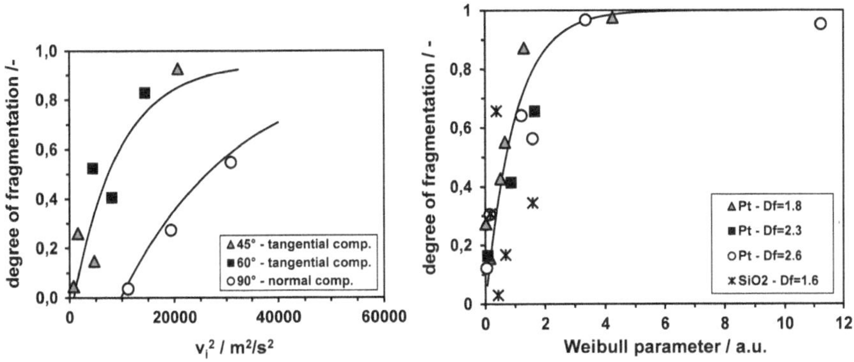

Fig. 26 Left: fragmentation curve for normal and oblique impact as a function of tangential (circles) and normal impaction velocity (squares). Right: degree of fragmentation versus Weibull parameter as defined in Eq. 34 (adapted from [10])

This behavior is similar for oblique impaction while the necessary impaction energies for fragmentation move towards smaller values. From the projection area, the corresponding degree of fragmentation was calculated by relating the measured area to the average projection area of a single primary particle. It becomes clear that for oblique impaction substantial fragmentation starts much earlier than for normal impaction, Fig. 25. In addition, higher degrees of fragmentation are reached for oblique impaction. Thus, the tangential component of the impaction velocity seems to play a major role in the fragmentation. Therefore, the degree of fragmentation is plotted against the tangential velocity component only for the oblique impaction and against the total velocity for the normal impaction Fig. 26 left. As shown in Fig. 26, the fragmentation under oblique angles is almost completed before the fragmentation for normal impaction even starts. The reason for this behavior seems to be the shear stressing of the agglomerates which is only realized under oblique impact. In comminution theory the Weibull statistics, which assumes that a chain breaks at the weakest unit, has been proven as valuable tool to generalize the breaking of solids. Similar attempts have been made for agglomerates [52]. Therefore, a master curve was constructed using the following equation for the degree of fragmentation FG Fig. 26 right:

$$\mathrm{FG} = \mathrm{FG}_0 \left(1 - \exp(-0.5 k_{agglo} \rho_{agglo} (v_{tang}^2 - v_{tang,thr}^2))\right) \quad (34)$$

where FG_0 is the maximum attainable fragmentation, k_{agglo} is a material-dependent parameter, ρ_{agglo} is the agglomerate density, v_{tang} the actual tangential velocity component and $v_{tang,thr}$ is the threshold value for the onset of fragmentation. Complete fragmentation may not be achieved due to several reasons. First, in many production processes growing primary particles encounter high enough temperatures to partially fuse together. This solid state bridge (necking) is usually too strong to be broken by mechanical forces. In addition, very small particles show a reduction of the melting point which further eases coalescence. For commercial silica nanoparticles, which are produced by flame synthesis, these smallest mechanical very stable clusters are called aggregates and have diameters of several tens of nanometers. Second, during impaction the distribution of the stress within agglomerates is not homogeneous. As already pointed out by the Thornton group [48], the primary particles at the point of first contact undergo reflection and heat release. Therefore, these particles may collide with still approaching particles which cannot even reach the wall. Such shielding effects can result in very non-uniform stressing of the different agglomerate parts.

Beside the probability of fragmentation given by Eq. 34, also the distribution of the fragments is important for a complete description of the fragmentation process. Therefore, a function similar to the one proposed by Vogel and Peukert [52] was adapted for the fragment size distribution B(x):

$$B(x) = [1 + \exp(-(x - x_M)/\Delta x)]^{-1} \tag{35}$$

where x is the fragment size, x_M is the mean fragment size and Δx is the width of the distribution. As shown in Fig. 27 for different materials and different agglomerate structures, with this function the fragment size distribution can be approximated very well. Since the fragmentation was achieved by only one agglomerate-wall collision the case of multiple impaction can simply be obtained by repeating the operation several times. Thus, a complete system of equations is presented for the first time to take into account the fragmentation process of nanoparticle agglomerates. The parameters x_M and Δx depend on the intensity and the distribution of stress within the agglomerate as well as on the distribution of inter-particle bond strength.

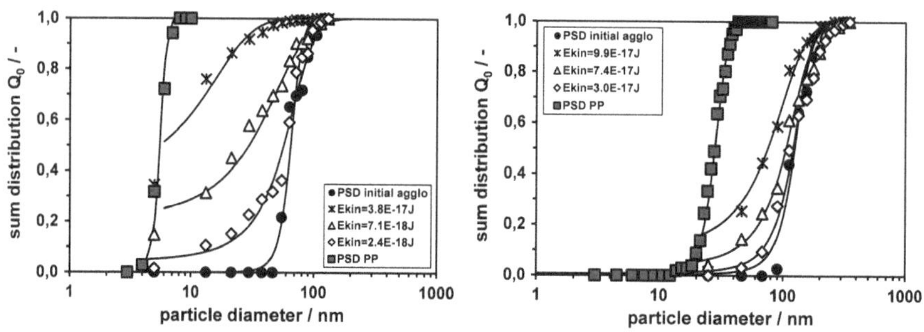

Fig. 27 Left: fragmentation of open Pt agglomerates ($Df = 1.8$) under 45 °C impact. Right: fragmentation of open ($Df = 1.6$) SiO_2 agglomerates under 45 °C impact: including the fit of Eq. 35 to the data (filled circles indicate original size distribution and filled squares indicate the primary particle size distribution) (adapted from [10])

Influence of Particle Size and Material as Well as Impact Velocity and Angle on the Bouncing and Fragmentation Behavior

The rebound of nanoparticle agglomerates has so far only marginally been investigated [13]. Besides the energy dissipation processes for single particles, fragmentation and internal restructuring can also occur during the impaction of agglomerates leading to additional energy loss [9]. In order to establish a more applicable approach than the one proposed by Ihalainen [13] the model of Dahneke [6], cf. Eq. 36 was employed where the agglomerate mass m_{aggl} is calculated by using the primary particle mass m_{PP} and the fractal expression for the number of primary particles N_{PP} in an agglomerate.

$$m_{aggl} = m_{pp} N_{pp} = m_{pp} k \left(\frac{x_m}{x_{pp}} \right)^{D_f} \tag{36}$$

By inserting Eq. 36 into Eq. 13, the following expression for the critical impact velocity for agglomerates is derived [51]. Here, k is the fractal prefactor and x_{PP} the primary particle size.

$$v_{cr}^2 = \frac{12 E_{adh}}{\pi \rho_{bulk}} \frac{x_{pp}^{D_f - 3}}{k} \frac{1 - e_n^2}{e_n^2} x_m^{-D_f} \tag{37}$$

When the adhesion energy for agglomerates is calculated by using the mobility diameter x_m, the following proportionality is obtained.

$$v_{cr}^2 \sim x_m^{\delta_i - D_f} \sim x_m^{-\beta} \tag{38}$$

where δ_i depends on the adhesion model and is 4/3 and 1 for the JKR and Hamaker model, respectively. The bounced fraction of single spherical Pt particles exhibit a similar evolution versus the impact velocity, Fig. 28 left, as the coefficient of restitution. When the critical velocity is reached, the bounced fraction rises with increasing impact velocity until a maximum is achieved. Since the rebound velocity was not measured, the coefficient of restitution could not be determined here. Nevertheless, the transition point v_{cr} from the sticking regime to the rebound regime and finally the transition point v_Y to the plastic deformation can be evaluated. The measured bounced fraction is given as a function of the impact velocity, which is calculated with the model of Rennecke and Weber [31] and refers to impact velocity in the center of the particle jet. While there is a slight decrease of the impact velocity with increasing radial distance which is even relatively more pronounced for higher impaction pressures, the measured bounced fraction is the superposition of the rebounding particles from all radial impaction positions. Thus, the critical velocity can be determined from the onset for rebound from the bounced fraction. However, with further increasing impact velocity an increasing cross section of the jet particles overcome

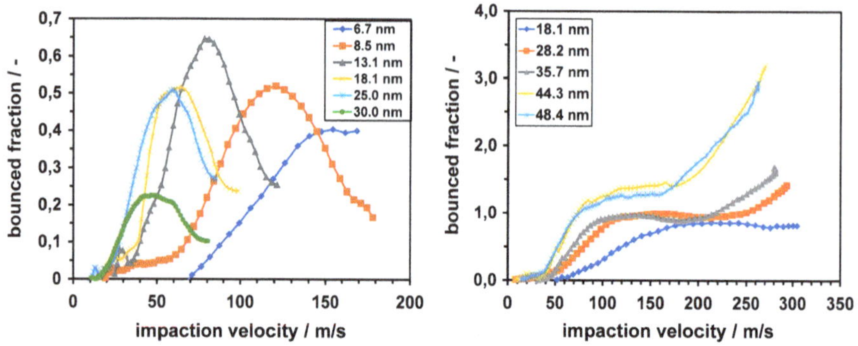

Fig. 28 Left: bouncing of spherical Pt particles on mica target. Right: bouncing of spherical SiO_2 particles on mica target (adapted from [10])

the critical velocity leading to a rising bounced fraction. For even higher velocities, the particles in the center reach the yield pressure and deform plastically which is accompanied by a reduced gain of the bounced fraction. Finally, more and more particles deform plastically and the bounced fraction decreases. Therefore, the exact velocity for the onset of plastic deformation may only be approximated as an average from the bounced fraction. The position of the maximum can be also affected by contact charging. During the contact charges are transferred as shown for particles in the micrometer range [53] and in the nanometer size range [33]. Using the insulator mica as substrate, the amount of transferred charges is negligible (estimated to less than 0.001 e [53], which is beyond the measurement resolution of the electrometers). Thus, the bounced fractions of platinum particles impacting on a mica substrate are not affected by contact charging.

The bounced fractions of all investigated particle sizes are shown in Fig. 28 for Pt particles on the left and silica particles on the right impacting on a mica substrate. For Pt particles, the maximum, which is caused by plastic deformation, as well as the critical velocity are shifted to lower impact velocities with increasing particle size. The bounced fraction of the silica particles also starts to increase monotonically above the critical velocity, until a value of around 1 is reached representing the rebound of all particles. At higher impact velocities, a second increase is observed indicating an additional charge transfer during the collision. The onset velocity for this contact charging depends on particle size where for the smallest particles of 18.1 nm, no additional charging is observed up to an impact velocity of 300 m s^{-1}. In contrast to the Pt nanoparticles, a maximum of the bounced fraction cannot be discerned for silica nanoparticles. The only reasonable explanation may be, that the solid state properties of the silica, which is usually amorphous, are changing. However, this point remains unresolved so far. The bounced fractions of differently structured Pt agglomerates and open SiO_2 agglomerates impacting on a mica surface are shown in Fig. 29. From a copper substrate, the agglomerates do not rebound and the bounced fraction remains zero (not shown here), whereas from mica substrate the

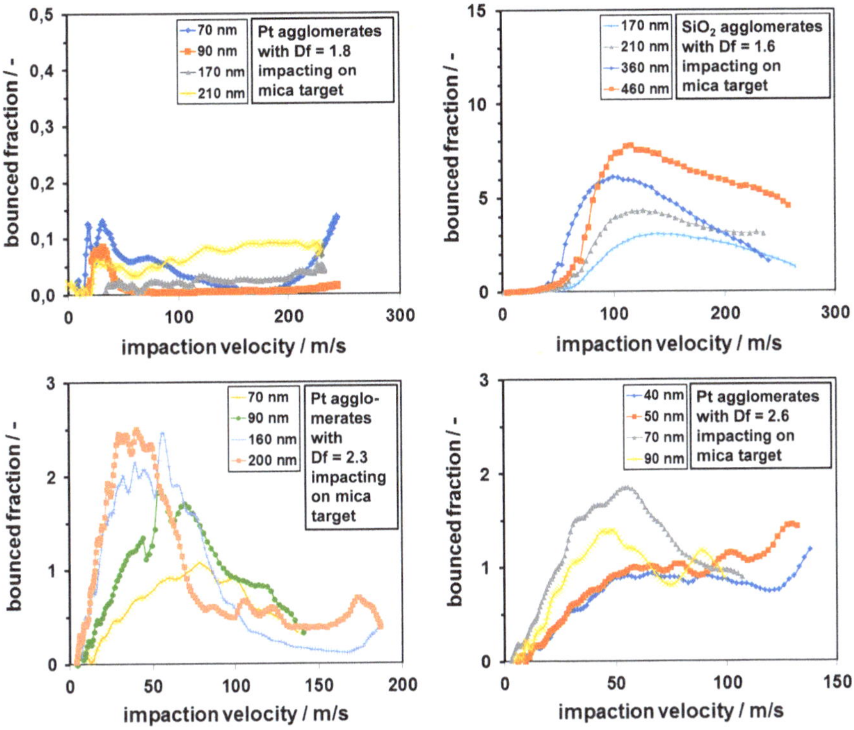

Fig. 29 Bounced fraction of differently structured Pt agglomerates and open silica agglomerates impacting on a mica target (bounced fraction based on particle charge measurements) (adapted from [10])

agglomerates show weak but clear rebounding. The absence of rebound (for copper substrate) and reduced bouncing (for mica substrate) is attributed to the additional energy dissipation due to fragmentation of the agglomerates, while internal restructuring starts at even much smaller velocities as discussed later. Fragmentation starts at impact velocities at around 55 m s^{-1} for the platinum agglomerates [9]. In contrast, the bounced fraction of openly structured silica agglomerates increases, when the critical velocity is exceeded, until a maximum is reached, Fig. 29. This behavior differs from the single silica particles. Bounced fractions with values higher than one are assumed to be caused by contact charging during the impaction process. Due to the bigger interaction area of larger agglomerates with the mica substrate, the acquired charges increase with increasing agglomerate size. However, the maximum, which is caused by fragmentation, occurs nearly at the same velocity, i.e. independent of the agglomerate size. Since the openly structured agglomerates behave aerodynamically transparent, the onset energy for fragmentation is determined by the primary particles.

Pt agglomerates with higher fractal dimensions ($D_f > 2$) were also impacted on a mica substrate. The bounced fractions of Pt agglomerates with fractal dimensions of

$D_f = 2.3$ and $D_f = 2.6$ show a similar evolution like the single spherical particles Fig. 28. After the onset for rebound is exceeded, the bounced fractions increase as long as most of the impaction energy is converted into rebound energy. When the impaction energy is high enough, additional energy dissipation channels (fragmentation and/or internal restructuring) occur leading to decreasing bounced fractions. The maxima of the bounced fractions reach values above one revealing contact charging of the agglomerates, where agglomerates with a fractal dimension of 2.3 acquire more charges than agglomerates with a fractal dimension of 2.6. For agglomerates with a fractal dimension of 2.6 with sizes of 40 and 50 nm, no pronounced maximum is observed. As for single particles, the contact charging may mask the onset for fragmentation. Therefore, the impact velocity, where the bounced fractions start to decrease, is compared to the onset velocity for fragmentation obtained from image analysis. Thereby, it is found that these two velocities are similar. Thus, the onset for fragmentation was determined from the bounced fractions neglecting possible charging effects. The following discussion focuses first on the onset of rebound v_{cr} and of plastic deformation v_Y. The measured critical velocities for single spherical Pt and silica particles are illustrated in Fig. 30 for a mica substrate, respectively. A decrease in the particle size leads to an increase of the critical impact velocity as expected from theoretical considerations. By fitting the measured values to the power law Eq. 38 a value of $\beta_{Pt-mica} = -1.07$ is obtained which is close to the theoretical predictions of $\beta_{elastic,Hamaker} = -1$, $\beta_{elastic,JKR} = -5/6$ and $\beta_{plastic} = -1$. However, on softer targets such as copper the value of β may vary more since the hardness affects the critical impact velocity according to Eq. 38 of [51].

The onset velocity for rebound for the openly structured Pt and silica agglomerates impacting on a mica surface could be determined from the bounced fractions Fig. 30. The critical velocity of the agglomerates does not depend on the mobility diameter in the investigated size range between 30 and 400 nm and is 25 m s^{-1} for Pt agglomerates impacting on a mica substrate, which is equal to the critical velocity for single Pt

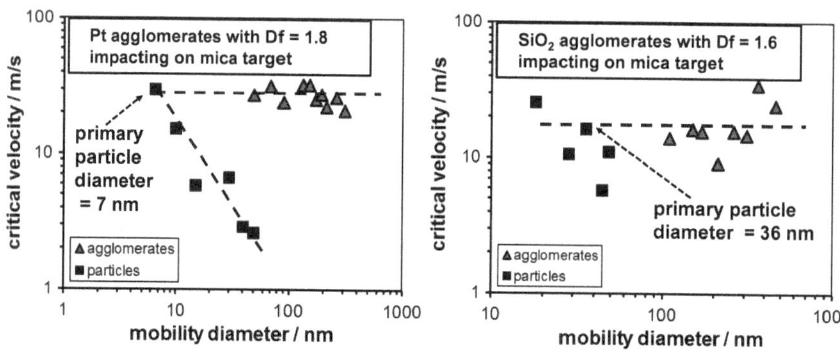

Fig. 30 Critical impact velocity of openly structured agglomerates impacting on a mica target as a function of the mobility diameter for Pt particles on the left and silica particles on the right. For comparison, the critical velocity of compact spherical particles is also shown (adapted from [10])

particles with the same size as the primary particles. The same result is obtained when the averaged critical velocity of the openly structured silica is compared to the critical velocity of the single particles with the size of the primary particles. Thus, it is concluded, that the critical velocity of openly structured agglomerates is determined by the critical velocity of the primary particles. A comparison with the critical velocity of single particles rebounding from copper substrate shows that the critical velocity for rebound is higher than the onset velocity for fragmentation, which supports the conclusion that the absence of rebound is caused by energy dissipation due to fragmentation. From the experimental investigations of single particles and the openly structured agglomerates ($D_f < 2$), it is evident that the critical velocity is determined by the primary particles as long as no other energy dissipation processes occur. Referring to the model proposed by Ihalainen et al. [13], the rebound of agglomerates depends on the ratio of the primary particles in contact with the substrate to the total number of primary particles in an agglomerate. From the investigations above, it can be concluded that all primary particles of openly structured agglomerates are in contact with the substrate. In the case of shielded particles, their mass would contribute to the kinetic energy of the agglomerates but no to the adhesion, which would reduce the critical impact velocity compared to single particles. Also for the closely structured agglomerates, v_{cr} was analyzed. The critical velocity decreases with increasing agglomerate size as shown in Fig. 31 left. A power law fit reveals the following proportionalities, $v_{Df2.3} \sim x^{-0.54}$ and $v_{Df2.6} \sim x^{-1.6}$. By using Eq. 38, δ can be determined to be $\delta_{Df2.3} = 1.22$ and $\delta_{Df2.6} = -0.7$, respectively. The value of the agglomerates with a fractal dimension of 2.3 is close to the theoretical prediction for elastic behavior ($\delta_{JKR} = 1.33$; $\delta_{Hamaker} = 1$), whereas the determined δ for agglomerates with a higher fractal dimension of 2.6 differs from the theoretical prediction indicating a size dependent coefficient of restitution, which can be related to changing adhesion energies with the agglomerate size. The adhesion energy during the impaction process depends on the ratio of the primary particles in

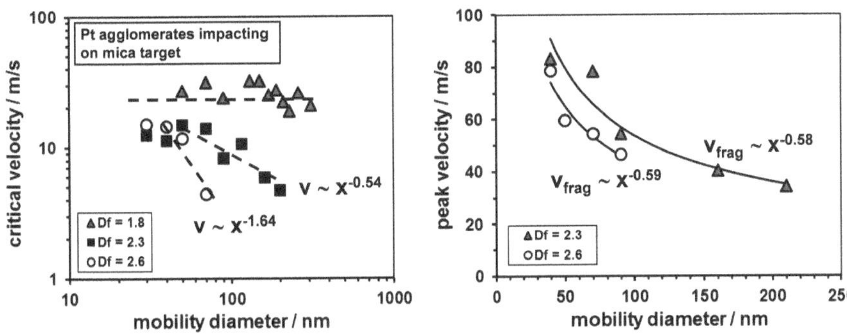

Fig. 31 Left: critical velocity for the investigated Pt agglomerate structures as a function of the mobility diameter for different agglomerate structures. Right: onset velocity for fragmentation of closely structured Pt agglomerates with fractal dimensions of $D_f = 2.3$ and $D_f = 2.6$ obtained from the bounced fractions (adapted from [10])

contact with the substrate to the total number of primary particles in an agglomerate [13]. This ratio decreases with increasing mobility diameter of the agglomerates (cf. Eq. (27)) where the effect may become only important for agglomerates of higher fractal dimensions.

$$\frac{N_{incontact}}{N_{total}} = \left[1 - \left(1 - \frac{x_{pp}}{x_m}\right)^{D_f}\right] \tag{39}$$

The decreasing bounced fraction at higher impact velocities are caused by internal restructuring or fragmentation. To evaluate which process is responsible, the projection area as a function of the impact velocity is measured by TEM analysis and compared to the peak of the bounced fraction. For fractal Pt agglomerates with $D_f = 2.3$, the flattening starts at relative low velocities indicating that a low amount of energy is needed for the movement of particles in the agglomerates. For agglomerates with a fractal dimension of $D_f = 2.6$, flattening is not observed up to an impact velocity of $45\,\mathrm{m\,s^{-1}}$. At this velocity, fragmentation starts indicated by the reduction of the projection area. The absence of restructuring is attributed to the higher coordination number of the more closely structured agglomerates [55]. The onset velocity for fragmentation determined by images analysis correlates to the peak velocity of the bounced fractions. Thus, the decrease of the bounced fraction is caused by fragmentation, i.e. the onset velocity for fragmentation can be determined from the bounced fractions and is shown in Fig. 31 right. The onset velocity for fragmentation decreases with increasing mobility diameter for closely structured agglomerates. For both fractal dimensions, nearly the same dependency on the mobility diameter ($v_{Df2.3} \sim x^{-0.58}$; $v_{Df2.6} \sim x^{-0.59}$) is observed.

Identification of Optimal Process Parameter for the Continuous Dry Dispersion of Nanopowders

In comparison to aerodynamic stresses such as shear flow and turbulent eddies, the impaction of nanoparticle agglomerates induces much higher strain rates in the order of $10^9\,\mathrm{s^{-1}}$. Therefore, impaction is most suitable for nanoparticle agglomerate dry dispersion by fragmentation and bouncing. However, for a continuous process the impact velocity needs to be high enough and the oblique impaction favors the fragment bouncing. According to Rennecke and Weber [32], the particles have to bounce far enough to reenter the gas stream to avoid recapturing by van der Waals attraction or diffusion, i.e. the stopping distance of the rebounding particles/fragments must be large enough compared to the characteristic dimensions of the substrate [32]. The stopping distance depends mainly on rebounding velocity, particle size and gas pressure, respectively. The effect of particle bouncing is well known from filtration and is called thermal rebound there [11, 54], which is considered as a consequence of the relatively high thermal velocity ($v_{therm} \sim T^{0.5}/x^{1.5}$). From an energy balance between kinetic impact energy and adhesion energy Wang and Kasper deduced a critical particle size below which bouncing becomes possible (typically < 10 nm), [54]. While Wang and Kasper considered only elastic deformation, Givehchi and Tan [11] took also plastic deformation and capillary forces into account where the additional capillary forces reduced the critical size from 37 to 9 nm. However, both studies were restricted to normal impaction. In contrast, Konstandopoulos investigated the oblique impact on filter fibers [16] and deduced a limiting impact angle below which bouncing occurs even when the normal velocity component is below the critical velocity. In summary, it has to be emphasized that particle can rebound due to their thermal velocity or to a fiber impact angle below the critical value even at atmospheric pressure where the mean free path of the carrier gas is typically in the order of 70 nm. In Fig. 32, an example of the bounced fraction of Pt agglomerates with a size of 70 nm under a geometric impact angle of 45 °C is shown. Bouncing starts at very low velocities which cannot be resolved with the experimental setup (resp. the evaluation model) used here. With increasing velocity, the bouncing fraction increases and continues to rise even after the onset of fragmentation (indicated by a dash-dotted line in Fig. 32). In particular, Fig. 32 shows that the agglomerate fragments contain enough energy for rebounding from the substrate surface. For the fragmentation on fibers, the agglomerates need to have sufficient collision energy. In the diffusive regime, the collision velocity can mainly be varied via the temperature according to the Maxwell-Boltzmann distribution. In the inertial regime, the collision velocity correlates with the gas velocity. As characteristic parameter for deposition of particles on fibers, the Stokes number may be employed (also called inertia parameter in filtration), which relates the stopping distance to a characteristic dimension (usually the fiber diameter). Particles with small Stokes numbers are deposited due to diffusion, while particles with high values of the Stokes number collide nearly undamped with the fibers. In Fig. 32 right, the Stokes numbers for open Pt agglomerates are calculated as a function of the fiber diameter for different agglomerate sizes. It becomes obvious that for fiber diameters below 500 nm, inertia parameters above 10 are reached. For smaller fiber diameters, another effect has to be taken into account, i.e. when the fiber diameter approaches the mean free path of the carrier gas (ca. 70 nm for air at standard conditions) the so called no-slip condition at the fiber surface is no longer fulfilled. Therefore, the gas stream is less deviated which results in a further increase of the inertia parameter [19]. For fiber Knudsen numbers $Kn_{fiber} > 10$, Maze et al. assume that the flow field is no longer influenced by the fiber [19]. In this case, the particle impact velocity would be the same as the gas velocity. These considerations imply that gasborne agglomerates can be continuously dispersed with net of nanofibers (assuming sufficient mechanical strength of the nanofiber array). On the one hand, the curved fiber surfaces lead to oblique impaction, which enhances fragmentation and bouncing. On the other hand, the stopping distance for fibers with diameters in the submicron range can be so small that particle rebound is even possible at elevated pressures, i.e. at atmospheric pressure. In addition, small fiber diameters allow the particle agglomerates to collide with the fiber surface with nearly gas velocity which can easily be adjusted.

In summary, the insights gained into the rapid collision of nanoparticles with walls using experiments and simulations offers a pathway to design the first continuous

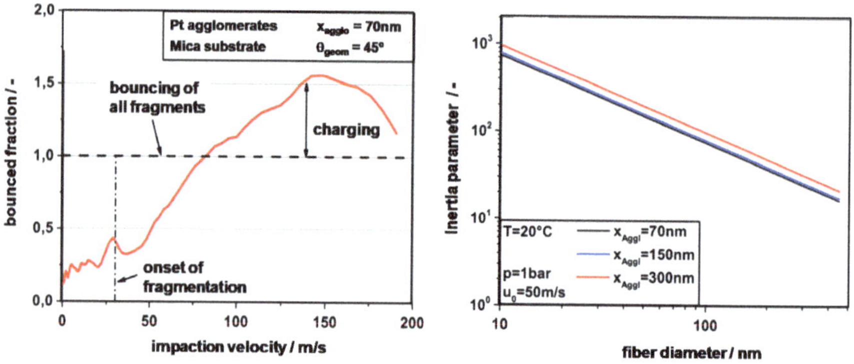

Fig. 32 Left: bounced fraction of Pt agglomerates with a size of 70 nm for a geometric impaction angle of 45 °C. Right: inertia parameter (~Stokes number) as a function of the fiber diameter for different agglomerate sizes ($D_f = 1.8$) (adapted from [10])

dry dispersion process for nanoparticle agglomerates. The achievable fragmentation/dispersion will be limited by the inter-particle bond strength. For instance, solid necks as they may originate from industrial flame synthesis [45] will possibly not be breakable with the approach outlined. However, for many applications such a dry disperser would be a significant step forward in the handling of nanopowders.

Acknowledgements We thank the German Research Foundation (Deutsche Forschungsgemeinschaft) for funding through the Cluster of Excellence "Engineering of Advanced Materials", the Collaborative Research Center SFB814, and Grants No. PO472/20 and WE 2331/12-1-3. We gratefully acknowledge the computing time granted by the John von Neumann Institute for Computing and provided on the supercomputer JUROPA at Jülich Supercomputing Centre.
We thank the colleagues involved in the SPP for intensive discussion and steady support.

References

1. Antony, S.J., Moreno-Atanasio, R., Musadaidzwa, J., Williams, R.A.: Impact fracture of composite and homogeneous nanoagglomerates. J. Nanomater. **2008**, 125,386 (2008). https://doi.org/10.1155/2008/125386
2. Armstrong, P., Peukert, W.: Size effects in the elastic deformation behavior of metallic nanoparticles. J. Nanoparticle Res. **14**, 1–13 (2012). https://doi.org/10.1007/s11051-012-1288-4
3. Bitter, J.G.A.: A study of erosion phenomena part 1. Wear **6**(1), 5–21 (1963). https://doi.org/10.1016/0043-1648(63)90003-6
4. Brilliantov, N.V., Albers, N., Spahn, F., Pöschel, T.: Collision dynamics of granular particles with adhesion. Phys. Rev. E **76**(5) (2007). https://doi.org/10.1103/PhysRevE.76.0513020
5. Brilliantov, N.V., Albers, N., Spahn, F., Pöschel, T.: Erratum: Collision dynamics of granular particles with adhesion [Physical Review e 76, 051302 (2007)]. Phys. Rev. E **87**(3) (2013). https://doi.org/10.1103/PhysRevE.87.039904
6. Dahneke, B.: The capture of aerosol particles by surfaces. J. Colloid Interface Sci. **37**(2), 342–353 (1971). https://doi.org/10.1016/0021-9797(71)90302-X

7. Daw, M.S., Baskes, M.I.: Embedded-atom method: derivation and application to impurities, surfaces, and other defects in metals. Phys. Rev. B **29**(12), 6443–6453 (1984). https://doi.org/10.1103/PhysRevB.29.6443
8. Froeschke, S., Kohler, S., Weber, A.P., Kasper, G.: Impact fragmentation of nanoparticle agglomerates. J. Aerosol Sci. **34**(3), 275–287 (2003). https://doi.org/10.1016/S0021-8502(02)00185-4
9. Gensch, M., Weber, A.: Fragmentierung von gasgetragenen Nanopartikel-Agglomeraten bei schräger Impaktion. Chem. Ing. Tech. **86**(3), 270–279 (2014). https://doi.org/10.1002/cite.201300134
10. Gensch, M.: Mechanische Stabilität von Nanopartikel-Agglomeraten bei mechanischen Belastungen. PhD Thesis TU Clausthal, Shaker Verlag, ISBN 978-3-8440-6110-9 (2018)
11. Givehchi, R., Tan, Z.: An overview of airborne nanoparticle filtration and thermal rebound theory. Aerosol Air Qual. Res. **14**(1), 46–63 (2014). https://doi.org/10.4209/aaqr.2013.07.0239
12. Halicioğlu, T., Pound, G.M.: Calculation of potential energy parameters form crystalline state properties. Phys. Status Solidi (a) **30**(2), 619–623 (1975). https://doi.org/10.1002/pssa.2210300223
13. Ihalainen, M., Lind, T., Ruusunen, J., Tiitta, P., Lähde, A., Torvela, T., Jokiniemi, J.: Experimental study on bounce of submicron agglomerates upon inertial impaction. Powder Technol. **268**, 203–209 (2014). https://doi.org/10.1016/j.powtec.2014.08.029
14. Kiener, D., Minor, A.M.: Source-controlled yield and hardening of Cu(100) studied by in situ transmission electron microscopy. Acta Mater. **59**(4), 1328–1337 (2011). https://doi.org/10.1016/j.actamat.2010.10.065
15. Kim, J.Y., Greer, J.R.: Tensile and compressive behavior of gold and molybdenum single crystals at the nano-scale. Acta Mater. **57**(17), 5245–5253 (2009). https://doi.org/10.1016/j.actamat.2009.07.027
16. Konstandopoulos, A.G.: Particle sticking/rebound criteria at oblique impact. J. Aerosol Sci. **37**(3), 292–305 (2006). https://doi.org/10.1016/j.jaerosci.2005.05.019
17. Kuninaka, H., Hayakawa, H.: Simulation of cohesive head-on collisions of thermally activated nanoclusters. Phys. Rev. E **79**(3) (2009). https://doi.org/10.1103/PhysRevE.79.031309
18. Marsaglia, G.: Choosing a point from the surface of a sphere. Ann. Math. Stat. **43**(2), 645–646 (1972). https://doi.org/10.1214/aoms/1177692644
19. Maze, B., Vahedi, H.T., Wang, Q., Pourdeyhimi, B.: A simulation of unsteady-state filtration via nanofiber media at reduced operating pressures. J. Aerosol Sci. **38**(5), 550–571 (2007). https://doi.org/10.1016/j.jaerosci.2007.03.008
20. Miles, R.E.: On random rotations in R^3. Biometrika **52**(3–4), 636–639 (1965). https://doi.org/10.1093/biomet/52.3-4.636
21. Müller, P., Pöschel, T.: Oblique impact of frictionless spheres: on the limitations of hard sphere models for granular dynamics. Granul. Matter **14**(2), 115–120 (2012). https://doi.org/10.1007/s10035-012-0324-5
22. Müller, P., Pöschel, T.: Event-driven molecular dynamics of soft particles. Phys. Rev. E **87**(3), 033,301 (2013). https://doi.org/10.1103/PhysRevE.87.033301
23. Müller, P., Krengel, D., Pöschel, T.: Negative coefficient of normal restitution. Phys. Rev. E **85**(4), 041,306 (2012). https://doi.org/10.1103/PhysRevE.85.041306
24. Ning, Z., Boerefijn, R., Ghadiri, M., Thornton, C.: Distinct element simulation of impact breakage of lactose agglomerates. Adv. Powder Technol. **8**(1), 15–37 (1997). https://doi.org/10.1016/S0921-8831(08)60477-X
25. Nowak, J.D., Beaber, A.R., Ugurlu, O., Girshick, S.L., Gerberich, W.W.: Small size strength dependence on dislocation nucleation. Scr. Mater. **62**(11), 819–822 (2010). https://doi.org/10.1016/j.scriptamat.2010.01.026
26. Ostendorf, F., Schmitz, C., Hirth, S., Kühnle, A., Kolodziej, J.J., Reichling, M.: How flat is an air-cleaved mica surface? Nanotechnology **19**(30), 305,705 (2008). https://doi.org/10.1088/0957-4484/19/30/305705
27. Pöschel, T., Müller, P.: Event-driven DEM of soft spheres. AIP Conf. Proc. **1542**, 149–152 (2013). https://doi.org/10.1063/1.4811889

28. Pöschel, T., Brilliantov, N.V., Formella, A., Heckel, M., Krülle, C., Müller, P., Salueña, C., Schwager, T.: Contact of granular particles and the simulation of rapid flows using event-driven molecular dynamics. Eur. J. Environ. Civ. Eng. **12**(7–8), 827–870 (2008). https://doi.org/10.1080/19648189.2008.9693051
29. Ramírez, R., Pöschel, T., Brilliantov, N.V., Schwager, T.: Coefficient of restitution of colliding viscoelastic spheres. Phys. Rev. E **60**(4), 4465–4472 (1999). https://doi.org/10.1103/PhysRevE.60.4465
30. Rennecke, S.: Kontaktphänomene bei Hochgeschwindigkeitskollisionen von Nanopartikeln mit Oberflächen. Ph.D. thesis, TU Clausthal (2015)
31. Rennecke, S., Weber, A.P.: A novel model for the determination of nanoparticle impact velocity in low pressure impactors. J. Aerosol Sci. **55**, 89–103 (2013). https://doi.org/10.1016/j.jaerosci.2012.07.014
32. Rennecke, S., Weber, A.P.: On the pressure dependence of thermal rebound. J. Aerosol Sci. **58**, 129–134 (2013). https://doi.org/10.1016/j.jaerosci.2013.01.006
33. Rennecke, S., Weber, A.: Charge transfer to metal nanoparticles bouncing from conductive surfaces. Aerosol Sci. Technol. **48**(10), 1059–1069 (2014). https://doi.org/10.1080/02786826.2014.955566
34. Richter, G., Hillerich, K., Gianola, D.S., Mönig, R., Kraft, O., Volkert, C.: Ultrahigh strength single crystalline nanowhiskers grown by physical vapor deposition. Nano Lett. **9**(8), 3048–3052 (2009). https://doi.org/10.1021/nl9015107
35. Sator, N., Hietala, H.: Damage in impact fragmentation. Int. J. Fract. **163**(1), 101–108 (2010). https://doi.org/10.1007/s10704-009-9406-8
36. Schmid, E., Boas, W.: Plasticity of Crystals. F. A. Hughes & Co Ltd., London (1950)
37. Schöner, C., Pöschel, T.: Orientation-dependent properties of nanoparticle impact. Phys. Rev. E **98**(2), 022,902 (2018). https://doi.org/10.1103/PhysRevE.98.022902
38. Schöner, C., Rennecke, S., Weber, A.P., Pöschel, T.: Introduction of a new technique to measure the coefficient of restitution for nanoparticles. Chem. Ing. Tech. **86**(3), 365–374 (2014). https://doi.org/10.1002/cite.201300132
39. Schwager, T.: Coefficient of restitution for viscoelastic disks. Phys. Rev. E **75**(5), 051,305 (2007). https://doi.org/10.1103/PhysRevE.75.051305
40. Schwager, T., Pöschel, T.: Coefficient of normal restitution of viscous particles and cooling rate of granular gases. Phys. Rev. E **57**(1), 650–654 (1998). https://doi.org/10.1103/PhysRevE.57.650
41. Schwager, T., Pöschel, T.: Coefficient of restitution and linear-dashpot model revisited. Granul. Matter **9**(6), 465–469 (2007). https://doi.org/10.1007/s10035-007-0065-z
42. Schwager, T., Pöschel, T.: Coefficient of restitution for viscoelastic spheres: The effect of delayed recovery. Phys. Rev. E **78**(5), 051,304 (2008). https://doi.org/10.1103/PhysRevE.78.051304
43. Schwager, T., Becker, V., Pöschel, T.: Coefficient of tangential restitution for viscoelastic spheres. Eur. Phys. J. E **27**(1), 107–114 (2008). https://doi.org/10.1140/epje/i2007-10356-3
44. Seipenbusch, M., Toneva, P., Peukert, W., Weber, A.P.: Impact fragmentation of metal nanoparticle agglomerates. Part. Part. Syst. Charact. **24**(3), 193–200 (2007). https://doi.org/10.1002/ppsc.200601089
45. Seipenbusch, M., Rothenbacher, S., Kirchhoff, M., Schmid, H.J., Kasper, G., Weber, A.P.: Interparticle forces in silica nanoparticle agglomerates. J. Nanoparticle Res. **12**(6), 2037–2044 (2010). https://doi.org/10.1007/s11051-009-9760-5
46. Shafiei Mohammadabadi, A., Dehghani, K.: A new model for inverse Hall-Petch relation of nanocrystalline materials. J. Mater. Eng. Perform. **17**(5), 662–666 (2008). https://doi.org/10.1007/s11665-008-9206-8
47. Subero, J., Ghadiri, M.: Breakage patterns of agglomerates. Powder Technol. **120**(3), 232–243 (2001). https://doi.org/10.1016/S0032-5910(01)00276-5
48. Subero, J., Ning, Z., Ghadiri, M., Thornton, C.: Effect of interface energy on the impact strength of agglomerates. Powder Technol. **105**(1), 66–73 (1999). https://doi.org/10.1016/S0032-5910(99)00119-9

49. Subero-Couroyer, C., Ghadiri, M., Brunard, N., Kolenda, F.: Analysis of catalyst particle strength by impact testing: the effect of manufacturing process parameters on the particle strength. Powder Technol. **160**(2), 67–80 (2005). https://doi.org/10.1016/j.powtec.2005.08.005
50. Tomsic, A., Marković, N., Pettersson, J.B.C.: Scattering of ice particles from a graphite surface: a molecular dynamics simulation study. J. Phys. Chem. B **107**(38), 10576–10582 (2003). https://doi.org/10.1021/jp030557b
51. Tsai, C.J., Pui, D.Y.H., Liu, B.Y.H.: Capture and rebound of small particles upon impact with solid surfaces. Aerosol Sci. Technol. **12**(3), 497–507 (1990). https://doi.org/10.1080/02786829008959364
52. Vogel, L., Peukert, W.: From single particle impact behaviour to modelling of impact mills. Chem. Eng. Sci. **60**(18), 5164–5176 (2005). https://doi.org/10.1016/j.ces.2005.03.064
53. Wang, H.C., John, W.: Dynamic contact charge transfer considering plastic deformation. J. Aerosol Sci. **19**(4), 399–411 (1988). https://doi.org/10.1016/0021-8502(88)90016-X
54. Wang, H.C., Kasper, G.: Filtration efficiency of nanometer-size aerosol particles. J. Aerosol Sci. **22**(1), 31–41 (1991). https://doi.org/10.1016/0021-8502(91)90091-U
55. Weber, A.P., Friedlander, S.K.: Relation between coordination number and fractal dimension of aerosol agglomerates. J. Aerosol Sci. **28**, S765–S766 (1997). https://doi.org/10.1016/S0021-8502(97)85381-5
56. Weir, G., McGavin, P.: The coefficient of restitution for the idealized impact of a spherical, nano-scale particle on a rigid plane. Proc. R. Soc. Lond. A: Math. Phys. Eng. Sci. **464**(2093), 1295–1307 (2008). https://doi.org/10.1098/rspa.2007.0289

Stochastic Nature of Particle Collisions and its Impact on Granular Material Properties

Nina Gunkelmann, Dan Serero, Aldo Glielmo, Marina Montaine, Michael Heckel and Thorsten Pöschel

Introduction

The dynamics of granular systems is driven by dissipative collisions of particles with each other and with the system walls. These interactions do not only depend on material parameters, e.g. elastic properties, particle sizes and impact velocities but are highly affected by the geometry of the particles. The degree of inelasticity of particle interactions may be quantified by the coefficient of normal restitution (COR), $\varepsilon < 1$, defined as

$$\varepsilon = -\frac{\mathbf{V}' \cdot \mathbf{n}}{\mathbf{V} \cdot \mathbf{n}}, \tag{1}$$

where $\mathbf{V}$ and $\mathbf{V}'$ are the relative velocities of the colliding particles before and after the collision and $\mathbf{n}$ is the unit vector normal to the impact plane in the instant of the collision. The COR is the most important characteristic of a granular system because it is the foundation of both granular hydrodynamics and Kinetic Theory of rapid granular flows [9, 16] as well as event-driven particle simulation of granular matter [43].

There are several techniques for experimentally measuring the coefficient of restitution, such as high-speed photography [13, 29, 34], pendulum systems [8, 28, 32], LASER Doppler velocimetry [57], particle tracking velocimetry [47] and others [45, 60]. The most prominent technique to measure the coefficient of restitution exploits

N. Gunkelmann (✉) · D. Serero · M. Montaine · M. Heckel · T. Pöschel
Institute for Multiscale Simulations, Friedrich-Alexander-Universität, Erlangen, Germany
e-mail: nina@gunkelmann.de

N. Gunkelmann
Institute of Applied Mechanics, Technische Universität Clausthal, Clausthal-Zellerfeld, Germany

A. Glielmo
Department of Physics, King's College London, Strand, London WC2R 2LS, United Kingdom

S. Antonyuk (ed.), *Particles in Contact*,
https://doi.org/10.1007/978-3-030-15899-6_18

the time lag between consecutive impacts of a particle bouncing repeatedly on a horizontal plane,

$$\begin{aligned} \varepsilon_i &= \left|\frac{V_i'}{V_i}\right| = \left|\frac{V_{i+1}}{V_i}\right| = \frac{\tau_{i+1}-\tau_i}{\tau_i-\tau_{i-1}}, \qquad i = 2, 3, \ldots . \\ V_i &= \frac{g}{2}(\tau_i - \tau_{i-1}) = \frac{g}{2}\Delta\tau_i . \end{aligned} \tag{2}$$

Here, V_i and V_i' are the normal pre- and post-collisional velocities of the ith impact occurring at time τ_i and $g = -9.81\ \mathrm{m/s^2}$ is the gravitational acceleration. This technique is extensively used for scientific as well as educational purposes, e.g. [1, 3, 11, 12, 14, 25, 26, 30, 36, 37, 41, 48, 54, 56, 58, 59, 61] and many others.

While in most references it is assumed that ε is a material constant, detailed analysis of the collision process as well as experiments show clearly that the coefficient depends on the impact velocity [21, 45].

Moreover, even for virtually perfect spheres like ball bearings, one observes significant scatter in experiments due to microscopic surface asperities [12, 23, 25, 32, 38]. This scatter may be expressed by considering the coefficient of restitution as a fluctuating quantity.

Performing large scale bouncing ball experiments, Montaine et al. [38] analyzed the fluctuations of the COR of more than 10^5 single impacts and found that besides the known dependence on impact velocity, the COR may be described as a fluctuating quantity whose probability distribution follows an asymmetric Laplace distribution

$$p_\varepsilon(t) = \begin{cases} e^{a\varepsilon+b} & \varepsilon < \varepsilon_{\max} \\ e^{c\varepsilon+d} & \text{else,} \end{cases} \tag{3}$$

where a, b, c, d, and $\varepsilon_{\max}$ are parameters depending on the impact velocity.

Collisions of particles that may be also considered as imperfect spheres, although of different type, were also studied in [15]. In this study, the coefficient of restitution for the repeated bounce of smooth but slightly eccentric particles of dumbbell type was investigated by means of numerical experiments. Assuming random angular orientation of the particle at the instant of collision it was shown that the variance was considerable, even for small eccentricities. The measured coefficient of restitution obeys a wide Laplacian probability density function including a finite probability for negative values of the coefficient of restitution.

The strong scatter observed in experiments [20, 38] may thus be attributed to both roughness and eccentricity of the spheres. Therefore, the assumption of a fluctuating coefficient of restitution is a possible way to extend both event-driven molecular dynamics and kinetic theory of dilute granular systems to slightly non-spherical particles.

The structure of the paper is as follows. In section "Stochastic Behavior of the Coefficient of Normal Restitution: Experimental Measurements and Numerical Models" it is shown that for measurements of the coefficient of restitution from the sound signal emitted by a sphere bouncing repeatedly off the ground, small deviations from the perfect shape of the sphere lead to large measurement errors. The coefficient of normal restitution may be described as a fluctuating quantity. To this end, in section "Stochastic Behavior of the Coefficient of Normal Restitution: Theoretical Derivation" we derive an analytic expression for the distribution p_ε by analyzing local impact properties of rough particles for a given collision geometry. Finally, in section "Impact of the Stochastic Nature on Cooling Properties of Rapid Granular Flows" a numerical and analytical analysis of the temperature behavior in the case of a stochastic, velocity-dependent coefficient of restitution is performed showing a decisive influence of the fluctuating coefficient of restitution to the dynamics of rapid granular flows.

Stochastic Behavior of the Coefficient of Normal Restitution: Experimental Measurements and Numerical Models

Experiment

The experimental setup is sketched in Fig. 1: A robot is used to move a small vertical glass tube to a desired position. In the beginning of each experimental trial, a stainless steel ball of radius $R = 3\,\text{mm}$ and mass $M = 0.9\,\text{g}$ is suspended at the end of the tube by means of a vacuum pump. The robot moves the sphere then to a desired

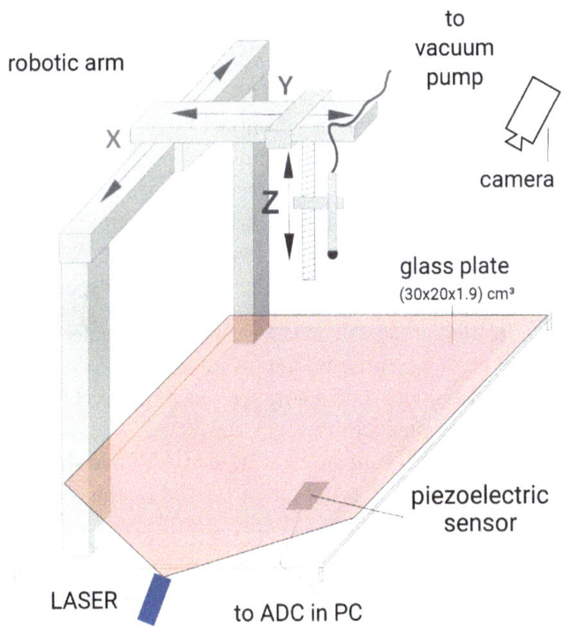

Fig. 1 Schematic diagram of the experimental setup

position $\{x, y, z\}$. By switching off the vacuum pump, the sphere is released to bounce repeatedly off the ground. When the ball eventually comes to rest, it is pushed to a defined position by a fan, from where the robot picks it up again. Sound emission caused by the impacts is recorded by a piezoelectric sensor mounted on the lower side of the plate, 5 cm away from the corner and recorded by a computer. A LASER sheet oriented parallel to the plate in distance of slightly more than the sphere diameter and a camera were used to determine the location of the impact, for more details see [20].

In each cycle, the initial horizontal (x, y)-position is chosen randomly within the central region of the ground plate. Edge-effects [55] are, therefore, not noticeable. The initial impact velocity is randomly chosen from $V \in (1.33, 1.4)$ m/s, corresponding to $z \in (9, 10)$ cm, leading to 90–100 bounces of the ball. This way, via Eq. (2) we perform about 6,000 measurements of $\varepsilon(V)$ per hour. To assure stationary conditions, the experiment was performed in a climate chamber to keep temperature and humidity of the experimental environment at constant values. The experiment was performed in darkness with the only illumination from the LASER sheet to conveniently determine the locations of the impacts from the reflections of the LASER at the sphere's surface. Figure 2 shows the coefficient of restitution as a function of the impact velocity, for the system described above. The figure shows about 280,000 data points. For the statistical analysis we neglected the first 10^4 bounces in order to suppress effects due to initial wear. Further information regarding the details of the experimental setup are given in [20]. Figure 2 (bottom) displays the normalized histograms of measurements of a certain value of ε for several small intervals of the impact velocity. The probability density is described by the asymmetric Laplacian in Eq. (3), first proposed in [38].

The results obtained from the experiment described above, are influenced by imperfections of different kind, resulting in experimental measurement errors. In [20], the authors identified the following measurement uncertainties as putative main sources of errors:

- uncertainty of the time of impact by measuring the emitted sound
- different location of consecutive impacts
- air drag
- wear of the plate and the sphere in long-time experiments
- imperfection of the spherical shape with respect to surface roughness
- imperfection of the spherical shape with respect to eccentricity.

The detailed analysis carried out in [20] revealed that the main contributions to the overall error are those related to imperfections of the spherical shape, due to either surface roughness or to a non-zero eccentricity. Only for large falling height, air drag may lead to considerable error, too.

When quantifying the error of the COR, one should keep in mind that the inelastic nature of a collision is not characterized by the coefficient of restitution, ε, but rather by the deviation of the (trivial) elastic case, $(1-\varepsilon)$. Therefore, an experimental error should be better related to $(1-\varepsilon)$. Below, we will see that an experimental error which appears acceptable when related to ε can render unacceptable when related to $(1-\varepsilon)$.

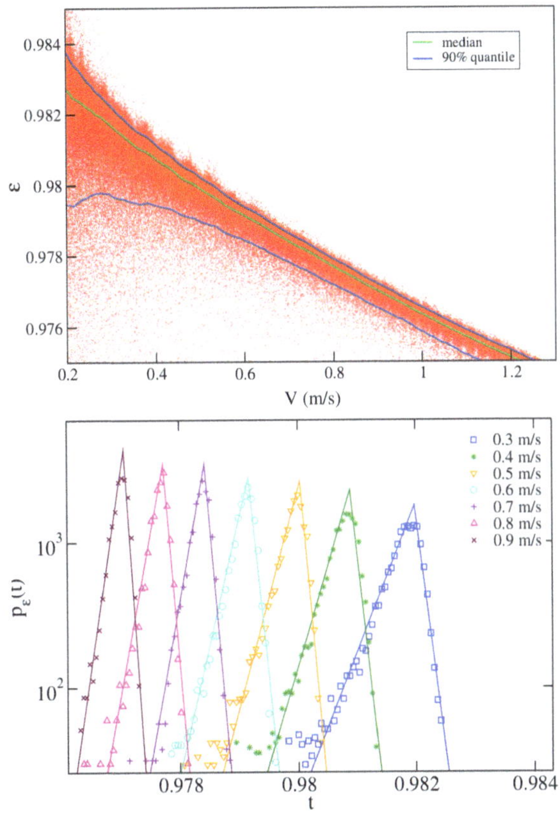

Fig. 2 Top: Coefficient of restitution as a function of the impact velocity. Each of the approximately 280,000 data points corresponds to an impact. The lines shows the median (full line) and the 90% quantile (dashed lines). Bottom: Histograms of the coefficient of normal restitution for impact velocities from small intervals. Each histogram represents a certain impact velocity. The lines are exponential fits

The coefficient of restitution describes the change of relative velocity of the particle and the plate in normal direction *at the point of contact*. Therefore, bouncing ball experiments as the one just described are only useful to determine the coefficient of restitution if the bouncing body is a sphere. Surprisingly, this simple but important detail is rarely mentioned in the literature. Deviations from the spherical shape lead to sizeable errors of the measured coefficient of restitution. In the following, we discuss the error due to imperfections of two different types: (a) surface roughness, that is, small scale impurities of the surface and (b) eccentricity, that is, deviations of the overall shape from the perfect sphere. Since there is no experimental means to individually control these deviations, we use a numerical simulation method to estimate the corresponding errors.

In the next subsection, we describe a mechanical model, well validated against experimental results [38], for the description of the dynamics of imperfect spheres. Numerical simulations of such a model are presented in section "Numerical Simulations: Fluctuations Due to Surface Roughness and Eccentricity". By varying the characteristics of the simulated particle we are able to isolate, and separately analyze, imperfections of type (a) and (b) from all other sources of error.

Mechanical Model of Collision: Microscopic and Macroscopic Coefficients of Restitution

In a model by Montaine et al. [38] a rough sphere was modeled by a central (perfect) sphere and a large number (several millions) microscopic spherical asperities located at random positions at the sphere's surface. Figure 3 (left) shows SEM pictures of the sphere after 10^6 impacts where tiny scratches are visible on the sphere's surface which can be modeled by tiny spheres attached to the surface (Fig. 3, right). The contact between the particle and the plane occurs, thus, via one of the asperities. The local change of velocity at the point of contact is then described by a microscopic coefficient of restitution, $\epsilon_n(v)$ at normal velocity v at the point of contact. The particle is dropped from a certain height and collides with a horizontal hard plate. We describe the inelastic collision of the particle with the plate as an instantaneous event leading to a change of linear and angular momentum of the particle due to the collisional impulse $\mathbf{P}$:

$$\begin{aligned} \Delta \mathbf{V} &\equiv \mathbf{V}' - \mathbf{V} = \frac{1}{M}\mathbf{P}, \\ \Delta \boldsymbol{\Omega} &\equiv \boldsymbol{\Omega}' - \boldsymbol{\Omega} = \frac{1}{J}\left(\mathbf{r} \times \mathbf{P}\right). \end{aligned} \tag{4}$$

where $\mathbf{V}$ and $\mathbf{V}'$ describe the pre- and post-collisional linear velocity of the particle, $\boldsymbol{\Omega}$ and $\boldsymbol{\Omega}'$ are the pre- and post-collisional angular velocities, J is the moment of inertia of the particle and $\mathbf{r}$ is the contact point, located at the surface of one of the asperities. The geometry of the impact is sketched in Fig. 4.

The *microscopic* impact velocity, that is, the velocity of the point where the asperity touches the floor at $\mathbf{r}$ is defined by

$$\mathbf{v} = (\mathbf{v} \cdot \mathbf{n})\,\mathbf{n} + (\mathbf{v} \cdot \mathbf{t}\,)\,\mathbf{t} = \mathbf{V} + \boldsymbol{\Omega} \times \mathbf{r}\,, \tag{5}$$

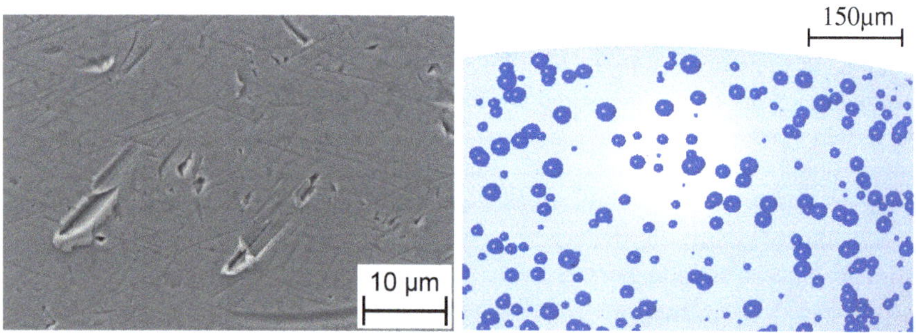

Fig. 3 Left: SEM picture at the end of the experiment after about 10^6 impacts, right: Numerical model of a rough sphere [38], where a large number of tiny spheres is attached to random positions at the surface of a large central (perfect) sphere

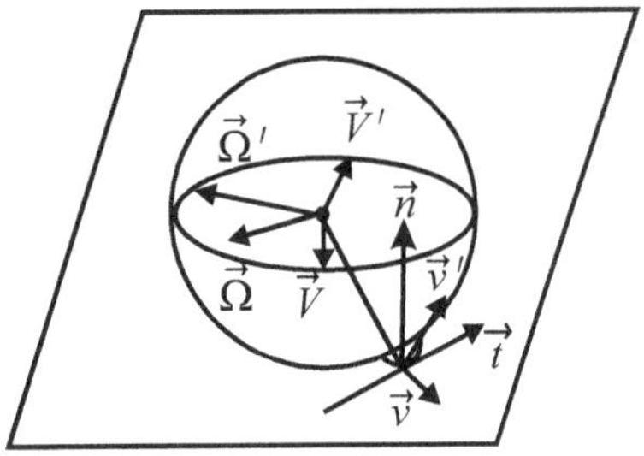

Fig. 4 Sketch of the collision of a rough particle with a flat surface. Only the asperity which is in contact with the plane is shown

where **n** and **t** are unit vectors in the plane pointing in directions normal and tangential to the velocity of the contact point.

The inelastic nature of the collision of the particle with the plate is taken into account by the *microscopic* coefficients of normal and tangential restitution, ϵ_n and ϵ_t, such that the post-collisional velocity at the point of contact are given by

$$\begin{aligned} \mathbf{v}' \cdot \mathbf{n} &= -\epsilon_n \left(\mathbf{v} \cdot \mathbf{n} \right) \\ \mathbf{v}' \cdot \mathbf{t} &= \epsilon_t \left(\mathbf{v} \cdot \mathbf{t} \right) . \end{aligned} \tag{6}$$

The third component of **v** which is perpendicular to both **n** and **t** does not change during the impact.

The change of the *microscopic* impact velocity of the particle is, therefore,

$$\begin{aligned} \mathbf{v}' - \mathbf{v} &= -\left(1 + \epsilon_n\right) \left(\mathbf{v} \cdot \mathbf{n}\right) \mathbf{n} + \left(-1 + \epsilon_t\right) \left(\mathbf{v} \cdot \mathbf{t}\right) \mathbf{t} \\ &= \Delta \mathbf{V} + \Delta \boldsymbol{\Omega} \times \mathbf{r} . \end{aligned} \tag{7}$$

The second line of Eq. (7) relates the change of the microscopic velocity at the point of contact to the change of the macroscopic velocities of the particle. By substituting $\Delta\mathbf{V}$ and $\Delta\boldsymbol{\Omega}$ from Eq. (4) into the right hand side of Eq. (7) we obtain the following system of linear equations describing the impact:

$$-\left(1 + \epsilon_n\right) \left(\mathbf{v} \cdot \mathbf{n}\right) \mathbf{n} + \left(-1 + \epsilon_t\right) \left(\mathbf{v} \cdot \mathbf{t}\right) \mathbf{t} = \frac{\mathbf{P}}{M} + \frac{1}{J} \left(\mathbf{r} \times \mathbf{P}\right) \times \mathbf{r} . \tag{8}$$

Using Eqs. (4) and (8) one can compute the dynamics of the bouncing particle provided the position of the asperity $\mathbf{r}_c$ and the *microscopic* coefficients of restitution, ϵ_n and ϵ_t, are given. Consequently, by means of Eq. (1) we can compute the *macroscopic* coefficient of restitution, ε, as observed in experiments. Obviously, as the position of the asperity at the instant of the impact is random, **v** and, thus, ε are fluctuating quantities too. Consequently, the stochastic properties of the coefficient of normal restitution, ε, are intimately related to the distribution of the asperities at the surface of the particle.

The value of the *microscopic* coefficient of restitution $\epsilon_n(v)$ was derived analytically for the viscoelastic collision of a perfect sphere with a plane [10, 51, 52] at normal velocity v at the point of contact and can be computed by the convergent Padé expansion of order [1/4], for details see [40, 46]:

$$\epsilon_n(v) \approx \frac{1 + c_1 v_*}{1 + b_1 v_* + b_2 v_*^2 + b_3 v_*^3 + b_4 v_*^4}; \quad v_* = \beta^{1/2} v^{1/10}, \tag{9}$$

with the universal (material independent) constants $c_1 = 0.501086$, $b_1 = 0.501086$, $b_2 = 1.15345$, $b_3 = 0.577977$, $b_4 = 0.532178$. The material constant $\beta = 0.0467$ was determined by fitting Eq. (9) to the experimental data; see [38]. The *microscopic* coefficient of tangential restitution, ϵ_t, depends on both bulk material properties and surface properties. Therefore, ϵ_t cannot be analytically derived from material properties, except for the limiting case of pure Coulomb friction, e.g. [2, 50]. Here we assume $\epsilon_t = 1$, that is, elastic no-slip interaction in tangential direction.

Numerical Simulations: Fluctuations due to Surface Roughness and Eccentricity

Surface Roughness

In numerical experiments we released the particle from a certain height just as in the experiment described above and simulated the sequence of collision in analogy to the experiment. When the particle contacts the plane through one of its asperities, we compute the local change of velocity due to the impact geometry and the microscopic coefficient of restitution, to obtain the impulse on the particle at each impact. These impulses govern the dynamics of the particle in an event-driven molecular dynamics simulation (see section "Mechanical Model of Collision: Microscopic and Macroscopic Coefficients of Restitution"). Micromechanically, when the particle collides with the plane, there act torques due to the asperities which cause transfer between the translational and rotational degrees of freedom. Since the rotation of the particle does not influence the coefficient of restitution determined via Eq. (1), the rotational degrees of freedom act like a reservoir of energy and at each collision a certain amount of energy is exchanged between the rotational and translational degrees of freedom. The amount depends on the details of the contact mechanics, that is, on the angular orientation of the particle at the instant of the contact and on the positions of the asperities which may be considered as random variables, thus appearing like an uncertainty of measurement. It was shown that the coefficient of restitution obtained from numerical simulations of the described model reveals the same characteristics as the experimental results. This concerns both the mean value $\bar{\varepsilon}(V)$ and the statistical characteristics of the data [17, 38]. Figure 5 shows the coefficient of restitution obtained from a numerical simulation. Here the central sphere of radius $R = 3$ mm was covered by 3×10^6 asperities of size $\rho_i \leq R \times 10^{-3}$.

Having the experimental data sufficiently closely reproduced, the error due to roughness can be determined from the standard deviation of the data, isolated from

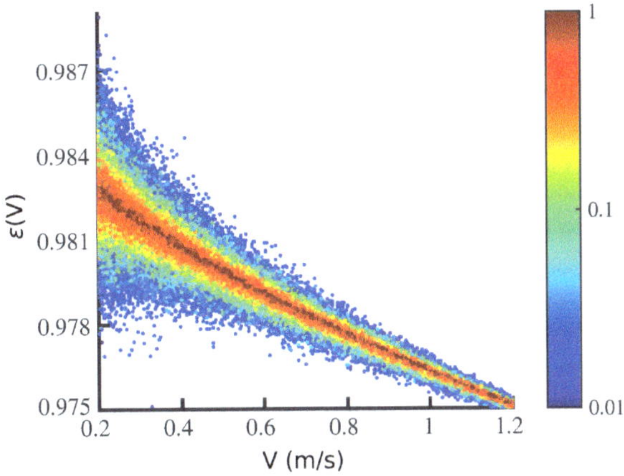

Fig. 5 Coefficient of restitution for a rough particle plotted against the impact velocity. The data are colored according to the normalized frequency of occurrence

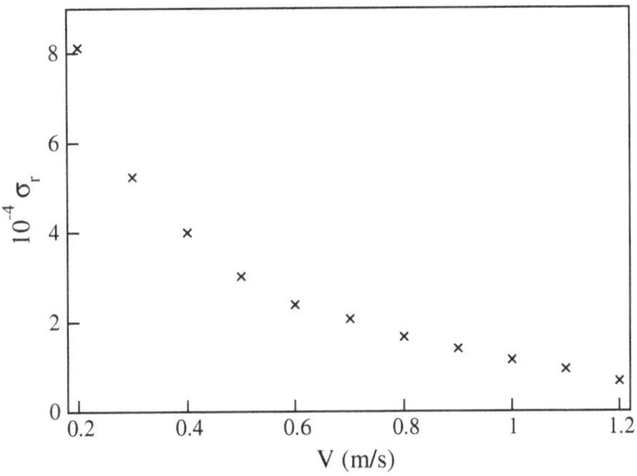

Fig. 6 Standard deviation of the data shown in Fig. 5 computed due to Eq. (10), as a function of impact velocity

other sources of experimental error. The error as a function of the impact velocity is determined by

$$\sigma_r(V) = \sqrt{\frac{1}{n(V)} \sum_{i=1}^{n(V)} [\varepsilon_i - \overline{\varepsilon}(V)]^2}$$
$$\overline{\varepsilon}(V) = \frac{1}{n(V)} \sum_{i=1}^{n(V)} \varepsilon_i \,, \tag{10}$$

where the sum is performed over all data points (ε_i, V_i) where $|V - V_i| < \delta_V$ and $n(V)$ is the number of such data points. The value of δ_V must be chosen with care since it compromises between good statistics (large δ_V) and good V-resolution (small δ_V); here we use $\delta_V = 0.01$ m/s. Figure 6 shows the standard deviation of the data shown in Fig. 5 due to Eq. (10), as a function of impact velocity, $\sigma_r(V)$. We observe that the standard deviation σ_r increases for small impact velocity.

Eccentricity

Complementary to surface roughness, that is, deviations from the perfect sphere on a microscopic scale, we consider deviations from the spherical shape on the scale of the particle size, disregarding roughness. To this end, the particle is modeled by two identical spheres of radius $\tilde{R}$ and distance L, with $L \ll \tilde{R}$, see Fig. 7. Similarly to the previous section, we perform bouncing-body simulations and determine the coefficient of restitution via Eq. (1), in analogy to the experiment [15]. Figure 8 shows the coefficient of restitution obtained from a numerical simulation of 4×10^4 bounces of the particle for two different values of the eccentricity, $L/\tilde{R} = 0.1$. The scatter increases steadily with decreasing impact velocity similar to our observations for rough particles. Again, the reason for the scatter is transfer between the translational and rotational degrees of freedom similar as discussed in the previous section. The amount of the transferred energy depends on the impact velocity and on the angular orientation of the particle at the instant of contact. The orientation may be considered as a random variable, except towards the end of the collision sequence when the height of the jumps approaches the order of L [15]. The randomness of the numerically determined coefficient of restitution characterizes the uncertainty of the measurement due to the deviation of the particle shape from the perfect sphere.

Figure 9 shows the uncertainty of the numerical measurement, quantified by the standard deviation due to Eq. (10), for eccentricity $L/\tilde{R} = 0.01$ and $L/\tilde{R} = 0.02$ corresponding to high precision glass spheres [33] as frequently used in bouncing ball experiments. Again, the standard deviation σ_r increases for small impact velocity. Even small deviations from the spherical shape lead to significant errors. In [20] it was shown that this error dominates the overall measurement error for moderate impact velocities considered in this paper. Even for high precision glass beads with typical asphericity of 1...2%, as frequently used in bouncing ball experiments the

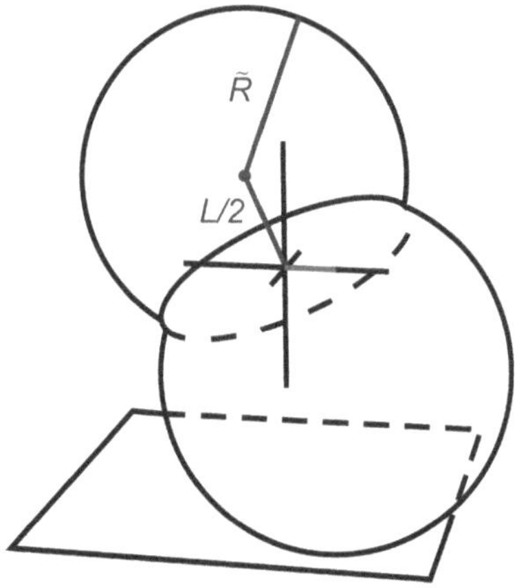

Fig. 7 Sketch of a dumbbell particle consisting of two identical spheres of radius $\tilde{R}$ and distance $L \ll R$. For better visibility, the value of L appears exaggerated

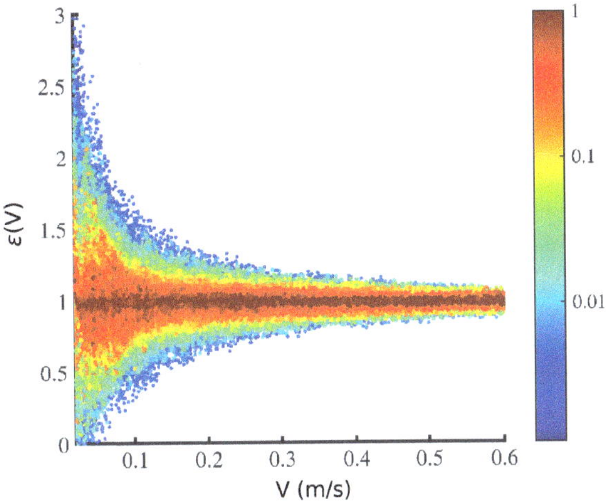

Fig. 8 The macroscopic coefficient of restitution as a function of velocity for eccentricity $L/\tilde{R} = 0.1$. The data are colored according to the normalized frequency of occurrence

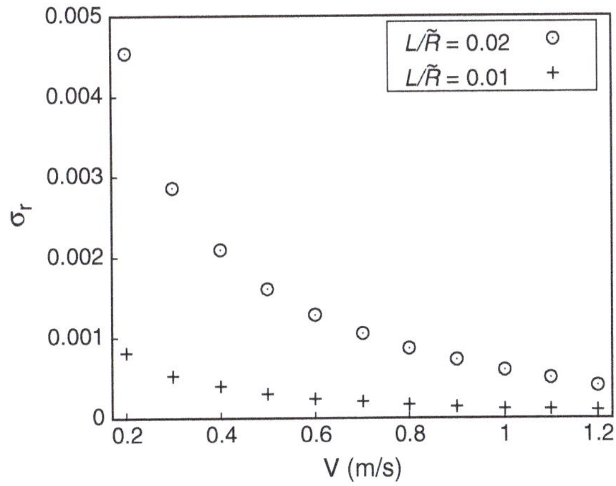

Fig. 9 Standard deviation of the numerically determined coefficient of restitution for eccentricity $L/\tilde{R} = 0.01$ and $L/\tilde{R} = 0.02$, as a function of impact velocity

standard deviation is $\sigma \approx 1 \ldots 5 \times 10^{-3}$. For larger impact velocity the error due to air drag is the largest contribution. Although errors of $\sigma \approx 1 \ldots 5 \times 10^{-3}$ seem to be small, the dissipation is quantified by the deviation from elasticity $(1 - \varepsilon)$. For spheres with asphericity $L/\tilde{R} = 0.02$, with $\sigma \approx 5 \times 10^{-3}$, the relative error of the inelasticity $(1 - \varepsilon)$ can easily exceed $20 \ldots 50$ %.

In this section, we presented experimental and numerical results suggesting to model the coefficient of restitution as a Laplacian random variable. In the next one, we show that a purely theoretical analysis leads to the same conclusion.

Stochastic Behavior of the Coefficient of Normal Restitution: Theoretical Derivation

As in section "Stochastic Behavior of the Coefficient of Normal Restitution: Experimental Measurements and Numerical Models" we consider a rigid particle consisting of a large sphere of radius R and mass M which is covered by a large number, $N \gg 1$, of tiny asperities at uniformly distributed random positions on its surface (see Fig. 3). Each of the asperities is represented by a sphere of radius $\rho \ll R$ and mass m. It is further assumed that the coverage of the surface of the central sphere is sufficiently dense such that each contact of the particle with a plane occurs through one of the asperities. For typical values $\rho/R \lesssim 10^{-4}$, thus $m/M \lesssim 10^{-12}$ and $N \sim 10^6$ we can safely neglect the contribution of the asperities to the mass and the moment of inertia of the particle. For the subsequent analysis we chose the coordinate system with the origin located at the center of the central sphere and the $(x - y)$-plane being in parallel to the plate.

Distribution of the Contact Vector r

Components of the Contact Vector

In order to compute the stochastic properties of the coefficient of normal restitution, ε, through Eqs. (1), (4) and (8), we consider the statistical distribution of the vector $\mathbf{r}$ indicating the point where the particle contacts the plane through one of its asperities, see Fig. 4. From geometry it is clear that the contacting asperity is the one which is closest to the south pole of the large central sphere, see Fig. 10.

Due to the random orientation of the particle and the random distribution of the asperities on its surface, the components of $\mathbf{r} = (r_x, r_y, r_z)$ obey probability densities

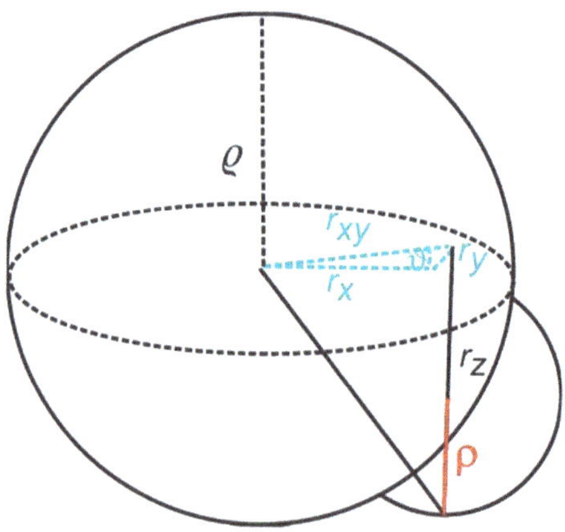

Fig. 10 Definition of variables describing the collision geometry

which will be computed in this section. The components r_x and r_y can be expressed by the length of the projection of $\mathbf{r}$ to the plane, $r_{xy} \equiv \sqrt{r_x^2 + r_y^2}$, and the angle ϑ (see Fig. 10) such that

$$\begin{aligned} r_x &= r_{xy} \cos \vartheta , \\ r_y &= r_{xy} \sin \vartheta . \end{aligned} \tag{11}$$

The remaining coordinate,

$$r_z = -\sqrt{R^2 - r_{xy}^2} - \rho . \tag{12}$$

In the following subsections, we compute the probability densities of r_{xy}, $\cos \vartheta$, $\sin \vartheta$ and derive then the probability densities of the components of $\mathbf{r}$.

Probability Density of r_{xy}

The probability to find k homogeneously distributed asperities in a small circle of area A obeys a Binomial distribution which can be approximated by a Poisson distribution for $N \gg 1$:

$$p_k(A) = \frac{e^{-\lambda A} (\lambda A)^k}{k!} , \tag{13}$$

where $\lambda = N / \left(4\pi R^2\right)$ follows from equating the total number of asperities N on the surface of the central sphere with the expectation value of k due to $p_k \left(4\pi R^2\right)$. We specify A as the area of a circle with radius r_{xy}, see Fig. 10. We approximated the area of the cap of the sphere by the area of a flat circle of the same radius leading to negligible errors for the parameters chosen in this work [17].

From these considerations follows the cumulative probability, $P_{r_{xy}}$, to find at least one asperity in a circle of radius r_{xy},

$$P_{r_{xy}}(t) = 1 - \exp\left(-\lambda \pi t^2\right), \quad t \geq 0, \tag{14}$$

and by differentiating we obtain the probability density to find exactly one asperity in the circle:

$$p_{r_{xy}}(t) = 2\pi t \lambda \exp\left(-\lambda \pi t^2\right), \quad t \geq 0. \tag{15}$$

Angular Location of the Contact Point in the Plane

Given the particle contacts the plane at the distance r_{xy} from the z-axis, the possible locations of the contact point are on a circle of radius r_{xy} where no angular orientation, ϑ, is preferred. Thus, the angle is homogeneously distributed in the interval $\vartheta \in [0, 2\pi)$. Consequently, the cumulative probabilities of $\sin \vartheta$ and $\cos \vartheta$ read

$$
\begin{aligned}
P_{\cos\vartheta}(t) &\equiv P(\cos\vartheta \le t) \\
&= 1 - [P(\vartheta \in [0, \arccos t)) \\
&\quad + P(\vartheta \in (2\pi - \arccos t, 2\pi))] \\
&= 1 - \frac{1}{\pi}\arccos t.
\end{aligned} \tag{16}
$$

Analogously, we find:

$$
\begin{aligned}
P_{\sin\vartheta}(t) &\equiv P(\sin\vartheta \le t) \\
&= 1 - \frac{1}{\pi}\arccos(t).
\end{aligned} \tag{17}
$$

Differentiating Eqs. (16) and (17) we obtain the probability densities

$$
\begin{aligned}
p_{\cos\theta}(t) &= \frac{1}{\pi}\frac{1}{\sqrt{1-t^2}}, \\
p_{\sin\vartheta}(t) &= \frac{1}{\pi}\frac{1}{\sqrt{1-t^2}}.
\end{aligned} \tag{18}
$$

According to Eq. (11), the components of $\mathbf{r}$ are products of r_{xy} and $\cos\vartheta$ or $\sin\vartheta$, respectively. Since the distributions of r_{xy} and ϑ (and, thus, $\cos\vartheta$ and $\sin\vartheta$) are statistically independent [31], we obtain the distribution of the product via

$$
p_{r_x}(t) = p_{r_y}(t) = p_{r_{xy}\cos\vartheta}(t) = p_{r_{xy}\sin\vartheta}(t) = \sqrt{\lambda}e^{-\pi\lambda t^2}. \tag{19}
$$

We see that the probability density distribution for r_x and r_y is a normal distribution. For more details of the calculation and numerical verifications see [17].

Distribution of the Coefficient of Normal Restitution

To obtain the distribution of the coefficient of normal restitution, we use Eq. (4) together with the definition, Eq. (1),

$$
\varepsilon = -1 - \frac{1}{M}\frac{\mathbf{P}\cdot\mathbf{n}}{\mathbf{V}\cdot\mathbf{n}}. \tag{20}
$$

The impulse $\mathbf{P}$ is found by solving the system Eq. (8):

$$
\mathbf{P} = -K\left(\mathbf{V}, \boldsymbol{\Omega}, \mathbf{r}\right) M \begin{pmatrix} r_x r_z \\ r_y r_z \\ I + r_z^2 \end{pmatrix}, \tag{21}
$$

with

$$K(\mathbf{V}, \boldsymbol{\Omega}, \mathbf{r}) \equiv [1 + \epsilon_n(v)] \frac{\left(\mathbf{V} \cdot \mathbf{n} + \Omega_x r_y - \Omega_y r_x\right)\left(I + r_z^2\right)}{\mathbf{r}^2 + I}, \tag{22}$$

with the components of the pre-collisional angular velocity $\boldsymbol{\Omega} = \left(\Omega_x, \Omega_y, \Omega_z\right)$ and the velocity at the contact point in z-direction, $V = (\mathbf{V} + \boldsymbol{\Omega} \times \mathbf{r}) \cdot \mathbf{n}$ and $I \equiv J/M$. Inserting in Eq. (20) yields

$$\varepsilon = -1 + \frac{K(\mathbf{V}, \boldsymbol{\Omega}, \mathbf{r})\left(I + r_z^2\right)}{\mathbf{V} \cdot \mathbf{n}}. \tag{23}$$

The moment of inertia of the particle is

$$J = \frac{2}{5} M R^2. \tag{24}$$

The solution of Eq. (8) for the post-collisional angular velocity reads

$$\boldsymbol{\Omega}' = \boldsymbol{\Omega} + K(\mathbf{V}, \boldsymbol{\Omega}, \mathbf{r}) \begin{pmatrix} r_y \\ -r_x \\ 0 \end{pmatrix}. \tag{25}$$

The post-collisional linear velocity reads

$$\mathbf{V}' = \mathbf{V} - K(\mathbf{V}, \boldsymbol{\Omega}, \mathbf{r}) \begin{pmatrix} r_x r_z \\ r_y r_z \\ I + r_z^2 \end{pmatrix}. \tag{26}$$

For the computation of ε according to Eq. (1) only the orientation of the z-axis of the coordinate system is important. The orientation of the y- and z-axis in the plane is arbitrary and can be chosen for each collision such that $\Omega_x = -\Omega_y$ which simplifies Eq. (22). The fluctuating quantity $r_x r_y$ has mean zero and a very small variance which suggests the approximation $r_x + r_y \approx \left(r_x^2 + r_y^2\right)^{1/2} = r_{xy}$, whose distribution is known, Eq. (15). Then Eq. (22) adopts the form

$$K(\mathbf{V}, \boldsymbol{\Omega}, \mathbf{r}) \equiv [1 + \epsilon_n(v)] \frac{\left(\mathbf{V} \cdot \mathbf{n} + \Omega_x r_{xy}\right)}{\mathbf{r}^2 + I}. \tag{27}$$

From geometry (see Fig. 10; $R \gg \rho$) it is clear that r_z has a much smaller fluctuation range than r_x and r_y, therefore, r_z can be assumed constant with $r_z = R + \rho$.

With these approximations, K and, thus, ε, Eq. (23), contain only a single fluctuating quantity which is r_{xy}. Hence, the distribution of $\varepsilon = g\left(r_{xy}\right)$ can be calculated by the transformation

$$p_\varepsilon(t) = p_{r_{xy}}\left[g^{-1}(t)\right] \frac{dg^{-1}(t)}{dt}, \tag{28}$$

where $p_{r_{xy}}$ is the probability density function of r_{xy}. From Eqs. (23) and (27) we obtain

$$p_\varepsilon(t) = -\frac{\pi\lambda\,[Q\Omega_x + f(t)]\,Q\,[h(t) + f(t)\Omega_x]}{2V_z^2(1+t)^3 f(t)} \exp\left[-\frac{\pi\lambda(h+f(t))^2}{4(1+t)^2V_z^2}\right] \quad (29)$$

$$f(t) \equiv \sqrt{[(1+\epsilon_n(v)]\,Q\Omega_x^2 + 4V_z^2\left[-t^2 + t\epsilon_n(v) - t + \epsilon_n(v)\right](I + r_z^2)} \quad (30)$$

$$h(t) \equiv 2(1+t)V_z^2 + Q\Omega_x^2 \quad (31)$$

$$Q \equiv [1+\epsilon_n(v)]\,(I + r_z^2). \quad (32)$$

The component Ω_x' can be expressed in terms of r_y:

$$\begin{aligned} \Omega_x' &= \Omega_x + \frac{r_z P_y - r_y P_z}{I} \\ &= \Omega_x + \frac{1}{I}\left[\frac{P_z r_z^2 r_y}{r_z^2 + I} - r_y P_z\right]. \end{aligned} \quad (33)$$

Consequently, as r_y obeys a Gaussian distribution and r_z is approximately constant, Ω_x' is Gaussian distributed with zero mean as well. Its standard deviation reads

$$\sigma = \frac{P_z}{I\sqrt{2\lambda\pi}}\left(1 + \frac{r_z^2}{r_z^2 + I}\right), \quad (34)$$

where $P_z = MV_z\,(-\varepsilon - 1)$ is the $z-$component of the collisional impulse. For more details see [17].

Coefficient of Restitution for a Rough Ball: Analytic Predictions

In bouncing ball experiments [38], the coefficient of restitution of a rough ball was characterized as a fluctuating quantity obeying an asymmetric Laplace distribution, Eq. (3). In order to compare the analytical result for the probability distribution of a rough sphere, Eq. (29) with the experiment, we perform a corresponding Monte Carlo simulation using a particle of radius $R = 3$ mm covered by 10^6 asperities of size $\rho = 5 \times 10^{-4}$ mm. We assume the initial velocity, $V_z = \mathbf{V} \cdot \mathbf{n} = 1$ m/s and angular velocity $\boldsymbol{\Omega} = \mathbf{0}$ and chose ε from the analytically obtained distribution Eq. (29) to obtain V_z' via Eq. (1) and we randomly generate Ω_x' according to Eqs. (33), (34) and update the impact velocity using Eq. (26). This procedure is iterated until the linear velocity decreases to 0.1 m/s. This process corresponds to the bounc-

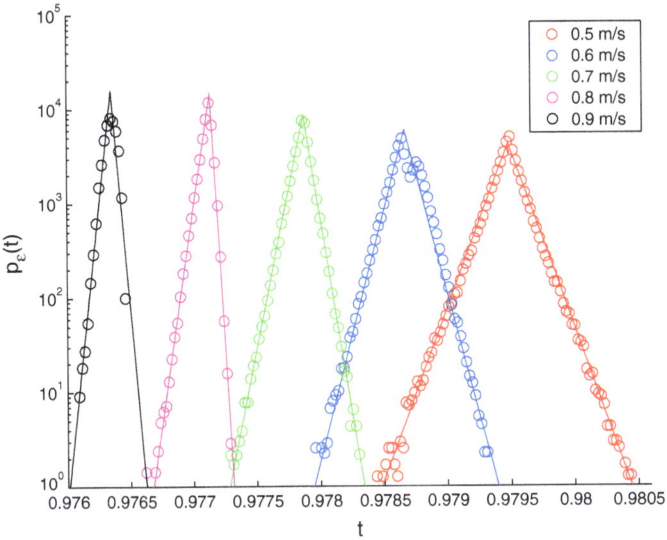

Fig. 11 Probability density of the coefficient of restitution, ε, due to Eq. (29) for different values of the impact velocity. For the Monte Carlo simulation we assume a particle of size $\varrho = 3\,\text{mm}$ covered by 10^6 asperities of size $\rho = 5 \cdot 10^{-4}\,\text{mm}$. The distribution is approximately Laplacian (see exponential fits, full lines) and shows good agreement with the experimental data, Fig. 2

ing ball experiment [38]. From the statistical data obtained by repeating the process 10^5 times, we can draw the histogram for the coefficient of restitution, ε. Figure 11 shows 5 histograms of ε for different impact velocities. The resulting probability density is approximately an asymmetric Laplace distribution of the form Eq. (3). Comparison with Fig. 2 shows good agreement with the experimental data.

In this section, the conjecture of the Laplacian shape in section "Stochastic Behavior of the Coefficient of Normal Restitution: Experimental Measurements and Numerical Models" could be justified by considering the statistical properties of the impact geometry of a rough particle. The origin of the fluctuations are thus imperfections of the sphere. In the next section, it is shown how these microscopic irregularities can change the dynamics of rapid granular flows.

Impact of the Stochastic Nature on Cooling Properties of Rapid Granular Flows

Decay of Temperature

In a dilute granular system the absolute velocities of the particles decay on average because of the dissipative nature of the collisions, characterized by $\varepsilon < 1$. This may be expressed by a decay of temperature. Using Eq. (1) the post-collisional velocity of two particles i and j can be obtained via

$$\mathbf{V}_i = \mathbf{V}_i' - \frac{(1+\varepsilon)}{2} V_{ij}' \mathbf{n}, \tag{35}$$

where

$$V_{ij}' = \left(\mathbf{V}_i' - \mathbf{V}_j'\right) \cdot \mathbf{n}. \tag{36}$$

If the coefficient of restitution is assumed independent of the impact velocity, $\varepsilon = \text{const}$, Haff's law is valid [19],

$$T(\tau) = \frac{T_0}{(1+\tau/\tau_0)^2} \quad \text{with} \quad \tau_0^{-1} \propto nd^2\left(1-\varepsilon^2\right)\sqrt{T_0}, \tag{37}$$

where T_0 is the initial temperature, that is asymptotically $T(\tau) \sim \tau^{-2}$ with time τ. Haff's law is the most simple expression for granular cooling. The asymptotic form has been confirmed experimentally, where gravity was compensated by diamagnetic levitation [35]. However, for realistic particles, the coefficient of restitution is a function of the impact velocity, $\varepsilon = \varepsilon(V)$. In this case, it can be shown that for viscoelastic particles, the decay of temperature reads

$$T(\tau) = \frac{T_0}{\left(1+\tau/\tau_0'\right)^{5/3}}, \tag{38}$$

that is, asymptotically $T(\tau) \sim \tau^{-5/3}$ [9].

Decay of Temperature—Kinetic Theory

At the kinetic level, the system is described by a distribution function for the velocities, $f(\mathbf{V})$, by mean of which one can express the number density field, and temperature, as:

$$n = \int f(\mathbf{V})\, d\mathbf{V} \tag{39}$$

$$T = \frac{1}{n}\int f(\mathbf{V})\frac{mV^2}{3} d\mathbf{V}. \tag{40}$$

The equation of motion for T is given by:

$$n\frac{\partial T}{\partial \tau} = -\gamma \tag{41}$$

where γ is the cooling coefficient, describing the rate of decay of the kinetic energy (multiplied by 2/3), due to the inelasticity of the collisions. Using the velocity transformations, Eq. (35), the change of energy ΔE in a collision between particles with velocities $\mathbf{V}_1$ and $\mathbf{V}_2$, along the direction spanned by the unit vector $\mathbf{e}$, and characterized by a coefficient of restitution ε, is given by:

$$\Delta E = -\frac{\left(1-\varepsilon^2\right)}{2} M \left(\mathbf{V}_{12} \cdot \mathbf{e}\right)^2 . \tag{42}$$

In addition, the collision frequency ν_{12} of such collisions is

$$\nu_{12} = d^2 f\left(\mathbf{V}_1\right) f\left(\mathbf{V}_2\right) \left|\mathbf{V}_{12} \cdot \mathbf{e}\right| d\mathbf{V}_1 d\mathbf{V}_2 d\mathbf{e}, \tag{43}$$

where d is the diameter of the particles. Therefore, the cooling coefficient is given in terms of the distribution function by:

$$\gamma = \frac{M}{12} d^2 \int\int\int\int p_\varepsilon(t) f\left(\mathbf{V}_1\right) f\left(\mathbf{V}_2\right) \left(1-\varepsilon^2\right) \left(\mathbf{V}_{12} \cdot \mathbf{e}\right)^3 \\ H\left(\mathbf{v}_{12} \cdot \mathbf{e}\right) d\mathbf{V}_1 d\mathbf{V}_2 d\mathbf{e}\, d\varepsilon \tag{44}$$

where a factor $1/2$ has been included to avoids double counting of collisions, $p_\varepsilon(t)$ denote the probability of the coefficient of restitution ε, depending on the normal relative velocity V, H is the Heaviside step function, and the integration range of ε is $[-\infty, +\infty]$. The distribution function depends on probability of the coefficient of restitution ε. In order to take into account this dependence, one needs to solve the Boltzmann equation pertaining to the system. Here, we will for sake of simplicity neglect this dependence and assume that the deviation of the velocity distribution function from the Maxwell distribution

$$f^{(0)} = n \left(\frac{M}{2\pi T}\right)^{3/2} e^{-\frac{MV^2}{2T}} \tag{45}$$

is small. For more details on this assumption see [18].

If we approximate the distribution function by a Maxwellian, $f(\mathbf{V}) \approx f^{(0)}(\mathbf{V})$, the cooling coefficient γ reads

$$\gamma = \frac{M}{12} d^2 \int\int\int \left(1-\langle\varepsilon^2\rangle\right) f^{(0)}\left(\mathbf{V}_1\right) f^{(0)}\left(\mathbf{V}_2\right) \left(\mathbf{V}_{12} \cdot \mathbf{e}\right)^3 \\ H\left(\mathbf{V}_{12} \cdot \mathbf{e}\right) d\mathbf{V}_1 d\mathbf{V}_2 d\mathbf{e} \tag{46}$$

where brackets denote an average over ε. It appears that the evolution of the temperature deviates from the decay corresponding to the case of non-fluctuating coefficient of restitution due both to the variance of the distribution, and to the asymmetry of the distribution function, $p_\varepsilon(t)$ (i.e, $\langle\varepsilon^2\rangle \neq \varepsilon_{max}^2$). While the former has negligible effect, the latter has a qualitative influence on the asymptotic evolution of the temperature.

We can reformulate the Laplacian distribution in Eq. 3 in the following way:

$$p_\varepsilon(t) = \begin{cases} \frac{1}{\sigma_p(V)+\sigma_m(V)} e^{-\frac{t-\varepsilon_{\max}(V)}{\sigma_p(V)}} & \text{if } t \geq \varepsilon_{\max}(V) \\ \frac{1}{\sigma_p(V)+\sigma_m(V)} e^{\frac{t-\varepsilon_{\max}(V)}{\sigma_m(V)}} & \text{else}\,. \end{cases} \tag{47}$$

By using the coefficients $\sigma_m(V)$ and $\sigma_p(V)$ obtained from simulation data and theory (cf. Sects. 2 and 3), the average coefficient of restitution corresponding to the distribution (47):

$$\langle\varepsilon\rangle(V) = \varepsilon_{\max}(V) - \left[\sigma_m(V) - \sigma_p(V)\right] \tag{48}$$

is found to be well described by a function of the form

$$\langle\varepsilon\rangle = 1 - \delta^* V^\eta \quad \text{with} \quad \delta^* = 0.05317\,; \quad \eta = \frac{4}{25}, \tag{49}$$

(for more details see [18]). A straightforward calculation then yields:

$$\gamma = \frac{288}{625} 2^{\frac{33}{50}} \sqrt{\pi}\,\Gamma\left(\frac{2}{25}\right) n^2 d^2 \frac{1}{m^{\frac{29}{50}}} T^{\frac{79}{50}} \delta^*, \tag{50}$$

i.e, solving Eq. (41):

$$T(\tau) = \frac{T_0}{\left[1 + \frac{4176}{15625}\left(2^{\frac{33}{50}}\right)\Gamma\left(\frac{2}{25}\right)\sqrt{\pi} n d^2 \delta^* \left(\frac{T_0}{m}\right)^{\frac{29}{50}} \tau\right]^{\frac{50}{29}}} \tag{51}$$

that is,

$$T = \frac{T_0}{(1+\frac{\tau}{\tau^\rho})^{50/29}} \xrightarrow{\tau\to\infty} \sim \tau^{-50/29} \tag{52}$$

with the constant τ^ρ defined by Eq. (51). Comparing this result with the decay of temperature of a gas of viscoelastic particles [51, 52], $T \sim \tau^{-5/3}$, Eq. (38), we obtain that the fluctuating coefficient of restitution lets the gas cool noticeable faster but slower than gases of particles interacting via $\varepsilon = \text{const}$, where $T \sim \tau^{-1/2}$, according to Haff's law [19].

Decay of Temperature—Numerical Simulation

Numerical Description of the Stochastic Coefficient of Restitution

For the numerical simulation in the following subsection we need random numbers distributed according to $p_\varepsilon(t)$ as given by Eq. (3) where $\varepsilon_{\max}$ in turn is given by

Eq. (9). Using the *inverse transformation method* [44], these random numbers have to be computed from standard random numbers, x, equally distributed in the interval $x \in [0, 1)$, available on the computer. The cumulative distribution function

$$\varrho_\varepsilon(t) = \int_{-\infty}^{t} p_\varepsilon(\tilde{t})\, d\tilde{t} \tag{53}$$

maps each value of ε to a number in the interval $[0, 1)$. So if ϱ_ε can be inverted, $\varrho_\varepsilon^{-1}(x)$, obeys the distribution p_ε if x is an equally distributed random number, $x \in [0, 1)$.

The inverse of the cumulative distribution function of Eq. (3) reads

$$\varrho_\varepsilon^{-1}(x) = \begin{cases} \frac{-b}{a} + \frac{1}{a}\ln(ax) & \varrho^{-1}(x) < \varepsilon_{\max}(V) \\ \frac{-d}{c} + \frac{1}{c}\ln\left(cx - \frac{c}{a}e^{a\varepsilon_{\max}+b} + e^{c\varepsilon_{\max}+d}\right) & \text{else.} \end{cases} \tag{54}$$

In this way, we obtain random numbers distributed according to $p_\varepsilon(t)$ from equidistributed random numbers $x \in [0, 1)$.

Note that the functions ϱ_ε^{-1} depend on the impact velocity through $\varepsilon_{\max}(V)$. We use the Padé approximation of order [1/4] as described in Eq. (9).

Direct Simulation Monte-Carlo Details

To study the decay of temperature in the homogeneous cooling state, we apply *Direct Simulation Monte Carlo* (DSMC) [4, 43]. DSMC simulations of dilute granular gases have been first performed by Brey et al. [7]. DSMC is by orders of magnitude faster than MD (event-driven as well as force controlled MD), however, its main advantage compared to Molecular Dynamics is that DSMC is *not* a particle simulation but a tool to integrate the Boltzmann (or Boltzmann-Enskog) equation by means of a Monte-Carlo procedure. This allows us to compare the results of a DSMC simulation directly with the results obtained from the Kinetic Theory. When investigating a granular gas in the homogeneous cooling state, MD is restricted to almost elastic particles, $\varepsilon \lesssim 1$, and low density, to avoid formation of correlations of different kind, conflicting the homogeneous state. In contrast, DSMC works for all values of the coefficient of restitution and density by avoiding correlations. We wish to remark here that large part of the literature confirms that influence of the correlation of particle velocities due to inelastic collisions on the cooling rate is small, e.g. [6, 22, 24, 42, 49], which has, however, also been questioned [5].

In our simulation, at each time step a pair of (quasi-) particles is selected randomly with equal probability. A collision according to the collision rule, Eq. (35), was executed if

$$|\mathbf{e} \cdot \mathbf{v}_{12}| > \text{RND}\, 7 v_T\,, \tag{55}$$

where RND is a uniformly distributed random number from the interval RND $\in$ [0, 1) and v_T is the thermal velocity. To compute the post-collisional velocities using Eq. (35) we determined a random unit-vector **e**. The term $7v_T$ is an estimate for the smallest upper bound of the average normal velocity, see [43] for justification of this term. The simulation was performed with 10^5 quasi-particles, initialized with a Maxwellian velocity distribution according to the thermal velocity $T_0 = 1$.

Decay of Temperatures Obtained From DSMC Simulations

We simulated a granular gas of particles interacting via a fluctuating coefficient of restitution due to Eq. (3) in the homogeneous cooling state and computed the granular temperature as a function of time via $T = \frac{1}{3}m\overline{v^2}$. For large times, the temperature decays according to a power law, $T \sim \tau^{-k}$, with the best fit $k = 1.71$ being in excellent agreement with the result obtained from Kinetic Theory, Eq. (52), $k = 50/29 \approx 1.7241$. Figure 12 shows the DSMC simulation data with a best fit of the function

$$T = \frac{T_0}{(1 + \tau/\tau_f)^{1.71}}, \tag{56}$$

with a fit value for τ_f together with the analytical result, Eq. (52). The asymptotic decay due to Haff's law [19] (for $\varepsilon = \text{const}$) and for a gas of viscoelastic particles [51] is also shown.

The result deviates from Haff's law ($k = 2$) [19] as well as from the evolution of temperature for a gas of viscoelastic particles ($k = 5/3$) [51]. Note that the devi-

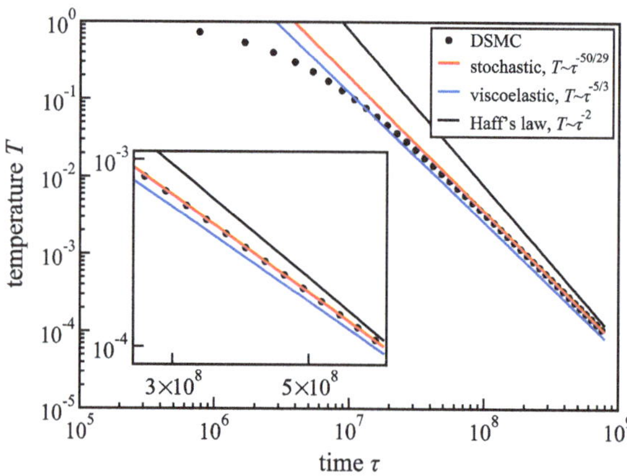

Fig. 12 Evolution of the granular temperature of a gas of 10^5 particles interacting via a fluctuating coefficient of restitution. The full line shows the analytical result, Eq. (52), the symbols show the simulation results. The dashed lines mark the asymptotic behavior for gases of viscoelastic particles, $T \sim \tau^{-5/3}$ [51], and particles interacting via a constant coefficient of restitution, Haff's law, $T \sim \tau^{-2}$ [19]

ation from the case of viscoelastic particles originates from the asymmetry of the probability function, $p_{\bar{\varepsilon}}$. Although the median of the coefficient of normal restitution equals the expression for viscoelastic spheres [51, 52], the mean value deviates considerably because of the skewness of the functions.

In a further study [53], the influence of the stochastic coefficient of restitution to dilute binary mixtures was examined by hydrodynamics and DSMC simulations. The stochasticity leads to quantitative differences in comparison to the non-fluctuating case depending on the shape of the distribution function. It was shown that differences in the shape of the probability distribution are sufficient to produce partial segregation in vertically vibrated systems.

Conclusion

A common technique to measure the coefficient of restitution exploits the analysis of the sound signal of a particle bouncing repeatedly off the ground. A detailed analysis of such large scale measurements reveals that the measured values of the coefficient of restitution show extraordinary large scatter; more than what could be expected from an error analysis. We have shown that this scatter can be attributed to small imperfections of the spherical shape of the particles.

Considering the coefficient of restitution as a fluctuating quantity, by means of experiments and theory, we have quantified the corresponding distribution function and related the characteristics of the distribution to the typical size of the surface imperfections. It turns out that for slightly non-spherical rough particles, the fluctuations are characterized by a rather uncommon asymmetric Laplace probability density.

Further, we applied the concept of a fluctuating coefficient of restitution to large assemblies of granular particles in a dilute state (rapid granular flow). Due to the non-trivial coupling of translational and rotational degrees of freedom substantial differences in the kinetic properties of rapid granular flows occur. The decay of temperature of rapid granular flows in the homogeneous cooling state deviates from Haff's law for gases of particles interacting via a constant coefficient of restitution also from the scaling law for gases of viscoelastic particles. Future research will cover impacts of the stochasticity on denser granular systems including pipe flow and sheared systems [16], granular jets [39] and vibrating systems [27].

Acknowledgements We acknowledge funding by Deutsche Forschungsgemeinschaft (DFG) through the Cluster of Excellence "Engineering of Advanced Materials" and through project PO 472/25. We thank the colleagues involved in the SPP for intensive discussion and steady support. NG acknowledges support by "Simulation Science Center Clausthal-Göttingen". AG acknowledges funding from the Engineering and Physical Sciences Research Council (EPSRC) through the Centre for Doctoral Training "Cross Disciplinary Approaches to Non-equilibrium Systems" (CANES, Grant No. EP/L015854/1) and thanks the Deutsche Akademische Austauschdienst for a RISE scholarship.

References

1. Aguiar, C.E., Laudares, F.: Listening to the coefficient of restitution and the gravitational acceleration of a bouncing balls. Am. J. Phys. **71**, 499–501 (2002)
2. Becker, V., Schwager, T., Pöschel, T.: Coefficient of tangential restitution for the linear dashpot model. Phys. Rev. E **77**, 011304 (2008)
3. Bernstein, A.D.: Listening to the coefficient of restitution. Am. J. Phys. **45**, 41–44 (1977)
4. Bird, G.A.: Molecular Gas Dynamics and the Direct Simulation of Gas Flows. Clarendon Press, Oxford (1994)
5. Brenig, L., Salazar, J.M.: Exact results for the homogeneous cooling state of an inelastic hard-sphere gas. J. Plasma Phys. **59**, 639–646 (1998)
6. Brey, J.J., Prados, A., García, M.I., Maynar, P.: Scaling and aging in the homogeneous cooling state of a granular fluid of hard particles. J. Phys. A Math. Theor. **40**, 14331–14342 (2007)
7. Brey, J.J., Ruiz-Montero, M.J., Cubero, D.: Homogeneous cooling state of a low-density granular flow. Phys. Rev. E **54**, 3664 (1996)
8. Bridges, F.G., Hatzes, A., Lin, D.N.C.: Structure, stability and evolution of saturn's rings. Nature **309**, 333–335 (1984)
9. Brilliantov, N., Pöschel, T.: Kinetic Theory of Granular Gases. Oxford University Press, Oxford (2003)
10. Brilliantov, N.V., Spahn, F., Hertzsch, J.-M., Pöschel, T.: A model for collisions in granular gases. Phys. Rev. E **53**, 5382–5392 (1996)
11. Ciocca, M., Wang, J.: Watching and listening to the coefficient of restitution. J. Ky. Acad. Sci. **72**, 100–104 (2002)
12. Falcon, E., Laroche, C., Fauve, S., Coste, C.: Behavior of one inelastic ball bouncing repeatedly off the ground. Eur. Phys. J. B **3**, 45–57 (1998)
13. Foerster, S.F., Louge, M.Y., Chang, H., Allia, K.: Measurements of the collision properties of small spheres. Phys. Fluids **6**, 1108–1115 (1994)
14. Foong, S.K., Kiang, D., Lee, P., March, R.H., Paton, B.E.: How long does it take a bouncing ball to bounce an infinite number of times? Phys. Educ. **39**(1), 40–43 (2004)
15. Glielmo, A., Gunkelmann, N., Pöschel, T.: Coefficient of restitution of aspherical particles. Phys. Rev. E **90**, 052204 (2014)
16. Goldhirsch, I.: Rapid granular flows. Annu. Rev. Fluid Mech. **35**, 267–293 (2003)
17. Gunkelmann, N., Montaine, M., Pöschel, T.: Stochastic behavior of the coefficient of normal restitution. Phys. Rev. E **89**, 022205 (2014)
18. Gunkelmann, N., Serero, D., Pöschel, T.: Temperature of a granular gas with regard to the stochastic nature of particle interactions. New J. Phys. **15**(9), 093030 (2013)
19. Haff, P.: Grain flow as a fluid-mechanical phenomenon. J. Fluid Mech. **134**, 401 (1983)
20. Heckel, M., Glielmo, A., Gunkelmann, N., Pöschel, T.: Can we obtain the coefficient of restitution from the sound of a bouncing ball? Phys. Rev. E **93**, 032901 (2016)
21. Hodgkinson, E.: On the effect of impact on beams. J. Franklin Inst. **19**(2), 140–141 (1835)
22. Huthmann, M., Orza, J.A.G., Brito, R.: Dynamics of deviations from the gaussian state in a freely cooling homogeneous system of smooth inelastic particles. Gran. Matter **2**, 189–199 (2000)
23. Jiang, Z., Liang, Z., Wu, A., Zheng, R.: Effect of collision duration on the chaotic dynamics of a ball bouncing on a vertically vibrating plate. Physica A Stat. Mech. Appl. **494**, 380–388 (2018)
24. Kawahara, R., Nakanishi, H.: Effects of velocity correlation on early stage of free cooling process of inelastic hard sphere system. J. Phys. Soc. Jpn. **73**, 68–75 (2004)
25. King, H., White, R., Maxwell, I., Menon, N.: Inelastic impact of a sphere on a massive plane: nonmonotonic velocity-dependence of the restitution coefficient. Europhys. Lett. **93**, 14002 (2011)
26. Koller, M.G., Kolsky, H.: Waves produced by the elastic impact of spheres on thick plates. Int. J. Solids Struct. **23**, 1387–1400 (1987)

27. Kollmer, J.E., Sack, A., Heckel, M., Zimber, F., Mueller, P., Bannerman, M.N., Pöschel, T.: Collective granular dynamics in a shaken container at low gravity conditions. AIP Conf. Proc. **1542**, 811–814 (2013)
28. Kuwabara, G., Kono, K.: Restitution coefficient in a collision between two spheres. Jpn. J. Appl. Phys. **26**, 1230–1233 (1987)
29. Labous, L., Rosato, A.D., Dave, R.N.: Measurements of collisional properties of spheres using high-speed video analysis. Phys. Rev. E **56**, 5717–5725 (1997)
30. Leconte, M., Garrabos, Y., Palencia, F., Lecoutre, C., Evesque, P., Beysens, D.: Inelastic ball-plane impact: an accurate way to measure the normal restitution coefficient. Appl. Phys. Lett. **89**(24), 243518 (2006)
31. Lewis, P., Shedler, G.: Simulation of nonhomogeneous poisson processes by thinning. Nav. Res. Logist. Quart. **26**, 403 (1979)
32. Lifshitz, J.M., Kolsky, H.: Some experiments on anelastic rebound. J. Mech. Phys. Solids **12**, 35–43 (1964)
33. Lorenz, A., Tuozzolo, C., Louge, M.Y.: Measurements of impact properties of small, nearly spherical particles. Exp. Mech. **37**, 292–298 (1997)
34. Louge, M.Y., Adams, M.E.: Anomalous behaviour of normal kinematic restitution in the oblique impacts of a hard sphere on an elastoplastic plate. Phys. Rev. E **65**, 021303 (2002)
35. Maaß, C.C., Isert, N., Maret, G., Aegerter, C.M.: Experimental investigation of the freely cooling granular gas. Phys. Rev. Lett. **100**, 248001 (2008)
36. Maynes, K.C., Compton, M.G., Baker, B.: Coefficient of restitution measurements for sport balls: an investigative approach. Phys. Teach. **43**(6), 352–354 (2005)
37. McLaskey, G.C., Glaser, S.D.: Hertzian impact: experimental study of the force pulse and resulting stress waves. J. Acoust. Soc. Am. **128**, 1087–1096 (2010)
38. Montaine, M., Heckel, M., Kruelle, C., Schwager, T., Pöschel, T.: Coefficient of restitution as a fluctuating quantity. Phys. Rev. E **84**, 041306 (2011)
39. Müller, P., Formella, A., Pöschel, T.: Granular jet impact: granular-continuum transition. J. Fluid Mech. (2013)
40. Müller, P., Pöschel, T.: Collision of viscoelastic spheres: compact expressions for the coefficient of normal restitution. Phys. Rev. E **84**, 021302 (2011)
41. Nunn, J.: The bounce meter. Phys. Educ. **49**(3), 303–309 (2014)
42. Pöschel, T., Brilliantov, N.V., Schwager, T.: Violation of molecular chaos in dissipative gases. Int. J. Mod. Phys. C **13**, 1263–1272 (2002)
43. Pöschel, T., Schwager, T.: Computational Granular Dynamics. Models and Algorithms. Springer, New York (2005)
44. Press, W.H., Teukolsky, S.A., Vetterling, W.T., Flannery, B.P.: Numerical Recipes in C, 2nd edn. Cambridge University Press, Cambridge (1992)
45. Raman, C.V.: The photographic study of impact at minimal velocities. Phys. Rev. E **15**, 442–447 (1918)
46. Ramírez, R., Pöschel, T., Brilliantov, N., Schwager, T.: Coefficient of restitution of colliding spheres. Phys. Rev. E **60**, 4465–4472 (1999)
47. Reagle, C.J., Delimont, J.M., Ng, W.F., Ekkad, S.V., Rajendran, V.P.: Measuring the coefficient of restitution of high speed microparticle impacts using a PTV and CFD hybrid technique. Meas. Sci. Technol. **24**, 105303 (2013)
48. Ribeiro, T.C., Romero, J.F.A., Romero, J.A.A.: Novel techniches for experimental determination of the restitution coefficient by means of acoustic signal analysis. ABCM Symp. Ser. Mech. **5**, 676–683 (2012)
49. Santos, A., Kremer, G.M., dos Santos, M.: Sonine approximation for collisional moments of granular gases of inelastic rough spheres. Phys. Fluids **23**, 030604 (2011)
50. Schwager, T., Becker, V., Pöschel, T.: Coefficient of tangential restitution for viscoelastic spheres. Eur. Phys. J. E **27**, 107 (2008)
51. Schwager, T., Pöschel, T.: Coefficient of restitution for viscous particles an cooling rate of granular gases. Phys. Rev. E **57**, 650–654 (1998)

52. Schwager, T., Pöschel, T.: Coefficient of restitution for viscoelastic spheres: the effect of delayed recovery. Phys. Rev. E **78**, 051304 (2008)
53. Serero, D., Gunkelmann, N., Pöschel, T.: Hydrodynamics of binary mixtures of granular gases with stochastic coefficient of restitution. J. Fluid Mech. **781**, 595621 (2015)
54. Smith, P.A., Spencer, C.D., Jones, D.E.: Microcomputer listens to the coefficient of restitution. Am. J. Phys. **49**, 136–140 (1981)
55. Sondergaard, R., Chaney, K., Brennen, C.E.: Measurements of solid spheres bouncing off flat plates. J. Appl. Mech. **57**, 694–699 (1990)
56. Stensgaard, I., Lægsgaard, E.: Listening to the coefficient of restitution—revisited. Am. J. Phys. **69**, 301–305 (2001)
57. Tabakoff, W.: Measurements of particles rebound characteristics on materials used in gas turbines. J. Prop. Pow. **7**, 805–813 (1993)
58. Wadhwa, A.: Measuring the coefficient of restitution using a digital oscilloscope. Phys. Educ. **44**(5), 517–521 (2009)
59. Wadhwa, A.: Measuring the rebound resilience of a bouncing ball. Phys. Educ. **47**(5), 620–626 (2012)
60. Wadlund, A.P.R.: A simple method for determining the coefficient of restitution. Am. J. Phys. **7**(3), 194–195 (1939)
61. Wang, J., Ciocca, M.: Watching and listening to the coefficient of restitution. Phys. Teach. **50**(8), 503–503 (2012)

Non-ohmic Properties of Particle Layers in Electrostatic Precipitators

Damian Pieloth, Helmut Wiggers and Peter Walzel

Nomenclature

A	Area (m^2)
E	Electric field strength (V/m)
I	Electric current (A)
j	Current density (mA/m^2)
R	Ohmic resistance (Ω)
s	Particle or dust layer thickness (m)
T	Temperature (°C)
T_d	Dew point (°C)
t	Time (s)
σ	surface charge density (C/m^2)
ρ	Resistivity (Ω cm)
φ	Relative humidity (%)
e	Euler-number (–)
ε_0	Electric permittivity of vacuum (As/Vm)
ε	Electric permittivity
η	The density of current carriers per unit volume (C/m^3)

D. Pieloth (✉)
Department of Applied Biosciences and Process Engineering, Anhalt University of Applied Sciences, Köthen, Germany
e-mail: damian.pieloth@hs-anhalt.de

H. Wiggers · P. Walzel
Laboratory of Solids Process Engineering, TU Dortmund University, Dortmund, Germany
e-mail: helmut.wiggers@tu-dortmund.de

P. Walzel
e-mail: peter.walzel@bci.tu-dortmund.de

S. Antonyuk (ed.), *Particles in Contact*,
https://doi.org/10.1007/978-3-030-15899-6_19

η_E Density of current carriers after exposing a material to a strong electric field (C/m^3)

κ Conductivity ($1/\Omega$ cm)

the κ_E Conductivity of the material after exposing a material to a strong electric field ($1/\Omega$ cm)

Indices

i Index of the initial condition like initial temperature, etc.

f Index of the final condition like final temperature, etc.

Introduction

The particle layers formed on the precipitation electrodes of electrostatic precipitators usually consist of particles with different composition, different sizes, and shapes. The particle layers have gas-filled voids and adsorbed liquid layers depending on the temperature and humidity of the gaseous environment. Electric charges emitted by the discharge electrodes under negative voltage accumulate to a certain amount within the particle layers despite of conduction to the grounded collection electrodes. The mechanical, thermodynamical and electrical properties of particle layers to a high degree influence the performance of electrostatic precipitators [1–3].

In agreement with other researchers, Tachibana characterizes a resistivity of particle layers between 10^4 and 10^{10} Ω cm for being uncritical to the performance of an electrostatic precipitator. In particle layers with low resistivity, charges collected on the particles pass very fast through the particle layer to the grounded collection electrode. The rapid discharge of the particle layers decreases the electric adhesion forces between the particles. Thus, particle layers with resistivity below 10^4 Ω cm tend to lose particles creating re-entrainment and decrease the separation efficiency of electrostatic precipitators. In contrast, high resistivity particle layers with resistivity above 10^{10} Ω cm impede the conduction of electric charges to the grounded collection electrode. The charges accumulate to a considerable amount within the particle layer creating a high electric potential with high gradients at the sharp edges of particles. A voltage breakdown can occur here. The so-called back-corona generates gas-ions opposite to the gas-ions induced during the corona discharging, which migrate into the process gas and neutralize the already charged particles. Also, particles deposited on the precipitation electrode disperse back into the process gas. Although the conduction mechanisms in particle layers are not fully understood, the resistivity of a particle layer is a measurable variable which can help to predict back-corona [4–7].

Surrounding process gas temperature and humidity influence the resistivity of particle layers and therefore the safe operation of electrostatic precipitators. Water molecules are adsorbed on particle surfaces depending on the temperature and humidity of the surrounding gas. The conductivity of the particle layer increases with the amount of water absorbed. With higher temperatures but unchanged dew point of the surrounding gas, water molecules start to desorb from the particle surface, which leads to a higher resistivity of the particle layer. To avoid back-corona, the effect can be counteracted by spraying water into the process gas. Masuda and Rothenberger studied the influence of temperature and humidity of the process gas on the collecting efficiency of electrostatic precipitators [8, 9]. With increasing temperature but unchanged water content of the gas, the resistivity of particle layers increases to reach a maximum between 100 and 140 °C for most particle species. With increasing the temperature beyond 140 °C, the resistivity decreases again.

Surface and volume conduction of the particle layers can explain the influence of temperate and humidity of the surrounding gas on the resistivity of particle layers. Within temperature range below 140 °C water adsorbed on the particle surfaces leads to conduction of electric charges on particles or through water bridges between the particles. With increasing temperature water desorbs from the particle surfaces and the conduction of electric charges through the volume of the particles gets dominant. The semiconductor properties of the particle's material explain the decrease of the resistivity of particle layers in the temperature range above 140 °C. The charge carriers in the semiconductors are free moving electrons, which are not present as a free electron gas but are released by external influences such as heating or light irradiation. With increasing temperature the electron density and thus the conductivity increases. In semiconductors, the electron energy belongs either to the valence band or the conduction band. A valence-conduction band gap separates the bands. Close to absolute zero, all available electron energy levels are occupied and belong to the valence band. The conduction band is empty. In this temperature regime, the semiconductor acts as an insulator. With a valence-conduction band gap not too high, the electrons can overcome the valence-conduction band gap due to a thermal extinction when the temperature of the surrounding gas increases. In this way, the conduction band is populated with electrons to a certain amount. In transferring electrons from the valence band to conduction band, the valence band is depopulated. Both, the electrons in the conduction band and the holes in the valence band contribute to the conductivity of the material, in that manner also the conductivity of the particle layer. In the further increase of the temperature more electrons succeed to overcome the valence-conduction band gap, the conductivity of the particle layers increases further. In comparison with metals, the density of electrons in the conduction band is lower in semiconductors. Thus, the conductivity of semiconductors differs several magnitudes from metals [10].

The influence of bulk properties like particle size distributions or porosity on the resistivity of dust layers was studied by Miller et al., Young et al., and Rose and Wood [3, 4, 11]. High-porous dust layers generate back corona more notably than low-porous dust layers [3, 4]. Due to the back corona, positive gas ions occur within the dust layer. When a negative voltage is applied for the separation process, the positive gas ions migrate in reverse direction to negative charges through the dust layer towards the main gas flow. This effect neutralizes the negatively charged particles in the main separation process and reduces the migration velocity of dust particles to the collection electrodes, which in turn decreases the overall collection efficiency [12].

The electric current I that flows through a conductor is proportional to the voltage drop U in the conductor, the constant of proportionality R, being the resistance of the conductor:

$$R = \frac{U}{I} \tag{1}$$

To get rid of the dependence of R on the cross-sectional area A of the conductor, the resistivity ρ of the conductor is introduced:

$$R = \rho \frac{s}{A} \tag{2}$$

Here s is the thickness of the conductor. In equate, Eqs. (1) and (2) the resistivity can be calculated by:

$$\rho = \frac{A \cdot U}{I \cdot s} \tag{3}$$

The definition of current density $j = I/A$ yields:

$$\rho = \frac{U}{j \cdot s} \tag{4}$$

The electric field strength E within the particle layer can be calculated as:

$$E = \frac{U}{s} \tag{5}$$

Using Eq. (4) it follows:

$$E = \rho j \tag{6}$$

The assumption of an ohmic behavior of particle layers fails as resistivity measurements indicate [9, 13, 14]. However, the resistivity of particle layers can be explained by the theory of electrets.

Electrets are insulators. They are fabricated from dielectric materials and are used, e.g., in microphones to transform sound waves into electric pulses. Per definition, electrets are dielectric materials that produce a permanent external electric field [15, 16]. The electric field results either from a permanent allocation of material dipoles or stable uncompensated surface and space charges [15, 16]. Electrets show a hysteresis behavior regarding the electric field strength E. As a consequence, the polarization P of an electret material placed in an electric field of strength E is not an explicit function of E. The fabrication process of nonporous electrets is much like the measurement process of the resistivity of particle layers. In most cases, a corona charging is applied to inject excess electrons into a dielectric material [15, 16]. The excess electrons are made responsible for the current conduction.

The current density j passing an electret material during a discharging process is equivalent to the change of surface charge density σ:

$$j = -\frac{d\sigma}{dt} = \kappa \cdot E \tag{7}$$

The change of surface charge density is associated with both the change of polarization charges and real charges [16]. In Eq. (7), κ denotes the conductivity of the material. The electric field strength E inside the electret material is a complex function of material properties like the space distribution of charges, bulk properties, the polarization of the material, boundary conditions like the distances of electrodes to the material, electrode substance in contact with the material, and an electric potential applied to the electrodes. However, to simplify the calculations, the electric field strength can be approximated by:

$$E = \frac{\sigma}{\varepsilon_0 \varepsilon}, \tag{8}$$

where ε_0 is electric permittivity of vacuum and ε is the electric permittivity of the electret material, respectively. Substituting Eq. (8) in (7) yields:

$$\frac{d\sigma}{dt} = -\frac{\kappa \cdot \sigma}{\varepsilon_0 \varepsilon} \tag{9}$$

Integrating Eq. (9) and assuming the initial surface charge density at $t = 0$ by σ_0 it follows:

$$\sigma = \sigma_0 \cdot \exp\left(-\frac{t}{\varepsilon_0 \cdot \varepsilon/\kappa}\right) = \sigma_0 \cdot \exp\left(-\frac{t}{\varepsilon_0 \cdot \varepsilon \cdot \rho}\right) \tag{10}$$

The period τ after which the surface charge density decreases to the 1/e primary value, e being the Euler's number, defines the time constant of the electret material:

$$\tau = \varepsilon_0 \cdot \varepsilon \cdot \rho \tag{11}$$

The resistivity ρ of the electret material defines, therefore, the lifetime of the electret.

Electrets made of PTFE material, one of the best insulators, have lifetimes up to 10^3–10^5 years [17, 18]. This implies a resistivity of the order 10^{22}–10^{24} Ω m, assuming the electric permittivity of the electret to be $\varepsilon = 2$ [16]. The specific resistivity of untreated PTFE material lies, however, in the order of 10^{19} Ω m [18]. This implies lifetimes of less than one year. Therefore it can be assumed that during the formation process, i.e., when exposed to a strong electric field for a long period, such as 20 kV mm^{-1} or more, the resistivity of the electret material increases drastically, thus

$$\rho_E \gg \rho \tag{12}$$

ρ_E defines the resistivity of the electret after the formation process. The condition (12) also holds for other electret materials [16]. Experimental data available for electrets indicate that condition (12) also holds for other dielectric materials [16].

The drastic increase in specific resistivity of dielectric materials placed in a strong electric field can be explained by the balance between the number of current carriers generated by thermal excitation, and the number of current carriers captured and successively stored in a potential trap within the valence band or conduction band of the material. When exposed to a strong electric field, the number of potential traps in the material increases. As a consequence, the number and the mobility of the current carriers decrease and so does the conductivity. The conductivity κ of a material is defined by

$$\kappa = \eta \cdot e \cdot u \tag{13}$$

with η the density of current carriers per unit volume and u the mobility of the carriers. The density of the current carriers η in the conduction band decreases in a material exposed to a strong electric field, i.e., $\eta_E \ll \eta$, with η_E being the density of the current carriers after exposing the material to a strong electric field [16]. If the number of traps in the conduction band is much higher than the number of current carriers in the conduction band, one can assume that:

$$\frac{\eta_E}{\eta} = \frac{\kappa_E}{\kappa} \tag{14}$$

with κ_E being the conductivity of the material after its former exposition to a strong electric field. Furthermore, the time t for the current carriers η_E to reach a constant value is equal to the time needed for the current carrier flow to pass a sample of thickness s:

$$t = \frac{s}{E \cdot u} \tag{15}$$

With Eq. (8) it follows:

$$t = \frac{\varepsilon_0 \cdot \varepsilon \cdot s}{\sigma \cdot u} \tag{16}$$

Inserting values for a dielectric material, e.g., PTFE film with 20 μm thickness, $\varepsilon = 2$, $\sigma = 10^{-4}$ C m^{-2} and $u = 10^{16}$ m^2/V s yields $t \approx 35{,}000$ s ≈ 10 h.

It can be assumed that comparable balancing times could be expected with other dielectrics like quartz glass (SiO_2) or ceramic materials as Al_2O_3. Typical examples of balancing periods up to 2 h in porous systems as Al_2O_3 powder layers are given in [19].

Materials and Methods

Experimental Setup

The measurement of the resistivity of particle layers was carried out in a newly designed chamber where the operating conditions like temperature and humidity of the process gas and the current density within the particle layer can be adjusted according to the operating conditions in an electrostatic precipitator; see Fig. 1.

The particle sample was placed on a carrier electrode. The point-plate and plate-plate-setup were used during the resistivity measurements; see Fig. 2.

Figure 3 shows a photograph of the measurement chamber.

A high voltage with negative polarity was applied to the upper electrode. In applying a high voltage to the upper electrode and measuring the electric current through the grounded lower electrode, the resistivity of the particle layer can be calculated according to Eqs. (1) and (2). Compressed air passes through a temperature controlled bath of deionized water. The dew point of the air entering the chamber was adjusted in that way. The air was heated up afterwards by an air heater. The air temperature

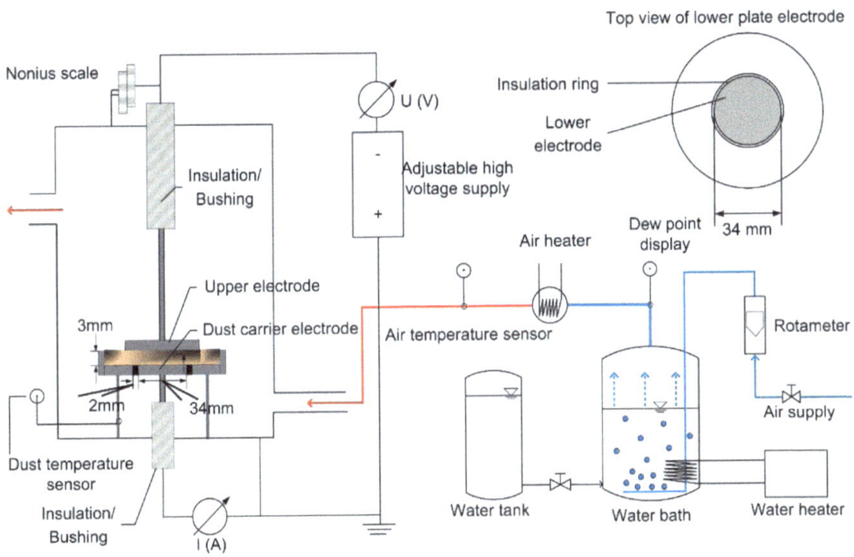

Fig. 1 Schematic of the setup to measure the resistivity of particle layers (Reprinted with permission from [14].)

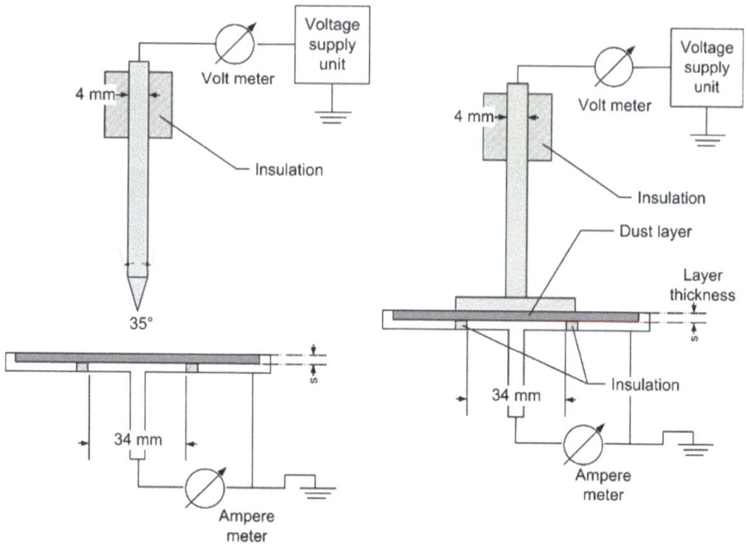

Fig. 2 Electrode arrangement for resistivity measurement of particle layers, point-plate left, plate-plate right (Reprinted with permission from [14].)

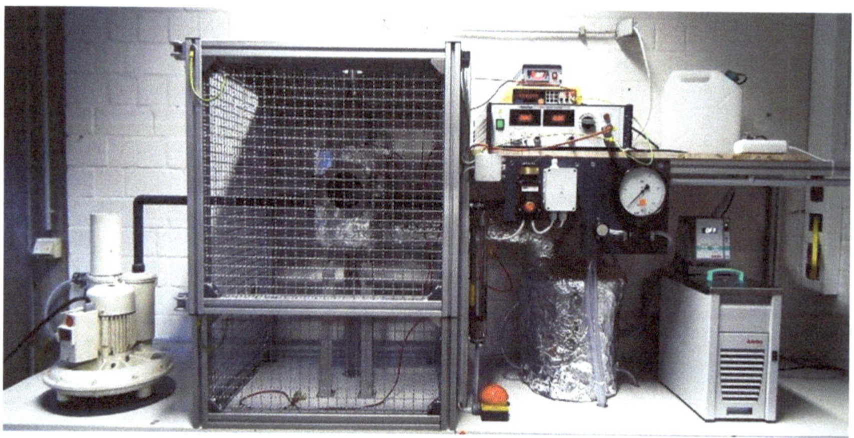

Fig. 3 Photograph of the resistivity measurement chamber

was measured with a digital thermometer. A thermocouple was used to measure the particle layer temperature. Below an air temperature of 90 °C, both air temperature and the dew point were additionally verified with the aid of a dew-point mirror (Michell S8000 Remote). Short-term and long-term measurements of the resistivity of particle layers were carried out. In short-term measurements, the chamber was pre-heated to the air temperature of 60 °C. The temperature was then raised successively with a ramp of 3 °C/min to ensure an acceptable exchange of humidity between the

Fig. 4 Scanning electron pictures of the test material, SiO_2 left Al_2O_3 right (Reprinted with permission from [20].)

particle layer and the air. The lower electrode used in the measurement is made of porous metal to facilitate the mass transfer between the air and the particle layer [17]. Short-term measurements were performed in the temperature range between 60 and 220 °C. In long-term measurements, the thermodynamic conditions in the measuring chamber, for instance, an air temperature of 100 °C and the dew point of 40 °C are maintained for at least 2 h. Only then the voltage is applied to the upper electrode. The upper electrode was connected to a laboratory high voltage unit (Heizinger PNC 40000). A multimeter was used to measure the voltage between the upper and lower electrode. The multimeter resolution accuracies are 10 V for the voltage applied between the electrodes. To register the current that flows through the particle layer a pico-amperemeter (Safelec M1501M respectively) was used. The particle layers examined had a resistivity $>10^{11}$ Ω cm. The samples were measured at a current density of 0.5 mA/m^2. This corresponds to typical current densities in electrostatic precipitators. The currents measured here are in the µA range, the pico-amperemeter resolution accuracies are in the range of ~100 pA.

Materials

The particle size distributions of the test particles (SiO_2) were measured by a CILAS 715 laser particle size analyzer after dispersion in water. Characteristic $d_{10,3}$, $d_{50,3}$ and $d_{90,3}$ values of test particles Al_2O_3 were provided by the distributor. Scanning electron pictures of the test particles are shown in Fig. 4.

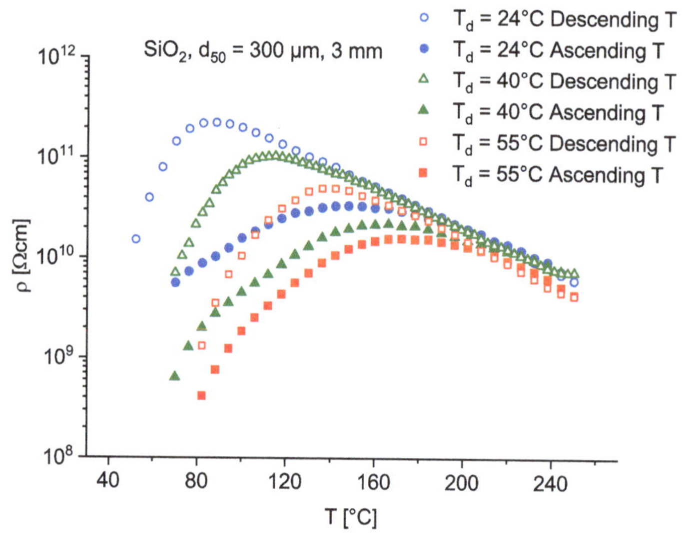

Fig. 5 Influence of temperature and humidity on the resistivity of particle layers (Reprinted with permission from [20].)

Experimental Results and Discussion

Influence of Temperature and Dew Point

The short-term measurement of the resistivity of a 3 mm SiO_2 particle layer are shown in Fig. 5.

Three different dew points $T_d = 24$, 40 and 55 °C were adjusted. The measurements were carried out with plate-plate-method and ascending temperature in the range of ~60–250 °C and afterward with descending temperatures from ~250–60 °C. The mean particle size was ~300 μm ($d_{50.3} = 300$ μm), the temperature ramp was set to ~3 °C/min. A voltage U needed to obtain a constant current density of 0.5 mA/m^2 was adjusted at every measuring point. The measurements show a characteristic hysteresis. The hysteresis is weakest at high humidity. Mass transfer between the surrounding air and the particle layer explains the distinction. With higher air humidity the adsorption and desorption rates of water at and from the particle layer decrease. The particle layer loses (or gains) only a small amount of water content during the increase (or decrease) of temperature. Thus the curve for $T_d = 55$ °C almost retraces itself. With smaller air humidity the adsorption and desorption rates of water increase. The particle layer loses (or gains) a higher amount of water content during the increase (or decrease) of temperature. Thus the curves do not retrace but show a more distinctive hysteresis. Figure 5 also shows the influence of air humidity on the maximum value of the resistivity of particle layers. For ascending temperatures, the maximum resistivity of 2×10^{10} Ω cm was measured under $T_d = 24$ °C at 140 °C.

For descending temperatures, the maximum resistivity of 2×10^{11} Ω cm was measured under $T_d = 24$ °C at 80 °C. For temperatures higher then 180 °C the resistivity of particle layers is almost independent of the water content of the surrounding gas. As mentioned before, the conduction of electric charges through the volume of the particles becomes dominant here.

Influence of Thermic and Electric Balancing Times

The influence of electric balancing time on the resistivity of a SiO_2 particle layer as a function of time under $T_d = 25$ °C and constant gas temperature T = 60 °C is presented in Fig. 6.

The resistivity was measured for 3, 7, and 9 mm particle layer thickness. The SiO_2 test particles had a mean particle size $d_{50,3} = 110$ μm. The plate-plate method was applied. The gas temperature in the measurement chamber was increased up to T = 60 °C. Like in the previous measurements, the voltage U was adjusted to gain a constant current density of 0.5 mA m^{-2} during the measurement. With constant gas temperature and dew point in the measurement chamber, the determination of dust resistivity was started. The tests were performed in 3-min steps for 200 min. In all three experiments, the resistivity did not reach a constant value even for a relatively long balancing time. All three curves show an increase of resistivity with time. Under the same ambient conditions and with the current density of 0.5 mA m^{-2}, a similar behavior of resistivity with time is observable for other particle size dis-

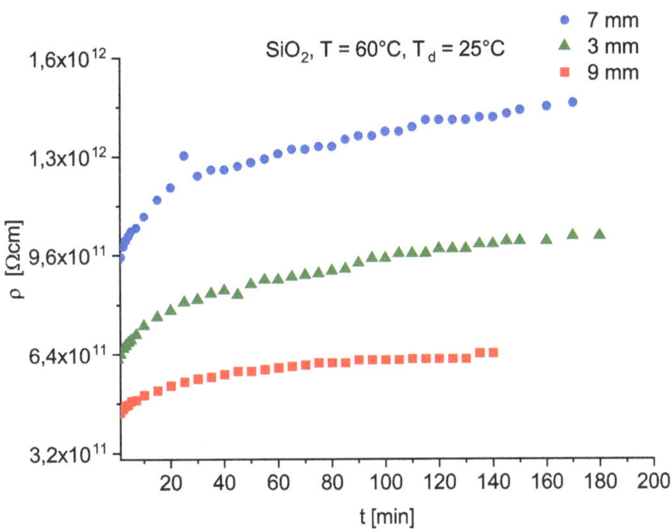

Fig. 6 Influence of electric balancing time on the resistivity of particle layers (Reprinted with permission from [20].)

tributions of SiO_2. Similar behavior of resistivity also occurred with Al_2O_3 particle layers as well with the point-plate method, thus favoring heat and mass exchange. No clear conclusions can be made about the electric balancing time and particle layer thickness or resistivity and particle layer thickness. Furthermore, there is no systematic order of the data to the particle layer thickness, which could be linked to a previous polarization or charging of the particle layer material. Keeping measurement conditions like ambient temperature and humidity, bulk density, and particle size distributions constant, particle resistivity always shown inconsistent behavior in long-time measurement.

The dependence of resistivity on particle layer thickness was measured for Al_2O_3 test particle layers with thicknesses between 1.5 and 3 mm under ambient gas temperature and a dewpoint of $T_d \approx -22$ °C with much lower current density by Riebel et al. [13]. The Al_2O_3 test particles had a mean particle size of $d_{50.3} = 95$ μm [13]. The resistivity measurements in [13] were carried out with variable current density. The resistivity increases with particle layer thickness in [13]. On the contrary, as can be seen in Fig. 6, the resistivity curve of the particle layer with 3 mm thickness lies above the resistivity curve of the particle layer with 9 mm thickness. Similarly, the Al_2O_3-particles provides no unique resistivity values for a given particle layer thickness under similar ambient conditions and a current density of 0.5 mA m^{-2}.

The resistivity curves shown in Fig. 6 cannot be fully explained by adsorption and desorption of water at and from the particle layers. The total time required for the diffusion of water vapor through the dust layers is less than a few minutes [14]. Rather, it can be assumed that with applying a voltage to the dust layer, electric processes are started within the dust layer, which overlap the mass and heat transfer and with balancing times much longer than 150 min.

To determine the balancing times, long-term measurements with constant gas temperature and dew point were performed. Initially, for the first 2–3 h, the measurements were carried out similarly as those described in Fig. 6. After that, the particle samples remain in the measurement chamber for the next 10 h where the ambient conditions, i.e., the voltage applied to obtain a constant current density of 0.5 mA m^{-2}, were kept at constant values. The measurements were continued after 10 h, in the same way as in the first 2–3 h. Measurement periods longer than 10 h are only feasible for temperatures $T < 100$ °C and dew points $T_d < 30$ °C. Measurements with higher temperatures and higher dew points are not reasonable due to excess deionized water consumption of the gas conditioning, respectively the tank capacity of the deionized water bath.

The resistivity was measured for 3, 7, and 9 mm particle layer thickness. Figure 7 illustrates the particle resistivity of Al_2O_3 test particle layers as a function of time after 16 h thermal conditioning and permanent application of voltage. The plate-plate method was applied. All measurements provide an almost constant value for resistivity as a function of time. After 16 h preconditioning time, the transient behavior of resistivity vanishes. Also, increasing resistivity with particle layer thickness in accordance with the results in [21, 13] was observed.

To clarify the contradictions between short-term and long-term resistivity measurements, tests were carried out after thermal preconditioning of the particle probe

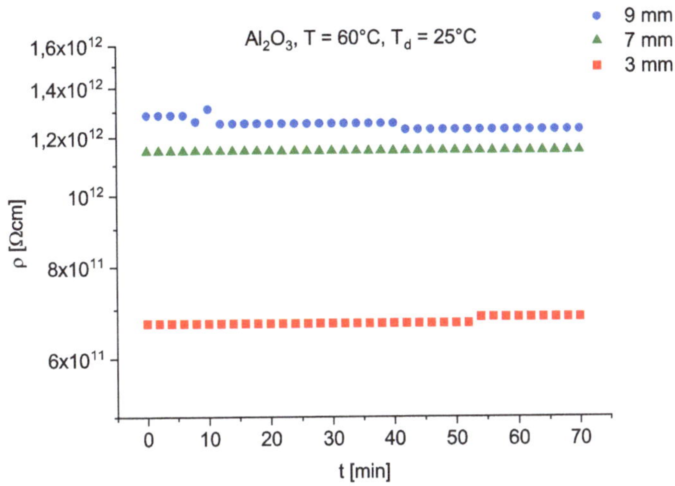

Fig. 7 Dust resistivity after thermal preconditioning and permanent application of voltage (Reprinted with permission from [20].)

for more than 16 h, with constant gas temperature T = 60 °C and dew point $T_d = 25$ °C, but no voltage applied during that time.

The particle sample was placed on the lower electrode. Afterward, the gas temperature in the measurement chamber was increased successively up to T = 60 °C. No voltage was applied to the upper electrode. With constant gas temperature and dew point, the particle sample remained in the measurement chamber for the next 16 h. Thereafter, the resistivity measurement started. A voltage between the electrodes is adjusted to gain a constant current density of 0.5 mA m^{-2} like in the tests performed earlier. Applying the plate-plate method again, with constant ambient conditions held for 16 h but no voltage applied, the particle resistivity increases and drifts again after starting the measurements; see Fig. 8. In long-term experiments with constant ambient conditions and voltage applied, the particle resistivity increases with particle layer thickness in accordance with [13]; see Fig. 7. In contradiction, long-term measurements with constant ambient conditions but no voltage applied show lower particle resistivity with higher particle layer thickness; see Fig. 8.

Similar results were obtained for both Al_2O_3 ($d_{50.3} = 106.5$ μm) and SiO_2 ($d_{50.3} = 110$ μm) test particle layers with similar ambient conditions, the point-plate method, and a constant current density of 0.5 mA m^{-2}. The results cannot be explained with adsorption and desorption of water at and from the surface of particles as Fig. 9 indicates. Figure 9 presents a sorption curve of H_2O on SiO_2 ($d_{50.3} = 110$ μm) particle surfaces determined in an adsorption chamber at gas temperature T = 70 °C and 280 mbar vapor pressure at the chair of Thermodynamics at TU Dortmund. The mass transfer to the particle surfaces is completed within 60 min.

Consequently, the increase and drifting of resistivity during measurement cannot be explained by thermodynamic or bulk properties only, but should also take electrical

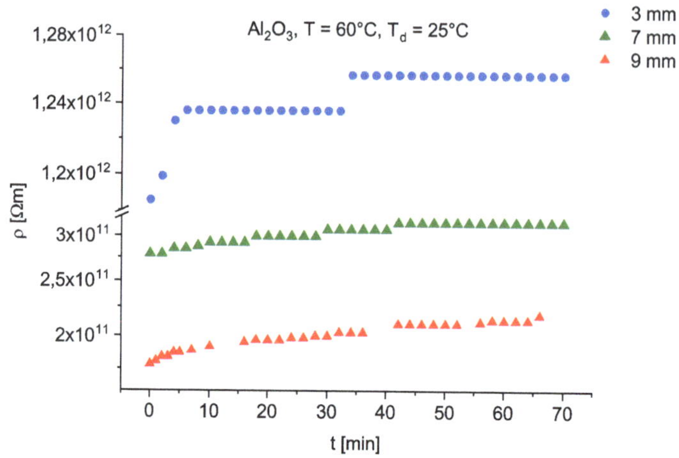

Fig. 8 Dust resistivity after thermal preconditioning but voltage application after 16 h (Reprinted with permission from [20].)

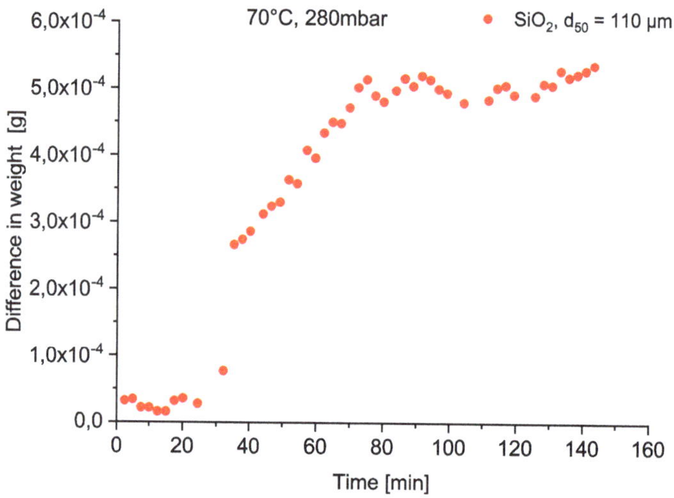

Fig. 9 Adsorption of H_2O on SiO_2 particle surfaces (Reprinted with permission from [20].)

material properties into account. Al_2O_3 and SiO_2 are insulators. When applying a strong electric field in connection with current flow to those dust layers, electric processes are started as well. Due to the strong electric field, the polarization of the particle material is initiated, similar to a dielectric material. Furthermore, due to the current flow, electric charges remain within the particle layer, which in turn increases the charge density within the layer. This changes the electric field within the particle layer and leads to higher dust resistivity similar to the increase of specific resistivity during electret formation from, e.g., PTFE material.

Conclusion

The resistivity of SiO_2 and Al_2O_3 particle layers shows non-ohmic and transient behavior. It also depends on the particle layer thickness, confirming the electret behavior as discovered in [13]. With higher particle layer thickness, higher ambient temperature and humidity, and higher current density than used in [13], the measurements indicate, however, very long electric balancing times of up to 16 h to reach a constant value of resistivity. Adsorption and desorption kinetics of H_2O into the particle layers cannot explain this effect. Sorption curves for H_2O at the test particle layer show balancing times of 1 h or less. Rather, it can be presumed that the thermal processes are overlapped by electric processes much like the polarization and charging of dielectric materials, e.g., during the formation of electrets. Balancing periods of the electric processes take a longer time by far than adjustment of thermodynamic equilibria. In measuring the resistivity of particle layers, the balancing periods and history of treatment should be considered to avoid misinterpretation of the results.

Acknowledgements The authors want to express their gratitude to the Deutsche Forschungsgemeinschaft (DFG) for the financial support of this project. We also acknowledge Prof. Dr.-Ing. U. Riebel for his valuable contributions to the scientific discussion of the subject.

References

1. Busby, T., Darby, K.: Efficiency of electrostatic precipitators as affected by the properties and combustion of coal. J. Inst. Fuel **36**(268), 184–197 (1963)
2. Watson, K., Blecher, K.: Further investigation of electrostatic precipitator for large pulverized fuel-fired boilers. J. Air Water Pollut. **9**(10), 577–583 (1966)
3. Miller, J., Hoferer, B., Schwab, A.J.: The impact of corona electrode configuration on electrostatic precipitator performance. J. Electrost. **44**, 67–76 (1988)
4. Young, R.P., Dubard, J.L., Sparks, L.E.: The onset of electrical breakdown in dust layers. JAPCA **38**(1412–1418), 1518–1522 (1988)
5. Lowe, H.J., Lucas, D.H.: The physics of electrostatic precipitation. Brit. J. Appl. Phys. Suppl. **2**, 540–547 (1953)
6. Lee, J.K., Hyun, O.C., Lee, J.I.: Relations between coal/fly ash properties and cohesive forces in electrostatically precipitated ash layers. KSME Int. J. **5**(15), 630–638 (2001)
7. Tachibana, N., Fujishima, H.: Application of electrostatic precipitation with the intermittent energization. In: Proceedings of the International Conference on Modern Electrostatics, Beijing, China (1988)
8. Masuda, S.: Einfluss von Temperatur und Feuchte auf die elektrische Leitfähigkeit hoch isolierender Stäube. Staub **25**, 175–179 (1965)
9. Rothenberg, S.J., Cheng, Y.S.: Coal combustion fly ash characterization—rates of adsorption and desorption of water. J. Phys. Chem. **8**, 1644–1649 (1980)
10. Meschede, D.: Gerthsen Physik, 25 Auflage. Springer (2015). ISBN 978-3-662-45977-5
11. Rose, H.E., Wood, A.J.: An Introduction to Electrostatic Precipitation in Theory and Practice. Constable & Company Ltd. (1966). ISBN 978-0094553408
12. Schubert, H.: Handbuch der Mechanischen Verfahrenstechnik, 1 Auflage. Wiley-VCH Verlag (2003). ISBN 978-3527305773
13. Riebel, U., Aleksin, Y., Vora, A.: Das elektrische Verhalten hochohmiger Stäube. Chem. Ing. Tech. **85**, 235–244 (2013)

14. Majid, M.: Dust resistivity and re-circulation in electrostatic precipitators. Ph.D. thesis, Technische Universität Dortmund (2012)
15. Erhard, D.: High performance polymer electrets. Ph.D. thesis, Universität Bayreuth (2010)
16. Hilczer, B., Malecki, J.: Electrets. Elsevier (1986). ISBN 978-0444995476
17. Sessler, G.M., West, J.E.: Self biased condenser microphone with high capacitance. J. Acoust. Soc. Am. **40**, 1433–1440 (1966)
18. Sessler, G.M., West, J.E.: Electret transducers: a review. J. Acoust. Soc. Am. **53**, 1589–1600 (1973)
19. Gauckler, L.J., Conder, K.: Ceramics II. Lecture Notes. ETH Zürich (2010)
20. Pieloth, D., Wiggers, H., Walzel, P.: Thermodynamic material and bulk property influence on electric resistivity of particle layers. Chem. Eng. Technol. **37**(4), 627–634 (2014)
21. Wiggers, H.: Measurement of dust resistivity—back corona in electrostatic precipitators. VGB Power Tech. **3**, 93–96 (2007)

Structure of Sheared Cohesive Granular Bulk

Lothar Brendel, Alexander Weuster, Dietrich E. Wolf, Harald Zetzener, Stephan Strege, Lutz Torbahn, Arno Kwade, Lisa Handl and Volker Schmidt

Introduction

Cohesive particle contacts have intrinsic kinetics, even if the particles do not move: Freshly formed contacts mature due to accommodation in the contact area of their rough surfaces, intensifying atomic binding and hence lowering the interfacial free energy. The underlying microscopic processes are very complex, because they can be influenced by various means, such as temperature, humidity, surface contaminants, or pressure. (Here we do not consider cementation in the sense that a foreign chemical substance forms a solid bridge between the particles, although the phenomenology should be similar.) All these parameters determine the characteristic time, on which the cohesion force between particles establishes itself. We call it *accommodation time* t_c.

The accommodation time can be probed in bulk powders by shearing: The idea is that shear localization, density inhomogeneities and anisotropic fabric reflect the microscopic competition between the accommodation time and the inverse shear rate. Hence a structural analysis of powder bulk sheared at different *rates*, will reveal properties of the contact mechanics. In this project review two model materials are considered, which represent opposite limiting cases. Glass spheres at room temperature have a very short accommodation time. Cohesion (due to van der Waals forces) is instantaneously present, when a contact forms, and does not increase noticeably,

L. Brendel · A. Weuster · D. E. Wolf (✉)
Faculty of Physics, Universität Duisburg-Essen, Duisburg, Germany
e-mail: dietrich.wolf@uni-due.de

H. Zetzener · S. Strege · L. Torbahn · A. Kwade
Institute Particle Technology, Technische Universität Braunschweig, Brunswick, Germany
e-mail: a.kwade@tu-bs.de

L. Handl · V. Schmidt
Institute of Stochastics, Universität Ulm, Ulm, Germany
e-mail: volker.schmidt@uni-ulm.de

S. Antonyuk (ed.), *Particles in Contact*,
https://doi.org/10.1007/978-3-030-15899-6_20

when it lasts (section "Complex Shear Localization in the Micro-shear Tester"). On the other hand, KCl-particles are a clear example for caking contacts, where the cohesion force becomes slowly stronger (section "Elastoplastic Contact Model for Potassium Chloride"). A completely caked powder resembles a porous solid rather than a granular medium. Instead of a shear band one expects a crack, which upon continued shear grinds the adjacent material, so that it becomes granular-like. Section "Rheology of Caking Granular Matter: A 2D Simulation Study" explores this process theoretically.

In order to investigate the bulk structure evolution, a novel experimental tool has been developed, the micro shear tester, combined with x-ray tomography, which allows grain-resolved rheological observations. The experiments are briefly reviewed in section "Micro Shear Tester and Tomography, Requirements for Model Materials". Statistical particle tracking tools have been developed to analyze these experiments (section "Particle Tracking and Statistical Analysis"). For connecting these results to contact mechanics, simulations with the discrete element method (DEM, see section "Discrete Element Method: Contact Force Models and Boundary Conditions") are indispensable. Such interdisciplinary efforts are needed to shed light on the question, what information about contact mechanics can be extracted from shear induced bulk structures.

Methods to Study Grain-Resolved Rheology

Micro Shear Tester and Tomography, Requirements for Model Materials

For the investigation of the particle motion and the subsequent change in powder structure a novel micro torsional shear tester was developed within the project. The shear tester was designed to be integrated into a high resolution XMT device (MicroXCT 400, Xradia, Inc., today Zeiss). This state of the art XMT scanner differs from the conventional scanners due to the implementation of special scintillators in combination with optical magnifying lenses (see e.g. [30]). The micro shear tester has a cylindrical shear cell with inner diameters from less than 2 mm to about 4 mm and allows a precise compaction and shear deformation of the small powder samples in the lower μL-range (typically around 10 μL) [30]. Another characteristic of this novel shear device is the decoupled and most precise determination of the acting forces and torque behavior whereas the whole system is theoretically frictionless using advanced technology of magnetic spring and air bearing mechanisms. Due to the implemented very precise position and force sensors the tester can be operated in strain or stress controlled mode. The possible axial deformations range from micrometres to several millimetres and the resulting normal and shear stresses from 0.5 to 30 kPa. By the combination of μST and XMT, Fig. 1 (left), a stepwise manipulation of the specimen can be performed in order to acquire high resolution 3D images in between these deformation steps. During the whole measurement the specimen remains in the scanning position and thus, the danger of potentially disturbing the fragile structure is minimized.

The cylindrical shear cell with the powder sample is exposed to the X-ray beam. The X-ray source was operated with an acceleration voltage of 50 kV and a constant current intensity of 200 μA. The specific absorption of the specimen due to its material and inner structure results in a characteristic radiograph, which the scintillator translates into the spectrum of visible light. The subsequent optical lens generates the required magnification of the image. In order to change the magnification the optical lens can be replaced using an objective revolver. For the usually applied shear cell diameter of 2 mm, a ten-fold optical magnification was adequate to image the entire cell with a resolution of 2.2 μm (2× binning). The 3D information is achieved by multiple projections while the sample rotates in small angle steps around its vertical axis. A record of 2000 two-dimensional projections provides the basis for a precise three-dimensional reconstruction of the powder microstructure.

In order to minimize the X-ray absorption the shear cell was build out of materials with low effective atomic number. Moreover, the thermal expansion coefficient should be as low as possible to avoid a geometrical change of the shear cell due to heat expansion. Therefore, for compression and shear tests the powder specimen is enclosed in a thin-walled borosilicate glass tube (wall thickness ≈50 μm, length 40 mm) with an effective atomic number of 10.7 and a thermal expansion coefficient of only 3.3×10^{-6}/K. The top and bottom sides of the powder sample are limited by the upper and lower pistons made out of PEEK (effective atomic number of about 6, thermal expansion coefficient of 47×10^{-6}/K, Fig. 1 (centre, right). While compaction experiments are performed in between two flat pistons, shear deformation can be induced with structured pistons. During the test the upper piston, which is firmly connected to the glass tube, performs a relative movement (i.e. translation or rotation) to the lower piston. In order to prevent agglomeration during the preparation

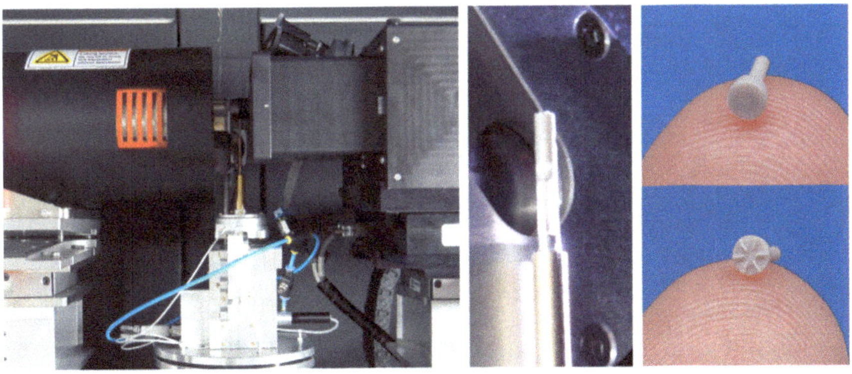

Fig. 1 Integration of the μST into the XMT system (left), powder specimen in front of the X-ray source (centre), compression and shear geometry on a fingertip (right) [30]

process, powder is sieved into the capillary. After the filling process, the sample is compressed to the desired normal load, usually 0.5 kPa. During shear deformation the sidewall and the upper piston rotates stepwise. Usually during the first 9.5° of shear deformation an angle increment of 0.5° is set. Afterwards, the angle increment is raised to usually 5° until the final shear deformation is achieved. In most experiments a total shear deformation of 39.5° was used. The shear deformation is performed in a quasi-static regime using an angle velocity ω of 0.1°/s. Between the shear steps tomographic data is recorded.

For μCT measurements the model particles have to fulfill specific requirements. On one side the effective atomic number should be high enough, at least in the range of 10. On the other side the particles should not be smaller than 10 μm in order to get good resolved pictures of the particles and, at the same time, a full image of the powder sample on particle level inside the 2 mm wide shear cell. As cohesive model material with very short accommodation times due to van der Waals forces massive borosilicate glass beads (BSGMS 27–32 μm, Cospheric, USA) are applied. Against that as model material for caking contacts potassium chloride (KCl) particles are used (K+S, Germany). As depicted in Fig. 2 the BSGMS micro beads are very spherical, whereas the KCl particles show a prismatic to irregular shape. Moreover, the BSGMS particles have a very narrow particle size distribution ($x_{10} \approx 20$ μm, $x_{50} \approx 30$ μm, $x_{90} \approx 40$ μm), while the size distribution of the specifically milled KCl particles is relatively wide ($x_{10} \approx 3$ μm, $x_{50} \approx 20$ μm, $x_{90} \approx 50$ μm) [31]. The mechanical characteristics of the single particles were determined by nanoindentation

Fig. 2 SEM images of model particles. Left: borosilicate glass micro spheres, right: potassium chloride particles [31]

(Triboindenter, Hysitron, Inc., USA). For BSGMS particles a Young modulus of 14.8 ± 7.2 GPa was achieved by fitting the deformation curve with the model of Hertz. The non-spherical KCl particles cannot be described by the model of Hertz, so that a contact stiffness of 9.8 ± 3.4 kN/m was determined [31]. The flowabilities of the two model powders were characterized within a Schulze ring shear tester RST-XS: The ff_c values of the BSGMS powder were above 10 for principal stresses higher than 8 kPa, so that the BSGMS powder can be characterized as free flowing. Against that for the KCl powder the ff_c values are in the range between 1.5 at low principal stresses and 3 at high principal stresses, so that the KCl powder has a very cohesive to cohesive flow behavior according to the definition of Jenike.

Particle Tracking and Statistical Analysis

Starting from a series of 3D tomographic images recorded in the micro shear-tester, we aimed to extract as much information on particle level as possible. For method development, we focused on a shear experiment with spherical particles (BSGMS 27–32 μm, CoSpheric LLC, USA; abbreviated BSGMS) with a diameter of roughly 30 μm. The size distribution and an SEM image of several particles are shown in Fig. 3. However, many of the developed methods are also applicable to other particle sizes, shapes or materials. The experiment we studied was conducted quasi-statically with a normal load of 0.5 kPa. Particles were sieved into the sample chamber and the

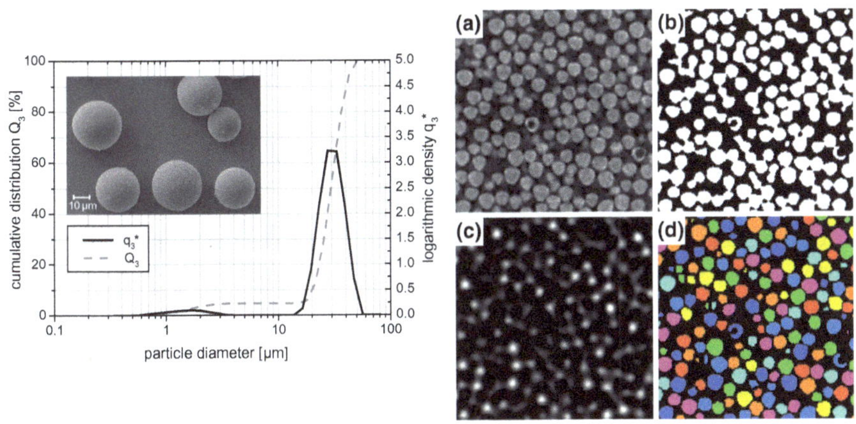

Fig. 3 Left: Particle size distribution (mass distribution, logarithmic density and cumulative distribution function) and SEM picture (inset) of BSGMS glass particles. Right: Visualization of the main steps of the segmentation process based on a small cutout of an image slice. Grayscale image obtained by XMT (**a**), binary image (**b**), convolution of grayscale image and particle mask used for marker selection (**c**) and final segmentation result after applying the marker-based watershed transform (**d**). Note that though the visualization is in 2D, all operations are carried out in 3D (Reprinted from [12] with permission from Elsevier.)

normal load was applied. Next, the sample was sheared, first in steps of 0.5° up to a cumulative shearing angle of 9.5°, then further in steps of 5° up to an overall angle of 39.5°. Directly after applying the normal load and after each step of shearing, a tomographic 3D image of the entire sample chamber was recorded. The experiment and analysis methods described in this section were previously published in [11, 12].

First, the particles were segmented by applying a marker-based watershed transformation. Markers were chosen as the local maxima of the convolution of the original image with an idealized particle mask as suggested in [39]. This approach is ideal for spherical particles with a narrow size distribution and also allowed for the correct segmentation of hollow particles in most cases, which were occasionally present in the sample. Based on the segmentation, we could extract the center of mass and other geometric characteristics (e.g. diameter, volume, sphericity) for each particle. The main steps of particle segmentation are visualized in Fig. 3.

Based on the extracted centers of mass, we computed tracks of particles over time. When shearing in steps of 0.5°, a comparatively simple algorithm was sufficient, which aims to minimize the sum of squared displacements in each step [4]. In these small steps, particles are expected to move at most 8.72 μm (at the upper outer edge of the cylindrical sample chamber), which is less than a particle radius. When larger angle increments of 5° are used between tomographic images, this approach is no longer feasible.

We solved this problem by deriving a first estimation of the average movement of particles in different areas of the shear cell directly from the 3D image data, extending an idea from [31]. Here, an image-based local shear deformation was computed by comparing image slices at two consecutive points in time and a fixed height, z. For a time step (t_1, t_2) the local angle of shear deformation $\Delta_\varphi^{t_1 \to t_2}$ at height z was then defined as the angle by which the first slice has to be rotated to best match the second slice. More precisely,

$$\Delta_\varphi^{t_1 \to t_2}(z) = \underset{\alpha \in [0°, 360°]}{\operatorname{argmax}} \left\{ \operatorname{corr}\left(\operatorname{rot}_\alpha\left(I_{t_1}^z\right), I_{t_2}^z\right)\right\}, \tag{1}$$

where I_t^z denotes the image slice at time t and height z, corr denotes the image cross-correlation, and $\operatorname{rot}_\alpha$ denotes the rotation around the image center by the angle α. The image rotations were implemented using bilinear interpolation and maximization was carried out in steps of 0.1°. We extended this approach to capture also radial variations of local shear deformation and to measure vertical compression or dilation along with the rotational deformation. For this purpose, we divided each image slice into k disjoint, concentric rings $R_1, \ldots, R_k$ with equal area and maximized image cross-correlation with the corresponding rings of the next time point while simultaneously applying a rotational and a translational deformation. Note that each ring is rotated independently and shifted vertically now, not the image slice as a whole. This yields the 2D optimization problem

$$\left(\Delta_{\varphi}^{t_1 \to t_2}(z, r_i), \Delta_{z}^{t_1 \to t_2}(z, r_i)\right) = \underset{\substack{\alpha \in [\alpha_{\min}, \alpha_{\max}] \\ h \in [h_{\min}, h_{\max}]}}{\operatorname{argmax}} \left\{ \operatorname{corr}\left(rot_{\alpha}\left(I_{t_1}^{z,R_i}\right), I_{t_2}^{z+h,R_i} \right) \right\}, \tag{2}$$

where $i \in \{1, \ldots, k\}$, r_i is the central radius of ring R_i, and I_t^{z,R_i} denotes ring i on the image slice at time t and height z. We used $k = 10$ rings and chose the range for the optimization of the vertical deformation based on the sizes of the 3D image stacks in z-direction, $z_{\max}^{t_1}$ and $z_{\max}^{t_2}$, as

$$[h_{\min}, h_{\max}] = \left[\min\left\{z_{\max}^{t_2} - z_{\max}^{t_1}, 0\right\}, \max\left\{z_{\max}^{t_2} - z_{\max}^{t_1}, 0\right\}\right]. \tag{3}$$

To speed up calculations, α was not optimized over the full range of 360° but was restricted to a realistic range, where $\alpha_{\min} = -1°$ and $\alpha_{\max}$ was set to the actual angle of shearing applied to the sample between times t_1 and t_2 plus 1°. Using bilinear interpolation in the coordinates z and r_i, this method can be used to obtain an estimate of the local shear deformation at any location in the sample. A similar approach can also be used to measure the rotation of upper and lower piston (see section "Complex Shear Localization in the Micro-shear Tester"). We will refer to the left-hand sides of Eqs. 1 and 2 as 1D and 2D image-based local shear deformation, respectively. An example for the results of the image-based local shear deformation is shown in Fig. 4.

We used the 2D image-based local shear deformation as prior information for tracking particles in shearing steps of 5°. For each particle at time t_1 we computed the hypothetical position where it would be expected at time t_2 if it perfectly followed the average movement summarized in the image-based local shear deformation. Then,

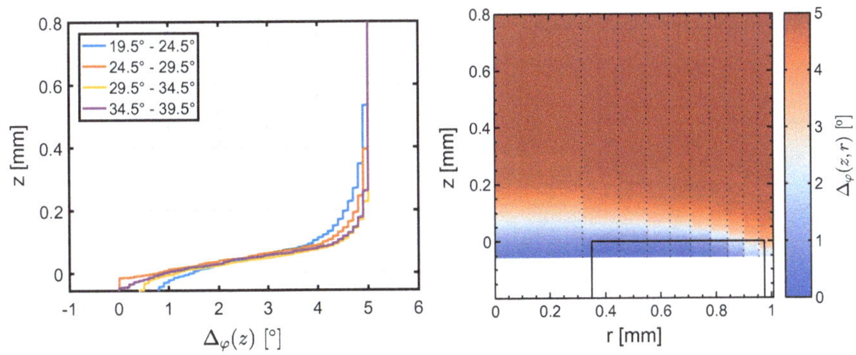

Fig. 4 Image-based local shear deformation in 1D and 2D in the steady state of the experiment. The 1D case (left) shows the shear deformation as a function of the height in the sample with almost congruent states. The 2D deformation, which is shown for the last time step as an example (right), reveals that the shear band geometry is curved close to the wall (Reprinted from [12] with permission from Elsevier.)

instead of minimizing the sum of squared displacements, we minimized the sum of squared distances between hypothetical and actual particle positions at time t_2. In both cases, with the simple algorithm for steps of 0.5° and the adapted one for steps of 5°, optimization was carried out on a reduced problem as suggested in [4], discarding possible assignments if the displacement (or distance between hypothetical and actual position) exceeded a predefined threshold.

Using this approach, we computed a particle tracking for all steps starting at $\omega t = 3.5°$. In previous time steps, the distance between upper and lower piston decreased considerably, from 2.12 to 1.93 mm, leading to large vertical displacements of particles. This densification was spatially inhomogeneous, even in φ-direction, so that it could not be estimated with the 2D image-based local shear deformation and reliable tracks could not be identified. In all other time steps, we achieved tracking efficiencies of at least 99.5% for angle increments of 0.5° and at least 97% for angle increments of 5°. Here, tracking efficiency is defined as the percentage of particles in the segmented image, which were successfully assigned to a particle track. We additionally validated our algorithm for tracking in steps of 5° using the first part of the experiment, where image data in steps of 0.5° is available. When comparing the results of the 5°-step tracking (without considering intermediate time points) and the 0.5°-step tracking (using all available time points), we could show that more than 98.5% of the identified tracks were identical.

Based on the particle tracking, we first computed absolute particle velocities, measured as the distance a particle traveled in a time step per degree of shearing (μm/°). Particle velocities on the same height but at two different points in time are visualized in Fig. 5. As clearly visible in this figure, the axis of rotation does not

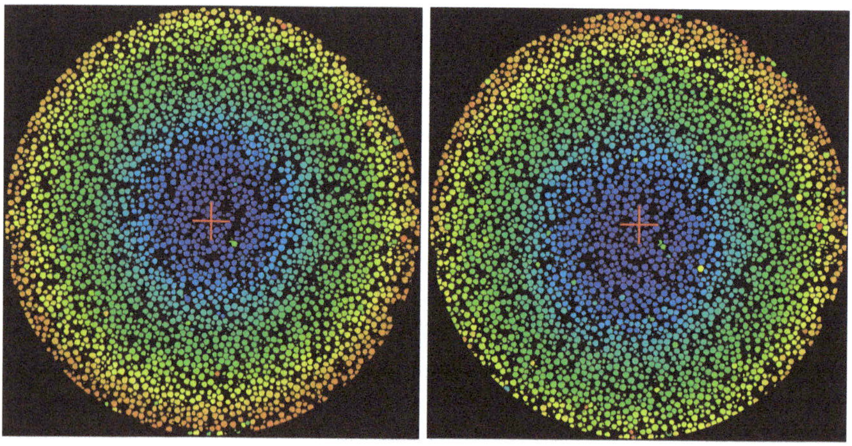

Fig. 5 Slice of the binary image at two different points in time. Each particle is colored according to the distance it travels in the subsequent time step, blue indicates small and red indicates large values. In the first time step (left) the center of rotation is slightly above and in the second time step (right) it is below and left of the center of the sample. The image centers are marked with a red cross (Reprinted from [12] with permission from Elsevier.)

always perfectly coincide with the central axis of the cylindrical sample chamber. This may be caused by small inaccuracies in the alignment of the upper and lower piston and the glass capillary. Since assuming a wrong axis of rotation would have led to significant errors in angular velocities, which are of major interest for shear band analysis, we estimated the actual axis of rotation from the data. This was achieved by fitting circles to particles with approximately the same height and absolute velocity at each time point using weighted least squares, and averaging the circle centers estimated at each z-coordinate. Note that the resulting estimated axis of rotation is not necessarily a straight line but may be curved. Since no rotation occurs at the bottom of the sample, all particles at this height have velocities close to zero and fall into the same velocity class, so the estimated axis of rotation will automatically be close to the center of the sample chamber in this region.

Discrete Element Method: Contact Force Models and Boundary Conditions

The simulations reported here are done with the discrete element method (DEM) using soft spherical particles, i.e. all particle trajectories are calculated by numerically solving Newton's equation of motion. Particles interact via pairwise forces and torques, dictated by the chosen contact model, which is a function of the particle diameters (d_i, d_j), their relative position $(\vec{r}_{ij} = \vec{r}_i - \vec{r}_j)$, velocity, angular velocity, and their contact history to account for friction. As usual, normal $\left(F_\text{n} = \vec{F}_{ij} \cdot \vec{r}_{ij} / \left|\vec{r}_{ij}\right|\right)$ and frictional forces are treated separately [13]. For the normal contact force three different models will be discussed in detail below. Coulomb friction, as well as rolling and torsional friction (in case of 3D simulations) are implemented with a model similar to the one presented in [13]: Each contact mobilization mode (sliding, rolling and twisting) is suppressed individually by using a linear spring-dashpot model until a particular threshold is exceeded. This threshold as well as the friction force and torque, when a contact is mobilized are proportional to the absolute value of normal plus adhesion force $|F_\text{n} + F_\text{adh}|$.

Contact Model for 2D Caking Simulations

To understand how the contact accommodation time t_c affects the bulk's rheology, we use 2D simulations and a simple linear spring-dashpot model. Hence, particles experience a viscoelastic force, which increases linearly with their overlap $\xi = (d_i + d_j)/2 - \left|\vec{r}_{ij}\right|$ and their relative normal velocity $\dot{\xi}$. A contact is assumed to form, when ξ becomes positive, and to last (even for negative overlap, see Fig. 6 (left)), until the normal force drops below a threshold $-F_\text{adh}(t_\text{k})$, which depends on the contact's age t_k,

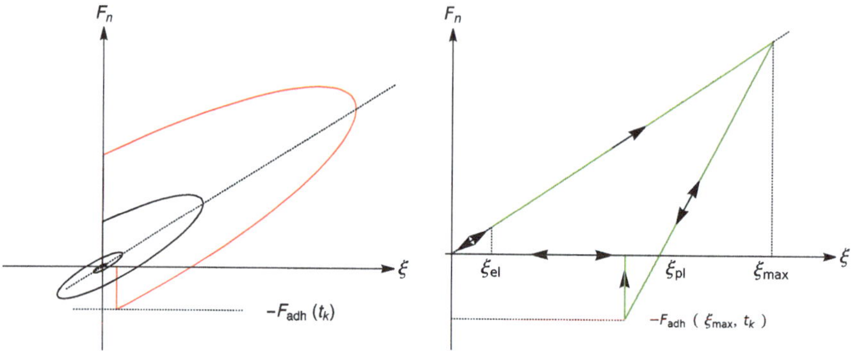

Fig. 6 Left: Normal force of the linear viscoelastic contact model with time dependent adhesion, see section "Contact Model for 2D Caking Simulations". Collision with adhesive capture (black), or with particle bounce back (red). Right: Elastoplastic contact model with time dependent adhesion, see section "Elastoplastic Contact Model for Potassium Chloride", suited for potassium chloride (Reprinted with permission from [42].)

$$F_{\rm adh}(t_{\rm k}) = F_{\rm adh}^{\rm (max)} \frac{4r_{\rm eff}}{\langle d \rangle}\left(1 - e^{-t_{\rm k}/t_{\rm c}}\right). \tag{4}$$

While the contact exists, the adhesion force grows from zero to the saturation value $F_{\rm adh}^{\rm (max)} 4r_{\rm eff}/\langle d \rangle$. The parameter $r_{\rm eff} = d_i d_j / 2\left(d_i + d_j\right)$ denotes the reduced radius, and d the average particle diameter. $F_{\rm adh}^{\rm (max)}$ is a given constant. This age dependence of the adhesion force is inspired by the formation of solid bridges between particles (see [35, 41]). With respect to the average contact time, which in case of quasi-static deformations is approximately given by the inverse shear rate $\dot{\gamma}^{-1}$ [42], two regimes can be identified. If $t_{\rm c} \ll \dot{\gamma}^{-1}$, the maximum adhesion force acts immediately as soon as the contact is established, defining an *adhesion dominated* flow regime. For $t_{\rm c} \gg \dot{\gamma}^{-1}$, no significant adhesion force is built up during the duration of a contact, and the bulk's behavior is determined by the preparation procedure prior to the shear, defining a *cementation dominated* regime [5, 40]. The theoretical results for the whole range of the dimensionless parameter $\dot{\gamma} t_{\rm c}$ will be discussed in section "Rheology of Caking Granular Matter: A 2D Simulation Study".

These simulations were done for plane shear under constant normal load P. The coordinate system was chosen such that x is the direction of shear and y the direction of normal pressure. Periodic boundary conditions were used in x-direction. In order to study shear banding in the bulk without perturbations by walls [20, 22, 26, 27], Lees-Edwards boundary conditions [3] were implemented in y-direction. However, in order to account for the non-affine nature of quasistatic shear [8, 34], we refrained from prescribing a homogeneous shear deformation. As we are interested in shear-banding rather than calculating linear transport coefficients, we induce the deformation purely by the boundaries in y-direction. Hence, these boundaries become well defined layers in space and break translational invariance in y-direction (see also [28]). The normal load P was regulated to a constant value by a variable system height. It responded

Table 1 Parameters for 2D simulations

Parameter	k_n/P	$\frac{\gamma_n}{\sqrt{k_n m_{eff}}}$	k_t, k_r	$\frac{\gamma_t}{\sqrt{k_t m_{eff}}}$	$F_{adh}^{(max)}/P\langle d\rangle$	μ	μ_r/r_{eff}
Value	10^5	1	$k_n/2$	2	10	0.5	0.1

to a deviation of the stress component σ_{yy} from the desired value P with a rate proportional to $(\sigma_{yy} - P)/P$. The corresponding control constant was set larger than $20 \cdot \dot{\gamma}$ to assure constant pressure during shear deformation. Following [14], we used a shear rate of $\dot{\gamma} = 5 \times 10^{-4}\sqrt{P/m}$ with m being the average particle mass, which in the adhesion dominated flow regime can be regarded as quasi-static deformation.

We used the normal load P, the particle diameter average $\langle d\rangle$ and the particle mass density ϱ as natural units. To avoid crystallization effects, a moderate polydispersity was utilized by evenly distributed particle diameters $0.5 \leq d/\langle d\rangle \leq 1.5$. Table 1 summarizes the parameter choice, where the same notation as in [13] was used.

Contact Model for Micro-sized Glass Beads

Micro-sized borosilicate glass particles at room temperature serve as a model material for the limit $\dot{\gamma} t_c \rightarrow 0$. Experimental results will be analyzed in section "Complex Shear Localization in the Micro-shear Tester". They compare favorably [37] with simulations done with the contact model described in the following:

Due to the spherical shape of the glass particles, modelling demands a Hertzian contact model. While literature offers a variety of sophisticated contact models for spherical particles, which include elastoplastic response [33, 36] or different formulations of the damping term [1, 2], we intentionally keep the model as well as the parameter choice as simple as possible. For the normal part of the contact force, we use the viscoelastic model combined with a non-zero, constant adhesion force F_{adh} for positive overlap,

$$F_n(\xi, \dot{\xi}) = \left(\frac{4E_{el}\sqrt{r_{eff}}}{3}\xi^{\frac{3}{2}} + \gamma_n\sqrt{\xi}\dot{\xi} - F_{adh}\frac{r_{eff}}{\langle r_{eff}\rangle}\right)\Theta(\xi), \tag{5}$$

with the effective elastic modulus E_{el}, a viscosity coefficient γ_n and the Heaviside function $\Theta(\xi)$. The particle stiffness and adhesion force are set according to the experimentally determined values. Tangential stiffness is set to $k_t = 2/7k_n$ with $k_n = 4/3E_{el}x_{50,2}/2$ (see [37] for details). In the quasi-static deformation regime, viscous forces play a minor role [14], so that one is free to enforce a high dissipation rate by setting $\gamma_n = \sqrt{4/3E_{el}m_{eff}}$, where m_{eff} is the reduced mass. The same argument applies to Coulomb friction implemented by damped linear springs with a damping constant of $2\sqrt{k_t m_{eff}}$. The torsion friction coefficient is linked to the Coulomb friction coefficient by setting $\mu_{tor} = r_{eff}\mu$, and we employ a load dependent rolling friction coefficient $\mu_{\text{rol}} = \sqrt{2r_{\text{eff}}\xi}$. The Coulomb friction coefficient is calibrated by comparing the macroscopic friction coefficient of plane shear simulations under a constant normal load of $\sigma = 5\,\text{kPa}$ with the experimental findings [37].

Elastoplastic Contact Model for Potassium Chloride

Micro-sized KCl particles serve as model material for the limit $\dot{\gamma}t_c \to \infty$. It shows load dependent caking on a time scale of 10–20 min. For this material we developed a protocol that allows a complete calibration (and independent validation) of an elastoplastic contact model, based on nanoindentation results and macroscopic shear experiments as described in [41]. Here, just a brief explanation of the main concepts will be given.

Although the KCl particles have facets with rounded edges, they are represented as spheres in the simulation. Nanoindentation experiments reveal an approximately piecewise linear relation [41] between the contact force and the particle deformation that is represented in the simulation model by the overlap ξ between the particles. Figure 6 (right) depicts the idealized normal force between two colliding particles as a function of their overlap. It consists of a loading branch, which apart from an initial small elastic part describes plastic deformation at yield stress σ_{pl}, and an elastic unloading/reloading branch governed by an effective elastic modulus E_{el}. For dimensional reasons, the two linear branches must have the form

$$\begin{aligned} F_{\text{n}}(\xi) &= \sigma_{\text{pl}} L_{\text{pl}} \xi && \text{for loading,} \\ F_n(\xi) &= E_{el} L_{el} (\xi - \xi_{\text{pl}}) && \text{for unloading,} \end{aligned} \tag{6}$$

where ξ_{pl} is the residual plastic deformation after complete unloading. The length scales L_{pl} and L_{el} characterize the geometry, while the stress scales are material constants, and the overlap stands for the deformation. For undeformed spheres the only length scale is the radius, hence $L_{\text{pl}} \propto r_{\text{eff}}$. However, the unloading takes place at an irreversibly flattened contact area after plastic deformation. Hence, in addition to r_{eff}, the contact geometry depends also on the maximum overlap ξ_{max}. Guided by the scaling of the radius of the circular contact area of overlapping spheres, we postulate $L_{\text{el}} \propto \sqrt{r_{\text{eff}}\xi_{\text{max}}}$. Because of the scaling of the unloading stiffness with the square root of the maximum overlap, a realistic velocity dependence of the restitution coefficient can be expected [32].

Due to adhesion, the elastic branch persists at tensile forces, as long as the threshold $-F_{\text{adh}}$ is not exceeded, at which the contact opens abruptly. The threshold combines a cementation as well as an adhesive component phenomenologically:

$$F_{\text{adh}}(t_{\text{k}}) = \sigma_{\text{adh}} A_{\text{adh}} e^{-t_{\text{k}}/t_{\text{c}}} + \sigma_{\text{cem}} A_{\text{cem}} \left(1 - e^{-t_{\text{k}}/t_{\text{c}}}\right), \tag{7}$$

i.e. contact crystallization leads to a replacement of the adhesive part $\sigma_{\text{adh}} A_{\text{adh}}$ by a cementation term $\sigma_{\text{cem}} A_{\text{cem}}$ on a timescale t_c. We find that the bulk's load dependence in the adhesive limit is best described by $A_{adh} = \chi_{adh} r_{eff}^2 (1 + \xi_{max}/r_{eff})$ and leave $\sigma_{adh}\chi_{adh}$ as a calibration parameter [41, 42]. The tensile strength in the cementation limit is given by the plastic yield stress of KCl and a load dependent contact area $A_{cem} = \chi_{cem} r_{eff} \xi_{max}$, where $\sigma_{cem}\chi_{cem}$ is used as a fit parameter.

The materials yield stress was determined with nanoindentation tests. Furthermore, we assumed the unloading and reloading branch to be viscoelastic and included a moderate damping coefficient. As in section "Contact Model for Micro-sized Glass Beads", torsion friction is linked to Coulomb friction. The influence of particle shape was taken into account by a rolling friction coefficient that was determined with SEM images and an approach presented in [7]. The Coulomb friction coefficient μ, as well as the parameters for tensile forces $\sigma_{adh}\chi_{adh}$ and $\sigma_{cem}\chi_{cem}$ where calibrated by comparing the stress response in plane shear experiments and simulations.

Boundary Conditions

Calibration of the contact models (sections "Contact Model for Micro-sized Glass Beads" and "Elastoplastic Contact Model for Potassium Chloride") was done by wall-driven plane shear simulations. Specifically, frictional walls structured by vanes (see Fig. 1 (right)) and periodic boundaries in the directions perpendicular to y were used. The motion of the upper wall was velocity controlled in shear direction and force controlled in y-direction.

Simulations of the micro-shear tester [37] were done with the same torsional shear cell geometry but a slightly smaller diameter (1.78 mm), i.e. a cylinder for particle confinement in horizontal direction and two structured pistons (one movable, one fixed) in vertical direction. Since we neglect plastic deformation, a small gap of 0.1 mm between the outer edge of the piston and the cylinder wall was introduced to avoid jamming of single particles. Particle-wall interaction is described by the viscoelastic part of the contact model in Eq. 5, using the same material parameters.

Rheology of Caking Granular Matter: A 2D Simulation Study

Here we answer the fundamental question, how a finite contact accommodation time t_c affects rheology. The simulations are performed in 2D with the contact force model described in section "Contact Model for 2D Caking Simulations". The results are compared with theoretical predictions obtained from the principle of minimal energy dissipation. Starting point is always a completely accommodated bulk, i.e. $t_k/t_c \to \infty$ in Eq. 4. Depending on the dimensionless parameter $\dot{\gamma} t_c$, qualitatively different shear heterogeneities emerge. As the average contact time is given by the inverse shear rate, $\dot{\gamma} t_c$ characterizes the competition between opening and forming of caked contacts. For $\dot{\gamma} t_c \ll 1$, we expect the behavior of a normal cohesive granular

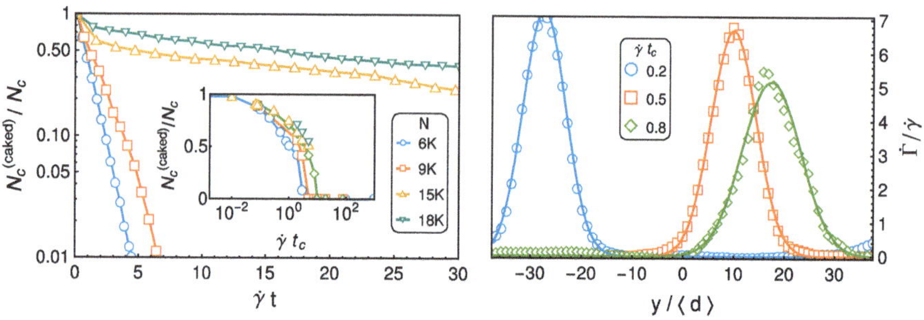

Fig. 7 Portion of caked contacts as a function of shear deformation for $\dot{\gamma} t_c = 80$ (left) and as a function of $\dot{\gamma} t_c$ after a total shear deformation of $\dot{\gamma} t = 30$ (inset). Snapshots of the local shear rate as a function of y, fitted by Eq. 8 (right) (Reprinted with permission from [42].)

bulk, while $\dot{\gamma} t_c \gg 1$ defines the *cementation dominated regime* where adhesion forces vanish, when a contact fails, and never recover to their former value, because the contacts don't live long enough. Therefore, shear banding in the cementation dominated regime is conceptually similar to fracture. Upon further shear, the initial crack is expected to widen, grinding up the adjacent caked material. This mechanism needs to be understood first, because it is at the core of the rheology of caking granular matter.

Inspired by the sample preparation in common shear tests [24, 25], we prepare a well-defined, reproducible initial state by driving the cohesive bulk, setting $t_c = 0$, into steady state shear [23, 43]. This choice avoids over-consolidation and hence the resulting dilatancy under shear, for all $t_c > 0$, as the same normal pressure P is applied for preparation and shear simulations. Different (average) system heights were implemented at a fixed system width ($=88\langle d\rangle$) in shear direction, in order to check finite size effects for the shear bands. System size is given in terms of total particle number N.

Figure 7 (left) illustrates for $\dot{\gamma} t_c = 80$ and different system sizes, how the initially caked configuration is ground up in the cementation dominated regime. We call contacts caked, if they have existed longer than the accommodation time t_c. Their number decreases roughly exponentially within the simulated time span. The two small systems ($L_y/L_x \lesssim 1$) loose all caked contacts and become a homogeneously sheared, weakly cohesive material on a timescale $\dot{\gamma}^{-1}$. There is a qualitative difference compared to the two larger systems, where the timescale of the exponential decay of caked contacts is approximately $t_c = 80\dot{\gamma}^{-1}$, instead.

We repeated this kind of simulation many times for a whole range of $\dot{\gamma} t_c$-values and system sizes. Evaluating only those runs, which reached a steady state by the time $t \approx 30\dot{\gamma}^{-1}$, the fraction of contacts with $t_k > t_c$ is shown in the inset of Fig. 7 (left). In the adhesion dominated regime ($\dot{\gamma} t_c \ll 1$), we observe no structural or mechanical change with respect to the initial configuration. Thus, the two limits of vanishing and infinite accommodation time tend towards the steady states of a cohesive or a non-cohesive bulk, respectively. In the intermediate regime between these limiting cases, the sheared system remains partially caked.

We examined this regime more closely by measuring the local shear rate $\dot{\Gamma}(y) = \partial v_x/\partial y$, which was calculated from the coarse-grained velocity field, averaged in x-direction. Shearing is localized between two caked slabs. The shear bands can be fitted by

$$\dot{\Gamma}(y) = \dot{\gamma} \frac{L_y}{\sqrt{2\pi} w_{sb} \operatorname{erf}\left(L_y/2\sqrt{2}w_{sb}\right)} e^{-\frac{(y-y_{sb})^2}{2w_{sb}}}, \tag{8}$$

as displayed in the right graph of Fig. 7. As the averaged contact time within these shear bands is less than t_c, adhesion forces are suppressed, which is why the shear bands turn out to be more densely packed than the average bulk [42]. Because of shear weakening due to reduced adhesion inside the shear band, its position y_{sb}, once established randomly, is rather stable. Only rare spontaneous jumps of the position were observed.

In the following we will present a qualitative explanation for these shear heterogeneities based on the principle of minimal energy dissipation [17, 18] which was previously used in the context of granular matter for the prediction of shear band configurations [15, 29, 38]. Under the assumption that the work done is fully dissipated, the dissipated power density equals the product of shear stress τ and shear rate $\dot{\gamma}$ [6]. For dense granular flow the shear stress depends on the shear rate in the following way:

$$\tau(\dot{\gamma}) = \tau_{min}(\eta) + b(\eta)\sqrt{m P}\dot{\gamma}, \tag{9}$$

where the threshold $\tau_{min}(\eta)$ as well as the coefficient $b(\eta)$ are functions of the dimensionless cohesion number $\eta = F_{adh}/Pd$ [21]. According to Eq. 4 one must consider a local cohesion number, which depends on the average lifetime of contacts

$$\eta = \eta_{max}\left(1 - e^{-\langle t_k\rangle/t_c}\right). \tag{10}$$

The combination of Eqs. 5 and 9 yields to first order in η

$$\tau(\dot{\gamma}) = \tau_0 + \tau_1\left(1 - e^{-\langle t_k\rangle/t_c}\right) + \left(b_0 + b_1\left(1 - e^{-\langle t_k\rangle/t_c}\right)\right)\dot{\gamma}. \tag{11}$$

The constants τ_0 and τ_1 were measured in simulations of the steady state flow of the adhesion dominated limit. b_0 and b_1 can be estimated with literature values [21]. To proceed further, we switch to the concept of local rheology. By neglecting the time dependence of the local shear rate, we assume translational invariance in shear direction and set $\langle t_k\rangle = \langle t_k\rangle(y) = 1/\dot{\Gamma}(y)$, hence the contact time of each layer within the bulk is proportional to the inverse local shear rate. Combining these equations, we get the dissipation rate

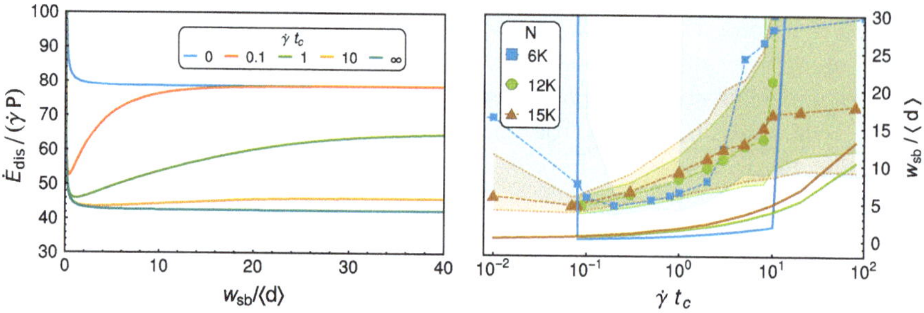

Fig. 8 Dissipation rate as a function of shear band width for different $\dot{\gamma}t_c$ (left). Median of measured shear band width together with interval between quantiles 0.1–0.9 as shaded region, and the theoretical prediction according to Eq. 12 as solid lines (right) (Reprinted with permission from [42].)

$$\dot{E}_{\text{dis}}[\dot{\Gamma}] = L_x \int_{-L_y/2}^{L_y/2} \tau(\dot{\Gamma}(y), \eta(\dot{\Gamma}(y), t_c))\dot{\Gamma}(y)dy, \tag{12}$$

as a functional of the local shear rate $\dot{\Gamma}$. Inserting Eq. 8 for $\dot{\Gamma}(y)$, the dissipation rate becomes a function of shear band width and accommodation time $\dot{E}_{\text{dis}}(w_{\text{sb}}, t_c)$. In the limits $t_c \to 0$ and $t_c \to \infty$ this function can be calculated analytically. For finite t_c the dissipation rate is determined by numerical integration (Fig. 8, left). In the cementation as well as in the adhesion dominated regime, the dissipation rate decreases monotonically as a function of shear band width. Hence, the theory predicts $w_{\text{sb}} \to \infty$ as the state of least dissipated power and therefore homogenous shear. However, for intermediate values of $\dot{\gamma}t_c$, a local minimum at finite w_{sb} exists, and the state of least energy dissipation predicts inhomogeneous flow and shear bands. The right graph in Fig. 8 shows the measured shear band width (median) as a function of $\dot{\gamma}t_c$ (symbols) as well as the predicted w_{sb} utilizing Eq. 12 for various system sizes L_y. (Shaded areas quantize fluctuations when fitting Eq. 8.) As predicted by theory, simulation results confirm shear zones with finite w_{sb}, which increases sub-linearly as a function of $\dot{\gamma}t_c$. As we argued in the framework of a local rheology, it is not surprising, that theory underestimates the measured shear band width (see e.g. [14] or [19] for a more detailed discussion). Still the functional dependence in the regime $\dot{\gamma}t_c \approx 1$ is captured qualitatively correct. Furthermore, within the framework of a local rheology, the finite shear-band width is predicted as another length scale besides the particle diameter.

Complex Shear Localization in the Micro-shear Tester

We analyzed the experiment described at the beginning of section "Particle Tracking and Statistical Analysis" to assess shear induced structural heterogeneity as well as shear band formation and evolution over time. This was a micro-sized torsional shear experiment with spherical glass particles and a constant normal load. As the cohesion forces between particles are small compared to the normal stress $\left(\eta = F_c/\left(\sigma x_{50,3}^2\right) \approx 0.2\right)$, they are expected to have only a minor influence. A total shear strain of $\omega t = 39.5°$ was analyzed based on tomographic images, which were taken in steps of 0.5° for the first 9.5° of shearing, and in steps of 5° for the remaining experiment. The results summarized in this section were previously published in [11, 12].

We first analyzed the spatial structure of the sample and its evolution over time. Particles were segmented in all tomographic images as detailed in section "Particle Tracking and Statistical Analysis" and the centers of mass were computed based on their voxel representations. Extremely small and extremely large particles were excluded since they likely occur due to artifacts or segmentation errors, though the volume of excluded particles was less than 0.1% of the sample volume at all time points. Based on the extracted centers of mass, we computed the number density, i.e., the expected number of particles per cubic millimeter, as a function of height (z) and radial coordinate (r). We coarse-grained the data with a Gaussian kernel with standard deviation w to obtain continuous fields and averaged over φ, assuming rotational invariance. Figure 9 shows the number density profiles for the initial configuration, a consolidated configuration and a configuration towards the end of the experiment. The profiles reveal that the sample is compressed during the first steps of shearing ($\boldsymbol{\omega t} = 0°$ to $\boldsymbol{\omega t} = 4°$) and that the number density is larger in the bottom of the sample than on top. The compression and low density at the top of the sample indicate that applying the uniaxial normal load after sieving the particles into the sample chamber is not sufficient to spread the particles on top out evenly. This process is completed during the first steps of shearing. Moreover, Fig. 9 shows that there is an almost cone-shaped, denser zone in the lower center of the shear cell. This might also be caused by the preparation procedure, e.g. by particles piling on top of each other at the beginning of sieving, which are later compressed by other particles on top. However, this interpretation needs further investigation. A dilation can only be observed after $\boldsymbol{\omega t} = 4°$ and occurs directly above the lower piston, suggesting that this is the location of the shear zone.

In addition, we computed particle tracks, estimated the axis of rotation from the data and computed angular velocities of all particles with respect to this estimated axis as described in section "Particle Tracking and Statistical Analysis". We also used the 1D image-based local shear deformation (see section "Particle Tracking and Statistical Analysis") to accurately measure the movements of upper and lower piston by maximizing image cross-correlation in slices where the pistons are visible using a very small step size of 0.01°. In principle, this movement should be exactly as dictated by the experiment, i.e., 0.5° or 5° per step for the upper piston and no

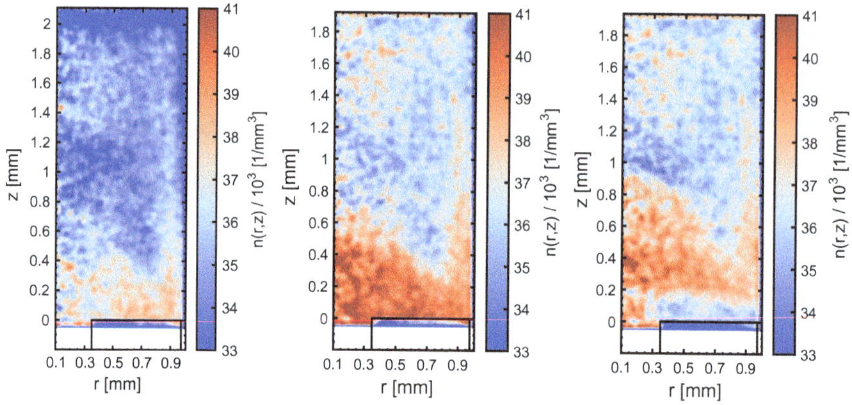

Fig. 9 2D density profiles (coarse-graining length $w = 14.9$ μm) of the initial configuration ($\omega t = 0°$; left), a compacted configuration ($\omega t = 4°$; center) and a configuration towards the end of the experiment $\omega t = 34.5°$; right). The black frame in the bottom right corner marks the location of the vanes on the lower piston (Reprinted from [12] with permission from Elsevier.)

movement for the lower piston. However, we measured deviations from this ideal, mostly at the lower piston. The mean absolute differences between target and actual movements were 0.02° and 0.19° at the upper and lower piston, respectively. The reason for this deviation in rotational movement probably lies in the experimental setup. The compensation unit tracks shear stress and is supposed to keep the lower piston in place, but particles interlocked between piston and wall might counteract this mechanism so that the lower piston gets partially dragged along. This problem could be solved by increasing the gap between lower piston and wall, yet this would simultaneously lead to more loss of sample material, which might be pushed out of the sample chamber through the gap. Moreover, only the rotation of the upper piston and glass capillary can be controlled in the micro shear-tester. Rotations of the lower piston can unfortunately not be observed with an accuracy of 0.01° during shear deformation. To ensure comparability of angular velocities between time steps, we normalized them to the interval of [0, 1] based on the movements of upper and lower piston measured in the image data.

We found that for a fixed radial distance from the axis of rotation, the normalized angular velocity profiles could well be described by the parametric function

$$v_\varphi(z) = \frac{1}{2} + \frac{1}{2}\mathrm{erf}\left(\frac{z - z_{sb}}{w_{sb}}\right), \tag{13}$$

where erf denotes the error function. The fitting parameters, z_{sb} and w_{sb}, can be determined with the least squares method and offer a quantitative description of the vertical location and width of the shear band (as a function of the radial coordinate). The parametric function given in Eq. 13 has previously been used in [8] to describe a symmetric shear zone and is attractive because of its simplicity. We defined the shear

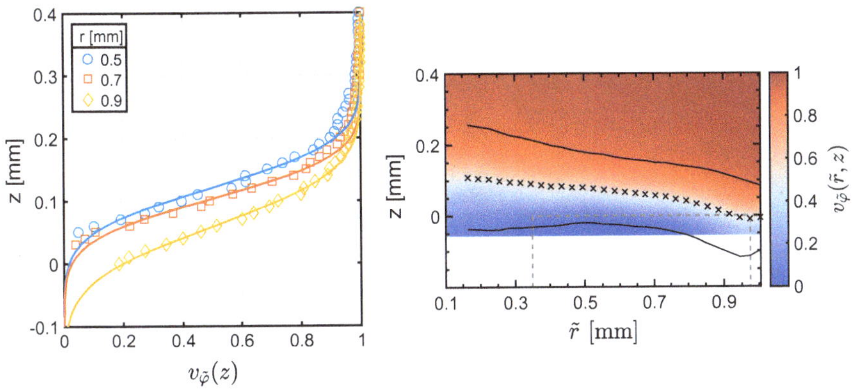

Fig. 10 Example of actual and fitted velocity profiles for one time step (29.5°–34.5° of shearing) and different distances from the axis of rotation (left), and 2D velocity profile after $\omega t = 10°$ with the fitted shear band indicated in black as $z_{sb} \pm w_{sb}$ (right) (Reprinted from [12] with permission from Elsevier.)

band as the interval $z_{sb} \pm w_{sb}$, which covers roughly 84% of the velocity range in the data. Different choices to define shear band width would be possible, yet since we mainly want to quantitatively compare the experimental shear band characteristics between different time points, we stick to this simple choice. Figure 10 shows examples of actual and fitted 1D velocity profiles for different radial coordinates and the time-averaged, coarse-grained ($w = 14.9$ μm) 2D velocity profile after $\omega t = 10°$ with fitted shear band.

The velocity profiles show that shear strain localizes directly above the lower piston, as was already indicated by the number density. Its center is at an average height of $z = 0.12$ mm. The 2D profile additionally reveals that the shear band is slightly wider in the center of the shear cell and curved downwards near the cylinder wall. The latter is probably due to the fact that the outer cylinder wall rotates together with the upper piston, and particles close to it are more easily dragged along and achieve higher angular velocities than particles in the center. We observe an average shear band width of $2w_{sb} = 250$ μm, which corresponds to roughly 8 median particle diameters and agrees well with values reported in literature [9, 10, 16, 31].

Using the parametric shear band representation, we analyzed the development of shear band width and location over time. A plot of z_{sb} and w_{sb} over time (averaged over the radial coordinate) is shown in Fig. 11. Shear band location and width fluctuate strongly at the beginning of the experiment, until a steady state seems to be reached at $\omega t \approx 10°$. Lower fluctuations are expected after this point, since we implicitly average over much larger time intervals (5° instead of 0.5° per step). In the initial phase of the experiment, the shear band is (on average) wider and rises from $z_{sb} \approx 0.15$ mm to $z_{sb} \approx 0.37$ mm before dropping back to its position right above the lower piston. Even in the more steady period after $\omega t \approx 10°$ the shear band width fluctuates perceptibly, taking values between $w_{sb} \approx 0.8$ mm to $w_{sb} \approx 0.16$ mm.

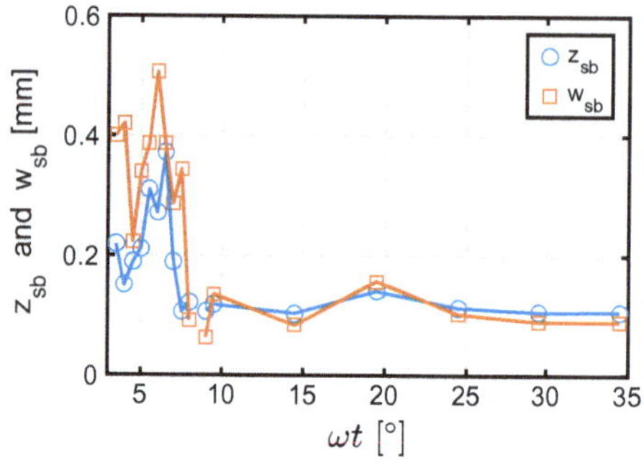

Fig. 11 Development of the fitted shear band width and location over time, averaged over the radial coordinate (Reprinted from [12] with permission from Elsevier.)

There could be a second drop in variability at $\omega t \approx 25°$ but with the present data we cannot judge if the steady state is actually only reached at this point or if the second drop is merely a coincidence. Therefore, based on our analysis, we assume that the steady state is reached at $\omega t \approx 10°$.

Conclusion and Outlook

The interplay between particle contact properties, bulk structure and mechanical behaviour of fine, adhesive granular matter is of wide interest. In particular, time consolidation effects, i.e. time dependent adhesion, are far from being well understood, although they are of eminent importance for the handling of powders after storage. This chapter reviewed some recent progress in this respect.

Recording as much experimental information as possible on the particle motion during a shear experiment is vital for gaining insight into the relationship between bulk structure and mechanical behaviour. Ideally, observing the trajectory of every single particle offers information on local densification, loosening or formation of failure areas. Before we started our project this was only possible virtually by discrete element method (DEM) simulations.

We came very close to this ideal by developing a novel micro torsional shear tester which enables to measure yield loci of extremely small powder samples in the range of only 10 μl with high accuracy in force and displacement measurements and which can be integrated into a high resolution X-ray tomography device (μCT). Moreover, we developed the necessary statistical analysis methods which allow the extraction of the trajectories of almost all particles. The processing of time-resolved data makes it possible to localize and track particles despite large angle increments of up to 5° between tomographic measurements.

With this device and analysis tool stepwise compression and shear of powder samples were monitored in detail using particles in the size range between 10 and 100 μm. In this chapter we reported results for two powder materials, borosilicate glass spheres and potassium chloride particles. The reason for this choice of model materials was that they represent very different particle contact properties, validated by simulations with the discrete element method (DEM). Contacts between the glass spheres could be described by a *viscoelastic* Hertz model shifted by a *constant adhesion* force. Using a geometry comparable to the shear tester, the width and position of shear bands and the radial and axial particle density distribution could be reproduced.

Potassium chloride, on the other hand can be described by an *elastoplastic* model with a *time dependent adhesion* force that approaches a load dependent maximal value on a time scale of $t_c \approx 15$ min, called accommodation time. A method to calibrate the DEM model was developed and proved. It is based on nanoindentation and shear tests. (This calibration strategy can also be applied to adhesion dominated materials, where $t_c \approx 0$.) After calibration, the simulation model could be independently validated.

For potassium chloride, accommodation happens very slowly compared to typical inverse shear rates, for the glass beads very fast. In this sense, the model materials represent extreme cases of cementation dominated, respectively adhesion dominated powders. In general, adhesive materials, when sheared, are expected to show behaviour in between these extreme cases. A systematic theoretical prediction of shear banding for finite accommodation times, and its qualitative support by DEM simulations has been given. The theory is based on the principle of minimal dissipation power and takes the shear weakening into account, which originates from the fact, that the average adhesion force reached during the lifetime of a particle contact decreases with increasing $\dot{\gamma} t_c$. Starting from a completely accommodated bulk, this leads to stable shear bands of finite width that increases with $\dot{\gamma} t_c$. Compared to the values obtained from simulation, the theory underestimates the shear band width. A possible reason is that the theory is based on local rheology. Creep effects are not taken into account. A nonlocal extension of the theory would be a promising future project.

References

1. Antypov, D., Elliott, J.A.: On an analytical solution for the damped Hertzian spring. Europhys. Lett. **94**, 50004 (2011)
2. Brilliantov, N.V., Spahn, F., Hertzsch, J.M., Pöschel, T.: Model for collisions in granular gases. Phys. Rev. E **53**, 5382 (1996)
3. Campbell, C.S.: Stress-controlled elastic granular shear flows. J. Fluid Mech. **539**, 273 (2005)
4. Crocker, J.C., Grier, D.G.: Methods of digital video microscopy for colloidal studies. J. Colloid Interface Sci. **179**, 298 (1996)
5. Cuccovillo, T., Coop, M.R.: On the mechanics of structured sands. Géotechnique **49**, 741 (1999)
6. Da Cruz, F., Emam, S., Prochnow, M., Roux, J.-N., Chevoir, F.: Rheophysics of dense granular materials: discrete simulation of plane shear flows. Phys. Rev. E **72**, 21309 (2005)
7. Estrada, N., Azéma, E., Radjaï, F., Taboada, A.: Identification of rolling resistance as a shape parameter in sheared granular media. Phys. Rev. E **84**, 11306 (2011)
8. Fenistein, D., van de Meent, J.W., van Hecke, M.: Universal and wide shear zones in granular bulk flow. Phys. Rev. Lett. **92**, 094301 (2004)

9. Gourlay, C.M., Dahle, A.K.: Dilatant shear bands in solidifying metals. Nature **445**, 70 (2007)
10. Guo, P.: Critical length of force chains and shear band thickness in dense granular materials. Acta Geotech. **7**, 41 (2012)
11. Handl, L., Torbahn, L., Spettl, A., Zetzener, H., Schmidt, V., Kwade, A.: Microstructural changes, particle tracking and shear localization of fine glass powders. In: Proceedings of the International Congress on Particle Technology, Nürnberg, paper-ID 2.29 (2016)
12. Handl, L., Torbahn, L., Spettl, A., Schmidt, V., Kwade, A.: Structural analysis and tracking of micron-sized glass particles during shear deformation: a study based on time-resolved tomographic data. Adv. Powder Technol. **28**, 1920 (2017)
13. Luding, S.: Cohesive, frictional powders: contact models for tension. Granul. Matter **10**, 235 (2008)
14. MiDi, G.D.R.: On dense granular flows. Eur. Phys. J. E **14**, 341 (2004)
15. Moosavi, R., Shaebani, M.R., Maleki, M., Török, J., Wolf, D.E., Losert, W.: Coexistence and transition between shear zones in slow granular flows. Phys. Rev. Lett. **111**, 148301 (2013)
16. Nemat-Nasser, S., Okada, N.: Radiographic and microscopic observation of shear bands in granular materials. Géotechnique **51**, 753 (2001)
17. Onsager, L.: Reciprocal relations in irreversible processes. I. Phys. Rev. **37**, 405 (1931)
18. Onsager, L.: Reciprocal relations in irreversible processes. II. Phys. Rev. **38**, 2265 (1931)
19. Ries, A., Brendel, L., Wolf, D.E.: Shearrate diffusion and constitutive relations during transients in simple shear. Comput. Part. Mech. **3**, 303 (2016)
20. Ries, A., Brendel, L., Wolf, D.E.: The cooperativity length in simple shear of dry granular media. Comput. Part. Mech. **4**, 379 (2017)
21. Rognon, P.G., Roux, J.-N., Naaim, M., Chevoir, F.: Dense flows of cohesive granular materials. J. Fluid Mech. **596**, 21 (2008)
22. Rognon, P.G., Miller, T., Metzger, B., Einav, I.: Long-range wall perturbations in dense granular flows. J. Fluid Mech. **764**, 171 (2015)
23. Rothenburg, L., Kruyt, N.P.: Critical state and evolution of coordination number in simulated granular materials. Int. J. Solids Struct. **41**, 5763 (2004)
24. Schwedes, J.: Review on testers for measuring flow properties of bulk solids. Granul. Matter **5**, 1 (2003)
25. Schulze, D.: Pulver und Schüttgüter. Springer, Berlin, Heidelberg (2009)
26. Shojaaee, Z., Roux, J.-N., Chevoir, F., Wolf, D.E.: Shear flow of dense granular materials near smooth walls. I. Shear localization and constitutive laws in the boundary region. Phys. Rev. E **86**, 11301 (2012)
27. Shojaaee, Z., Brendel, L., Török, J., Wolf, D.E.: Shear flow of dense granular materials near smooth walls. II. Block formation and suppression of slip by rolling friction. Phys. Rev. E **86**, 11302 (2012)
28. Soddemann, T., Dünweg, B., Kremer, K.: Dissipative particle dynamics: a useful thermostat for equilibrium and nonequilibrium molecular dynamics simulations. Phys. Rev. E **68**, 046702 (2003)
29. Stegmann, T., Török, J., Brendel, L., Wolf, D.E.: Minimal dissipation theory and shear bands in biaxial tests. Granul. Matter **13**, 565 (2011)
30. Strege, S., Weuster, A., Zetzener, H., Brendel, L., Kwade, A., Wolf, D.E.: Approach to structural anisotropy in compacted cohesive powder. Granul. Matter **16**, 401 (2014)
31. Strege, S.: Röntgenmikrotomographische Analyse der Verdichtung und Scherung feiner und kohäsiver Pulver. Ph.D. Thesis, TU Braunschweig (2014). ISBN 978-3868446647
32. Thornton, C., Cummins, S.J., Cleary, P.W.: On elastic-plastic normal contact force models, with and without adhesion. Powder Technol. **315**, 339 (2017)
33. Thornton, C., Ning, Z.M.: A theoretical model for the stick/bounce behaviour of adhesive, elastic-plastic spheres. Powder Technol. **99**, 154 (1998)
34. Tighe, B.P.: Critical viscoelastic response in jammed solids. arXiv:1205.2960 (2012)
35. Tomas, J.: The consolidation of particulate solids—microprocesses of caking and kinetic models. Chem. Ing. Tech. **69**, 455 (1997)

36. Tomas, J.: Adhesion of ultrafine particles—a micromechanical approach. Chem. Eng. Sci. **62**, 1997 (2007)
37. Torbahn, L., Weuster, A., Handl, L., Schmidt, V., Kwade, A., Wolf, D.E.: Mesostructural investigation of micron-sized glass particles during shear deformation—an experimental approach vs. DEM simualtion. In: EPJ Web of Conferences, vol. 140, p. 03027 (2017)
38. Unger, T., Török, J., Kertész, J., Wolf, D.E.: Shear band formation in granular media as a variational problem. Phys. Rev. Lett. **92**, 214301 (2004)
39. Wenzl, J., Seto, R., Roth, M., Butt, H.-J., Auernhammer, G.K.: Measurement of rotation of individual spherical particles in cohesive granulates. Granul. Matter **15**, 391 (2013)
40. Weuster, A., Brendel, L., Wolf, D.E.: Simulation of sheared, caking powder. In: AIP Conference Proceedings, vol. 1542, p. 515 (2013)
41. Weuster, A., Strege, S., Brendel, L., Zetzener, H., Wolf, D.E., Kwade, A.: Shear flow of cohesive powders with contact crystallization: experiment, model and calibration. Granul. Matter **17**, 271 (2015)
42. Weuster, A.: Planare Scherung kohäsiver und zeitverfestigender granularer Materie mit der Diskrete-Elemente-Methode. Ph.D. thesis, Universität Duisburg-Essen (2017). https://duepublico2.uni-due.de/receive/duepublico_mods_00044560
43. Wood, D.M.: Soil Behaviour and Critical State Soil Mechanics. Cambridge University Press (1990)

Improved Flowability of Ultrafine, Cohesive Glass Particles by Surface Modification Using Hydrophobic Silanes

Zinaida Todorova, Steffi Wünsche and Werner Hintz

Introduction

In the past decades, fine (d < 100 μm), ultrafine (d < 10 μm) and nano-sized (d < 100 nm) particulate materials have attracted a huge interest for many industrial branches, such as chemical, pharmaceutical, food, automotive, construction materials industry [1]. The reason for the increasing usage of these particle systems is in their special material properties; one of them is their large specific surface area. In addition, ultradispersed particles often provide controlled functionality and increased reactivity, making them advantageous over other conventionally used coarse grain materials [2]. However, their large specific surface areas lead to an intensification of the interparticle attractive forces that exceed the gravity force by several orders of magnitude. It results in a high adhesion potential of these particle systems leading to a bad flowability, which significantly impairs their handling and processing compared to chemically identical, coarse materials. Such materials are cohesive. Typical problems of cohesive powders in shipping and storage containers or processing apparatuses related to the high interparticle attraction are bridging, channeling, oscillating mass flows, feeding and dosing problems.

There are different approaches aimed to overcome the interparticle adhesion and facilitate the handling of cohesive powders. One of them implies the introduction of external energies by agitators, air streams, mechanical vibrations or acoustic waves [3]. However, such technical tools do not have a lasting effect on the flowability of the treated powders since their effect fades after the end of the treatment. Another approach is to reduce the adhesion forces between cohesive particles directly by

Z. Todorova · S. Wünsche · W. Hintz (✉)
Institute for Process Engineering, Otto-von-Guericke Universität Magdeburg, Magdeburg, Germany
e-mail: werner.hintz@ovgu.de

Z. Todorova
e-mail: zinaida.todorova@ovgu.de

S. Antonyuk (ed.), *Particles in Contact*,
https://doi.org/10.1007/978-3-030-15899-6_21

changing their surface properties, which are known to depend strongly on the surface morphology and chemical composition. The adhesion reduction through the modification of the surface morphology is a widely used technique and is most commonly achieved through coating with small nanoparticles. While this approach is well-known and it is applied for decades for improving the flowability and facilitating the handling of cohesive powders [4–8], the approach of adhesion reduction through a modification of the chemical composition of the surface is not being well studied.

This paper investigates the effect of a changed chemical surface composition on the flowability of fine, cohesive particles. As a model particle system, microscopic spherical glass particles are used. The chemistry modification of the surface is achieved by the surface hydrophobization through a coating with hydrophobic silanes. This leads to a decreased surface energy, and thus to a decreased work of adhesion, and it also reduces the effect of capillary forces. Silanes are widely available and well suited to modify the surfaces of metals, metal oxides and glass [9–11]. They react with the hydroxyl groups that are on the particle surfaces and replace them with low polar organic rests, thus effecting an altering of the physical interactions of the treated substrates and rendering them hydrophobic. Five hydrophobic silanes with different hydrophobic groups—alkyl, fluoroalkyl and phenyl—and with different chain lengths were selected for the surface hydrophobization. The impact of the silane coating on the powder flow behavior of the particle systems was investigated by shear testing. By a model-based reverse micro-macro transition, the micromechanical adhesion and contact properties of the surface-modified particles were determined from the shear tests. This will contribute to the better understanding of the flow properties of the modified powders on macroscopic level.

Experimental Set-Up

Preparation of the Surface-Modified Particles

Glass particles were selected as a reference material due to a number of reasons. First, they find a very broad industrial application as fillers, reflective coating, polymer reinforcement and in a lot of other applications. Secondly, they possess reactive hydroxyl (OH) groups on the surface that can be easily subjected to surface modifications. The used—commercially available product Spheriglass 5000 CP00—was purchased from Potters Europe GmbH. Spheriglass particles are smooth, nonporous (see Fig. 1) and have a relatively high sphericity (>91%).

All particles were classified through a 20 μm sieve using an ALPINE Air Jet Sieve e200 LS. The particle size distribution $Q_3(d)$ of the glass particles after sieving, measured with the laser diffraction spectroscopy (Mastersizer 2000, Malvern Instruments), is presented in Fig. 2. Particles are relatively narrow distributed and possess a median particle size of 7.8 μm.

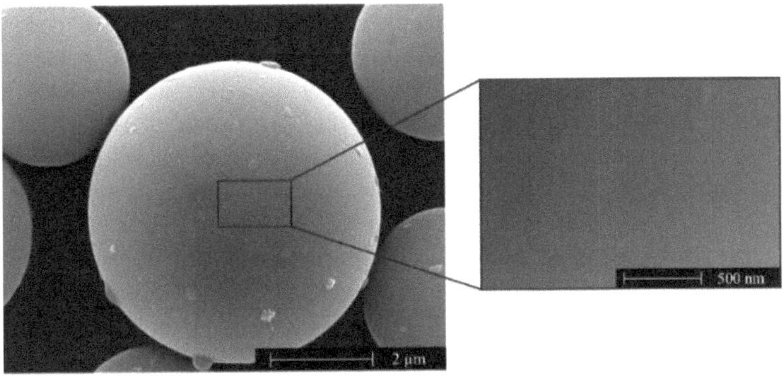

Fig. 1 Scanning electron micrograph (SEM) of Spheriglass CP00 5000

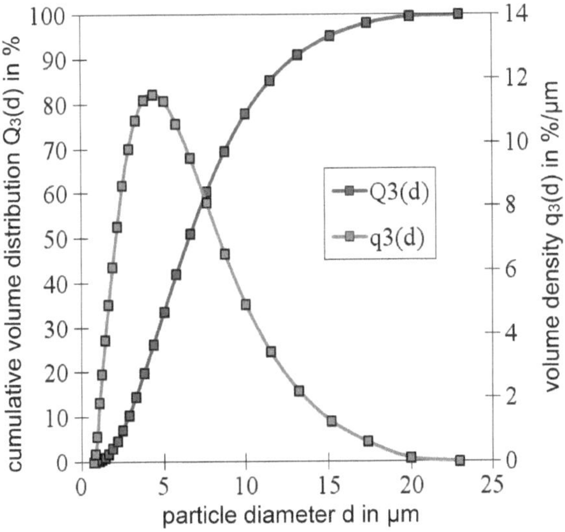

Fig. 2 Particle size distribution $Q_3(d)$ and particle size frequency distribution $q_3(d)$ of the Spheriglass 5000 system, quantiles of $Q_3(d)$: $d_{5,3} = 3.15\ \mu m$, $d_{50,3} = 7.8\ \mu m$, $d_{95,3} = 18\ \mu m$

Prior to modification, the glass particles were cleaned and activated, leading to a hydroxylated glass surface. This was necessary to let the silanes react with the surface hydroxyl groups and substitute them with hydrophobic rests. Since the OH-groups are the sites at which the silanes are linked to the surface, the presence of those groups constitutes a crucial prerequisite for a successful surface hydrophobization. In the current study peroxymonosulfuric acid (Piranha solution) was used which is well suited to clean the glass surface from any organic contaminants and add OH-groups. The Piranha solution was prepared as a mixture of concentrated sulfuric acid and hydrogen peroxide in ratio 3:1. The charring effect of the sulfuric acid is complemented by the formation of elemental oxygen which can dissolve and oxidize elemental carbon, thus to remove organic residues completely by oxidizing them to CO_2. After cleaning with Piranha solution the treated surfaces are extremely hydrophilic (the contact angle after cleaning is lower than 10° [12]). For the cleaning 100 g particles were mixed with 100 ml freshly prepared Piranha solution; the reaction mixture was agitated for 12 h to remove all organic contaminations completely. Subsequently, the particles were separated with a frit, washed several times with distilled water and then dried at 100 °C immediately before silanization.

Silanization was carried out in a wet process chemically with dry toluene as a solvent. Prior to coating, the particles were mixed with the solvent and sonicated for 10 min. After the sonication procedure, the silane was added to the solvent-particle suspension. To modify the thickness of the produced silane coating three different silane concentrations were applied: 0.010 molar, 0.025 molar and 0.035 molar.

Used silanes and their structural formulas are given in Table 1. The compounds Chlorotrimethylsilane (CTMS), Chlorodimethylphenylsilane (CDMPS), Chlorodimethyloctylsilane (CDMOS) and Triethoxyoctylsilane (TEOS) were purchased from Sigma Aldrich, and 3,3,3-Trifluoropropyltrimethoxysilane (FPTS), 1H,1H,2H,2H-Perfluorooctyltriethoxysilane (PFOTES) from Alfa Aesar. All silanes were applied without further purification.

Table 1 Used silanes with their structural formulas and molecular weights

Silane	Structural formula	Molecular weight (mol/g)
1H,1H,2H,2H-Perfluorooctyltriethoxysilane (PFOTES)	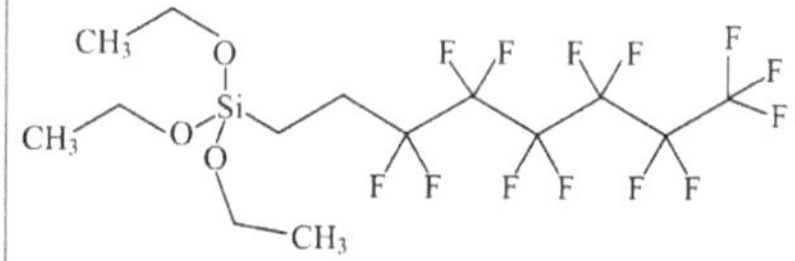	510.36
Chlorodimethyloctylsilane (CDMOS)	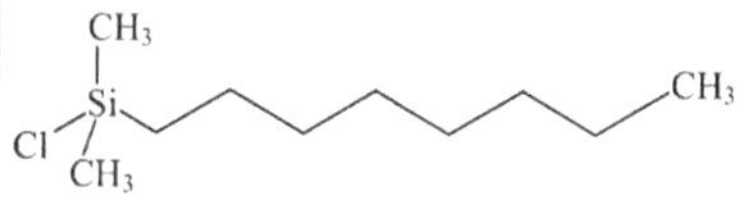	206.83
3,3,3-Trifluoropropyltrimethoxysilane (FPTS)	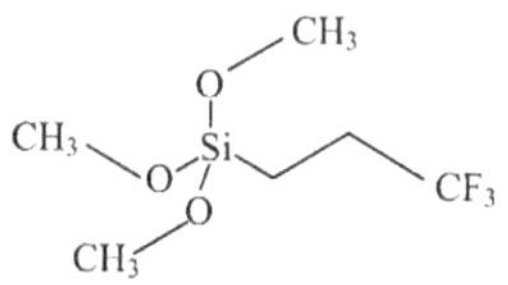	218.25
Trimethylchlorsilane (CTMS)	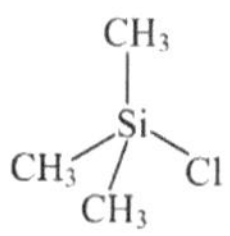	108.64
Chlorodimethylphenylsilane (CDMPS)	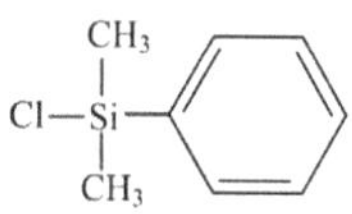	170.71

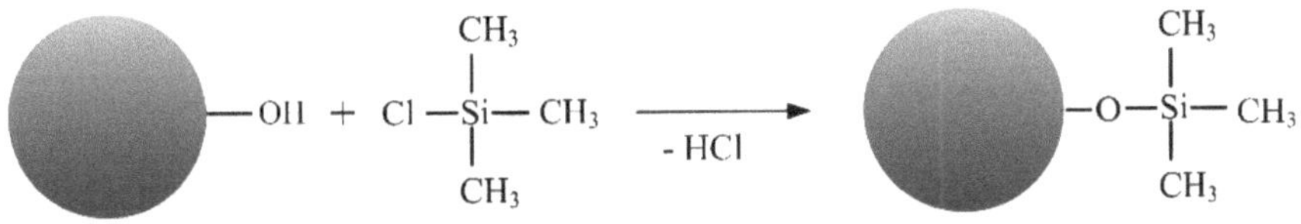

Fig. 3 Theoretical bonding mechanism of Trimethylchlorsilane (CTMS) to the glass surface

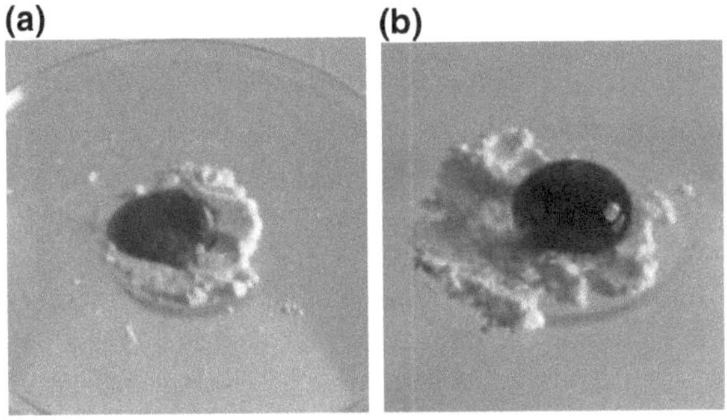

Fig. 4 Hydrophobicity of Spheriglass 5000 micro particles, empirically tested with the help of a water drop colored with blue food dye on a layer of **a** unmodified particles and of **b** silanized particles

The silanization process was carried out in a Teflon beaker in order to avoid a potential silanization of the laboratory glassware. The glass particles were allowed to react with the silane at room temperature for 24 h, while continuously stirred with a magnetic stirrer. Subsequently, the solid phase was separated from the liquid using a ceramic filter, washed several times with toluene, followed by chloroform and ethanol to remove the non-converted silane species. At the end, the particles were air-dried under a fuming hood for 12 h and heated in an oven at 100 °C for 2 h in order to insure a covalent bonding of the silane molecules to the glass surface. A schematic representation of the theoretical surface bonding mechanism of the silanes in a non-aqueous solvent for CTMS is shown in Fig. 3.

To prove that the silanization was successful and the particles have really obtained hydrophobic properties, an empirical hydrophobicity test was carried out. For this purpose, a water drop colored with blue food dye was placed on a thin layer of unmodified and silanized particles (see Fig. 4). If the droplet beads up, the silanization process is considered successful (see Fig. 4b).

Characterization of the Surface-Modified Particles

Hydrophobicity and Surface Free Energy

The characterization of the induced hydrophobicity is an important aspect for the generated hydrophobic surfaces. Typically, in the field of surface engineering wettability tests are used for this characterization. These tests measure the contact angle

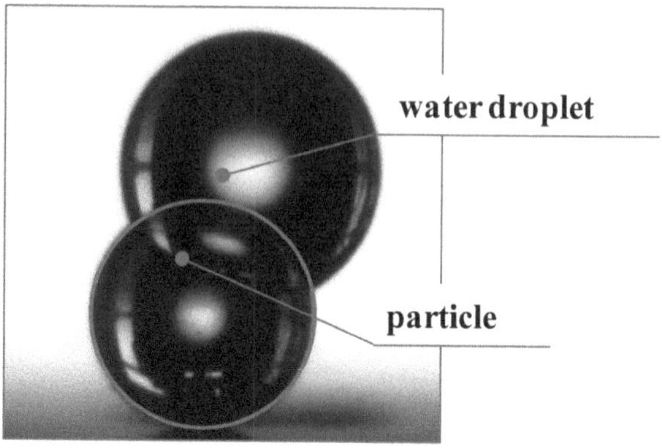

Fig. 5 Contact angle measurement of water droplet placed on a model glass sphere ($d_{50,3}$ = 491 μm) modified with a PFOTES concentration of 0.01 M

formed between the liquid drop and the solid surface. Surfaces on which a water droplet forms a contact angle above or equal to 90° are regarded as hydrophobic. If the contact angle is below this value, the surface is considered hydrophilic. The contact angle (see Fig. 5) represents the interactions between the molecules in the liquid and solid phase directly. However, the roughness of the solid surface can impact the contact angle by either increasing it if the surface is hydrophobic or decreasing it when it is hydrophilic [13]. Thus, to avoid any misinterpretation of the contact angle measurements, model glass beads made of soda-lime glass were used for the characterization of the hydrophobicity induced after the modification with the selected silanes. These model glass beads as well as the Spheriglass 5000 particles were made of the same material and exhibited the same chemical composition, but the median diameter $d_{50,3}$ of the model glass beads amounted to 491 μm. This allowed to measure the contact angle after silanization directly on the surface of the glass bead. For the contact angle determination an OCA 20 device from DataPhysics Instruments was used. For each chemical surface modification, 20 measurements on different particles were conducted.

In addition to measure the degree of hydrophobicity, the OCA 20 instrument was also applied to determine the surface free energy using the Owens, Wendt, Rabel and Kaelble (OWRK) model [14]. The free surface energy of solids is determined by their forces acting between the surface molecules in the outermost surface layer. These forces are of different origins, such as Coulomb interaction between two permanent dipoles (Keesom force), the interactions between a permanent and an induced dipole (Debye force), hydrogen bonds, Lewis acid-base interactions, metal bonds or London dispersive force [15, 16].

The determination of the free surface energy by the QWRK model requires the contact angle measurement with two standard liquids: one polar and one purely dispersive. As a polar liquid water and as a dispersive liquid 1-Bromonaphthalene were applied.

Effective Silane Coverage

The effective silane coverage (ESC) with silanes was estimated as a degree of substitution of the hydroxyl-groups with silane molecules by measuring the water vapor adsorption prior and after the surface modification. The polar water vapor adsorbs onto the glass surface due to the presence of the OH-groups. If nonpolar organosilane rests, depending on the degree of substitution, substitute these groups, a diminished to no water vapor adsorption will take place. Water vapor adsorption was measured gravimetrically using a dynamic water vapor sorption (DVS) device from POROTEC. The water adsorption isotherm was determined at room temperature while gradually increasing the relative humidity in 5% steps to a maximum value of 95%. To adjust the desired relative humidity, dry nitrogen was added to the water vapor. Using the BET adsorption theory, and provided that 1 g of water equals to 1 cm^3 of water, the amount of water required for a monolayer formation was calculated. Based on findings of Farrar et al. [17], 1 g of a monomolecular water layer covers an area of 3500 m^2, the specific surface area of the particle before and after the silanization was determined and the effective silane coating was calculated.

Silane Coating Thickness

However, the effective silane coverage (ESC) provides no information about what amount of silane is bonded to the surface and what is the thickness of the silane coating. To find this out, the silanized glass particles were additionally subjected to a thermogravimetric analysis (TGA) using a STA 449 C instrument from Netzsch. For this analysis, the particles were first gradually heated under a nitrogen atmosphere to a temperature of 600 °C with a rate of 5 K/min. After reaching this temperature, the heating rate was decreased to 1 K/min and the particles were heated up under an oxygen atmosphere to 650 °C in order to oxidize any pyrolysis residues and remove them from the glass surface. The amount of bonded silane was determined from the mass loss at the end of the measurement. The cleaned glass particles served as a reference. Higher temperatures during the thermogravimetric analysis were avoided as they would approach the softening point of the glass particles that starts at 704 °C. By combining the dynamic water vapor sorption and TGA analysis, a theoretical layer thickness of the silane coatings was calculated.

In order to calculate the layer thickness of the silanes from the thermogravimetric analysis, analogous to a surfactant molecule, the silanes were considered as consisting of a head part and a tail part. A silicon tetrahedron together with hydrolysable groups and the first group of the hydrophobic tail linked directly to the tetrahedron were assumed as head of the PFOTES and FPTS silanes. In the case of CDMOS, chlorine and a methyl group bound to the tetrahedron were assumed as head group. In the case of CDMPS and CTMS, the whole molecule was assumed as head group and approximated as a sphere. To calculate the cross-sectional area δ of the head part

Table 2 Theoretical number of silane molecules per unit area and length of silane molecules

Silane	Number of silane molecules per nm^2 (SM/nm^2)	Molecular length (nm)
PFOTES	2.48	1.20
FPTS	2.69	0.63
CDMPS	2.13	0.69
CDMOS	2.25	1.13
CTMS	2.58	0.37

of a silane molecule Eq. (1) was applied, which is used for the determination of the cross-sectional area of an adsorptive molecule on the surface of an adsorbent [18]

$$\delta = 1.1 \cdot \left(\frac{M_A}{\rho_{l,A} \cdot N_A}\right)^{\frac{2}{3}} \tag{1}$$

In Eq. (1) N_A is the Avogadro constant, M_A the molar mass and $\rho_{l,A}$ the density of the liquid adsorbent. The quotient of M_A and $\rho_{l,A}$ is the molar volume of the adsorbent $V_{m,l}$.

Using the cross-sectional area of the silane molecules, the maximum number of silane molecules (SM) that cover one square nanometer surface area was calculated for each silane (see Table 2). From this theoretical occupancy density of the glass surface and the weight loss, determined by TGA, the number of silane layers after the modification process was calculated. For the silane molecules, a hexagonal dense spherical packing with a packing density of 0.7405 was assumed. To calculate the layer thickness of the silanes from the calculated number of layers, the maximum molecular lengths of the silanes were calculated using the software ChemSketch of the ACD Labs. Results are given in Table 2.

Assuming a homogeneous distribution of the silane layer thickness, the theoretical thickness of the layers was calculated using the effective silane coverage determined by the water vapor adsorption and the theoretical thicknesses of the produced silane.

Particle Flowability

The powder flowability, expressed though the Jenike flow function ff_c [19], was experimentally tested with a Schulze ring shear tester RST-XS.s which allows an automatic, computer-controlled test procedure [20]. For the shear testing the shear cell type XS-SV3 with a specimen volume of approximately 3.5 cm^3 was used. The shear tests were made according to the international standard shear test method for bulk solids using the Schulze ring shear tester [21]. For an extensive powders characterization, four yield loci (YL) for each sample were measured. The pre-shear stresses were chosen based on the bulk density ρ_b of the powder that was roughly estimated before the shear tests. The pre-shear stresses amounted to following values: YL 1: $\sigma_{pre} = 2{,}000$ Pa; YL 2: $\sigma_{pre} = 4{,}000$ Pa; YL 3: $\sigma_{pre} = 8{,}000$ Pa and YL 4: $\sigma_{pre} = 16{,}000$ Pa. The normal stresses at shearing for YL 1 and YL 2 were set to 20% of the normal stress at pre-shearing. For measurement of YL 3 and YL 4, the instrument automatically selected the shear stresses. For reliability and repeatability of the results, each yield locus was measured three times. Before the start of the shear testing all samples were sieved using a 32-μm meshed sieve to eliminate the presence of large agglomerates.

Results and Discussion

Hydrophobicity of Surface-Modified Particles

Figure 6 illustrates the contact angles measured on the surface of the model glass beads ($d_{50,3} = 491.3$ μm) as a function of silane concentration.

A general trend shows that the contact angle increases with the increase of silane concentration, however the particles coated with CDMOS and CTMS do not entirely follow this general tendency. Nevertheless, as the mean contact angles differ only slightly for different silane concentrations and the standard deviation relative to these differences is very high, no clear conclusions about the influence of these silanes can be drawn. Furthermore, Fig. 6 shows a trend for the degree of hydrophobicity after the modification with different silanes. The particles modified with the long chained fluoroalkylsilane PFOTES appear to be the most hydrophobic ones followed by the particles modified with the long chained alkylsilane CDMOS. The particles silanized with phenylsilane CDMPS are slightly hydrophilic because, regardless of the concentration, their contact angle lies below the threshold of 90°. Modifications with the short-chained silanes FPTS and CTMS show moderate hydrophobicity.

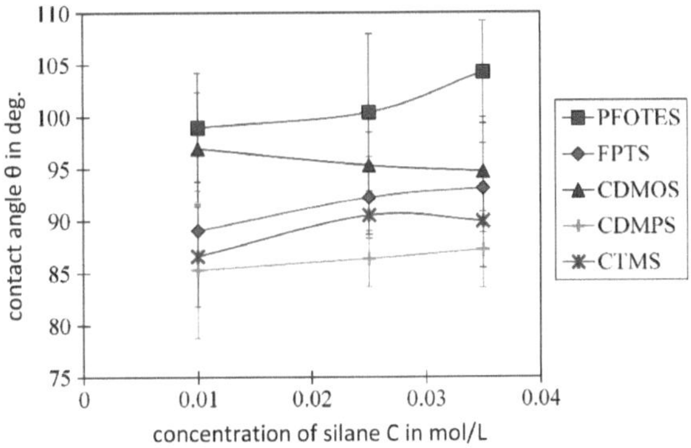

Fig. 6 Contact angles of sessile water droplet versus silane concentration measured on surfaces of silanized model glass particles with an average diameter $d_{50,3} = 491.3$ μm

The low wettability of the glass particles after modifying with the long-chained silanes can be attributed to a shielding effect of those chains. The chains protrude from the surface, thus preventing a direct contact between the water droplet and glass surface. As a result, the surface wettability decreases and the contact angle increases. In the case of the short-chained silanes, this shielding effect is less pronounced which results in lower hydrophobicity. The low contact angle occurring after the silanization with the phenylsilane CDMPS can be attributed to the planar conformation of the benzene ring. After the modification, the ring lies parallel to the particle surface, which results in a weaker shielding. If during the silanization process silane molecules substitute not all OH-groups, these OH-groups could form hydrogen bonds with the water molecules, which leads to lower contact angles.

Figure 7 shows the free surface energy of the modified particles calculated from contact angle measurements with water and 1-bromonaphthalene using the OWRK-model as a function of the silane concentration. The surface energy of the unmodified particles was 53.4 ± 7.2 mJ/m^2 and of the particles chemically cleaned with Piranha solution amounted to 75.5 ± 0.57 mJ/m^2. The huge standard deviation determined for the unmodified particles results from surface contaminations, such as organic impurities adsorbed from the atmosphere or during the manufacturing process. These contaminations cause surface inhomogeneities and lead to strong fluctuations of the surface energy. As mentioned, the treatment with Piranha solution removes all organic contaminations and hydroxylates the glass surface. The surface hydroxylation and the removal of organic contaminations contribute to the intensification of the interactions between the water molecules and the OH species. This results in an increase of the surface energy of the particles after cleaning with the Piranha solution.

In case of the modified particles the lowest free surface energy is achieved after the modification using the long-chained fluoroalkyl silane PFOTES. The reason is the high electronegativity of fluorine, which leads to a reduced polarizability of this atom. It results in reduced charge fluctuations (instant dipole induction) and, as a consequence, in a lower London dispersion force [22] and a lower free surface energy. The surface energy of the particles modified with the short-chained fluoroalkyl silane FPTS is considerably higher. Even though its hydrophobic chain also contains flu-

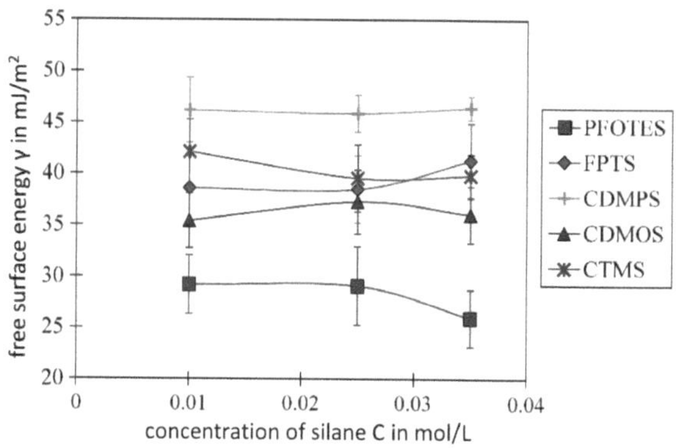

Fig. 7 Free surface energy of the modified model glass particles ($d_{50,3} = 491.3$ μm) versus silane concentration

orine atoms, the chain length is significantly shorter than those of PFOTES, which according to Lindner und Arias [23] plays a crucial role for the magnitude of the surface energy. Lindner und Arias [23] have discovered that the dispersive forces are inversely proportional to the length of the fluorocarbon chain. Thus, the particles coated with the short-chained FPTS have higher surface energy than those coated with PFOTES having a longer silane rest. Modification with phenylsilane CDMPS leads to the smallest reduction of the free surface energy compared to the unmodified particles. This could come from the planar conformation of the benzene ring, which allows the surrounding molecules to interact with the surface and thus to increase the interaction potential. Another reason for the high magnitude of surface energy of CDMPS modified particles might be π-electron interactions, occurring between the phenyl ring of the silane and the aromatic rings of the 1-bromonaphthalene, which leads to an increase of the London dispersion force.

As plotted in Fig. 7, an increase of the silane concentration causes a slight decrease of the free surface energy for the surface-modified particles. Only the particles coated with FPTS exhibit a peak of the surface energy at the highest silane concentration. This sudden increase of the surface energy might be due to irregular silane coating i.e. due to decrease of effective silane coverage.

As already introduced, the effective silane coverage (ESC) was evaluated using water vapor adsorption isotherms. H_2O vapor adsorption isotherms are shown as a function of the relative humidity for different silane concentrations in Figs. 8, 9, 10 and 11; the numerical values for the effective silane coverage are listed in Table 3.

Figure 8 presents the adsorption isotherms for these particles modified with the lowest silane concentration of 0.01 mol/L. As reference, the isotherm of the unmodified, Piranha cleaned particles is shown. All adsorption isotherms can be classified as IUPAC isotherm type II. This means that in the range of low relative humidity (RH) an adsorption monolayer is formed. This monolayer grows with increased relative

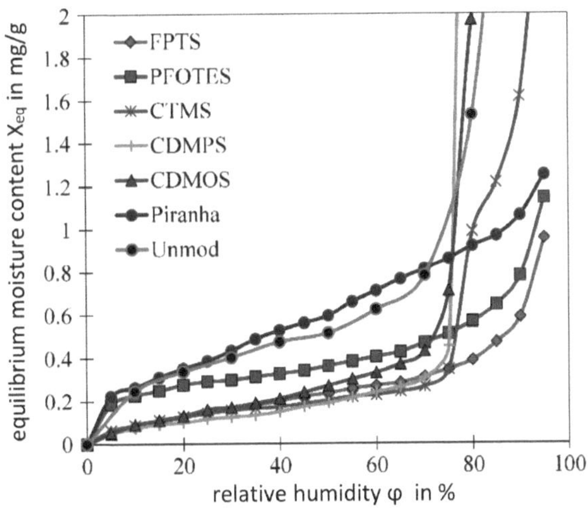

Fig. 8 Water vapor adsorption isotherms of the modified glass particles modified with a silane concentration of 0.01 mol/L

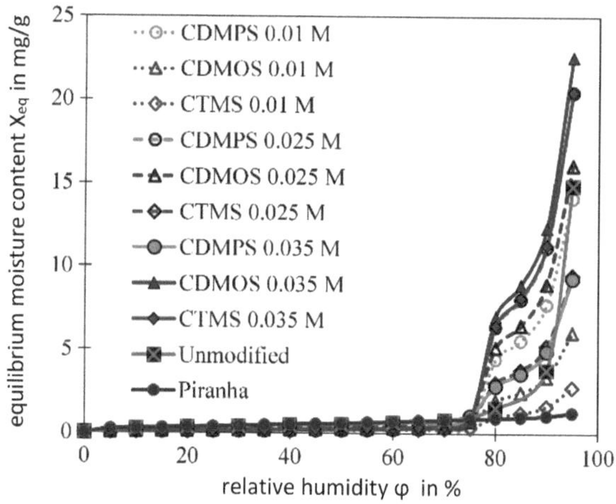

Fig. 9 Water vapor adsorption isotherms of Spheriglass particles modified with different concentrations of chlorosilanes versus relative humidity

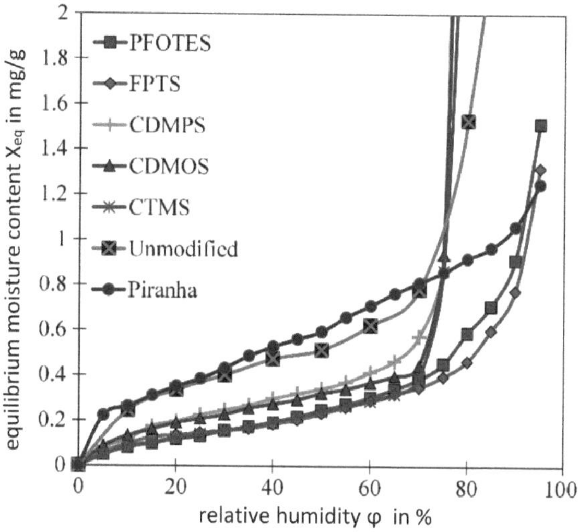

Fig. 10 Water vapor adsorption isotherms of modified glass particles at a silane concentration of 0.025 mol/L

water vapor pressure and forms multilayers. As expected, the purified and unmodified glass particles adsorb the highest amount of water vapor. This adsorption takes place due to the hydroxyl groups present on the surface. These polar groups form hydrogen bonds with water molecules leading to a high water uptake.

It should be mentioned that the unmodified particles show only a slightly smaller water absorption compared to Piranha cleaned particles. There could be two possible reasons for this. On the one hand, the surface of the unmodified glass already might be in a hydroxylated state due to water adsorbed from the atmosphere. On the other hand, the surface of the unmodified particles might be contaminated with impurities that favor the absorption of water.

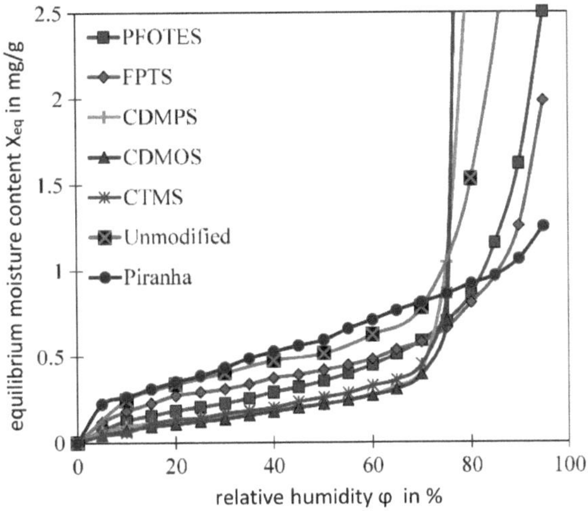

Fig. 11 Water vapor adsorption isotherms of the modified glass particles at a silane concentration of 0.035 mol/L

Table 3 Effective silane coverage at different silane concentrations

Concentration (mol/L)	Effective silane coverage (ESC) (%)				
	PFOTES	FPTS	CDMPS	CDMOS	CTMS
0.010	37	60	71	64	64
0.025	61	64	37	44	60
0.035	36	26	65	55	51

Looking at the adsorption isotherms of the silanized particles, there can be seen that the coated glass particles show a reduced water adsorption compared to the cleaned ones. During the silanization process, a large portion, but not all hydroxyl groups, were substituted with silane groups. All silanized particles except those modified with PFOTES show a similar water adsorption behavior in a BET range between 5–35%. The explanation for this behavior is in the reactivity of PFOTES. The hydrolysis reaction is the rate-controlling step for the silanization process and, therefore, has a decisive effect on the efficiency of the silanization [24]. Among the used silanes, the highest hydrolysis rates have those compounds which have chlorine as a leaving group (CTMS, CDMOS, CDMPS), followed by the silanes with a methoxy leaving group (FPTS). PFOTES reacts most slowly with the glass surface due to an ethoxy-leaving group [25].

Considering the adsorption in the relative humidity range of above 70%, the unmodified particles and particles coated with the chlorosilanes CDMPS, CTMS and CDMOS drastically increase their water absorption. Forny et al. [26] also report this peculiar behavior. They have found that this effect is caused by the formation of sodium chloride NaCl crystals during silanization. Hydrochloric acid (HCl) is formed as a by-product of the silanization with chlorosilanes, since the soda-lime glass contains a comparatively high amount of sodium oxide Na_2O. The steep increase in the water absorption at a humidity range above 70% is caused by the dissolution of the hygroscopic NaCl crystals. After the dissolution of these crystals, a supersaturated salt solution is achieved. Thus, with increasing humidity, the particles adsorb more water to reach the equilibrium concentration of the salt solution at the corresponding relative humidity. Forny et al. [26] have also discovered that the influence of the NaCl crystals on the water adsorption is noticeable only after their dissolution at a relative humidity RH > 70%. The adsorption behavior before reaching this point of inflection is attributed solely to the glass particles. The unmodified glass particles show a similar uptake behavior as the glass particles treated with chlorosilanes. The high water absorption of the unmodified and chlorosilane-treated particles compared to the particles cleaned with Piranha solution is shown in Fig. 9. The water adsorption increases with increasing silane concentration due to an increasing concentration of formed NaCl crystals.

Figure 10 shows the water adsorption isotherms of glass particles after a modification with a silane concentration of 0.025 mol/L. Glass particles modified with PFOTES absorb the smallest amount of water. In contrast, water absorption of particles silanized with CDMOS and CDMPS increases.

The water vapor adsorption isotherms of the glass particles silanized with a silane concentration of 0.035 mol/L is plotted in Fig. 11. An increase in the water adsorption for silane modified glass particles that have an alkoxy-leaving group (PFOTES, FPTS) is found.

The steep increase in water adsorption at higher water vapor pressures (relative humidity RH > 70%) is caused by capillary condensation [27]. In this case, the water molecules link together to form a two-dimensional structure, resulting in a water film on the adsorbent surface.

In case of hydrophobic particles no to little capillary condensation should occur due to the hindrance effects of the hydrophobic silane groups attached to the glass surface. However, an increase of the water absorption is measured for the glass particles treated with PFOTES and FPTS. The reason for this is presumably an irregular distribution of the silane coating, as well as an incomplete coverage of the particle surface with silanes. When comparing the adsorption isotherms at different silane concentrations (see Figs. 8, 9, 10 and 11), a significant increase of the amount of condensed water with an increasing silane concentration is observed. One possible reason for that might be the presence of unreacted alkoxy groups of the silanes, which can hydrolyze with water vapor to form OH species and thus increase the amount of available adsorption sites. As the silane concentration increases, more irregular silane structures are formed on the glass surface. As a result the concentration of non-hydrolyzed alkoxy-groups also increases, and the probability of water condensation rises. This can also explain why the effective silane coverage with PFOTES and FPTS at 0.035 mol/L is so low (see Table 3).

Table 3 gives the effective silane coverage at the various silane concentrations. It is evident, that the maximum surface coverage for the investigated silanes is between 60 and 70%. Therefore, during the modification process with silanes no continuous silane coating layer was produced on the glass surface; and the surface is only partially coated with silanes. Furthermore, the maximum silane coverage of chlorosilanes

(CDMPS, CDMOS and CTMS) was achieved at the lowest reaction concentration of 0.01 mol/L. The increment of chlorosilane concentration to 0.025 mol/L has led to a decreased silane coverage. With a further increase of the concentration to 0.035 mol/L the effective silane coverage shows a higher value. The silanes with an alkoxy group as a hydrolysable unit (FPTS and PFOTES) exhibit a different behavior. The highest surface coverage with these silanes was observed at a concentration of 0.025 mol/L and the lowest coverage was measured at the highest silane concentration of 0.035 mol/L. The reason for the low surface coverage with PFOTES at the lowest silane concentration of 0.01 mol/L is in the low reactivity of this silane.

Thickness of the Produced Silane Layers

In order to draw further conclusions about the produced silane coating thermogravimetric analysis (TGA) was applied (see Fig. 12).

Looking at the mass losses of the silanized particles, an interesting pattern can be observed. All recorded mass losses have a maximum value at the middle silane concentration of 0.025 mol/L. At the highest concentration of 0.035 mol/L the lowest mass loss has been recorded. One possible reason for this behaviour is the high moisture sensitivity of the silanes, which even in the presence of water traces can lead to a dimerization of the monofunctional (CTMS, CDMPS and CDMOS) or to a polymerization of the trifunctional silanes (FPTS and PFOTES). Despite the fact that the modification was carried out in dry toluene, the presence of water traces cannot be excluded since the silanes were not purified before the modification and the silanization was not carried out under an inert atmosphere. At the lowest concentration of 0.01 mol/L few monomer units are present in the reaction solution. Most

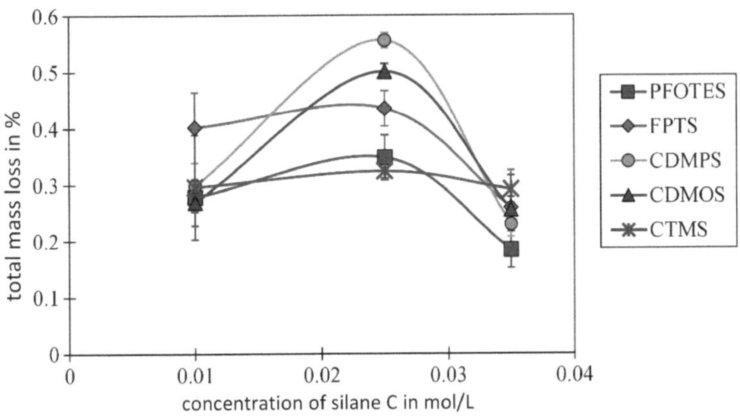

Fig. 12 Thermogravimetric determination of the silane bound to the glass particle surface at different silane concentrations

likely, the silane moieties preferably react with the OH groups on the glass surface and the polymerization of the silane units is not well pronounced. However, when the concentration is increased to 0.025 mol/L the probability increases to build-up silane polymers. This polymer structures are most likely adsorbed on the glass surface, resulting in an increased mass loss. This effect is less pronounced in case of the less reactive alkoxysilanes FPTS and PFOTES compared to the silanes, which have chlorine as the leaving group. Therefore, with increased silane concentration the mass loss of the chlorosilanes shows a sudden increase with simultaneous decrease in the effectiveness of the coating (see Table 3).

At the highest silane concentration of 0.035 mol/L even larger silane polymer structures are formed which are probably too large and thus adsorb only partially on the surface. The non-adsorbed silane polymers are believed to be removed during the filtration process together with the solvent. Therefore, the mass loss, e.g. the amount of bound silane decreases.

Looking at the layer thicknesses in Table 4 it can be seen that the silanization has led to a formation of relatively thick layers ranging from several nanometers to several tens of nanometers. The massive increase of the layer thickness of CDMPS and CDMOS at the middle silane concentration of 0.025 mol/L results from the low effective silane coverage and the high mass loss.

The macroscopic hydrophobicity, determined with contact angle measurements, seems to increase with increasing silane concentration, while at the microscopic level the ability to absorb water i.e. the hydrophilicity increases. One possible explanation for this phenomenon gives the structure of the formed layers. The high contact angles probably result from an interaction with the thick silane layers that prevent the direct contact between water droplets and the rest OH-groups being at the glass surface. Moreover, the silane structures increase the roughness of the glass surface. As a result, the interactions between glass surface and water droplets are reduced and the contact angle increases. If, however, the water vapor adsorption is measured, the individual molecules of the water vapor are freely movable and can adsorb to the OH species.

The morphology of the particle surface after coating was visualized by scanning electron microcopy (SEM). In Figs. 13 and 14 SEM images of glass particles modified with PFOTES and FPTS respectively using a silane concentration of 0.01 mol/L are shown. As seen in Fig. 13, some small roughnesses with a diameter of about 40 nm are present on the glass surface after coating with PFOTES. This structure of the

Table 4 Theoretical layer thickness of the silane coatings calculated from DVS and TGA data

Concentration (mol/L)	Theoretical layer thickness DVS + TGA (nm)				
	PFOTES	FPTS	CDMPS	CDMOS	CTMS
0.010	7.03	6.96	5.71	9.99	6.92
0.025	5.34	7.07	20.41	27.06	8.08
0.035	6.38	10.36	4.80	11.04	8.56

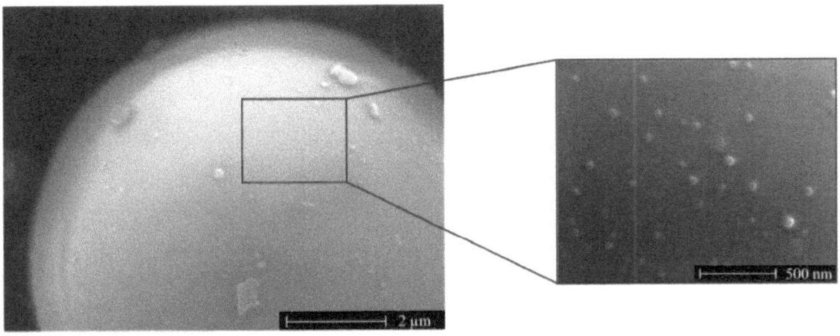

Fig. 13 SEM image of the glass particles coated with PFOTES (concentration 0.01 mol/L)

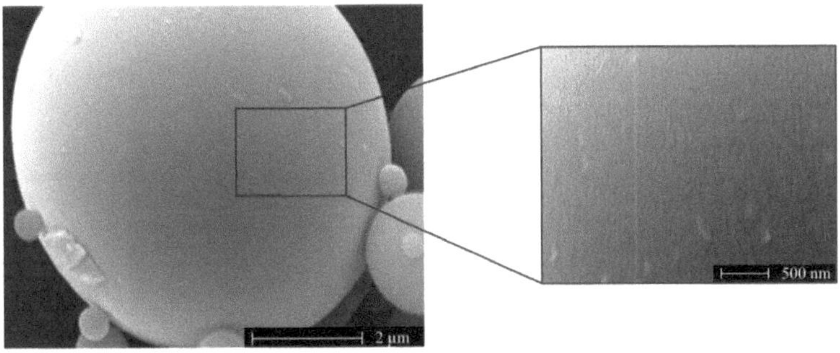

Fig. 14 SEM-image of the glass particles coated with FPTS (concentration 0.01 mol/L)

coating is believed to be the reason for the low effective silane coverage, which is only 37% at this concentration.

On the other hand, particles modified with FPTS (see Fig. 14) show only a small-pronounced surface roughness, suggesting a more even distribution of the silane coating which is consistent with the results for the effective silane coverage (ESC) of 60%.

Consequently, the applied method for the assessment of an effective silane coverage by measuring water vapor adsorption can be considered as successful.

Flowability of Surface-Modified Particles

Figure 15a illustrates the macroscopic flow function of the glass particles modified with a silane concentration of 0.01 mol/L. As a reference, values are shown for the flow functions of the unmodified particles and for particles cleaned with Piranha acid. The flow function $\sigma_C = f(\sigma_1)$ correlates characteristically the uniaxial compressive strength σ_C (called also unconfined yield strength) and the major principal stress σ_1 which corresponds to the consolidation stress. The ratio ff_c as a quotient of the consolidation stress σ_1 and the unconfined yield strength σ_C is applied for the

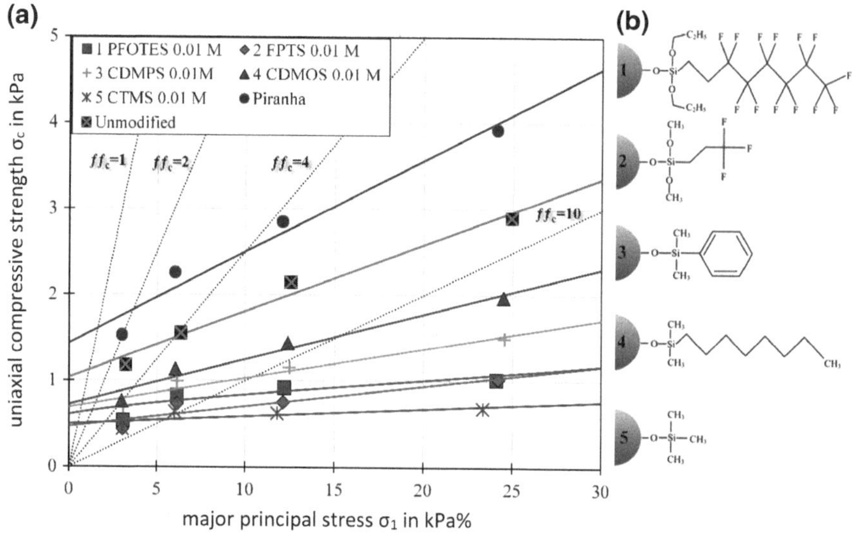

Fig. 15 **a** Macroscopic flow function of silanized particles with silane concentration of 0.01 mol/L and **b** structural formula of bound silane chains

numerical characterization of the bulk solid behavior. Table 5 gives a classification of bulk solid flowability based on the value of ff_c according to Jenike [19] with a small addition by Tomas [28]. According to this classification, the larger the value of ff_c is the better is the flowability of the bulk solid.

Figure 15a clearly shows that the modification of the glass surface with the hydrophobic silanes leads to an improved flowability of the glass particles; the flowability ff_c of the silanized particles lays in the range of easy flowing to free-flowing powders (see Table 5).

The reduced compressive strength σ_C of the particles after modification with organic silanes could be attributed to a reduced surface energy, which according to JKR [29] and DMT [30] theories leads to a reduction of the interparticle interactions. Because of the reduced adhesion, the flowability of the silanized particles is improved. However, it must be noted that the surface energy is not the only factor that influences the flowability.

Table 5 Classification of the flowability of powders according to Jenike [19] and Tomas [28]

Flowability ff_c	Bulk solid evaluation
$ff_c < 1$	Hard flowing, hardened
$1 < ff_c < 2$	Very cohesive
$2 < ff_c < 4$	Cohesive
$4 < ff_c < 10$	Easy flowing
$ff_c > 10$	Free flowing

According to the results shown in Fig. 7, the particles modified with PFOTES, have the lowest surface energy. Nevertheless, these particles do not have the best flowability. The best flowability among all modifications is observed for those particles treated with CTMS. Moreover, the worst flowability is determined for particles modified with CDMOS, although according to Fig. 7 the CDMPS-treated powder has the highest free surface energy.

To explain this behavior the calculated theoretical layer thicknesses from Table 4 have to be considered. According to these results, the low flowability after modification with CDMOS can be explained with the high thickness of the silane coating. However, the results for the theoretical layer thickness (see Table 4) do not explain the best flowability of particles coated with CTMS since they have almost identical layer thickness as the particles modified with PFOTES and FPTS. Thus, for a further explanation of the flowability the effective silane coverage (see Table 3) and the morphology of the coating shown in Figs. 13 and 14 have to be considered. They reveal that due to the low effective silane coverage (ESC) of the particles coated with PFOTES, their morphology differs significantly from that of the particles modified with FPTS. Thus, it can be concluded that the thickness and morphology of the coating play a significant role for the particles flowability. Due to the acting normal load during the shearing process in the flowing particle packing, the hydrophobic silane chains (Fig. 15b) can interfere with each other. This leads probably to mutual penetration and shear induced interlocking of the organic chains and decreases the flowability. Concluded from Fig. 15a, this interlocking effect is less pronounced in the range of low normal stresses, as at this range the silanes CTMS, FPTS and PFOTES are characterized by an approximately equal flowability. If, however, the normal stress is increased, the particles are compressed and pressed more tightly against each other. This causes an overlapping of the silane-polymer grains on the surface of the particles and entanglement of the silane chains occurs.

Furthermore, it can be seen that, in addition to the coating thickness, the length of the hydrophobic chain also seems to play an important role for the described interlocking effect. Despite the similar coating thicknesses and effective silane coverage (ESC) of the particles modified with CTMS and FPTS, particles coated with CTMS have a better flowability compared with those silanized with FPTS. This effect may be caused by the length of the hydrophobic chain of the silane, which in case of CTMS is C1 (one carbon atom) and in case of FPTS is C3 (three carbon atoms).

Glass particles cleaned with Piranha solution have the worst flowability among all powders. This is probably due to the high free surface energy of these particles, which intensifies the interparticle forces. Another effect that contributes to the reduction of the flowability of these particles is their dense surface coverage with polar hydroxyl groups. In contact with the surrounding atmosphere, water molecules are easily attracted to their surface, which leads to the formation of water adsorption layers. These layers interact with each other via hydrogen bonds, which also leads to an increase of the adhesion forces. However, there is no significant difference in the water absorption capacity measured for the Piranha cleaned and the unmodified particles (see Fig. 8). Thus, the deteriorated flow behavior of the particles cleaned with Piranha solution is caused mainly by the increased surface energy.

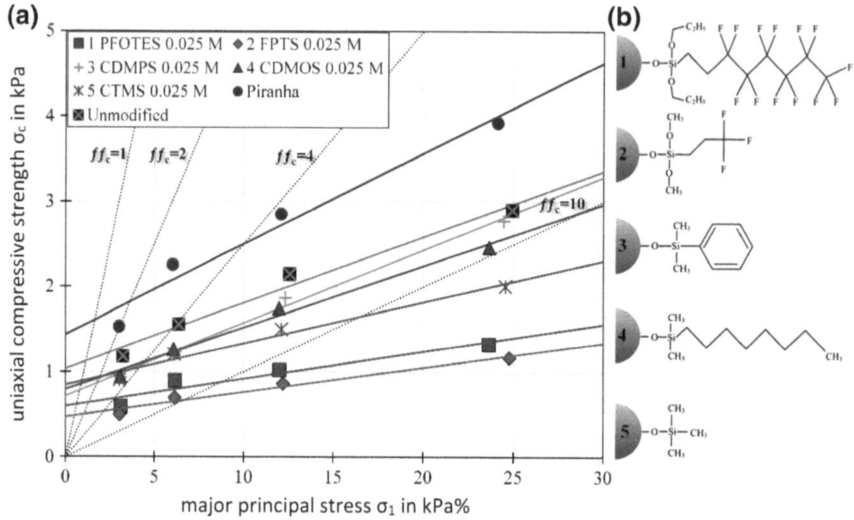

Fig. 16 **a** Macroscopic flow function of silanized particles at a silane concentration of 0.025 mol/L and **b** structural formula of bound silane chains

Considering the flow behavior of the particles that had been modified with the higher silane concentration of 0.025 mol/L, shown in Fig. 16a, a different behavior can be seen compared to the lower silane concentration of 0.01 mol/L. On one hand, a substantial reduction of the flowability of the CTMS modified particles is measured at this concentration. On the other hand, the powder silanized with CDMPS exhibit the worst flow behavior. At this silane concentration, the glass particles modified with FPTS have the best flowability. This clear change in the flow behavior after an increase of the used silane concentration cannot be explained by the effect of the surface energy, since it has remained almost unchanged (see Fig. 7). In this case, the thickness of the modification layers and the effective silane coverage of the glass surface appear to have a decisive influence on the flowability.

As mentioned, when the silane concentration is increased to 0.025 mol/L, a radical change in the flow behavior of the particles modified with CTMS can be seen. Although the theoretical coating thickness has increased only by about 1.0 nm, and the effective silane coverage (ESC) is reduced only by 4% (Tables 3 and 4), these particles show a drastic decrease in their flowability. The reason for this behavior could lie in the distribution of the coating thickness. As already explained, the silane thicknesses listed in Table 4 are only theoretical values, which do not take into account a coating distribution. In the produced silane coating, however, inhomogeneous distribution is very probable. Furthermore, substitution of one OH group by only one silane molecule cannot be excluded. Thus, it could be possible that, although the amount of the bound silane remained almost unchanged, its distribution at the glass surface is different and very uneven, which causes the seen change of the flowability.

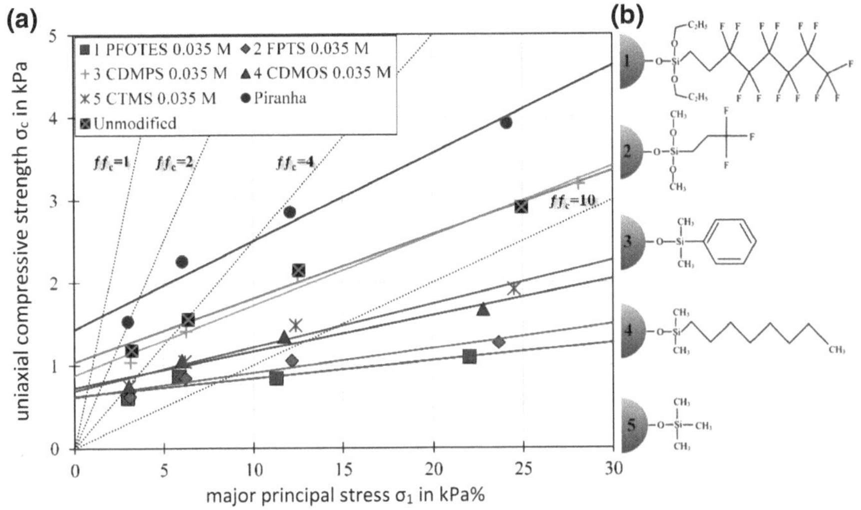

Fig. 17 **a** Macroscopic flow function of the silanized particles at a silane concentration of 0.035 mol/L and **b** structural formula of bound silane chains

As for the particles modified with CDMPS and CDMOS, their deteriorated flow behavior is attributed to the large increase of the amount of silane bound to the glass surface and the enlarged thickness of the produced coating. The increased silane concentration seems to have no impact on the flow behavior of the particles coated with the fluoroalkyl silanes PFOTES and FTPS, which remains practically unchanged.

Figure 17a illustrates the flow function of these glass particles treated with the highest silane concentration of 0.035 mol/L. In this case, some changes in the general picture can be seen. Particles modified with FPTS have a partially reduced flowability and the best flow behavior is seen after the modification with PFOTES. Particles treated with CDMPS are flowing as poorly as the unmodified particles, although they have a significantly lower surface energy. This shows ones again that the presence of the aforementioned interlocking effect of the silane organic chains has a crucial impact on the powders flowability, which despite the reduced surface energy does not increase.

To summarize, the surface energy decrease is not the only factor affecting the flowability of the modified particles. Interlocking of the hydrophobic chains takes place, which diminish or even eliminate the flow-improving effect of the hydrophobic silane coating. Consequently, a high effective silane coverage and a thin silane coating are advantageous for the flowability. This means that it is possible to improve the flowability of cohesive powders by coating their surface with hydrophobic silanes, however low silane concentrations are advisable for the modification.

To ensure a better comparability of the impact of the silane modification on the glass particles flowability and its dependence on the silane concentrations, a mutual

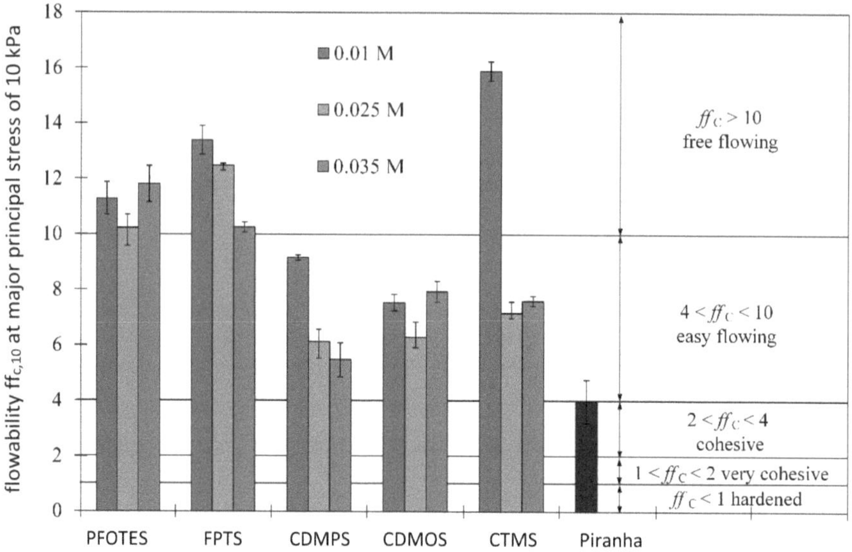

Fig. 18 Flowability $ff_{c,10}$ of the silanized particles at the major principal stress of 10 kPa as a function of the silane concentration

flow parameter is used. This flow parameter, referred to as $ff_{c,10}$, was selected in accordance with Schumann [31] and reflects the flowability of the particles at a consolidation stress of $\sigma_1 = 10$ kPa. The flow parameter $ff_{c,10}$ was determined by a linear interpolation between the flowability at the second and the third yield locus. The flow parameter $ff_{c,10}$ for all silanized particles is presented in Fig. 18. As a reference, the flowability of the particles cleaned with Piranha solution (starting material for the silanization process) has also been plotted in Fig. 18. The influence of the silane concentration on the flow behavior of glass particles as well as the type of silane applied can be identified by this flow parameter. All used silanes have changed the flowability of the starting material from cohesive to easy or free flowing.

The highest flowability $ff_{c,10}$ was calculated for glass particles modified with 0.01 mol/L CTMS with $ff_{c,10}$ is around 16. However, it should be noted that the flow-improving effect of CTMS depends strongly on the silane concentration and decreases substantially with its increase. Interestingly, for this silane both the effective silane coverage (ESC) and the theoretical coating thickness remain almost constant at all concentrations. Thus, no significant difference in the coating configuration with increasing silane concentration was found using these characterization approaches. Therefore, this flow behavior can be explained with a distribution of the layer thickness.

In Fig. 18 it can be seen that the second best flow-improving effect is achieved through silanization with FTPS, the third best one with PFOTES. In contrast to CTMS, the flow-improving effect of the fluoroalkyl silanes FPTS and PFOTES has a moderate dependence on the concentration and the particles remain free flowing,

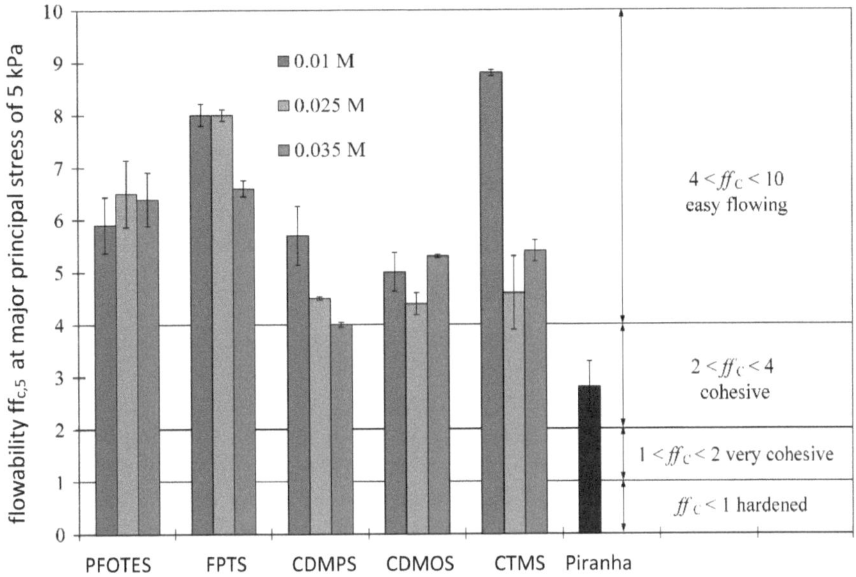

Fig. 19 Flowability $ff_{c,5}$ of the silanized particles at major principal stress of 5 kPa

even if the silane concentration is increased almost four times. The other two silanes (CDMPS and CDMOS) do not improve the flowability of the glass particles to a significant extent, even at low silane concentrations. Thus, they are not rather suitable for flow enhancers.

An additional flow parameter $ff_{c,5}$ is used (see Fig. 19) to compare the behavior of the silanized particles at lower consolidation stresses, which normally occur at silo outlets. This flow parameter represent the particles flowability at the major principal stress (consolidation stress) of $\sigma_1 = 5$ kPa. It can be seen in Fig. 19, at this stress level all modified particles, regardless the silane type and concentration, are located in the easy flowing range, while the particles cleaned with Piranha acid, which has been plotted as a reference, behave cohesive. The impact of the silane concentration on the flowability, seen at a higher consolidation stress of 10 kPa, is not so strong. An explanation could be that at higher stresses the silane coating is subjected to a greater deformation. Thus, at higher stress levels, the interlocking effect of the silane chains, which is assumed to weaken the flowability, is more pronounced.

Flow characteristics of the particles at the steady state flow [32–35] are listed in Table 6. The first flow parameter—the isostatic tensile strength σ_0—is a direct consequence of the interparticle attraction forces when no external force is applied. Thus, this parameter can be used to measure the interparticle attraction forces [36]. In Table 6, it can be seen that for the majority of the used silanes the tensile strength σ_0 exhibits its lowest value at the lowest silane concentration and increases directly proportional to the silane concentration. A slightly different behavior can be found for those particles modified with CDMPS and CTMS. In this case, the tensile strength has its peak at the middle silane concentration and decreases when the silane concentration reaches its maximum. The lowest isostatic tensile strength σ_0 among all silanized powders was calculated for the particles modified with the short chained fluoroalkyl silane FPTS. This implies that after silanization with this silane the glass particles exhibit the lowest cohesion forces. That is why the particles modified with this silane are expected to be the most free flowing (at least at silane concentrations of 0.01 and 0.025 mol/L). However, if we look at the flow function of the particles after modification with the lowest silane concentration of 0.01 mol/L (see Fig. 15a), it can be noticed that the best flowability belongs to those particles modified with CTMS. This behavior can be attributed to a microscopic repulsion effect, which turns the adhesion to a repulsion force when the normal load is increased. Further information about this effect will be provided in the next chapter.

The second flow parameter—stationary angle of internal friction φ_{st}—is a measurement of the friction between the particles in the stationary flowing bulk solid. This parameter is analogous to the sliding friction between solids and represents the equilibrium friction between the failing and the new developing particle contacts [36]. This parameter does not depend on the pressure and is a material characteristic. The highest friction is calculate for the unmodified glass particles (see Table 6), whereas,

Table 6 Comparison of selected flow characteristics for the glass particles modified with different silanes

Silane	Concentration (mol/L)	Isostatic tensile strength of unconsolidated powder, σ_0 (kPa)	Stationary angle of internal friction, φ_{st} (deg)
PFOTES	0.010	0.37	25.7
	0.025	0.65	24.8
	0.035	0.72	22.0
FPTS	0.010	0.31	25.7
	0.025	0.42	25.5
	0.035	0.71	24.6
CDMPS	0.010	0.48	26.5
	0.025	0.70	28.4
	0.035	0.49	28.8
CDMOS	0.010	0.57	26.4
	0.025	0.73	27.4
	0.035	0.83	24.1
CTMS	0.010	0.56	23.7
	0.025	0.72	27.1
	0.035	0.65	27.1
Piranha	–	0.72	26.0
unmodified	–	0.52	30.0

after cleaning with Piranha solution the friction is slightly reduced. This is probably an indication that after the chemical treatment with Piranha solution the glass surface is not only free from organic contaminations, but also becomes smoother. During the manufacturing process of the origin glass particles asperities caused from small splinters or fragments can be found on the surface of the untreated glass beads. These asperities are probably etched away during the cleaning procedure. Generally, Piranha acid does not attack glass, but due to the high surface area of these roughnesses, it cannot be excluded that they are leached by Piranha solution. Thus, friction between the chemically cleaned particles is reduced. Regarding the friction of the glass particles after the modification with silanes, following behavior can be seen. Particles modified with the CTMS and CDMPS exhibit the lowest friction at the lowest silane concentration. Their friction raises with increasing concentration and becomes higher than the friction between unmodified particles. This is probably due to an increase of the surface roughness, which results from the silanization process. The reversed effect is found for the glass particles modified with FPTS, PFOTES and CDMOS. In this case, friction between the particles decreases with increasing silane concentration. The length of the hydrophobic rest probably causes this effect. According to Zisman [37], who investigated the influence of the hydrocarbon chain length on the friction between a glass slide modified with a lubricant monolayer and a stainless steel surface, friction decreases slowly as the chain length of the lubricant increases. Even though the glass particles are coated with more than one monolayer of silane, the effect of reduced friction can also be seen in this case. Although the silane FPTS has a short hydrophobic rest (C3 three carbon atoms), it exhibits almost the same angle of internal friction as the CDMOS modified glass particles with a longer hydrophobic rest (C8 eight carbon atoms). This can be attributed to the fluoroalkyl chain of this silane. This is the reason why the particles coated with the long chained (C8) fluoroalkyl silane PFOTES exhibit the lowest angle on internal friction. It is structurally similar to the polytetrafluoroethylene (PTFE), which has a very low coefficient of friction [38]. Thus, it can be concluded that modification with PFOTES leads to a certain lubrication.

Microscopic Contact and Adhesion Properties

A model based calculation of the microscopic contact properties for the modified glass particles was carried out using Tomas' contact model "stiff particles with soft contacts" [12, 39, 40]. Using this model it is possible to back-calculate the characteristic adhesion force F_{H0}, the elastic-plastic contact consolidation coefficient κ and the linear adhesion function $F_H(F_N)$ based on shear tests making a reverse micro-macro transition. The determination of micro mechanical contact and adhesion properties should contribute to a better understanding of the flow behavior of the surface-modified glass particles on the macroscopic scale. The contact model as well as the reverse micro-macro transition are published by Tomas in previous publications [12, 39, 40].

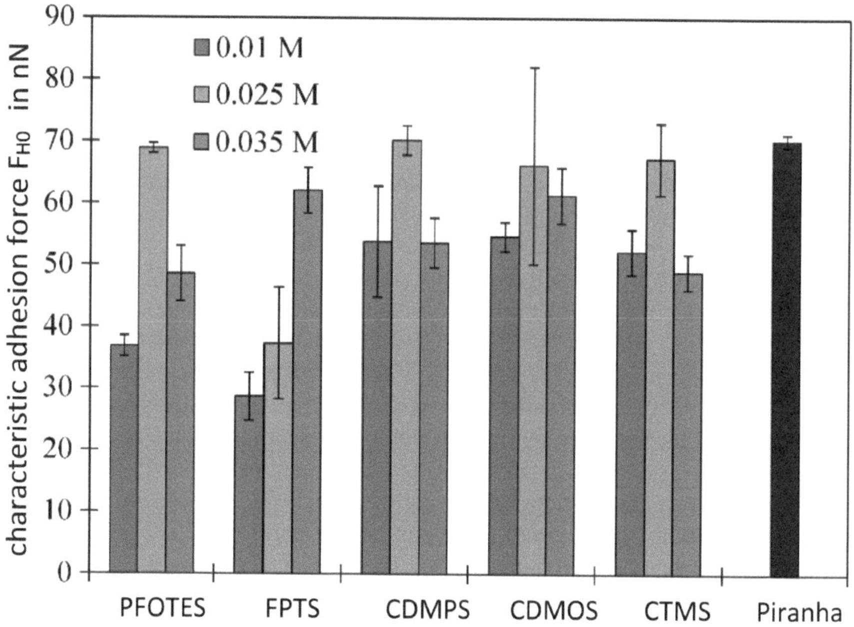

Fig. 20 Characteristic adhesion force between two particles from the unconsolidated bulk solid F_{H0} back-calculated from the shear-test results using Tomas' contact model "stiff particles with soft contacts"

Figure 20 illustrates the characteristic adhesion force F_{H0} between two representative particles from unconsolidated bulk solids. Only the acting surface forces (in this case van der Waals forces) influence the extent of the adhesion force. No external exerted forces that could contribute to the adhesion force by causing a contact deformation and thus increasing the contact area, are taken into account. For the particles modified with 0.01 mol/L silane, it can be seen that the lowest adhesion force F_{H0} of ca. 29 nN is achieved after treatment with FPTS followed by PFOTES ($F_{H0} = 38$ nN). The particles treated with the other three silanes CDMPS, CDMOS and CTMS exhibit almost the same adhesion force of around 50 nN. The particles treated with Piranha solution have the highest adhesion force of around 70 nN. When the silane concentration was increased from 0.01 to 0.025 mol/L, the adhesion force of the particles after the modification with PFOTES, CDMPS, CDMOS and CTMS was increased and equals the adhesion force of Piranha treated particles. The particles treated with FPTS exhibit only a slightly increased adhesion force in comparison to the lower silane concentration, which is lower than those of the other silanized particles are. A further increase of the silane concentration to its maximum value of 0.035 mol/L changed again the adhesion behavior of the particles. In this case, the particles modified with FTPS turned out to be the most adhesive ones approaching the adhesion force of the Piranha treated particles. In this case, the lowest adhesion force is determined for the PFOTES treated particles.

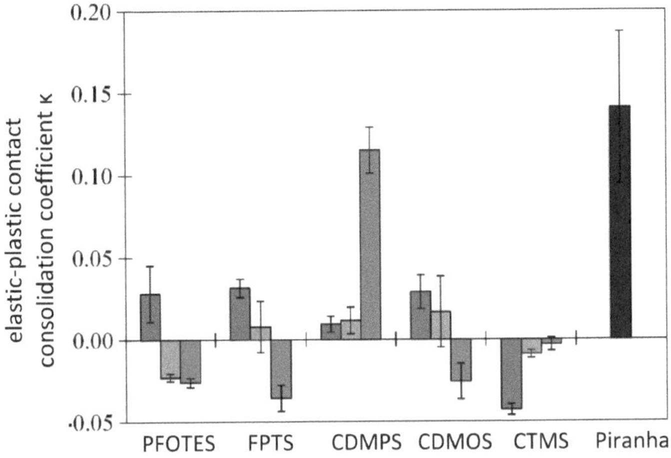

Fig. 21 Elastic-plastic contact consolidation coefficient κ of the silanized glass particles back-calculated from the shear-test results using Tomas' contact model "stiff particles with soft contacts"

Figure 21 shows the dimensionless elastic-plastic contact consolidation coefficient κ [34]. The κ-coefficient characterizes the instantaneous contact stiffness as well as the increase in the adhesion force F_H as a result of an external compressive normal force F_N. One contact exhibits a stiff behavior when it does not deform or undergoes only a slight deformation when an external compressive normal force F_N is applied. The κ-coefficient characteristic for such contact behavior varies between zero and 0.1. On the contrary, a soft contact can be easily deformed or flattened, as a result of which the contact area increases. The enlargement of the contact area leads to the intensification of the adhesion force, since its extent depends on the size of the contact area. An evaluation of the contact behavior based on the contact consolidation coefficient κ can be seen in Table 7.

When considering the elastic-plastic contact consolidation coefficient κ of the silanized particles, plotted in Fig. 21, an interesting effect can be seen. After the modification with almost all silanes (especially at high silane concentrations), the particles exhibit a negative contact consolidation coefficient κ. At a microscopic level, this means the contact between two modified particles reacts to a compressive normal force with repulsion. Brown et al. [42] have observed a similar repulsion behavior during the deformation of surfactant micelles, which they have attributed

Table 7 Contact behavior according to the contact solidification coefficient κ [28, 41]

Contact consolidation coefficient κ	Contact behavior
0–0.1	Stiff
0.1–0.3	Soft
0.3–0.8	Very soft
>0.8	Extremely soft

to a steric repulsion due to elastic deformation of these micelles. The silanes used in experiments are extremely moisture sensitive and tend to partially polymerize and form multilayers. Irregular, micelle-like agglomerates are formed and adsorbed on the surface. The observed repulsion effect between the silanized particles is also presumably caused by steric repulsion during the deformation of the silane polymer structures.

At the lowest silane concentration of 0.01 mol/L such repulsion behavior is found only for those particles silanized with CTMS, probably because of its very high reactivity. With an increasing silane concentration, more irregular silane structures are formed; this behavior is also observed for the other silanes. Thus, it is possible to explain why the particles treated with CDMOS first show a decrease in the flowability when increasing the silane concentration to 0.025 mol/L and then with a further concentration increase to 0.035 mol/L suddenly the flowability rises. This also explains the fact that, while the tensile strength (Table 6) of the particles modified with PFOTES increases monotonically with the concentration increase, the flowability of these particles does not change. Only the particles modified with the phenyl silane CDMPS do not show such repulsion behavior. The reason for this is the planar shape of the benzene molecule, which does not lead to the formation of three dimensional micelle-like silane structures. The effect of the extent of the elastic-plastic contact consolidation coefficient κ (contact behavior) on the interparticle attraction is plotted in Fig. 22.

Figure 22 shows the load-dependent adhesion force of a representative particle-particle contact F_H as a function of the total acting normal contact force $F_{N,H}$ according to Tomas [12, 34, 39, 40] for particles silanized with the lowest silane concentration of 0.01 mol/L. As a reference, the adhesion functions of the unmodified particles

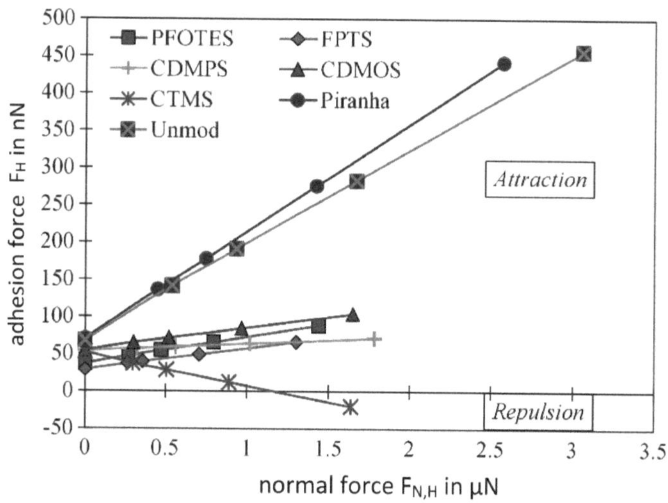

Fig. 22 Linear load-dependent adhesive force function F_H ($F_{N,H}$) of the glass particles, silanized with 0.01 mol/L silane

and those cleaned with Piranha solution are plotted. The slope of the adhesion force characterizes the contact consolidation coefficient κ. A large slope of the adhesion force F_H indicates a compliant (soft) contact with a high adhesion potential. On the contrary, a low slope is typical for a stiff contact behavior with a low adhesion potential.

As evident from Fig. 22, the particles cleaned with Piranha solution have the most compliant contact behavior and the highest adhesion potential, and hence they have the lowest flowability. The normal force $F_{N,H}$ acting between two Piranha treated or two unmodified particles is higher than for the silanized particles, although they were subjected to nearly same stresses during shear testing. There are several reasons for this behavior. On the one hand, the contact stiffness and the characteristic adhesion force F_{H0} play an important role for the extent of the normal load $F_{N,H}$ [12, 39, 40]. A soft contact behavior (high κ-values) and a high characteristic adhesion force F_{H0} lead to an increase of the contact normal force $F_{N,H}$. On the other hand, the porosity of the particle packing has an effect on the normal stress distribution among the contact points of the particles. With the bulk density of the particle system, determined by shear testing, and the applied contact model, it was possible to calculate the porosity of the particle packing prior to and after modification. Results showed an increase of the bulk density and a decrease of porosity after the silanization. Since unmodified and Piranha cleaned particles have a lower bulk density than the silanized particles, the normal stress is distributed over fewer contact surfaces. Because of this, and due to the soft contact behavior and high characteristic adhesion force F_{H0}, the compressive normal force $F_{N,H}$ acting between two Piranha treated or unmodified particles is higher than between two silanized particles.

If we consider the adhesion function of the silanized particles, it can be seen that there is a similar behavior for almost all modifications. Only the particles modified with CTMS exhibit a different behavior. Due to the negative value of the contact consolidation coefficient κ, with increasing normal load the adhesion force does not increase but decreases until, at high load values, it becomes a microscopic repulsion force. This explains the flow behavior of these particles (see Fig. 15) which despite the relatively high adhesion (see Fig. 20) have the best flowability among all modified particles and which progressively increases with compressive stress. When increasing the silane concentration such behavior of the adhesion function (transition to repulsion) is observed also for particles modified with the other silanes (except CDMPS).

Conclusion

The focus is laid on the effect for the reduction of the surface free energy through a surface hydrophobization and on the flow behavior of ultrafine, cohesive particles. As a model particle system, microscopic glass particles with a median diameter $d_{50,3}$ of 7.8 μm were selected. An extensive analysis of the effect of the surface hydrophobization on the flowability of the glass particles was accomplished by coating them using five different silanes with three concentrations applied in order to achieve variable coating thicknesses. After the silanization, the particles were subjected to an extensive characterization. A combination of water vapor sorption (DVS) and thermogravimetric analysis (TGA) allowed the determination of the produced coating thickness and homogeneity. The flow behavior of the modified particles was assessed by shear testing using an automatic ring shear tester.

Results showed the modification with all tested silanes led to an enhanced powder flowability, which is attributed to the decreased van der Waals forces due to reduced surface energy. However, the flow enhancing effect of the majority of the used silanes exhibited a certain concentration dependence. The most effective flow improving effect was observed at a lowest silane concentration of 0.01 mol/L and decreased with increasing concentration. This was attributed to an increased coating thickness and a decreased coating homogeneity due to the formation of thick silane polymer structures at higher silane concentrations. During the shear tests an overlapping of the thick silane polymer structures takes place that causes a shear induced interlocking of the hydrophobic chains and reduces the particle flowability. Additionally, the surface energy is not the only parameter that affects the flow behavior of the silanized particles due to this interlocking effect of the hydrophobic chains. This is clearly visible in the case of the particles modified with the long-chained silane PFOTES, which despite the very low surface energy have worse flowability than particles coated with the short-chained silane FPTS. Thus, the flow improving effect of the silanes stems from a combination between several parameters, such as surface energy, coating thickness and chain length of the hydrophobic chain.

Furthermore, the application of the micromechanical contact model "stiff particles with soft contacts" allowed to determine the characteristic adhesion force F_{H0}, the adhesion as a function of the normal load F_H as well as the elastic-plastic contact consolidation coefficient which characterizes the contact behavior. This contributed to a better understanding of the macroscopic flow behavior of the modified particles. An interesting effect of microscopic repulsion between the particles with increasing deformation was observed for almost all silanes. Despite the facts that at high silane concentrations the adhesion force between the particles was almost as high as in case of the unmodified particles, the silanized particles still had a better flowability.

Generally, all silanes are applicable for the reduction of particles adhesion and for the improvement of flowability and handling of cohesive powders. However, for the most effective flow effect silanes with short hydrophobic chains and low silane concentrations should be selected.

Acknowledgements We would like to acknowledge the financial support of the German Research Foundation (DFG) through the priority program 'PiKo—Particles in Contact'.

References

1. Tomas, J.: Adhesion of ultrafine particles—a micromechanical approach. Chem. Eng. Sci. **62**, 1997–2010 (2007)
2. Valverde, J.M., Castellanos, A.: Fluidization of nanoparticles: a simple equation for estimating the size of agglomerates. Chem. Eng. J. **140**, 296–304 (2008)
3. Xu, C.C., Zhang, H., Zhu, J.: Improving flowability of cohesive particles by partial coating on the surfaces. Can. J. Chem. Eng. **87**, 403–414 (2009)
4. Yang, J., Sliva, A., Banerjee, A., Dave, R.N., Pfeffer, R.: Dry particle coating for improving the flowability of cohesive powders. Powder Technol. **158**, 21–33 (2005)
5. Kojima, T., Elliott, J.A.: Effect of silica nanoparticles on the bulk flow properties of fine cohesive powders. Chem. Eng. Sci. **101**, 315–328 (2013)
6. Zimmermann, I., Eber, M., Meyer, K.: Nanomaterials as flow regulators in dry powders. Nano Online, De Gruyter (2016)
7. Tomas, J., Kleinschmidt, S.: Improvement of flowability of fine cohesive powders by flow additives. Chem. Eng. Technol. **32**, 1470–1483 (2009)
8. Ghoroi, C., Gurumurthy, L., McDaniel, D.J., Jallo, L.J., Davé, R.N.: Multi-faceted characterization of pharmaceutical powders to discern the influence of surface modification. Powder Technol. **236**, 63–74 (2013)
9. Arkles, B.: Tailoring surfaces with silanes. ChemTech **7**(12), 766–778 (1977)
10. Dow Corning Corporation, Organosilane Technology in Coating Applications: Review and Perspectives
11. Gelest, I., Arkles, B.C., Larson, G.L.: Silicon Compounds: Silanes and Silicones: A Survey of Properties and Chemistry, 3rd edn. Gelest Inc. (2013). ISBN 978-0-578-12235-9
12. Mader-Arndt, K., Kutelova, Z., Tomas, J.: Microscopic particle contact adhesion models and macroscopic behavior of surface modified particles. In: Mittal, K., Jaiswal, R. (eds.) Particle Adhesion and Removal, pp. 105–155. Wiley, Hoboken, NJ, USA (2015)
13. Cassie, A.B.D., Baxter, S.: Wettability of porous surfaces. Trans. Faraday Soc. **40**, 546 (1944)
14. Owens, K.D., Wendt, R.C.: Estimation of the surface free energy of polymers. J. Appl. Polym. Sci. **13**, 1741–1747 (1969)
15. London, F.: The general theory of molecular forces. Trans. Faraday Soc. **33**, 8b–26 (1937)
16. Schwuger, M.J., Findenegg, G.H., Haegel, F.H.: Lehrbuch der Grenzflächenchemie. Wiley-VCH (1996). ISBN 978-3131375018
17. Farrar, D.M.: The use of vapour-pressure and moisture content measurements to deduce the internal and external surface area of soil particles. J. Soil Sci. **14**, 303–321 (1963)
18. Kast, W.: Adsorption aus der Gasphase, Ingenieurwissenschaftliche Grundlagen und Technische Verfahren. VCH Verlagsgesellschaft, Weinheim (1988). ISBN 3-527-26719-0
19. Jenike, A.W.: Storage and flow of solids. Bulletin No. 123 **53**(26) (1964)
20. Schulze, D.: Ringschergerät RST-XS.s - noch kleiner - noch mehr Funktionen
21. D18 Committee: Test Method for Bulk Solids Using Schulze Ring Shear Tester. ASTM International, West Conshohocken, PA (2008)
22. Lemal, D.M.: Perspective on fluorocarbon chemistry. J. Org. Chem. **69**, 1–11 (2004)
23. Lindner, E., Arias, E.: Surface free energy characteristics of polyfluorinated silane films. Langmuir **8**, 1195–1198 (1992)
24. Lork, K.D., Unger, K.K., Kinkel, J.N.: Role of the functional group in n-octydimethylsilanes in the synthesis of C8 reversed-phase silica packings for high-performance liquid chromatography. J. Chromatogr. A **352**, 199–211 (1986)
25. Butt, H.-J., Graf, K., Kappl, M.: Physics and Chemistry of Interfaces. Wiley-VCH (2003). ISBN 978-3-527-40629-6
26. Forny, L., Pezron, I., Saleh, K., Guigon, P., Komunjer, L.: Peculiar absorption of water by hydrophobized glass beads. Colloids Surf. A **270–271**, 263–269 (2005)
27. Fuji, M., Fujimori, H., Takei, T., Watanabe, T., Chikazawa, M.: Wettability of glass-bead surface modified by Trimethylchlorosilane. J. Phys. Chem. B **102**, 10498–10504 (1998)

28. Tomas, J.: Zur Produktgestaltung kohäsiver Pulver – Mechanische Eigenschaften, Kompressions- und Fließverhalten. Chem Ing Tech **75**, 651–661 (2003)
29. Johnson, K.L., Kendall, K., Roberts, A.D.: Surface energy and the contact of elastic solids. Proc. R. Soc. A: Math. Phys. Eng. Sci. **324**, 301–313 (1971)
30. Derjaguin, B., Muller, V., Toporov, Y.: Effect of contact deformations on the adhesion of particles. J. Colloid Interface Sci. **53**, 314–326 (1975)
31. Schumann, M.: Einfluss von Adsorptionsschichten auf die Fließeigenschaften feiner Schüttgüter. PhD thesis, Freiberg (2005)
32. Molerus, O.: Schüttgutmechanik: Grundlagen und Anwendungen in der Verfahrenstechnik. Springer (1985). ISBN 978-3-540-15428-0
33. Molerus, O.: Theory of yield of cohesive powders. Powder Technol. **12**, 259–275 (1975)
34. Tomas, J.: Fundamentals of cohesive powder consolidation and flow. Granul. Matter **6**, 75–86 (2004)
35. Tomas, J.: Assessment of mechanical properties of cohesive particulate solids. Part 2: Powder flow criteria. Part. Sci. Technol.**19**, 111–129 (2001)
36. Tomas, J.: Produkteigenschaften ultrafeiner Partikel: Mikromechanik, Fliess- und Kompressionsverhalten kohäsiver Pulver. Hirzel (2009). ISBN 978-3777616810
37. Zisman, W.A.: Friction, durability and wettability properties of monomolecular films solids: friction and Wear. In: Davies, R. (ed.) Friction and Wear. Elsevier (1959)
38. Biswas, S.K., Vijayan, K.: Friction and wear of PTFE—a review. Wear **158**, 193–211 (1992)
39. Tomas, J.: Flow properties of cohesive nanopowders. China Particuol. **1**, 231–241 (2003)
40. Tomas, J.: Particle adhesion fundamentals and bulk powder consolidation. KONA **18**, 157–169 (2000)
41. Mader-Arndt, K., Kutelova, Z., Fuchs, R., Meyer, J., Staedler, T., Hintz, W., Tomas, J.: Single particle contact versus particle packing behavior: model based analysis of chemically modified glass particles. Granul. Matter **16**, 359–375 (2014)
42. Brown, S.C., Rabinovich, Y.I., Vakarelski, I.U., Sharma, P., Moudgil, B.M.: Evaluation of the mechanical and tribological properties of self-assembled surfactant, nanostructures using atomic force microscopy. In: Strarov, V.M. (ed.) Nanoscience: Colloidal and Interfacial Aspects. CRC Press Books (2017). ISBN 978-1138198944

MIX
Papier | Fördert
gute Waldnutzung
FSC® C083411

Kolibri 360 GmbH
Ferdinand-Jühlke-Straße 7
99095 Erfurt, Deutschland
produktsicherheit@kolibri360.de